ELECTRICAL SAFETY HANDBOOK

ELECTRICAL SAFETY HANDBOOK

Dennis K. Neitzel, C.P.E., C.E.S.C.P.
Mary Capelli-Schellpfeffer, M.D., M.P.A.
Al Winfield, C.E.S.C.P.

Fifth Edition

Mc Graw Hill

New York Chicago San Francisco Lisbon London
Madrid Mexico City Milan New Delhi San Juan
Seoul Singapore Sydney Toronto

Library of Congress Control Number: 2019936594

Electrical Safety Handbook, Fifth Edition

1 2 3 4 5 6 7 8 9 LCR 24 23 22 21 20 19

ISBN 978-1-260-13485-8
MHID 1-260-13485-7

Sponsoring Editor
Lara Zoble

Editorial Supervisor
Donna M. Martone

Acquisitions Coordinator
Elizabeth Houde

Project Manager
Poonam Bisht, MPS Limited

Copy Editor
Mohammad Taiyab Khan, MPS Limited

Proofreader
Heather Mann

Indexer
Jerry Ralya

Production Supervisor
Pamela A. Pelton

Composition
MPS Limited

Art Director, Cover
Jeff Weeks

To my wife, Brenda Neitzel, who always believed in me and encouraged me to continue my education and strive to be the best that I could be; to the U.S. Air Force for giving me my start in an electrical career in 1967; to all of my employers, who gave me countless opportunities to learn and progress; and to John Cadick, who believed in me enough to ask me to contribute to this book.

Dennis K. Neitzel

In dedication to Michael Allen, Sarah, Benjamin, Amelia, and Natalie, with all my love.

Mary Capelli-Schellpfeffer

I dedicate this effort to Gerry, my wife and best friend, for her endless love, support, encouragement, and belief in me. It is my honor to have John Cadick as a cherished friend and coworker. The confidence John Cadick has shown in me by inviting my contribution to this esteemed project is deeply appreciated.

Al Winfield

ABOUT THE AUTHORS

Dennis K. Neitzel, C.P.E., C.E.S.C.P., has over 50 years in the electrical industry in various capacities, specializing in electrical equipment and systems maintenance, testing, engineering, inspection, and safety. He is a past Chair of the IEEE-IAS Electrical Safety Workshop (2012). He is an IEEE Senior Member, and is also a member of ASSE, AFE, IAEI, and NFPA. Mr. Neitzel is a Certified Plant Engineer (C.P.E.), Certified Electrical Safety Compliance Professional (C.E.S.C.P.), and a Certified Electrical Inspector-General. He served as a Principal Committee Member and Special Expert for the NFPA 70E, *Standard for Electrical Safety in the Workplace* from 1992 to 2018; served as Working Group Chair of IEEE Std. 3007.1-2010, *Recommended Practice for the Operation and Management of Industrial and Commercial Power Systems*, IEEE Std. 3007.2-2010, *Recommended Practice for the Maintenance of Industrial and Commercial Power Systems*, IEEE Std. 3007.3-2012, *Recommended Practice for Electrical Safety in Industrial and Commercial Power Systems*, and IEEE Std. 45.5-2014, *Recommended Practice for Electrical Installations on Shipboard-Safety Considerations*; and is a co-author of the *Electrical Safety Handbook*, McGraw-Hill Publishers. Mr. Neitzel earned his Bachelor's degree in Electrical Engineering Management and his Master's degree in Electrical Engineering Applied Sciences. He has authored, published, and presented numerous technical papers and magazine articles on electrical safety, maintenance, and technical training.

Mary Capelli-Schellpfeffer, M.D., M.P.A., delivers outpatient medical care services to employees in occupational health service centers. Board-certified as a physician in general preventive medicine and public health, she is also a consultant to both the NJATC (National Joint Apprenticeship and Training Committee of the National Electrical Contractors Association and International Brotherhood of Electrical Workers) and IEEE (Institute of Electrical and Electronics Engineers) Standards Committees. She lives in Chicago, Illinois.

Al Winfield has more than 50 years of hands-on electrical construction, repair, system operations, and training experience. Mr. Winfield started his career in the electrical industry in 1960. During his career in the public utility industry, his experience included hands-on electrical work as a high-voltage lineman, operations experience in system operations, and several years as the supervisor of training—system operations. He has specialized in providing technical and safety training for electrical system operations personnel and electrical construction and maintenance personnel for the past three decades. Over the past two decades, Mr. Winfield has also provided electrical consulting services for several manufacturing, mining, and petrochemical corporations around the world. He is currently the director of safety and training for Cadick Corporation.

CONTENTS

Chapter 13. Safety Management and Organizational Structure 579

Chapter 14. Safety Training Methods and Systems 613

FOREWORD

Electrical power makes modern life easier. It provides energy for appliances and factory processes that simplify life and industry. Electricity can be manipulated to carry signals that are easy to interpret, and it can be easily converted to other forms of energy. However, the hazards associated with electricity can also cause injuries and fatalities.

In most instances, electricity must be converted to another form before it can be used. For example, electrical energy must be converted to thermal energy before it can be used to cook food, heat water, or warm a room. When electrical energy is converted to a more useful form, the conversion process must be a controlled event. When controlled, the conversion process is desirable, the result is good, and the process is safe.

Normally, a user of electricity does not think about the conversion process. It is not necessary for process operators to think about converting electrical energy into mechanical energy for the rotation of a motor when they push a start button. Electrical workers and electrical manufacturers have provided the necessary electrical equipment to convert the electrical energy safely to make the motor run.

Consensus electrical standards provide adequate guidance for manufacturers, engineers, employees, and employers. When workers are trained to understand and follow the guidance provided by consensus standards, operators can push a start button without concern for their safety. However, when they do not understand the guidance, workers sometimes create hazard exposures.

Electrical hazards exist in many different forms. Direct contact with an energized conductor exposes workers to current flow through their body. Current flowing through body tissue produces heat and damages or destroys the tissue, sometimes resulting in death. An arcing fault is electrical energy that is being converted to another form of energy, such as heat or pressure, by an *uncontrolled* process. An arcing fault might expose a worker to injury from the thermal hazard or from the effects of the accompanying pressure wave.

To avoid injury from an electrical hazard, workers must avoid exposure to the hazard or use adequate protective equipment and safe work practices. The most important safe work practice is to *remove* all electrical energy and *eliminate* any chance that the energy might reappear. If the energy cannot reappear, the equipment or circuit is considered to be in an electrically safe work condition. Consensus standards discussed in this book provide guidance about how to establish an electrically safe work condition.

Each worker must be trained to recognize how exposure to each electrical hazard might exist and how to avoid that exposure. Workers are exposed to electrical hazards in many different ways, including the following:

- Electrical equipment, devices, and components have a normal lifetime. Control devices sometimes wear out and malfunction with age or lack of maintenance. When a failure occurs, a worker is expected to identify the problem, repair the problem, and restore the equipment to normal service.
- Electrical equipment must be maintained. Although the electrical energy sometimes is removed before a worker begins a maintenance task (best practice), those tasks often are executed while the source of electricity is energized.
- Equipment and circuits sometimes are modified to add new devices or circuits. Short-term employees might be expected to work in an environment that includes exposure to energized

electrical circuits and components. Consultant and service employees are frequently exposed to energized electrical equipment and circuits.

- When a problem exists that causes a process to malfunction, a worker might open a door or remove a cover and expose an energized electrical conductor or component. In many cases, the worker might troubleshoot while the circuit is energized. Components and conductors might be added within a piece of equipment while the equipment or parts of the equipment remain energized.
- After correcting a problem, workers sometimes create further hazardous conditions by leaving an equipment door ajar, leaving latches open, replacing covers with a minimum number of screws, and removing devices that create holes in a door or cover.

When workers understand that these conditions expose themselves or others to possible injury, they are more likely to avoid the hazard exposure. Training must build and reinforce that understanding. This book provides guidance to help trainers and other workers develop the necessary understanding.

Workers must understand the limits of their knowledge and ability. They should not accept and perform a work/task unless they have been trained and have the experience necessary to avoid all hazards, including electrical hazards. When workers are trained to understand electrical hazards and how to avoid them, then they become a valuable asset to the employer.

Ray Jones, P.E. (Retired)
IEEE Fellow and Former NFPA 70E Technical Committee Chair

PREFACE

This fifth edition of the *Electrical Safety Handbook* comes during an avalanche of changes in the world of electrical safety.

Since the fourth edition was published, the National Fire Protection Association (NFPA) released the 2015 and 2018 editions of the *Standard for Electrical Safety in the Workplace* (NFPA 70E). Both documents include numerous changes that both add to and further explain the practical aspects of electrical safety. NFPA 70E has been adopted by a multitude of facilities, companies, and organizations around the world as their standard for electrical safety.

Labor unions such as the International Brotherhood of Electrical Workers have widely promoted the electrical safety portions of their apprenticeship programs. Colleges and universities such as Murray State University have added electrical safety as part of their environmental safety and health (ES&H) degree programs.

Intensive research is ongoing in areas such as the following:

- Electrical shock hazard in systems as low as 30 volts
- Electrical arc flash hazards in systems of 208 volts and below
- Field testing and measurement of arc flash energies—a collaboration between NFPA and the Institute of Electrical and Electronics Engineers (IEEE)
- Calculation of incident arc flash energies in dc systems

The many vendors who write and supply the software packages used for performing engineering studies such as arc flash incident energy analysis have frequently updated their software to give the engineering community better and faster tools to perform the necessary calculations. In addition, the U.S. Occupational Safety and Health Administration (OSHA) has revised the 29 CFR 1910.269 regulation, *Electric Power Generation, Transmission, and Distribution*, to include the requirements for an arc flash analysis and associated arc-rated clothing and personal protective equipment. The construction equivalent, 29 CFR 1926, Subpart V, *Electric Power Transmission and Distribution*, has been revised to be consistent with 1910.269. As part of the revisions of 1910.269 and 1926, Subpart V, OSHA also revised 1910.137 and 1926.97, Electrical Protective Equipment, to include class 00, 500-volt ac gloves.

IEEE C2, the *National Electrical Safety Code*, in the 2007 edition required an arc flash analysis and arc-rated clothing and personal protective equipment. This standard has also been revised for 2017 and expands on and clarifies the existing requirements.

In 2008, the Canadian Standards Association published the first edition of CSA Z462-08, *Workplace Electrical Safety*, which is essentially the Canadian version of NFPA 70E, and continues to update to the latest information on electrical safety, currently the 2018 edition.

The fourth edition of the *Electrical Safety Handbook* (ESH) has continued to be widely accepted and used throughout the electrical industry. In fact, the authors have noted that many copies of the ESH are appearing on booksellers' sales lists from all over the world.

Because of the nationality of the authors, the ESH has always used North American regulatory standards for the purpose of example and identifying regulatory needs. While we continue to use the U.S. and Canadian regulations and standards as our guideline, we have modified some of the text to be more inclusive.

Chapters 1, 3, and 4 continue to serve as the central core of the book by presenting the case for electrical safety (Chapter 1), a broad coverage of electrical safety equipment (Chapter 3), and detailed coverage of electrical safety procedures (Chapter 4). In this fifth edition, we have updated and improved each of these chapters. Chapter 1 has been augmented by inclusion of some information on arc-related hazards such as toxic materials and acoustic injuries. Chapter 3 has been generally edited and new information added on such topics as arc-fault circuit-interrupters. Finally, Chapter 4 has also been edited and now includes sections on remote operating devices to be used for enhanced safety when operating switchgear.

Chapter 2 addresses the fundamental physics underlying the various electrical hazards. The material is presented in a much more technical format than Chapter 1 and uses advanced mathematics and citation of high-level research. The authors' purpose in including this chapter is not to move away from the practical information provided in all previous editions. Rather, we are presenting some of the more technical data used as the foundation for all electrical safety research—whether theoretical or practical. In making this information available in a public way, we hope that others will add their voices and efforts to the ongoing work in basic research in electrical safety.

Chapter 5 provides a detailed and updated overview of the general requirements for grounding and bonding electrical systems and equipment. The fifth edition features many updated, improved diagrams to help clarify the subject of electrical grounding and bonding. Further, the information in the chapter has been edited and rewritten to help with a subject that many find very difficult to understand. As with all of the chapters in this handbook, Chapter 5 is not intended to replace or be a substitute for the requirements of the current NEC or OSHA regulations. Always use the most current standards and regulations when designing, installing, and maintaining the grounding systems within a facility.

Chapter 6 has been extensively edited and contains newly written material. In addition to the information first introduced in the third edition, Chapter 6 has been enhanced with a new section dealing with shock and arc flash risk assessments. As always, readers of the fifth edition should refer to other references for more detailed information on electrical maintenance and risk assessments. One good source of detailed maintenance information is the InterNational Electrical Testing Association (NETA), whose website is http://www.netaworld.org. The best source for information concerning electrical hazard risk assessments is the current edition of NFPA 70E, Standard for Electrical Safety in the Workplace.

Chapter 7 updates the fourth edition coverage of the consensus and mandatory standards and regulations in the workplace. The specific information reprinted from OSHA has been updated to the most recent versions as of the date of this publication. As before, readers should always refer to OSHA publications, available at www.osha.gov, for the most recent information.

Chapter 8 has been generally updated.

Chapter 9 provides recent injury and fatality statistics and updated medical evaluation and treatment information.

Chapters 10 and 11 continue to be a valuable synopsis of low-voltage (Chapter 10) and medium- and high-voltage (Chapter 11) safety. The reader may refer to these chapters for quick coverage of key safety issues in electrical systems. Of course, Chapters 3, 4, and 5 provide detailed information.

Chapter 12 includes additional references to standards addressing human factors considerations, as well as information about electrical industry resources regarding ergonomics and human performance.

Chapter 13 includes general edits and some minor error corrections. Electrical safety, like any human activity, has developed its own share of anecdotes, legends, and so-called urban myths. This culture is often based on assumptions that are not valid. Chapter 13 provides some information on how to change that culture.

Chapter 14 contains in-depth information about how adult learners should be trained. We provide a comparison of the four most common ways of training adults—classroom presentations, computer-based training (CBT), Internet (Web-based or online) training (WBT), and simple video training. The other sections of the chapter have been edited and clarified in some cases.

Dennis K. Neitzel, C.P.E., C.E.S.C.P.
Mary Capelli-Schellpfeffer, M.D., M.P.A.
Al Winfield, C.E.S.C.P.

ACKNOWLEDGMENTS

The authors gratefully acknowledge contributions, permissions, and assistance from the following individuals and organizations:

AVO Training Institute, Inc. (a Subsidiary of Megger); Erico, Inc. (Cadweld); Ronald P. O'Riley, *Electrical Grounding* (Delmar Publishers); National Fire Protection Association (NFPA), *2017 National Electrical Code Handbook* (NEC), and NFPA 70E-2018, *Standard for Electrical Safety in the Workplace.*

Dennis K. Neitzel

Dr. Capelli-Schellpfeffer's research was supported in part by grant R01 OH04136-02 from the U.S. Centers for Disease Control and Prevention (CDC) and the National Institute of Occupational Safety and Health (NIOSH). Her comments do not represent official agency views.

Mary Capelli-Schellpfeffer

CHAPTER 1
HAZARDS OF ELECTRICITY

INTRODUCTION

Modern society has produced several generations that have grown accustomed to electricity. This acclimatization has been made easier by the fact that electricity is silent, invisible, odorless, and has an "automatic" aspect to it. In the late 1800s, hotels had to place signs assuring their guests that electricity is harmless. By the late 1900s, signs had to be hung to remind us that electricity is a hazard. In fact, the transition of electricity from a silent coworker to a deadly hazard is a change that many cannot understand until it happens to them. Because of these facts, the total acceptance of an electrical safety procedure is a requirement for the health and welfare of workers.

Understanding the steps and procedures employed in a good electrical safety program requires an understanding of the nature of electrical hazards. Although they may have trouble writing a concise definition, most people are familiar with electric shock. This often painful experience leaves its memory indelibly etched on the human mind. However, shock is only one of the electrical hazards. Others include arc, blast, acoustic, light, and toxic gases. This chapter describes each of these hazards and explains how each affects the human body.

Understanding the nature of the hazards is useless unless protective strategies are developed to protect the worker. This chapter also includes a synopsis of the types of protective strategies that should be used to safeguard the worker.

HAZARD ANALYSIS

The division of the electrical power hazard into three components is a classic approach used to simplify the selection of protective strategies. The worker should always be aware that electricity is the single root cause of all of the injuries described in this and subsequent chapters. That is, the worker should treat electricity as the hazard and select protection accordingly.

SHOCK

Description

Electric shock is the physical stimulation that occurs when electric current flows through the human body. The distribution of current flow through the body is a function of the resistance of the various paths through which the current flows. The final trauma associated with the electric shock is usually determined by the most critical path called the *shock circuit.* The symptoms may include a mild tingling sensation, violent muscle contractions, heart arrhythmia, or tissue damage. Detailed descriptions of electric current trauma are included in Chap. 9. For the purposes of this chapter, tissue damage may be attributed to at least two major causes.

Burning. Burns caused by electric current are almost always third degree because the burning occurs from the inside of the body. This means that the growth centers are destroyed. Electric-current burns can be especially severe when they involve vital internal organs.

Cell Wall Damage. Research funded by the Electric Power Research Institute (EPRI) has shown that cell death can result from the enlargement of cellular pores due to high-intensity electric fields.[1] This research has been performed primarily by Dr. Raphael C. Lee and his colleagues at the University of Chicago. This trauma, called electroporation, allows ions to flow freely through the cell membranes, causing cell death.

Influencing Factors

Several factors influence the severity of electrical shock. These factors include the physical condition and responses of the victim, the path of the current flow, the duration of the current flow, the magnitude of the current, the frequency of the current, and the voltage magnitude causing the shock.

Physical Condition and Physical Response. The physical condition of the individual greatly influences the effects of current flow. A given amount of current flow will often cause less trauma to a person in good physical condition. Moreover, if the victim of the shock has any specific medical problems such as heart or lung ailments, these parts of the body will be severely affected by relatively low currents. A diseased heart, for example, is more likely to suffer ventricular fibrillation than a healthy heart.

Current Duration. The amount of energy delivered to the body is directly proportional to the length of time that the current flows; consequently, the degree of trauma is also directly proportional to the duration of the current. Three examples illustrate this concept:

1. Current flow through body tissues delivers energy in the form of heat. The magnitude of energy may be approximated by

$$J = I^2Rt \tag{1.1}$$

where J = energy, joules
I = current, amperes
R = resistance of the current path through the body, ohms
t = time of current flow, seconds

If sufficient heat is delivered, tissue burning and/or organ shutdown can occur. Note that the amount of heat that is delivered is directly proportional to the duration of the current (t).

2. Some portion of the externally caused current flow will tend to follow the current paths used by the body's central nervous system. Since the external current is much larger than the normal nervous system current flow, damage can occur to the nervous system. Note that nervous system damage can be fatal even with relatively short durations of current; however, increased duration heightens the chance that damage will occur.
3. Generally, a longer duration of current through the heart is more likely to cause ventricular fibrillation. Fibrillation seems to occur when the externally applied electric field overlaps with the body's cardiac cycle. The likelihood of this event increases with time.

Frequency. Table 1.1 lists the broad relationships between frequency and the harmful effects of current flow through the body. Note that at higher frequencies, the effects of Joule (I^2t) heating become

TABLE 1.1 Important Frequency Ranges of Electrical Injury

Frequency	Regimen	Applications	Harmful effects
DC–10 kHz	Low frequency	Commercial electrical power; soft tissue healing; transcutaneous electrical stimulation	Joule heating; destructive cell membrane potentials
100 kHz–100 MHz	Radio frequency	Diathermy; electrocautery	Joule heating; dielectric heating of proteins
100 MHz–100 GHz	Microwave	Microwave ovens	Dielectric heating of water
10^{13}–10^{14} Hz	Infrared	Heating; CO_2 lasers	Dielectric heating of water
10^{14}–10^{15} Hz	Visible light	Optical lasers	Retinal injury; photochemical reactions
10^{15} Hz and higher	Ionizing radiation	Radiotherapy; x-ray imaging; UV therapy	Generation of free radicals

less significant. This decrease is related to the increased capacitive current flow at higher frequencies.

It should be noted that some differences are apparent even between DC (zero Hz) and standard power line frequencies (50 to 60 Hz). When equal current magnitudes are compared (DC to AC rms), anecdotally, DC seems to exhibit two significant behavioral differences:

1. Victims of DC shock have indicated that they feel greater heating from DC than from AC. The reason for this phenomenon is not totally understood; however, it has been reported on many occasions.
2. The DC current "let-go" threshold seems to be higher than the AC "let-go" threshold.

Despite the slight differences, personnel should work on or near DC power supplies with the same level of respect that they use when working on or near AC power supplies. This includes the use of appropriate protective equipment and procedures.

Note: Unless otherwise specifically noted, the equipment and procedures suggested in this handbook should be used for all power frequencies up to and including 400 Hz.

Voltage Magnitude. The magnitude of the voltage affects electric shock in one or more of the following three ways:

1. At voltages above 400 volts (V), the electrical pressure (voltage) may be sufficient to puncture the epidermis. Since the epidermis provides the major part of the resistance of the human body, the current magnitude will increase dramatically and lethally when this puncture occurs.
2. The degree of electroporation is higher for greater voltage gradients. That is, the higher voltages cause more intense fields, which in turn increase the severity of the electroporation.
3. Higher voltages are more likely to cause electrical arcing. While this is not a shock hazard per se, it is related to the shock hazard since arcing may occur at the point of contact with the electrical conductor.

Although current regulatory and consensus standards use 50 V as the lower limit for the shock hazard, recent research has shown that harmful or even fatal shocks can result from contact with circuits as low as 30 V.[2]

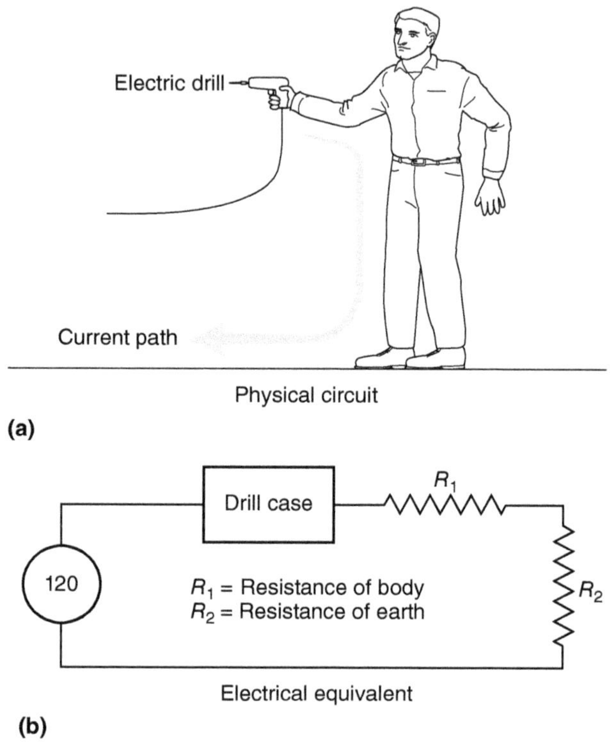

FIGURE 1.1 Electric shock current path.

Current Magnitude. The magnitude of the current that flows through the body obeys Ohm's law, that is,

$$I = \frac{E}{R} \tag{1.2}$$

where I = current magnitude, amperes (A)
E = applied voltage, volts (V)
R = resistance of path through which current flows, ohms (Ω)

In Fig. 1.1 the worker has contacted a 120-V circuit when an electric drill short-circuits internally. The internal short circuit impresses 120 V across the body of the worker from the hand to the feet. This creates a current flow through the worker to the ground and back to the source. The total current flow in this case is given by the formula

$$I = \frac{E}{R_1 + R_2} \tag{1.3}$$

Variable R_2 is the resistance of the earth and for the purposes of this analysis may be ignored. Variable R_1 is the resistance of the worker's body and includes the skin resistance, the internal body resistance, and the resistance of the shoes where they contact the earth.

Typical values for the various components can be found in Tables 1.2 and 1.3. Assume, for example, that the worker shown in Fig. 1.1 is wearing leather shoes and is standing in wet soil. This person is perspiring heavily and has an internal resistance of 200 Ω. From Tables 1.2 and 1.3 the total resistance can be calculated as

$$500\ \Omega\ \text{(drill handle)} + 200\ \Omega\ \text{(internal)} + 5000\ \Omega\ \text{(wet shoes)} = 5700\ \Omega$$

TABLE 1.2 Nominal Resistance Values for Various Parts of the Human Body

Condition (area to suit)	Resistance Dry	Resistance Wet
Finger touch	40 kΩ–1 MΩ	4–15 kΩ
Hand holding wire	10–50 kΩ	3–6 kΩ
Finger-thumb grasp*	10–30 kΩ	2–5 kΩ
Hand holding pliers	5–10 kΩ	1–3 kΩ
Palm touch	3–8 kΩ	1–2 kΩ
Hand around $1\frac{1}{2}$-inch (in) pipe (or drill handle)	1–3 kΩ	0.5–1.5 kΩ
Two hands around $1\frac{1}{2}$-in pipe	0.5–1.5 kΩ	250–750 Ω
Hand immersed	—	200–500 Ω
Foot immersed	—	100–300 Ω
Human body, internal, excluding skin	—	200–1000 Ω

*Data interpolated.
Source: This table was compiled from Kouwenhoven and Milner. Permission obtained from estate of Ralph Lee.

TABLE 1.3 Nominal Resistance Values for Various Materials

Material	Resistance*
Rubber gloves or soles	>20 MΩ
Dry concrete above grade	1–5 MΩ
Dry concrete on grade	0.2–1 MΩ
Leather sole, dry, including foot	0.1–0.5 MΩ
Leather sole, damp, including foot	5–20 kΩ
Wet concrete on grade	1–5 kΩ

*Resistances shown are for 130-cm² areas.
Source: Courtesy Ralph Lee.

From this information the total current flow through the body for a 120-V circuit is calculated as

$$I = \frac{120}{5700} = 21.1 \text{ milliamperes (mA)} \tag{1.4}$$

Table 1.4 lists the approximate effects that various currents will have on a 68-kilogram (kg) human being. The current flow of 21.1 mA is sufficient to cause the worker to go into an "electrical hold." This is a condition wherein the muscles are contracted and held by the passage of the electric current—the worker cannot let go. Under these circumstances, the electric shock would continue until the current was interrupted or until someone intervened and freed the worker from the contact. Unless the worker is freed quickly, tissue and material heating will cause the resistances to drop, resulting in an increase in the current. Such cases are frequently fatal.

The reader should note that the values given in this example are for illustration only. Much lower values can and do occur, and many workers have been electrocuted in exactly this same scenario.

TABLE 1.4 Nominal Human Response to Current Magnitudes

Current (60 Hz)	Physiological phenomena	Feeling or lethal incidence
<1 mA	None	Imperceptible
1 mA	Perception threshold	Mild sensation
1–3 mA		Painful sensation
3–10 mA		
10 mA	Paralysis threshold of arms	Cannot release hand grip; if no grip, victim may be thrown clear (may progress to higher current and be fatal)
30 mA	Respiratory paralysis	Stoppage of breathing (frequently fatal)
75 mA	Fibrillation threshold 0.5%	Heart action discoordinated (probably fatal)
250 mA	Fibrillation threshold 99.5% (≥5-s exposure)	
4 A	Heart paralysis threshold (no fibrillation)	Heart stops for duration of current passage; for short shocks, may restart on interruption of current (usually not fatal from heart dysfunction)
≥5 A	Tissue burning	Not fatal unless vital organs are burned

Notes: (1) This data is approximate and based on a 68-kg (150-lb) person. (2) Information for higher current levels is obtained from data derived from accident victims. (3) Responses are nominal and will vary widely by individual.

Source: Courtesy Ralph Lee.

Parts of the Body. Current flow affects the various bodily organs in different manners. For example, the heart can be caused to fibrillate with as little as 75 mA. The diaphragm and the breathing system can be paralyzed, which possibly may be fatal without outside intervention, with less than 30 mA of current flow. The specific responses of the various body parts to current flow are covered in later sections.

ARC

Caution: The calculations and formulas in this section are shown to illustrate the basic concepts involved in the calculation of arc parameters including current, voltage, and energy. The calculation of actual values for specific field conditions is a complex, safety-related procedure and should be done only under the direction of experienced engineers.

Definition and Description

ANSI/IEEE Std 100-1988 defines arc as: "A discharge of electricity through a gas, normally characterized by a voltage drop in the immediate vicinity of the cathode approximately equal to the ionization potential of the gas."[3]

A similar definition, perhaps more useful in the discussion of electrical safety, is given in the glossary of this handbook as: "The heat and light energy release that is caused by the electrical breakdown of and subsequent electrical discharge through an electrical insulator such as air."

Electric arcing occurs when a substantial amount of electric current flows through what previously had been air. Since air is a poor conductor, most of the current flow is actually occurring through the vapor of the arc terminal material and the ionized particles of air. This mixture of superheated, ionized materials, through which the arc current flows, is called a *plasma*.

TABLE 1.5 Ignition Temperatures and Characteristics of Clothing Fibers

Fiber	Melt temperature	Decomposition temperature	Ignition temperature	Burning temperature
Cotton/Rayon*	N/A	554/581	752/788	1562
Polyester	482	734	1040	1337
Wool	N/A	446	1112	Unknown
Nylon 6,6	490	653	990	1607
PBI	N/A	860	N/A	N/A
NOMEX	N/A	900	N/A	N/A

*FR treatment of cotton or rayon does not affect ignition temperatures.

All temperatures are expressed in °F. Please note that polyester ignites at a higher temperature and burns at a lower temperature than cotton. This shows the fallacy of using untreated cotton as an FR garment.

Source: Courtesy Bulwark Protective Clothing.

Arcs can be initiated in several ways:

- When the voltage between two points exceeds the dielectric strength of the air. This can happen when overvoltages occur due to lightning strikes or switching surges.
- When the air becomes superheated with the passage of current through some conductor. For example, if a very fine wire is subjected to excessive current, the wire will melt, superheating the air and causing an arc to start.
- When two contacts part while carrying a very high current. In this case, the last point of contact is superheated and an arc is created because of the inductive flywheel effect.

Electric arcs are extremely hot. Temperatures at the terminal points of the arcs can reach as high as 50,000 kelvin (K). Temperatures away from the terminal points are somewhat cooler but can still reach 20,000 K. Although the specific results of such temperatures will vary depending on factors such as distance from the arc, ambient environmental conditions, and arc energy, anecdotal evidence supported by experimental results developed by the Institute of Electrical and Electronics Engineers (IEEE) clearly shows the following:

- The heat energy of an electrical arc can kill and injure personnel at surprisingly large distances. For example, second-degree burns have been caused on exposed skin at distances of up to 12 feet (ft) or (3.6 meters [m]) and more.
- Virtually all types of clothing fibers can be ignited by the temperatures of electrical arcs. Clothing made of non-flame-resistant fibers will continue to burn after the arc source has been removed and will continue to cause serious physical trauma. Table 1.5 shows the ignition temperature of various fabrics and identifies those that will support combustion after the arc energy is gone.

The amount of energy, and therefore heat, in an arc is proportional to the maximum available short circuit volt-amperes in the system at the point of the arc. Calculations by Ralph Lee indicate that maximum arc energy is equal to one-half the available fault volt-amperes at any given point.[4] Later research by Neal, Bingham, and Doughty shows that while the maximum may be 50 percent, the actual value will usually be somewhat different depending on the degree of distortion of the waveform, the available system voltage, and the actual arc power factor.[5] The same research also shows that enclosing the arc to create a so-called arc in the box focuses the incident arc energy and increases its effect by as much as threefold.[5,6]

The arc energy determines the amount of radiated energy and, therefore, the possible degree of thermal injury from radiation effects, including convective, infrared, and ultraviolet radiation. The arc energy will be determined by the arc voltage drop and the arcing current. After the arc is established, the arc voltage tends to be a function of arc length; consequently, the arc energy is less dependent on the system voltage and more dependent on the magnitude of the fault current. This means that *even low-voltage systems have significant arc hazard* and appropriate precautions must be taken. Figures 1.2 and 1.3 show the results of two experiments that were conducted with manikins exposed to electric arcs. As can be seen, both high and low voltages can create significant burns.

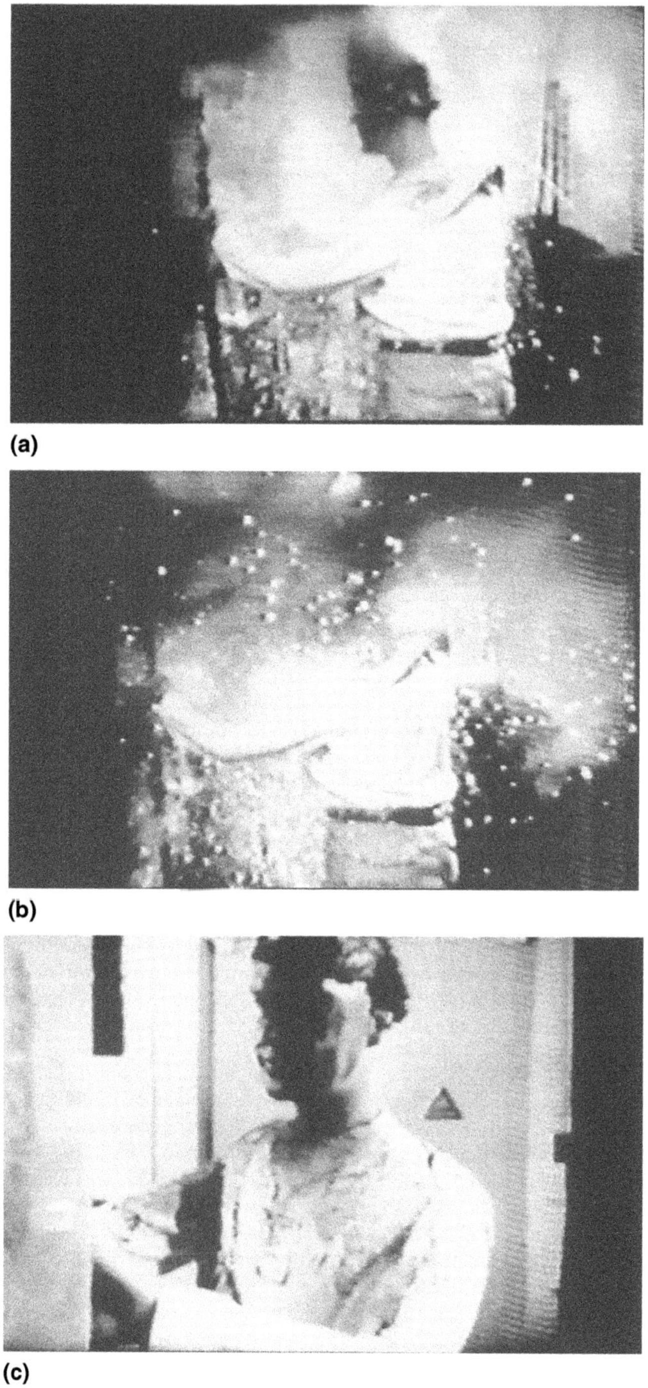

FIGURE 1.2 Electric arc damage caused by a medium-voltage arc. (*Courtesy Brosz and Associates.*)

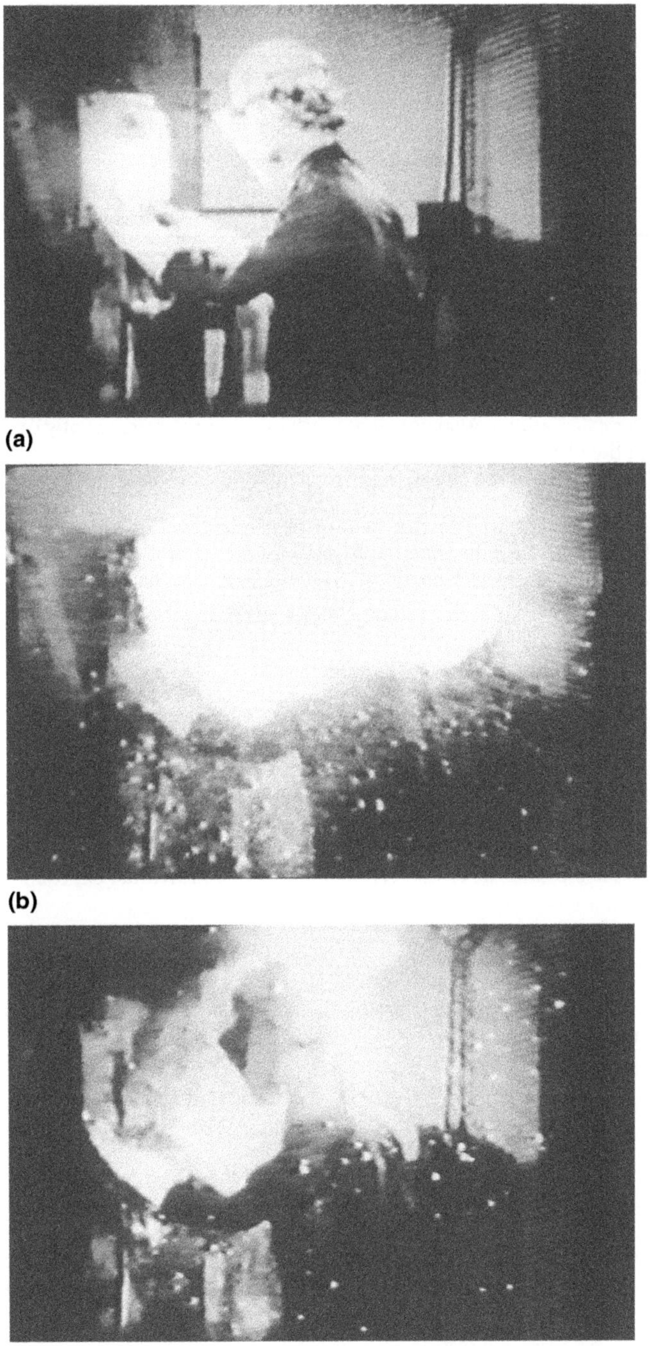

(a)

(b)

(c)

FIGURE 1.3 Electric arc damage caused by 240-volt arc. (*Courtesy Brosz and Associates.*)

TABLE 1.6 Electrical Arc Injury, Energy Sources

Energy	Nature of injuries
Light	Principally eye injuries, although severe burns can also be caused if the ultraviolet component is strong enough and lasts long enough
Heat	Severe burns caused by radiation and/or impact of hot objects such as molten metal
Mechanical	Shrapnel injuries from flying objects or blast debris; concussive injuries to brain/central nervous system; air pressure wave injuries to ears, lungs, and gastrointestinal organs

Arc Energy Release

Arc energy is released in multiple forms, including electrical, thermal, mechanical, and photonic or light energy. Table 1.6 describes the nature of these energy releases and the injuries that they cause. Note that light and heat tend to cause similar injuries and will, therefore, be treated as one injury source in later calculations. Also note that mechanical injuries are usually categorized as blast injuries, even though the ultimate cause is the electric arc.

To be conservative in arc energy release calculations, two assumptions must be made:

1. All arc energy is released in the form of heat measured in cal/cm^2 or J/cm^2. *The reader should remember that this assumption is made solely for the purpose of analyzing electric arc thermal injury. Other hazards such as shock and blast are considered separately.*
2. Every arc is fed by a sinusoidal source, thereby creating the maximum amount of energy release.

Arc Energy

Several major factors determine the amount of energy created and/or delivered by an electric arc. Table 1.7 lists the major factors and their qualitative effect. The quantitative effects of electric arc are the subject of many ongoing studies.

An individual's exposure to arc energy is a function of the total arc energy, the distance of the subject from the arc, and the cross-sectional area of the individual exposed to the arc.

Arc Energy Input

The energy supplied to an electric arc by the electrical system, called the arc input energy, may be calculated using the formula

$$J_{arc} = \int_0^t V_{arc} \times I_{arc} \times \cos\theta \times dt \qquad (1.5)$$

where J_{arc} = arc energy, joules
V_{arc} = arc voltage, volts
I_{arc} = arc current, amperes
t = time, seconds

Research has shown that electric arcs are rarely perfect sinusoids; however, the perfect sinusoid creates the greatest arc power. Therefore, Eq. 1.5 can be solved as

$$J_{arc} = V_{arc} \times I_{arc} \times t \times \cos\theta \qquad (1.6)$$

where θ = the angle between current and voltage

TABLE 1.7 Factors That Affect the Amount of Trauma Caused by an Electric Arc

Distance	The amount of damage done to the recipient diminishes by approximately the square of the distance from the arc. Twice as far means one-fourth the damage. (Empirical evidence suggests that the actual value may be somewhat different because of the focusing effect of the surroundings.)
Temperature	The amount of energy received is proportional to the difference between the fourth power of the arc temperature and the body temperature ($T_a^4 - T_b^4$).
Absorption coefficient	The ratio of energy received to the energy absorbed by the body.
Time	Energy received is proportional to the amount of time that the arc is present.
Arc length	The amount of energy transmitted is a function of the arc length. For example, a zero-length arc will transmit zero energy. Note that for any given system, there will be an optimum arc length for energy transfer.
Cross-sectional area of body exposed to the arc	The greater the area exposed, the greater the amount of energy received.
Angle of incidence of the arc energy	Energy is proportional to the sine of the angle of incidence. Thus, energy impinging at 90° is maximum.

Arcing Current. The actual arcing current varies as a function of several variables and has been calculated or estimated in different ways. IEEE Standard (Std) 1584-2002, for example, gives two equations that may be used to calculate the arcing current.[7] Equation 1.7 is the formula used for electrical arcs in systems with voltages less than 1000 V, and Eq. 1.8 is used for systems with voltages equal to or greater than 1000 V.

$$\log_{10}(I_a) = K + 0.662\log_{10}(I_{bf}) + 0.0966V + 0.000526G + 0.5588V\log_{10}(I_{bf}) - 0.00304G\log_{10}(I_{bf}) \tag{1.7}$$

$$\log_{10}(I_a) = 0.00402 + 0.983\log_{10}(I_{bf}) \tag{1.8}$$

where I_a = arcing current (kA)
K = a constant (−0.153 for open configurations or −0.097 for box configurations)
I_{bf} = the bolted, RMS symmetrical, three-phase fault current (kA)
V = system phase-to-phase voltage (kV)
G = the gap between the arcing conductors (mm)

Note that these equations are based on a specific model as developed for Std 1584-2002. The model includes the following:

- Three-phase voltages in the range of 208 to 15,000 V phase-to-phase
- Frequencies of 50 or 60 Hz
- Bolted fault current in the range of 700 to 106,000 A
- Grounding of all types and ungrounded
- Equipment enclosures of commonly available sizes
- Gaps between conductors of 13 to 152 mm
- Faults involving all three phases

Arcing Voltage

Arc voltage is somewhat more difficult to determine. Values used in power system protection calculations vary from highs of 700 V/ft (22.72 V/cm) to as low as 300 V/ft (9.84 V/cm). Two things are well understood:

1. Arc voltages start low and tend to rise. Periodically, the arc voltage will drop if the arc lasts long enough.[8]
2. Arc voltage is proportional to arc length. Therefore, from Eq. 1.6, arc power and energy are proportional to arc length.

Modern software programs used to calculate incident arc energy take different approaches to determine arc voltage. It should be noted that, at best, arc voltage calculation is only approximate for any given scenario.

Arc Surface Area

While the actual shape of an electrical arc may vary, all classic, realistic solutions start by assuming that an arc causes an approximately cylindrical plasma cloud with length L and radius r. This cylindrical structure will have a lateral surface area equal to $2\pi rL$. The areas of the ends of the cylinder are ignored in this calculation since they are so small relative to the side of the arc. To simplify the calculation of energy density, the arc is assumed to form a sphere with a surface area equal to the cylinder (Fig. 1.4). Thus, the arc sphere will have a radius of

$$r_s = \frac{1}{2}\sqrt{2rL} \tag{1.9}$$

where r_s = radius of equivalent sphere
r = radius of arc cylinder
L = length of arc

Incident Energy

The most commonly used arc calculation for personnel protection is the one that determines the radiant energy transfer from the electrical arc source to a body or equipment surface at an assumed distance away from the electrical arc. This is called the incident energy. This information can be used

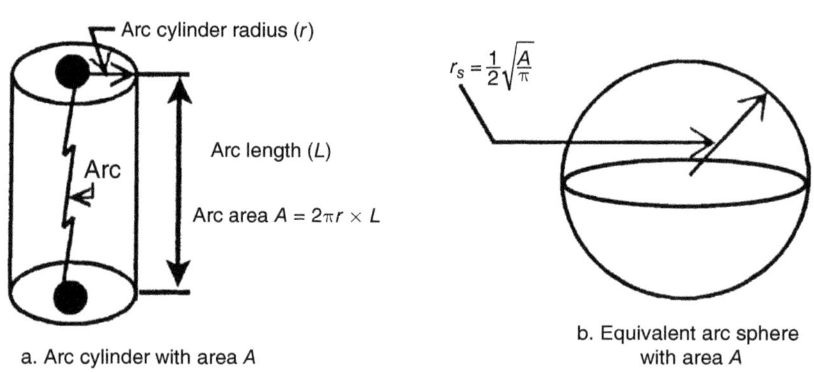

FIGURE 1.4 Arc cylinder and equivalent arc sphere.

to determine the necessary level of protective clothing required and can also be used in the performance of a risk analysis.

The estimation for tissue injury from electric arc depends on the temperature change and duration of the tissue exposure. However, the estimation starts with the calculation of the amount of energy, or heat flux (measured in calories per square centimeter), delivered to the skin. Many methods have been developed to calculate the incident energy—some more conservative than others. The following sections describe some of the methods that have been determined either empirically or theoretically. The reader should be aware that research into these areas is continuing at a frantic pace. Always refer to the most recent industry literature for the most up-to-date information.

The Lee Method. Ralph Lee has predicted that the heat energy received by an object (or worker) can be calculated using Eq. 1.10.

$$Q_o = \frac{Q_s \times A_s}{4\pi \times r^2} \times t \tag{1.10}$$

where Q_o = heat flux received by the object (cal/cm²)
Q_s = heat flux generated by source (cal/s/cm²)
A_s = surface area of arc sphere
r = distance from center of source to object (cm)
t = length of arc exposure

Using Eq. 1.8 as a starting point, Lee determined that the energy received by the worker is calculated using Eq. 1.11.

$$E = 2.142 \times 10^6 VI_{bf}\left(\frac{t}{D^2}\right) \tag{1.11}$$

where E = incident energy in J/cm²
V = system voltage (phase-to-phase)
t = arcing time (seconds)
D = distance from arc point to person or object (mm)
I_{bf} = bolted fault current (kA)

Other Research. Research by Bingham and others[4,5,9] has yielded a slightly different result based primarily on empirical results. Using an experimental setup,[9] the researchers measured energy received from an electric arc at various distances. The arc was created using a 600-V source, and different configurations were used to simulate a completely open-air arc versus the so-called arc-in-a-box. Using these experiments, they developed two equations to model the amount of energy received.

$$E_{MA} = 5271D_A^{-1.9593}t_A(0.0016F^2 - 0.0076F + 0.8938) \tag{1.12}$$

$$E_{MB} = 1038.7D_B^{-1.4738}t_B(0.0093F^2 - 0.3453F + 5.975) \tag{1.13}$$

where E_{MA} = maximum open-arc incident energy (cal/cm²)
E_{MB} = maximum arc-in-a-box incident energy (cal/cm²)
D = distance from the arc electrodes in inches ($D \geq 45.7$ cm [18 in])
F = bolted fault current available in kA (16 to 50 kA range)
t_A = duration of the open-air arc (seconds)
t_B = duration of the arc-in-a-box (seconds)

Note that these equations were developed with three constraints:

1. System voltage = 600 V
2. System available fault current F – 16,000 A < F < 50,000 A
3. Electrode distance $D \geq 45.72$ cm (18 in)

Using the model previously described (see "Arcing Current"), the IEEE Std 1584-2002 calculates incident energy by first calculating the normalized incident energy. The normalized energy is calculated for an arc time of 0.2 second and a distance of 610 mm. The empirically developed formula for this calculation is shown in Eq. 1.14.

$$\log_{10}(E_n) = K_1 + K_2 + 1.081[\log_{10}(I_a)] + 0.0011G \tag{1.14}$$

where I_a = arc current calculated from Eq. 1.7
E_n = incident energy (J/cm^2) normalized for time and distance
K_1 = constant and is equal to −0.792 for open configurations and −0.555 for enclosed configurations
K_2 = constant and is equal to 0 for ungrounded and high-resistance systems and −0.113 for grounded systems
G = gap between conductors (mm)

After the $\log_{10} E_n$ is calculated from Eq. 1.14, Eq. 1.15 is used to calculate E_n, and Eq. 1.16 is used to calculate the actual incident energy.

$$E_n = 10^{\log_{10} E_n} \tag{1.15}$$

$$E = 4.184 C_f E_n \left(\frac{t}{0.2}\right)\left(\frac{610^x}{D^x}\right) \tag{1.16}$$

where E = incident energy (J/cm^2)
C_f =calculation factor equal to 1.0 for voltages above 1 kV and 1.5 for voltage equal to or below 1 kV
E_n = normalized incident energy calculated from Eqs. 1.14 and 1.15
t = arcing time in seconds
D = distance from the arc point to the exposed worker (mm)
x = distance exponent whose value is dependent on voltage (see IEEE Std 1584-2002, Table 4 for values of x)

Extensive research continues to be performed on the subject of incident arc energy. Practical applications used in the selection of protective equipment are covered in detail in Chap. 3.

Arc Burns

Arc burns are thermal in nature and, therefore, fall into one of four categories:

1. *First-degree burns.* First-degree burning causes painful trauma to the outer layer (epidermis) of the skin. Little permanent damage results from a first-degree burn because all the growth areas survive. Healing is usually prompt and leaves no scarring.
2. *Second-degree burns.* Second-degree burns are sometimes called *partial thickness* burns and cause damage to the epidermis and dermis layers. The second-degree burn is usually very painful and typically causes blistering of the skin; however, this type of burn will heal without skin grafts. Second-degree burns are often further classified into two subcategories:
   a. *Superficial:* This type of second-degree burn affects the epidermis and the upper layer of the dermis called the papillary dermis.
   b. *Deep:* The deep secondary burn involves the epidermis and extends through the papillary dermis into the reticular dermis.
3. *Third-degree burns.* Third-degree burns (sometimes called *full thickness* burns) destroy the epidermis and the dermis and usually cause damage to the subcutaneous layer. These types of burns result in complete destruction of the growth centers. If the burn is small, healing may

occur from the edges of the damaged area; however, extensive third-degree burns require skin grafting.

4. *Fourth-degree burns.* Fourth-degree burns cause severe damage to all three skin layers and extend into the muscle, nerve, tendon, ligament, vascular, organ, and bone tissues below the skin. Most severe electrical burns are fourth degree.

Refer to Chap. 9 for more detailed coverage of electrical arc trauma.

BLAST

When an electrical arc occurs, the vaporization of solid metal conductors into a gas is an exothermic or heat-releasing event that leads to rapid superheating of the surrounding air. The metallic vapor can be toxic exposure to respiratory or lung tissue because of its chemical composition and high heat. The superheating of the surrounding air can create a blast effect leading to acoustic trauma or tissue destruction from explosion.

This rapid expansion of the air creates a wavefront that can reach pressures of 100 to 200 lb per square foot (lb/ft2) (4.79 to 9.58 kPa). Such pressure is sufficient to explode switchgear, turn sheet metal into shrapnel, turn hardware into bullets, push over concrete walls, and propel molten metal and superheated plasma at extremely high velocities.

Blasts do not always occur. Sometimes an arc is not accompanied by a blast, but when it is, it can be lethal. Figure 1.5*a* to *c* shows physical evidence of the pressure exerted by an electric blast.

In Fig. 1.5*a* the interior of a medium-voltage cubicle can be seen. The severe scorching on the right-hand side of the interior and exterior of the cubicle is clear evidence of a significant arc-flash event. By looking closely at the cable terminations, evidence of the contact points for the electric arc as it occurred can be seen.

Figure 1.5*b* shows the aisle between the switchgear where the arc-flash occurred (right side) and the adjacent gear across the aisle. The metal panel in the aisle is the one that covered the cubicle where the arc-flash event occurred. Note that the fully secured panel was blown completely off the faulted switchgear across the aisle and smashed into the adjacent switchgear.

Figure 1.5*c* is a close-up view of the adjacent switchgear showing where the metal panel, seen in the lower right side of the photo, is dented and crumpled in the cubicle cover.

Taken together, these three photos clearly illustrate the following two key points:

- Workers may not assume that they are safe from electrical arc-flash events even though the access doors and panels are fully secured.
- Unless it is specially designed arc-resistant switchgear, metal-clad equipment will probably not withstand the explosive force of an electrical blast.

AFFECTED BODY PARTS

General

Detailed information on the medical aspects of electrical trauma is provided in Chap. 9. The following sections are for overview only.

Skin

Definition and Description. Skin is the outer layer that completely encloses and envelops the body. Each person's skin weighs about 4 lb, protects against bacterial invasion and physical injury of

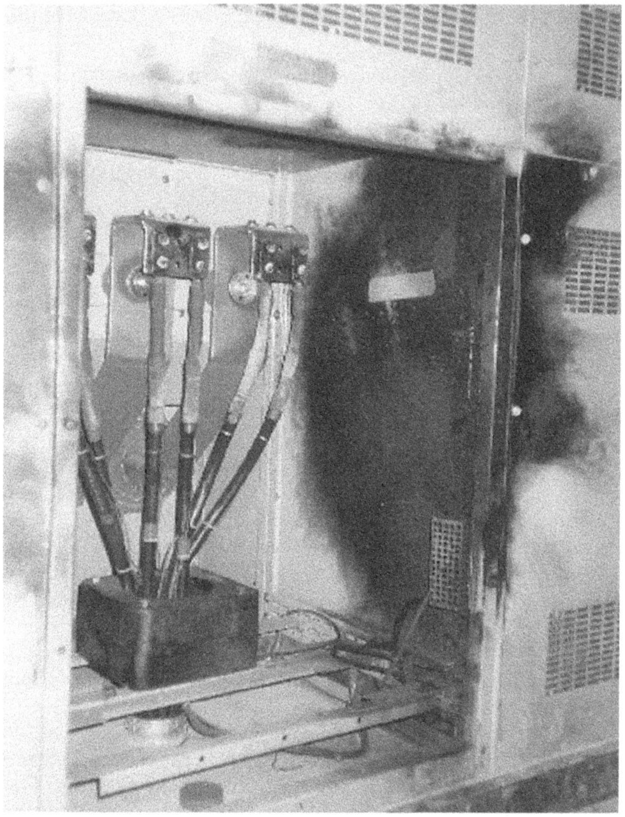

(a)

(b)

FIGURE 1.5 (a) Interior of a medium-voltage cubicle showing the results of an electrical arc and accompanying electrical blast; (b) external view of an aisle and adjacent switchgear for arc-flash event shown in Fig. 1.5*a*; (c) close-up view of adjacent switchgear showing metal covers damaged by impact of panel blown across aisle by arc-flash event.

(c)
FIGURE 1.5 *(Continued).*

underlying cells, and prevents water loss. It also provides the body with sensation, heat regulation, and excretion (sweat) and absorbs a few substances. There are about 20 million bacteria per square inch on the skin's surface as well as a forest of hairs, 50 sweat glands, 20 blood vessels, and more than 1000 nerve endings. Figure 1.6 is a cross section of the upper layers of skin tissue.

The main regions of importance for electrical purposes are the epidermis, the dermis, and the subcutaneous layers of the skin. For severe electrical burns, the underlying muscles and bone tissues may be involved as well.

The epidermis, the topmost layer of skin, is 0.1 to 1.5 millimeters thick; however, it is made up of five layers including the basal cell layer, the squamous cell layer, the stratum granulosum, the stratum lucidum, and the outermost layer called the *stratum corneum* or "horny layer."

The stratum corneum comprises 10 to 30 thin layers of dead cells that have been "pushed" up from the lower layers in the process of the normal growth process. It is called the horny layer because its cells are toughened like an animal's horn.

The stratum corneum is composed primarily of a protein material called *keratin*. Of all the skin layers, keratin exhibits the highest resistance to the passage of electricity. When areas of the epidermis such as the hands or feet are subjected to friction, the horny layer becomes thickened and toughened. Areas that are toughened in this manner are called corns or calluses.

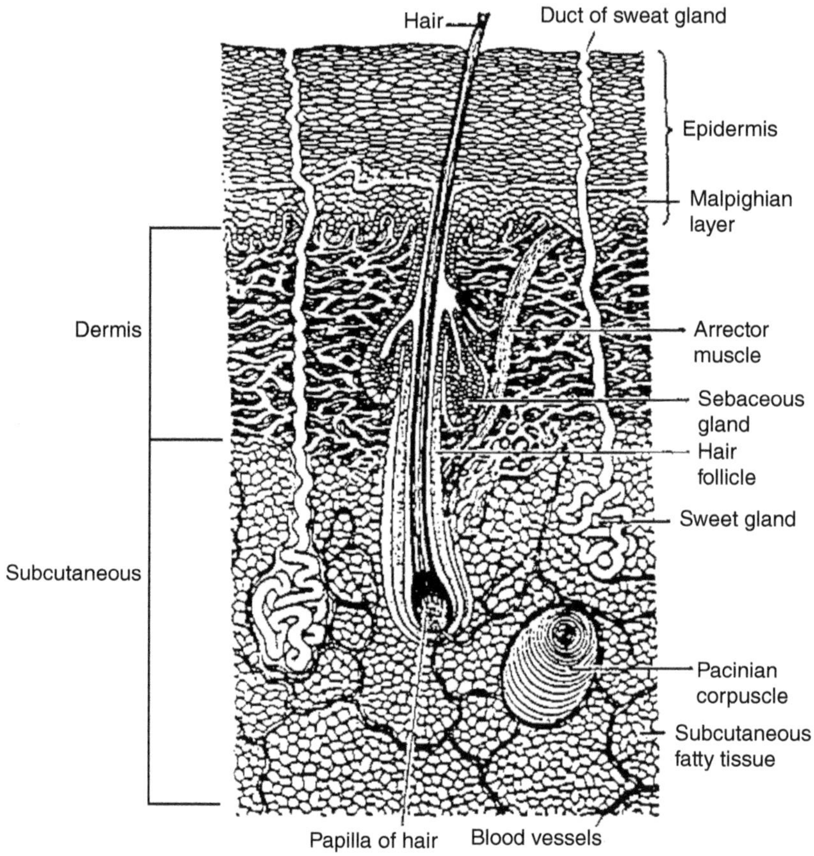

FIGURE 1.6 Typical skin cross section.

The sweat glands and the blood vessels have relatively low resistance to the passage of electricity and provide a means of electrical access to the wet and fatty inner tissues. Most of the electrical resistance exhibited by the human body (see Table 1.2) comes from the stratum corneum. Internal resistance is typically in the area of 200 Ω.

Effects on Current Flow. Since the body is a conductor of electricity, Ohm's law applies as it does to any other physical substance. The thicker the horny layer, the greater the skin's electrical resistance. Workers who have developed a thick stratum corneum have a much higher resistance to electricity than a child with an extremely thin layer. However, as Table 1.2 shows, even high skin resistance is not sufficient to protect workers from electric shock.

Skin resistance is also a function of how much skin area is in the circuit. Therefore, grasping a tool with the entire hand gives a much lower resistance than touching the tool with a finger. Also, any cut or abrasion penetrates the horny layer and significantly reduces the total resistance of the shock circuit. Moisture, especially sweat, greatly reduces the skin's resistance.

A remarkable thing occurs to the skin insulation when voltages above 400 V are applied. At these voltages the epidermis is punctured like any film insulation and only the low-resistance inner layers are left. This is a major consideration for the many 480-volt distribution systems commonly used

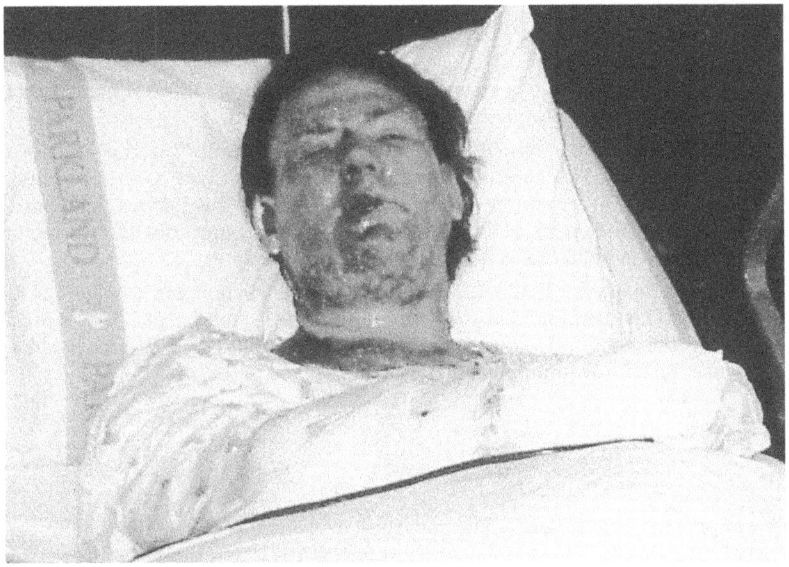

FIGURE 1.7 Thermal burns caused by high-voltage electric arc.

today. Note that the epidermis may not puncture, but if it does, the current flow increases and shock injury is worse.

Burns. Electrically caused burns can come from at least four different sources:

- Physical contact with conductors, tools, or other equipment that have been heated by the passage of electrical current flow. These types of burns are no different than burns received from any hot object.
- Current flow through the skin can cause burns from the I^2R energy. Since such burns occur from the inside-out, they are frequently third degree.
- Thermal or radiation burns are caused by the radiant energy of the electrical arc.
- Contact with superheated plasma caused by the vaporization of solid materials in the vicinity of an electrical arc.

Figure 1.7 shows a victim of burns caused by both the radiant arc energy and contact with the superheated plasma.

The Nervous System

Definition and Description. The nervous system comprises the electrical pathways that are used to communicate information from one part of the human body to another. To communicate, electric impulses are passed from one nerve to another. For example, the heart beats when an electric impulse is applied to the muscles that control it. If some other electric impulse is applied, the nervous system can become confused. If the current is high enough, the damage can be permanent.

Shock. As far as the nervous system is concerned, at least three major effects can occur when current flows through the body:

1. *Pain.* Pain is the nervous system's method of signaling injury. When current flows through the nerves, the familiar painful, tingling sensation can result.
2. *Loss of control.* An externally applied current can literally "swamp" the normal nervous system electric impulses. This condition is similar to electrical noise covering an information signal in a telemetering or other communications system. When this happens, the brain loses its ability to control the various parts of the body. This condition is most obvious during the electrical paralysis, or electrical hold, that is described later in this chapter.
3. *Permanent damage.* If allowed to persist, electric current can damage the nervous system permanently. This damage takes the form of destroyed neurons and/or synapses. Since the nervous system is the communications pathway used to control the muscles, such damage can result in loss of sensation and/or function depending on the type of injury.

Muscular System

Definition and Description. The muscular system provides motor action for the human body. When the nervous system stimulates the muscles with electric impulses, the muscles contract to move the body and perform physical activity. Skeletal muscles are either flexor or extensor muscles. When a flexor muscle, such as the biceps, contracts, it bends or closes the joint. When an extensor muscle, such as the triceps, contracts, it unbends or opens the joint.

The heart and pulmonary system are also muscle related, and they will be covered in a later section.

Shock. Electrical shock can affect muscles in at least three significant ways:

1. *Reflex action.* Muscular contractions are caused by electric impulses. Normally these impulses come from the nervous system. When an externally induced current flows through a muscle, it can cause the muscle to contract, perhaps violently. This contraction can cause workers to fall off ladders or smash into steel doors or other structures.
2. *Electrical paralysis.* Current magnitudes in excess of 10 mA may be sufficient to block the nervous system signals to the muscular system. Thus, when such an external current is flowing through the body, the victim may be unable to control his or her muscles. This means that the victim cannot let go—he or she is caught in an electrical hold. As the current continues, the heating and burning action can lower the path resistance and cause an increase in the current. If the current is not cut off or if the victim is not freed from the circuit, death will occur.
3. *Permanent damage.* If the current is high enough, the muscle tissue can be destroyed by burning. Even very low currents will cause tissue destruction if they last long enough. Because such burning destroys the growth areas in tissue, the damage can be extremely slow to heal. Physical therapy and other extraordinary methods may be required to restore muscular function.

The Heart

Definition and Description. The heart is a fist-size pump that beats more the 2.5 billion times in a 75-year lifetime. Figure 1.8 shows the structural layout of the heart. The atria and ventricles work together, alternately contracting and relaxing to pump blood through your heart. The electrical system of your heart is the power source that makes this possible. Normally a heartbeat starts in the sinus

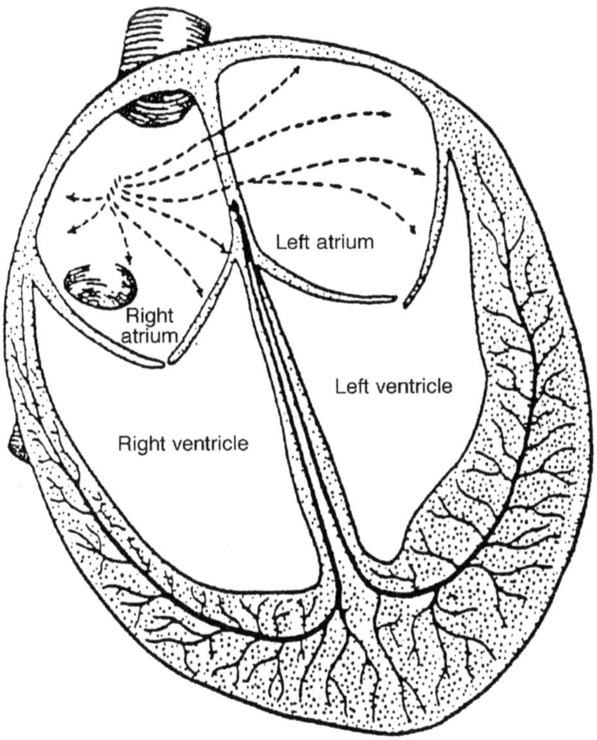

FIGURE 1.8 The heart.

node, travels at approximately 2.1 m/s (7 ft/s) through the AV node, "HIS bundle," and right and left bundle branches. The resulting contraction sends blood flowing from the heart.

This sequence occurs with every beat (usually 60 to 100 times per minute). If the path is interrupted for any reason, even for a few minutes, changes in the heart rate and rhythm occur that can be fatal.

Shock. When the heart's electrical system is disturbed for any reason, such as an outside current from an electric shock, changes in the heart's rate and rhythm occur. Such disruptions result in a large percentage of heart deaths.

The electric impulses in the heart must be coordinated to give a smooth, rhythmic beat. An outside current of as little as 60 to 75 mA can disturb the nerve impulses so that there is no longer a smooth, timed heartbeat. Instead, the heart fibrillates—that is, it beats in a rapid, uncoordinated manner. When a heart is fibrillating, it flutters uselessly. Prolonged exposure to an outside current exceeding 75 mA is likely to result in death.

Like any muscle, the heart will become paralyzed if the current flowing through it is of sufficient magnitude. Oddly, paralysis of the heart is not often fatal if the current is removed quickly enough. In fact, such paralysis is used to an advantage in defibrillators. A defibrillator intentionally applies heart-paralyzing current. When the current is removed, the heart is in a relaxed state ready for the next signal. Frequently the heart restarts.

Burns. Any internal organ, such as the heart, can be burned by current flows in excess of 5 A. Such burns are often fatal.

The Pulmonary System

Definition and Description. With the exception of the heart, the pulmonary system is the most critical to human life. If breathing stops, which can happen with as little as 30 mA, brain damage can occur in as little as four minutes. When the lower diaphragm moves down, it creates a partial vacuum in the chest chamber. This in turn draws air into the sacs in the lungs. The oxygen is then passed to the bloodstream through the tiny capillaries. At the same time, carbon dioxide is returned to the air in the lungs. When the lower diaphragm moves up, the air is forced out of the lungs, thus completing the breathing cycle.

Shock. Current flow through the midsection of the body can disrupt the nervous system impulses that regulate the breathing function. This disruption can take the form of irregular, sporadic breathing, or—if the current flow is sufficient—the pulmonary system may be paralyzed altogether. When such stoppage occurs, first aid is often required.

SUMMARY OF CAUSES—INJURY AND DEATH

Shock Effect

Table 1.4 summarizes the effects that electric shocks of varying amounts of current will have on a 68-kg (150-lb) person. Note that these effects are only approximate and vary among individuals.

Arc-Flash Effect

Tables 1.6 and 1.8 summarize the effects that electric arc-flash energy have on personnel. Note that these effects are still being studied and the full range of injury potential is not yet fully evaluated and documented.

Causes of Injury

Injury from electrical hazard can come from both direct and indirect sources:

- The reflex action caused by the passage of current flow can cause falls resulting in cuts, abrasions, or broken limbs.
- Nerve damage from shock or burns can cause loss of motor function, tingling, and/or paralysis.
- Burns, both thermal- and current-induced, can cause extremely long duration and intensely painful suffering. Third- and fourth-degree burns may require skin grafting to heal.
- The light intensity, molten metal, and/or burns to the eyes can cause blindness.
- The concussion of a blast can cause partial or complete loss of hearing.
- Current-induced burns to internal organs can cause organ dysfunction.
- The superheated plasma may be inhaled, causing severe internal burns.
- Metal vapors may be inhaled, filling the lungs with toxic residues.

Causes of Death

If the electrical injury is severe enough, death can result.

- An electric shock–induced fall can cause fatal physical injuries.
- When the skin is severely burnt, large quantities of liquid are brought to the burnt areas to aid in the healing process. This creates a stress on the renal system and could result in kidney failure.

- Severe trauma from massive burns can cause a general systemic failure.
- Burnt internal organs can shut down—causing death. Thus, the more critical the organ that is burnt, the higher the possibility of death.
- The pressure front from the blast can cause severe injury to the lungs, called *blast-lung,* resulting in death.
- Heart failure can result from fibrillation and/or paralysis.

PROTECTIVE STRATEGIES

The types of strategies that may be employed to protect from each of the three electrical hazards are remarkably similar. Table 1.8 summarizes the types of protective strategies that may be used. Note that the information given in Table 1.8 is general. Specific equipment and procedures are covered in Chaps. 3 and 4.

Be aware that any given strategy may not be applicable in a given situation. For example,

- When troubleshooting equipment, de-energization may not be possible.
- De-energization may create an additional, unacceptable hazard. For example, if de-energization shuts down ventilation equipment in a hazardous area, workers may opt for working with energized equipment.
- Shutdown of an entire continuous process plan to work on or around one small auxiliary circuit may not be economically feasible.

TABLE 1.8 Equipment and Procedural Strategies for Protection from the Three Electrical Hazards (See Chaps. 3 and 4 for detailed information.)

Hazard	Equipment strategy	Procedural strategy for all three hazards
Shock	• Rubber insulating equipment including gloves with leather protectors, sleeves, mats, blankets, line hose, and covers. • Insulated tools when working near energized conductors.	• De-energize all circuits and conductors in the immediate work area. • Develop and follow a lockout/tagout procedure. • Maintain a safe working distance from all energized equipment and conductors.
Arc	• Approved flash/flame-resistant work clothing. • Approved flash suits when performing work with a high risk of arcing. • Use hot sticks to keep as much distance as possible. • Wear eye protection. • Wear rubber gloves with leather protectors and/or other flashproof gloves.	• Use all specified safety equipment. • Follow all safety procedures and requirements. • Carefully inspect all equipment before placing it into service. This includes tools, test equipment, electrical distribution equipment, and safety equipment.
Blast	• Approved flash/flame-resistant work clothing. This may protect from splashed molten material. • Approved flash suits when performing work with a high risk of arcing. This may protect from splashed molten material. • Wear face shields.	• Make certain that all nonenergized equipment is properly grounded. This applies to both normal system grounding and temporary safety grounds. • Design or redesign systems to be intrinsically safe.

If electrical equipment cannot be de-energized, a hierarchy of electrical safe work practices and procedures must be implemented as follows:

- Document the reason(s) for having to work on or near energized electrical conductors or circuit parts. Identify that de-energizing is infeasible.
- Obtain responsible manager's (employer's designated person) signature, verifying knowledge and recognizing the need to work on or near exposed energized electrical conductors or circuit parts.
- Perform a job briefing that includes a hazard-risk analysis.
- Prepare and follow a documented plan of work processes and tasks.
- If worker(s) must enter closer to the potential arc source(s) than the flash protection boundary, then ensure that the worker is a "qualified person" and is wearing appropriate personal protective clothing (PPC) and using other personal protective equipment (PPE).

REFERENCES

1. Lee, Raphael C., "Injury by Electrical Forces: Pathophysiology, Manifestations and Therapy," *Current Problems in Surgery*, 34 (9): 1997.
2. Morse Michael S., *How Much Damage Can a Low Voltage Shock Do?—Traditional Theory, Modern Theory, and Pervasive Myths*, IEEE paper ESW2008-07 presented at the IEEE Electrical Safety Workshop in Dallas, 2008.
3. Institute of Electrical and Electronics Engineers (IEEE), *Standard Dictionary of Electrical and Electronics Terms*, ANSI/IEEE Std 100-1988, p. 240.
4. Lee, Ralph H., "The Other Electrical Hazard: Electric Arc Blast Burns," *IEEE Trans. Industrial Applications*, 1A-18 (3): p. 246, 1982.
5. Neal, Thomas E., Bingham, Allen H., and Doughty, Richard I., *Protective Clothing Guidelines for Electric Arc Exposure*, IEEE Paper PCIC-96-34 presented at the Petroleum Chemical Industry Conference in Philadelphia, 1996.
6. Doughty, Richard L., Neal, Thomas E., Dear, Terrence A., and Bingham, Allen H., *Testing Update on Protective Clothing & Equipment for Electric Arc Exposure*, IEEE Paper PCIC-97-35, 1997.
7. Institute of Electrical and Electronic Engineers (IEEE), *IEEE Guide for Performing Arc-Flash Calculations*, IEEE Std 1584-2002.
8. Blackburn, J. Lewis, *Protective Relaying Principles and Applications*, 2nd ed., Marcel Dekker, New York, 1998.
9. Doughty, Richard L., Neal, Thomas E., and Floyd, H. Landis, *Predicting Incident Energy to Better Manage the Electric Arc Hazard on 600-V Power Distribution Systems*, IEEE Paper PCIC-98-36, 2000.

CHAPTER 2
BASIC PHYSICS OF ELECTRICAL HAZARDS

INTRODUCTION

Electrical phenomena are based on fundamental principles of physics. The same foundational physical principles serve as the basis for recognition of electrical hazards, the implementation of safety precautions, and requirements for safety equipment used around electricity. For the reader with sufficient background in math and science, this chapter will help to provide a scientific baseline to help in the struggle to achieve electrical safety.

ELECTROMAGNETISM

Introduction

This chapter briefly describes electromagnetism and explains how it affects our lives and electrical safety. Electromagnetism is one of the four fundamental forces of nature. Playing a crucial role in everyday life, electromagnetism is the foundation of electrical phenomena and, by extension, electrical safety.

The Four Fundamental Forces (Interactions) of Nature

Table 2.1 lists a few of the important characteristics of the four fundamental forces.

The strong and weak forces play no known role in the hazards of electricity. Only the electromagnetic force and gravity (to a lesser extent) affect electrical safety.

Gravity. Any two masses have a force of attraction given by

$$F = G\frac{m_1 m_2}{r^2} \tag{2.1}$$

where F = force between the two masses in newtons (N)
m_1 and m_2 = masses of the two objects in kilograms (kg)
r = distance between the objects in meters (m)
G = gravitational constant = $6.674 \times 10^{-11}\,\mathrm{N \cdot m^2/kg^2}$

Gravity is the centripetal force that keeps orbiting objects from continuing on in a straight line. It is also the force that causes a rock to fall when it is dropped. And—of greatest importance to the topics in this handbook—gravity is the force that causes workers to hit the ground when they fall from a ladder or other elevated location.

Because mass is always a positive quantity, gravity is positive and is a force of attraction.

TABLE 2.1 The Four Fundamental Forces and Their Characteristics

Name	Relative strength	Range (meters)	Particle	Comments
Strong[1]	1	1 to 2×10^{-15}	Gluon p (nucleon)	Binds quarks
Residual strong[2]	1			Binds nucleons (protons and neutrons)
Weak	10^{-6}	10^{-18}	Vector bosons W^+, W^-, Z_0	Radioactive decay
Gravity	6×10^{-39}	Infinite	Not yet discovered Posited to be the *graviton*	Only significant for large masses
Electromagnetic	7.3×10^{-3}	Infinite	Photon	Responsible for all large-scale interactions not caused by gravity

[1] The strong force is also called the color force.
[2] *Residual strong force* is that portion of the strong force that extends beyond the nucleons and binds them together.

Electromagnetic Force. The electromagnetic force can be thought of as two different aspects—electrostatic force and magnetostatic force.

The *electrostatic force* is the force between two electrically charged objects. Electric charge is measured in coulombs and can be either positive or negative. Two like charges will repel each other, and two unlike charges will attract each other. This is why:

- Protons attract electrons.
- Protons repel other protons.
- Electrons repel other electrons.

Charles-Augustin de Coulomb (1736–1806) developed the formula (named Coulomb's law) for calculating the strength of the electrostatic force.

$$F = k\frac{q_1 q_2}{r^2} \tag{2.2}$$

where F = force between the two charged objects in newtons (N)
q_1 and q_2 = charges on the two objects in coulombs (C)
r = distance between the objects in meters (m)
k = Coulomb's constant = 9×10^9 Nm^2/C^2

Unlike the force of gravity, electrostatic force can be either positive or negative. This means that a positive electrostatic force (two like charges) is a force of repulsion. For the same reason, a negative electrostatic force is a force of attraction.

The *magnetostatic force* is the force between two magnetic poles. Magnetic poles are similar to electric charges in that they come in two varieties—north poles and south poles. Hundreds of years ago, it was noted that one end of a magnet *pointed* toward the earth's North Pole. This end of the magnet is called the *north seeking pole,* which has been reduced to just *north pole.*

Magnetic poles have another similarity to charged particles. Like poles attract and unlike poles repel. This is why:

- North magnetic poles attract south magnetic poles.
- North magnetic poles repel other north magnetic poles.
- South magnetic poles repel other south magnetic poles.

TABLE 2.2 Effects of Electromagnetic Force

• Matter is mostly space. The electrostatic force between electrons is so great that it keeps separate objects from falling through each other. Gravity attracts us to the earth; the electromagnetic force keeps us from falling through the earth. Almost all physical experiences that we experience are caused by electromagnetic force. This includes touch, pressure, the motion of trees caused by wind, the light that we see, and many others.
• Electrical arcs occur when the outer electrons in air molecules become superexcited by either heat or strong electrostatic force (i.e., high-voltage gradient). These electrons are pushed out of their normal orbits and cause the air to become a conductor. This process releases great amounts of heat and light. When an electrical arc occurs, the released electrons are extremely energetic and cause the plasma to expand by pushing against each other. The path of the plasma flow is influenced by electromagnetic force. Electromagnetic fields can act on water molecules in the body in a detectable manner, which is the basis for medical diagnostic imaging with a magnetic resonance imaging (MRI) scanner.
• Electromagnetic fields can induce electric field effects within the body, which explains in part the effects of lightning on the body's nervous and cardiac (heart) systems.

In his research, Coulomb determined that the force between two magnetic poles could be calculated by a formula that is similar to the force of gravity and the electrostatic force.

$$F = \frac{\mu(q_{m1}q_{m2})}{4\pi r^2} \tag{2.3}$$

where F = force between the two magnetic poles in newtons (N)
q_{m1} and q_{m2} = strengths of each pole in ampere meters (Am)
r = distance between the objects in meters (m)
μ = permeability of the medium in which the poles are located (N/A^2)

Positive force is a repelling force and negative force is attractive. Equation 2.3 is not used because magnetic poles cannot be located as precisely as electrical charges. However, it is a good comparison to Eq. 2.2.

The Effects of Electromagnetic Force. The effect that gravity has on everyday life is well known and understood. Gravity holds humans and other objects firmly onto the surface of Earth. Gravity causes apples to fall from trees. It keeps the moon in orbit around Earth and Earth in orbit around the sun. Gravity causes dying stars to collapse into white dwarfs, neutron stars, or black holes. How about electromagnetic force? Table 2.2 lists a few of the effects of electromagnetic force on our daily lives.

The Electromagnetic Spectrum

The frequency of an electromagnetic wave determines its nature and how it affects us. Radio waves, gamma waves, infrared light, and microwave radiation are all types of electromagnetic radiation and differ only by the frequency of their oscillation. Consider Table 2.3, for example.

Sometimes the wavelength of the waves is used rather than the frequency. The relationship is

$$\lambda = c/f$$

where λ = wavelength in meters
f = frequency in hertz (cycles per second)
c = speed of light in free space (2.9979×10^8 m/s)

Figure 2.1 shows a graphic representation of the electromagnetic spectrum and how the various frequencies interact with our daily lives.

TABLE 2.3 Comparison of Various Frequencies in the Electromagnetic Spectrum

Type	Frequency range (Hz)*	Description
Radio waves	$3 \times 10^3 – 300 \times 10^9$	With the exceptions of the microwave frequencies (>1 $\times 10^9$ Hz), radio waves pass harmlessly through the human body unless they are very high energy levels. Radio waves are used for all types of communications applications including line of sight, long distance, wireless data networks, and a host of other applications.
Infrared light	$1 \times 10^{12} – 100 \times 10^{12}$	Perceived as heat when it falls on human tissue. Over time, infrared light falling on the eye can cause a slow, irreversible opacity of the lens. Is useful in photographic astronomy by allowing warm bodies not visible in the light spectrum to appear.
Visible light	$400 \times 10^{12} – 790 \times 10^{12}$	Reflected by solid objects. Captured by the optic nerve allowing us to see objects. Frequencies near ultraviolet in high concentrations pass through the cornea and dazzle or damage the optic nerve complex.
Ultraviolet light	$7.5 \times 10^{14} – 3 \times 10^{16}$	Help the body to produce vitamin D. Used as a therapy for psoriasis. Used to disinfect fish tanks and sterilize medical equipment. Cause sunburn. Can be extremely dangerous if not controlled. Cause photoaging in high doses. Are the cause of tissue damage in the human cornea leading to a condition known as *arc eye*.
X-rays†	$3 \times 10^{16} – 3 \times 10^{19}$	Created by electrons' energy state change when outside of the nucleus. Will pass through human soft tissue with relatively little reflection. Reflected by bone and other types of dense tissues.
Gamma rays†	Greater than 3×10^{19}	Created by decay of high energy states in atomic nuclei. Fairly easily absorbed by human tissue. Absorption causes extreme damage to living tissue when absorbed.

*Frequency ranges are approximate.
†The frequencies of X-rays and gamma rays may overlap. They are distinguished more by the cause than the frequency.

ELECTRICAL PROPERTIES OF MATERIALS

Conductors

A conductor is a material that presents very little resistance to current flow. Conductors have resistivities within the range of 10^{-6} to 10^{-2} ohm meters (ohm m) (conductivity of 10^2 to 10^6 siemens per meter (S/m)). Conductors may be made of silver, copper, aluminum, gold, or many other metals. Silver is the best naturally occurring conductor.

Conductors are very conductive because their valence electrons are held to their atoms relatively loosely, so it takes less electromagnetic force to move them.

Metals. A variety of metals are used as electrical conductors. The four most common in order of their conductivity are silver, copper, gold, and aluminum, based on a temperature of 20°C. The conductivity values of these metals are derived from the *CRC Handbook of Chemistry and Physics*, 95th edition.

Silver: Silver has an electrical conductivity of 6.30×10^7 S/m. This makes it the most conductive of all naturally occurring metals. Silver is a soft, ductile, and malleable metal and is used for making very low-resistance contacts in electrical circuits. Silver is relatively expensive and somewhat rare. Silver is chemically active, and its various compounds (silver-oxide for example) are nonconductors.

Copper: The electrical conductivity of copper (5.96×10^7 S/m) makes it almost as conductive as silver. The low cost and relative abundance of this metal have made it the most widely used electrical conductor in the world. It is also malleable and ductile. It is somewhat less chemically active than silver; however, over time it, too, will corrode.

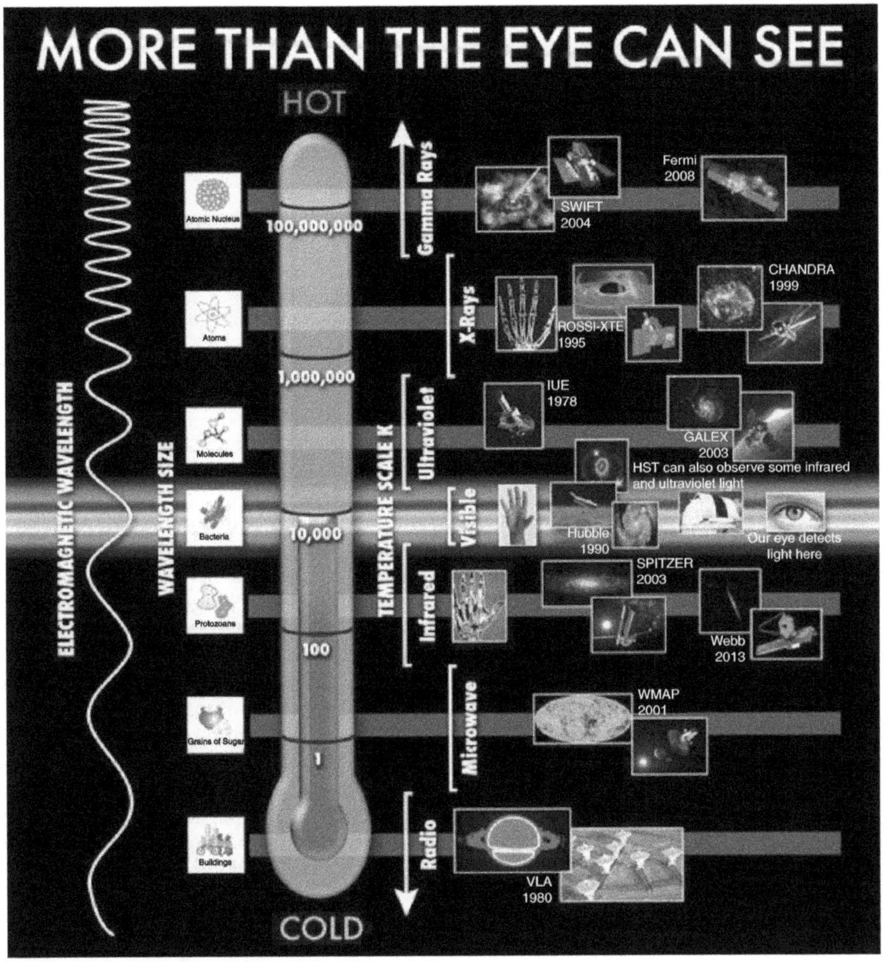

FIGURE 2.1 The electromagnetic spectrum.

Gold: Gold is not as conductive (4.52×10^7 S/m) as silver or copper, and it is very expensive relative to silver, copper, and aluminum. Because gold is nonreactive with most other substances, it does not corrode and its conductivity is generally not affected by corrosive or chemically active atmospheres. It is commonly used for contacts and connectors on low-voltage/low-current connections, such as those found on electronic printed circuit boards.

Aluminum: With a conductivity of only 3.77×10^7 S/m, aluminum seems like a poor choice for an electrical conductor when compared with the other four. In addition, aluminum forms a very strong, nonconductive oxide when exposed to air. It also has a tendency to *cold-flow* when it is compressed improperly. However, because it is much lighter and generally less expensive than copper, it is used successfully as a conductor by electric utilities, primarily in long-distance transmission lines. An aluminum wire can be made as conductive as a copper wire by increasing its cross-sectional area. Even with the added material in aluminum wire construction (material such as steel messenger wires for added strength), it is less expensive to use than copper wire.

Aluminum was used extensively in industrial and residential wiring between the late 1950s and early 1970s, leading to many house fires and industrial fires attributed to improper application in these locations and installations. Aluminum should not be installed unless proper installation methods are used for safe operational conditions. Aluminum is no longer allowed for residential installations.

TABLE 2.4 Material Characteristics of Common Conductors (Metal and Nonmetal)

Type of characteristic	Metals	Nonmetals
Chemical	• Form cations • Reactive as basic oxides • Reducing agents	Form anions Reactive as acidic oxides Oxidizing agents
Physical	• High electrical conductivity • High thermal conductivity • High luster • Malleable under stress • High to moderate density	Poor electrical conductivity Poor thermal conductivity Dull luster in solid form Brittle under stress Low to moderate density
Mechanical	• Very ductile—can be stretched into wire • Reversible elasticity	Poor ductility

Table 2.4 shows the general chemical, physical, and mechanical characteristics of many electrical conductors.

The Human Body. While we may not think of our bodies as behaving like a roll of copper or aluminum electrical cable, as is true for these metals, the human body is a good conductor of electrical charges. In copper and aluminum, electrical charge is carried by electrons. In the human body, charge moves via ions like sodium, potassium, and chloride. Charge movement can cause heating.

Some tissues in the body are especially adept at ion charge movements. We call these tissues *excitable*. In particular, our sense organs, muscles, nervous system, and heart are excitable tissues. Ion charge movement in the body can affect the actual structures of individual tissue cells in very short periods of time, measured in thousandths of a second, or milliseconds.

Charge movement in metals like copper or aluminum or in the body is also called *current flow*. The words *movement* or *flow* suggest a "path" where the charge might go. The basic principle that governs current flow of electricity is the tendency for the movement of charge to follow all paths, especially the path of least resistance. Thus, if charged particles are given multiple paths, the amount of charge flow through each path will be inversely proportional to the resistance of that path. In electrical terms, resistance is a measurable characteristic of engineered and natural materials, including the different tissues in the body.

While the force that causes electrical current flow in engineered operating equipment is usually measured in volts, hundreds of volts, or thousands of volts (kilovolts), in the human body the force of electrical current flow is typically measured in fractions of a volt, in the order of a certain number of thousandths of a volt.

Whereas we typically see that in an electrical installation, like a range light, the force of electrical current flow needed to illuminate a lightbulb may be more than 100 volts, in the human body basic excitation of a nerve cell membrane that might be experienced as a "tingle" may be as little as 15 to 20 mV or 15 to 20 thousandths of 1 volt.

Nonconductors

In an insulator, the valence electrons are held very tightly. Thus an insulator presents very high opposition to electric current flow, a characteristic referred to as *nonconductive*. Resistivities of insulators range from 10^8 to 10^{20} ohm m (conductivity of 10^{-20} to 10^{-8} S/m). Most insulators are nonmetals. Following are resistivities of naturally occurring insulators:

- Glass: 1.00×10^{11} ohm m
- Mica: 1×10^{13} ohm m
- Fused quartz: 7.5×10^{17} ohm m

Most insulators used in the early history of electricity were made of naturally occurring materials such as rubber and those items just mentioned. As electrical knowledge grew, complex insulators such as varnish cambric or oil-impregnated paper (usually lead covered) were developed. In modern times, synthetic materials, such as polyethylene, provide excellent electrical characteristics. They also are more impervious to corrosion or other chemical reaction and are easy to extrude into cable.

Of course, the safety application of insulators is to isolate workers from the inner metal conductors, reducing the risk of electric shock and greatly reducing the possibility of exposure to electric arc.

Insulation is inherently safe unless it is improperly applied, used in the wrong environment, or physically damaged by external or electrical action.

PHYSICS CONSIDERATIONS IN ELECTRICAL FAULT CONDITIONS

Risks

Electric fault conditions are widely recognized and well-known risks of energized electrical systems.

Electrical malfunction was the fire cause for 6.1 percent of reported building fires and fires with losses and 9.1 percent of reported building fires and fires with injuries, according to 2006 data compiled by the U.S. Fire Administration in a reference on electrical malfunction as a cause of nonresidential fires.

The recognition of fire hazard in electrical systems operations dates to the late 19th and early 20th Centuries, when the electrification of cities and optimization of fire safety evolved through the organizations that ultimately formed the National Fire Protection Association (NFPA). Today, the NFPA publishes the National Electrical Code, or NEC, and the *NFPA 70E, Standard for Electrical Safety in the Workplace*. NEC and NFPA 70E code and standards implementation is the bedrock foundation of U.S. risk reduction for electrical hazards. During delivery of electrical energy in an electrical power system, electrical current is expected by design to flow within its intended path via metallic conductors. When electrical current flows outside of its intended path, an electrical fault is said to occur.

Bolted Fault

When electrical current flows outside its intended path along solidly connected metallic conductors, the electrical fault is referred to as a "bolted fault." The ease of the electrical current flow is a function of the conductor's material properties as discussed earlier in this chapter.

Under a bolted fault scenario, the presence of current is an electrical shock hazard. As is discussed more fully in the following chapters, to mitigate the electrical shock hazard, routine establishment of an electrically safe working condition is essential. A written electrical safety program, lockout-tagout policies and procedures, and use of barriers and electrically protective personal protection are all required for successful mitigation.

When an employee comes into unprotected contact with a bolted fault or an unrecognized energized conductor like a metal wire in a light installation, electrical energy can flow from the conductor across the skin into the human body. At the body's contact point with an energized electrical conductor, the physical chemistry is such that an "electrochemical" reaction occurs as electron charge movement converts to ion charge movement. This reaction may leave no visible mark on the outer body. Or this reaction can be accompanied by a vaporizing of the outer layer of skin so that a wound or puncture site is seen.

Whether skin change at the body site of contact with electricity is visible to an onlooker depends on multiple factors, including the following:

- Voltage at the point of contact
- The duration of the current flow
- The completeness of the coupling or connection between the body and current source during current flow

- The environmental conditions at the time of coupling, like humidity, temperature, presence of contaminants, or fire ignition
- Surrounding presence of other conductors

Once electricity flows into the body, it continues to flow through all available paths, with most of the current flowing through the paths with the lowest resistance. This means that current flow path may "exit" or "reconnect" with other conductors that the body is touching, like a floor or nearby walls or rails, again depending on multiple factors as suggested in the previous paragraph.

Arcing Fault

Description. When electrical current flows outside its intended path, and not along metallic conductors, but instead across space, the fault condition is referred to as an electric arc or arcing fault. An electric arc propagating into work spaces can harm nearby employees by creating multiple physical forms of energy that transfer to human tissues, leading to negative health effects. This is why an electric arc occurring as a fault in an electrical power system is an occupational as well as operational hazard.

Available Energy. The total amount of energy of all types released by an electric arc is limited to no more than some fraction of the energy supplied by the power system. All theoretical models of arc energy start by determining the maximum power transfer from the power system to the electrical arc. Figure 2.2 shows the equivalent circuit that is used to determine the maximum power that the system can transfer to the arc.

The figure shows an equivalent circuit of a power system:

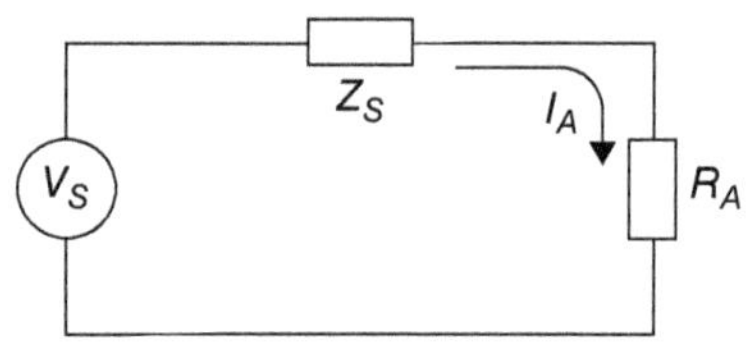

FIGURE 2.2 Simplified diagram of an arcing power system.

where V_S = power system voltage in volts
Z_S = power system equivalent impedance in ohms
R_A = arc resistance in ohms
I_A = arc current in amperes

Since arcs are largely resistive, R_A is modeled as a pure resistance, and Z_S is modeled as complex impedance because power systems supplies are highly inductive with X/R ratios ranging from 2 to 1 in lower-voltage systems to as high as 40 to 1 in higher-voltage systems.

In rectangular form:

$$\bar{Z}_S = R_S + jX_S \tag{2.4}$$

where

$$j = \sqrt{-1} = 1\angle 90° \tag{2.5}$$

From electrical theory, we know that

$$\bar{S} = (\overline{VI}^*) = \bar{Z} \times \bar{I} \times \bar{I}^* \tag{2.6}$$

where S = apparent power in volt-amperes
V = complex voltage drop of the circuit in volts
I = complex current in the circuit in amperes
I^* = complex conjugate of the current flow, also in amperes

From Ohm's law, the current in the circuit of Fig. 2.2 is given as

$$\bar{I}_A = \frac{\bar{E}_S}{R_A + R_S + jX_S} \tag{2.7}$$

Since $\bar{E}_S$ is the supply voltage and the phase angle reference, $\bar{E}_S = \left|\bar{E}_S\right|\angle 0$, which we will treat as a real (noncomplex) number.

Noting this and rearranging:

$$\bar{I}_A = \frac{E_S}{(R_A + R_S) + jX_S} \times \frac{(R_A + R_S) - jX_S}{(R_A + R_S) - jX_S} = \frac{E_S(R_A + R_S) - jE_S X_S}{(R_A + R_S)^2 + X_S^2} \tag{2.8}$$

By definition, the conjugate of a complex number $a + jb$ is equal to $a - jb$.

$$\therefore \quad \bar{I}_A^* = \frac{E_S(R_A + R_S) + jE_S X_S}{(R_A + R_S)^2 + X_S^2} \tag{2.9}$$

And

$$\bar{I}_A \times \bar{I}_A^* = \left[\frac{E_S(R_A + R_S) - jE_S X_S}{(R_A + R_S)^2 + X_S^2}\right] \times \left[\frac{E_S(R_A + R_S) + jE_S X_S}{(R_A + R_S)^2 + X_S^2}\right] = \frac{[E_S(R_A + R_S)]^2 + (E_S X_S)^2}{\left[(R_A + R_S)^2 + X_S^2\right]^2} \tag{2.10}$$

Expanding

$$\bar{I}_A \times \bar{I}_A^* = \frac{[E_S(R_A + R_S)]^2 + (E_S X_S)^2}{\left[(R_A + R_S)^2 + X_S^2\right]^2} = E_S^2 \times \frac{\left[(R_A + R_S)^2 + X_S^2\right]}{\left[(R_A + R_S)^2 + X_S^2\right] \times \left[(R_A + R_S)^2 + X_S^2\right]} \tag{2.11}$$

Which simplifies to

$$\bar{I}_A \times \bar{I}_A^* = \frac{E_S^2}{\left[(R_A + R_S)^2 + X_S^2\right]} \tag{2.12}$$

And from Eq. 2.6

$$\bar{S} = \bar{I}_A \times \bar{I}_A^* \times \bar{Z}_S = \frac{E_S^2}{\left[(R_A + R_S)^2 + X_S^2\right]} \times (R_A + R_S + jX_S) \tag{2.13}$$

Since we are trying to maximize the power transferred to the arc, we can focus on only the arc energy term, which yields

$$P_A = \frac{E_S^2 R_A}{\left[(R_A + R_S)^2 + X_S^2\right]} \tag{2.14}$$

Note the switch to P (real power) instead of S. Since R_A is resistance, there will be no reactive power present.

We wish to determine the value of R_A that will allow the maximum power transfer. To do this, we must perform a partial derivative of P_A with respect to R_A and set the result equal to zero. The partial derivative of P_A is calculated as

$$\frac{\partial P_A}{\partial R_A} = \frac{E_S^2(R_S^2 + X_S^2 - R_A^2)}{\left(R_S^2 + 2R_S R_A + X_S^2 + R_A^2\right)^2} \tag{2.15}$$

This expression will be zero when the numerator is zero. Thus

$$R_S^2 + X_S^2 - R_A^2 = 0 \Rightarrow R_A^2 = R_S^2 + X_S^2 \Rightarrow R_A = \pm\sqrt{R_S^2 + X_S^2} \tag{2.16}$$

The maximum power will be transferred to the arc when the magnitude of the system impedance is equal to the magnitude of the arc resistance.

Two clarifying points need to be made:

1. Since a negative resistance is meaningless in this context, the negative solution in Eq. 2.16 $\left(R_A = -\sqrt{R_S^2 + X_S^2}\right)$ need not be considered.
2. Since P_A is zero for both $R_A = 0$ and $R_A = \infty$, the solution in Eq. 2.16 must be a maximum not a minimum.

Some form of this procedure has been used by all researchers in developing their theoretical calculations of maximum incident energy. Unfortunately, R_A is rarely equal to $\sqrt{R_S^2 + X_S^2}$. This causes all theoretical calculations to yield unrealistically high incident energies. This has been shown by the empirical tests that have been and are being performed in the industry.

One of the authors of this handbook is performing some statistical analysis and research to allow better theoretical calculations. More will be reported on the research during the next few years. When final, the results will be published in the next edition of the *Electrical Safety Handbook*.

Review of Foundational Approaches to Interpreting Arcing Phenomena

In presenting a schematic view of a simplified energy balance of an arc fault in a cubicle, in 1982, F. Lutz and G. Pietsch noted that although the physical processes of energy transfer from the arc to surroundings were well known, these processes were difficult to describe in quantitative detail.[1]

While emphasizing it was not a "physical constant," Lutz and Pietsch used a working assumption that 40 to 50 percent of arc fault power led to pressure rise for air-insulated switchgear installations, rationalizing that the remaining part of the electrical energy is likely emitted by radiation and transferred to the terminals, so as not to contribute to heating of the air.

With regard to nearby employees, equipment, and structures, damage from an electric arc fault can develop on several different levels including the following:

- Electrical disruption
- Conductor heating
- Combustion
- Radiation
- Overpressure

According to the Law of Energy Conservation, represented in Eq. 2.17, energy cannot be created or destroyed, merely transformed from one variety to another.

$$\Delta E_{\text{universe}} = 0 \tag{2.17}$$

In relation to electric arcs in electrical power systems, this physical law suggests that the electrical energy input to an arc fault can be expected to equal the sum of the energy output or energy released in its various physical forms. Conceptually, this can be represented as follows:

$$\text{Energy input (joules)} = \text{Energy output (joules)} \tag{2.18}$$

or

$$\begin{aligned}\text{Electrical power in (watts)} = {} & \text{Electrical power out (watts)} \\ & + \text{Plasma formation (watts)} + \text{Heat flow (watts)} \\ & + \text{Mechanical work (watts) (including acoustic waves} \\ & \quad \text{and shock waves)} + \text{Light power (watts) (including} \\ & \quad \text{optical, infrared, UV, and associated heat)}\end{aligned} \tag{2.19}$$

Note conversion rules:

One joule per second is equal to one watt:

$$1 \text{ J/s} = 1 \text{ W} \tag{2.20}$$

One joule is equal to 0.239 calorie, and one calorie is equal to 4.184 joules:

$$1 \text{ cal} = 4.184 \text{ J} \tag{2.21}$$

In commercial and industrial power supplies, while the energy input may be 60 Hz alternating current or direct current, that is, a specific form of electromagnetic energy (referred to as extremely low frequency, or ELF), the energy output may span the electromagnetic spectrum to include electrical energy and other energy forms.

The electromagnetic spectrum can be thought of as a continuum, with different forms of energy characterized by their frequency and wavelength. The percentage distribution of the physical forms of energy in the output consequent to an electrical arc is variable, described by physics and influenced by environmental considerations.

Following are environmental considerations:

- Geography
- Altitude
- Humidity
- Geology
- Meteorological factors (such as ambient pressure, temperature, and winds)

The output terms in Eq. 2.19 can be linked to their hazard potential or ability to harm nearby employees. In brief:

Electrical power out refers to the continued flow of electrical energy through the installation metallic conductors and conductive gases. As long as input watts are available to output from the electrical fault condition, maximum hazard is present in the situation. The hazard associated includes the possibility of electrical shock.

In addition, the magnetic field effects associated with current flow (and consistent with Maxwell's equations) can be substantial, creating electromagnetic interference with certain electronic devices, bending installation metals or moving nearby metallic structures, and creating structural instability in the installation.

The magnetic fields can also induce electric field effects, even at distance from the arc initiation. If of sufficient magnitude, induced electric field effects on people can interfere with pacemaker function or other implanted electrical devices.

Plasma is a term that describes electrical charges in a gas, which is why the defining feature of plasma is ionization. The term was first used by Langmuir in his studies of mercury vapor lighting. Today, plasmas have industrial uses and are common in industrial consumer products and processes. In devices like televisions or in plasma processes, plasmas can be employed for productive gain.

In an electric arc fault, ionization is driven by the input electrical energy available to act on the space and metals in the faulting electrical installation. Before an electrical arc fault, oxygen, nitrogen, and other gaseous or fluid elements surround the installation. The installation has charges present and flowing. Some of these charges may collect on metallic surfaces, creating a "corona" or "luminescence."

When enough charge accumulates to bridge the dielectric of the space or dielectric gap, an alternate route for electrical energy movement exists. Via an electrochemical-thermal reaction, metallic ions and electrically charged particles in the gaseous space of the installation can thus provide an electrical path for current to flow. This path can be characterized by terms including *arc current*, *arc electrical conductivity*, *arc velocity of distribution*, and *arc duration*.

A plasma can range in temperature from low to high (we talk about cold and hot plasmas) and so can vary in its physical effects. A simple flame at the top of a candle is an example of a low-temperature partial plasma. A white "fireball" from ionized copper electrical equipment is an example of a higher-temperature plasma, in which the white color signals the optical feature of vaporized copper. The sun's core is an example of a very high temperature plasma.

Plasmas are conductive, and their physical phenomena, or how they behave, are very complex. The complexity of plasma behavior is an important source of the variability in data recorded in arc experiments related to electrical power failures.

Because plasmas are electrically conductive, their physical nature is associated with its own magnetic field. In low-voltage conditions, for example up to 600 V, as explained by Land, the Lorenz force law gives insight to arc movement as the magnetic fields will push the arc away from the current source.[2]

In low- and higher-voltage conditions, arc behavior is also influenced by environmental and structural geometric interactions that may alter arc movements. Computations to describe and predict these interactions are physics and engineering based using higher-order mathematics. In applications environments, mathematical complexity is reduced via use of estimates and data collection for equations based on specific experience.

The characteristics of plasmas, in terms of their size, duration, electrical density, temperature, and magnetic field effects, can scale by orders of magnitudes. As a consequence, so can their destructive potential.

Heat refers to the form of energy transfer that results when there is a temperature difference. Heat energy is measured in calories, where 1 calorie equals 4.184 joules (J). With temperatures rising in and around an arc, burn hazard is present from ohmic heating; ignition and combustion of nearby materials, notably including worn clothing and adjacent equipment; and sprayed or blown hot or melting installation elements.

Radiation is another major source of heat from of an electric arc. Radiation is energy traveling through space in waves or particles. With reference to the electromagnetic spectrum, visible light, invisible infrared, and invisible ultraviolet light and their associated heat are forms of nonionizing radiation.

Research suggests some fraction of the electric arc input energy, perhaps as much as one-half, does not heat the vapor or gas in the compartment directly but is emitted as radiation, latent heat of melting, and vaporization of metal components.

The vaporization of metal components refers to the sublimation of solid metallic conductors. Sublimation occurs when the arc drives the phase transition of installation metals to vapor without the metals going through a liquid stage. This electrochemical reaction is highly exothermic, so it yields enormous heat. The reaction also is highly uncontrolled, meaning it spreads spatially from the arc ignition event in an extremely fast and unpredictable manner.

Visible Light. People will naturally avoid too intense a visible light, with the aversion response of closing one's eyes taking about 0.25 second to occur.

Infrared Radiation. Under controlled circumstances, when two physical objects are at different temperatures, infrared radiation will flow between them from the hotter to the cooler in an amount following the Stefan-Boltzmann law, in which the total radiation is expressed in watts per square centimeter (W/cm^2). The object's temperature determines how much radiation is emitted and the specific wavelength of the emission.

The infrared portion of the electromagnetic spectrum at wavelengths ranging from 780 nm to 1 mm is divided into subregions. By convention of the International Commission on Illumination (CIE), these include IRA, IRB, and IRC.

In industrial situations, the most common injury from infrared radiation exposure is eye injury to the cornea, with the formation of cataracts (e.g., glass blower's cataract or furnaceman's cataract), retinal burns, and corneal burns.

As for hazards to the skin, infrared threshold effects are time dependent. An irradiation of 10 kWm^{-2} may cause painful sensation within 5 seconds.

To protect against infrared exposures, international organizations, including the World Health Organization (WHO), the International Labour Organization (ILO), the International Commission on Non-Ionizing Radiation Protection (ICNIRP), and the American Conference of Governmental Industrial Hygienists (ACGIH), have suggested exposure limits to protect the structures of the eye, including the cornea, retina, and lens.

Apart from direct health effects, the infrared radiation from an electric arc likely plays a role in the spread of fire at the scene. When nearby surface materials are not reflective, these will absorb infrared (IR). As IR drives up the material's temperature beyond its ignition value, combustion occurs, propagating the spread of flames.

Ultraviolet Radiation. The ultraviolet hazard associated with arcs is known, for example, with metal halide arc lamps. Metal halide arc lamps that lose their protective quartz shields are highly efficient sources of ultraviolet light and are known to severely injure vulnerable organs, including eyes and skin.

The UV hazard is also commonly recognized in arc processes like welding, and explains the welder's required routine use of a protective hood that can filter the UV light emitted from the arc during work.

Following are common eye injuries from UV exposure:

- Corneal "flash burn"
- "Ground glass" eye
- "Welder's flash"
- Photokeratitis

The ACGIH has promulgated exposure limits to radiation exposure from UV. Limits are generally expressed as 1 milliwatt per square centimeter (1 mW/cm^2) for durations longer than 1000 seconds, and as a total energy exposure of 1 joule per square centimeter (1 J/cm^2) for exposure times less than 1000 s. Specific threshold limit values (TLVs) are available from the ACGIH.

Mechanical work refers to electric arc mechanical effects, including those effects that push, pull, collapse, shift, or propel via acoustic or mechanical shock. Primary and secondary hazards are associated with mechanical work.

A primary hazard exists directly, while a secondary hazard exists indirectly. For example, a primary hazard is when the acoustic effect of an arc ruptures a nearby worker's eardrum.

An example of a secondary hazard is when the acoustic effect of an arc ruptures the metal sheeting on a cable raceway, and the sheeting hits the employee in the head: the shrapnel of the metal sheeting is a secondary hazard to the primary hazard of the arc effects on the electrical component's metal.

Estimates of the potential for mechanically destructive effects from electrical current flow can be roughly calculated. One stick of dynamite weighs about 1/3 lb TNT and in energy terms yields 0.7 MJ (about 1 MJ).

The power of a stick of dynamite in joules per second or watts can be estimated when the time course for ignition is known. Thus we can use a rough rule of thumb that one stick of dynamite (about 1 MJ) is approximately equivalent to 1 megawatt (1 MW) when ignited across 1 second (1 MJ/s).

Depending on the speed of an electrical arc event, which may be knowable from electrical system monitoring, coordination studies, test lab conditions, or operational control records, we can estimate the equivalent destructive potential of an electric arc with electrical or mechanical terms.

We can also assess an electrical explosion in terms comparable to the destructive potential of dynamite and TNT. Using an equivalency to dynamite and TNT allows an estimate of the explosive yield of an electric arc. With this information, engineering knowledge about building tolerances to pressure waves can be incorporated into management of electrical installations. Furthermore, researchers using staged tests have repeatedly documented that with higher available arc energies, increasingly prominent blast effects are observed.

SUMMARY

The principles of physics apply to our understanding of electrical hazards. Physical characterizations of metal and conductors compared with nonmetals and nonconductors influence the material composition of electrical installations. Notably, the features of nonconductors can be used to create insulation against electrical current flow in component design as well as the design and use of personal protective technologies. The improved understanding of the physics of electrical current behavior under fault conditions in electrical systems has led to wider recognition of the human effects of exposure to electrical shock as well as unintentional electric arcing with release of destructive energy. Quantification of the physical relations among the forces of nature permits increasingly more accurate estimations and calculations of the magnitude of electrical, thermal, mechanical, and chemical effects in electrical systems.

REFERENCES

1. Lutz, F., and Pietsch, G., The Calculation of overpressure in metal-enclosed switchgear due to internal arcing. *IEEE Tiunsactions on Power Apparatus and Systems.* PAS-101(11): pp. 4230–4236, 1982.
2. Land, H. B., Sensing switchboard arc faults. *IEEE Power Engineering Review.* 22(4): pp. 18–27, 2002.

CHAPTER 3
ELECTRICAL SAFETY EQUIPMENT

INTRODUCTION

The safety aspects of any job or procedure are greatly enhanced by the use of proper tools, meters, apparel, and other such equipment. This chapter outlines the construction and use of a variety of electrical safety equipment. Some of the equipment is used to actually perform work—items such as insulated tools or voltage-measuring devices fall into this category. Other safety products are used strictly to protect the worker, for example, flash suits and rubber goods.

Each specific piece of safety equipment is used to protect the worker from one or more of the numerous electrical safety hazards, and each piece of equipment should be employed when performing various types of jobs in the electric power system.

Always be certain that the clothing or apparel and the meters or tools you are using are designed and tested to match or exceed the incident energy level and the voltage level for the application to which you will be exposed.

GENERAL INSPECTION AND TESTING REQUIREMENTS FOR ELECTRICAL SAFETY EQUIPMENT

Each of the types of electrical safety equipment described in this chapter has specific inspection and testing requirements. These requirements are identified in each of the individual sections. In addition to specific requirements, the following precaution should always be observed:

Always perform a detailed inspection of any piece of electrical safety equipment before it is used. Such an inspection should occur at a minimum immediately prior to the beginning of each work shift, and should be repeated any time the equipment has had a chance to be damaged.

Where possible, the guidelines for wearing and/or using the various types of safety equipment discussed in this chapter are based on existing industry standards. In any event, the guidelines used in this book should be considered *minimum.* Requirements for specific locations should be determined on a case-by-case basis using current industry standards.

ARC-FLASH AND THERMAL PROTECTION

The extremely high temperatures and heat content of an electric arc can cause extremely painful and/or lethal burns. Since an electric arc can occur at any time in electrical equipment that has not been placed in an electrically safe work condition, the worker must take precautions and wear protection when exposed to potential arc hazards. Note that these sections address equipment for electrical hazard. Fire protection equipment has slightly different requirements and is not covered.

Table 3.1 itemizes the type of equipment required to protect the worker from the thermal hazards of electric arc. The next sections describe the type of equipment used and will identify when and how to use that equipment.

A Note on When to Use Thermal Protective Clothing

The usage directions given in this chapter should be used as minimum guidelines only. Modern technology has enabled the calculation of actual incident arc energies. When these arc energies are compared to the Arc Thermal Performance Value (ATPV) discussed later in this chapter, the exact weight and type of thermal clothing can be determined.

If workers are required to place any part of their body within the *flash boundary* distance of an energized electrical component, they must wear thermal protective clothing with an ATPV or E_{BT} equal to or greater than the amount of arc energy to which they might be exposed. Chapter 4 of this handbook provides specific methods for these calculations. The ATPV or E_{BT} for any given material is calculated using the procedures defined in ASTM standard F 1959/F 1959M.[1]

Thermal Performance Evaluation

Arc-Rated. Virtually all clothing will ignite when exposed to a sufficient heat source. Most fabric, clothing, and personal protective equipment (PPE) manufacturers still use the term "flame-resistance (FR)," but the correct term as per NFPA 70E is arc-rated. There must be no mistake when purchasing protective clothing and PPE for arc flash protection. When the heat source is removed, normal clothing will continue to burn. Arc-rated clothing and PPE may burn and char when it is exposed to a heat source, but it will not continue to burn after the heat source is removed.

The most common test for arc-rating is defined in Method 5903.1 of Federal Test Standard 191A (Flame Resistance of Cloth: Vertical). This test suspends a 12-inch-long specimen of fabric vertically in a holder. The fabric is enclosed and subjected to a controlled flame on the bottom edge of the fabric for 12 seconds. Table 3.2 lists the three sets of data that are recorded in this test.

Note that the results are gathered *after* the flame source has been removed. Note also that the *afterglow* is not included in most of the industry standards that reference this method.

TABLE 3.1 Equipment Used to Protect Workers from Arc Hazard

Area of body to be protected	Equipment used
Torso, arms, legs	Thermal work uniforms, flash suits
Eyes	Face shields, goggles, safety glasses
Head	Insulating hard hats, flash hoods
Hands	Rubber gloves with leather protectors, thermally resistant gloves

TABLE 3.2 Test Data Drawn from Method 5903.1 of Federal Test Standard 191A

Test result measured	Description
Afterflame	Number of seconds (in tenths) during which there is a visible flame remaining on the fabric
Afterglow	Number of seconds (in tenths) during which there is a visible glow remaining on the fabric
Char length	Length of the fabric in tenths of an inch destroyed by the flame that will readily tear by application of a standard weight

Arc Thermal Performance Value (ATPV). Research by Stoll and Chianta[2] developed a curve (the so-called Stoll curve) for human tolerance to heat. The curve is based on the minimum incident heat energy (in kJ/m^2 or cal/cm^2) that will cause a second-degree burn on human skin. Modern standards that define the level of thermal protection required are based on the Stoll curve. That is, clothing must be worn that will limit the degree of injury to a second-degree burn. This rating is called the *Arc Rating* or the *Arc Thermal Performance Value* (ATPV).

Energy Break-Through (E_{BT}). Some types of flame-resistant materials become brittle when exposed to very high temperatures. The combination of the brittleness and the concussion from the electrical blast can cause the material to fail and break open. ASTM has defined this value as the average of the five highest incident energy exposure values below the Stoll curve where the specimens do not exhibit breakopen.

Since the E_{BT} is defined for values below the Stoll curve, garments made from such materials can be safely used when exposures will be less than the energy break-through level.

ASTM and Other Standards. The American Society for Testing and Materials (ASTM) has three standards that apply to the thermal protective clothing to be worn by electrical workers. Table 3.3 identifies these standards.

ASTM standard F 1506 specifies three requirements for workers' clothing in Part 6:

1. Thread, bindings, and closures used in garment construction shall not contribute to the severity of injuries to the wearer in the event of a momentary electric arc and related thermal exposure.
2. Afterflame is limited to 2.0 s or less and char length is limited to 6.0 in or less. Afterglow is mentioned but is not judged a serious hazard.
3. Garments must be labeled with the following information:
   a. Tracking identification code system
   b. Meets requirements of ASTM F 1506
   c. Manufacturer's name
   d. Size and other associated standard labeling
   e. Care instructions and fiber content

Standard F 1959 defines the technical specifications of measuring the Arc Thermal Performance Value. Note that garments only pass the 5903.1 method with 6.0 in or less char length and 2.0 s or less afterflame. It is important to note that ASTM 1959 is a test procedure for testing garment material, not the garment itself.

Standard F 2178 defines the technical specifications for measuring the arc rating for protective face shields used in arc protective clothing. The standard provides methods for arc thermal ratings only and specifically does not address molten materials, fragmentation materials, and other such flying debris.

Usage Standards. The principal standards for electrical worker thermal protection are OSHA 1910.269 and ANSI/NFPA 70E. Of these two, 70E is the most rigorous and provides the best level of protection, and it defines user thermal protection requirements on the basis of the ATPV. The usage requirements described in this handbook are based primarily on ANSI/NFPA 70E.

TABLE 3.3 ASTM Standards Defining Electrical Worker Thermal Clothing

Standard number	Title
F 1506	Standard Performance Specification for Textile Materials for Wearing Apparel for Use by Electrical Workers Exposed to Momentary Electric Arc and Related Thermal Hazards
F 1959	Standard Test Method for Determining the Arc Thermal Performance Value of Materials for Clothing
F 2178	Standard Test Method for Determining the Arc Rating of Face Protective Products

Clothing Materials

Materials used to make industrial clothing fall into two major categories, with several subcategories under each as follows:

1. Non-arc-rated materials. When these materials are treated with a flame-retardant chemical, they become flame resistant or arc-rated.
   a. Natural fibers such as cotton and wool
   b. Synthetic fibers such as polyester, nylon, and rayon
2. Arc-rated materials
   a. Non-arc-rated materials that have been chemically treated to be made flame resistant or arc-rated
   b. Inherently arc-rated materials such as PBI, Kermel, and Nomex

The following sections describe some of the more common fibers and identify their general capabilities with respect to thermal performance.

Non-Arc-Rated Materials

Contrary to some misunderstandings, natural fibers such as cotton and wool *are not arc-rated.* In fact, the only advantages that natural fibers exhibit over synthetics such as polyester is that they do not melt into the burn, and the gases that they generate while burning are generally much less toxic than those of the synthetics. Do not use natural fibers and expect to get the type of protection afforded by true arc-rated materials.

Cotton. Cotton work clothing made of materials such as denim and flannel is a better choice than clothing made from synthetic materials. Cotton does not melt into the skin when heated; rather, it burns and disintegrates, falling away from the skin. Thick, heavy cotton material provides a minimal barrier from arc temperatures and ignites quickly. At best, cotton provides only minimal thermal protection.

Wool. Wool clothing has essentially the same thermal properties as cotton clothing.

Synthetic Materials. Untreated synthetic clothing materials such as polyester and nylon provide extremely poor thermal protection and should *never be used* when working in areas where an electric arc may occur. Some synthetic materials actually increase the danger of exposure to an electric arc. Synthetic materials have a tendency to melt into the skin when exposed to high temperatures. This melting causes three major difficulties.

1. The melted material forms a thermal seal that holds in heat and increases the severity of the burn.
2. Circulation is severely limited or cut off completely under the melted material. This slows healing and retards the flow of normal nutrients and infection-fighting white blood cells and antibodies.
3. The removal of the melted material is extremely painful and may increase the systemic trauma already experienced by the burn victim.

Synthetic-Cotton Blends. Synthetic-cotton blends such as polyester-cotton are used to make clothing that is easier to care for. Although slightly less vulnerable to melting than pure polyester, the blends are still extremely vulnerable to the heat of an electric arc and the subsequent plasma cloud. Such blends provide poor thermal protection and should not be used in areas where the hazard of electric arc exists.

Arc-Rated Materials

Chemically Treated Materials. Both natural and synthetic fibers can be chemically treated to render them arc-rated. Such materials are sometimes used in disposable, coverall-type clothing. While some chemical treatments (such as Borax and boric acid–salt combinations) may be temporary in nature, others are quite satisfactory and may last for the life of the garment.

Historically, chemically treated natural fibers did not exhibit as high an ATPV as synthetic materials when compared by weight. Over the past several years, the methods of treating natural fibers have erased any advantage previously exhibited by the synthetic FR materials, and some workers report that the natural fibers are more comfortable to wear in the climatic extremes of heat and cold.

Heavy weights of chemically treated natural fibers may provide superior protection against certain molten metals.

Nomex* IIIA. Nomex is an aramid fiber made by the DuPont Company. It has a structure that thickens and carbonizes when exposed to heat. This unique characteristic allows Nomex to provide excellent thermal protection.

Nomex has been modified in the years since it was first introduced. Nomex IIIA is made with an antistatic fiber and is, therefore, suitable for use in hazardous environments such as those with high concentrations of hydrocarbon gas.

Since the characteristics of Nomex are inherent to the fiber, and not a chemical treatment, the thermal protection capabilities of Nomex are not changed by repeated laundering.

Polybenzimidazole (PBI).[†] PBI is a product of Performance Products Inc., a subsidiary of Hoechst Celanese Corporation. It is similar to Nomex in that it is a synthetic fiber made especially to resist high temperatures. PBI is nonflammable, chemically resistant, and heat stable. This heat stability makes it less prone to shrinking or embrittlement when exposed to flame or high temperatures.

PBI does not ignite, melt, or drip in Federal Vertical Flame Tests FSTM 5903 and FSTM 5905. PBI's characteristics are permanent for the life of the garment. Hoechst Celanese performed tests that indicate that PBI has heat protection characteristics that are equal to or superior to other materials.

Kermel.** Kermel is a synthetic polyamide imide aramid fiber manufactured in France by Rhone-Poulenc. Kermel fiber is only offered in fabrics blended with other fibers. Kermel is blended with wool for dress uniforms, sweaters, and underwear, and with high-tenacity aramid for bunker gear and gloves. In the professional firefighter and work wear areas, Kermel is offered in a 50/50 blend with FR viscose rayon.

Like other synthetic flame-retardant materials, Kermel is flame resistant and does not drip or melt when heated. This material is suitable only if it has an arc rating.

Material Comparisons. Table 3.4 illustrates the relative properties of the various types of clothing materials. The information given in Table 3.4 is drawn from general industry experience and/or manufacturer experiments.

Work Clothing

Construction. Work clothing used for routine day-to-day electrical safety may be employed as secondary flash protection. Arc-rated cotton, arc-rated synthetic-cotton blend, Nomex, PBI, or other

*Nomex is a registered trademark of the E. I. Dupont de Nemours Company.
**Kermel is a registered trademark of Rhone-Poulenc.
[†]PBI is a trademark of the Hoechst Celanese Corporation.

TABLE 3.4 Clothing Material Characteristics

Generic name	Fiber	Manufacturer	Moisture regain*	Tenacity g/den†	Comments
Aramid (meta)	Nomex	DuPont	5.5	4.0–5.3	• Long-chain synthetic polyamide fiber. Excellent thermal stability. Will not melt and drip. Excellent chemical and abrasion resistance. Fair colorfastness to laundering and light exposure.
Aramid (para)	Kevlar Twaron Technora	DuPont Akzo (Netherlands) Teijin (Japan)	4.3 4.0	21–27 22.6	• Blended with Nomex for fabric integrity in high-temperature exposures. Fair abrasion resistance. Sensitive to chlorine bleach, light, and strong mineral acids.
Polyamide imide	Kermel (France)	Rhone-Poulenc	3.4	4.0–4.5	• Long-chain synthetic polyamide fiber. Excellent thermal stability. Will not melt and drip. Excellent chemical and abrasion resistance. Fair colorfastness to laundering and light exposure.
Basofil	Basofil	BASF	5.0	2.0	• A melamine fiber formed when methylol compounds react to form a three-dimensional structure of methylene ether and methylene bridges. Resistant to many solvents and alkalis. Moderately resistant to acids. Will not shrink, melt, or drip when exposed to a flame.
Modacrylic	Protex	Kaneka (Japan)	2.5	1.7–2.6	• Long-chain synthetic polymer fiber containing acrylonitrile units modified with flame retardants. Excellent chemical resistance. Fair abrasion resistance. High thermal shrinkage.
FR Acrylic	Super Valzer	Mitsubishi (Japan)	2.5	1.7–2.6	• Long-chain synthetic polymer fiber containing acrylonitrile units modified with flame retardants. Excellent chemical resistance. Fair abrasion resistance. High thermal shrinkage.
PBI	PBI Gold	Celanese	15.0	2.8	• Polymer is a sulfonated poly (2.2-m-phenylene-5,5 bibenzim idazole). Will not ignite, does not melt. Excellent chemical resistance. Dyeable in dark shades only.
Polyimide	P84	Imitech (Austria)	3.0*	4.3	• Long-chain synthetic polyimide fiber. High thermal shrinkage. Thermal properties inferior to Nomex.
FR Viscose	PFR Rayon	Lenzing (Austria)	10.0	2.6–3.0	• Man-made cellulosic, properties similar to cotton. Fiber contains flame retardants.
FR Cotton	FR Cotton	Natural Fiber	8.0	2.4–2.9	• Flame retardant treated in fabric form. Poor resistance to acids. Fair abrasion resistance. Relatively poor colorfastness to laundering and light exposure. Wear properties similar to untreated cotton.
Vinyl	Vinex FR9B®	Westex	3.0	3.0	• Fabric blended of 85% vinyl/15% rayon. Fiber composed of vinyl alcohol units with acetal crosslinks. Sheds aluminum splash. Very sensitive to shrinkage from wet and dry heat.
FR Polyester	Trevira FR Polyester	Trevira	0.4	4.5	• Polyester with proprietary organic phosphorus compound incorporated into the polymer chain. Properties similar to regular polyester except as modified by flame retardants. Melt point 9°C lower than regular polyester.
Polyamide	Nylon	DuPont Monsanto	6.0	6.0–8.0	• Long-chain synthetic polyamide in which less than 85% of the aramide linkages are attached. Blended with FR cotton to improve abrasion resistance. Wear properties significantly better than untreated cotton.

*A measure of ability to absorb moisture. (Percent by weight of moisture gained from a bone dry state at 65% relative humidity.)
†A measure of strength and durability. (Tenacity is defined as force per unit linear density to break a known unit of fiber.)
Source: Courtesy Bulwark

arc-rated materials are preferred. The clothing should meet the following minimum requirements:

1. Long sleeves to provide full arm protection
2. Sufficient weight for both thermal and mechanical protection
3. Sufficient arc rating (ATPV) for the locations where used

The suggested minimum is 4 ounces per square yard (oz/yd^2) if synthetics such as Nomex or PBI are used and 7 to 8 oz/yd^2 if flame-retardant cloth or blends are used. Figure 3.1 shows several examples of arc-rated work clothing.

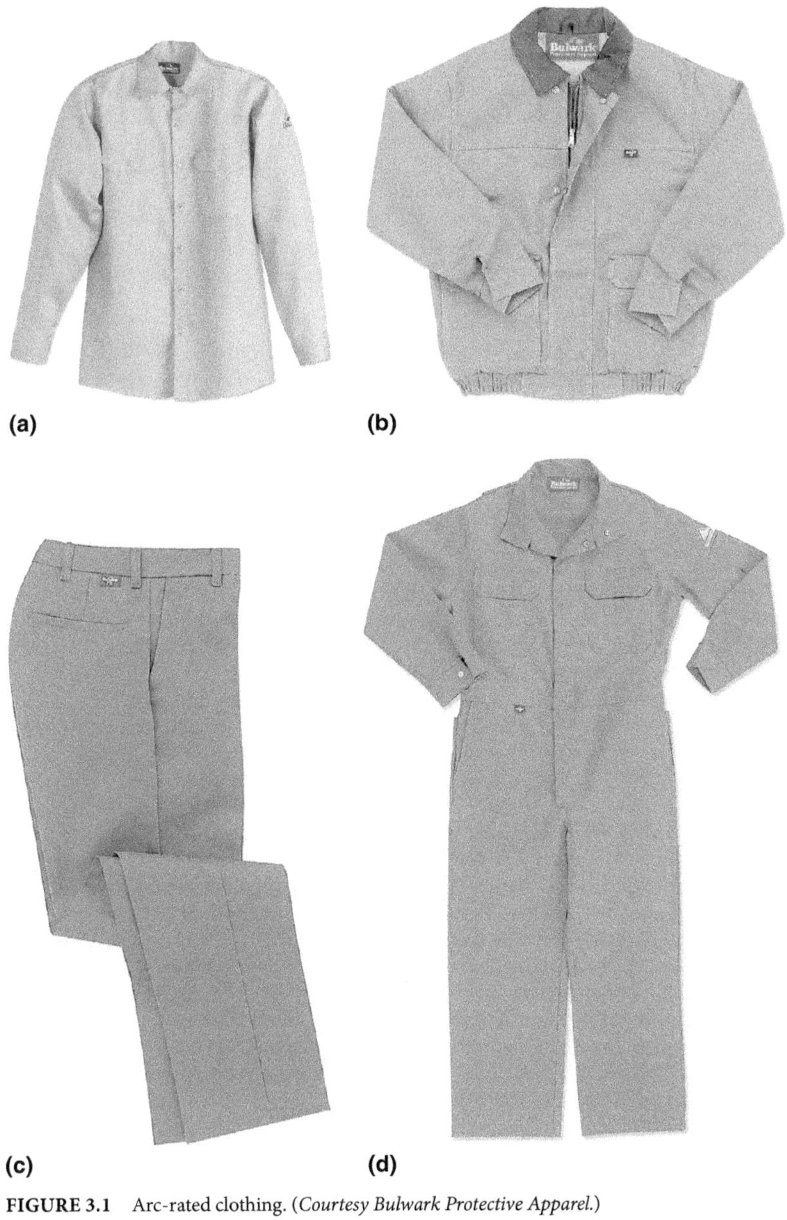

FIGURE 3.1 Arc-rated clothing. (*Courtesy Bulwark Protective Apparel.*)

TABLE 3.5 Care and Use Guidelines for Thermal Protective Clothing

- Clothing should not be allowed to become greasy and/or impregnated with flammable liquids.
- Launder according to manufacturer's instructions. Generally, home laundering in hot water with a heavy-duty detergent will be effective.
- Do not mix flame-resistant garments with items made of other materials in the same wash.
- Do not use bleaches or other treatments unless recommended by the manufacturer.
- Remember that laundering may degrade the chemical treatment on some flame-retardant materials. Observe manufacturer's recommendations as to how many washes constitute the life of the garment.
- Inspect work uniforms and flash suits before each use. If they are contaminated, greasy, worn, or damaged in any way, they should be cleaned or replaced as required.

When to Use Thermally Protective Work Uniforms. Thermally protective uniforms should be required for all workers who are routinely exposed to the possibility of electric arc and/or arc flash. This applies especially to workers in the industries that have the added hazard of flash fire. At a minimum, all employees who are routinely exposed to 480 V and higher should use the thermally protective materials. See Chap. 4 for more specific information.

Thermal protective clothing should be worn every time the worker works inside the arc flash boundary. The clothing should be selected with an ATPV or E_{BT} sufficient for the maximum incident energy that may be created in the work area. See Chap. 4 for information on determining the arc flash boundary.

Care of Thermally Protective Work Uniforms. The following information is necessarily general in nature. Always refer to the manufacturer's care and laundering instructions for specific information. Work uniforms should be kept clean and free of contaminants. Contaminated work clothing can be extremely hazardous. Table 3.5 lists typical care and use precautions for thermal or arc-rated work clothing and arc-rated arc flash suits.

Arc Flash Suits

Construction. An arc flash suit is a thermal protective garment primarily intended for use in locations where the calculated incident energy exceeds 12 cal/cm^2. The arc flash suit provides whole-body protection. The arc flash suit shown in Fig. 3.2 is made of 12-oz twill and 88 percent cotton. The entire garment is arc-rated and has an ATPV rating of 40 cal/cm^2.

Arc flash suits are composed of a minimum of three parts—the face shield/hood (Fig. 3.3), the jacket (Fig. 3.4), and coveralls (Fig. 3.5). Some older arc flash suits may not include coveralls. In these cases, the jacket must be long enough to completely protect the legs. As a matter of good safety practice, an arc flash suit without a coverall or leggings should be avoided.

Using Arc Flash Suits. Arc flash suits should be worn any time the worker works inside the arc flash boundary. The suit should be selected with an ATPV or E_{BT} sufficient for the maximum incident energy that may be created at the working distance. Work should never be performed inside the working distance. See Chap. 4 for information on determining the arc flash boundary. The procedures listed in Table 3.6 are typical of those in which many companies require the use of arc flash suits; however, specific rules should be developed for each company. *Arc flash suits should always be used in conjunction with adequate head, eye, and hand protection. See NFPA 70E Section 130.7 for required PPE.* Note that all workers in the vicinity of the arc flash potential should be wearing an arc flash suit.

Table 3.6 lists tasks that typically require the use of flash suits. Note that this table is not exhaustive. The ultimate decision on exactly what PPE is required to perform any task safely is based on the incident energy inside the flash boundary.

FIGURE 3.2 Arc flash suit. (*Courtesy Bulwark Protective Apparel.*)

Head, Eye, and Hand Protection

When wearing arc flash suits, or whenever exposed to an arc flash hazard, employees must wear full protection for the head, eyes, and hands. Head and eye protection will be provided if the employee is equipped with an arc flash suit. When not in an arc flash suit, however, employees must wear hard hats and face shields plus safety glasses or goggles. Hand protection must be provided by electrical insulating rubber gloves covered with leather protectors, especially if a shock hazard also exists.

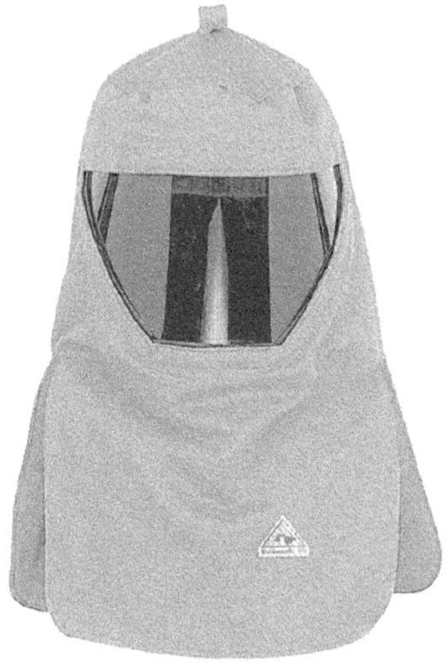

FIGURE 3.3 Face shield for arc flash suit. (*Courtesy Bulwark Protective Apparel.*)

FIGURE 3.4 Arc flash suit jacket. (*Courtesy Bulwark Protective Apparel.*)

FIGURE 3.5 Arc flash suit coveralls. (*Courtesy Bulwark Protective Apparel.*)

TABLE 3.6 Procedures That Require the Use of Arc Flash Suits

- Operating open-air switches on circuits of 480 V and higher
- Open-door switching and racking of circuit breakers—480 V and higher
- Removing and installing motor starters in motor control centers—208 V and higher
- Applying safety grounds—480 V and higher
- Measuring voltage in any circuit that is uncertain or has exhibited problems—208 V and higher
- Working on or near any exposed, energized conductors or circuit parts—208 V or higher
- Any time the incident energy may exceed a level that can be protected by day-to-day arc-rated work clothing

HEAD AND EYE PROTECTION

Hard Hats

Construction and Standards. In addition to wearing protection from falling objects and other blows, electrical workers should be equipped with and should wear hard hats that provide electrical insulating capabilities. Such hats should comply with the latest revision of the American National Standards Institute (ANSI) standard Z89.1, which classifies hard hats into three basic classes.

1. Class G hard hats are intended to reduce the force of impact of falling objects and to reduce the danger of contact with exposed low-voltage conductors. They are proof-tested by the manufacturer at 2200 V phase-to-ground.
2. Class E* hard hats are intended to reduce the force of impact of falling objects and to reduce the danger of contact with exposed higher-voltage conductors. They are proof-tested by the manufacturer at 20,000 V phase-to-ground.
3. Class C hard hats are intended to reduce the force of impact of falling objects. They offer no electrical protection.

*Prior to 1998, the class G hat was labeled class A and class E hats were labeled class B.

Figure 3.6 shows two examples of class E (formerly class B) hard hats. Note that a hard hat must be a class G or class E hat to be used in areas where electrical shock may occur. The label for a class B (now class E) hard hat is shown in Fig. 3.7.

Use and Care. Electrically insulating class G or E hard hats should be worn by workers any time there is a possibility they will be exposed to electrical shock, arc flash, arc blast, mechanical blows, or injuries. Table 3.7 lists typical working conditions in which workers should be wearing such protection. Note that this table is not comprehensive. Many other electrical activities may expose workers to incident energy levels above 8 cal/cm^2.

All components of the hard hat should be inspected daily, before each use. This inspection should include the shell, suspension, headband, sweatband, and any accessories. If dents, cracks, penetrations, or any other damage is observed, the hard hat should be removed from service. Class G and E hard hats may be cleaned with warm water and soap. Be sure to thoroughly rinse and dry the hat after washing. Solvents and other harsh cleaners should be avoided. Always refer to the manufacturer's instructions for specific cleaning information.

FIGURE 3.6 Hard hats suitable for use in electrical installations. *(Courtesy Mine Safety Appliances Company.)*

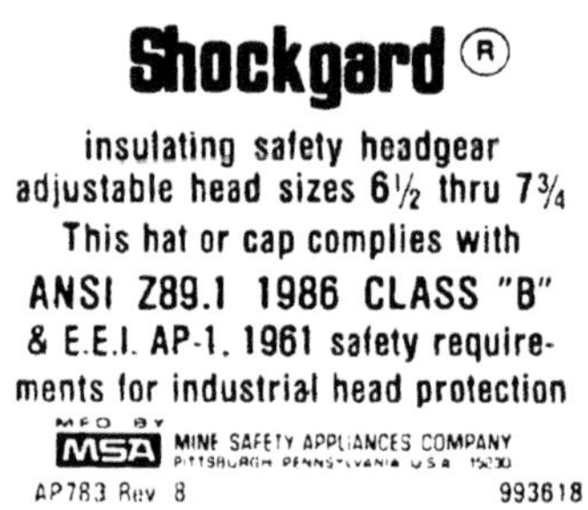

FIGURE 3.7 Hard hat label for a class E hard hat. *(Courtesy Mine Safety Appliances Company.)*

TABLE 3.7 Work Situations That Require Nonconductive Head Protection and Eye Protection

- Working close to exposed, overhead energized lines
- Working in switchgear, close to exposed energized conductors
- Any time that an arc flash suit is recommended (see Table 3.6)
- When any local rules or recognized standards require the use of nonconductive hard hats or eye protection
- Any time there is danger of head, eye, or face injury from electric shock, arc flash, or arc blast

Safety Glasses, Goggles, and Face Shields

The plasma cloud and molten metal created by an electric arc are projected at high velocity by the blast. If the plasma or molten metal enters the eyes, the extremely high temperature will cause injury and possibly permanent blindness. Electrical workers exposed to the possibility of electric arc flash and arc blast must be equipped with and must wear eye and face protection. Such protection should comply with the latest revision of ANSI standard Z87.1 and should be nonconductive when used for electric arc flash and arc blast protection.

Arc flash suit face shields (Fig. 3.3) will provide excellent face protection from molten metal and the plasma cloud. Goggles or safety glasses that reduce the ultraviolet light intensity (Fig. 3.8) must also be used. Figure 3.9 is a photograph of a worker with an insulating hard hat and protective goggles.

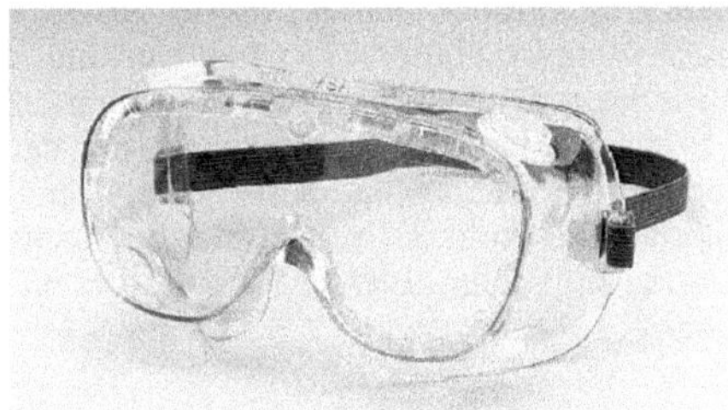

FIGURE 3.8 Ultraviolet-resistant safety goggles. (*Courtesy Mine Safety Appliances Company.*)

FIGURE 3.9 Hard hat and goggles. (*Courtesy Mine Safety Appliances Company.*)

Use and Care. Eye and face protection should be worn by workers any time they are exposed to the possibility of electric arc flash and arc blast. Table 3.7 lists typical situations where such protection might be required.

Face protection should be cleaned before each use. Soft, lint-free cloths and warm water will normally provide the necessary cleaning action; however, most manufacturers supply cleaning materials for their specific apparatus.

RUBBER INSULATING EQUIPMENT

Rubber insulating equipment includes rubber gloves, sleeves, line hose, blankets, covers, and mats. Employees must use such equipment when working in an area where the hazard of electric shock exists. This means any time employees are working on or near an energized, exposed conductor or circuit parts, they should be using rubber insulating equipment.

Rubber goods provide an insulating shield between the worker and the energized conductors. This insulation will save the workers' lives should they accidentally contact the conductor. The American Society for Testing and Materials (ASTM) publishes recognized industry standards that cover rubber insulating goods.

Rubber Gloves

Description. A complete rubber glove assembly is composed of a minimum of two parts—the rubber glove itself and a leather protective glove. In service, the leather protector fits over the outside of the rubber glove and protects it from physical damage and puncture. Sometimes the glove set will include a sheer, cotton insert that serves to absorb moisture and makes wearing the gloves more pleasant. Figure 3.10 shows a typical set of rubber gloves with leather protectors. *Caution:* Rubber gloves should never be used without their leather protectors except in certain specific situations as described later in this chapter.

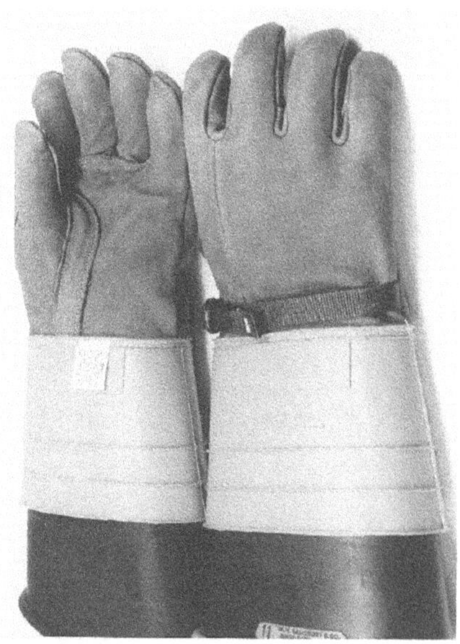

FIGURE 3.10 Typical rubber glove set. (*Courtesy W.H. Salisbury and Co.*)

Construction and Standards. The ASTM publishes four standards that affect the construction and use of rubber gloves.

1. Standard D 120 establishes manufacturing and technical requirements for the rubber glove.
2. Standard F 696 establishes manufacturing and technical requirements for the leather protectors.
3. Standard F 496 specifies in-service care requirements.
4. Standard F 1236 is a guide for the visual inspection of gloves, sleeves, and other such rubber insulating equipment.

Rubber gloves are available in six basic voltage classes from class 00 to class 4, and two different types: types I and II. Table 3.8 identifies each class, its maximum-use voltage, and the root-mean-square (rms) and direct current (dc) voltages that are used to proof-test the gloves. Figure 3.11 shows the general design and dimensions of rubber gloves. Rubber gloves are available in the following lengths:

For class 00:	278 mm (11 in), 356 mm (14 in)
For class 0:	280 mm (11 in), 360 mm (14 in), 410 mm (16 in), and 460 mm (18 in)
For classes 1, 2, 3:	360 mm (14 in), 410 mm (16 in), 460 mm (18 in)
For class 4:	410 mm (16 in), 460 mm (18 in)

Note that all lengths have an allowable tolerance of ±13 mm (±1⁄2 in).

Table 3.9 lists minimum and maximum thicknesses for the six classes of gloves.

In addition to the voltage classes, rubber gloves are available in two different types: type I, which is *not* ozone resistant, and type II, which is ozone resistant.

All rubber goods must have an attached, color-coded label subject to the minimum requirements specified in Table 3.10.

When to Use Rubber Gloves. Rubber gloves and their leather protectors should be worn any time there is danger of injury due to contact between the hands and energized parts of the power system. Each of the work situations described in Table 3.7 should require the use of rubber gloves and their leather protectors.

Rubber gloves should also be worn whenever a worker's hands are closer to an energized conductor than the distance specified by the restricted approach boundary. Some companies require

TABLE 3.8 Rubber Insulating Equipment Classifications, Use Voltages, and Test Voltages

Class of insulating blankets	Nominal maximum-use voltage* phase-phase, ac, rms, max	AC proof-test voltage, rms, V	DC proof-test voltage, avg, V
00	500	2500	10,000
0	1000	5000	20,000
1	7500	10,000	40,000
2	17,000	20,000	50,000
3	26,500	30,000	60,000
4	36,000	40,000	70,000

**Note:* The ac voltage (rms) classification of the protective equipment designates the maximum nominal design voltage of the energized system that may be safely worked. The nominal design voltage is equal to (1) the phase-to-phase voltage on multiphase circuits or (2) the phase-to-ground voltage on single-phase grounded circuits. Except for class 0 and 00 equipment, the maximum-use voltage is based on the following formula: Maximum-use voltage (maximum nominal design voltage) = 0.95 ac proof-test voltage − 2000.

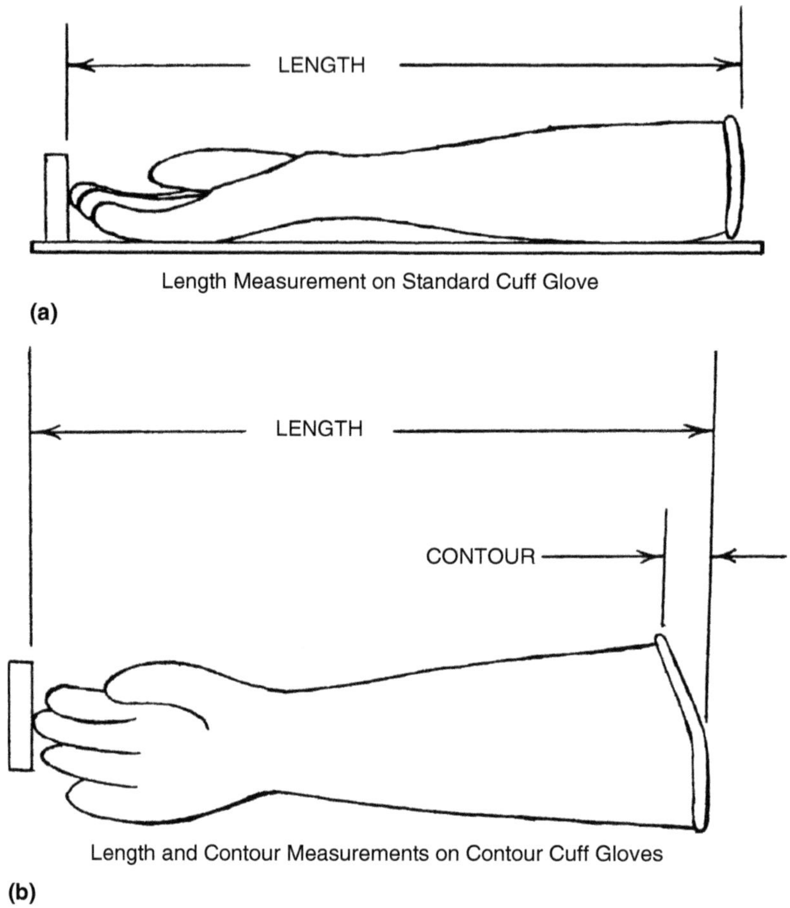

FIGURE 3.11 Dimension measurements for standard and contour cuff rubber gloves. (*Courtesy ASTM.*)

TABLE 3.9 Rubber Glove Thickness Standard

Class of glove	Minimum thickness				Maximum thickness	
	In crotch		Other than crotch			
	mm	in	mm	in	mm	in
00	0.20	0.008	0.25	0.010	0.75	0.030
0	0.46	0.018	0.51	0.020	1.02	0.040
1	0.63	0.025	0.76	0.030	1.52	0.060
2	1.02	0.040	1.27	0.050	2.29	0.090
3	1.52	0.060	1.90	0.075	2.92	0.115
4	2.03	0.080	2.54	0.100	3.56	0.140

TABLE 3.10 Labeling Requirements for Rubber Goods

- Color-coded label according to voltage class: class 00–beige, class 0–red, class 1–white, class 2–yellow, class 3–green, class 4–orange
- Manufacturer's name
- Voltage class (00, 0, 1, 2, 3, 4)
- Type
- Size (gloves only)

their use when the worker crosses the limited approach boundary. Although this is ultraconservative, doing so provides enhanced protection. See Chap. 4 for the definition and calculation methods for the restricted approach boundary.

When to Use Leather Protectors. As stated earlier in this chapter, leather protectors should always be used over rubber gloves to provide mechanical protection for the insulating rubber. Furthermore, leather protectors should *never* be used for any purpose other than protecting rubber gloves.

Sometimes the need for additional dexterity may require that the leather protectors not be used. The various industry standards allow such an application in only three situations.

- Class 00 Up to and including 250 V, leather protectors may be omitted for class 00 gloves. Such omission is permitted only under limited-use conditions when small-parts manipulation requires unusually good finger dexterity.
- Class 0 Leather protectors may be omitted under limited-use conditions when small-parts manipulation requires unusually good finger dexterity.
- Classes 1, 2, 3, 4 Under limited-use conditions, the leathers may be omitted. *However: when the leathers are omitted for these classes, the user must employ gloves rated at least one (1) voltage class higher than normal.* For example, if working in a 4160-V circuit without leather protectors, the worker must use class 2 gloves in lieu of class 1 gloves.

Leather protectors should never be omitted if there is even a slight possibility of physical damage or puncture. Also, rubber gloves previously used without protectors shall not be used with protectors until given a thorough inspection *and electrical retest.*

How to Use Rubber Gloves. Rubber gloves should be thoroughly inspected and air-tested before each use. They may be lightly dusted inside with pure talcum powder or, better yet, manufacturer-supplied powder. This dusting helps to absorb perspiration and eases putting them on and removing them. *Caution:* Do not use baby powder on rubber gloves. Some baby powder products contain additives that can damage the glove and reduce its life and effectiveness. Cotton glove liners worn under the rubber gloves are highly recommended for worker comfort and convenience because they absorb perspiration.

Rubber gloves should be applied before any activity that exposes the worker to the possibility of contact with an energized conductor. As previously stated, any time the worker's hands must be closer to an energized conductor than the restricted approach boundary, rubber gloves must be worn. Be certain to wear the leather protectors with the gloves. Always check the last test date marked on the glove and do not use it if the last date is more than six months earlier than the present date.

Rubber Mats

Description. Rubber mats are used to cover and insulate floors for personnel protection. Rubber insulating mats should not be confused with the nonelectrical rubber matting used to help prevent slips and falls. This type of mat is sold by many commercial retail outlets and is not intended for electrical insulation purposes. Rubber insulating mats will be clearly marked and labeled as such.

Insulating rubber matting has a smooth, corrugated, or diamond design on one surface and may be backed with fabric. The back of the matting may be finished with cloth imprint or other slip-resistant material.

Construction and Standards. The ASTM standard D 178 specifies the design, construction, and testing requirements for rubber matting.

Rubber mats are available in five basic voltage classes, from class 0 to class 4, in two different types, and in three different subcategories. Table 3.8 identifies each class, its maximum-use voltage, and ac rms and dc voltages that are used to test them. Table 3.11 identifies each of the types and special properties for insulating mats. Table 3.12 identifies the thickness requirements for insulating mats. Table 3.13 identifies the standard widths for insulating rubber matting.

Rubber mats must be clearly and permanently marked with the name of the manufacturer, type, and class. ASTM D 178 must also appear on the mat. This marking is to be placed a minimum of every 3 ft (approximately 1 m).

When to Use Rubber Mats. Employers should use rubber mats in areas where there is an ongoing possibility of electric shock. Because permanently installed rubber mats are subject to damage, contamination, and embedding of foreign materials, they should not be relied upon as the sole or primary source of electrical insulation.

How to Use Rubber Mats. Rubber mats are usually put in place only when required or on a permanent basis to provide both electrical insulation and slip protection. Mats should be carefully inspected before work is performed that may require their protection. Rubber mats should be used only as a backup type of protection. Rubber blankets, gloves, sleeves, and other such personal apparel should always be employed when electrical contact is likely.

TABLE 3.11 Types and Special Property Specifications of Insulating Rubber Matting

	Type I	Type II
Composition	Made of any elastomer or combination of elastomeric compounds, properly vulcanized	Made of any elastomer or combination of elastomeric compounds with one or more of the special properties listed by subcategory
Subcategories	None	A: Ozone resistant B: Flame resistant C: Oil resistant

TABLE 3.12 Thickness Specifications for Insulating Rubber Mats

	Thickness		Tolerance	
Class	mm	in	mm	in
0	3.2	0.13	0.8	0.03
1	4.8	0.19	0.8	0.03
2	6.4	0.25	0.8	0.03
3	9.5	0.38	1.2	0.05
4	12.7	0.50	1.2	0.05

Source: Courtesy ASTM.

TABLE 3.13 Standard Widths for Insulating Rubber Matting

(610 ± 13 mm)	24 ± 0.5 in
(760 ± 13 mm)	30 ± 0.5 in
(914 ± 25 mm)	36 ± 1.0 in
(1220 ± 25 mm)	48 ± 1.0 in

Rubber Blankets

Description. Rubber blankets as shown in Fig. 3.12 are rubber insulating devices used to cover conductive surfaces, energized or otherwise. They come in a variety of sizes and are used any time employees are working in areas where they may be exposed to energized conductors or have a need to isolate themselves from grounded surfaces.

Construction and Standards. The ASTM publishes three standards that affect the construction and use of rubber blankets.

1. Standard D 1048 specifies manufacturing and technical requirements for rubber blankets.
2. Standard F 479 specifies in-service care requirements.
3. Standard F 1236 is a guide for the visual inspection of blankets, gloves, sleeves, and other such rubber insulating equipment.

Rubber blankets are available in five basic voltage classes (0 to 4), two basic types (I and II), and two styles (A and B). Table 3.8 identifies each class, its maximum-use voltage, and the ac rms and dc voltages that are used to proof-test the blankets. Table 3.14 lists standard blanket sizes, and Table 3.15 lists standard blanket thicknesses.

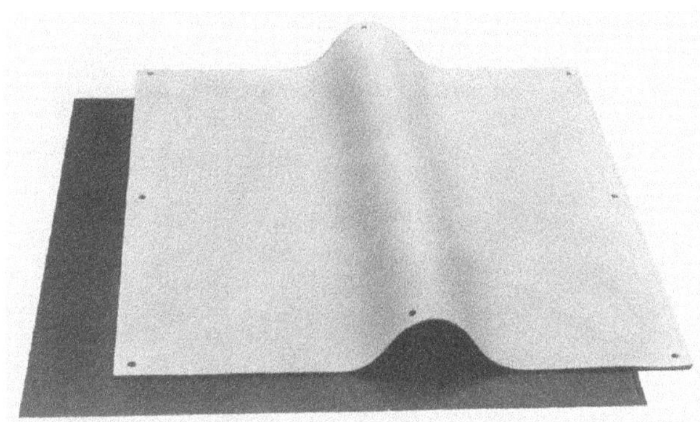

FIGURE 3.12 Typical insulating rubber blanket.

TABLE 3.14 Standard Blanket Sizes—Length and Width

Without slot mm (in)	
457 by 910	(18 by 36)
560 by 560	(22 by 22)
690 by 910	(27 by 36)
910 by 910	(36 by 36)
910 by 2128	(36 by 84)
1160 by 1160	(45.5 by 45.5)
With slot mm (in)	
560 by 560	(22 by 22)
910 by 910	(36 by 36)
1160 by 1160	(45.5 by 45.5)

Source: Courtesy ASTM.

TABLE 3.15 Rubber Blanket Thickness Measurements

Class	Thickness	
	mm	in
0	1.6 to 2.2	0.06 to 0.09
1	2.6 to 3.6	0.10 to 0.14
2	2.8 to 3.8	0.11 to 0.15
3	3.0 to 4.0	0.12 to 0.16
4	3.2 to 4.3	0.13 to 0.17

Source: Courtesy ASTM.

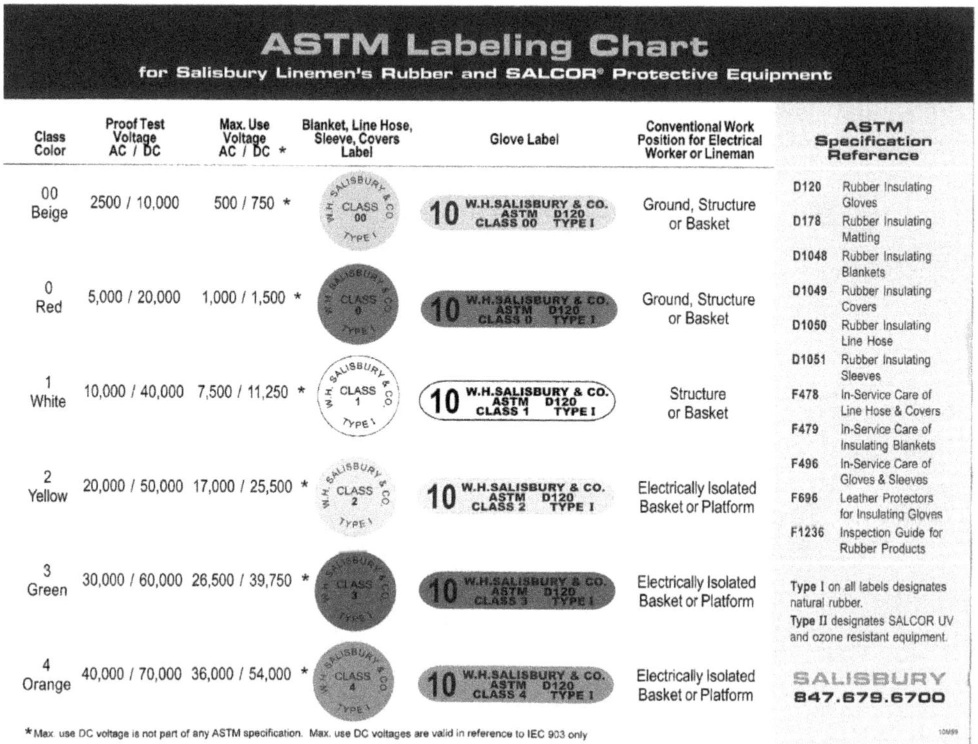

FIGURE 3.13 ASTM labeling chart. (*Courtesy W.H. Salisbury and Co.*)

Type I blankets are made of an elastomer that is not ozone resistant. Type II blankets are ozone resistant. Both type I and type II blankets are further categorized into style A and style B. Style A is a nonreinforced construction, and style B has reinforcing members built in. The reinforcing members may not adversely affect the insulating capabilities of the blanket.

Blankets have a bead around the entire periphery. The bead cannot be less than 0.31 in (8 mm) wide nor less than 0.06 in (1.5 mm) high. Blankets may have eyelets to facilitate securing the blanket to equipment; however, the eyelets must not be metal.

Rubber blankets must be marked either by molding the information directly into the blanket or by means of an attached, color-coded label. The labeling is subject to the minimum requirements specified in Table 3.10. Figure 3.13 summarizes the voltage ratings for all rubber goods and illustrates the labels that are applied by one manufacturer.

When to Use Rubber Blankets. Rubber blankets should be used any time there is danger of injury due to contact between any part of the body and energized parts of the power system. Rubber blankets may be used to cover switchgear, lines, buses, or concrete floors. They differ from mats because they are never permanently installed.

How to Use Rubber Blankets. Rubber blankets should be thoroughly inspected before each use. They may then be wrapped around or draped over metal conductors or buses. When used in either of these ways, they are held in place by using nonconductive blanket buttons or nonconductive "clothespins" that are specially designed for this use. Magnetic blanket buttons, specially designed for this purpose, may be used to hang blankets on metal switchgear frames or switchboards to create an insulating barrier. Rubber blankets are never secured by using electrical plastic or friction tape because when the tape is removed, the face of the blanket is damaged, thereby rendering the blanket unfit for service.

Blankets should be applied before any activity that exposes the worker to the possibility of contact with an energized conductor. Always check the last test date marked on the blanket and do not use it if the last test date was more than one year earlier than the present date.

Rubber Covers

Description. Rubber covers are rubber insulating devices that are used to cover specific pieces of equipment to protect workers from accidental contact. They include several classes of equipment such as insulator hoods, dead-end protectors, straight line hose or connector line hose equipped with serrations on the outside of the male end and serrations on the inside of the female connector end designed to hold them together, cable end covers, and miscellaneous covers. Rubber covers are molded and shaped to fit the equipment for which they are intended. None of these devices should ever be taped with electrical plastic or friction tape to hold them in place. If they are taped on, the insulating rubber is damaged or destroyed when the tape is removed. Always secure them with the proper equipment to prevent damage.

Construction and Standards. The ASTM publishes three standards that affect the construction and use of rubber covers.

1. Standard D 1049 specifies manufacturing and technical requirements for rubber covers.
2. Standard F 478 specifies in-service care requirements.
3. Standard F 1236 is a guide for the visual inspection of blankets, gloves, sleeves, and other such rubber insulating equipment.

Rubber covers are available in five basic voltage classes (0 to 4), two basic types (I and II), and five styles (A, B, C, D, and E). Table 3.8 identifies each class, its maximum-use voltage, and the ac rms and dc voltages that are used to proof-test the covers. Many varieties of rubber covers are available (Fig. 3.14). Their size and shape are determined by the equipment that they are designed to cover. Refer to ASTM standard D 1049 for complete listings of the various standard covers.

Type I covers are made of a properly vulcanized, cis-1,4-polyisoprene rubber compound that is not ozone resistant. Type II covers are made of ozone-resistant elastomers. Both type I and II covers are further categorized into styles A, B, C, D, and E. Table 3.16 describes each of the five styles.

Rubber covers must be marked either by molding the information directly into the cover or by means of an attached, color-coded label. The labeling is subject to the minimum requirements specified in Table 3.10.

When to Use Rubber Covers. Rubber covers should be used any time there is danger of an injury due to contact between any part of the body and energized parts of the power system.

FIGURE 3.14 Typical insulating rubber covers. (*Courtesy W.H. Salisbury and Co.*)

TABLE 3.16 Styles of Rubber Covers

Style	Description
A	Insulator hoods
B	Dead-end protectors
C	Line hose connectors
D	Cable end covers
E	Miscellaneous covers

How to Use Rubber Covers. Rubber covers should be thoroughly inspected before each use. They may then be applied to the equipment that they are designed to cover. Any covers that appear to be defective or damaged should be taken out of service until they can be tested.

Covers should be applied before any activity that exposes the worker to the possibility of contact with an energized conductor. Covers that are used to connect line hose sections should always be used when multiple line hose sections are employed.

Line Hose

Description. Rubber insulating line hoses are portable devices used to cover exposed power lines and other energized or grounded equipment, to protect workers from accidental contact with energized or grounded surfaces such as guy wires, transformer riser taps, or miscellaneous equipment in

TABLE 3.17 Characteristics of the Four Styles of Line Hose

Style	Description
A	Straight style, constant cross section
B	Connect end style; similar to straight style with connection at one end
C	Extended lip style with major outward-extending lips
D	Same as style C with a molded connector at one end

the immediate work area. Line hose segments are molded and shaped to completely cover the equipment to which they are affixed.

Construction and Standards. The ASTM publishes three standards that affect the construction and use of rubber line hose.

1. Standard D 1050 specifies manufacturing and technical requirements for rubber line hose.
2. Standard F 478 specifies in-service care requirements.
3. Standard F 1236 is a guide for the visual inspection of blankets, gloves, sleeves, and other such rubber insulating equipment.

Rubber line hose is available in five basic voltage classes (0 to 4), three basic types (I, II, and III), and four styles (A, B, C, and D). Table 3.8 identifies each class, its maximum-use voltage, and the ac rms and dc voltages that are used to proof-test them.

Type I line hose is made of a properly vulcanized, cis-1,4-polyisoprene rubber compound that is not ozone resistant. Type II line hose is made of ozone-resistant elastomers. Type III line hose is made of an ozone-resistant combination of elastomer and thermoplastic polymers. Type III line hose is elastic. All three types are further categorized into styles A, B, C, and D. Table 3.17 lists the characteristics of each of the four styles.

Rubber line hose must be marked either by molding the information directly into the hose or by means of an attached, color-coded label. The labeling is subject to the minimum requirements specified in Table 3.10.

When to Use Rubber Line Hose. Rubber line hose should be used any time personnel are working on or close to energized lines or grounded surfaces, or any equipment that could become energized or provide an accidental path to ground.

How to Use Rubber Line Hose. Line hoses should be thoroughly inspected before each use. They may then be applied to the equipment that they are designed to cover. Any line hose that appears to be defective or damaged should be taken out of service until it can be tested.

Line hose should be applied before any activity that exposes the worker to the possibility of contact with an energized conductor. When more than one section of line hose is used, connecting line hoses should be employed. The line hose should completely cover the line or equipment being isolated.

Rubber Sleeves

Description. Rubber sleeves (Fig. 3.15) are worn by workers to protect their arms and shoulders from contact with exposed energized conductors. They fit over the arms and complement the rubber gloves to provide complete protection for the arms and hands. They are especially useful when work must be performed in a cramped environment.

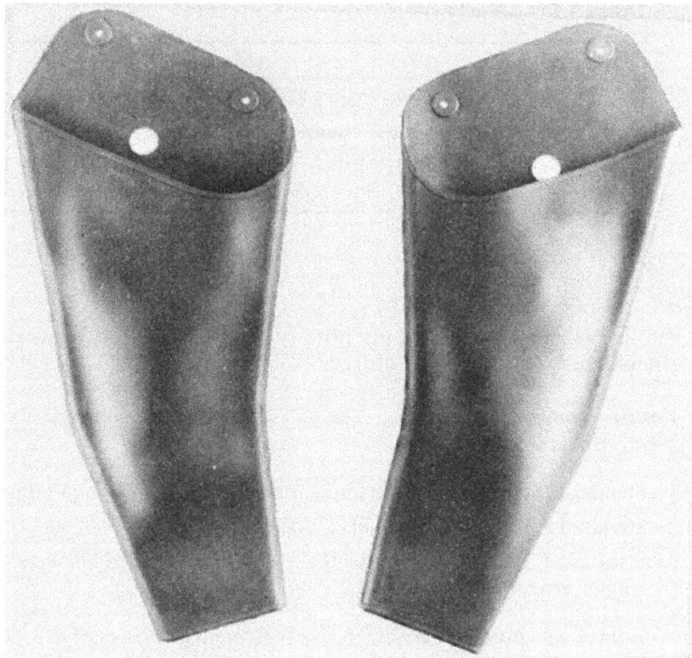

FIGURE 3.15 Insulating rubber sleeves. (*Courtesy W.H. Salisbury and Co.*)

Construction and Standards. The ASTM publishes three standards that affect the construction and use of rubber sleeves.

1. Standard D 1051 specifies manufacturing and technical requirements for rubber sleeves.
2. Standard F 496 specifies in-service care requirements.
3. Standard F 1236 is a guide for the visual inspection of blankets, gloves, sleeves, and other such rubber insulating equipment.

Insulating sleeves are available in five basic voltage classes (0 to 4), two basic types (I and II), and two styles (A and B). Table 3.8 identifies each class, its maximum-use voltage, and the ac rms and dc voltages that are used to proof-test them.

Type I sleeves are made of properly vulcanized, cis-1,4-polyisoprene rubber compound that is not ozone resistant. Type II sleeves are made of ozone-resistant elastomers.

Style A sleeves are made in a straight, tapered fashion (Fig. 3.16). Type B sleeves are of a curved elbow construction (Fig. 3.15).

Rubber sleeves are manufactured with no seams. They have a smooth finish and self-reinforced edges. Sleeves are manufactured with holes used to strap or harness them onto the worker. The holes are nominally 5/16 in (8 mm) in diameter and have nonmetallic, reinforced edges.

Rubber sleeves must be marked clearly and permanently with the name of the manufacturer or supplier, ASTM D 1051, type, class, size, and which arm they are to be used on (right or left). Such marking shall be confined to the shoulder flap area and shall be nonconducting and applied in such a manner as to not impair the required properties of the sleeve.

A sleeve shall have a color-coded label attached that identifies the voltage class. The labeling is subject to the minimum requirements specified in Table 3.10. Table 3.18 shows standard thicknesses for rubber sleeves, and Table 3.19 lists standard dimensions and tolerances.

FIGURE 3.16 Style A, straight taper rubber insulating sleeves. (*Courtesy ASTM.*)

TABLE 3.18 Standard Thickness for Rubber Insulating Sleeves

	Minimum		Maximum	
Class of sleeve	mm	in	mm	in
0	0.51	0.020	1.02	0.040
1	0.76	0.030	1.52	0.060
2	1.27	0.050	2.54	0.100
3	1.90	0.075	2.92	0.115
4	2.54	0.100	3.56	0.140

Source: Courtesy ASTM.

When to Use Rubber Sleeves. Rubber sleeves should be used any time personnel are working on or close to energized lines or lines that could be energized. They should be considered any time rubber gloves are being worn and should be required for anyone working around or reaching through energized conductors.

How to Use Rubber Sleeves. Rubber sleeves should be inspected before each use. They may be worn to protect the worker from accidental contact with energized conductors. Be certain to check the last test date marked on the sleeve. If the date is more than 12 months earlier than the present date, the sleeve shall not be used until it has been retested.

In-Service Inspection and Periodic Testing of Rubber Goods

Field Testing. Rubber goods should be inspected before each use. This inspection should include a thorough visual examination and, for rubber gloves, an air test. Table 3.20 is a synopsis of the

TABLE 3.19 Standard Dimensions and Tolerances for Rubber Insulating Sleeves

		Dimensions*							
		A		B		C		D	
Style	Size	mm	in	mm	in	mm	in	mm	in
Straight taper	regular	667	$26\frac{1}{4}$	394	$15\frac{1}{2}$	286	$11\frac{1}{4}$	140	$5\frac{1}{2}$
	large	724	$28\frac{1}{2}$	432	17	327	$12\frac{7}{8}$	175	$6\frac{7}{8}$
	extra large	762	30	483	19	337	$13\frac{1}{4}$	175	$6\frac{7}{8}$
Curved elbow	regular	673	$26\frac{1}{2}$	394	$15\frac{1}{2}$	311	$12\frac{1}{4}$	146	$5\frac{1}{4}$
	large	705	$27\frac{1}{2}$	406	16	327	$12\frac{7}{8}$	175	$6\frac{7}{8}$
	extra large	749	$29\frac{1}{2}$	445	$17\frac{1}{2}$	327	$12\frac{7}{8}$	178	7

* Tolerances shall be as follows:
A—±13 mm (±$\frac{1}{2}$ in)
B—Minimum allowable length
C—±13 mm (±$\frac{1}{2}$ in)
D—±6 mm (±$\frac{1}{4}$ in)

Source: Courtesy ASTM.

TABLE 3.20 Inspection Techniques for Rubber Insulating Equipment

Gloves and sleeves	Inspect glove and sleeve surface areas by gently rolling their entire outside and inside surface areas between the hands. This technique requires gently squeezing together the inside surfaces of the glove or sleeve to bend the outside surface areas and create sufficient stress to inside surfaces of the glove or sleeve to highlight cracks, cuts, or other irregularities. When the entire outside surface area has been inspected in this manner, turn the glove or sleeve inside-out and repeat the inspection on the inside surface (now the outside). If necessary, a more careful inspection of suspicious areas can be achieved by gently pinching and rolling the rubber between the fingers. Never leave a glove or sleeve in an inside-out condition. Stretch the thumb and finger crotches by pulling apart adjacent thumb and fingers to look for irregularities in those areas.
Blankets and mats	Place rubber blankets on a clean, flat surface and roll up tightly starting at one corner and rolling toward the diagonally opposite corner. Inspect the entire surface for irregularities as it is rolled up. Unroll the blanket and roll it up again at right angles to the original direction of rolling. Repeat the rolling operations on the reverse side of the blanket.
Line hose	Examine the inside surfaces of the insulating line hose by holding the hose at the far end from the lock and placing both hands side-by-side palms down around the hose. With the slot at the top and the long free end of the hose on the left, slowly bend the two ends of the hose downward while forcing the slot open with the thumbs. The hose should be open at the bend, exposing the inside surface. Slide the left hand about a foot down the hose and then, with both hands firmly gripping the hose, simultaneously move the left hand up and the right hand down to pass this section over the crown of the bend for inspection. Slide the right hand up the hose to the left hand. Hold the hose firmly with the right hand while the left hand again slides another foot down the hose. Repeat the inspection and, in this way, the entire length of hose passes through the hands from one end to the other.

Source: This material is reproduced verbatim from ASTM standard F 1236. It appears with the permission of the American Society for Testing and Materials.

inspection procedures defined in ASTM standard F 1236. Rolling is a procedure in which the rubber material is gently rolled between the hands or fingers of the inspector. This procedure is performed on both the inside and outside of the material. Figures 3.17 and 3.18 illustrate two types of rolling techniques.

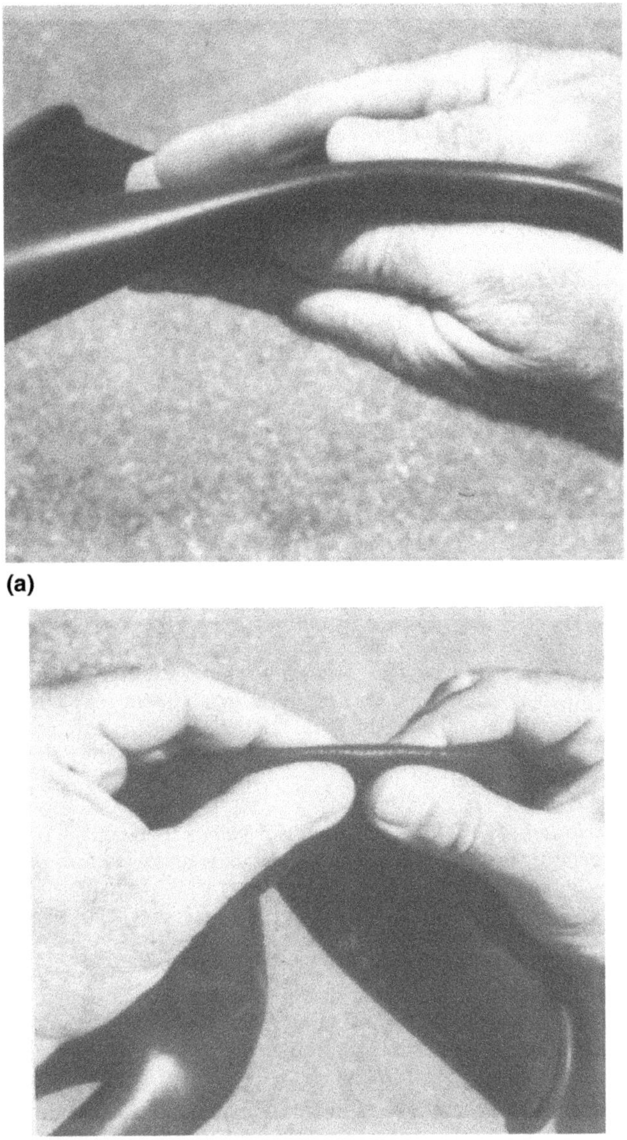

(a)

(b)

FIGURE 3.17 Inspection of gloves and blanket by rolling. (a) Hand rolling; (b) pinch rolling. (*Courtesy ASTM.*)

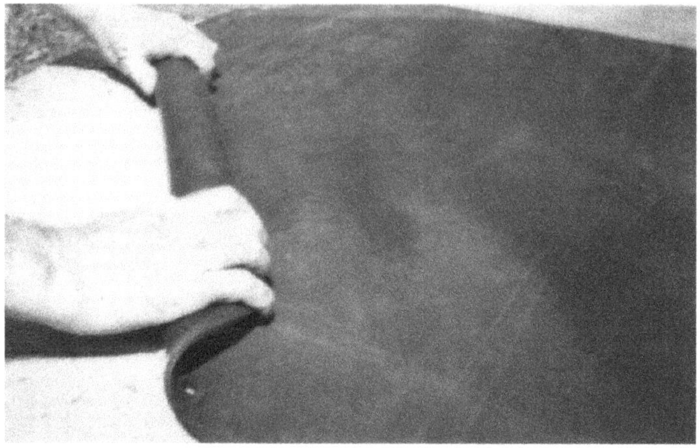

FIGURE 3.18 Inspection of rubber blanket by rolling. (*Courtesy ASTM.*)

Rubber gloves should be air-tested before each use. First, the glove should be inflated with air pressure and visually inspected, and then it should be held close to the face to feel for air leaks through pinholes. Note that when air-testing rubber gloves, never inflate them to more than 150 percent of normal glove size. Overinflation will stretch and stress the rubber, thereby destroying its insulating quality. Rubber gloves can be inflated by twirling, by rolling, or by using a mechanical glove inflater. Note that "rolling" in this situation is not the same as the rolling discussed earlier in the description of rubber goods inspection.

To inflate a glove by twirling, grasp the side edges of the glove (Fig. 3.19*a*), gently stretch the glove until the end closes and seals (Fig. 3.19*b*), and then twirl the glove in a rotating motion using the rolled edges of the glove opening as an axis (Fig. 3.19*c*). This will trap air in the glove and cause it to inflate.

Gloves that are too heavy to inflate by twirling may be inflated by rolling. To do this, lay the glove on a flat surface, palm up, and press the open end closed with the fingers (Fig. 3.20*a*). Then while holding the end closed, tightly roll up about $1\frac{1}{2}$ in of the gauntlet (Fig. 3.20*b*). This will trap air in the glove and cause it to inflate. Gloves may also be inflated by commercially available mechanical inflaters (Fig. 3.21). After the gloves are inflated, they may be visually inspected (Fig. 3.22) and then checked by holding them close to the face to listen and feel for air leaks through pinholes (Fig. 3.23).

Description of Irregularities in Rubber Goods. If damage or imperfections are discovered during inspections, the rubber goods may need to be removed from service. The following are typical of the types of problems that may be discovered.

1. *Abrasions and scratches* (Fig. 3.24). This is surface damage that normally occurs when the rubber material makes contact with an abrasive surface.
2. *Age cracks* (Fig. 3.25). Surface cracks that may look like crazing of glazed ceramics and may become progressively worse. Age cracks result from slow oxidation caused by exposure to sunlight and ozone. They normally start in areas of the rubber that are under stress.
3. *Chemical bloom* (Fig. 3.26). A white or yellowish discoloration on the surface. It is caused by migration of chemical additives to the surface.
4. *Color splash* (Fig. 3.27). This is a spot or blotch caused by a contrasting colored particle of unvulcanized rubber that became embedded in the finished product during the manufacturing process.
5. *Cuts* (Fig. 3.28). These are smooth incisions in the surface of the rubber caused by contact with sharp-edged objects.

(a)

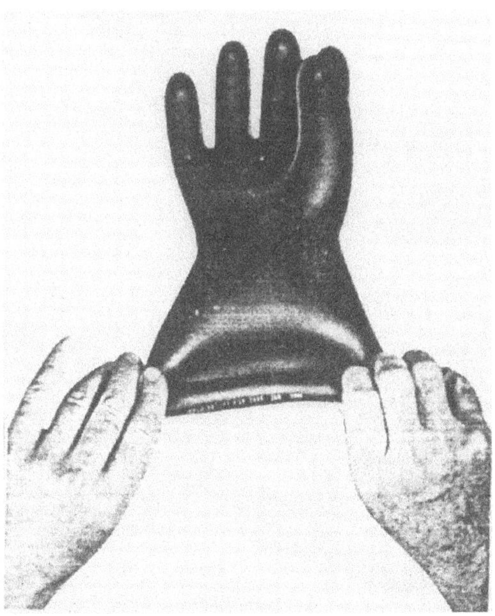

(b)

FIGURE 3.19 Inflating rubber gloves by twirling. (a) Grasping; (b) stretching; (c) twirling. (*Courtesy ASTM.*)

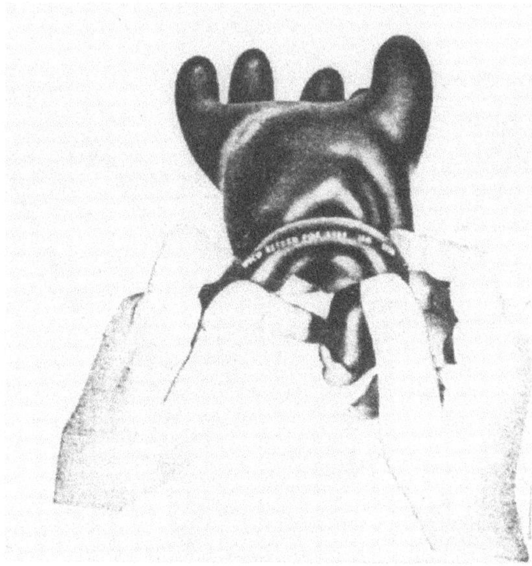

(c)

FIGURE 3.19 (*Continued*)

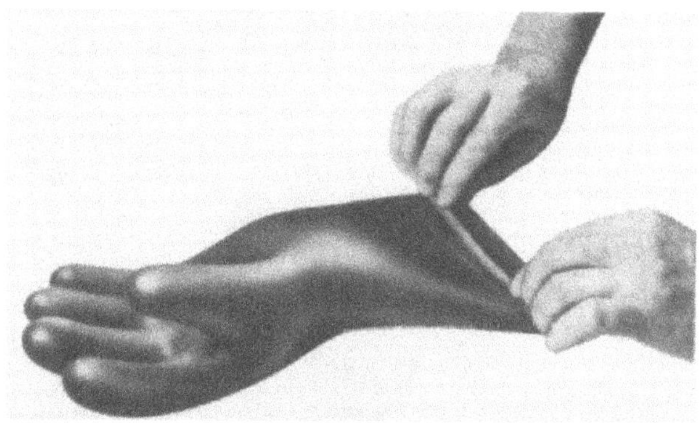

(a)

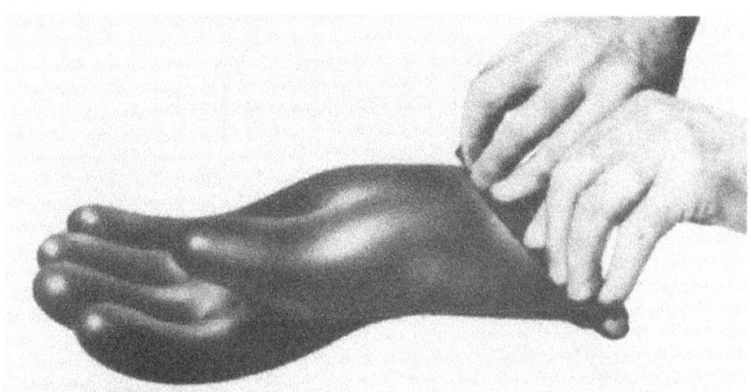

(b)

FIGURE 3.20 Inflating rubber gloves by rolling. (a) Pressing; (b) rolling. (*Courtesy ASTM.*)

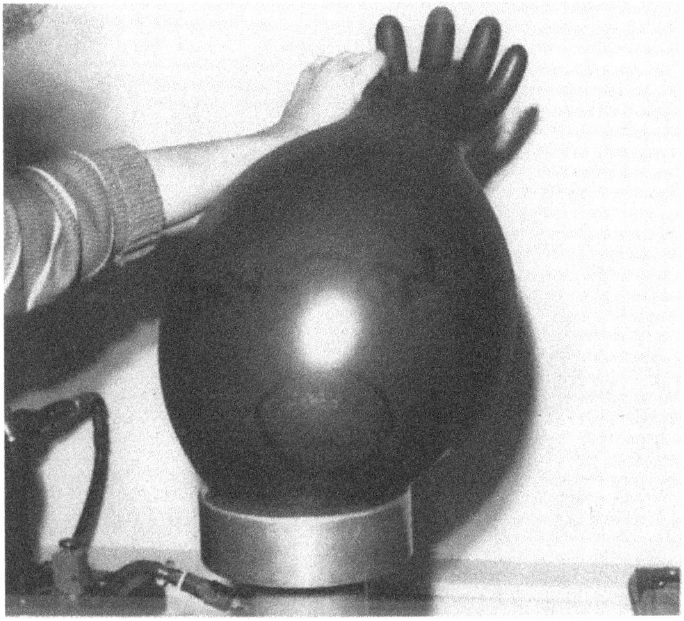

FIGURE 3.21 Mechanical inflation. (*Courtesy ASTM.*)

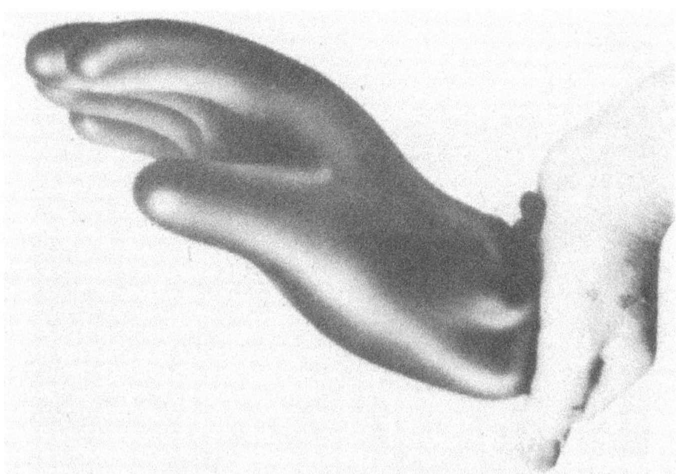

FIGURE 3.22 Visually inspecting inflated rubber gloves. (*Courtesy ASTM.*)

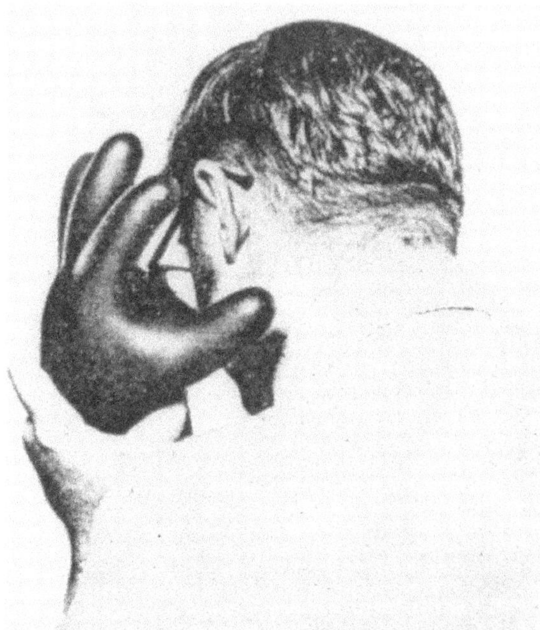

FIGURE 3.23 Inspecting rubber gloves by listening and feeling for air leaks through pinholes. (*Courtesy ASTM.*)

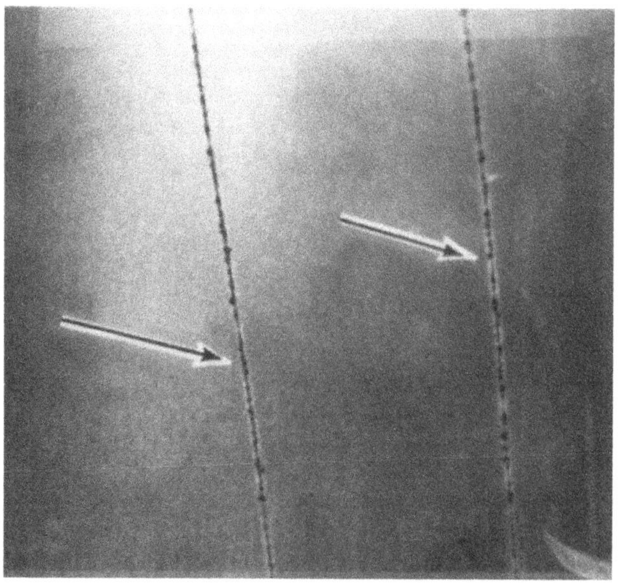

FIGURE 3.24 Scratches. (*Courtesy ASTM.*)

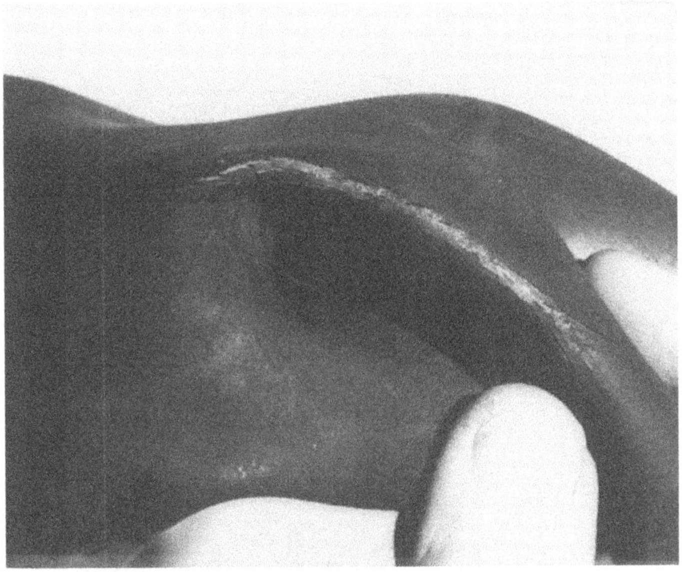

FIGURE 3.25 Age cracks. (*Courtesy ASTM.*)

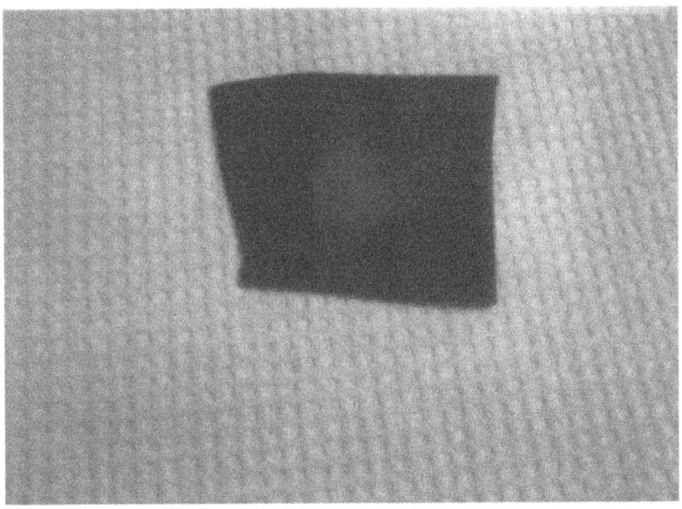

FIGURE 3.26 Chemical bloom. (*Courtesy ASTM.*)

FIGURE 3.27 Color splash. (*Courtesy ASTM.*)

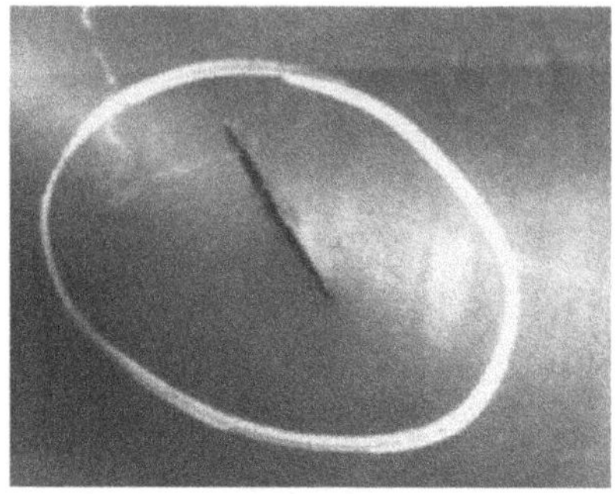

FIGURE 3.28 Cuts. (*Courtesy ASTM.*)

FIGURE 3.29 Depression or indentations. (*Courtesy ASTM.*)

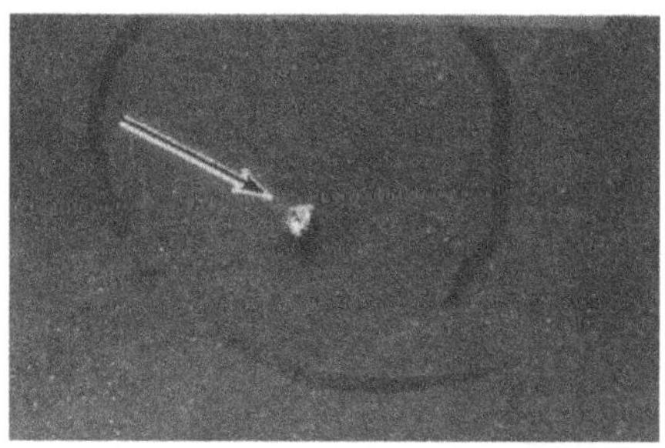

FIGURE 3.30 Embedded foreign material. (*Courtesy ASTM.*)

6. *Depressions or indentations* (Fig. 3.29). A shallow recess that exhibits a thinner rubber thickness at the bottom of the depression than in the surrounding areas.
7. *Detergent cracks.* Cracks that appear on the inside surface of the glove or sleeve. The cracks form around a spot of detergent residue that was not removed during the cleaning and rinsing of the form prior to the dipping process.
8. *Embedded foreign matter* (Fig. 3.30). This is a particle of nonrubber that has been embedded in the rubber during the manufacturing process. It normally shows up as a bump when the rubber is stretched.
9. *Form marks.* This is a raised or indented section on the surface of the rubber. It is caused by an irregularity in the form that was used to mold the product.
10. *Hard spot* (Fig. 3.31). A hardened spot caused by exposure to high heat or chemicals.

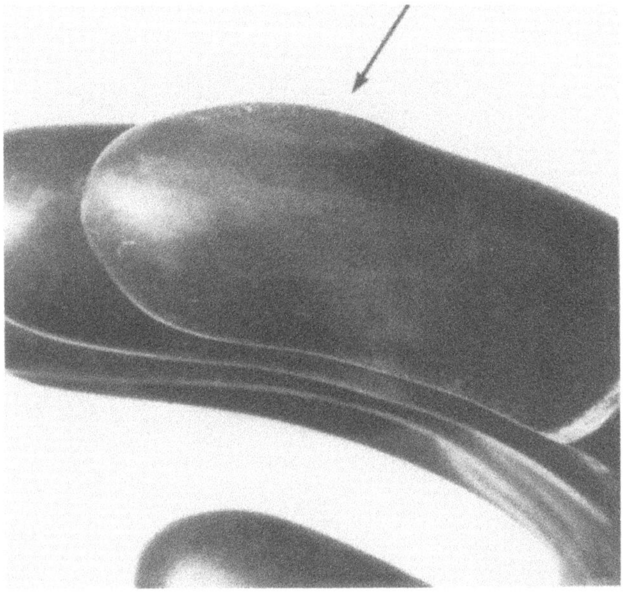

(a)

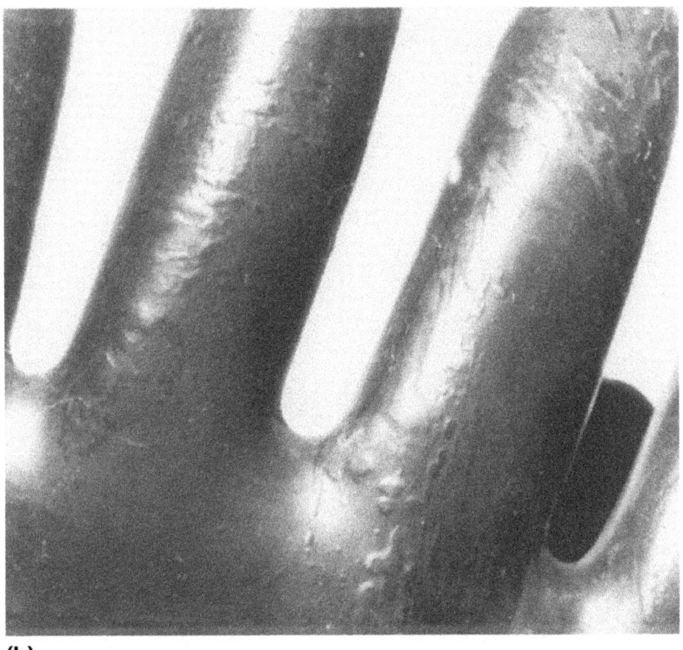

(b)

FIGURE 3.31 Hard spot. (*Courtesy ASTM.*)

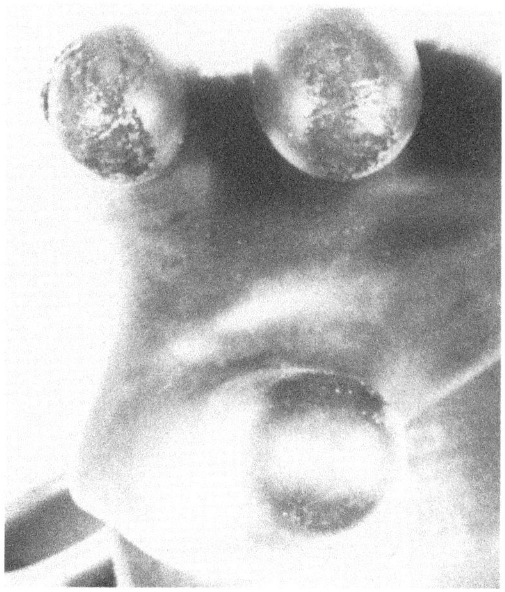

(c)

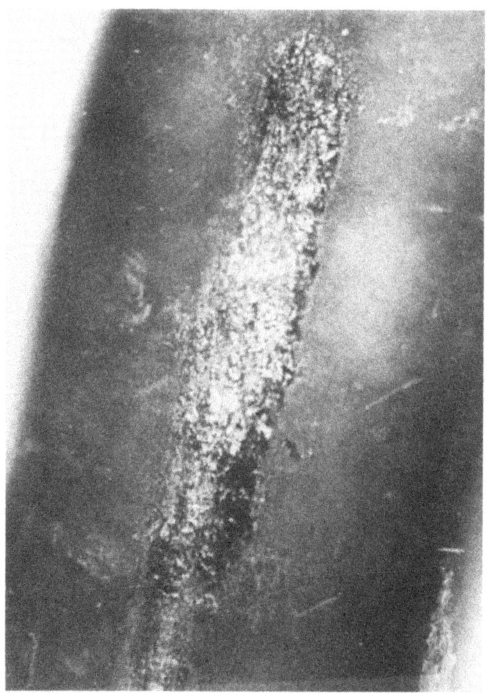

(d)

FIGURE 3.31 (*Continued*)

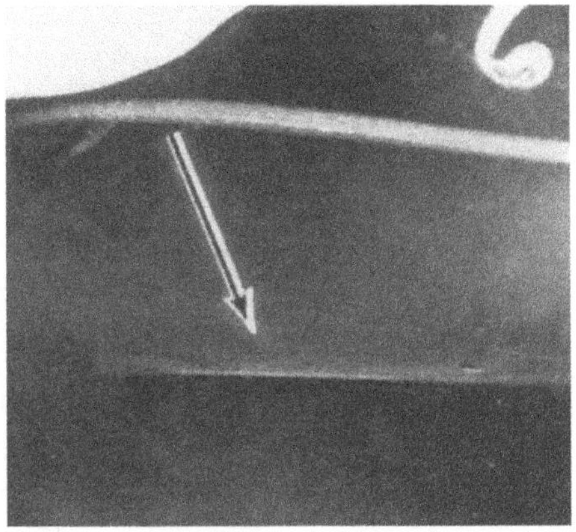

(a)

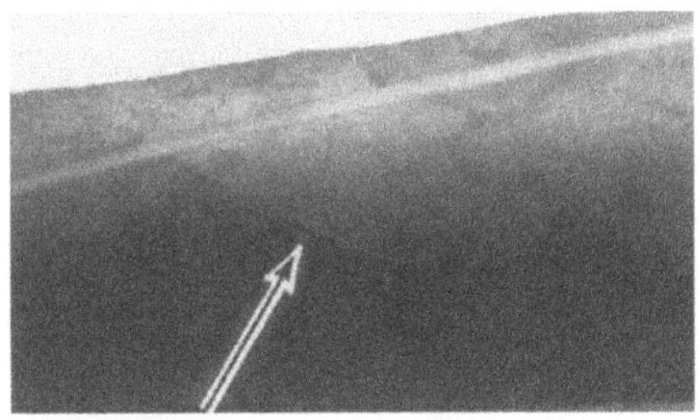

(b)

FIGURE 3.32 Mold mark. (*Courtesy ASTM.*)

11. *Mold mark* (Fig. 3.32). A raised or indented section caused by an irregularity in the mold.
12. *Scratches, nicks, and snags* (Figs. 3.24, 3.33, and 3.34). Angular tears, notches, or chiplike injuries in the surface of the rubber caused by sharp objects such as wire, pointed tools, staples, or other similar sharp-edged hazards.
13. *Ozone cracks* (Fig. 3.35). A series of interlacing cracks that start at stress points and worsen as a result of ozone-induced oxidation.
14. *Parting line or flash line.* A ridge of rubber left on finished products. They occur at mold joints during manufacturing.
15. *Pitting* (Fig. 3.36). A pockmark in the rubber surface. It is often created by the rupturing of an air bubble close to the surface during manufacturing.
16. *Protuberance* (Fig. 3.37). A bulge or swelling above the surface of the rubber.
17. *Puncture* (Fig. 3.38). Penetration by a sharp object through the entire thickness of the product.
18. *Repair marks* (Fig. 3.39). An area of the rubber with a different texture. Usually caused by repair of the mold or form.

FIGURE 3.33 Nick. (*Courtesy ASTM.*)

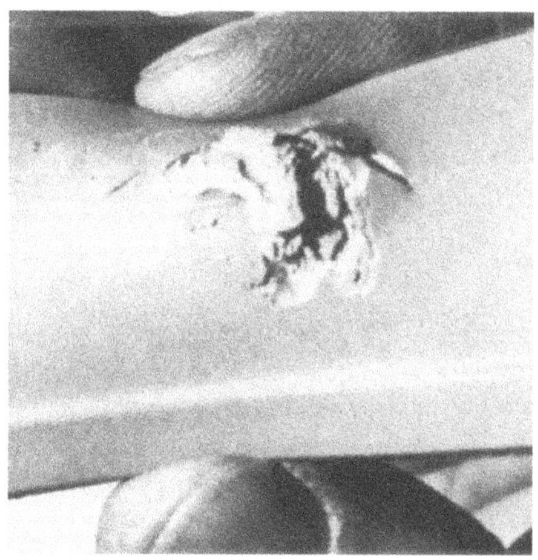

FIGURE 3.34 Snag. (*Courtesy ASTM.*)

19. *Runs.* Raised flow marks that occur on rubber glove fingers during the dipping process.
20. *Skin breaks.* Cavities in the surface of the rubber. They have filmy ragged edges and smooth interior surfaces. They are usually caused by embedded dirt specks during the manufacturing process.
21. *Soft spots.* Areas of the rubber that have been soft or tacky as a result of heat, oils, or chemical solvents.
22. *Tears* (Fig. 3.40). A rip through the entire thickness of the rubber. Usually caused by forceful pulling at the edge.

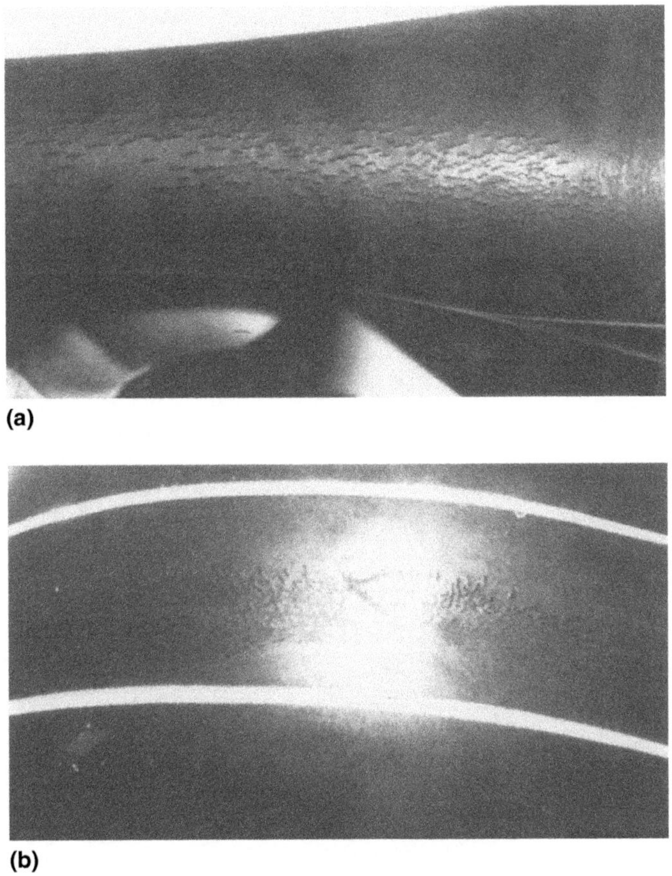

FIGURE 3.35 Ozone cracks. (*Courtesy ASTM.*)

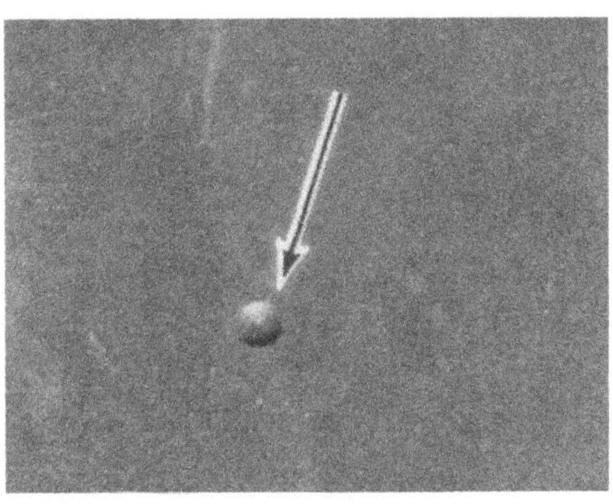

FIGURE 3.36 Pitting. (*Courtesy ASTM.*)

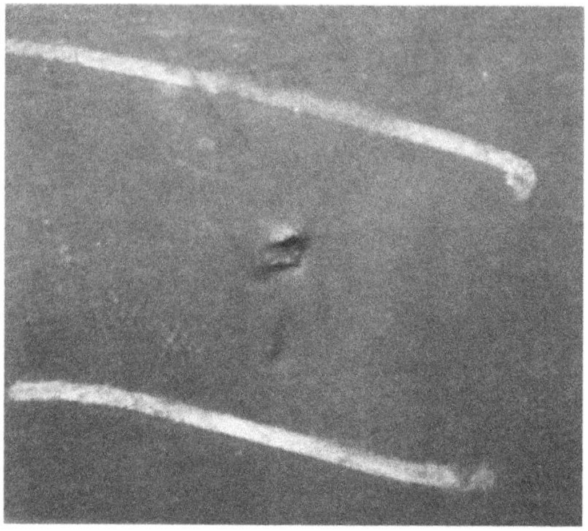

FIGURE 3.37 Protuberance. (*Courtesy ASTM.*)

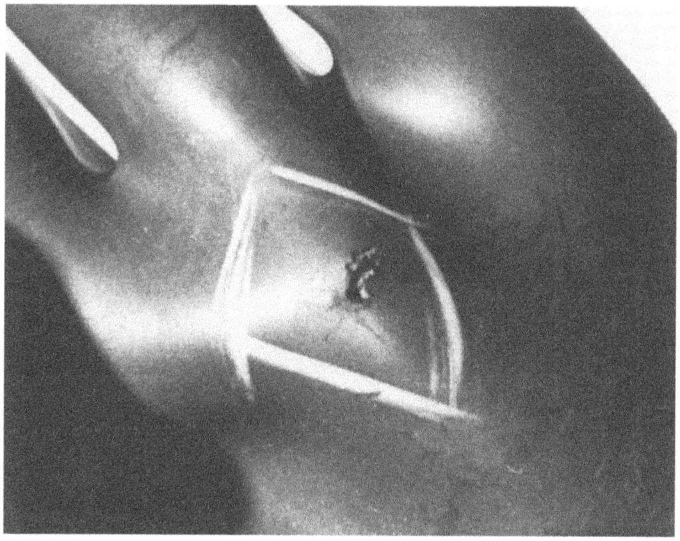

FIGURE 3.38 Puncture. (*Courtesy ASTM.*)

FIGURE 3.39 Repair mark. (*Courtesy ASTM.*)

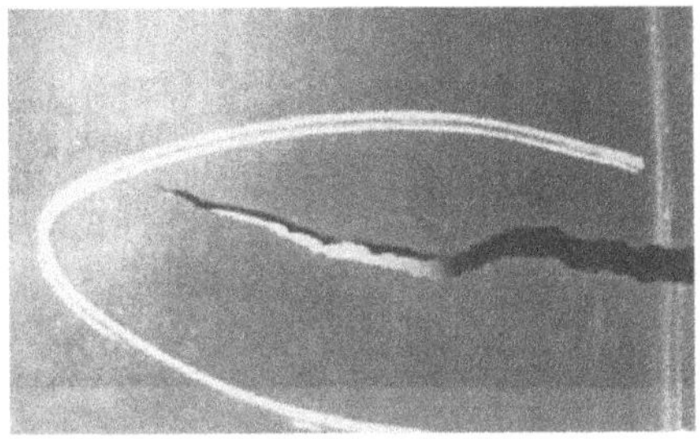

FIGURE 3.40 Tear. (*Courtesy ASTM.*)

TABLE 3.21 In-service Electrical Tests for Rubber Insulating Goods

Product	Maximum test interval, months	ASTM standard	Notes
Gloves	6	F 496	Tested, unused gloves may be placed into service within 12 months of the previous tests without being retested.
Mats	—	D 178	There is no regulatory requirement for in-service electrical testing of mats. They are tested by the manufacturer when new.
Blankets	12	F 479	Tested, unused blankets may be placed into service within 12 months of the previous tests without being retested.
Covers	—	F 478	Covers should be retested when in-service inspections indicate a need.
Line hose	—	F 478	Line hose should be retested when in-service inspections indicate a need.
Sleeves	12	F 496	Tested, unused sleeves may be placed into service within 12 months of the previous tests without being retested.

New insulating equipment is not permitted to be placed into service unless it has been electrically tested within the previous 12 months. Insulating equipment that has been issued for service is not new and is required to be retested in accordance with the intervals in this table.

Electrical Testing. Rubber insulating equipment should be electrically tested periodically. Table 3.21 summarizes the requirements and/or recommendations for such testing. Electrical testing of rubber goods is a relatively specialized procedure and should be performed only by organizations with the necessary equipment and experience. For detailed testing information, refer to the ASTM standards referenced in Table 3.21. Note that all rubber goods are tested at the same voltages dependent upon their class (00, 0, 1, 2, 3, or 4). See Table 3.8 for a listing of ASTM-required test voltages.

HOT STICKS

Description and Application

Hot sticks are poles made of an insulating material. They have tools and/or fittings on the ends that allow workers to manipulate energized conductors and equipment from a safe distance. Hot sticks vary in length depending on the voltage level of the energized equipment and the work to be performed. Modern hot sticks are made of fiber glass and/or epoxy glass. Older designs were made of

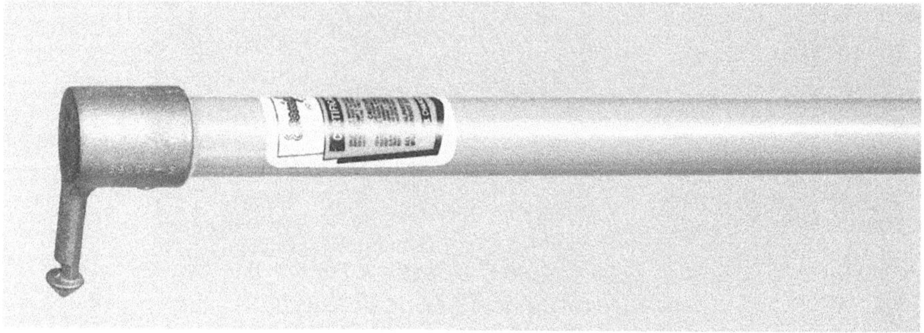

FIGURE 3.41 Typical hot stick. (*Courtesy AB Chance Corp.*)

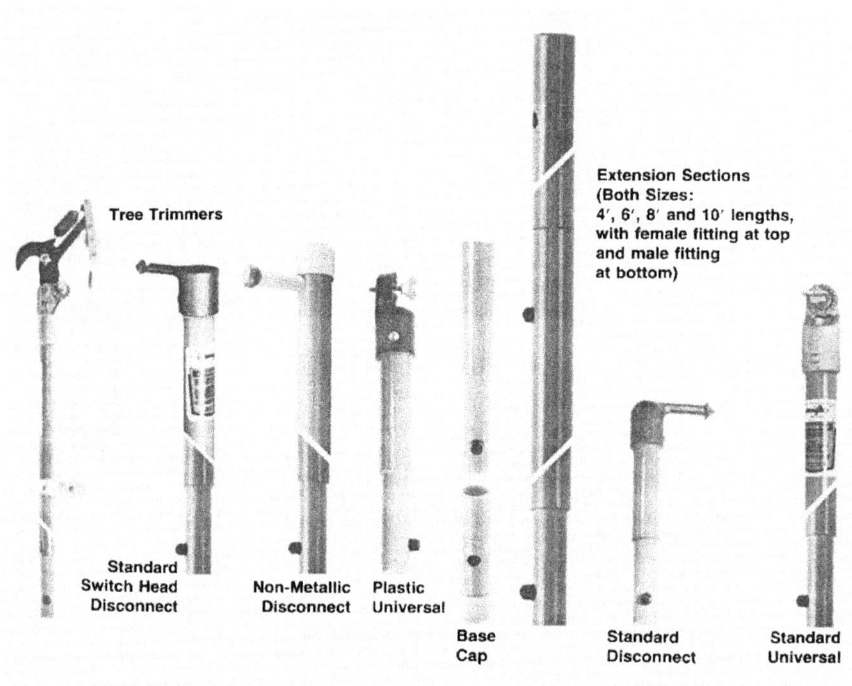

FIGURE 3.42 Various fittings, couplings, and tools for hot sticks. (*Courtesy AB Chance Corp.*)

wood that was laminated or treated and painted with chemical-, moisture-, and temperature-resistant materials. Figure 3.41 is an example of a simple hot stick fitted with a tool suitable for operation of open-air disconnect switches.

Hot sticks can be fitted with a variety of tools and instruments. The most common fitting is the NEMA standard design shown in Fig. 3.42 as the *standard universal* fitting. This fitting allows a variety of tools and equipment to be connected to the hot stick. Figure 3.43 shows a voltage tester attached to a hot stick using a standard fitting.

Figure 3.42 also shows other attachments and extensions that can be used to increase the usefulness of hot sticks. In addition to the equipment shown in Fig. 3.42, hot sticks can also be equipped with wrenches, sockets, screwdrivers, cutters, saws, and other such tools.

FIGURE 3.43 Voltage-measuring tool attached to a hot stick. (*Courtesy W.H. Salisbury and Co.*)

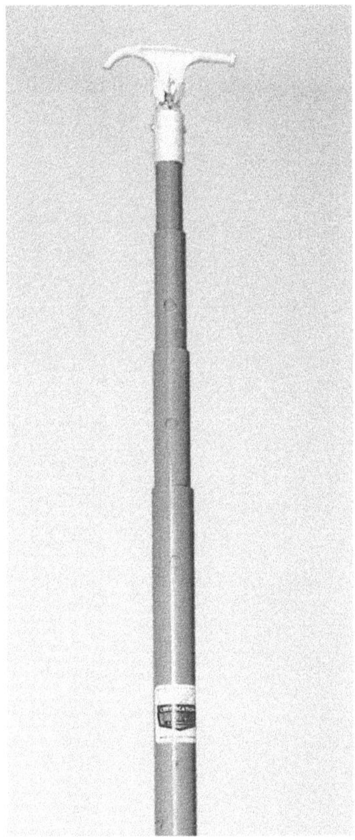

FIGURE 3.44 Telescoping hot stick—shown collapsed. (*Courtesy AB Chance Corp.*)

Hot sticks can also be purchased in telescoping models (Fig. 3.44) and so-called shotgun models (Fig. 3.45). The telescoping type of hot stick is composed of several hollow, tubular sections that nest inside of each other. The topmost section is first extended and locked in place by means of a spring-loaded button that snaps into a hole. The user of the hot stick extends as many of the sections as are required to accomplish the job at hand. The telescoping hot stick makes very long hot stick lengths available, which then collapse to a small, easy-to-carry assembly.

The shotgun hot stick (Fig. 3.45) has a sliding lever mechanism that allows the user to open and close a clamping hook mechanism at the end. In this way, the user can attach the stick to a disconnect ring and then operate it. After the switch is operated, the shotgun mechanism is operated to open the hook. The shotgun stick gets its name from its similarity to the pump-action shotgun.

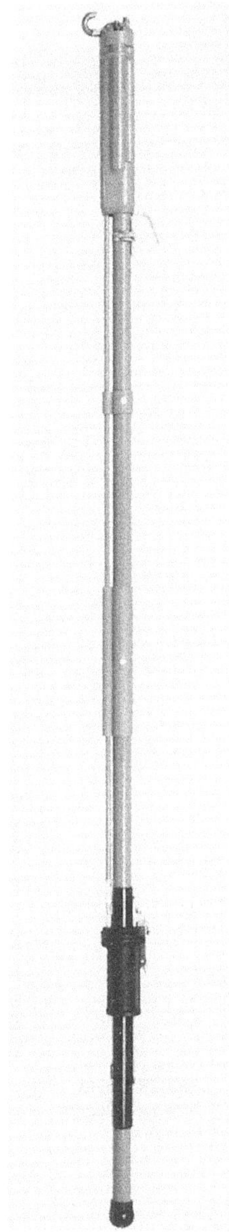

FIGURE 3.45 Shotgun-type hot stick. (*Courtesy AB Chance Corp.*)

Figure 3.46 shows a hot stick kit with several sections and various tools. This type of package provides a variety of configurations that will satisfy most of the day-to-day needs for the electrician and the overhead line worker. The kit includes the following components:

1. Six 4-ft sections of an epoxy glass snap-together hot stick
2. Aluminum disconnect head for opening and closing switches and enclosed cutouts

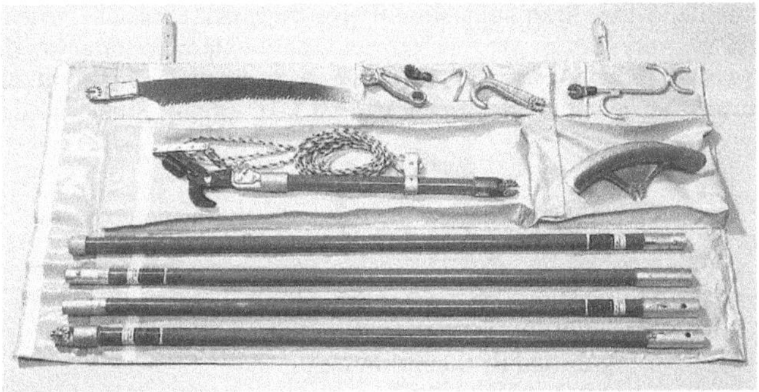

FIGURE 3.46 Typical hot stick kit for electricians and line workers. (*Courtesy AB Chance Corp.*)

TABLE 3.22 Typical Procedures Requiring Use of Hot Sticks

Medium voltage and higher
• Voltage measurement
• Any repairs or modifications to energized equipment
All voltages
• Operation of disconnects and cutouts
• Application of safety grounds

3. Nonmetallic disconnect head for use in indoor substations where buswork and switches are in close proximity
4. Clamp stick head for use with 6-in-long eye-screw ground clamps; used to apply and remove safety grounds
5. Tree trimmer attachment with 1½ ft of additional stick and pull rope used to close jaws of the trimmer
6. Pruning saw
7. Pistol-grip saw handle for use when tree limbs can be reached and insulation is not required
8. Heavy-duty vinyl-impregnated storage case

Electricians involved primarily in indoor work might wish to substitute other tools for the tree trimming and pruning attachments.

When to Use

Hot sticks should be used to insulate and isolate the electrician from the possibility of electric shock, arc, or blast. Table 3.22 identifies the types of procedures for which hot sticks are recommended.

How to Use

The specifics of hot stick use will depend on the task being performed and the location in which the worker is positioned. As a general rule, if hot sticks are being used, the worker should also wear other protective clothing. At a minimum, rubber gloves and face shields should be employed. However, many recommend that flash suits should also be worn, especially when safety grounds are being applied and the worker must approach closer than the flash boundary to do the job.

Before each use, the hot stick should be closely inspected for signs of physical damage, which may affect its insulating ability. If the hot stick is cracked, split, or otherwise damaged, it should be taken out of service.

Testing Requirements

ASTM standard F 711 requires that manufacturers test hot sticks to very stringent standards before they are sold. Additionally, OSHA standards require that hot sticks be inspected and/or tested periodically. The following should be the minimum:

1. Hot sticks should be closely inspected for damage or defects
   a. prior to each use.
   b. at least every two years.
2. If any damage or defects are noted, the hot stick should be repaired or replaced.
3. Hot sticks should be electrically tested according to ASTM standard F 711
   a. any time an inspection reveals damage or a defect.
   b. every two years.

INSULATED TOOLS

Description and Application

Insulated tools, such as those shown in Fig. 3.47, are standard hand tools with a complete covering of electrical insulation. Every part of the tool is fully insulated. Only the minimum amount of metallic work surface is left exposed. Such tools are used to prevent shock or arc in the event that the worker contacts the energized conductor.

ASTM standard 1505 defines the requirements for the manufacture and testing of insulated hand tools. Such tools are to be used in circuits energized 50 V ac to 1000 V ac, or 50 V dc to 1500 V dc. The insulated hand tools shown in Fig. 3.47 are covered with two layers of material. The inner layer provides the electrical insulation and the outer layer provides mechanical protection for the electrical insulation. Insulated tools can also be purchased that have a single, thicker layer of insulation. Either type meets the OSHA and NFPA 70E requirements because all are rated for 1000 V ac and 1500 V dc and are tested at 5000 V ac by the manufacturer.

When to Use

Insulated tools should be used any time work is being performed on or near exposed, energized conductors energized at any voltage exceeding 50 V but less than 1000 V ac. They should be inspected before each use.

How to Use and Care For

Insulated tools are used in the same way that ordinary hand tools are used, and all the same precautions should be observed. Avoid using the tools in any application that may damage the insulation. Insulated tools should be kept clean and stored in such a manner as to not cause damage to the insulation. Electrical workers should have a designated tool pouch for insulated tools, so the insulation is not damaged by carrying and storing them in the same pouch as their noninsulated tools are stored and carried in.

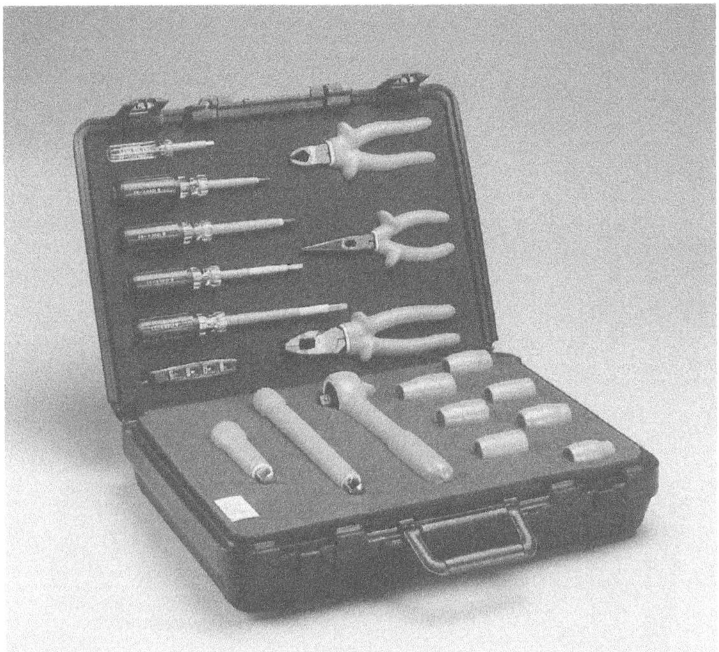

FIGURE 3.47 Insulated tool set. (*Courtesy Cementex.*)

BARRIERS AND SIGNS

Whenever work is being performed that requires the temporary removal of normal protective barriers such as panels or doors, barriers and signs should be used to warn personnel of the hazard.

Barrier Tape

Barrier tape is a continuous length of abrasion-resistant plastic tape. It should be a minimum of 2 in wide and should be marked or colored to clearly indicate the nature of the hazard to which employees will be exposed if they cross the tape line. Figure 3.48 shows a type of barrier tape suitable for marking and barricading an area where an electrical hazard exists.

Such tape should be yellow or red to comply with Occupational Safety and Health Administration (OSHA) standards. Generally red tape, or red tape with a white stripe, is the preferred color for such applications. Whatever color is chosen should be a standard design that is used consistently for the same application.

Signs

Warning signs should be of standardized design and easily read. They should be placed in such a way as to warn personnel of imminent hazard. Figure 3.49 shows a type of sign suitable for use as an electrical hazard warning.

FIGURE 3.48 Barrier tape styles suitable for electrical hazards. (*Courtesy Direct Safety Supply Co.*)

FIGURE 3.49 Typical electrical hazard sign. (*Courtesy Ideal Industries, Inc.*)

When and How to Use

General Requirements. Table 3.23 lists the general procedures for using signs, barriers, and attendants to warn personnel of electrical hazards. Signs should be placed so that they are easily seen and read from all avenues of access to the hazardous area.

Temporary Hazard Barricades. Temporary hazard barricades should be constructed with a striped barrier tape using the following procedure:

1. The tape should be placed so that it completely encloses the hazardous area.
2. The tape should be clearly visible from all directions of approach.
3. The tape should be at a level such that it forms an effective barrier. Approximately 3 ft is suitable.
4. Allow an area of sufficient size to give adequate clearance between the hazard and any personnel outside the hazardous area.

TABLE 3.23 Summary of the Use of Warning Methods

Type of warning	When to use
Signs	Signs should be placed to warn employees about electrical hazards that may harm them.
Barricades	Barricades should be used to prevent and limit employee access to work areas where they may be exposed to electrical hazards. Such barriers should be made of nonconductive material such as plastic barrier tape.
Attendants	If signs and/or barricades cannot provide a level of safety sufficient to protect employees, attendants shall be placed to guard hazardous areas. Attendants shall be familiar with the nature and extent of the hazard so as to adequately warn other employees.

FIGURE 3.50 Typical tags suitable for tagout purposes. (*Courtesy Ideal Industries, Inc.*)

5. If test equipment is being used on equipment inside the hazardous area, the tape should be arranged so that the equipment can be operated outside the area.
6. Do not use the same style and color of tape for any purpose other than marking temporary hazards.
7. Such a barricaded area should be considered to be the same as a metal enclosure; that is, *access is not possible*.
8. After the hazard has been eliminated, remove the tape.

SAFETY TAGS, LOCKS, AND LOCKING DEVICES

Safety tags, locks, and locking devices are used to secure and mark equipment that has been taken out of service. They are applied in such a way that the equipment cannot be reenergized without first removing the tags and/or locks. For more information about lockout and tagout procedures, refer to Chap. 4.

Safety Tags

Safety tags are applied to equipment to indicate that the equipment is not available for service. They are tags constructed of a durable, environment-proof material. They should be of standardized construction and include an "action statement" warning that says Do Not Start, Do Not Open, Do Not Close, Do Not Operate, or other such warning. The tag must also indicate who placed it on the equipment and the nature of the problem with the equipment. Figure 3.50 shows tags that are suitable for such an application.

FIGURE 3.51 Typical padlocks suitable for lockout purposes. (*Courtesy Ideal Industries, Inc.*)

Some manufacturers supply tags with individual employee photographs on them. This type of tag helps to further identify who placed the tag and to personalize the installation.

Tags are to be applied using strong, self-locking fasteners. Nylon cable wraps are suitable for such an application. The fastener must have a breaking strength of not less than 50 lb.

Locks and Multiple-Lock Devices

Locks are used to prevent operation of equipment that has been de-energized. They must be strong enough to withstand all but the most forceful attempts to remove them without the proper key. If a lock can be removed by any means other than a bolt cutter or the key that fits it, the lock should not be used.

Standard padlocks are normally applied for lockout purposes (Fig. 3.51). Each employee should have a set of padlocks that can be opened only by his or her key. See Chap. 4 for situations in which a safety lock may be removed by someone other than the individual who placed it.

Departments may also have "group" padlocks that are placed by shift personnel and that are keyed to a departmental key. For example, the operations department may have shift operators who remove and restore equipment from service. Equipment may be removed from service during one shift and later returned to service during another shift. In such situations, the group lock will be placed by one individual and removed by another. Thus, the group locks will have master keys so that any authorized shift operator can place or remove them.

Multiple-Lock Devices. Sometimes several workers will need to place a lock on one piece of equipment. This often happens when several crafts are working in the area secured by the lock. In these circumstances, a multiple-lock device is used (Fig. 3.52). To lock out a piece of equipment, the multiple-lock device is first opened and applied as though it were a padlock. Then the padlock is inserted through one of the holes in the device. The padlock prevents opening of the multiple-lock device, which, in turn, prevents operation of the equipment. The devices shown in Fig. 3.52*a* can accommodate up to six locks; however, multiple-lock devices can be "daisy chained" to allow as many locks as required.

The multiple-lock devices in Fig. 3.52*b* combine the multiple-lock device with a safety tag. The information required on the tag is written on the tag portion of the device, which is then applied to the equipment.

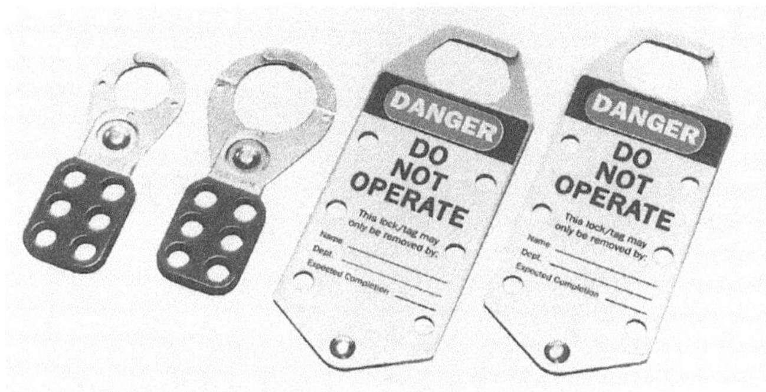

(a) (b)

FIGURE 3.52 Multiple-lock devices. (*Courtesy Ideal Industries, Inc.*)

Locking Devices

Some equipment, such as wall switches and molded-case circuit breakers, do not readily accommodate locks. In these instances, when lockout is required, a locking device must be used. Figure 3.53*a* shows locking devices that may be placed over the handle of a molded-case circuit breaker and clamped in place. The lock is then installed through the hole left for that purpose. The breaker cannot be operated until the device is removed, and the locking device cannot be removed until the padlock is open.

Figure 3.53*b* is a similar device that mounts on a standard wall switch. The locking device is first attached to the wall switch with the switch faceplate mounting screws. The switch is moved to the off position and the hinged cover of the device is closed. A padlock is placed through the flange supplied for that purpose.

Devices like the wall switch lockout device can be left in place permanently for switches that are frequently locked out. Although the device shown in Fig. 3.53*b* can be used to lock the switch in the on position, this practice is not a good safety practice.

When and Where to Use Lockout-Tagout

Equipment should be locked out and tagged out when it is being serviced or maintained and an unexpected start-up could injure personnel who are working in the area. Thus tags and locks should be placed any time an employee is exposed to contact with parts of the equipment that have been de-energized.

The locks and tags must be applied to all sources of power for the affected equipment. They must be applied in such a way that the equipment cannot be reenergized without first removing the locks and/or tags. See Chap. 4 for more complete information on lockout-tagout procedures. Figure 3.54 shows locks, tags, and multiple-lock devices being applied to electrical switching equipment.

VOLTAGE-MEASURING INSTRUMENTS

Safety Voltage Measurement

Safety voltage measurement actually involves measuring for zero voltage. That is, a safety measurement is made to verify that the system has been de-energized and that no voltage is present.

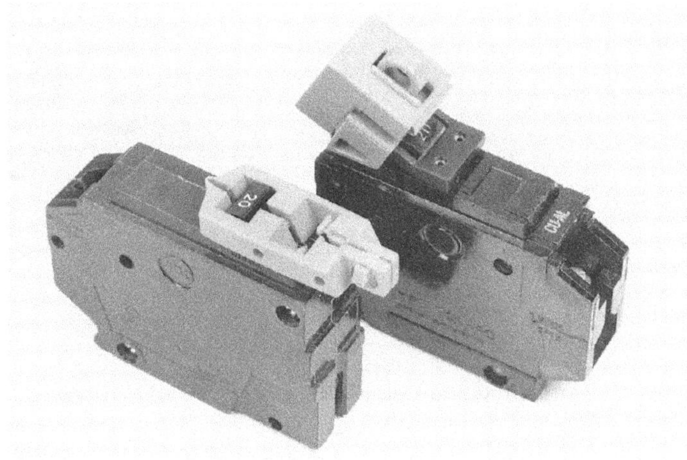

(a)

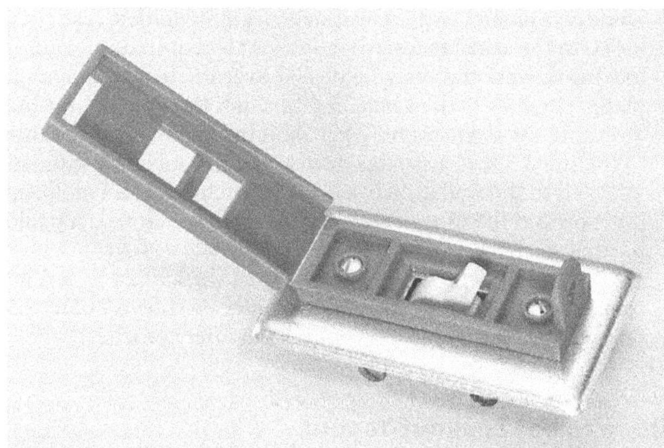

(b)

FIGURE 3.53 Locking devices. (*Courtesy Ideal Industries, Inc.*)

(See Chap. 4 for more details.) Because of this, the instruments that are used for safety voltage measurement need only be accurate enough to determine whether or not the nominal system voltage is present at the point of worker exposure.

The instruments discussed in the following sections are intended primarily for safety voltage measurements. Some of them are quite accurate. Contact the individual manufacturers for more specific information.

For various reasons, some authorities prefer a proximity tester over a contact tester, while others prefer a contact tester over a proximity tester. Whichever type is selected, it must be capable of indicating the presence of dangerous voltages and discriminating between nuisance static voltages and actual hazardous voltage sources. See Chap. 4 for more information.

FIGURE 3.54 Typical application of locks, tags, and multiple-lock devices. (*Courtesy Ideal Industries, Inc.*)

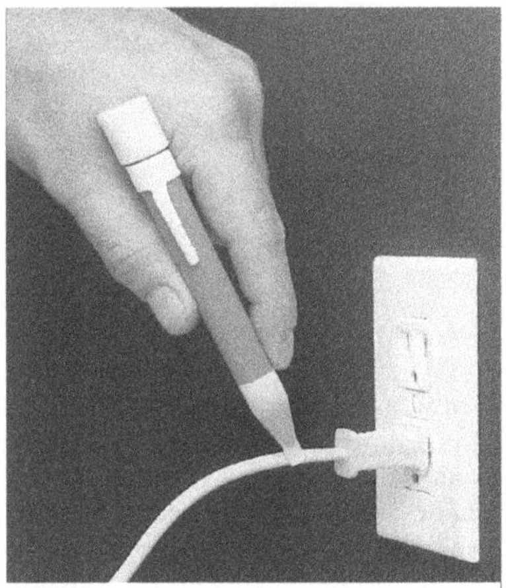

FIGURE 3.55 "AC Sensor" proximity voltage sensor for use on circuits up to 600 V alternating current. (*Courtesy Santronics, Inc., Sanford, NC.*)

Proximity Testers

Proximity testers do not require actual metal-to-metal contact to measure the voltage, or lack of voltage, in a given part of the system. They rely on the electrostatic field established by the electric potential to indicate the presence of voltage. Proximity testers will indicate voltage levels through insulation. They may not provide accurate results when cable is shielded.

Proximity testers do not indicate the actual level of the voltage that is present. Rather they indicate the presence of voltage by the illumination of a light and/or the sounding of a buzzer. Figures 3.55 through 3.57 are three different types of proximity detectors for various voltage levels.

Figure 3.55 shows a simple neon light proximity tester. The end of the unit is plastic and sized to fit into a standard 120-V duplex receptacle. It requires two AAA flashlight cells to operate. When placed in proximity to an energized circuit, the red neon, located in the white plastic tip, glows.

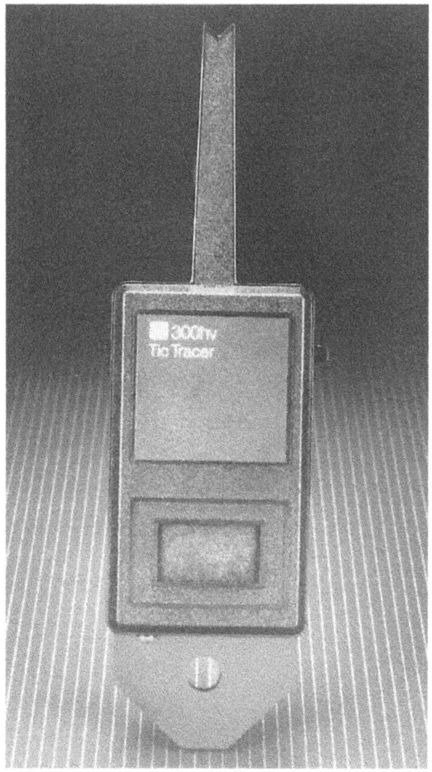

FIGURE 3.56 "TIC" tracer proximity voltage sensor for use on circuits up to 35,000 V. (*Courtesy TIF Electronics.*)

Figure 3.56 shows a proximity tester that uses a combination of audio and visual indicators. When turned on, the unit emits a slow beeping sound that is synchronized to a small flashing light. If the unit is placed close to an energized circuit, the number of beeps per second increases. The light flashes increase in frequency along with the beeps. The higher the voltage and/or the closer the voltage source, the faster the beeps. When the voltage is very high, the beeping sound turns into a steady tone. This unit has a three-position switch. One position turns the unit off, and the two others provide low- and high-voltage operation, respectively. If high voltages are being measured, the unit should be attached to a hot stick. The manufacturer supplies a hot stick for that purpose.

Figure 3.57 shows a proximity tester that is similar in some ways to the one shown in Fig. 3.56. It has both a light and a tone; however, the light and tone are steady when a voltage is sensed above the unit's operational threshold. The unit in Fig. 3.57 also has a multiposition switch that has an off position, a 240-V test position, a battery test position, and multiple voltage level positions from 4200 V up to the unit's maximum. This unit is shown attached to a hot stick in Fig. 3.43. It should not be used to measure voltages without the use of a hot stick and/or appropriate rubber insulating equipment.

Contact Testers

Some personnel prefer the use of testers that make actual metal-to-metal contact with the circuit being energized. Such instruments are called contact testers. *Contact testers* may be simple indicators, but more often they are equipped with an analog or digital meter that indicates actual voltage level. Figures 3.58 through 3.61 illustrate various styles of contact testers that may be used for safety-related measurements.

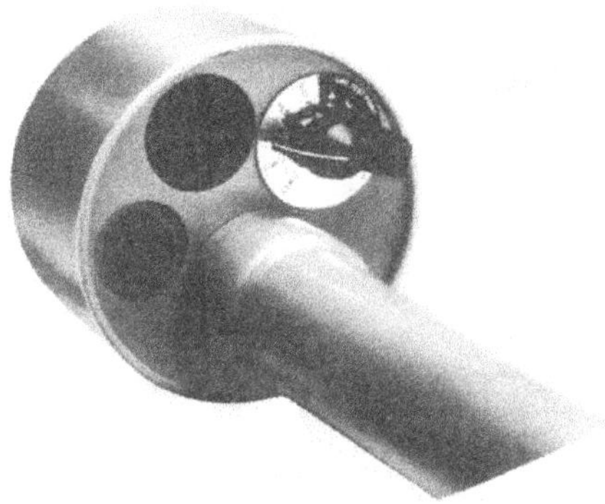

FIGURE 3.57 Audiovisual proximity voltage tester for use in circuits from 240 V to 500 kV. (*Courtesy W.H. Salisbury and Co.*)

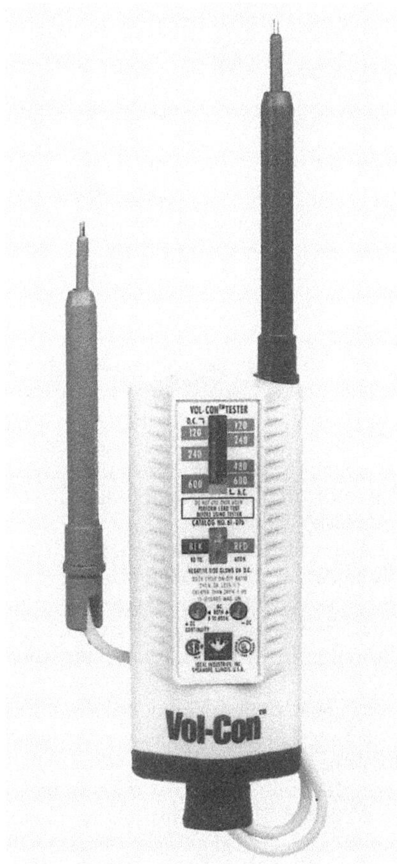

FIGURE 3.58 Safety voltage and continuity tester. (*Courtesy Ideal Industries, Inc.*)

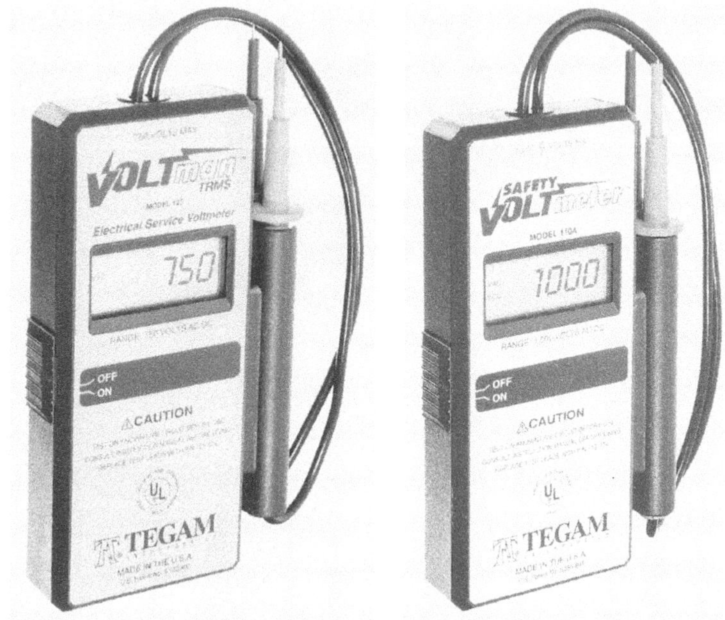

(a) **(b)**

FIGURE 3.59 Digital readout contact-type safety voltmeter. (*Courtesy Tegam, Inc.*)

Figure 3.58 shows one of the more popular models used for voltages up to 600 V alternating current or direct current. This unit is a solenoid type of instrument. That is, a spring-loaded solenoid plunger is connected to an indicator that aligns with a voltage scale. The distance that the plunger travels is proportional to the voltage level of the measured circuit. The voltage scale is read in volts. The instrument shown in Fig. 3.58 also indicates continuity and low voltage. It switches automatically between those functions. Note that this style of unit is one of the older, but still popular styles in use for low-voltage measurements.

Caution: Because of their solenoid mechanism, these styles are subject to small arcs when contact is made with the measured circuit. This problem is mitigated in some instruments by using test leads equipped with fused resistors. Such resistors not only eliminate the arcing problem but also open in the event that an internal short circuit occurs in the meter. Because of the possible arcing problem, these testers are not allowed in many petrochemical plants or other locations with potentially explosive atmospheres.

Figure 3.59 shows a modern, digital readout safety voltmeter. This instrument is suitable for circuits up to 1000 V and is tested to 2300 V. It has an inherently high impedance and often has fused leads; therefore, it is not prone to arcing when the leads make contact. The meter should only be used for voltage measurements. It has no continuity or ammeter scales.

Figure 3.60 shows a typical digital multimeter. These instruments are in common use by virtually all electricians and electrical and electronic technicians. Such instruments often have voltage ranges well above 1000 V; however, their use in power circuits with voltages above 600 V is not recommended. Care should be taken when using this style of instrument in an electric power circuit. When making a voltage measurement, the user must carefully check to be sure that the instrument is in the voltage mode and that the leads are plugged into the correct jacks.

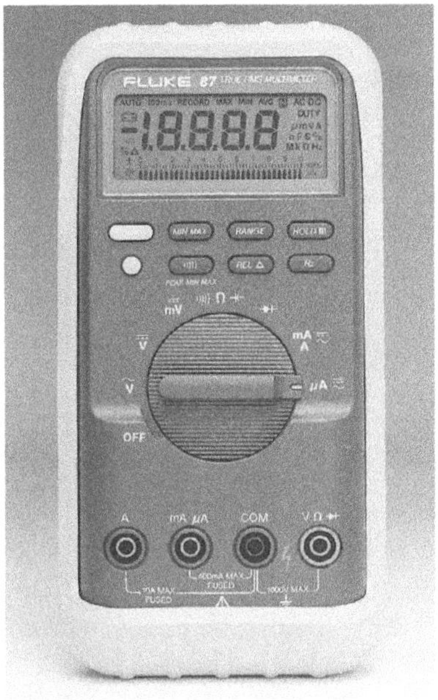

FIGURE 3.60 Digital readout multimeter. (*Courtesy Fluke.*)

The instrument shown in Fig. 3.61 is called a *phasing tester* because it is often used to phase two circuits—that is, to check that *A* phase in circuit 1 is the same as *A* phase in circuit 2 and so on. Such units are also used for safety-related voltage measurements. The instrument is composed of two high-resistance elements in series with an analog instrument. The resistances are selected in such a way that the meter tracks accurately up to 16 kV. To extend the range of the instrument up to 48 or 80 kV, extension resistors can be added.

Selecting Voltage-Measuring Instruments

Voltage-measuring instruments must be selected based on a variety of criteria. The following sections describe each of the steps that should be used in the selection of voltage-measuring instruments.

Voltage Level. The instrument used must have a voltage capability at least equal to the voltage of the circuit to be measured. Make certain that the manufacturer certifies the instrument for use at that level.

Application Location. Some instruments are designed for use solely on overhead lines or solely in metal-clad switchgear. Make certain that the manufacturer certifies the instrument for the application in which it will be used.

Internal Short-Circuit Protection. If the measuring instrument should fail internally, it must not cause a short circuit to appear at the measuring probes. Instruments with resistance leads and/or internal fuses should be employed. If an internal fuse opens, always replace it with the identical fuse or one that is recommended by the manufacturer.

Sensitivity Requirements. The instrument must be capable of reading the lowest voltage that can be present. This is from all sources such as backfeed as well as normal voltage supply.

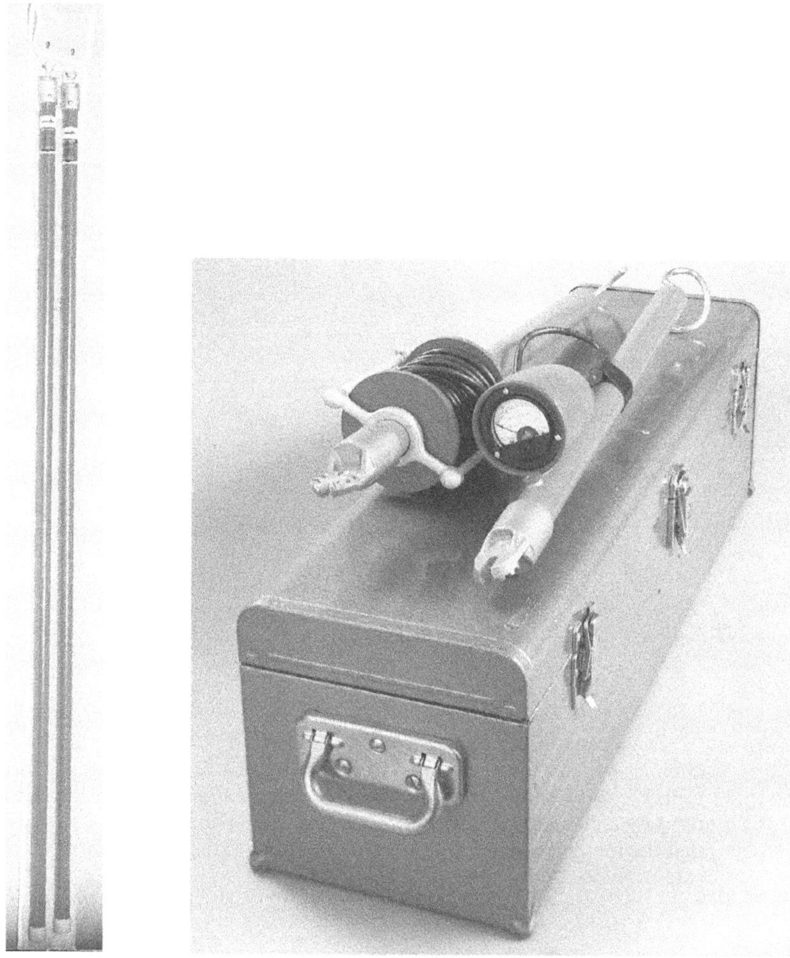

(a) (b)

FIGURE 3.61 Phasing tester extension handles. (*Courtesy AB Chance Corp.*)

Circuit Loading. The instrument must be capable of measuring voltages that are inductively or capacitively coupled to the circuit. Therefore, it must have a high enough circuit impedance so that it does not load the circuit, thereby reducing the measured voltage to apparently safe levels.

Instrument Condition

Before performing voltage measurements, the measuring instrument must be carefully inspected to ensure that it is in good mechanical and electrical condition.

Case Physical Condition. The case and other mechanical assemblies of the instrument must be in good physical condition and not cracked, broken, or otherwise damaged. Any instrument with a broken case should be taken out of service and repaired or replaced.

TABLE 3.24 Summary of IEC Meter Categories

Overvoltage category	Location of usage	Examples
Category I	Electronics and other types of cord connected equipment locations	• Any cord-connected equipment where the power supply or the equipment itself has built-in transient suppression capabilities • In equipment that is inherently low energy
Category II	Single-phase receptacle connected loads	• Appliances • Portable tools • Long branch circuits
Category III	Three-phase or single-phase distribution systems that are isolated from outdoor electric utility supplies by transformers, distance, or other specific types of surge protection	• Indoor lighting circuits • Bus and feeders in industrial facilities • Motors, switchgear, and other such industrial equipment
Category IV	Facilities with direct paths to outdoor utility circuits and feeders	• Outdoor substation facilities • Electric meter equipment connected directly to the utility • Service entrance equipment • Overhead lines to isolated facilities

Probe Exposure. Only the minimum amount of lead should be exposed on contact-type instruments. This minimizes the chance of accidentally causing a short circuit when the lead contacts more than one conductor at a time. Low-voltage instruments such as those shown in Figs. 3.58 and 3.59 have spring-loaded plastic sleeves. Spring loaded sleeves are no longer allowed and have been replaced by fixed plastic sleeves. Probes with spring-loaded sleeves should be replaced.

Lead Insulation Quality. The lead insulation should be closely inspected. If the insulation is frayed, scored, or otherwise damaged, the leads should be replaced before the instrument is used to measure voltage.

Fusing. If the instrument you are using is fused, the fuse should be checked to make certain that it is of the right size and capability. Instruments used to measure power system voltages should be equipped with high interrupting capacity fuses that will safely interrupt 200,000 A at rated voltage.

Operability. Make certain that the instrument is operable. If it uses batteries, they should be checked before the instrument is placed in service. Check the instrument on a known hot source before it is taken into the field. If it does not work, take it out of service until it can be repaired or replaced.

Low-Voltage Voltmeter Safety Standards

Background. IEC 61010 is a standard that has been developed to define design and usage safety requirements for low-voltage (less than 1000 V ac) meters. The standard applies primarily to contact-type instruments and establishes requirements for meter categories based on the meter's ability to withstand voltage surges such as those experienced in a modern low-voltage power system.

Categories. Table 3.24 shows the four categories as defined in the standard. For industrial and utility applications, the minimum category used should be category III. If the work to be performed is on the incoming service, incoming substation, or other electrical systems close to the utility system, category IV should be selected.

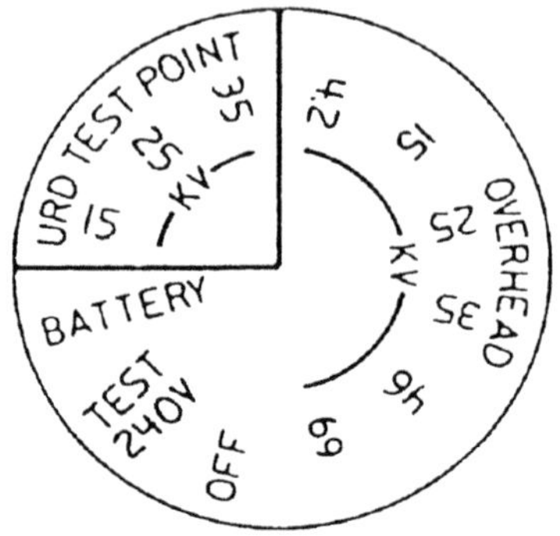

FIGURE 3.62 Selector switch positions for the meter shown in Fig. 3.57. (*Courtesy W.H. Salisbury and Co.*)

It is also important to note that many meters are rated for 600 V while others are rated at 1000 V. The 1000 V rated meters are always the preferred choice from a safety standpoint. In many applications, a category III meter rated at 1000 V is superior in surge withstand to a category IV rated at 600 V.

Three-Step Voltage Measurement Process

Since a safety-related voltage measurement is normally made to make certain that the circuit is dead, the measuring instrument must be checked both before and after the actual circuit is read. This ensures that a zero reading is, in fact, zero and not caused by a nonfunctional instrument. Note that this before-and-after check should be made in addition to the check that is given when the instrument is placed into service from the tool room or supply cabinet. This before-and-after measurement process is called the *three-step process.*

Step 1—Test the Instrument Before the Measurement. The measuring instrument should be applied to a source that is known to be hot. Ideally this source would be an actual power system circuit; however, this is not always possible because a hot source is not always readily available—especially at medium voltages and higher.

Because of this problem, manufacturers often have alternative means to check their instruments in the field. Some provide low-voltage positions on their instruments. Figure 3.62 shows the switch settings for the instrument of Fig. 3.57. This instrument may be verified in three different ways:

1. With the switch in the test-240v position, place the instrument head close to a live circuit in excess of 110 V.
2. With the switch in the test-240v position, rub the instrument head on cloth or clothing to obtain a static charge. The unit should indicate periodically.
3. Set the switch to the 35-KV overhead position and place the head close to a spark plug of a running engine.

After the instrument is verified, the battery position can be used to verify the battery supply circuitry and the battery condition.

Figure 3.63*b* shows a test device and setup used to verify the type of instrument shown in Fig. 3.61. One lead from the test device (Fig. 3.63*a*) is plugged into a special jack mounted on the instrument. The other end is clipped to each of the measuring probes, one at a time. The reading on the instrument meter is used to determine the operability of the test instrument.

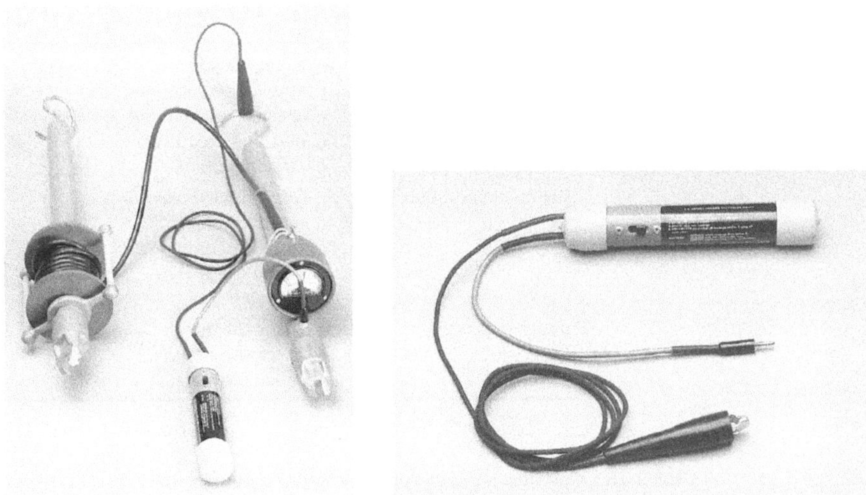

(a) (b)

FIGURE 3.63 Field test unit for voltage testers similar to Fig. 3.61. (*Courtesy AB Chance Corp.*)

Of course, the best way to test an instrument is to check it on a known hot source using the scale position that will be used for the actual measurement.

Step 2—Measure the Circuit Being Verified. The instrument should then be used to actually verify the presence or absence of voltage in the circuit. See Chap. 4 for a detailed description of how to make such measurements.

Step 3—Retest the Instrument. Retest the instrument in the same way on the same hot source as was used in step 1. Table 3.25 summarizes the voltage-measuring steps discussed in this section. Because of the extreme importance of voltage measurement, this procedure is repeated in more detail in Chap. 4.

General Considerations for Low-Voltage Measuring Instruments

Most general-purpose multimeters are not designed for power system use. This is certainly true of many of the small units that are available from some major retail companies. Table 3.26 lists specific requirements for test instruments used in power system voltage measurement.

SAFETY GROUNDING EQUIPMENT

The Need for Safety Grounding

Even circuits that have been properly locked and tagged can be accidentally energized while personnel are working on or near exposed conductors. For example,

- If capacitors are not discharged and grounded, they could accidentally be connected to the system.
- Voltages could be induced from adjacent circuits. Such voltages can be extremely high if the adjacent circuit experiences a short circuit.

TABLE 3.25 Voltage-Measurement Instrument Selection and Use*

Instrument selection	• Voltage level • Application location • Internal short-circuit protection • Sensitivity requirements • Circuit loading
Instrument condition	• Case physical condition • Probe exposure • Lead insulation quality • Fusing • Operability
Three-step process	• Check the meter • Check the circuit • Check the meter

*See text for details.

TABLE 3.26 Checklist for Low-Voltage Measuring Equipment

Instrument should be certified for use in electric power systems
Leads should be of proper design
Secured permanently to instrument
Fully insulated probes with minimum metallic exposure
Insulation rated for voltage of circuits to be measured
Internal fusing capable of interrupting fault current if short circuit occurs
High-resistance measurement to minimize arcing at probe tips
Single-function (voltage readings) instrument preferred
If a multimeter is used, the resistance and current measurement circuits must be adequately fused
Auto-voltage detection is desirable
Instrument insulation levels should be rated for voltage of the circuits to be measured
Instrument should be the correct category for the application per IEC 61010

- Switching errors could result in reenergizing of the circuit.
- An energized conductor can fall or be forced into the de-energized circuit, thereby energizing it and causing injury.
- Lightning strikes could induce extremely high voltages in the conductors.

Employees who are working on or near such exposed conductors could be severely injured from shock, arc, or blast that results when the conductors are accidentally energized. Because of this, safety grounding equipment should be employed as one additional safety measure when employees must work near exposed conductors.

Whenever possible, safety grounds are applied to create a zone of equal potential, or "equipotential" zone of protection, around the employee. This means that the voltage is equal on all components within reach of the employee. Correct methods for applying safety grounds are discussed more fully in Chap. 4.

Safety Grounding Switches

Grounding switches are specially manufactured units designed to replace a circuit breaker in medium-voltage metal-clad switchgear. The switch rolls into the space normally occupied by the breaker, and the switch stabs connect to the bus and line connections that normally connect to the breaker stabs. Figures 3.64 through 3.66 show various views of a typical grounding switch.

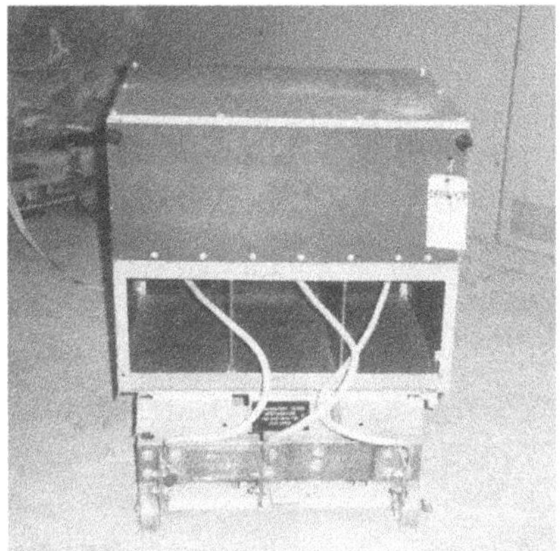

FIGURE 3.64 Front view of a 15-kV grounding switch. (*Courtesy UNO-VEN Refinery, Lemont, IL.*)

FIGURE 3.65 Rear view of a 15-kV grounding switch. (*Courtesy UNO-VEN Refinery, Lemont, IL.*)

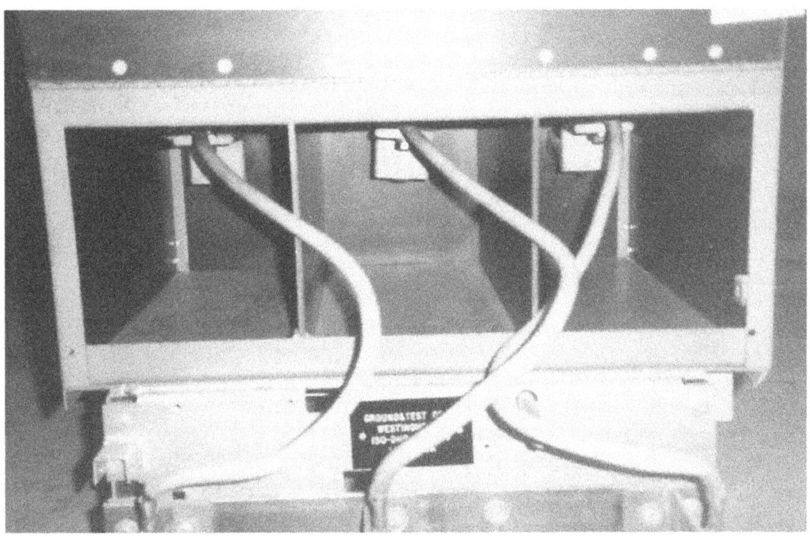

FIGURE 3.66 Close-up of jumper connections for 15-kV grounding switch. (*Courtesy UNO-VEN Refinery, Lemont, IL.*)

Two different varieties of grounding switches are in common use. One type of switch has two sets of disconnect stabs: one set connects to the bus phase connections, and the other connects to the line connections. The other type of switch has only one set of stabs, which can be moved to connect to either the bus or the line side of the switchgear connections.

For either variety of grounding switch, internal connections can be made to electrically ground the bus, the line, or both sides of the switch. Thus the switch grounds the circuit at the source.

Ground switches should be used very carefully and should not be relied on for 100 percent personal safety protection. Except in metal-clad switchgear, grounding switches cannot create a zone of equalized potential. Safety grounds (see next section) should be employed when secure personal protection is required.

Safety Grounding Jumpers

General Description. Safety grounding jumpers (also called safety grounds) are lengths of insulated, highly conductive cable with suitable connectors at both ends. They are used to protect workers by short-circuiting and grounding de-energized conductors. So if a circuit is accidentally energized, the safety grounds will short-circuit the current and protect the workers from injury. Safety grounds also drain static charges and prevent annoying or dangerous shocks. Figure 3.67 illustrates a typical safety ground.

Notice that the safety ground is composed of three major component parts: the clamp, the ferrule, and the cable. The entire assembly must be capable of withstanding the enormous thermal and magnetic forces that are applied when the circuit to which the jumper is applied becomes energized. The construction of safety grounds and their component parts is regulated by ASTM standard F 855.

Clamps. Clamps serve to connect the grounding jumper to the cable or bus system that is being grounded. ASTM standard F 855 defines three types of clamps (I, II, and III). Each of the three

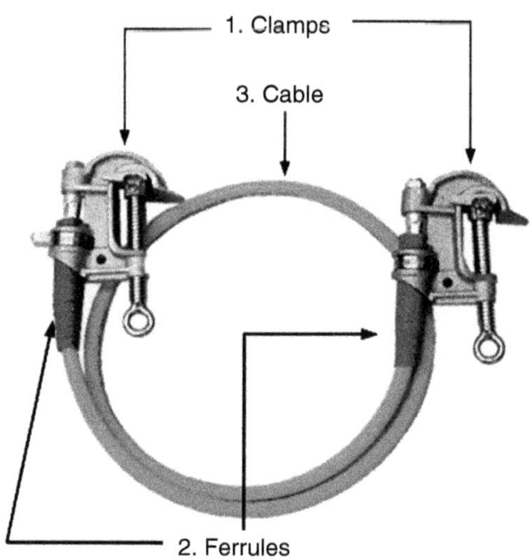

FIGURE 3.67 A typical safety ground with component parts marked. (*Courtesy W.H. Salisbury and Co.*)

types is available in two basic classes (A and B). Table 3.27 describes the basic types and classes of clamps.

Types I and II are used to connect the grounding jumper to the actual de-energized conductor. Type II has hot sticks that are permanently mounted to the clamp, and type I has only rings to which shotgun-type hot sticks can be connected. Type III clamps are used for connection to ground bus, tower steel, grounding clusters, and other such permanently grounded points. Figure 3.68*a* to *c* illustrates the three types of grounding clamps.

Each type of clamp can be purchased with either nonserrated (class A) or serrated (class B) jaws. Regardless of which type of jaw is purchased, the conductor surfaces should be thoroughly cleaned before application of the clamp. The class B jaws provide additional assurance that

TABLE 3.27 Types and Classes of Grounding Jumper Clamps

Type I	Clamps for installation on de-energized conductors equipped with eyes for installation with removable hot sticks
Type II	Clamps for installation on de-energized conductors equipped with permanently mounted hot sticks
Type III	Clamps for installation on permanently grounded conductors of metal structures; these clamps are equipped with tee handles, eyes, or square or hexagon head screws, or both
Class A	Clamps equipped with smooth contact surfaces
Class B	Clamps equipped with serrated, cross-hatched, or other means used to abrade or bite through corrosion products on the conductor to which the jaw is clamped

Source: From ASTM standard F 855.

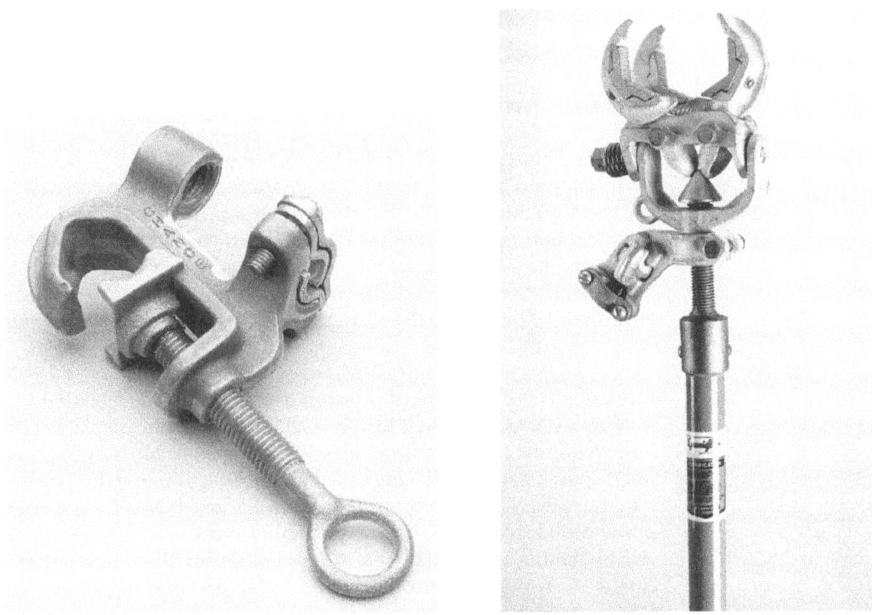

(a) **(b)**

FIGURE 3.68 ASTM type I, II, and III clamps. (a) Type I clamp; (b) type II clamp; (c) type III clamp. (*Courtesy AB Chance Corp.*)

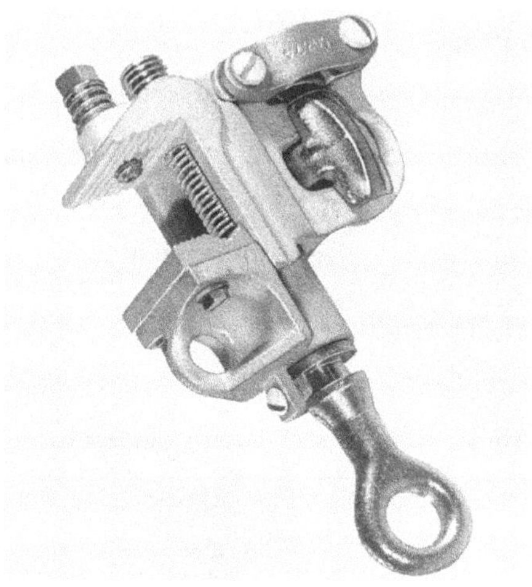

(c)

FIGURE 3.68 (*Continued*)

adequate electric contact will be made by biting through surface corrosion or dirt.

Clamps are available in seven ASTM grades depending on the short-circuit capability of the circuit to which they will be connected. Table 3.28 is taken from ASTM standard F 855 and identifies the seven grades and their associated electrical and mechanical characteristics.

Ferrules. Ferrules are used to connect the jumper cable to the clamp. ASTM standard F 855 identifies six basic types of ferrule as described in Table 3.29. Figure 3.69*a* through *d* illustrates each of the six basic types.

In addition to types, ferrules are manufactured in seven grades depending upon the short-circuit currents that they are designed to withstand. Table 3.30 shows the electrical and mechanical characteristics for the seven grades of ferrule. Table 3.31 shows the physical specifications for grounding cable ferrules. Of course, ferrules must be attached to the clamp with a mechanically and electrically secure type of termination. Table 3.32 lists the ASTM standard types of terminations between ferrules and grounding clamps.

Cables. The cables used for grounding jumpers are specified in three basic types (I, II, and III) as illustrated in Table 3.33. Note that the insulation can be of any color; however, most manufacturers use yellow, black, or clear insulation. Clear insulation has the advantage of allowing easy visual inspection to detect signs of conductor fusing after severe short circuits. Clear insulation, however, is sometimes sensitive to ultraviolet degradation. Yellow insulation has the advantage of being easily visible and indicating a potentially hazardous condition. Black insulation, while acceptable, can be easily mistaken for a normal conductor. Clear or yellow insulation is the preferred choice for grounding jumper cable.

Grounding jumper cable sizes are determined by the amount of available short-circuit current that is present in the part of the system to which they are being

TABLE 3.28 Grounding Clamp Ratings

	Grounding clamp torque strength, min				Short-circuit properties*									
	Yield†		Ultimate		Withstand rating, symmetrical kA rms, 60 Hz			Ultimate rating/capacity,‡ symmetrical kA rms, 60 Hz						
Grade	lbf·in	n·m	lbf·in	n·m	15 cycles (250 ms)	30 cycles (500 ms)	Copper cable size	6 cycles (100 ms)	15 cycles (250 ms)	30 cycles (500 ms)	60 cycles (1 s)	Maximum copper test cable size	Continuous current rating, A rms, 60 Hz	Minimum cable size with ferrule installed equal or larger than
1	280	32	330	37	14.5	10	#2	30	19	13	9	2/0	200	#2
2	280	32	330	37	21	15	1/0	48	30	21	15	4/0	250	1/0
3	280	32	330	37	27	20	2/0	60	38	27	19	4/0	300	2/0
4	330	37	400	45	36	25	3/0	76	48	34	24	250 kcmil	350	3/0
5	330	37	400	45	43	30	4/0	96	60	43	30	250 kcmil	400	4/0
6	330	37	400	45	54	39	250 kcmil or 2 2/0	113	71	50	35	350 kcmil	450	250 kcmil or 2 2/0
7	330	37	400	45	74	54	350 kcmil or 2 4/0	159	100	71	50	550 kcmil 550 kcmil	550	350 kcmil or 2 4/0

Note: lbf·in = foot-pound–inch, n·m = newton-meter, ms = millisecond, kcmil = thousand circular mil.
*Withstand and ultimate short-circuit properties are based on performance with surges not exceeding 20% asymmetry factor.
†Yield shall mean no permanent deformation such that the clamp cannot be reused throughout its entire range of application.
‡Ultimate rating represents a symmetrical current that the clamp shall carry for the specified time.
Source: Courtesy ASTM.

applied. Table 3.34 lists the maximum current ratings and the size of cable required by ASTM standards. Table 3.34 also identifies the current ratings for commercially manufactured, complete jumpers described in the next section.

Complete Jumpers. While some users make their own safety ground jumpers using the ASTM specifications and buying the individual components, most prefer to buy a completely preassembled jumper set as illustrated in Fig. 3.70. The jumper illustrated in Fig. 3.70 is designed for use on live-front pad-mount underground distribution (UD) switchgear and transformers. It is rated at 15,000 A for 30 cycles and 21,000 A for 15 cycles. It has a three-way terminal block assembly, four bronze body grounding clamps, and three 6-ft lengths of 2/0 clear jacketed copper cable.

TABLE 3.29 Types of Ferrules for Use on Safety Ground Jumpers

Type I	Compression ferrule; cylindrical and made for installation on cable stranding by compression
Type II	Solder ferrule is tubular and made for installation on cable stranding with solder
Type III	Plain stud shrouded compression ferrule with a stepped bore that accepts the entire cable over the jacket
Type IV	Threaded stud shrouded compression ferrule with stepped bore that accepts entire cable over jacket and has male threads at forward end
Type V	Bolted shrouded compression ferrule with internal threads and a bolt at forward end
Type VI	Threaded stud compression ferrule with male threads at forward end

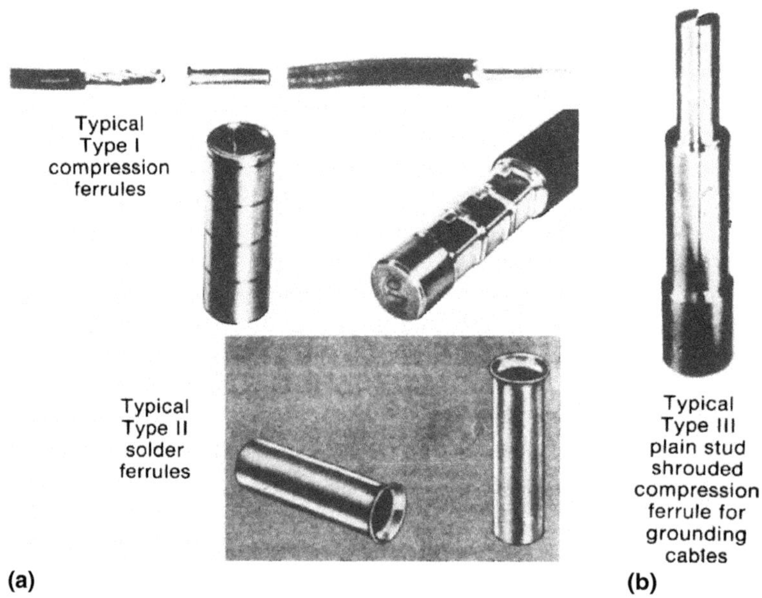

FIGURE 3.69 ASTM types I, II, III, IV, V, and VI ferrules. (*Courtesy ASTM.*)

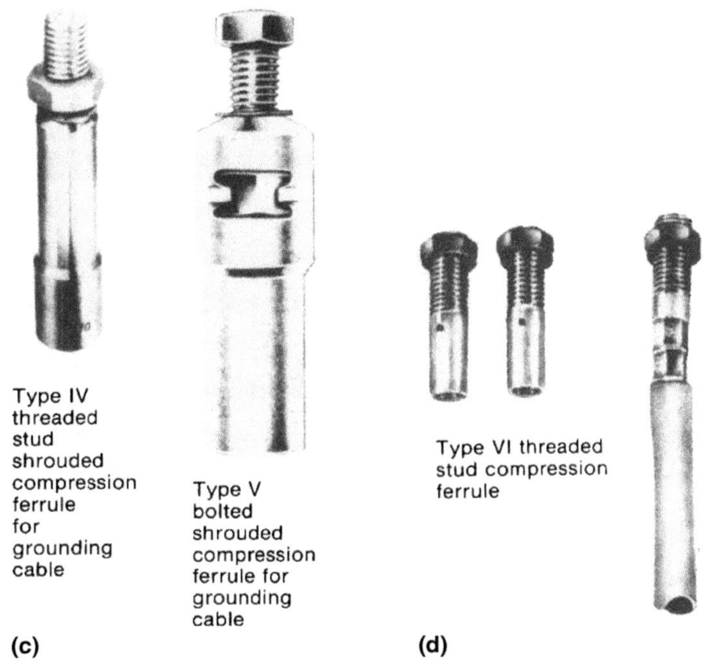

FIGURE 3.69 *(Continued)*

TABLE 3.30 Grounding Cable Ferrule Ratings

		Short-circuit properties*—Symmetrical kA rms, 60 Hz						Continuous current rating, rms 60 Hz
		Withstand rating		Ultimate rating/capacity†				
Grade	Cable size	15 cycles (250 ms)	30 cycles (500 ms)	6 cycles (100 ms)	15 cycles (250 ms)	30 cycles (500 ms)	60 cycles (1 s)	
1	2	14.5	10	30	19	13	9	200
2	1/0	21	15	48	30	21	15	250
3	2/0	27	20	60	38	27	19	300
4	3/0	36	25	76	48	34	24	350
5	4/0	43	30	96	60	43	30	400
6	250 kcmil	54	39	113	71	50	35	450
7	350 kcmil	74	54	159	100	71	50	550

*Withstand and ultimate short-circuit properties are based on performance with surges not exceeding 20% asymmetry factor.

†Ultimate rating represents a symmetrical current that the ferrule shall carry for the time specified.

Source: Courtesy ASTM.

TABLE 3.31 Grounding Cable Ferrule Physical Specifications

Type	Description	Shape	Minimum Specifications to be Supplied by Manufacturers in Addition to Cable Capacity and Material
I	Compression	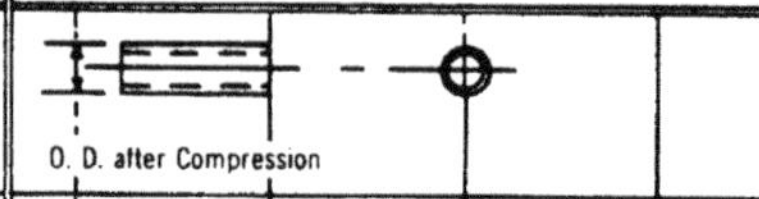	Installing Die Code O. D. after Installation
II	Solder	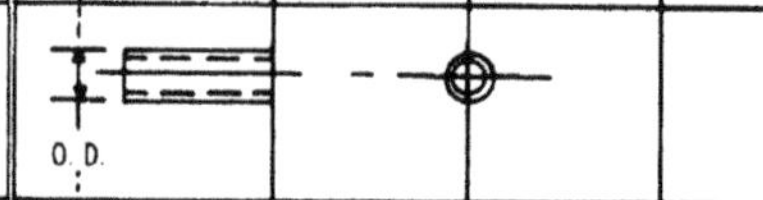	O. D.
III	Plain Stud Shrouded Compression	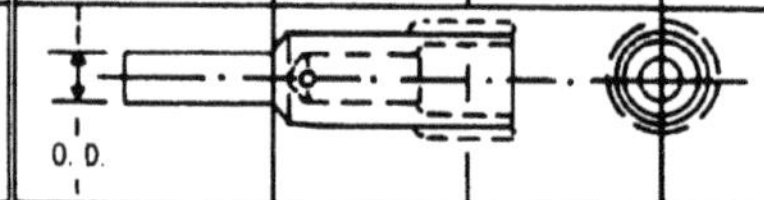	Installing Die Codes Stud O. D.
IV	Threaded Stud Shrouded Compression	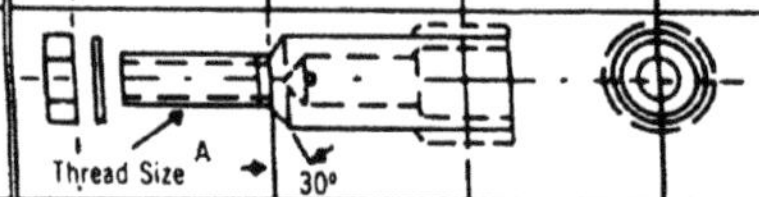	Installing Die Codes Thread Size A
V	Bolted Shrouded Compression	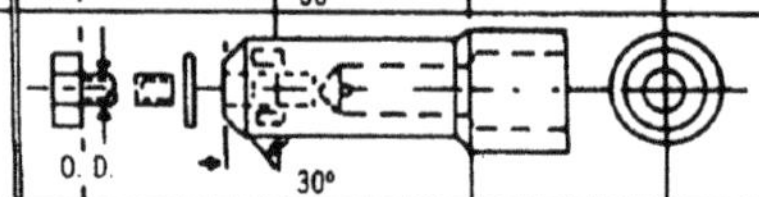	Installing Die Codes Bolt O. D. or Thread Size
VI	Threaded Compression	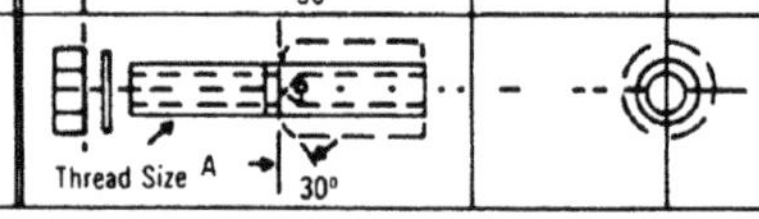	Installing Die Codes Thread Size A

A Standard thread sizes are as follows:
½ in.–13 UNC ≃ M12 × 1.75, ⅝ in.–11 UNC ≃ M16 × 2.00, ¾ in.–10 UNC ≃ M20 × 2.50.

Note: Inspection or vent holes are optional for types III, IV, V, and VI.

Source: Courtesy ASTM.

TABLE 3.32 Grounding Cable Ferrules and Compatible Grounding Clamp Terminations

CABLE FERRULES ASTM			MATCHING GROUNDING CLAMP CABLE TERMINATION ASTM		
ASTM Ferrule Type	Ferrule Description	Illustration of Ferrule Including Essential Interface Application Data Required	ASTM Termination Style	Termination Description	Essential Interface Application Data Required
I	Compression Ferrule	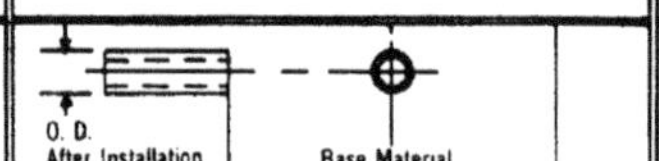	1.	Cable Retaining Eyebolt	
II	Solder Ferrule	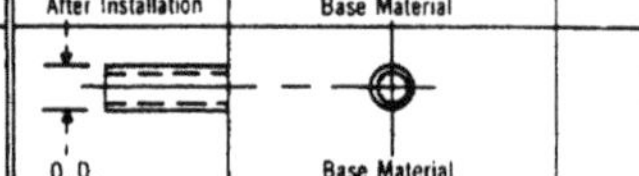	2. 3.	Eyebolt & Cable Support Plain Bore Clamp and Cable Support	Ferrule or Stud O. D. Accepted
III	Plain Stud Shrouded Compression Ferrule	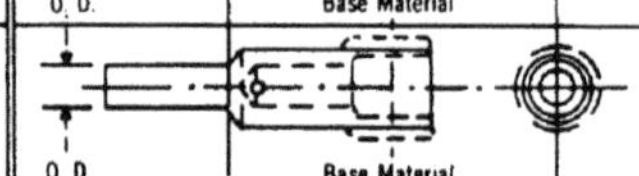	4.	Plain Bore Tubular With Screws and Cable Support	
IV	Threaded Stud Shrouded Compression Ferrule	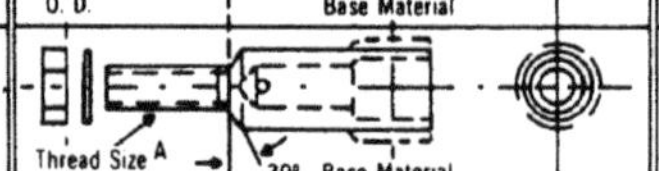	5. 6. 7.	Plain Bore Boss Plain Bore Boss and Cable Support Threaded Bore Boss	Tap Boss I. D. or Thread Size A and Material
V	Bolted Shrouded Compression Ferrule	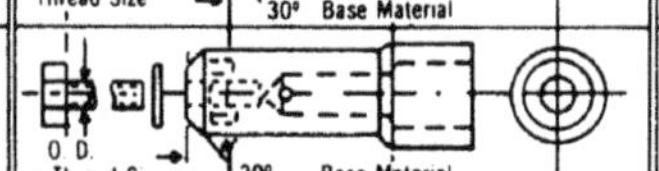	5. 6.	Plain Bore Boss Plain Bore Boss and Cable Support	Tap Boss I. D and Material
VI	Threaded Compression Ferrule	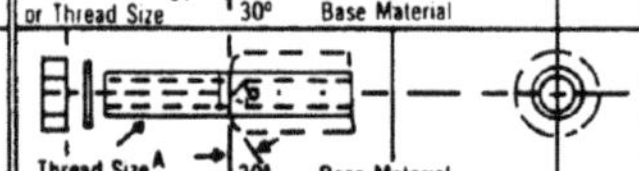	6. 8. 9.	Plain Bore Boss and Cable Support Threaded Bore Boss and Cable Support Threaded Bore Clamp and Cable Support	Tap Boss Thread Size A and Material

A Standard thread sizes are as follows:
½ in.–13 UNC ≃ M12 × 1.75, ⅝ in.–11 UNC ≃ M16 × 2.00, ¾ in.–10 UNC ≃ M20 × 2.50.

Note: Inspection or vent holes are optional for types III, IV, V, and VI.

Source: Courtesy ASTM.

TABLE 3.33 Cable Types for Use on Safety Ground Jumpers

Type I	Cables with stranded, soft drawn copper conductor with 665 or more strands of #30 American Wire Gauge (AWG) (0.0100 in) or #34 AWG (0.0063 in). Flexible elastomer jackets rated for installation and continuous service at temperatures from −40°F (−40°C) through +194°F (+90°C).
Type II	Cables with stranded, soft drawn copper conductor with 133 wires or more for size #2 or 259 wires or more for size 1/0 and larger. Flexible elastomer jackets rated for installation and continuous service at temperatures from −13°F (−25°C) to +194°F (+90°C).
Type III	Cables with stranded, soft drawn copper conductor with 665 or more #30 AWG (0.0100 in). Flexible thermoplastic jackets rated for installation and continuous service at temperatures ranging from +14°F (−10°C) to +140°F (+60°C). Caution: The use of type III jacketed cables is restricted to open areas or spaces with adequate ventilation so that fumes that could be produced by overheating the jacket during a short-circuit fault on the cable can be dispersed.

TABLE 3.34 Grounding Jumper Ratings

	Short-circuit properties*						
	Withstand-rating symmetrical kA rms, 60 Hz			Ultimate capacity symmetrical kA rms, 60 Hz			
Grounding cable size (copper)	15 cycles (250 ms)	30 cycles (500 ms)	6 cycles (100 ms)	15 cycles (250 ms)	30 cycles (500 ms)	60 cycles (1 s)	Continuous current rating, rms, 60 Hz
#2	14.5	10	29	18	13	9	200
1/0	21	15	47	29	21	14	250
2/0	27	20	59	37	26	18	300
3/0	36	25	74	47	33	23	350
4/0	43	30	94	59	42	29	400
250 kcmil	54	39	111	70	49	35	450
350 kcmil	74	54	155	98	69	49	550

*Withstand and ultimate short-circuit properties are based on performance with surges not exceeding 20% asymmetry factor.
Source: Courtesy ASTM.

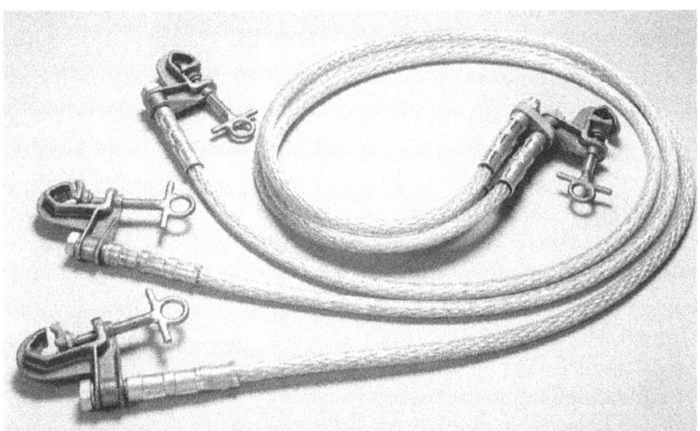

FIGURE 3.70 Commercially manufactured safety ground jumper. (*Courtesy AB Chance Corp.*)

Many types of jumpers are available from the manufacturer's stock, and many more types can be quickly assembled by the manufacturer for specific applications.

Selecting Safety Grounding Jumpers

The selection of grounding jumpers may be performed with a relatively simple four-step process. The following four sections are provided courtesy of W.H. Salisbury and Company.

Short-Circuit Capacity. Determine the maximum short-circuit capability at each typical grounding location. From this information, the size of the cable can be determined from Table 3.34. (Table 3.34 is the same as Table 5 in ASTM standard F 855.) Note that in selecting the cable size, it is perfectly acceptable to use smaller sizes of cable such as 2/0 and then double the cables when required for higher short-circuit duties.

Grounding Clamps. Grounding clamp types and styles must be determined for each conductor size and shape where grounding jumpers may be applied. If cable and/or bus is expected to be corroded and difficult to clean, serrated jaw clamps should be specified. Remember to specify clamps for both the conductor and the ground end of the grounding set.

Grounding Ferrules. Selection of the grounding ferrules is based on the cable size and material and the cable termination style of the grounding clamp(s). Aluminum or copper-plated ferrules should be used with an aluminum clamp and copper cable. Bronze clamps should be used only with copper ferrules and copper cables. Some manufacturers simplify the ferrule selection by employing only one ferrule and cable termination design.

Complete Jumper Specification. To simplify the selection of grounding sets, the user may first determine the short-circuit capacity and conductor size and shape, and then refer the design of the jumper to the manufacturer. Most manufacturers have a variety of stock grounding jumpers and will design and build special systems on an as-needed basis.

Installation and Location

The actual procedures and methods for the installation of safety grounds are discussed in detail in Chap. 4. In general the following rules apply:

1. Always make a three-step voltage measurement, where the tester is tested on a live circuit before and after the actual measurement, to verify the system is dead before installing safety grounds.
2. Grounds should be installed in such a way that they equalize the voltage on all equipment within reach of the worker. This equipotential grounding method is illustrated in Chap. 4.
3. When possible, safety grounds should be applied on both sides of the work area. Generally, the more grounding points, the better.
4. Application of safety grounds is considered to be a procedure with more than a normal risk of electric arc. Always wear face shields, rubber gloves, and arc protection rated for the maximum incident energy present at the location where the grounds are being installed.
5. Safety grounds should be applied on systems of any voltage level to add additional protection for personnel when the system is out of service and conductors are exposed. Systems of 480 V and higher should always be safety-grounded when locked and tagged. Lower-voltage systems may be safety-grounded on an as-needed basis.
6. Of course, if the nature of the work precludes the use of safety grounds, they should not be applied. For example, the measurement of insulation resistance cannot be done when safety grounds are installed.

GROUND-FAULT CIRCUIT INTERRUPTERS

Operating Principles

Most 120-V circuits are fed from standard thermal-magnetic molded-case circuit breakers. The circuit is similar to that shown in Fig. 3.71. In this configuration, the only protection from overcurrent is the circuit breaker, which requires a minimum of 15 A to even begin operation. Clearly, if a human being contacts the 120-V circuit, the circuit breaker will not operate unless a minimum of 15 A flows. Even then, it will operate slowly. Since current levels as low as 10 to 30 mA can be fatal to human beings, such installations are clearly not effective for personnel protection.

Figure 3.72 shows a diagram of a safety device that has been in use since the late 1960s. This device, called a *ground-fault circuit interrupter* (GFCI), has a current transformer that encloses the hot and the neutral lead. The resulting output of the current transformer is proportional to the difference in the current between the two leads.

To understand its operation, consider what happens when a normal load is attached to the duplex outlet of Fig. 3.72. Under such circumstances, all the current that flows through the hot wire will go to the load and return to the source on the neutral wire. Since the currents on the hot wire and the neutral wire are equal in magnitude and out of phase by 180°, the output from the GFCI current transformer will be zero and the GFCI will not operate.

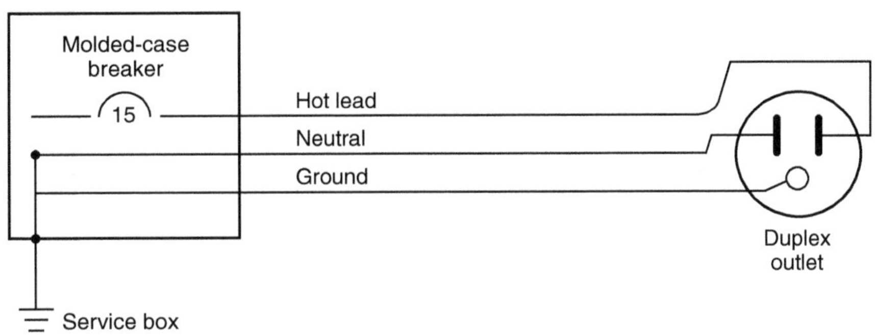

FIGURE 3.71 Standard protection schemes for 120-V circuits.

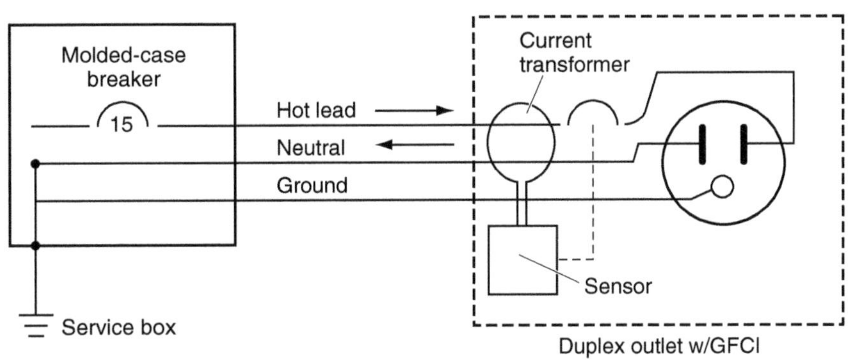

FIGURE 3.72 Standard 120-V supply with ground-fault circuit interrupter. (*Courtesy AB Chance Corp.*)

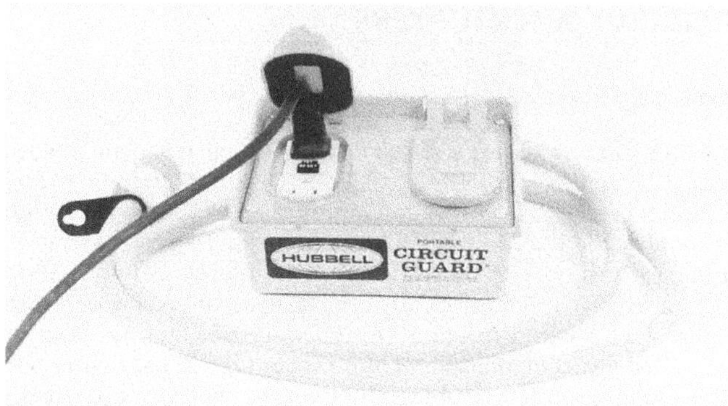

FIGURE 3.73 Portable ground-fault circuit interrupter. (*Courtesy Direct Safety Supply Co.*)

Now consider what happens when a grounded person contacts the hot wire downstream of the GFCI. Under these circumstances, current will flow from the hot wire through the person and will return on the ground wire. Because of this, the currents on the hot wire and the neutral wire will not be equal. The current transformer will produce an output to the sensor that will, in turn, cause the GFCI breaker contacts to open.

This operation occurs instantaneously, that is, with no intentional time delay. The circuit is disconnected very quickly and the person is spared the long-duration shock. While a GFCI does not guarantee the complete safety of personnel, it is set sensitively enough to save the life of the person an overwhelming percentage of the time.

Ground-fault circuit interrupters are set to trip when the difference between the hot lead and the neutral lead exceeds 5 mA $\pm$ 1 mA. They open typically in less than 25 milliseconds. GFCIs are sometimes built into low-voltage circuit breakers. They may also be equipped with AFCI capabilities. For more information on AFCIs, see the Arc-Fault Circuit Interrupters section later in this chapter.

Applications

The National Electrical Code (NEC) specifies locations that require the use of GFCIs. These locations include bathrooms, swimming pool outlets, and temporary power supplies for construction work. Refer to the most recent edition of the NEC for a detailed list of these requirements.

At a minimum, GFCI receptacles should be used in portable power cords used for temporary construction power during electrical system and other types of industrial work. Figure 3.73 shows a typical GFCI suitable for such use.

Ground-fault circuit interrupters should be tested before each use. In addition to being visually inspected to check for frayed insulation and a damaged case, the GFCI circuit should be tested for proper operation. This test can be easily performed using a simple test device like that shown in Fig. 3.74. This device plugs into the duplex receptacle. When the receptacle is energized, the user can observe the lights on the end, which will display in a certain pattern if the wiring to the receptacle is correct. The user then presses the test button and an intentional 5-mA ground is placed on the hot wire. If the GFCI is functional, it will operate and open the circuit. The user can then reset the GFCI and put it into service.

Most GFCIs have built-in test buttons. Pressing this button places a resistor between the hot lead and the neutral. The hot lead connection is made ahead of the internal CT and the neutral lead is made after it. This unbalances the current flow and causes the GFCI to trip.

Permanently installed GFCIs should be tested at least once per month. Temporary GFCIs, such as those in extension cords, should be tested prior to each use.

FIGURE 3.74 Receptacle and GFCI tester. (*Courtesy Ideal Industries, Inc.*)

ARC-FAULT CIRCUIT INTERRUPTERS

Thousands of fires are caused each year by electrical wiring failures. Arcing faults are among the major sources of these failures. Arc-fault circuit interrupters (AFCI) are used to detect the existence of an arcing fault and quickly open to clear the problem. Most AFCIs are included in the low-voltage circuit breakers that are found in most homes in the United States, Canada, and many other countries. Figure 3.75 is a photo of an AFCI breaker.

Such breakers include both of the normal overcurrent elements found in standard circuit breakers. They also have electronic circuitry that analyzes the current waveform—usually by looking for arc-indicating harmonics or spikes typical to arcing faults or through other methods.

Some AFCIs also include GFCI circuitry and thus provide standard overcurrent, arcing fault, and ground-fault protection.

The National Electrical Code (NFPA 70) requires AFCIs to be used in most rooms in new construction. Major changes in an electrical system will also require that a standard circuit breaker be replaced with an AFCI.

It should be noted that AFCIs alone do not provide much additional personnel protection above that afforded by a regular circuit breaker. True personnel protection is best provided by GFCIs.

SAFETY ELECTRICAL ONE-LINE DIAGRAM

One-line diagrams are used in electric power systems for a variety of purposes including engineering, planning, short-circuit analysis, and—most important—safety. The safety electrical one-line diagram (SEOLD) provides a road map for the electric power system. Figure 3.76 is an example of a typical SEOLD. Safety electrical one-line diagrams are used to ensure that switching operations are carried out in a safe, accurate, and efficient manner. *An accurate, up-to-date, legible SEOLD should be*

FIGURE 3.75 AFCI breaker

available for all parts of the electric power system. Safety electrical one-line diagrams should be accurate, concise, and legible.

1. *Accuracy.* The SEOLD should be accurate and up to date. Regular revision reviews should be performed on the SEOLD, and any required changes or modifications should be implemented immediately. Personnel should be aware of the review process and should have easy access to suggesting changes on the SEOLD.
2. *Concise.* Because one-line diagrams are used for a variety of purposes, some facilities put a large amount of information on one diagram. Items such as cable lengths, sizes, and impedances; current transformer ratios; transformer impedances; protective relay logic circuits; and metering circuits are often found on one-line diagrams used for safety and operations. This practice is not acceptable. Table 3.35 lists the only items that should be included on a one-line diagram. Note that the use of computer-aided drafting systems makes the development of job-specific one-line diagrams quite simple. All equipment shown on the SEOLD should be clearly identified with symbols or nomenclature that is identical to what is physically on the equipment.
3. *Legibility.* All too often, SEOLDs are allowed to become illegible. They may be sun bleached, or they may simply be the product of an original diagram that has been allowed to become too old. The SEOLD should be easy to read even in subdued lighting.

The SEOLD is the road map for the field electrician. Each electrical worker should have access to the SEOLD for day-to-day switching activities.

THE ELECTRICIAN'S SAFETY KIT

Each electrician should be supplied with a minimum list of electrical safety equipment, as shown in Table 3.36. This represents the minimum requirement for an electrician's electrical safety. Other equipment should be added on an as-needed basis.

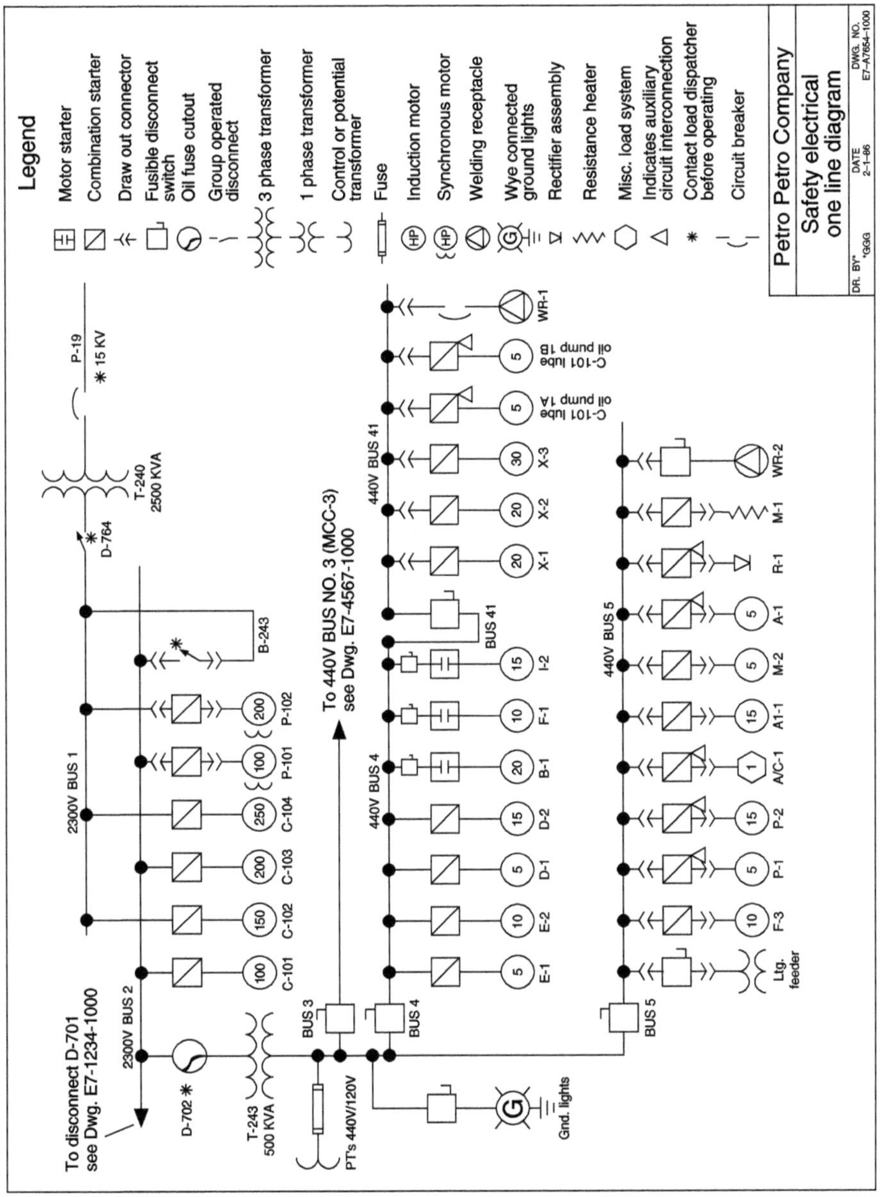

FIGURE 3.76 Typical safety electrical one-line diagram (SEOLD).

TABLE 3.35 Information Required to Be Included on Safety Electrical One-Line Diagrams

Energy sources	All sources of energy in the power system should be included on the SEOLD. This includes the normal sources such as power company connections and generators, as well as possible sources of backfeed such as instrument transformers and foreign voltages brought in for remote-control purposes.
Disconnect devices	The circuit breakers, switches, cutouts, fused disconnects, and any other devices that can be used to control the power system energy.

Note: All energy sources and disconnect devices should be clearly marked with a unique identifying name and/or number. The device itself should be clearly marked with an identical name and/or number.

TABLE 3.36 Minimum Safety Equipment for Electricians

- Rubber insulating gloves with leather protectors and protective carrying bags. Glove voltage classes should be consistent with the voltages around which the electrician is expected to work. Recommend one set of class 0 or 00 gloves and one set suitable for the highest voltage that the electrician will encounter.
- Rubber insulating sleeves. Quantities and voltage classes consistent with the rubber gloves described above.
- Rubber insulating blankets. Quantities and voltage classes consistent with the systems around which the electrician will be working.
- Safety voltage tester. One each low-voltage plus one medium- or high-voltage unit.
- Insulated tools for below 1000 V.
- Padlocks, multiple-lock devices, and lockout tags.
- Hard hats, ANSI standard Z89.1 class E.
- Nonconductive safety glasses with full side shields, ANSI standard Z87.1 glasses suitable for thermal protection.
- Warning signs—"Danger—High Voltage"—and others as required—red barrier tape with white stripe.
- Safety ground cables as required.
- Flame-retardant work clothing—minimum ATPV rating of 8 cal/cm^2.
- Flame-retardant flash suit—ATPV rating in excess of greatest arc-flash hazard the electrician will encounter, with full head, body, and leg protection along with in-the-ear-canal hearing protection.
- Safety electrical one-line diagram.

Note: This listing is for electrical hazards only. Other protection such as fire extinguishers and flashlights should be supplied as required.

Since electrical protective devices such as circuit breakers and fuses have ratings and capabilities that must be taken into account when installing or replacing them, field personnel should also have complete information about the protective devices in the system. Table 3.37 identifies some of those ratings for circuit breakers, and Table 3.38 covers fuses. A complete coverage of engineering design is beyond the scope of this handbook. However, field personnel should be very familiar with these ratings and be able to properly select the protective device when a replacement is required. In general, protective devices should always be replaced by an identical device.

TABLE 3.37 Circuit Breaker Ratings

Rating	Meaning	Comments
Voltage rating	The maximum voltage of the circuit where the circuit breaker is used	A breaker installed in a system with too high a voltage will likely fail. Additionally, the maximum amount of current that the breaker can interrupt is based on the system voltage.
Continuous current rating	The maximum continuous current that a circuit breaker can carry without overheating	Also referred to as *frame size*.
Minimum trip rating	The minimum current required to initiate operation of the protective circuitry	Continuous current and minimum trip rating may be two entirely different values. For example, a breaker with a 600-A frame size may be equipped with a 50-A protective device.
Interrupting rating	The maximum short-circuit current that an overcurrent protective device can safely interrupt under standard test conditions	This is the rating that manufacturers put on the circuit breaker nameplate.
Interrupting capacity	The maximum current that a circuit breaker can interrupt at the specified voltage level and system X/R ratio	Note that for any given installation, depending on the type testing methodology, this value may be different from the interrupting rating. This should be taken into account when the circuit breaker is being selected for use in a specific circuit.

TABLE 3.38 Fuse Ratings

Rating	Meaning	Comments
Voltage rating	The maximum voltage of the circuit where the fuse is applied	Fuses are usually manufactured in different sizes and configurations depending on the voltage level. For example, a fuse intended for use in a 250-V circuit will normally be shorter than one that is intended for use in a 600-V circuit.
Fuse rating	The maximum continuous current that a fuse can carry without tripping	Also referred to as *continuous current rating*. This is the current that is referred to when we speak of a 10-A fuse or a 600-A fuse.
Interrupting rating	The maximum short-circuit current that an overcurrent protective device can safely interrupt under standard test conditions	This is the rating that manufacturers put on the fuse label.
Interrupting capacity	The maximum current that a fuse can interrupt at the specified voltage level and system X/R ratio	Type testing of fuses is such that the interrupting rating and the interrupting capacity are usually the same value.

REFERENCES

1. ASTM Std F 1959/F 1959M B 99, "Standard Test Method for Determining the Arc Thermal Performance Value of Materials for Clothing," American Society for Testing and Materials (ASTM), Conshohocken, PA.
2. Stoll, A. M., and Chianta, M. A., "Method and Rating System for Evaluation of Thermal Protection," *Aerospace Medicine*, 40: pp. 1232–1238, 1968.
3. Doughty, Richard L., Neal, Thomas E., Dear, Terrence A., and Bingham, Allen H., *Testing Update on Protective Clothing & Equipment for Electric Arc Exposure*, IEEE Paper PCIC-97-35, 1997.
4. Doughty, Richard L., Neal, Thomas E., and Floyd, Landis H., *Predicting Incident Energy to Better Manage the Electric Arc Hazard on 600-V Power Distribution Systems*, IEEE Paper PCIC-98-36, 2000.

CHAPTER 4
SAFETY PROCEDURES AND METHODS

INTRODUCTION

The way work is performed in or around an electrical power system is just as important as the safety equipment that is used. Proper voltage measurement can mean the difference between life and death. Standing in the right place during switching operations can mitigate or eliminate the effects of an electric arc or blast, and proper application of safety grounds can prevent an accidental reenergization from becoming a fatality. Electrical hazard risk assessments must be performed in order to identify the electrical hazards that are present or are likely to be present, along with the safe work practices and personal protective equipment (PPE) that would be required to protect personnel.

This chapter summarizes many industry-accepted practices performing the electrical hazard risk assessments and for working on or around energized electric power circuits. The methods and techniques covered in this section should be used as guidelines. Local work rules and regulatory standards always take precedence. Note that any and all safety procedures should be reviewed at least annually. System changes, employee reassignments, or accidents should be considered excellent reasons for modification of existing procedures or implementation of new procedures.

Safety is the one truly personal concern in an electric power system. In the majority of electrical accidents, the injured victim was the so-called last link in the chain. The use of proper procedures and/or proper safety equipment could have prevented the accident. Equipment and procedures can be provided, but only the individual employee can make the decision to use them. Employees must be impressed with the knowledge that only they can make this final decision—a decision that may result in life or death.

Caution: This chapter presents several procedures that require technical calculations using a variety of mathematical and engineering applications. These calculations should be performed or evaluated only by qualified, experienced engineers or other technical personnel with the requisite knowledge and skills.

ELECTRICAL HAZARD RISK ASSESSMENTS

Before any procedure development or personnel training can occur, an electrical hazard risk assessment must be performed. The information in procedures, as well as training programs, must be based on all potential electrical hazards in the workplace. The risk assessment procedures must address employee exposure to electrical hazards and must identify the process to be used by the employee, before work is started, to accomplish the following:

1. Identify the hazard(s)
2. Assess the risks of the identified hazards
3. Implement risk control according to the Hierarchy of Risk Control Methods

NFPA 70E, Informative Annex F *Risk Assessment and Risk Control*, in F.3 Hierarchy of Risk Control, tells us that the purpose of specifying and adhering to the hierarchy of risk control methods is to identify the most effective protective measures needed to reduce the associated risk and to protect employees from the electrical hazards. Each method of risk control is considered less effective than the one before it. NFPA 70E Section 110.1(H)(3) lists the hierarchy of risk control identified in this and other safety standards and provides examples of each.

NFPA 70E, Table F.3 *The Hierarchy of Risk Control Methods* examples are as follows:

1. Elimination—Conductors and circuit parts in an Electrically Safe Working Condition
2. Substitution—Reduce energy by replacing 120 V control circuitry with 24 V
3. Engineering Controls—Install barricades to guard energized conductors and parts of equipment to reduce the likelihood of inadvertent contact or exposure to arc flash
4. Awareness—Warning signs alerting of the potential presence of hazards
5. Administrative Controls—Procedures and job planning tools
6. Personal Protective Equipment (PPE)—Shock and arc flash PPE

NFPA 70E, Section 110.1(H)(3) Hierarchy of Risk Control Methods identifies in order the most effective risk control method down to the least effective method for risk control for each risk assessment. Section 105.4 Priority states that "elimination" must be the first priority when implementing safety-related work practices. Once the hazards have been identified, it must be determined how the hazard(s) can be eliminated. Elimination means the removal of the hazard as though it never existed, which in turn removes the potential for human error when operating or maintaining the electrical equipment. Completely eliminating the hazard is often not a viable option for existing installed electrical equipment. Although elimination can be achieved by applying other controls such as administrative, as in establishing an electrically safe work condition, one caution that must be stated here is that other controls may introduce a potential for human error. We must start with elimination and work our way down the list of risk control methods; only after this we can resort to using PPE. Keep in mind that elimination is the FIRST choice and PPE is the LAST choice.

Section 110.1(H)(3) Informational Note No. 1 provides additional information to aid in the understanding of the hierarchy of risk control by stating that "elimination, substitution, and engineering controls are the most effective methods to reduce risk as they are usually applied at the source of possible injury or damage to health and they are less likely to be affected by human error. Awareness, administrative controls, and PPE are the least effective methods to reduce risk as they are not applied at the source and they are more likely to be affected by human error."

Informational Note No. 3 states that the risk assessment procedure could also include identifying when a second person could be required and the training and equipment that person should have. As an example of this, OSHA 29 CFR 1910.269(l) *Working* on or near *exposed energized parts* requires two (2) qualified workers when exposed to possible contact with over 600 volts.

As noted above, elimination by establishing an electrically safe work condition, which includes energy control such as lockout/tagout, is the first choice. One thing to keep in mind here is that the process of establishing an electrically safe work condition is still energized work and all required PPE must be used until the hazard is controlled and proven by testing. Essentially, until the worker is certain that no energy is present, it must be assumed that energy is still present.

Performing a proper risk assessment under the conditions that permit normal operation, establishing an electrically safe work condition should remove the hazard before the final voltage measurement is made. Although the process of establishing an electrically safe work condition must be considered energized work, it should not expose the worker to an energized circuit or condition. The final voltage measurement should always be zero. If the voltage measurement is other than zero, then there is a possible backfeed from another source or induced voltage from an adjacent source. All sources must be identified and controlled, or it is energized work. After establishing an electrically safe work condition, it is then possible to remove the PPE worn to protect from electrical hazards. The approach boundaries for shock and arc flash no longer exist, and other employees in the vicinity should not be at risk.

Working While Exposed to Electrical Hazards

Safety-related work practices must be used to protect employees from injury while they are exposed to electrical hazards. The specific safety-related work practices must be consistent with the electrical hazards and the associated risk. The required safety-related work practices must be determined before any person is exposed to the electrical hazards by using the shock risk assessment and the arc flash risk assessment. Remember that only qualified persons are permitted to work on electrical conductors or circuit parts that have not been put into an electrically safe work condition.

Shock Risk Assessment. A shock risk assessment is a process that:

- Identifies exposure to the potential electrical shock hazards
- Estimates the potential severity of a shock injury
- Estimates the likelihood of occurrence of this injury
- Determines if protective measures are required
- Determines the appropriate protective measure to use

Risk of electric shock is determined by a person's distance to an exposed energized electrical conductor or circuit part. As the person's distance decreases the risk of making contact increases and as the person's distance increases the risk decreases. When conducting a shock risk assessment, the employee must determine the voltage of all exposed energized electric conductors or circuit parts that are in the vicinity of the person's body or any tools that may be used. The shock approach boundaries provide the trigger for added protection for all qualified employees.

As a general rule the shock risk assessment is performed to:

1. Identify all potential shock hazards
2. Estimate the likelihood of occurrence of injury or damage to health
3. Estimate the potential severity of injury or damage to health
4. Determine if additional protective measures are required, including the use of PPE

If additional protective measures are required, they must be selected and implemented according to the hierarchy of risk control discussed earlier. When the additional protective measures include the use of PPE, the following must be determined:

1. The voltage to which personnel will be exposed
2. The shock boundary requirements
   a. Limited Approach Boundary
   b. Restricted Approach Boundary
3. The personal and other protective equipment required by NFPA 70E and OSHA to protect personnel against the electric shock hazard

Arc Flash Risk Assessment. Whenever there are exposed energized electrical conductors or circuit parts and a person is working within the Arc Flash Boundary or the Limited Approach Boundary, a potential arc flash hazard is presumed to exist. Also, when an employee is interacting with the energized electrical equipment, even though the equipment doors are closed and latched or covers are installed and secured, a potential arc flash hazard may exist. The risk of an arc flash occurring under normal operating conditions is greatly reduced if:

1. The equipment is properly installed
   a. Installed in accordance with applicable industry codes and standards and the manufacturer's recommendations

2. The equipment is properly maintained
   a. The equipment has been maintained in accordance with the manufacturer's recommendations and applicable industry codes and standards
3. The equipment is used in accordance with instructions included in the listing and labeling and in accordance with manufacturer's instructions
4. The equipment doors are closed and secured
5. All equipment covers are in place and secured
6. There is no evidence of impending failure
   a. Evidence such as arcing, overheating, loose or bound equipment parts, visible damage, deterioration, the smell of ozone, presence of smoke, sound of arcing, visible damage, or warning lights. Other signs to look for include unsealed openings in equipment through which snakes or rodents could pass through and get into the equipment.

Performing an arc flash risk assessment:

1. Identify all potential arc flash hazards
2. Estimate the likelihood of occurrence of injury or damage to health
   a. When evaluating the likelihood of occurrence there are several considerations, which include: When racking-out or racking-in a power circuit breaker, a conductor springs loose and contacts another energized conductor or ground, a tool drops into an energized circuit creating a phase-to-phase or phase-to-ground fault, an animal contacts an energized circuit part, a disconnect is opened or closed, or a starter operates automatically. Most arcing faults start with human interaction.
   b. When using hand tools for work on or near exposed energized conductors or circuit parts, the tools must be insulated for the voltage. Tools used where exposure is less than 1,000 Vac the tools must be insulated and approved by ASTM F1505, *Standard Specification for Insulated and Insulating Hand Tools*. Wrapping the tool with electrical tape or using rubber dipping, heat-shrink tubing, or tygon tubing is not insulated and must never be used.
3. Estimate the potential severity of injury or damage to health if the arc flash occurs
4. If an arc flash has the potential to occur, take steps to determine if additional protective measures will be required. This includes the use of PPE, which is the last choice. Always evaluate the hierarchy of risk control and start with elimination, as noted above.

When estimating the likelihood of occurrence of injury or damage to health, as well as estimating the potential severity of injury or damage to health should an arc flash occurs, we must take into consideration the following:

1. The design of the electrical equipment. This especially includes the overcurrent protective device(s) and the operating time
2. The electrical equipment operating condition and condition of maintenance

The above two considerations are very critical to the arc flash risk assessment. If there is an unintentional time delay in the operation of a circuit breaker, due to a dirty, corroded, or sticky operating mechanism, the incident energy of the arc flash could increase to dangerous levels in a fraction of a second.

When additional protective measures are required they must be selected and implemented according to the hierarchy of risk control identified above. If these additional protective measures include the use of PPE, the following must be determined:

1. Develop and implement the appropriate safety-related work practices for the work to be performed and for the electrical hazards
2. Identify the arc flash boundary and use safety signs or tags, barricades, and if necessary an attendant to keep people at a safe distance from the potential arc flash
3. Identify and select the PPE to be used within the arc flash boundary

Remember that only after all other risk control methods have been exhausted should PPE be selected. Selection of PPE is not the first, but it is the LAST choice after all other options have been considered. PPE is the least effective and lowest level of risk control for employee protection and should not be the first or only method used.

The results of the shock and arc flash risk assessments must be documented.

THE SIX-STEP SAFETY METHOD

Table 4.1 lists six important steps to practicing safe behavior and will serve as the foundation for a personnel safety philosophy. If implemented, this method will greatly enhance the safe and efficient performance of electrical work.

Individuals are responsible for their own safety. The steps listed in Table 4.1 are all individual steps that can be taken by anyone who works on or around electric power circuits and conductors.

Think—Be Aware

Many accidents could have been prevented if the injured victim had concentrated on the safety aspects of the job. Thinking about personal or job-related problems while working on or near energized conductors is a one-way ticket to an accident. Always stay alert to the electrical hazards around the work area.

Understand Your Procedures

Every company has defined safety procedures that are to be followed. Workers should be thoroughly familiar with all the safety procedures that affect their jobs. Knowledge of the required steps and the reasons for those steps can save a life. All employees should go through extensive safety training.

Follow Your Procedures

In the past, some facilities have allowed the violation of safety procedures in the name of production. Such actions are in violation of the law and have invariably proven to be costly in terms of human injury and/or death. Violation of safety procedures without good cause should be a discharge offense. What constitutes "good cause" must be decided on a local basis; however, excuses of lesser significance than immediate danger to life should not be acceptable.

Use Appropriate Safety Equipment

No matter how meticulous workers are, accidents do occasionally happen. Equipment failures, lightning strikes, switching surges, and other such events can cause shock, arc, or blast. Also, sometimes it

TABLE 4.1 Six-Step Safety Method

- Think—be aware.
- Understand your procedures.
- Follow your procedures.
- Use appropriate safety equipment.
- Ask if you are unsure, and do not assume.
- Do not answer if you do not know.

becomes necessary for employees to work on or very close to energized conductors, which increases the chance of accidental contact.

Because of these reasons, appropriate safety equipment should be used any time workers are exposed to the possibility of electrical hazards. Remember that nothing is sadder than an accident report that explains that the dead or injured worker was not wearing safety equipment.

Ask If You Are Unsure, and Do Not Assume

Ignorance kills and injures many people each year. No one should ever get fired for asking a question—especially if it is a safety-related question. Anyone who is uncertain about a particular situation should be encouraged to ask questions, which should then be answered by a qualified person immediately and to the fullest extent possible.

Do Not Answer If You Do Not Know

No one should answer a question if they are not certain of the answer. Self-proclaimed experts should keep their opinions to themselves.

JOB BRIEFINGS

Definition

A job briefing (sometimes called a "tailgate meeting") is a meeting that informs all workers of the job requirements. In particular, a job briefing is used to alert workers to potential safety hazards. A job briefing need not be a formal gathering; however, it is mandatory that all workers involved attend, and worker attendance should be documented.

What Should Be Included?

OSHA rules require that a job briefing discuss, at a minimum, the following issues:

- **S**pecial precautions to be taken
- **H**azards associated with the job
- **E**nergy control procedures
- **P**rocedures and policies
- **P**ersonal protective equipment (PPE)

Note that the first letters of each of these bulleted items form the acronym ***SHEPP,*** which can be used to help remember the important issues that need to be discussed.

Job briefings should be proactive meetings in which workers are informally quizzed to make certain that they fully understand the safety issues they will face.

When Should Job Briefings Be Held?

- At the beginning of each shift
- At the beginning of any new job
- Any time that job conditions change
- When new personnel are introduced to an ongoing job

ENERGIZED OR DE-ENERGIZED?

The Fundamental Rules

All regulatory standards are quite clear in their requirements to de-energize a circuit before employees work on or near it. Stated simply:

All circuits and components to which employees may be exposed should be de-energized before work begins.

Some references, such as NFPA 70E, refer to this process as *making a work area electrically safe*. This phrase is preferred over *de-energized* since it specifically encompasses turning equipment off, locking and tagging, voltage measurement, and so on.

A few basic points will clarify this requirement:

- Production or loss of production is *never* an acceptable, sole reason to work on or near an energized circuit.
- If work can be rescheduled to be done de-energized, it should be rescheduled.
- De-energized troubleshooting is always preferred over energized troubleshooting.
- The qualified employee doing the work must *always* make the final decision as to whether the circuit is to be de-energized. Such a decision must be free of any repercussions from supervision and management.

A Hot-Work Decision Tree

Figure 4.1 illustrates a method that may be used to determine the need to work on a circuit when it is energized. The numbers in the following explanation refer to the numbers assigned to each of the decision blocks shown in Fig. 4.1.

1. Work performed on or near circuits of less than 50 V to ground may *usually* be considered to be de-energized work. Note that if the circuit has high arcing capability, decision 1 should be answered as "Yes."
2. If de-energizing simply changes the hazard from one type to another, or if it actually increases the degree of hazard, this decision should be answered "Yes." Table 4.2 lists the types of additional hazards that should be considered in answering this decision.
3. The need to keep production up is common to all industries—manufacturing, petrochemical, mining, steel, aluminum, electric utilities, shipyards, and ships. However, many employers abuse the concept that "production must continue." The following points should clarify when production issues may be allowed to influence the decision to de-energize.
    a. Shutdown of a continuous process that will add extraordinary collateral costs may be a signal to work on the circuit energized. Table 4.3 lists examples of these types of collateral costs.
    b. Shutdown of a simple system that does not introduce the types of problems identified in Table 4.3 should always be undertaken rather than allowing energized work.
4. In some cases, the very nature of the work or the equipment requires that the circuit remain energized. Table 4.4 shows three of the most common examples of such work. Note, however, that this work should still be de-energized if it is possible to do it that way. For example, troubleshooting a motor starter may be faster with the circuit energized; however, if it can be done de-energized it should be, even at the cost of a little more time.
5. If decision 2 or 3 leads in the direction of energized work, the next decision should be rescheduling. If energized work can be done de-energized on a different shift or at a later time, it should be postponed. Many companies miss this elegantly simple alternative to exposing their personnel to hazardous electrical energy.
6. The final, and arguably the most important, decision of all is to determine whether the work can be done safely. If, in the opinion of the qualified personnel assessing the job, the work is simply too dangerous to do with the circuits energized, then it *must* be de-energized.

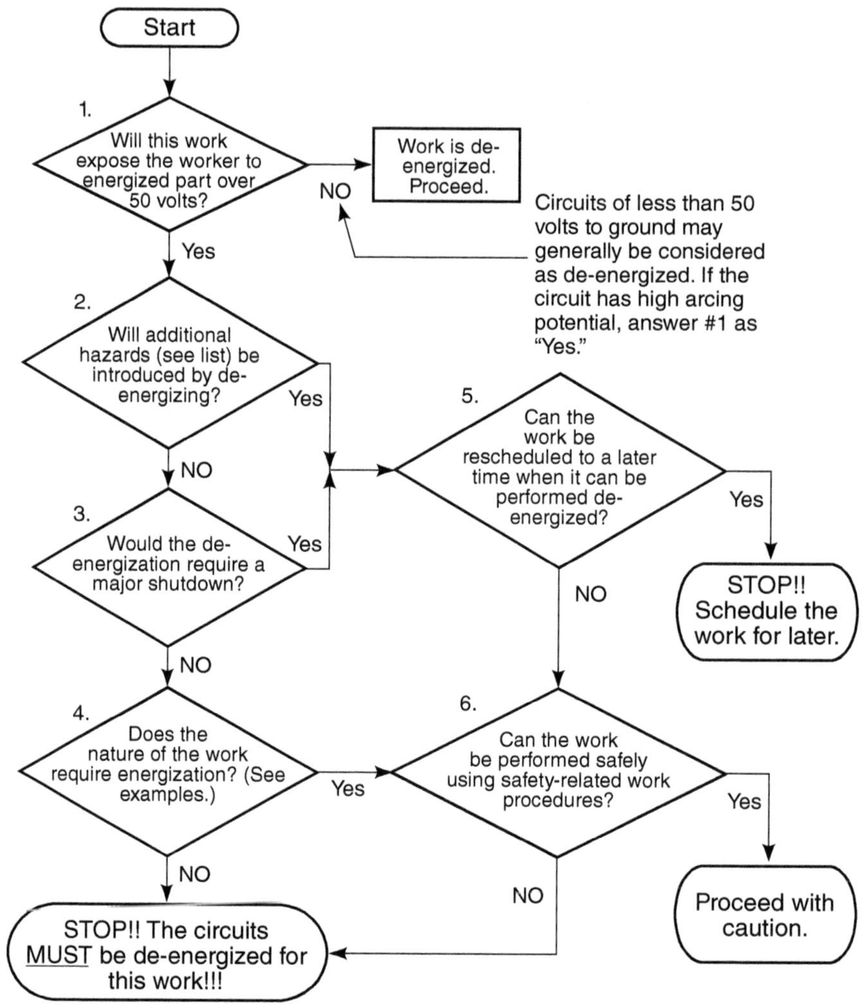

FIGURE 4.1 Hot-work flowchart.

TABLE 4.2 Examples of Additional Hazards

- Interruption of life-support systems
- Deactivation of emergency alarms
- Shutdown of ventilation to hazardous locations
- Removal of illumination from the work area

TABLE 4.3 Examples of Collateral Costs That May Justify Energized Work

- Excessive restart times in continuous process systems
- High product loss costs (in polyethylene process, for example, the product has to be physically dug out of process equipment after an unscheduled outage)

TABLE 4.4 Examples of Work That May Require Energization of Circuits

- Testing electrical circuits (to verify de-energization, for example)
- Troubleshooting complex controls
- Infrared scan

TABLE 4.5 Steps Required before De-energized Work May Commence

1. All energy control devices feeding the work area must be opened.
2. Locks and tags shall be placed on the energy control devices.
3. Voltage measurements shall be made at the point(s) of exposure to verify that the circuit is de-energized.
4. Safety grounds (if required) shall be placed to ensure the existence of an equipotential work zone.
5. The work area must be closely inspected by a qualified person to make certain that no energized parts remain. This critical step is often missed.

After the Decision Is Made

If the work must be done energized, all employees who work on or near energized conductors must be qualified to do the work, must use appropriate PPE, and must use appropriate safety-related work practices.

If the circuits are to be de-energized, the steps listed in Table 4.5 must be followed. Note that proper, safe procedures for each of the items in Table 4.5 are discussed later in this chapter.

SAFE SWITCHING OF POWER SYSTEMS

Introduction

The most basic safety procedure is to de-energize the parts of the system to which workers may be exposed. This procedure virtually eliminates the hazards of shock, arc, and blast. De-energizing, also called *clearing*, involves more than simply turning the switches off. To ensure maximum safety, de-energizing procedures that are precise for each situation should be written.

The following sections discuss the proper safety techniques for operation of various types of equipment and provide de-energizing and reenergizing procedures that may be used as the basis for the development of site-specific procedures. Please note that specific procedures may vary depending on the application and type of equipment. Refer to manufacturer's and/or local facility procedures for specific information. The methods given in these sections should be considered minimum requirements. These procedures assume that the device is being operated when one or both sides are energized.

Caution:

- Switching of electric power should be done only by qualified personnel who are familiar with the equipment and trained to recognize and avoid the safety hazards associated with that equipment.
- Non-load-interrupting devices that are not intended to interrupt any current flow should never be used to interrupt current flow. They should be operated only after all loads have been disconnected from the circuit by other means.
- Circuit breaks should *never* be closed after they interrupt a fault until the cause of the fault has been determined and corrected. Most circuit breakers are tested to operate only once at their maximum interrupting rating.

Remote Operation

There are at least three different ways that electrical switchgear such as circuit breakers and switches can be operated safely.

1. Operating from a remote control room
2. Operating using a remote operating device
3. Operating manually from the device panel

Operating from a Remote Control Room. Many companies are opting to install remote switching centers to better protect their workers. These types of installations are being used extensively in new construction and are being retrofitted into many existing facilities. This type of system requires that the devices to be operated can be operated electrically. Circuit breakers, for example, are equipped with electrical close, open, and/or racking accessories. When the operator in the remote control room operates the remote control handle, an electrical signal is sent to the breaker, causing it to open, close, or rack in or out.

These types of installations provide a very high level of safety because the worker is completely isolated from the gear. If the gear arcs or explodes, the worker is protected by brick or concrete walls.

All such control rooms have the ability to remotely open or close the breaker. However, since the most dangerous part of the operation is racking the breaker (moving it to connect or disconnect from the electrical bus), optimal safety is not realized unless the breaker can be racked remotely as well.

Operating Using a Remote Operating Device. Several companies are manufacturing and selling remote operating appliances. These devices have an operating mechanism that is temporarily (or permanently if desired) connected to the gear. The mechanism is designed so that, when triggered by the operator, it will physically manipulate the toggle switch, push button, or pistol-grip switch and close, open, charge, rack, or otherwise operate the circuit breaker or switch.

The control box is held by the operator and connected to the operating mechanism by an umbilical cable. The cable is long enough to allow the operator to stand well outside the flash hazard boundary. Of course, standing as far away as possible is the safest approach.

Figure 4.2 shows the control box of one such device in place to open or close a medium-voltage circuit breaker. Figure 4.3 is an *operator's-eye view* of the operating mechanism, umbilical cable, and control box for the same device. Note that the control box has a switch for open and close as well as a toggle switch to control the amount of movement made by the operating mechanism.

Figure 4.4 shows another manufacturer's operating mechanism attached to a CH Magnum DS low-voltage circuit breaker. Figure 4.5 shows a different operating mechanism by the same manufacturer. This one is designed for use on ABB (previously ITE) K-line low-voltage circuit breakers. Low-voltage circuit breakers sometimes require manual charging of closing springs. The units shown in Figs. 4.4 and 4.5 will not only open and close the breakers but also charge the closing springs.

Most electrical accidents involving circuit breakers occur when the breaker is being inserted into or removed from its operating cradle. The action, called *racking*, has a very critical moment when the circuit breaker connections engage the bus (*racking in*) or when they disengage the bus (*racking out*).

Because of the way the racking mechanism is designed, workers usually cannot stand to the side of the breaker cubicle where they are safest. This means that should there be a failure that causes an electrical arc and blast, the worker would be directly in the path of the heat and/or pressure.

Figure 4.6 shows one manufacturer's solution to this problem. The equipment shown in the figure is designed to rack breakers in or out remotely. The mechanism can be attached to many different breaker types, both low- and medium-voltage, using couplings supplied by the manufacturer. The breaker in this photo is a Siemens RL breaker.

Figure 4.7 is a view of the system showing the operator and the umbilical cable. After the unit is connected properly to the breaker, the operator moves away until he or she is at least outside of the flash hazard boundary. With the control panel, the operator starts the racking operation. When the breaker has racked in or out completely, the equipment shuts off automatically.

FIGURE 4.2 Operating mechanism of a remote open-close appliance. (*Courtesy www.chickenswitch.com*)

FIGURE 4.3 View of operating control and umbilical cable for device shown in Fig. 4.2. (*Courtesy www.chickenswitch.com*)

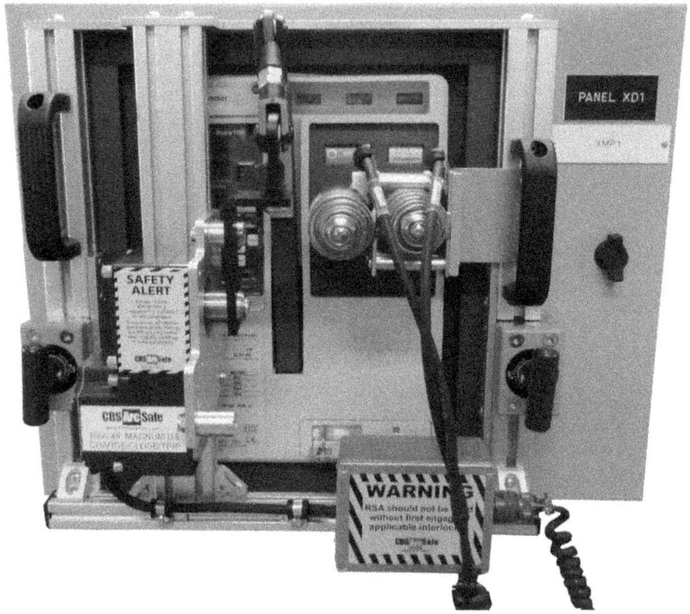

FIGURE 4.4 Operating mechanism designed for use on a CH Magnum DS. *(Courtesy CBSArcSafe, www.CBSArcSafe.com)*

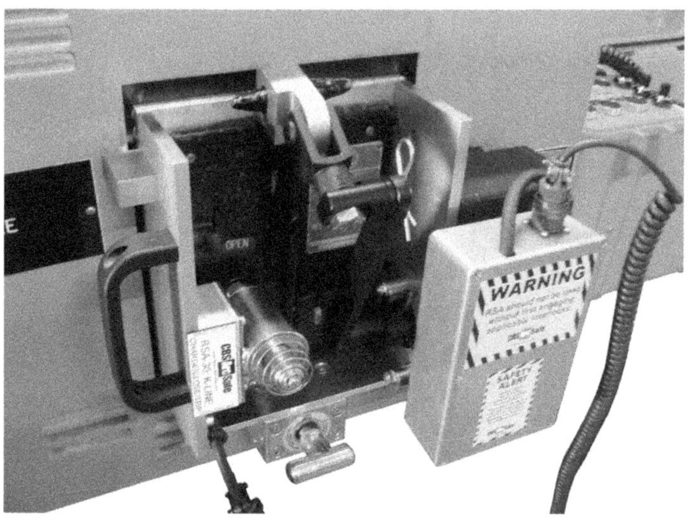

FIGURE 4.5 Operating mechanism designed for use on K-line circuit breakers (old ITE now ABB). (*Courtesy CBSArcSafe, www.CBSArcSafe.com*)

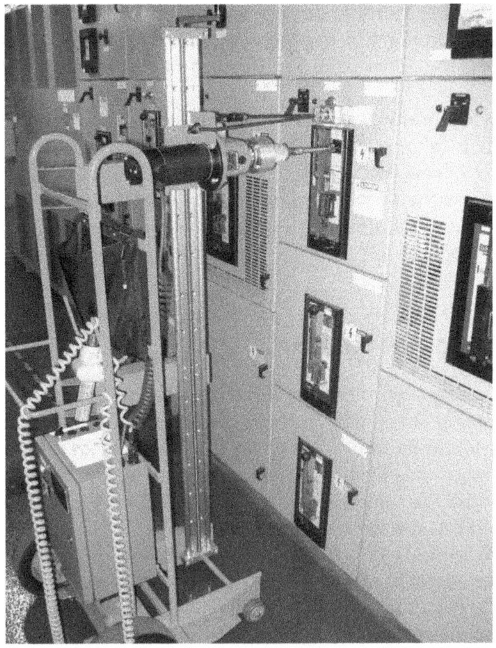

FIGURE 4.6 A device designed to remove and install circuit breakers remotely. *(Courtesy CBSArcSafe, www.CBSArcSafe.com)*

FIGURE 4.7 View of operator and umbilical cable for device shown in Fig. 4.6. *(Courtesy CBSArcSafe, www.CBSArcSafe.com)*

Although these remote devices will usually allow the worker enough distance to be outside of the flash hazard boundary, it is generally recommended that rated protective clothing be worn anyway. This protects the operator in the event of a catastrophic failure.

Operating Manually from the Device Panel. Although remote operation is being used more and more often, local operation is still the most widely used. Thankfully, this is starting to change, but it will be many years or even decades before remote operation becomes the norm. In addition, some equipment is difficult to set up for remote operation.

The following paragraphs describe various types of electrical gear and how to operate them locally and safely.

Operating Medium-Voltage Switchgear

General Description. Figures 4.8 through 4.14 illustrate various types of medium-voltage switchgear and the breakers used in them. In this style of gear, the circuit breaker rolls into the switchgear on wheels as shown in Fig. 4.9 or on a sliding type of racking mechanism as shown in Fig. 4.12. The opening and closing of the breaker is performed electrically using a front-panel-mounted, pistol-grip type of control. Turning the grip in one direction (usually counterclockwise) opens the breaker, and turning it in the other direction closes it.

The breaker connects to the bus and the line via a set of disconnects, visible at the top of Fig. 4.9 and on the right side of Fig. 4.12. When the breaker is open, it can be moved toward the front of the cubicle so it disconnects from the bus and line. This action is referred to as *racking* the breaker. The breaker may be completely removed from the switchgear, or it may be put into two or more auxiliary positions. Racking a circuit breaker is accomplished using removable cranks. These cranks may be of a rotary type or a lever bar that is used to "walk" the breaker from the cubicle.

Most switchgear have two auxiliary positions: the *test* position and the *disconnected* position. In the test position, the breaker is disconnected from the bus; however, its control power is still applied through a set of secondary disconnects. This allows technicians to operate the breaker for maintenance purposes. In the disconnected position, the breaker is completely disconnected; however, it is still in the switchgear.

For some types of switchgear, the front panel provides worker protection from shock, arc, and blast. This means that the switchgear is designed to contain arc and blast as long as the door is properly closed and latched. Such gear is called *arc-resistant* switch gear.

Closed-Door Operation. Table 4.6 lists the recommended safety equipment to be used by operators when performing both closed-door and open-door switching on medium-voltage switchgear. Figure 4.13 is a top view of such equipment showing alternative positions for the control handle and the preferred positions for the primary and backup operators. Note that both the primary operator and backup operator should be wearing the recommended clothing. The primary operator is the worker who actually manipulates the handle that opens and/or closes the circuit breaker. The backup operator's responsibility is to support the primary operator in the event there is a problem. The backup operator should be positioned several feet outside of the arc-flash boundary so he or she does not become an additional casualty in the event of a catastrophic failure of the gear. The backup operator may be optional in some facilities.

To operate the switchgear with a closed door, the following steps apply:

1. The primary operator stands to the side of the cubicle containing the breaker to be operated. The side to which he or she stands should be determined by which side the operating handle is on. If the handle is in the middle, the operator should stand on the hinge or the handle side of the door depending on which side is stronger. (Refer to the manufacturer.)
2. The primary operator faces away from the gear. Some workers prefer to face the gear to ensure better control. In either case, the primary operator should turn his or her face away from the breaker being operated.

FIGURE 4.8 Medium-voltage metal-clad switchgear. (*Courtesy Westinghouse Electric.*)

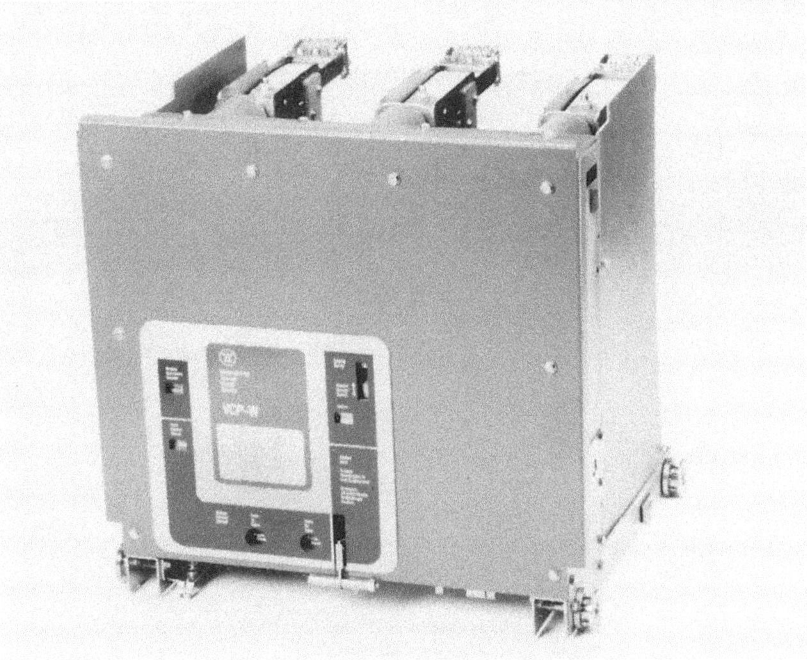

FIGURE 4.9 Circuit breaker used in switchgear shown in Fig. 4.8. (*Courtesy Westinghouse Electric.*)

FIGURE 4.10 Switchgear of Fig. 4.8 with door open showing circuit breaker cubicle. (*Courtesy Westinghouse Electric.*)

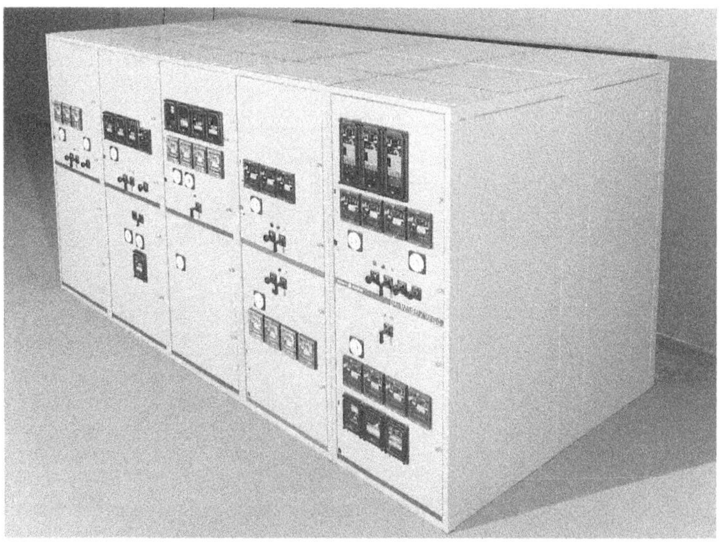

FIGURE 4.11 Metal-clad, medium-voltage switchgear for vacuum-type circuit breakers. (*Courtesy General Electric.*)

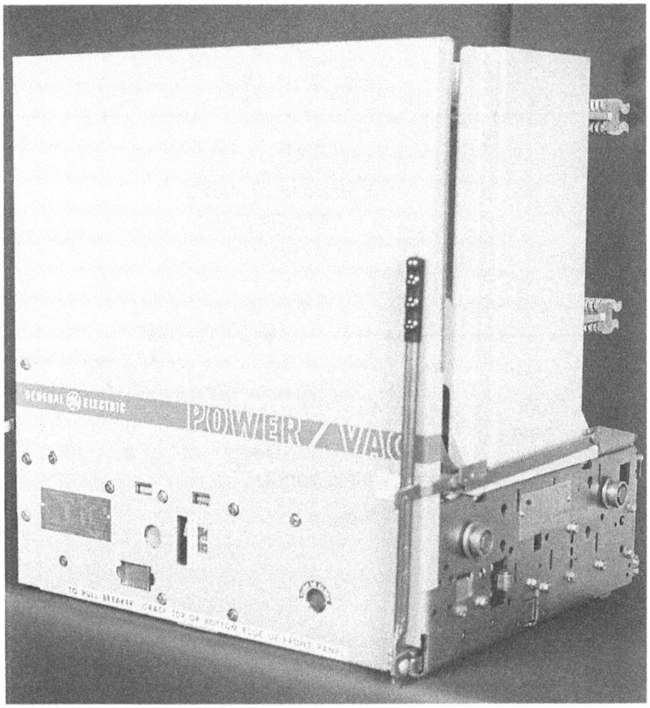

FIGURE 4.12 Medium-voltage, vacuum interrupter circuit breaker used in switchgear of Fig. 4.11. (*Courtesy General Electric.*)

3. The backup operator stands even farther from the cubicle and outside the arc-flash boundary, facing the primary operator.
4. The primary operator reaches across to the operating handle and turns it to open or close the breaker. Note that the primary operator continues to keep his or her face turned away from the gear. Some operators prefer to use a hot stick or a rope for this operation. This keeps the arms as far as possible from any hazard.
5. If the breaker can be racked with the door closed, and if the breaker is to be racked away, the primary operator inserts the racking handle. In this operation, the primary operator may have to face the breaker cubicle.
6. If lockout-tagout procedures are required, the primary operator places the necessary tags and/or locks.

Open-Door Operation. Refer to Table 4.6 for a minimum listing of safety equipment for this operation. If the door must be open for racking the breaker away from or onto the bus, the following steps should be observed:

1. The breaker is opened as described earlier under closed-door operation.
2. The primary operator opens the cubicle door and racks the breaker to the desired position.
3. If lockout-tagout procedures are required, the primary operator places the necessary tags and/or locks.

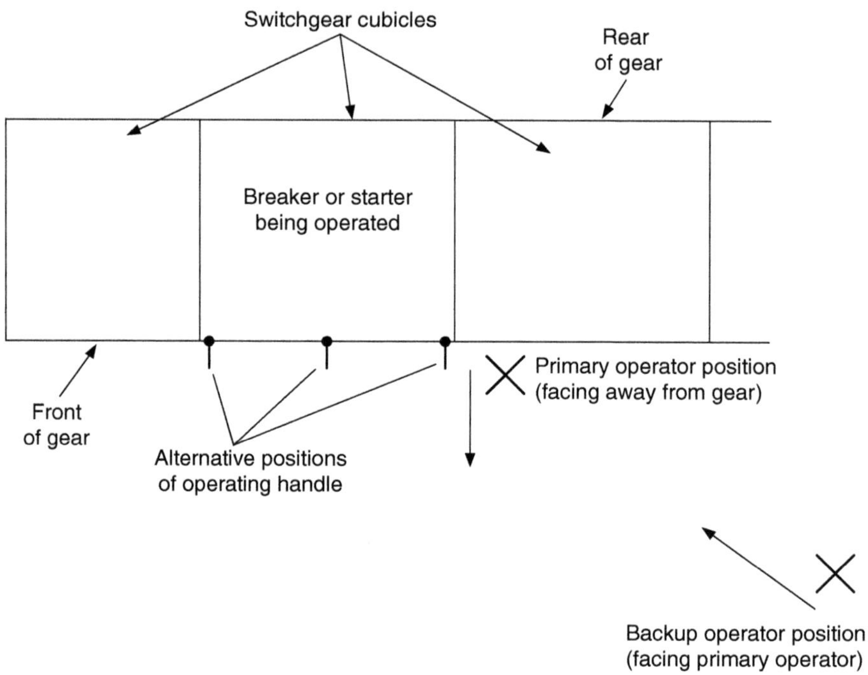

FIGURE 4.13 Proper position for operating electrical equipment (top view).

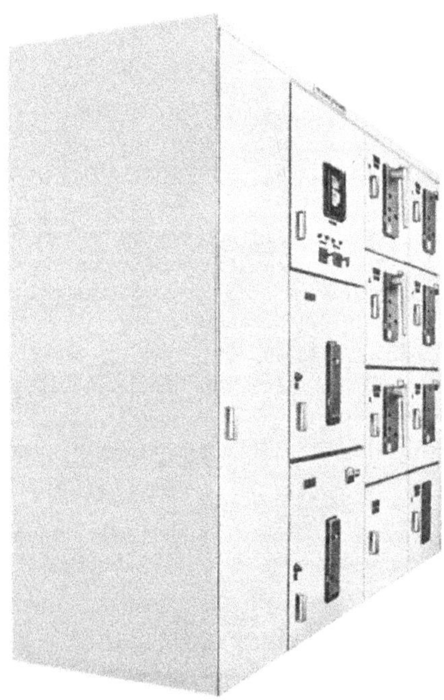

FIGURE 4.14 Low-voltage metal-clad switchgear. (*Courtesy General Electric.*)

TABLE 4.6 Recommended Minimum Safety Equipment for Operating Metal-Clad Switchgear

Closed door (arc-resistant gear only) • Hard hat—ANSI Z89.1 class E or G (as required by voltage level) • ANSI Z87.1 Safety glasses with side shields • Flame-resistant work clothing (select using arc-flash calculations)
Open-door or non-arc-resistant gear • Hard hat—ANSI Z89.1 class E or G (as required by voltage level) • ANSI Z87.1 Safety glasses with side shields • Rubber gloves with leather protectors (class according to voltage level) • Flame-resistant work clothing (select using arc-flash calculations)

Note: Closed door means that the front safety panels are closed and latched. If the front panels are not designed to contain blast and arc (such as with arc-resistant switchgear), then *open-door* protective equipment must be worn.

Operating Low-Voltage Switchgear

General Description. With some exceptions, the operation of low-voltage switchgear is very similar to the operation of medium-voltage gear. Figures 4.14 through 4.17 illustrate various types of low-voltage switchgear and the breakers used in them.

In this style of gear, the circuit breaker rolls into the switchgear on a sliding type of racking mechanism as in Fig. 4.17. The circuit breaker is tripped by the release of a powerful set of springs. Depending on the breaker, the springs are released either manually and/or electrically using a front panel button and/or control switch. The springs may be charged manually and/or electrically, again depending on the breaker.

The breakers may be closed either with another set of springs or by means of a manual closing handle. The springs for spring-operated breakers may be charged either manually or electrically. For example, the breaker shown in Fig. 4.15*c* is a manually operated breaker. The large gray handle is cranked several times to charge the closing spring. When the spring is fully charged, the breaker may be closed by depressing a small, mechanical button on the face of the breaker. Tripping the breaker is accomplished by depressing another push button. Other breakers have different means of opening and closing.

Low-voltage breakers connect to the bus and the line via a set of disconnects, visible on the right side of Fig. 4.15*b*. When the breaker is open, it can be moved toward the front of the cubicle so that it disconnects from the bus and line. This action is referred to as *racking* the breaker. Racking the larger-size low-voltage breakers is accomplished using removable cranks. These cranks are usually of a rotary type. The breaker may be completely removed from the switchgear, or it may be put into two or more auxiliary positions. Smaller low-voltage breakers are racked by simply pulling or pushing them.

Like medium-voltage switchgear, most low-voltage breakers have two auxiliary positions (*test* and *disconnected*). In the test position, the breaker is disconnected from the bus; however, its control power and/or auxiliary switches are still applied through a set of secondary disconnects. This allows technicians to operate the breaker for maintenance purposes. In the disconnected position, the breaker is completely disconnected; however, it is still in the switchgear.

For some types of switchgear, the front panel provides worker protection from shock, arc, and blast. This means that the switchgear is designed to contain arc and blast as long as the door is properly closed and latched. Such gear is called *arc resistant.*

Closed-Door Operation. The closed-door operation of low-voltage breakers is virtually identical to the closed-door operation of medium-voltage circuit breakers. Table 4.6 lists the recommended safety equipment to be used by operators when performing both closed-door and open-door switching on

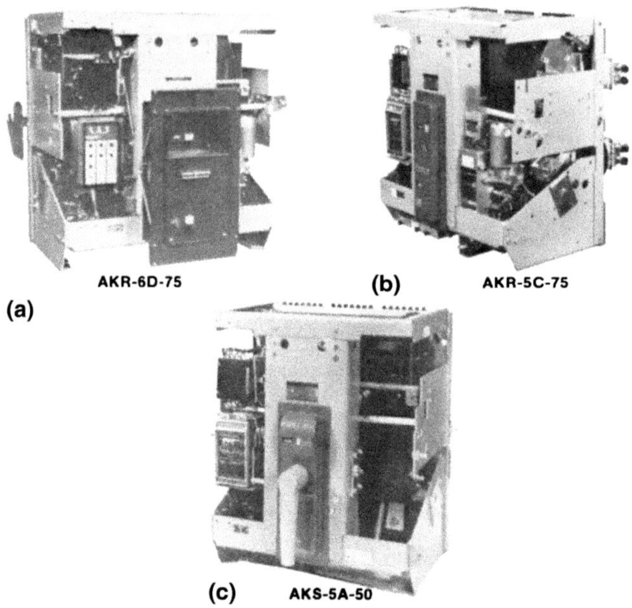

FIGURE 4.15 Low-voltage circuit breaker used in the switchgear shown in Fig. 4.14. (*Courtesy General Electric.*)

FIGURE 4.16 Low-voltage metal-enclosed switchgear. (*Type R switchgear manufactured by Electrical Apparatus Division of Siemens Energy and Automation, Inc.*)

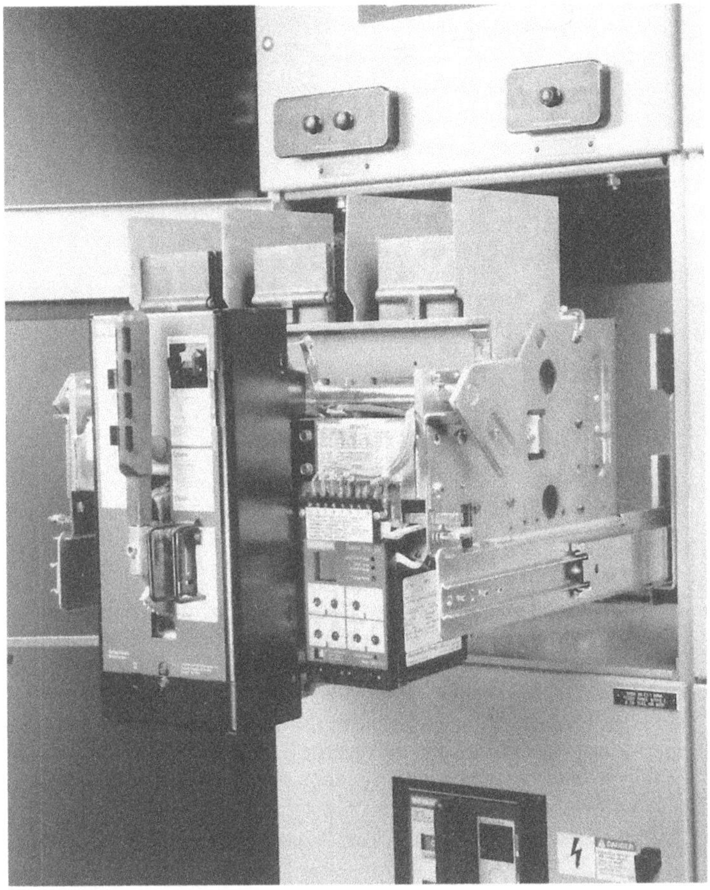

FIGURE 4.17 Low-voltage power circuit breaker racked completely away from the bus. (*Type RL circuit breaker manufactured by Electrical Apparatus Division of Siemens Energy and Automation, Inc.*)

low-voltage switchgear. Note that both the primary operator and backup operator should be wearing the recommended clothing.

The primary operator is the worker who actually manipulates the handle that opens and/or closes the circuit breaker. The backup operator's responsibility is to back up the primary operator in the event there is a problem. The backup operator may be optional in some facilities.

To operate the switchgear with a closed door, the following steps apply:

1. If the breaker requires manual spring charging, the primary operator may face the breaker to obtain the necessary leverage on the cranking handle.
2. After the springs are charged, the primary operator stands to the side of the cubicle containing the breaker to be operated. The side to which he or she stands should be determined by which side the operating handle is on. If the handle is in the middle, the operator should stand on the hinge or the handle side of the door, depending on which side is stronger. (Refer to the manufacturer.)
3. The primary operator faces away from the gear (preferred).
4. The backup operator stands even farther from the cubicle and outside the arc-flash boundary, facing the primary operator.

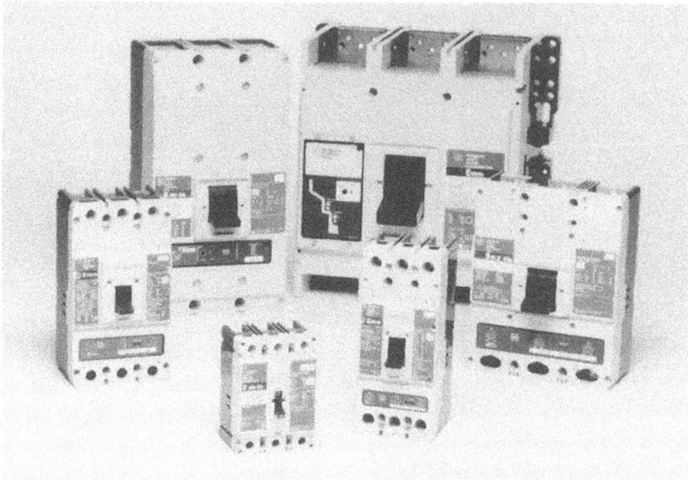

FIGURE 4.18 Various molded-case circuit breakers. (*Courtesy Westinghouse Electric.*)

5. The primary operator reaches across to the operating buttons or handle and operates them to open or close the breaker. Note that the primary operator continues to keep his or her face turned away from the gear. Some operators prefer to use a hot stick or a rope for this operation. This keeps the arms as far as possible from any hazard.
6. If the breaker can be racked with the door closed, and if the breaker is to be racked in or out, the primary operator inserts the racking handle and turns it. Note that breakers that are racked manually cannot be racked with the door closed. In this operation, the primary operator may have to face the breaker cubicle.
7. If lockout-tagout procedures are required, the primary operator places the necessary tags and/or locks.

Open-Door Operation. Refer to Table 4.6 for a minimum listing of safety equipment for this operation. If the door must be open for racking the breaker, the following steps should be observed:

1. The breaker is opened as described earlier under closed-door operation.
2. The primary operator opens the cubicle door and racks the breaker to the desired position.
3. If lockout-tagout procedures are required, the primary operator places the necessary tags and/or locks.

Operating Molded-Case Breakers and Panelboards

General Description. Molded-case circuit breakers are designed with a case that completely contains the arc and blast of the interrupted current, as shown in Fig. 4.18. Such breakers are permanently mounted in individual enclosures or panelboards (Fig. 4.19) along with many other such breakers.

Molded-case breakers have a three-position operating handle—open, closed, and tripped. When the operator opens the breaker, he or she does so by moving the operating handle to the open position. Likewise, the close operation is accomplished by moving the handle to the close position.

When the breaker trips via its internal automatic protective devices, the handle moves to the tripped position. The trip position is normally an intermediate position between the full-closed and full-open positions. After a trip operation, the breaker cannot be operated until the handle is moved

FIGURE 4.19 Panelboards equipped with molded-case circuit breakers. (*Courtesy McGraw-Hill, Inc.*)

TABLE 4.7 Recommended Minimum Safety Equipment for Operating Molded-Case Circuit Breakers

- Hard hat—ANSI Z89.1 class E or G
- ANSI Z87.1 Safety glasses with side shields
- Flame-resistant work clothing (select using arc-flash calculations)
- Hand protection—leather or flame-resistant gloves (do not need to be insulating)

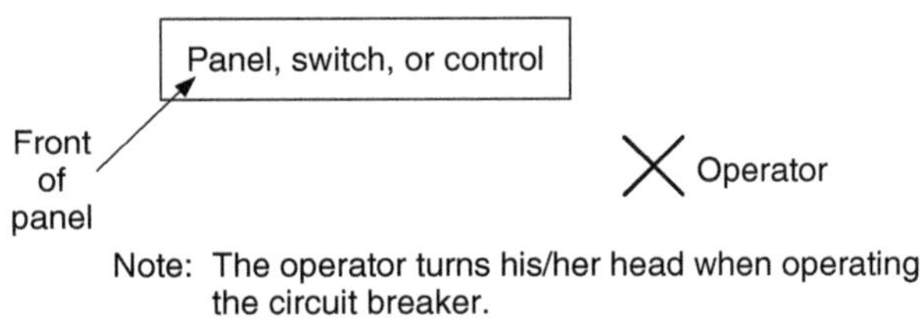

FIGURE 4.20 Proper position for operating molded-case circuit breakers, enclosed switches, and motor controls (top view).

forcefully to the open position. This action resets the internal tripping mechanism and reengages the manual operating mechanism.

Operation. *Caution:* Circuit breakers should not be used for the routine energizing and de-energizing of circuits unless they are manufactured and marked for such purpose. They may be used for occasional or unusual disconnect service.

Table 4.7 lists the minimum recommended safety equipment to be worn when operating molded-case circuit breakers, and Fig. 4.20 illustrates the proper position for the operation. Notice that a backup operator is not required for this procedure; however, secondary assistance is always a good practice. The procedure can be summarized as follows:

1. The operator stands to the side of the breaker and/or panel, facing the panel. The operator may stand to either side, depending on the physical layout of the area.
2. The operator grasps the handle with the hand closest to the breaker.
3. The operator turns his or her head away from the breaker and then firmly moves the operating handle to the desired position.
4. If locks or tags are required, they are placed on the breaker using the types of equipment as described in Chap. 3.

Operating Enclosed Switches and Disconnects

General Description. Figure 4.21 shows several basic types of enclosed switches. These devices are used to connect and/or disconnect circuits. Such devices may be load interrupting or non-load-interrupting. If they are non-load-interrupting, they must not be operated when current is flowing in the circuit. If you are uncertain as to whether the switches are load interrupting or not, look on the nameplate or check with the manufacturer. The presence of arc interrupters, such as those in Fig. 4.21*d*, is a good indication that the device is intended to interrupt load current.

These switches are operated by moving the handle. In some units, the handle is bolted or locked in place to prevent inadvertent operation. Enclosed switches have a mechanical interlock that prevents the case from being opened until the handle is in the open position. Electrically qualified personnel may temporarily defeat the interlock if needed for maintenance or troubleshooting purposes. Such switches should not be operated with the door open when load current if flowing. Be very cautious when opening medium-voltage switches.

Operation. The basic operating procedure for such switches is similar to the procedure given for molded-case circuit breakers. *Caution:* Switches should not be used to interrupt load current unless they are intended for that purpose. Refer to the manufacturer's information.

Table 4.8 lists the minimum recommended safety equipment to be worn when operating enclosed switches, and Fig. 4.20 illustrates the proper position for the operation. Notice that a backup operator is not required for this procedure; however, secondary assistance is always a good practice. The procedure can be summarized as follows:

1. The operator stands to the side, of the switch and/or panel, facing the panel. The operator may stand to either side, depending on the physical layout of the area.
2. The operator grasps the handle with the hand closest to the switch.
3. The operator turns his or her head away from the switch and then firmly moves the operating handle to the desired position.
4. If locks or tags are required, they are placed on the switch using the types of equipment described in Chap. 3.

Operating Open-Air Disconnects

General Description. Open-air disconnects may be manually operated or mechanism operated. Mechanism types (Fig. 4.22) are normally installed as overhead devices in medium-voltage substations or on pole lines. The switch has an operating handle close to the ground that is used to open or close the contacts. At ground level, a metal platform is often provided for the operator to stand on. This platform is bonded to the switch mechanism and to the ground grid or ground rod to create an equipotential zone of protection for the switch operator. The operator stands on the grid; thus, the operator's hands and feet are at the same electric potential. Note that the operation of some switches

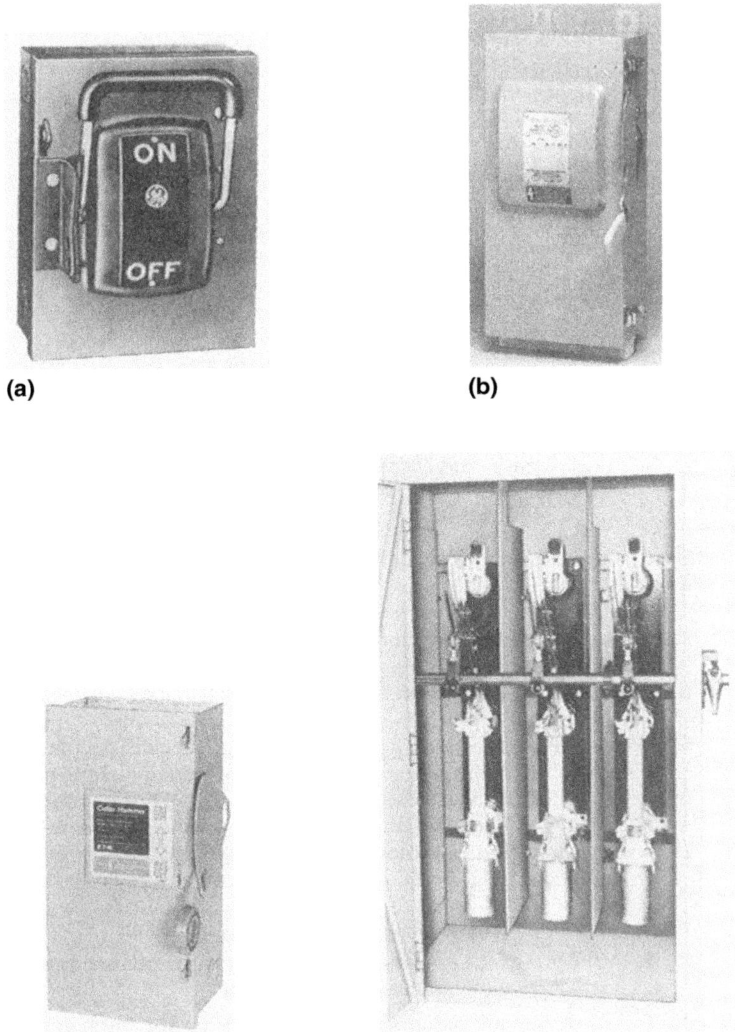

(c) (d)

FIGURE 4.21 Various types of low- and medium-voltage, enclosed disconnect switches. (*Courtesy General Electric, Crouse-Hinds Co., and Eaton Corporation, Cutler Hammer Products.*)

TABLE 4.8 Recommended Minimum Safety Equipment for Operating Enclosed Switches and Disconnects

- Hard hat—ANSI Z89.1 class E or G (as required by voltage level)
- ANSI Z87.1 Safety glasses with side shields
- Flame-resistant work clothing (select using arc-flash calculations)
- Rubber gloves with leather protectors (class according to voltage level)

FIGURE 4.22 Mechanism-operated, three-phase, open-air switch. (*Courtesy Alan Mark Franks.*)

is accomplished by moving the handle in the horizontal plane, while others are moved in a vertical direction.

Manually operated switches are operated by physically pulling on the blade mechanism. In some cases, such as the one shown in Fig. 4.23, the switch blade is a fuse. The manual operation is accomplished by using a hot stick. Manually operated switches may be located overhead in outdoor installations, or they may be mounted indoors inside metal-clad switchgear.

Operation. Recommended protective clothing depends on the type and location of the switch. Table 4.9 lists the minimum recommended safety equipment for operating overhead, mechanism-operated switches such as those shown in Fig. 4.22. Figure 4.24 shows the correct operating position for operating such a switch. *Caution:* Not all open-air switches are designed to interrupt load current. Do not use a non-load-interrupting switch to interrupt load current.

The basic operating procedure can be summarized as follows:

1. The operator stands on the metal platform (if available).
2. He or she grasps the operating handle firmly with both hands and moves it rapidly and firmly in the open or close direction as required.
3. If locks or tags are required, they are placed on the mechanism using the types of equipment described in Chap. 3.

Table 4.10 lists the recommended equipment for operation of manually operated open-air switches. Note that the use of flash suits is dependent on the location of the switch. If the switch is on an overhead construction that is outdoors, the worker may opt to wear flame-resistant clothing instead of a flash suit. Because the operator must use a hot stick to operate this type of switch, he or she must stand directly in front of the switch.

The general operating procedure is as follows:

1. Stand in front of the switch.
2. Carefully insert the hot stick probe into the switch ring.
3. Look away from the switch and pull it open with a swift, firm motion.
4. Since one side of the switch may be hot, locks and tags are not always applied directly to the switch. If the switch is in an indoor, metal-clad enclosure, the lock and tag may be applied to the door of the gear.

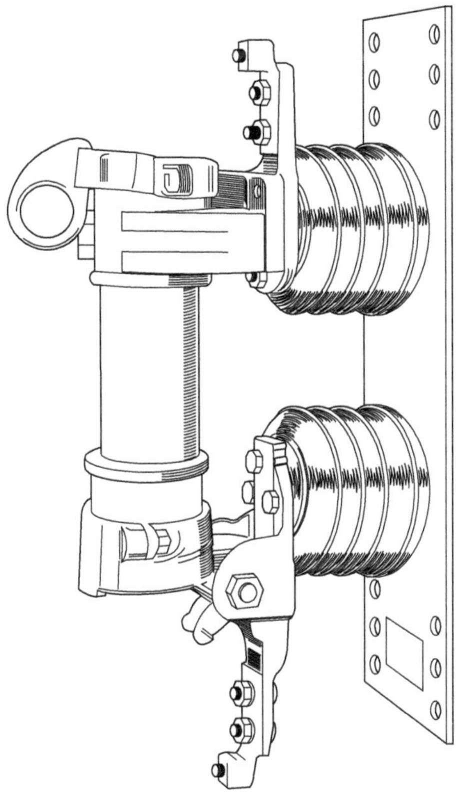

FIGURE 4.23 Open-air, fused disconnect switch. (*Courtesy General Electric.*)

TABLE 4.9 Recommended Minimum Safety Equipment for Operating Overhead, Mechanism-Operated Switches

- Hard hat—ANSI Z89.1 class E or G (as required by voltage level)
- ANSI Z87.1 Safety glasses with side shields
- Flame-resistant work clothing (select using arc-flash calculations)
- Rubber gloves with leather protectors (class according to voltage level)

Operating Motor Starters

General Description. With some exceptions, the operation of motor starters is very similar to the operation of low- and medium-voltage gear. Figure 4.25 shows a single motor starter in a cabinet suitable for mounting on a wall.

In the motor control center, the starter mounts on a mechanism specially designed for the purpose. In either type of construction, the motor is stopped and started by depressing the appropriate button. The starter also has a fused disconnect or a molded-case circuit breaker that is used to disconnect the motor and its circuitry from the power supply.

Motor starters used in motor control centers connect to the bus and the line via a set of disconnects. When the starter is open, it can be moved toward the front of the cubicle so that it disconnects

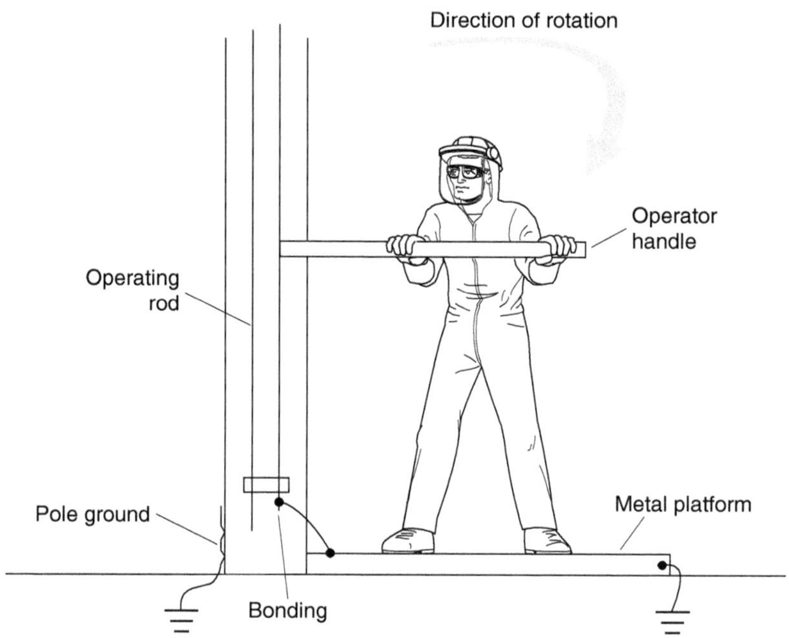

FIGURE 4.24 Correct position for operating an overhead, mechanism-operated switch.

TABLE 4.10 Recommended Minimum Safety Equipment for Operating Manually Operated, Open-Air Disconnect Switches

- Hard hat—ANSI Z89.1 class E or G (as required by voltage level)
- ANSI Z87.1 Safety glasses with side shields
- Flame-resistant work clothing (select using arc-flash calculations)
- Rubber gloves with leather protectors (class according to voltage level)
- Hot stick of proper length with proper fittings

from the bus and line. This action is referred to as *racking*. Racking starters is usually accomplished manually. The starter may be completely removed from the motor control center. Note that large medium-voltage contactors may be operated like medium-voltage circuit breakers.

For many types of motor control centers, the front panel provides worker protection from shock, arc, and blast. This means the motor control center is designed to contain arc and blast as long as the door is properly closed and latched.

Closed-Door Operation. The closed-door operation of motor starters is virtually identical to the closed-door operation of low-voltage circuit breakers. Table 4.6 lists the recommended safety equipment to be used by operators when performing both closed-door and open-door switching on motor starters. Note that both the primary operator and backup operator should be wearing the recommended clothing. The primary operator is the worker who actually manipulates the handle that opens and/or closes the motor starter. The backup operator's responsibility is to support the primary operator in the event there is a problem. The backup operator may be optional in some facilities.

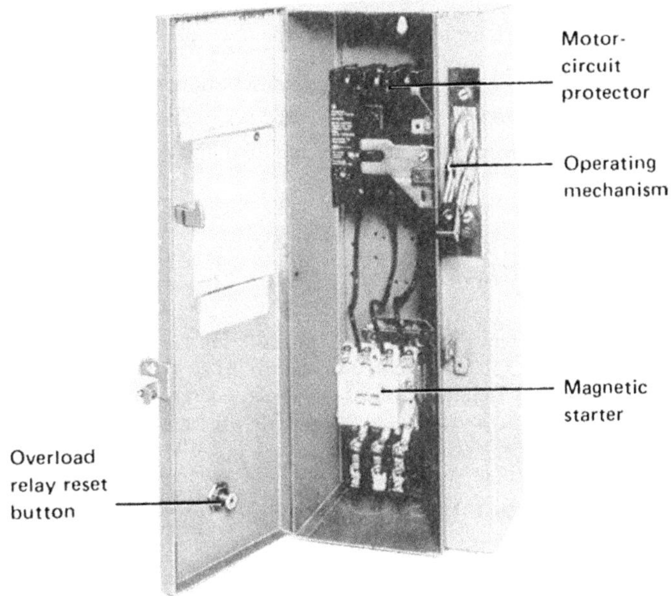

FIGURE 4.25 Combination motor starter in individual metal-clad cabinet.

To operate the starter with a closed door, the following steps apply:

1. While wearing proper hand protection (either rubber insulating gloves with leather protectors or other suitable protection), stand to the side of the motor starter and depress the stop button to properly interrupt the load current.
2. After the motor is stopped, the primary operator stands to the side of the cubicle containing the starter to be operated. The side to which he or she stands should be determined by which side the operating handle is on. If the handle is in the middle, the operator should stand on the hinge side or the handle side of the door, depending on which side is stronger. (Refer to the manufacturer.)
3. The primary operator faces away from the gear. *Note:* If the operating handle of the disconnect has a very tight operating mechanism, the primary operator may face the motor starter to obtain the necessary leverage on the cranking handle.
4. The backup operator stands outside the arc-flash hazard boundary with line of sight to the operator.
5. The primary operator reaches across to the operating handle and operates it to open or close the starter disconnect. Note that the primary operator continues to keep his or her face turned away from the gear. Some operators prefer to use a hot stick or a rope for this operation. This keeps the arms as far as possible from any hazard.
6. If the starter can be racked with the door closed (an unusual configuration), and if the starter is to be racked in or out, the primary operator inserts the racking handle and turns it. Note that starters that are racked manually cannot be racked with the door closed. In this operation, the primary operator will undoubtedly have to face the breaker cubicle.
7. If lockout-tagout procedures are required, the primary operator places the necessary locks and/or tags.

Open-Door Operation. Refer to Table 4.6 for a minimum listing of safety equipment for this operation. If the door must be open for racking the starter, the following steps should be observed:

1. The motor is stopped as described earlier under closed-door operation.
2. The primary operator opens the cubicle door and racks the starter to the desired position.
3. If lockout-tagout procedures are required, the primary operator places the necessary locks and/or tags.

ENERGY CONTROL PROGRAMS

An energy control program is a procedure for the proper control of hazardous energy sources. It should include a listing of company-approved steps for the proper and safe energizing and de-energizing of energy isolation devices as well as general company policy statements on preferred methods of operation. Energy control programs fall into two categories—general and specific.

General Energy Control Programs

Overview. A general energy program is one that is inherently generic in nature. Its steps are broad-based and designed in such a way that the program can be used as a procedure for a wide variety of equipment types. General energy control programs should be used only when the equipment being isolated meets all the following criteria:

- The equipment can be disabled, so it has no potential for the release of stored or residual energy after shutdown.
- The equipment is supplied from a single energy source that can be readily identified and isolated.
- The equipment is completely de-energized and deactivated by the isolation and locking-out procedure.
- No employees are allowed to work on or near the equipment until it has been tagged and locked. (See the lockout-tagout section later in this chapter.)
- A single lockout device will achieve a locked-out condition.
- The isolating circuit breakers or switches are under the exclusive control of the employee(s) who placed the lock and tag.
- De-energizing and working on the equipment does not create a hazard for other employees.
- There have been no accidents involving unexpected activation or reenergization of the equipment during previous servicing.

Specific Energy Control Programs

When a part of the system or piece of equipment does not meet all the criteria laid out in the overview to this section, a specific energy control program should be written. Although the procedures will vary depending on the specifics of the installation, at a minimum the program should include the following information:

- The description of the system and/or equipment that will be de-energized
- Any controls, such as motor starter push buttons, that exist on the equipment
- The voltages and short-circuit capacities of the parts of the system that will be de-energized
- The circuit breakers, switches, or contactors that are used to de-energize the system

- The steps that must be used to de-energize the system. The steps should include:
  1. The methods and order of operation of the circuit breakers, switches, and so on.
  2. Any special requirements for the lockout-tagout procedure.
  3. Special notifications and safety requirements.
- Reenergizing requirements and procedures.

Basic Energy Control Rules

- The safest and most secure method to protect personnel from the electrical hazard is to de-energize the conductors that they must work on or near. De-energization is the preferred method.
- If conductors cannot be de-energized, safety equipment and safety-related work practices must be used to protect personnel exposed to the energized conductors.
- Before personnel are allowed to work on or near any exposed, de-energized conductors, the circuit breakers and/or disconnect switches must be locked and tagged to prevent their inadvertent operation.
- All personnel should be instructed to never operate or attempt to operate any circuit breaker and/or disconnect switch when it is tagged and/or locked.
- Only authorized, qualified, and trained personnel should be allowed to operate electrical equipment.
- Locks and tags should be removed only by the personnel that placed them. Two exceptions may apply under the following situations:
  1. If the worker who placed the lock and tag is not available, his or her lock and tag may be removed by a qualified person who is authorized to perform such an action. This procedure is often called *bypassing control* as the person who removes the lock and tag is, in fact, bypassing the authority (control) of the person whose tag is being removed. If this operation is performed, the employees that had the lockout-tagout applied must be properly informed that their lockout-tagout has been violated and removed before they are allowed to return to work on the property.
  2. Some facilities may authorize the concept of a group lock. A group lock is placed by an authorized shift worker, such as the shift operator, and may be removed by another authorized shift worker. This activity should not be used to prevent any employee from placing his or her tag and lock on energy-isolating devices that may feed conductors that they must work on or near.

De-energizing Equipment. The general energy program for de-energizing equipment should include the following steps:

1. Before beginning the process, carefully identify the voltage levels and incident energy levels of the portion of the system that will be de-energized. This information serves to establish the level of the hazard to all personnel.
2. Notify all employees who will be affected by the de-energization that the system is to be de-energized.
3. Perform necessary checks and inspections to ensure that de-energizing the equipment will not introduce additional safety hazards, for example, de-energizing safety-related ventilation systems.
4. Using properly rated load-interrupting devices and proper operating instructions, shut down all processes being fed by the electric system that is to be de-energized.

5. Open the appropriate circuit breaker and/or switch.
6. Rack the circuit breaker away from the bus if it is of the type that can be manipulated in this manner.
7. Release stored energy from springs, hydraulic systems, or pneumatics.
8. Discharge and ground any capacitors located in the de-energized portions of the system.
9. Apply tags and/or locks.
10. Attempt to operate the breaker and/or switch to make certain that the locks are preventing operation. If a motor starter is involved, press the start button to make certain the motor will not start.
11. Measure the voltage on the conductors to which employees will be exposed, at the point where they will be exposed.
12. Notify personnel that the system is safely de-energized, locked, and tagged.

Reenergizing Equipment. Reenergization of some systems is more hazardous than de-energization. While the equipment has been out of service, personnel have grown used to its de-energized voltage status. In addition, tools and/or other equipment may have been inadvertently left on or near exposed conductors.

Because of these factors, the same type of rigorous steps should be followed during reenergization.

1. All personnel should be notified that the system is to be reenergized and warned to stay clear of circuits and equipment.
2. A trained, qualified person should conduct all tests and visual inspections necessary to verify that all tools, electric jumpers, shorts, grounds, and other such devices have been removed and that the circuits are ready to be reenergized.
3. Close and secure all cabinet doors and other safety-related enclosures.
4. Because the tests and inspections may have required some time, the personnel warnings should be repeated.
5. Locks and tags should be removed by the personnel that placed them.
6. If breakers were racked into disconnected positions, they should be racked in the connected position.
7. Make final checks and tests, and issue final warnings to all personnel.
8. Reenergize the system by reconnecting and closing circuit breakers and switches. These operations are normally carried out in the reverse order of how they were opened.

Procedures Involving More Than One Person. When more than one person is required to lock and tag equipment, each person will place his or her own personal lock and tag on the circuit breakers and/or switches. The placement of multiple locks and tags on equipment is often called *ganging*. Since few circuit breakers or switches have the ability to accept multiple locks and tags, this procedure can take one of two common approaches.

1. A multiple-lock hasp may be applied to the breaker or switch. Such hasps will accept up to six locks. If more than six locks are required, multiple-lock hasps may be cascaded. See Fig. 3.52 for examples of such hasps.
2. A lockbox may be used. In such an operation, the lock is applied to the breaker or switch and the key is then placed inside the lockbox. The lockbox is then secured by the use of a multiple-lock hasp. This approach is used when the presence of many locks on the switch or breaker might cause operational problems.

After the work has been completed, each employee removes his or her lock from the lockbox. The key for the lock is retrieved and the lock can then be removed by the person who placed it.

LOCKOUT-TAGOUT

Definition and Description

Tags are used to identify equipment that has been removed from service for maintenance or other purposes. They are uniquely designed and have clear warnings printed on them instructing personnel not to operate the equipment. Locks are applied to de-energized equipment to prevent accidental or unauthorized operation. Locks and tags are normally applied together; however, some special circumstances may require the use of a tag without a lock and/or a lock without a tag. See Chap. 3 for a detailed description of the construction of safety locks and tags.

Employers should develop a written specification that clearly defines the lockout-tagout rules for the facility. This specification should be kept on file and reviewed periodically to ensure that it is kept up to date. The following sections define the key elements that should be included in the specification.

When to Use Locks and Tags

Locks and tags should be applied to open circuit breakers, switches, or contactors whenever personnel will be exposed to the conductors that are normally fed by those devices. The application of the tags will warn and inform other employees that the equipment is not available for service, who applied the tag, and why the tag was applied. The lock will prevent the operation of the breaker, switch, or contactor so that the circuit cannot be accidentally reenergized.

Minor inspections, adjustments, measurements, and other such servicing activities that are routine, repetitive, and integral to the use of the equipment do not require the placement of locks and tags unless one of the following conditions exist:

1. Guard, insulation, or other safety devices are required to be removed or bypassed.
2. An employee is required to place his or her body into close proximity with an exposed, energized electric conductor. Remember that only qualified personnel are allowed to approach such locations any closer than the limited approach boundary.

Locks and tags do not need to be used on plug- and cord-connected equipment as long as the cord and plug stay under the exclusive control and within sight of the employee who is exposed to the electrical hazard.

Locks without Tags or Tags without Locks

Tags may be used without locks under both of the following conditions:

1. The interrupting device is not designed to accept a lock.
2. An extra means of isolation is employed to provide one additional level of protection. Such an extra procedure might take the form of an additional open point such as removing a fuse or disconnecting a wire or the placement of safety grounds to provide an equipotential work area.

Application of a lock is preferred by OSHA. OSHA 1910.147 Control of Hazardous Energy mandates: *"Energy isolation devices for machinery or equipment installed after January 2, 1990, shall be capable of accepting a lockout device."* When equipment is removed from service that has no means of accepting the hasp of a lock, the equipment should be retrofitted, so a lock can be applied and proper lockout-tagout can be routinely performed.

Locks may be used without tags under both of the following conditions:

1. The de-energization is limited to only one circuit or piece of equipment.
2. The lockout lasts only as long as the employee who places the lock is on-site and in the immediate area.

Rules for Using Locks and Tags

All electric equipment with the capability to be reenergized and harm employees shall be safely isolated by means of a lock and tag during maintenance, repair, or modification of the equipment.

When two or more crafts must both have access to the equipment, authorized employees from both crafts shall place locks and tags on behalf of the members of their craft. This practice is referred to as *ganging*. This should not be construed to limit the right of any employee to place his or her individual lock and tag on the equipment.

Responsibilities of Employees

Employees who are authorized to place locks and tags have certain responsibilities that they must exercise when placing those tags.

- The system must be surveyed to ensure that all sources of power to the system have been identified.
- All the isolating equipment (circuit breakers, switches, etc.) must be identified and correlated with the portions of the system to which they apply.
- The voltage level and incident energy magnitude for each part of the system to be de-energized must be determined. This step helps to assess the hazard to individuals who will be exposed to the de-energized system parts.
- All personnel who will be affected by the outage must be notified. This includes employees who may be served by the electric power or who may work on or around the equipment that will be affected by the outage.
- The employee(s) who place the locks and tags must maintain knowledge and control of the equipment to which they have affixed their locks and tags.

When locks and tags are removed, authorized, qualified employees must perform certain tasks, including the following:

1. Notify all affected personnel the system is going to be reenergized.
2. If appropriate, in a gang lock situation, make certain that all authorized employees are prepared to remove their locks and tags.
3. Inspect and/or test all parts of the system to make certain that they are ready to be reenergized.

Note: See the "Energy Control Programs" section earlier in this chapter for more information about these requirements.

Sequence

The following steps should be followed when shutting down an energized electric system:

1. Motors and other operational equipment should be shut down using normal or emergency procedures as required. During the shutdown process, personnel safety must be the prime consideration.
2. All isolating equipment (circuit breakers, switches, and/or contactors) should be opened.
3. Isolating equipment that is capable of being racked out should be racked to the disconnected position.
4. Stored energy, such as closing/tripping springs, hydraulic pressure, pneumatic pressure, or other such mechanisms should be discharged and released.
5. Discharge and ground capacitors.
6. Lockout-tagout should be performed as described elsewhere in this handbook.

Lock and Tag Application

Isolating equipment is capable of being locked out if either of the following conditions is met:

- The equipment has an attachment that is an integral part of the equipment, through which the locking device may be passed in such a way as to prevent operation of the isolating equipment.
- The lock can be attached in some other way to prevent operation without dismantling, rebuilding, or replacing the energy-isolating equipment. This might apply to a breaker that cannot be locked open but can be locked in the racked or disconnected position.

Locks and tags should be applied to all isolating equipment that is capable of being locked. Locks and tags should never be applied to selector switches, push buttons, or other such control devices as the sole means of control.

Isolation Verification

After the locks and tags have been applied, the following steps should be employed to verify that the de-energization was successful:

1. Make certain that all employees are clear of the de-energized conductors.
2. Attempt to reenergize the system by operating the circuit breaker control handles, pushing the switch to close it, or whatever other procedure is appropriate.
3. Using proper procedures and test equipment, measure the voltage on conductors at the point(s) where employee exposure will take place. The voltage must be zero.

Removal of Locks and Tags

Normal Removal. When the work is finished and an employee is ready to remove his or her locks and tags, the following general approaches should be used:

- The work area should be inspected to ensure that nonessential items have been removed and all components are operationally intact.
- The work area should be inspected to ensure that all employees have been safely removed or positioned.
- Remove any specialized equipment such as safety grounds or spring tension blocks.
- Notify all affected employees that locks and tags are to be removed.
- Locks and tags should be removed by the personnel that placed them.

Control Bypass. If the employee who placed the locks and tags is absent and not available to remove them, and if the locks and tags absolutely must be removed, another authorized employee should assume control of the equipment and remove the absent employee's locks and tags. The following steps must be employed:

1. The employee who is assuming control must make an assessment of the situation and determine that a genuine, critical need exists to remove the absent employee's locks and tags. Examples of such critical needs include:
   a. Immediate operation and/or production requires the equipment to resume operation and avoid general shutdown.
   b. Personnel safety requires restoration of power to the de-energized systems.
   c. Equipment must be temporarily returned to service to allow testing or other such evaluation.

2. The employee assuming control must make every reasonable attempt to contact the absent employee to obtain his or her assessment of the equipment.
3. The employee assuming control must contact other employees who may have knowledge of the availability of the equipment and the advisability of removing the absent employee's locks and tags.
4. The employee assuming control must develop and document a formal conclusion with regard to the decision to remove the absent employee's locks and tags.
5. If the employee assuming control intends to apply his or her own locks and tags, he or she should do this, following all of the steps normally used for lockout-tagout—including voltage measurement.
6. The employee assuming control must follow the normal removal steps to remove the locks and tags and either reinstall his or her own locks and tags or reenergize the system.
7. The area where the absent employee's locks and tags were removed should be posted with large, easy-to-read signs indicating that the locks and tags have been removed.
8. The employee assuming control must contact the absent employee immediately upon his or her return. The returning employee must be told that his or her locks and tags have been removed and completely updated as to the status of the equipment. The absent employees shall remove the signs posted in step 7.

Figure 4.26 shows a form that may be used to document the control bypass procedure. In this figure the employee who is assuming control is referred to as the *authorized controller*. The absent employee is referred to as the *authorized employee*. Notice that each one of the major steps is documented on this form.

Temporary Removal. Locks and tags may be temporarily removed by the employee who placed them. When doing so, the same steps used in normal removal should be observed.

Safety Ground Application

Safety grounds should be applied as an additional safety measure when equipment is removed from service. The only exception to this is when the nature of the work precludes the use of safety grounds. Procedures like insulation measurement cannot be performed when safety grounds are applied; therefore, they may not be used, or must be temporarily disconnected for the measurement process.

Control Transfer

Lockout-tagout control may be transferred from one employee to another as long as both employees are present. The following steps should be used:

1. The employee assuming control first applies his or her locks and tags using all of the steps described elsewhere in this handbook.
2. The employee relinquishing control then removes all of his or her locks and tags.

All logs or other such records should be immediately updated to show the new lockout-tagout condition.

Nonemployees and Contractors

Nonemployees such as contractors should be required to use a lockout-tagout program that provides the same or greater protection as that afforded by the facility's procedure. Contractors should be

LOCKOUT/TAGOUT CONTROLLER'S BYPASS FORM

DATE ________________ **TIME** ______________

LOCATION ____
____

AUTHORIZED CONTROLLER ____

DEPARTMENT ____

AUTHORIZED EMPLOYEE ____

TO BE BY-PASSED DEPARTMENT ____________________________________

DESCRIPTION OF EQUIPMENT OR DEVICE(S) TO BE BY-PASSED __________
____

NEED FOR BY PASS PROCEDURE ____________________________________
____
____
____
____
____
____

EFFORTS TO CONTACT AUTHORIZED EMPLOYEE _________________________
____
____
____
____
____
____

EFFORTS TO CONTACT OTHER PERSONNEL THAT MAY BE AWARE OF THE APPLIED LOCKOUT/TAGOUT DETAILS ______________________________
____
____
____
____
____
____

CONTROLLER EVALUATION OF THE SITUATION _________________________
____
____
____
____
____
____

ACTION TAKEN BY THE CONTROLLER TO RELEASE EQUIPMENT FOR USE
____
____
____
____
____
____

ACTIONS TAKEN TO PROPERLY MARK AND POST EQUIPMENT AND AREA AS BEING RE-ENERGIZED ______________________________________
____
____
____
____
____
____

FIGURE 4.26 Lockout-tagout control bypass form.

ACTIONS TAKEN TO NOTIFY EFFECTED EMPLOYEES ______________

AUTHORIZED CONTROLLER ______________
DEPARTMENT ______________ DATE ______________

*AUTHORIZED EMPLOYEE ______________
DEPARTMENT ______________ DATE ______________
* To be use after notification of By-Pass

FIGURE 4.26 *(Continued)*

required to submit their procedure for review and approval. No work should be allowed until the contractor's lockout-tagout program has been reviewed and approved by qualified management.

If the contractor's lockout-tagout program is more stringent than the owner's program, the contractor's program should be utilized. If the contractor's program is not as stringent as the owner's program, the contractor must be trained in and required to use the owner's lockout-tagout program. OSHA requires a meeting of the minds in this situation.

If contractors are allowed to use their lockout-tagout procedures, all employees should be familiarized with the contractors' procedure and required to comply with it. Copies of the respective standards should be made available for both contractors and employees. Likewise, all contractors must be familiar with the facility's procedure and must comply with it.

If the contractor is working on machines or equipment that have not yet been turned over to the facility, the contractor shall be required to lock and tag that equipment if it might operate and endanger employees. The contractor must guarantee the safety of the locked and tagged installation. As an alternative to this procedure, the contractor should be required to allow gang lockout and tagout with employees and contractor personnel.

Lockout-Tagout Training

All employees should be trained in the use, application, and removal of locks and tags. See Chap. 14 for more information on general employee safety training.

Procedural Reviews

The entire lockout-tagout procedure should be reviewed at intervals no longer than one year. A review report should be issued that identifies the portions of the procedure that were reviewed, changes that were considered, and changes that were implemented.

VOLTAGE-MEASUREMENT TECHNIQUES

Purpose

No circuit should ever be presumed dead until it has been measured using reliable, prechecked test equipment. Good safety practice and current regulatory standards require that circuits be certified de-energized by measurement as the last definitive step in the lockout-tagout procedure.

Instrument Selection

Voltage-measuring instruments should be selected for the category, voltage level, the application location, short-circuit capacity, sensitivity requirements, and the circuit loading requirements of the circuit that is to be measured.

During voltage measurement, the instrument may be subjected to voltage transients. This is especially critical in low-voltage (below 1000 V) applications. When selecting a low-voltage-measuring instrument, be certain to select the correct category of instrument as defined in IEC 61010. See Chap. 3 for a description of the categories.

Voltage Level. The instrument must be capable of withstanding the voltage of the circuit that is to be measured. Use of underrated instruments, even though the circuit is dead, is a violation of good safety practice.

Application Location. Some instruments are designed for outdoor use only and should not be used in metal-clad switchgear. Always check the manufacturer's information and verify that the instrument is designed for the location in which it is being used.

Internal Short-Circuit Protection. Industrial-grade, safety-rated voltage-measuring instruments are equipped with internal fusing and/or high-resistance elements that will limit the short-circuit current in the event the instrument fails internally. Be certain the instrument being used has internal protection with an interrupting/limiting rating that is at least equal to the short-circuit capacity of the circuit that is being measured. Also, if a fuse must be replaced, always replace it with the exact type used by the manufacturer or with a different one that is approved by the manufacturer.

Sensitivity Requirements. The instrument must be capable of measuring the lowest normal voltage that might be present in the circuit that is being measured. If the instrument has two ranges (high and low, for example), be certain to set it on the correct lowest range that applies to the voltage being measured.

Circuit Loading. A real voltmeter can be represented by an ideal meter in parallel with a resistance. The resistance represents the amount of loading the meter places on the circuit (Fig. 4.27). In the diagram, *RS* represents that resistance of the system being measured and *RM* is the internal resistance of the voltmeter. The voltmeter will read a voltage given by the formula:

$$V_M = V \times \frac{R_M}{R_M + R_S} \tag{4.1}$$

If the meter internal resistance is very high compared to system resistance, the voltage across the meter will be close to the actual system voltage. If the meter resistance is very low compared to the system resistance, the meter will read a lower voltage than actually exists because most of the voltage drop will be across the system resistance instead of the meter resistance. In this second situation, the meter is "loading" the circuit.

If a very low-resistance meter is used, the meter may not read a potentially lethal static or inductively coupled voltage. This situation is unacceptable since an inductively coupled or static voltage can be lethal. The meter that is selected for a safety-related voltage measurement must have an internal resistance that is high enough to avoid this problem.

Instrument Condition

Before each use, an instrument must be closely inspected to ensure it is in proper working order and that insulation systems have not been damaged. The case physical condition, probe exposure, lead insulation, fusing, and the operability of the instrument should be verified.

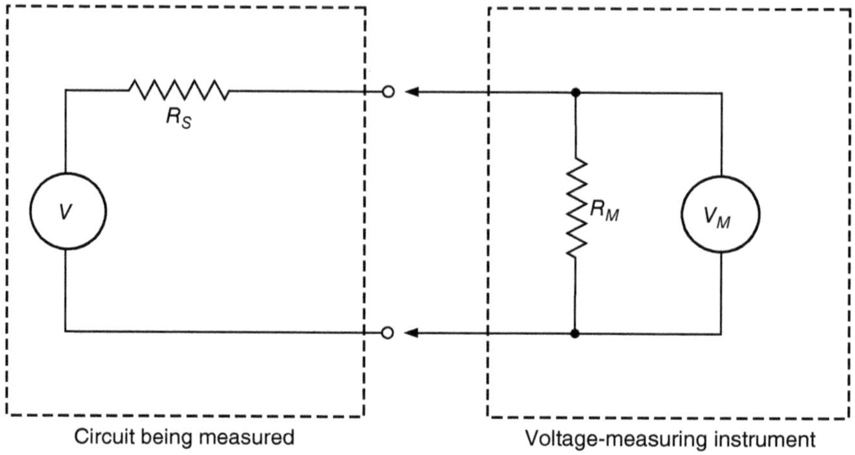

FIGURE 4.27 Equivalent circuit of voltage measurement.

Case Physical Condition. The case must be free of breaks, cracks, or other damage that could create a safety hazard or misoperation. Broken instruments should be taken out of service until they can be repaired or replaced.

Probe Exposure. Modern instrument probes have sleeves that cover all of the probe except the tip. The older spring-loaded sleeve types are no longer allowed. Check to make certain that only the minimum amount of probe required to do the job is exposed.

Lead Insulation Quality. Carefully inspect the lead insulation to make certain it is not damaged in any way.

Fusing. Accessible fuses should be checked to be certain that they are correctly installed and have not been replaced by incorrect units.

Operability. Before each usage (at the beginning of each shift, for example), the instrument should be checked to make certain that it is operable. Do not substitute this check for the instrument checks required in the three-step process.

Three-Step Measurement Process

Step 1—Test the Instrument. Immediately before each measurement, the instrument should be checked on a source that is known to be energized. This step confirms that the instrument is working before the actual circuit verification is made.

The preferred method is to use an actual power system conductor of the same voltage as that being verified. Finding such a circuit is easier when low-voltage measurements are being made; however, even then an energized circuit may not be available. Some manufacturers supply a device that creates a voltage that is sufficient for testing the instrument.

Some instruments intended for measurement of medium voltages have a low-voltage switch setting on them that allows them to be checked on low voltage. This is not a preferred method since a problem with the switch could give erroneous, dangerously false readings.

Step 2—Measure the Circuit Being Verified. After the instrument is tested, the worker then measures the circuit being verified to make certain it is de-energized. The actual wires that should be measured and the techniques to be used are discussed later in this chapter.

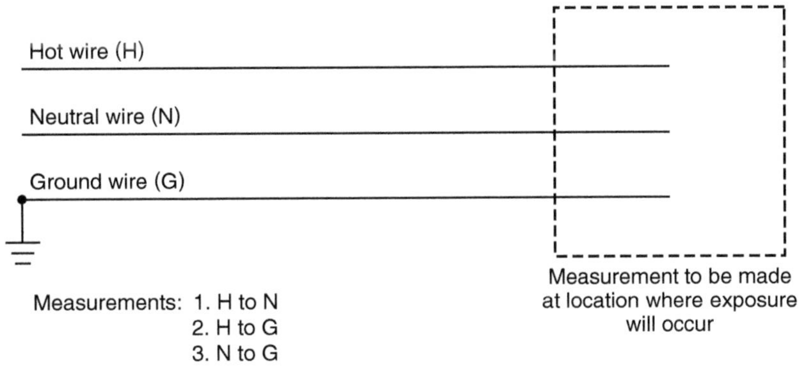

FIGURE 4.28 Measurement points for a single-phase system (one hot wire).

Step 3—Retest the Instrument. After the circuit has been verified, the instrument should be rechecked as it was in step 1. This ensures that the instrument was operable both before and after the measurement, thus affirming that a zero measurement is zero and not caused by an inoperable instrument.

What to Measure

As a general rule of thumb, all normally energized conductors should be measured to ground and to each other. The readings to ground should be made whether the system is grounded or not. Note that all readings should be made as close to the point of exposure as possible.

The procedures given in the following paragraphs will apply only to contact-type instruments. Proximity instruments measure the electrostatic field set up around an energized conductor rather than the actual voltage between two conductors. If the proximity indicator shows an unexpected voltage, the circuit should be rechecked, possibly with a contact tester.

Single-Phase Systems. The hot wire of single-phase systems should be measured both against neutral and ground. Figure 4.28 shows the points that should be measured. The voltage between neutral and ground should also be checked.

Two-Phase Systems. A voltage measurement should be taken between the two hot phases, between each hot phase and neutral (one at a time), and between each hot phase and ground. If the neutral is remotely grounded, a measurement between neutral and local ground should also be made. Figure 4.29 shows the measurement locations for a typical two-phase system.

Three-Phase Systems. Measure between each of the hot wires, two at a time, between each hot wire and neutral, and between each hot wire and ground. If neutral is remotely grounded, a measurement should be made between neutral and local ground. Figure 4.30 shows the measurement locations for a typical three-phase system.

Using Voltage Transformer Secondaries for Voltage Measurements. Many medium-voltage power systems are equipped with step-down voltage transformers that are used for metering or telemetering purposes. Because of the lower voltage on the secondary, some workers wish to measure the secondary winding voltage to verify that the primaries are de-energized. Such transformers may be used for safety-related voltage measurements under the following conditions:

1. The transformers can be visually traced and are known to be connected to the system where exposure will occur.
2. The transformers are located close to the part of the system where exposure will occur.

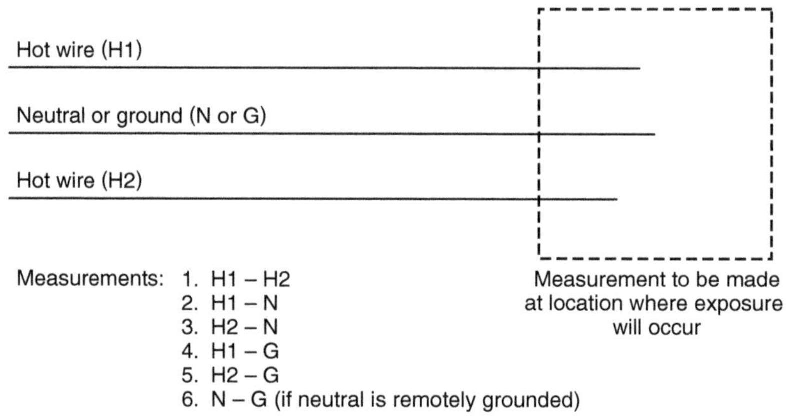

FIGURE 4.29 Measurement points for a two-phase system (two hot wires).

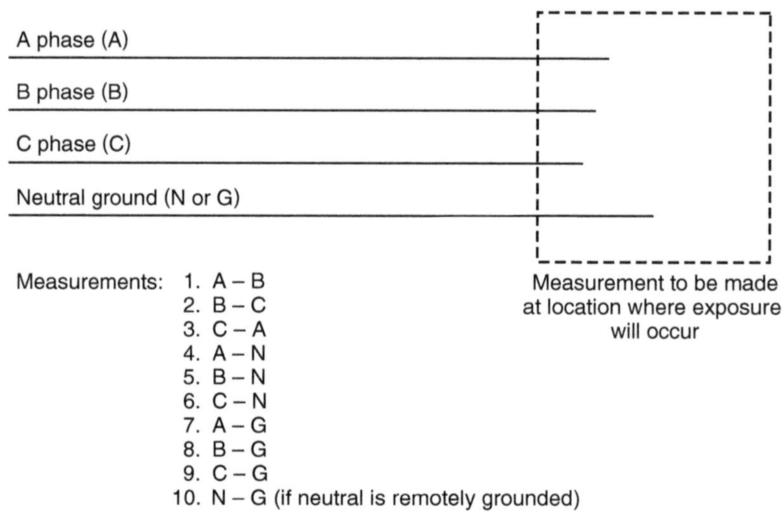

FIGURE 4.30 Measurement points for a three-phase system.

3. The transformer secondaries must be measured both before and after the circuit is de-energized. This verifies that the transformer fuses are not blown, resulting in an erroneous zero voltage reading.
4. A safety ground is applied to the primary circuit of the transformer after the measurement but before personnel contact occurs.
5. All other safety-related techniques should be used as described in the other parts of this chapter.

Caution: This procedure is considered a secondary, nonpreferred method. Do not use this approach unless absolutely necessary.

Using Panel Voltmeters for Voltage Measurements. Panel voltmeters may be used as a general indication of system energization; however, they **should not be used** for safety-related voltage measurements.

TABLE 4.11 Voltage Measurement Preparations

- Clear area of unnecessary personnel.
- Wear appropriate safety equipment.
- Expose conductors to be measured.
- Position measuring instrument.

TABLE 4.12 Recommended Minimum Safety Equipment for Making Low-Voltage Safety Measurements

- Hard hat—ANSI Z89.1 class E or G (as required by voltage level)
- ANSI Z87.1 Electrical safety glasses with side shields
- Flame-resistant work clothing (select using arc-flash calculations)
- Class 00 or higher rubber gloves with leather protectors
- Class 00 or higher rubber sleeves (if the measurement requires proximity to energized conductors)

TABLE 4.13 Recommended Minimum Safety Equipment for Making Medium- and High-Voltage Safety Measurements

- Hard hat—ANSI Z89.1 class E
- ANSI Z87.1 Electrical safety glasses with side shields
- Flame-resistant work clothing (select using arc-flash calculations)
- Class 00 or higher rubber gloves with leather protectors (class according to voltage level)
- Class 00 or higher rubber sleeves, if the measurement requires proximity to energized conductors (class according to voltage level)
- When sleeves and gloves are worn simultaneously, the voltage class of each should match.

How to Measure

Preparation. Table 4.11 identifies the steps that should precede the actual measurement procedure. The area should be cleared of unnecessary personnel. This will prevent their exposure to arc or blast in the event that a problem occurs. The person making the measurement should wear and use appropriate safety equipment. (Safety equipment information will be covered later in this chapter.)

After the safety equipment is on, the panels, doors, or other access means should be opened to expose the conductors that are to be measured. The measuring instrument is then carefully positioned. When making medium-voltage measurements with hot sticks, be certain that the hot stick is not contacting your body. The instrument should be securely positioned so that it does not fall to the floor if the leads are inadvertently overextended.

Safety Equipment. Table 4.12 lists the minimum recommended safety equipment to be worn when making low-voltage measurements. Table 4.13 lists equipment for medium-voltage measurements. The flash suit is identified as optional for outdoor, open-air measurements. Although many line personnel do not wear flash suits when performing overhead work, the use of these uniforms is strongly recommended.

Measurement. After all preparations are made and the safety equipment is put on, the measuring instrument should be applied to the conductors. If a measurement to ground is being made, one lead should be connected to the ground first and then the phase connection made. When measuring between two hot wires, the order of connection is unimportant. If a contact instrument is being used, each lead should be carefully placed on the appropriate conductor. The meter or readout is then observed to see if the circuit is energized. If a proximity instrument is being used, it should be moved gradually toward the conductor until it indicates or until the conductor is touched.

PLACEMENT OF SAFETY GROUNDS

Safety Grounding Principles

Safety grounds are conductors that are connected to a known earth connection and temporarily applied to de-energized system conductors in a grounded and short-circuited configuration. They are used to provide an equipotential zone of protection for personnel working on or around de-energized conductors and to ensure that an accidental reenergization will not cause injury. Power system components should be considered energized until safety grounds and short circuits are in place. Safety grounds should never be placed until the conductors on which they are to be installed have been measured and verified to be de-energized.

See Chap. 3 for a detailed description of the design and description of safety grounds. *Caution:* Safety grounds must be properly designed and sized for the conductors and short-circuit capacity to which they may be subjected. Refer to Chap. 3 and/or manufacturer's information for specifics.

In the event that the system is accidentally reenergized, enormous magnetic forces are exerted on the safety ground wires. The forces can cause the grounds to whip violently and cause injury to personnel. To minimize this effect, the safety grounds should be as short as possible, and the conductors should be restrained. Also, the grounds should be installed away from the work area and barricaded to provide safety for unqualified and affected personnel.

Safety Grounding Location

Equipotential Grounding. Safety grounds should be applied in such a way that a zone of equal potential is formed in the work area. This equipotential zone is formed when fault current is bypassed around the work area by metallic conductors. Figure 4.31 shows the proper location of safety grounds for three different work situations. In each of these situations, the worker is bypassed by the low-resistance metallic conductors of the safety ground.

Assume the worker contacts the center phase in Fig. 4.31*b*. With a fault-current capacity of 10,000 A, safety ground resistance (R_j) of 0.001 Ω, and worker resistance (R_w) of 500 Ω, the worker will receive only about 20 mA of current flow.

Figure 4.32*a* through *d* shows two nonpreferred or incorrect locations for the application of safety grounds. In these diagrams, the worker's body is in parallel with the series combination of

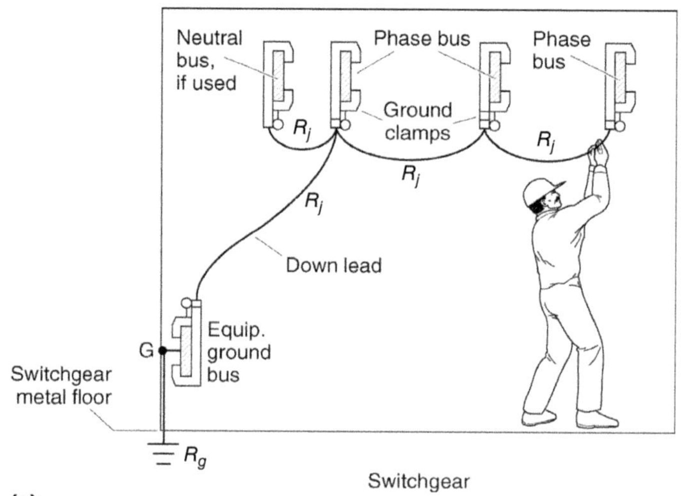

(a)

FIGURE 4.31 Application of safety grounds to provide a zone of equal potential.

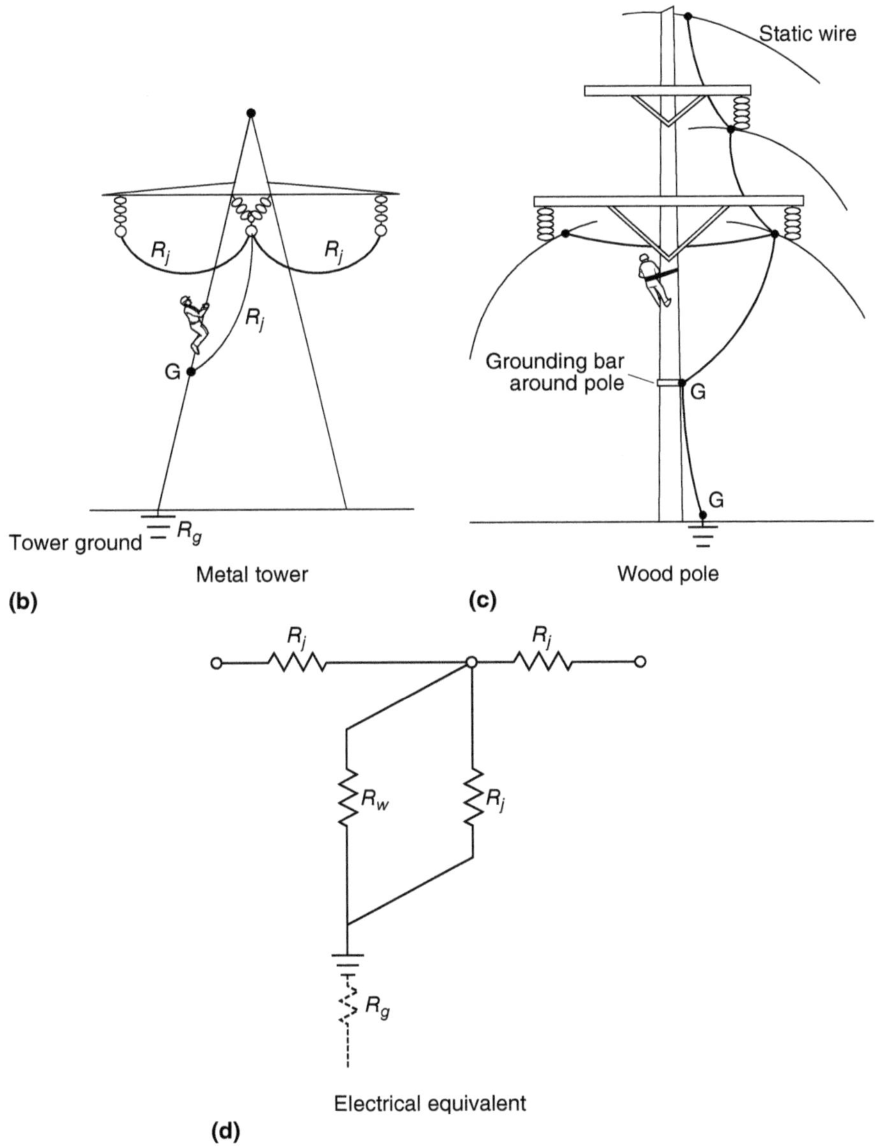

FIGURE 4.31 *(Continued)*

the jumper resistance (R_j) and the ground resistance (R_g). The placement of grounds in this fashion greatly increases the voltage drop across the worker's body in the event the circuit is reenergized. Such placements should not be used.

Figures 4.31 and 4.32 show ground sets installed on the structures where line workers are performing work. This is for graphic demonstration only. Normally, grounds are installed on both ends of the de-energized, overhead line section being worked on. Personal ground sets are installed on either side of the actual work site no more than two spans away from the work site and within sight of the work site. The only installation on the structure being worked on may be a grounding cluster (a grounding cluster can be seen in Fig. 4.33) attached to the pole or wooden structure below the work area with a grounding conductor connected to earth (by use of a driven ground rod), a jumper to the system

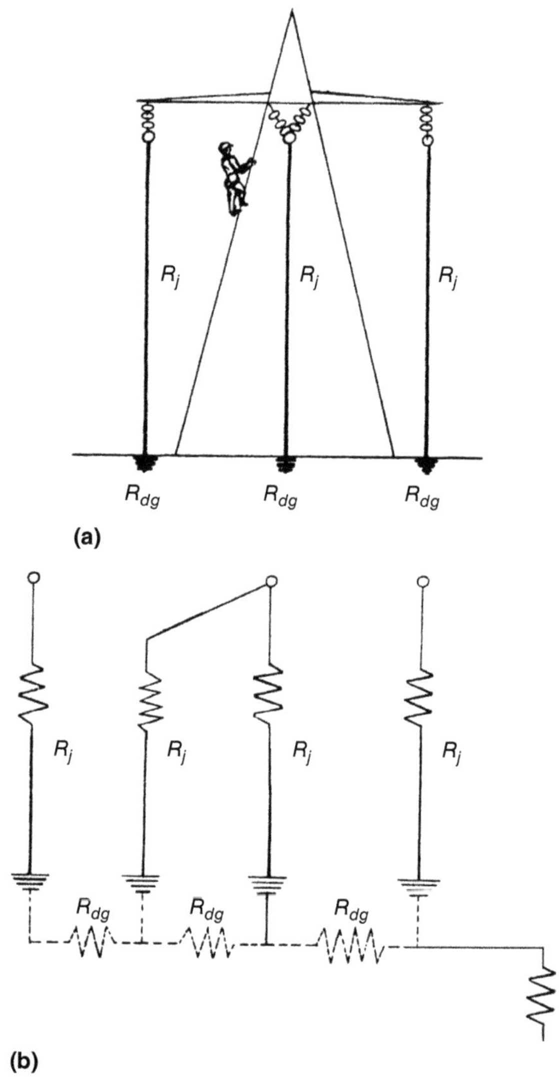

FIGURE 4.32 Two incorrect safety ground configurations.

neutral in a wye configuration or a jumper to one of the phase conductors in a delta configuration. All grounding jumpers are installed so the line worker's safety is not jeopardized in the event of a ground set failure or an accidental reenergization of the line through a switching error or contact with a foreign energized line.

Single-Point versus Two-Point Grounding. Single-point grounding is the placement of only one safety ground set. In this procedure, the safety grounds are placed near the point where the work is to be performed, but far enough away so the electrical worker's safety is not jeopardized in the event of a ground set failure or the violent whipping action of the ground set if the circuit is accidentally reenergized with the grounds in place. When possible, the grounds are placed between the worker and the source of electric energy.

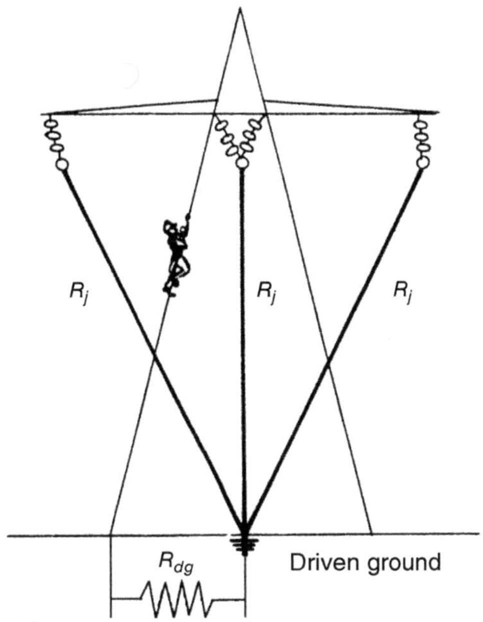

(c)

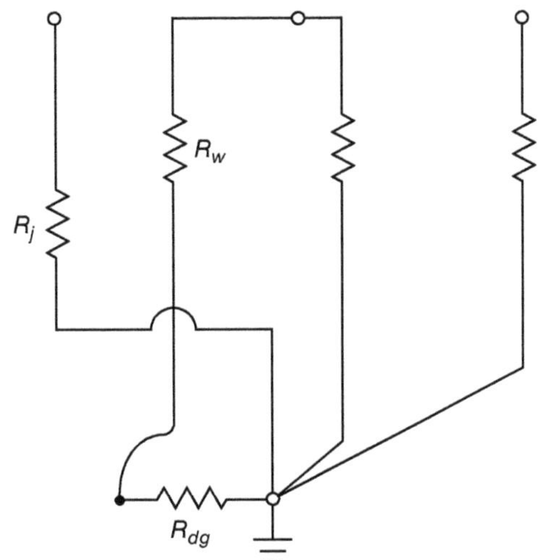

(d)

FIGURE 4.32 (*Continued*)

Two-point grounding is the placement of two safety ground sets. They are usually placed on opposite sides of the work area; that is, one set is placed "upstream" and one set is placed "downstream" from the workers.

In general, more safety ground sets are better than fewer; however, the safety ground system must provide a zone of equal potential as described in the previous section.

FIGURE 4.33 Worker applying safety grounds. (*Courtesy McGraw-Hill, Inc.*)

Application of Safety Grounds

Safety Equipment. Table 4.14 lists the minimum recommended safety equipment that should be worn and used when applying safety grounds. Full shock, arc, and blast protective clothing is required. No matter how carefully the work area is prepared, the possibility still exists for error and/or inadvertent reenergization during the application process.

Notice also that hot sticks are recommended for the actual application process. The use of hot sticks distances the worker from the point of contact and, therefore, minimizes the possibility of injury caused by arc or blast.

Procedure. The actual application procedure will be different for each specific application. The following general steps are recommended.

1. Thoroughly inspect the safety ground set that is to be used. Points to check include
   a. Insulation quality
   b. Condition of conductors
   c. Condition of clamps
   d. Condition of ferrule
   e. Condition of cable-to-ferrule connection
2. Identify the point at which each ground clamp will be connected to the system. Be certain to select points that minimize the amount of slack in the safety grounds. This will minimize the whipping action in the event the system is reenergized.
3. Put on required safety equipment.
4. Measure the system voltage to make certain that the system to be grounded is de-energized. (See the previous sections on voltage measurement.)

TABLE 4.14 Recommended Minimum Safety Equipment for Application of Safety Grounds

- Hard hat—ANSI Z89.1 class E
- ANSI Z87.1 Electrical safety glasses with side shields
- Flame-resistant work clothing with full head protection (select using arc-flash calculations)
- Class 0 or higher rubber gloves with leather protectors (class according to voltage level)
- Hot stick with suitable fittings for application of safety ground clamps*

*Hot sticks may not be usable in cramped switchgear locations. Workers should exercise best safety judgment in such situations.

5. Make certain that all unnecessary personnel have been cleared from the area.
6. Clean the connection to remove galvanic corrosion with a brush designed for that purpose or use ground clamps with serrated jaws to bite through the corrosion to ensure a low-resistance connection between the conductor and the ground clamp.
7. Apply the ground end of the safety ground sets first. (These points are labeled with the letter "G" in Fig. 4.31.)
8. Connect the phase-end safety ground clamp to the hot stick.
9. Firmly contact the de-energized conductor with the phase end of the safety ground.
10. Tighten the grounding clamp firmly. Remember that the amount of resistance in the clamp connection can make the difference between a safe connection and a hazardous one.
11. Repeat steps 7 through 9 for each of the phases to be grounded. **Note:** If using individual jumpers on overhead line construction, attach the second jumper to the conductor already grounded and then to the second phase conductor. For the third jumper, connect the first end to the second grounded conductor and then attach the other end to the third phase conductor. This configuration satisfies the "short-circuited" part of OSHA mandate to install ground sets so as to provide a "short-circuited and grounded" configuration. The short circuiting provides the most rapid clearing time because it causes maximum incident energy to flow instantaneously.
12. Record the placement of each safety ground by identification number. (See the section titled "Control of Safety Grounds" in the next section.)

Figure 4.33 shows a worker applying safety grounds to a de-energized system.

The Equipotential Zone

When a proper equipotential zone is established, there will be no lethal potential differences that the worker can reach in the work area. Figure 4.31 illustrates configurations that can be used to accomplish this end.

Note, however, that some situations make the establishment of such a zone difficult or impossible. Consider, for example, the employee who must stand on the earth when he or she is working. Since the earth has a relatively high resistivity, the worker's feet will be at a different potential from that of the metallic elements that he must contact.

Two approaches can be used in such circumstances:

1. A metal platform can be laid on the ground and bonded to the grounded metal of the electrical system. Figure 4.24 is an example of this approach. In this situation, as long as the worker stays on the metal pad, he or she will remain in an equipotential zone.
2. The worker can be insulated from the earth or other high-resistivity conductors. Rubber mats, gloves, or blankets can be placed so that the worker is completely insulated from electrical contact.

Of course, the best approach is always to establish the equipotential zone; however, these two "work-around" approaches may be used when absolutely necessary.

Removal of Safety Grounds

Safety Equipment. The removal of safety grounds is no less hazardous than applying them. All the safety equipment listed in Table 4.14 should be worn. Safety grounds should be removed using hot sticks.

Procedure. Safety grounds should be removed as follows:

1. Put on all required safety equipment.
2. Remove each of the phase connections one at a time.
3. Remove the ground connection.
4. Check the ground off as being removed.

Remember that when safety grounds are not present, the system should be considered to be energized.

Control of Safety Grounds

Safety grounds must be removed before the power system is reenergized. They must also be inspected periodically and before each use. The following sections describe two methods of controlling the safety ground sets.

Inventory Method. Each safety ground set should be identified with its own unique serial number. The serial number should be etched or impressed on a metal tag that is permanently attached to the safety ground set.

A safety officer should be appointed to control the inventory of safety ground sets. This person will control the use of the safety grounds and is responsible for keeping lists of where the grounds are applied during an outage.

As each safety ground set is applied, the safety officer notes its placement on a placement control sheet. As each ground set is removed, the officer notes its removal on the sheet. No reenergization is allowed until the safety officer is satisfied that all safety grounds have been removed. Figure 4.34 is a typical safety ground placement control sheet.

The sheet shown in Fig. 4.34 has the minimum required information. Other columns—such as where the ground set is installed, who installed it, and who removed it—may be added as needed.

Visual Method. Some small facilities may not be able to justify the complexity of a control system such as that described in the previous section. A visual control system may suffice for facilities that have only one or two safety grounding sets. In such a system, a brightly colored rope is permanently attached to each grounding set. Nylon ski rope is ideal for this application. The length of the rope should be determined by the applications; however, 3 meters (m) (10 ft) is a good starting point.

At the remote end of the rope, attach a brightly colored warning sign stating "Grounds Are in Place" or "This Equipment Is Temporarily Grounded." After the safety grounds are attached to the system, the rope should be arranged on the switchgear so that it is easily seen. If it can be arranged, the rope should be placed so the gear cannot be closed up until the rope is removed. The green sign should be placed on the breaker control handle, start push button, or other such device so that it is obviously visible.

When this procedure is used, it is very difficult to reenergize the system with the grounds in place. However, the inventory method is the preferred and recommended method.

ARC-FLASH HAZARD CALCULATIONS AND APPROACH DISTANCES

Caution: All of the formulas, methods, and calculations discussed in this chapter are based upon regulatory standards and engineering practices that are in use as of July 2011. A great deal of research

SAFETY GROUND PLACEMENT CONTROL SHEET			
Safety Officer:		Outage Date:	
Ground Set ID Number	Date Applied:	Date Removed:	Safety Officer's Initials

FIGURE 4.34 Safety ground placement control sheet.

work has been and is currently under way; consequently, current literature and a competent, qualified engineer with sufficient background and experience should be consulted to verify that these methods are still the recognized approach.

Introduction

How close to an electrical hazard can I get? Obviously, distance provides a safety barrier between a worker and an electrical hazard. As long as workers stay far enough away from any electrical energy sources, there is little or no chance of an electrical injury. This section describes methods that can be used to determine the so-called approach distances. Generally, if work can be carried on outside the approach distances, no PPE is required. If the worker must cross the approach distances, appropriate PPE must be worn and appropriate safety procedures must be used.

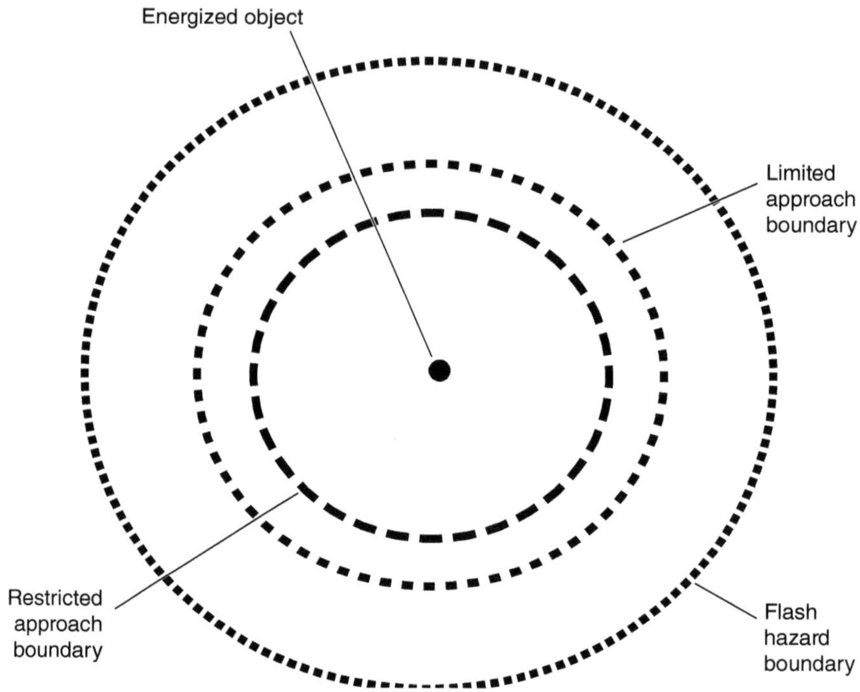

FIGURE 4.35 Minimum approach distance definitions.

Approach Distance Definitions

Approach distances generally take two forms—*shock hazard distance* and *arc-flash hazard distance.* Note that the flash or arc hazard distance also includes the *blast hazard distance.* Figure 4.35 illustrates the approach distances that are defined by the NFPA in its publication *NFPA 70E Standard for Electrical Safety in the Workplace.* Note that the space inside any given boundary is named for that boundary. Thus the space inside the restricted approach boundary is referred to as the *restricted space.*

The Restricted Approach Boundary and the Arc-Flash Boundary may not be crossed without meeting the following general requirements:

1. The employee must be qualified to cross the boundary.
2. The employee must be wearing appropriate PPE.
3. Proper planning must be carried out to prepare the employee for the hazards he/she may face.

Detailed requirements for each boundary will be given in later sections.

Determining Shock Hazard Approach Distances

Table 4.15 may be used to determine the minimum approach distances for employees for shock hazard purposes. Note that the shock approach distances are all based on voltage levels. In general, the higher the voltage level, the greater the approach distance.

Unqualified Persons. Unqualified persons may not approach exposed energized conductors any closer than the values specified in columns (2) and (3) of Table 4.15. Although unqualified persons

TABLE 4.15 Approach Boundaries to Live Parts for Shock Protection

(All dimensions are distance from live part to employee.)

(1)	(2)	(3)	(4)
	Limited approach boundary		Restricted approach boundary; includes inadvertent movement adder
Nominal system voltage range, phase to phase	Exposed movable conductor	Exposed fixed circuit part	
0 V to 50 V	Not specified	Not specified	Not specified
51 V to 300 V	10 ft 0 in	3 ft 6 in	avoid contact
301 V to 750 V	10 ft 0 in	3 ft 6 in	1 ft 0 in
751 V to 15 kV	10 ft 0 in	5 ft 0 in	2 ft 2 in
15.1 kV to 36 kV	10 ft 0 in	6 ft 0 in	2 ft 7 in
36.1 kV to 46 kV	10 ft 0 in	8 ft 0 in	2 ft 9 in
46.1 kV to 72.5 kV	10 ft 0 in	8 ft 0 in	3 ft 3 in
72.6 kV to 121 kV	10 ft 8 in	8 ft 0 in	3 ft 2 in
138 kV to 145 kV	11 ft 0 in	10 ft 0 in	3 ft 7 in
161 kV to 169 kV	11 ft 8 in	11 ft 8 in	4 ft 0 in
230 kV to 242 kV	13 ft 0 in	13 ft 0 in	5 ft 3 in
345 kV to 362 kV	15 ft 4 in	15 ft 4 in	8 ft 6 in
500 kV to 550 kV	19 ft 0 in	19 ft 0 in	11 ft 3 in
765 kV to 800 kV	23 ft 9 in	23 ft 9 in	14 ft 11 in

Notes:
for SI units: 1 in = 25.4 mm; 1 ft = 0.3048 m.
Source: *Courtesy National Fire Protection Association.*

cannot approach any closer than the limited boundary when they are unaccompanied, they may cross the boundary temporarily as long as they are continuously accompanied and supervised by a qualified person. This allows an unqualified person to transit through the zone.

Note that "movable" conductors are those that are not restrained by the installation. Overhead and other types of suspended conductors qualify as movable conductors. Fixed-circuit parts include buses, secured cables, and other such conductors.

Qualified Persons. Qualified personnel approach distances are determined from columns (4) and (5) of Table 4.15. No one may cross the restricted or the prohibited approach boundaries without meeting the requirements, which are defined later in this section. Note that the minimum requirements for crossing either of these boundaries include:

1. The worker must be qualified and fully trained for the work and the hazards that will be encountered inside the boundary.
2. The worker must wear all appropriate personnel protective clothing and use all required safety procedures and tools such as insulated hand tools inside the restricted approach boundary.

Specific Requirements for Crossing the Restricted Approach Boundary. To cross the restricted approach boundary, the following criteria must be met:

- The worker must be qualified to do the work.
- There must be a plan in place that is documented and approved by the employer.

TABLE 4.16 Human Tissue Tolerance to Heat, Second-Degree Burn

Exposure time	Heat flux		Total heat		Calorimeter* equivalent		
s	kW/m²	cal/cm²s	kWs/m²	cal/cm²	ΔT°C	ΔT°f	ΔmV
1	50	1.2	50	1.20	8.9	16.0	0.46
2	31	0.73	61	1.46	10.8	19.5	0.57
3	23	0.55	69	1.65	12.2	22.0	0.63
4	19	0.45	75	1.80	13.3	24.0	0.69
5	16	0.38	80	1.90	14.1	25.3	0.72
6	14	0.34	85	2.04	15.1	27.2	0.78
7	13	0.30	88	2.10	15.5	28.0	0.80
8	11.5	0.274	92	2.19	16.2	29.2	0.83
9	10.6	0.252	95	2.27	16.8	30.2	0.86
10	9.8	0.233	98	2.33	17.3	31.1	0.89
11	9.2	0.219	101	2.41	17.8	32.1	0.92
12	8.6	0.205	103	2.46	18.2	32.8	0.94
13	8.1	0.194	106	2.52	18.7	33.6	0.97
14	7.7	0.184	108	2.58	19.1	34.3	0.99
15	7.4	0.177	111	2.66	19.7	35.4	1.02
16	7.0	0.168	113	2.69	19.8	35.8	1.03
17	6.7	0.160	114	2.72	20.2	36.3	1.04
18	6.4	0.154	116	2.77	20.6	37.0	1.06
19	6.2	0.148	118	2.81	20.8	37.5	1.08
20	6.0	0.143	120	2.86	21.2	38.1	1.10
25	5.1	0.122	128	3.05	22.6	40.7	1.17
30	4.5	0.107	134	3.21	23.8	42.8	1.23

*Iron/constantan thermocouple.

Source: Stoll, A. M., and Chianta, M. A., "Method and Rating System for Evaluation of Thermal Protection," *Aerospace Medicine*, vol. 40: pp. 1232–1238, 1968.

- The worker must be certain that no part of the body crosses the prohibited approach boundary.
- The worker must work to minimize the risk that may be caused by inadvertent movement by keeping as much of the body out of the restricted space as possible. Allow only protected body parts to enter the restricted space as necessary to complete the work.
- PPE appropriate for the hazards of the exposed energized conductor must be used.

Calculating the Arc-Flash Hazard Minimum Approach Distance (Arc-Flash Protection Boundary)

Introduction. The concept of a minimum approach boundary for arc-flash protection is based on the amount of tolerance that human tissue has to heat. The current industry standards use the so-called Stoll Curve developed by Stoll and Chianta in the 1960s.[1] The arc-flash boundary then represents the closest distance that an unprotected (meaning not wearing thermal protective clothing beyond normal cotton garments) worker may approach an electrical arcing source. Theoretically, at the arc-flash boundary distance, if the worst-case electrical arc occurs, the unprotected worker has a 50 percent chance of a *just-curable burn*—that is, a second-degree burn. Table 4.16 is a reproduction of the Stoll Curve table.

TABLE 4.17 Methods for Calculating Flash Boundary Approach Distances below 600 V

#	Conditions	Method/Formulas	Notes
1	$I_{SC} \times t \leq 5000$ Ampere-seconds	$D_c = 4$ feet (1.22 meters)	I_{SC} = Maximum fault current (amps) t = Duration of fault in seconds D_c = Flash boundary distance
2	$I_{SC} \times t > 5000$ Ampere-seconds	$D_c = \sqrt{2.65 \times MVA_{bf} \times t}$ $D_c = \sqrt{53 \times MVA \times t}$	I_{SC} = Maximum fault current (amps) MVA_{bf} = Max fault MVA* MVA = Supply transformer MVA†¶ t = Duration of fault in seconds D_c = Flash boundary distance (in feet)
3	16,000 A < I_{SC} < 50,000 A	See Eq. 4.2‡	I_{SC} = Maximum fault current (kA) t = Duration of fault in seconds Dc = Flash boundary distance
4	16,000 A < I_{SC} < 50,000 A	See Eq. 4.3§	I_{SC} = Maximum fault current (kA) t = Duration of fault in seconds Dc = Flash boundary distance
5	700 A ≤ I_{SC} ≤ 106 kA 208 V ≤ V ≤ 15 kV	IEEE Std 1584 guide for performing arc-flash hazard calculations	I_{SC} = Maximum fault current V = System phase to phase voltage

*The bolted three-phase fault MVA at the point of exposure. Note that for a three-phase system $MVA_{bf} = kV_{LL} \times I_{SC} \times \sqrt{3} \times 1000$.
†This is the maximum self-cooled, full-load MVA of the supply transformer.
‡This formula applies to open-air short circuits.
§This formula applies to enclosed short circuits—the so-called arc-in-a-box.
¶For transformers with ratings less than 0.75 MVA, multiply the MVA rating times 1.25.

Low-Voltage Calculations (Below 600 V). Table 4.17 lists five methods that may be used for calculating the minimum approach distances for electrical arc-flash when the circuit voltage is below 600 V.

$$D_C = \left(\frac{5271 \times t \times (0.0016 I_{SC}^2 - 0.0076 I_{SC} + 0.8938)}{1.2}\right)^{1/1.9593} \tag{4.2}$$

$$D_C = \left(\frac{1038.7 \times t \times (0.0093 I_{SC}^2 - 0.3453 I_{SC} + 5.9675)}{1.2}\right)^{1/1.4738} \tag{4.3}$$

Note that methods 3 and 4 in Table 4.17 are actually solutions to the Doughty, Neal, and Floyd equations for incident energy = 1.2 cal/cm^2. This means that methods 3 and 4 must be used only within the constraints under which the empirical equations were developed—circuits with fault currents between 16 and 50 kA, 600 V or less, 18 in or more from the arc source. The "arc-in-a-box" method is based on a test method using a box 20 in on a side.

If we assume a 480 V circuit with a maximum fault current of 20,000 A and a tripping time of 0.1 s, the four methods generate the following results:

Method 1: 4 ft
Method 2: 2.1 ft (top formula only)
Method 3: 2.19 ft
Method 4: 3.44 ft
Method 5: 2.28 ft

The NFPA method 2; the Doughty, Neal, and Floyd equations for open-air arcs (3); and method 5, the IEEE calculator, agree very closely. Method 4 gives a larger arc-flash boundary, since the "arc-in-a-box" method assumes that the energy is focused toward the worker. That is, most of the energy goes in one direction.

Method 5 employs the IEEE Flash Hazard calculator provided with the IEEE Std 1584 2018. This calculator is a Microsoft Excel spreadsheet application that calculates both incident energy and flash boundary. It is a relatively "easy to use" application; however, as with all of the calculation methods given in this chapter, it should be employed only by qualified, experienced engineers with sufficient background and experience to use the tool correctly.

Medium- and High-Voltage Calculations. Much of the research that has been done on arc-flash boundary calculations has concentrated on the low-voltage spectrum (600 V and less); consequently, the availability of empirical support applied to medium- and high-voltage systems for the various derived formulas is sparse.

Method 5 is the best choice for calculating the arc-flash boundaries for medium-voltage systems up to 15 kV. However, in the absence of the IEEE standard, the engineer may use method 2 with certain cautions.

Calculating medium- and high-voltage flash boundaries using method 2 (Table 4.17) appears to give conservatively high results when compared to software programs derived using industry-standard heat transfer calculations.

To calculate the arc-flash boundary for medium-voltage systems, use the top formula given in method 2. Then enter the result into one of the commercially available (or freeware) software programs and verify that the received heat flux density is on the order of 1.2 cal/cm^2.

CALCULATING THE REQUIRED LEVEL OF ARC PROTECTION (ARC-FLASH HAZARD CALCULATIONS)

Note: The formulas and methods used in the following sections are based on the most current information available at the time of this writing (July 2011).

Ongoing research is being performed in an attempt to improve these formulas and methods. Always refer to the latest editions of NFPA 70E and IEEE Standard 1584 for the most current information.

Introduction

Arc-flash protection beyond normal cotton or wool work clothing is not required as long as all parts of the worker's body stay outside the arc-flash boundary as calculated above. Outside the arc-flash boundary, the minimum recommendation is either:

1. arc-rated clothing with an ATPV of 4.5 cal/cm^2 or higher
2. Natural fabric (cotton or wool) work clothing of 7 oz/yd^2 or more

Remember that these are *minimum* recommendations. Many organizations (including the author's company) require their electrical personnel to wear 8 cal/cm^2 at all times when in the field.

If, however, the worker must cross the arc-flash boundary and is qualified to do so, he or she must wear arc-rated clothing to protect against the arc incident energy levels that may be encountered.

To provide arc protection, the worker must select and wear arc-rated clothing with an ATPV or E_{BT} equal to or greater than the incident energy level. Note that the incident energy level is calculated based on a given distance between the arc source and the worker, for example 24 in. The worker will not be adequately protected if he or she gets any closer than the distance used in the calculation for the specific situation.

Arc-flash boundaries can be calculated from the formulas in this section by setting the incident energy level to 1.2 cal/cm^2 (5.02 J/cm^2) and then solving the appropriate equation for the distance.

Caution: The following sections present several complex calculation procedures using a variety of mathematical and engineering applications. These calculations should be performed or evaluated only by qualified, experienced engineers or other technical personnel with the requisite knowledge and skills.

The Lee Method

This method is named after its developer, Ralph Lee, P.E. (deceased). Lee was one of the early and most successful researchers into the theoretical models for the amount of energy created by an electrical arc and the amount of incident energy received by workers in the vicinity of the arc.

The Lee method is a theoretical model and is based on two major assumptions:

1. The maximum energy transfer from the power system to an electric arc is equal to one-half of the bolted short-circuit power available at the point where the arc occurs.
2. All of the arc electrical energy will be converted to the incident heat energy.

Starting with these two assumptions, Lee and subsequent researchers have refined the Lee method to the form shown in Eq. 4.4

$$E = K \times VI_{bf}\left(\frac{t}{D^2}\right)\times 10^6 \tag{4.4}$$

where E = incident energy (J/cm² or cal/cm² depending on the value of K)
K = a constant; if $K = 2.142$ then E is in J/cm²; if $K = 0.512$ then E is in cal/cm²
V = system phase-to-phase voltage (kV)
I_{bf} = bolted fault current (kA)
t = arcing time (seconds)
D = distance from the arcing point to the worker (mm)

The Lee method is a very conservative method and was developed without the aid of the empirical results available today from the extensive research in this field. While it can be used for all situations, its principal application is in those cases that have not been modeled based on actual experimental measurements. It is most often used in systems where the system voltage is greater than 600 V and/or where fault currents, arc lengths, or other such parameters are beyond the ranges that have been tested using field measurements.

Although the Lee method does not take the focusing effect of equipment into account, the so-called arc-in-a-box, it is nonetheless very conservative. The use of multipliers should be required only in the most extreme applications.

Note: All of the NFPA and IEEE methods described in the following sections are based on empirical field testing using a configuration similar to what is shown in Fig. 4.36. A manikin outfitted with heat sensors is placed inside a Faraday cage. Electrical terminals are placed in proximity to the manikin and an arc is started. The heat energy is captured on instruments.

Multiple experiments are performed using different distances and arc times. The resulting data is mathematically manipulated into a curve fit. The resulting equations (such as Eq. 4.5) are then used to calculate arc energies at various distances.

Methods Outlined in NFPA 70E

In addition to the Lee method, *NFPA 70E, Standard for Electrical Safety in the Workplace,* gives two different methods for calculating incident energy levels: the method based on empirical formulas developed by Doughty, Neal, and Floyd[2] and the method developed by the IEEE research in the development of IEEE Standard 1584.

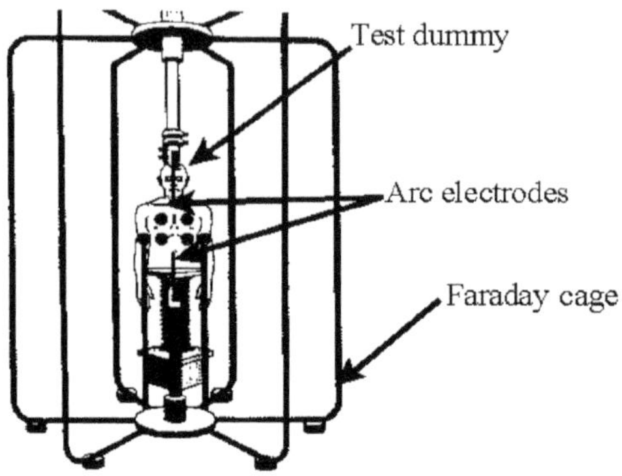

FIGURE 4.36 Electrical arc test setup used to determine Arc Thermal Performance Value (ATPV).

Both of these methods rely on curve-fitting data collected using a system similar to the one shown in Fig. 3.30. The manikin has sensors embedded in it, primarily in the chest region. When the arc is initiated, the sensors capture the amount of energy released. Multiple runs are made and the data from the runs is curve-fitted to create the empirical formulas included in this chapter.

Method #1. This method is based on a research by Richard L. Doughty, Thomas E. Neal, and H. Landis Floyd.[2] They performed research using the thermal manikin approach described in Chap. 1. Their research was bounded by the following conditions:

1. Systems with voltage levels of 600 V and below
2. Systems with maximum available short-circuit currents between 16 kA and 50 kA
3. Working distances of equal to or greater than 18 in

After running numerous tests by creating electric arcs in open air, they took their results and performed a curve fit to predict the incident energy. Equation 4.5 is the formula that was developed and that most closely models the energies they measured.

$$E_{MA} = 5271D_A^{-1.9593} \times t_A(0.0016I^2_{SC} - 0.0076I_{SC} + 0.8938) \tag{4.5}$$

where E_{MA} = maximum open-air arc incident energy (cal/cm^2)
D_A = distance from the electrodes (in)—note that $D_A \geq 18$ in
t_A = duration of the arc (seconds)
I_{SC} = short-circuit current (kA)—note that I_{SC} is the actual arc current, not the maximum bolted fault current

In the second part of their research, they created electrical arcs in a cubic box with one open side. The box was 20 in on a side and the measurements were taken at various distances from the open end. They then performed a curve fit to model the results. Equation 4.6 is the resulting formula.

$$E_{MB} = 1038.7D_B^{-1.4738} \times t_A(0.0093I^2_{SC} - 0.3453I_{SC} + 5.9675) \tag{4.6}$$

where E_{MB} = maximum arc-in-a-box incident energy (cal/cm^2)
D_B = distance from the electrodes (in)—note that $D_B \geq 18$ in

All other variables are as described for the open-air formula (Eq. 4.5).

Using either Eq. 4.5 or 4.6, incident energies can be calculated, then protective clothing can be purchased with the necessary ratings.

TABLE 4.18 System Limits for Calculation of Incident Energy Using IEEE Standard 1584-2002

System phase-to-phase voltage	0.208 to 15 kV
System frequency	50 to 60 Hz
Short-circuit current range	700 to 106 kA
Conductor gap (arc length)	13 to 152 mm

Method #2. This method is actually the method taken from the IEEE Standard 1584-2002.[3] This method is described in the next section.

IEEE Std 1584-2018

System Model Limitations. Institute of Electrical and Electronics Engineers (IEEE) working group P1584 has performed research and theory development over several years to expand the work done by Doughty, et al.[2] As shown in Table 4.18, the working group has significantly expanded the applicable range for the application of their methods.

Calculation of Arcing Current. One of the principal drawbacks of the Lee method is that the bolted fault current, which is used in that method, is always somewhat greater than the actual current flow when an electrical arc is formed. At voltages below 1000 V, the arcing current is often substantially less than the bolted fault current. (As mentioned in Chap. 2, the Lee method also assumes maximum power transfer, which is not usually the case in actual applications.)

Using both theoretical and empirical data, Standard 1584 develops a formula to calculate the arcing current that will be present in a system when the bolted fault current is known. Equations 4.7 and 4.8 are used to calculate the logarithm to the base 10 of arcing current ($\log_{10} I_a$). Equation 4.7 is used for systems below 1 kV, and Eq. 4.8 is used for systems that are equal to or greater than 1 kV.

$$\log_{10} I_a = K + 0.662(\log_{10} I_{bf}) + 0.0966V + 0.000526G + 0.5588V(\log_{10} I_{bf}) - 0.00304G(\log_{10} I_{bf}) \tag{4.7}$$

$$\log_{10} I_a = 0.00402 + 0.983(\log_{10} I_{bf}) \tag{4.8}$$

where I_a = arcing current (kA)
K = a constant; $K = -0.153$ for open-air arcs and $K = -0.097$ for arc-in-a-box
I_{bf} = maximum, symmetrical rms, three-phase bolted short circuit current (kA)
V = system three-phase voltage (kV)
G = conductor gap (arc length) (mm)

After $\log_{10} I_a$ has been calculated from Eq. 4.7 or 4.8, Eq. 4.9 is used to calculate I_a.

$$I_a = 10^{(\log_{10} I_a)} \tag{4.9}$$

Calculating Incident Energy. The calculation of actual incident energy is done in two steps using the empirically derived arc current (I_a). In step one, normalized incident energy (E_n) is calculated. E_n is normalized for an arcing time of 0.2 second and a distance from the arc of 610 mm. Equations 4.10 and 4.11 are used for this calculation.

$$\log_{10} E_n = k_1 + k_2 + 1.081(\log_{10} I_a) + 0.0011G \tag{4.10}$$

$$E_n = 10^{(\log_{10} E_n)} \tag{4.11}$$

TABLE 4.19 Distance Factors (X) Used in Eq. 4.12

System voltage (kV)	Type of equipment	Typical conductor gap (mm)	
0.208 to 1	Open-air	10–40	2.000
	switchgear	32	1.473
	MCCs and panelboards	25	1.641
	cables	13	2.000
>1 to 5	Open-air	102	2.000
	switchgear	13–102	0.973
	cables	13	2.000
>5 to 15	Open-air	13–153	2.000
	switchgear	153	0.973
	cables	13	2.000

where E_n = normalized incident energy (J/cm^2)
$k_1 = -0.792$ for open-air arcs and -0.555 for arcs-in-a-box
$k_2 = 0$ for ungrounded and high-resistance ground systems and -0.113 for grounded systems
G = conductor gap (arc length) (mm)

The actual incident energy at any distance D is then calculated using Eq. 4.12.

$$E = k_{CJ} C_f E_n \left(\frac{t}{0.2}\right)\left(\frac{610}{D}\right)^X \tag{4.12}$$

where E = incident arc energy at distance D in cal/cm^2 ($k_{CJ} = 1$) or J/cm^2 ($k_{CJ} = 4.184$)
$k_{CJ} = 1$ for E_n is cal/cm^2 and 4.184 for E_n in J/cm^2
$C_f = 1$ for systems above 1 kV or 1.5 for systems equal to or less than 1 kV
E_n = normalized incident energy calculated from Eq. 4.11
t = arcing time in seconds
D = distance from the arc (mm)
X = a distance exponent taken from Table 4.19

In Eq. 4.12, the arcing time (t) will be determined by how long the protective devices require to completely interrupt the short circuit. If the arcing current decreases, the arcing time will generally increase. Because of this, NFPA 70E suggests recalculating Eqs. 4.7 through 4.12 using 0.85 I_a and the resulting arcing time. The standard then suggests using the larger incident energy from the two calculations.

Software Solutions

Several software products on the market allow the calculation of incident energy and/or flash boundaries. For example, IEEE Standard 1584-2018 comes with a complete set of Microsoft Excel spreadsheet applications. Users need only enter the values for their particular system to determine the incident energy and the flash boundary using the IEEE method.

Also, virtually all of the commercially available engineering software packages, such as SKM Systems Analysis, Inc.—PowerTools for Windows and ESA, Inc. EasyPower, have added arc-flash calculation packages to their short-circuit analysis and coordination study packages.

Required PPE for Crossing the Arc-Flash Hazard Boundary

To select the level of arc-rated protection required, the following procedure may be used:

1. Calculate the incident arc energy value as shown previously.
2. Select clothing that provides an ATPV* or E_{BT}† that is less than the incident energy value previously calculated. Double layers of protective clothing may be required at the higher energy levels. Refer to the manufacturers for recommendations.

Caution: Current regulatory standards strongly suggest that locations with incident energy in excess of 40 cal/cm^2 should be evaluated very closely. Generally, workers should not be required to work in areas where the arc energy exceeds 40 cal/cm^2.

A Simplified Approach to the Selection of Protective Clothing

The National Fire Protection Association provides a simplified approach to the selection of protective clothing. While this method is convenient and economical, it should be used with extreme caution. Many have used this method without paying attention to the boundary conditions set up by the tables.

The method assumes certain short-circuit and operating time values. Users must be absolutely certain that their power system falls within the assumed values.

Note: The tables used in this section are taken from the 2009 edition of NFPA 70E. The user should always refer to the current version of that document to make sure that no changes have occurred.

1. Identify the Hazard/Risk category from Table 4.20. Note that this is based on the type of work that will be performed. Note also that Table 4.20 identifies whether insulating gloves and/or tools are required.
2. Use Table 4.21 to select the various types of PPE required for the hazard determined in step 1.
3. Use Table 4.22 to select the weight of flame-resistant clothing required for the task.

Although this procedure is quite simple and straightforward, it should be used carefully for at least two reasons:

- Because it is conservative, it tends to result in substantial amounts of clothing for the employee. Workers may tend to disregard necessary equipment out of frustration.
- The standard is task based rather than location based. Using the quantitative methods described previously will provide, ultimately, a more easily applied set of rules.
- In some rare cases, the simplified approach may give values that do not provide adequate protection. This is especially critical if the limits imposed by the method are ignored.
- This method is dependent on the boundary conditions given at the bottom of Tables 4.20 and 4.21. If the location does not comply with these footnotes, this method should not be used.

---

*Arc Thermal Performance Value—material rating provided by the manufacturer. This is the amount of heat energy that will just cause the onset of a second-degree burn (see Chap. 1).

†The average of the five highest incident energy values that did not cause the fabric to break open. Test value supplied by the manufacturer and developed per ASTM F 1959/F 1959M-99.

TABLE 4.20 Hazard Risk Category Classifications

Task (Assumes Equipment Is Energized, and Work Is Done within the Flash Protection Boundary)	Hazard/Risk Category	V-rated Gloves	V-rated Tools
Panelboards rated 240 V and below—Notes 1 and 3	—	—	—
Circuit breaker (CB) or fused switch operation with covers on	0	N	N
CB or fused switch operation with covers off	0	N	N
Work on energized parts, including voltage testing	1	Y	Y
Remove/install CBs or fused switches	1	Y	Y
Removal of bolted covers (to expose bare, energized parts)	1	N	N
Opening hinged covers (to expose bare, energized parts)	0	N	N
Panelboards or Switchboards rated > 240 V and up to 600 V (with molded-case or insulated case circuit breakers)—Notes 1 and 3	—	—	—
CB or fused switch operation with covers on	0	N	N
CB or fused switch operation with covers off	1	N	N
Work on energized parts, including voltage testing	2*	Y	Y
600 V Class Motor Control Centers (MCCs)—Notes 2 (except as indicated) and 3	—	—	—
CB or fused switch or starter operation with enclosure doors closed	0	N	N
Reading a panel meter while operating a meter switch	0	N	N
CB or fused switch or starter operation with enclosure doors open	1	N	N
Work on energized parts, including voltage testing	2*	Y	Y
Work on control circuits with energized parts 120 V or below, exposed	0	Y	Y
Work on control circuits with energized parts >120 V exposed	2*	Y	Y
Insertion or removal of individual starter "buckets" from MCC—Note 4	3	Y	N
Application of safety grounds, after voltage test	2*	Y	N
Removal of bolted covers (to expose bare, energized parts)	2*	N	N
Opening hinged covers (to expose bare, energized parts)	1	N	N
600 V Class Switchgear (with power circuit breakers or fused switches)—Notes 5 and 6	—	—	—
CB or fused switch operation with enclosure doors closed	0	N	N
Reading a panel meter while operating a meter switch	0	N	N
CB or fused switch operation with enclosure doors open	1	N	N
Work on energized parts, including voltage testing	2*	Y	Y
Work on control circuits with energized parts 120 V or below, exposed	0	Y	Y
Work on control circuits with energized parts >120 V exposed	2*	Y	Y
Insertion or removal (racking) of CBs from cubicles, doors open	3	N	N
Insertion or removal (racking) of CBs from cubicles, doors closed	2	N	N
Application of safety grounds, after voltage test	2*	Y	N
Removal of bolted covers (to expose bare, energized parts)	3	N	N
Opening hinged covers (to expose bare, energized parts)	2	N	N
Other 600 V Class (277 V through 600 V, nominal) Equipment—Note 3	—	—	—
Lighting or small power transformers (600 V, maximum)	—	—	—
Removal of bolted covers (to expose bare, energized parts)	2*	N	N
Opening hinged covers (to expose bare, energized parts)	1	N	N
Work on energized parts, including voltage testing	2*	Y	Y
Application of safety grounds, after voltage test	2*	Y	N
Revenue meters (kW-hour, at primary voltage and current)	—	—	—
Insertion or removal	2*	Y	N
Cable trough or tray cover removal or installation	1	N	N
Miscellaneous equipment cover removal or installation	1	N	N
Work on energized parts, including voltage testing	2*	Y	Y
Application of safety grounds, after voltage test	2*	Y	N

(Continued)

TABLE 4.20 Hazard Risk Category Classifications *(Continued)*

Task (Assumes Equipment Is Energized, and Work Is Done within the Flash Protection Boundary)	Hazard/Risk Category	V-rated Gloves	V-rated Tools
NEMA E2 (fused contactor) Motor Starters, 2.3 kV through 7.2 kV	—	—	—
Contactor operation with enclosure doors closed	0	N	N
Reading a panel meter while operating a meter switch	0	N	N
Contactor operation with enclosure doors open	2*	N	N
Work on energized parts, including voltage testing	3	Y	Y
Work on control circuits with energized parts 120 V or below, exposed	0	Y	Y
Work on control circuits with energized parts >120 V, exposed	3	Y	Y
Insertion or removal (racking) of starters from cubicles, doors open	3	N	N
Insertion or removal (racking) of starters from cubicles, doors closed	2	N	N
Application of safety grounds, after voltage test	3	Y	N
Removal of bolted covers (to expose bare, energized parts)	4	N	N
Opening hinged covers (to expose bare, energized parts)	3	N	N
Metal Clad Switchgear, 1 kV and above	—	—	—
CB or fused switch operation with enclosure doors closed	2	N	N
Reading a panel meter while operating a meter switch	0	N	N
CB or fused switch operation with enclosure doors open	4	N	N
Work on energized parts, including voltage testing	4	Y	Y
Work on control circuits with energized parts 120 V or below, exposed	2	Y	Y
Work on control circuits with energized parts >120 V, exposed	4	Y	Y
Insertion or removal (racking) of CBs from cubicles, doors open	4	N	N
Insertion or removal (racking) of CBs from cubicles, doors closed	2	N	N
Application of safety grounds, after voltage test	4	Y	N
Removal of bolted covers (to expose bare, energized parts)	4	N	N
Opening hinged covers (to expose bare, energized parts)	3	N	N
Opening voltage transformer or control power transformer compartments	4	N	N
Other Equipment 1 kV and above	—	—	—
Metal clad load interrupter switches, fused or unfused	—	—	—
Switch operation, doors closed	2	N	N
Work on energized parts, including voltage testing	4	Y	Y
Removal of bolted covers (to expose bare, energized parts)	4	N	N
Opening hinged covers (to expose bare, energized parts)	3	N	N
Outdoor disconnect switch operation (hookstick operated)	3	Y	Y
Outdoor disconnect switch operation (gang-operated, from grade)	2	N	N
Insulated cable examination, in manhole or other confined space	4	Y	N
Insulated cable examination, in open area	2	Y	N

Legend:

V-rated Gloves are gloves rated and tested for the maximum line-to-line voltage upon which work will be done.

V-rated Tools are tools rated and tested for the maximum line-to-line voltage upon which work will be done.

2* means that a double-layer switching hood and hearing protection are required for this task in addition to the other Hazard/Risk Category 2 requirements of Table 3-3.9.2 of Part II.

Y = yes (required)

N = no (not required)

Notes:

1. 25 kA short circuit current available. 0.03 second (2 cycle) fault clearing time.
2. 65 kA short circuit current available. 0.03 second (2 cycle) fault clearing time.
3. For <10 kA short circuit current available, the Hazard/Risk Category required may be reduced by one number.
4. 65 kA short circuit current available, 0.33 second (20 cycle) fault clearing time.
5. 65 kA short circuit current available, up to 1.0 second (60 cycle) fault clearing time.
6. For <25 kA short circuit current available, the Hazard/Risk Category required may be reduced by one number.

Source: *Courtesy National Fire Protection Association.*

TABLE 4.21 Protective Clothing and Personal Protective Equipment (PPE) Matrix

Protective clothing and equipment	Protective systems for hazard/risk category					
Hazard/Risk category number	−1 (Note 3)	0	1	2	3	4
Untreated natural fiber	—	—	—	—	—	—
a. T-shirt (short-sleeve)	X			X	X	X
b. Shirt (long-sleeve)		X				
c. Pants (long)	X	X	X (Note 4)	X (Note 6)	X	X
FR Clothing (Note 1)	—	—	—	—	—	—
a. Long-sleeve shirt			X	X	X (Note 9)	X
b. Pants			X (Note 4)	X (Note 6)	X (Note 9)	X
c. Coverall			(Note 5)	(Note 7)	X (Note 9)	(Note 5)
d. Jacket, parka, or rainwear			AN	AN	AN	AN
FR protective equipment	—	—	—	—	—	—
a. Flash suit jacket (2-layer)						X
b. Flash suit pants (2-layer)						X
Head protection	—	—	—	—	—	—
a. Hard hat			X	X	X	X
b. FR hard hat liner					X	X
Eye protection		—	—	—	—	—
a. Safety glasses	X	X	X	AL	AL	AL
b. Safety goggles				AL	AL	AL
Face protection (double-layer switching hood)				AR (Note 8)	X	X
Hearing protection (ear canal inserts)				AR (Note 8)	X	X
Leather gloves (Note 2)			AN	X	X	X
Leather work shoes			AN	X	X	X

Legend:
AN = As needed
AL = Select one in group
AR = As required
X = Minimum required

Notes:

1. See Table 3-3.9.3. (ATPV is the Arc Thermal Performance Exposure Value for a garment in cal/cm^2.)
2. If voltage-rated gloves are required, the leather protectors worn external to the rubber gloves satisfy this requirement.
3. Class 1 is only defined if determined by Notes 3 or 6 of Table 3-3.9.1 of Part II.
4. Regular weight (minimum 12 oz/yd^2 fabric weight), untreated, denim cotton blue jeans are acceptable in lieu of FR pants. The FR pants used for Hazard/Risk Category 1 shall have a minimum ATPV of 5.
5. Alternate is to use FR coveralls (minimum ATPV of 5) instead of FR shirt and FR pants.
6. If the FR pants have a minimum ATPV of 8, long pants of untreated natural fiber are not required beneath the FR pants.
7. Alternate is to use FR coveralls (minimum ATPV of 5) over untreated natural fiber pants and T-shirt.
8. A double-layer switching hood and hearing protection are required for the tasks designated 2* in Table 3-3.9.1 of Part II.
9. Alternate is to use two sets of FR coveralls (each with a minimum ATPV of 5) over untreated natural fiber clothing, instead of FR coveralls over FR shirt and FR pants over untreated natural fiber clothing.

Source: *Courtesy National Fire Protection Association.*

TABLE 4.22 Protective Clothing Characteristics

Hazard/Risk Category	Typical protective clothing systems: Clothing description (typical number of clothing layers is given in parentheses)	Required minimum arc rating of PPE [J/cm^2 (cal/cm^2)]
1	FR shirt and FR pants or FR coverall (1)	16.74(4)
2	Cotton underwear—conventional short sleeve and brief/shorts, plus FR shirt and FR pants (1 or 2)	33.47(8)
3	Cotton underwear plus FR shirt and FR pants plus FR coverall, or cotton underwear plus two FR coveralls (2 or 3)	104.6(25)
4	Cotton underwear plus FR shirt and FR pants plus multilayer flash suit (3 or more)	167.36(40)

Source: *Courtesy National Fire Protection Association.*

BARRIERS AND WARNING SIGNS

Barriers and warning signs should be placed to control entrance into a work area where there are exposed energized conductors. The installation of such equipment will vary depending on the layout of the work area. The following general criteria should be applied:

- The signs should be distinctive, easy to read, and posted at all entrances to the work area. The signs should clearly warn personnel of the hazardous or energized condition.
- Barriers and barrier tape should be placed at a height that is easy to see. Three feet or so is a good starting point. Adjust the height as dictated by the specific installation.
- Barriers and barrier tape should be placed so that equipment is not reachable from outside of the barrier. This will prevent the accidental or intentional operation of equipment by unauthorized personnel.
- If sufficient work room is not available when barriers are placed, attendants should be used to warn employees of the exposed hazards.

Figure 4.37*a* shows a sample arrangement for a work area in a switchgear room. The layout given here is for example only; however, the general principles may be used. Key points for this installation are

1. Stanchions are used to provide a firm structure for stringing the barrier tape.
2. Warning signs are placed at both entrances warning personnel that hazards exist inside.
3. Warning signs are also posted on the switchgear itself.
4. Five to ten feet of work clearance is allowed between the tape and the switchgear. This number may vary depending on the space available and the work to be performed.

Figure 4.37*b* shows a similar barrier system for a hallway lighting panel. The clearance distance here is less because of the need to allow passage down the hallway. Warning signs are used to alert personnel to the hazard.

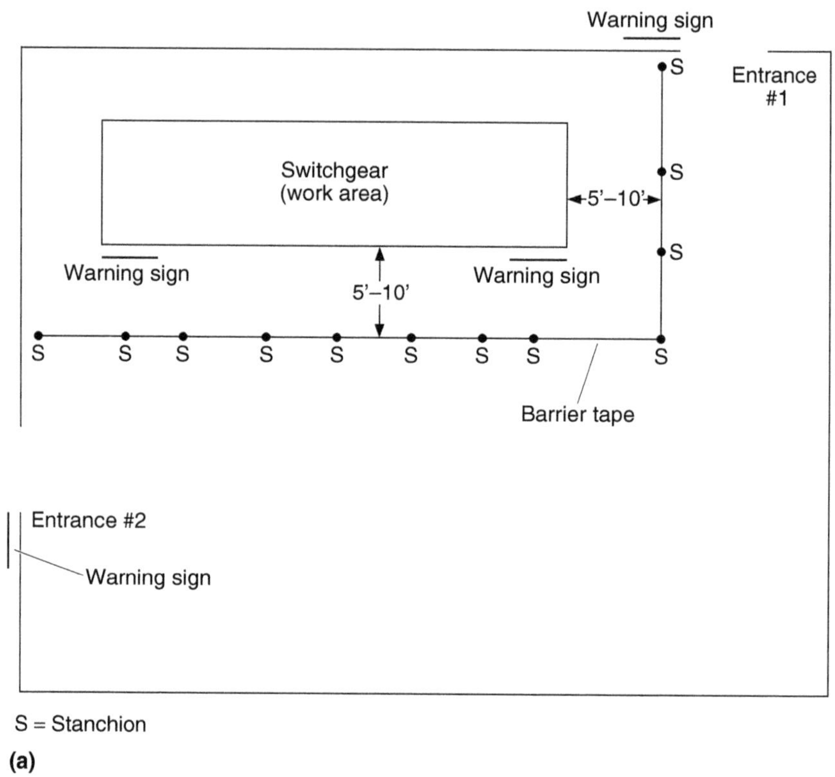

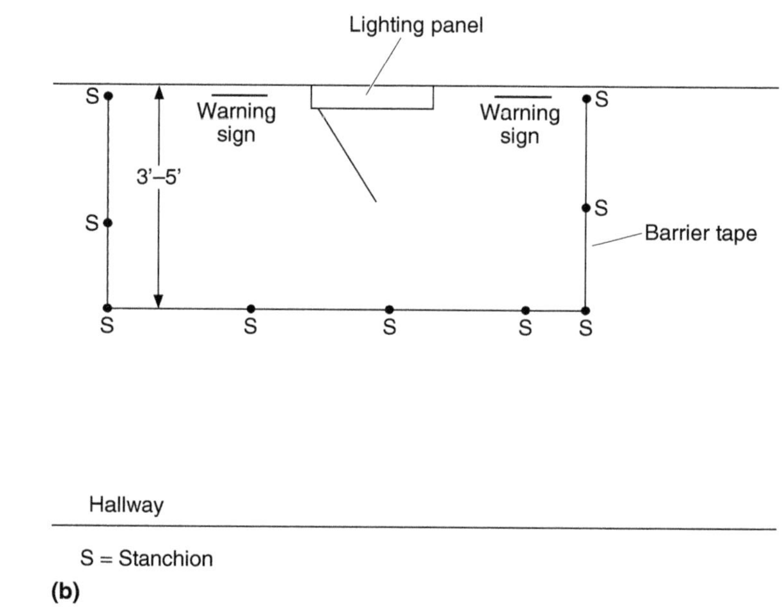

FIGURE 4.37 *(a)* Barriers and warning signs around a work area in a switchgear room with two entrances. *(b)* Barriers and warning signs around a lighting panel that is open for work.

Illumination

Personnel must not reach into or work in areas that do not have adequate illumination. This statement makes it mandatory to set up temporary lighting for many jobs. In areas where lighting is poor, the installation of additional permanent lighting fixtures is the indicated correct remedy. Although de-energization is the preferred method of eliminating electrical hazards, if de-energization also eliminates illumination, alternative safety measures must be employed.

Conductive Clothing and Materials

When working on or around energized conductors, personnel should remove watches, rings, keys, knives, necklaces, and other such conductive items they have on their bodies because (1) they conduct electricity and could possibly cause an arcing fault, and (2) in the event of an arcing fault, they conduct heat.

Conductive materials such as wire, tools, and ladders should be handled in such a way that they do not come into contact with energized conductors. If such material cannot be kept at a safe clearance distance (see "Approach Distances" section), they should be wrapped or insulated. Metal ladders should never be used in electrical installations.

Confined Work Spaces

A complete coverage of confined work spaces is beyond the scope of this handbook. Always refer to the current version of 29CFR1910.146. Electrical safety hazards in confined work spaces should be avoided. The following steps should be employed:

1. If safe approach distances cannot be maintained, energized conductors should be covered or barricaded so personnel cannot contact them.
2. Doors, hatches, and swinging panels should be secured so that they cannot swing open and push personnel into energized conductors.
3. Confined work spaces should be well ventilated to prevent the concentration of gases that may explode in the presence of an electric arc.
4. Exits should be clearly marked. Employees should be familiarized with the exits before entering the confined work space.
5. Confined work spaces must be well illuminated so that all hazards are clearly visible.

TOOLS AND TEST EQUIPMENT

General

Many electrical accidents involve failures of electric tools or test equipment. Such failures take a number of different forms including insulation failure, open ground return wires, internal shorts to ground, and overheating. These problems can be aggravated or even caused by using tools or test equipment improperly. For example, if extension cords not designed for use in wet areas are used in standing water, severe electric shock can result. The following general procedures should be employed when using electric tools or test equipment:

1. Tools and test equipment should be closely inspected before each use. (See the "Visual Inspections" section.)
2. Cord-connected tools should never be lifted or handled by their power cord. If tools must be lifted, a rope should be attached.

3. Grounded tools and equipment must have a continuous metallic connection from the tool ground to the supply ground.
4. If a grounded supply system is not available, double-insulated tools should be used.
5. Three-wire connection plugs should never be altered to fit into two-wire sockets.
6. So-called cheater plugs should not be used unless the third wire of the plug can be securely connected to the supply ground.
7. Locking types of plugs should be securely locked before the tool is energized.
8. Cords and tools should not be used in a wet environment unless they are specifically designed for such an application.

Authorized Users

Only authorized, trained persons should be allowed to use electric tools and test equipment. Training should include all the necessary inspection techniques for the tools that will be employed, plus recognition of the common types of safety hazards that are specific to the tool being used.

Visual Inspections

Tools and test equipment should be closely inspected before each use. This inspection should occur at the beginning of each work shift and again any time a problem is suspected. The inspector should look for the following types of problems:

- Missing, corroded, or damaged prongs on connecting plugs
- Frayed, worn, or missing insulation on connecting cords and/or test leads
- Improperly exposed conductors
- Bent or damaged prongs or test probes
- Excessive exposure on test prongs
- Loose screws or other poorly made electrical connections
- Missing or mis-sized fuses
- Damaged or cracked cases
- Indication of burning, arcing, or overheating of any type

If any of these problems are noted, the tool or test equipment should be removed from service until the problem can be repaired. If repair is not feasible, the tool or test equipment should be replaced.

Electrical Tests

Tools and extension cords should be electrically tested on a monthly basis. The following tests should be performed:

1. *Ground continuity test.* A high current (25 A minimum) should be applied to the tool's ground circuit. The voltage drop on the ground circuit should be no more than 2.5 V.
2. *Leakage test.* This test determines how much current would flow through the operator in the event that the tool's ground circuit were severed.
3. *Insulation breakdown.* This test applies a high voltage (up to 3000 V) to the tool's insulation system and then measures the amount of leakage current.
4. *Operational test.* This test applies rated voltage to the tool and determines how much current the tool draws.

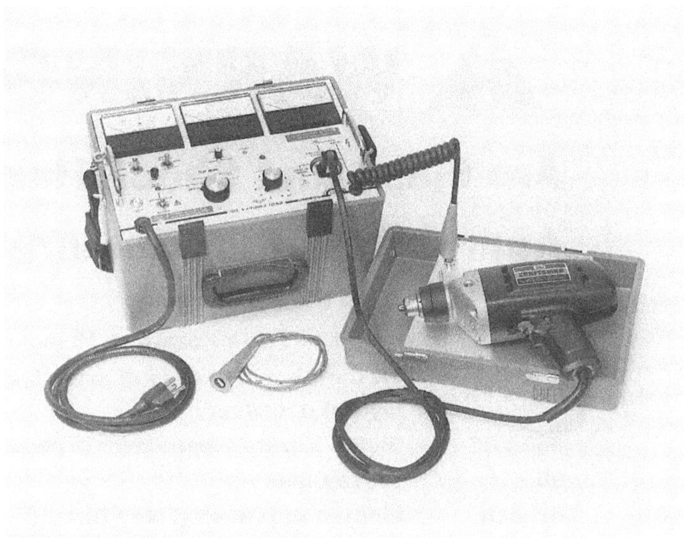

FIGURE 4.38 Tool and appliance tester. *(Courtesy of Megger.)*

If the tool fails any of the above tests, it should be removed from service until it can be repaired. If repair is not possible, the tool or cord should be replaced. Figure 4.38 is a test set designed especially to check the power circuits for cord-connected tools and extension cords.

Wet and Hazardous Environments

Tools and test equipment should be used only in the environments for which they have been designed. If the work area is wet, only tools that are rated for wet work should be used. Fully insulated, waterproof cords should be used if they will be exposed to water or conductive liquids.

If work must be performed in explosive environments, the tools used should be sealed or otherwise designed so that electric arcs will not ignite the explosive materials.

FIELD MARKING OF POTENTIAL HAZARDS

The type and degree of hazard determined by the various procedures outlined in this chapter should be clearly posted at each piece of electrical equipment in the field. These warning labels should clearly identify the type and degree of the hazard and should include the type and amount of PPE required to work in, on, or around the equipment. Figure 4.39 is an example of such a label.

Warning labels of this type should be placed on switchboards, panelboards, industrial-control panels, metal-clad switchgear, meter-socket enclosures, motor-control centers, transformers, motors, generators, and all other such equipment where employees may be exposed to one of the electrical hazards.

Some equipment may have more than one such label. For example, the primary of a transformer may have a different arc-energy level than the secondary. Depending on the mechanical configuration, a label may be required for the primary feeder section of the transformer and the secondary transition cabinet into the switchgear that it feeds.

⚠ WARNING

Arc Flash and Shock Hazard

Appropriate PPE Required

81 inch	**Flash Hazard Boundary**
14.1	**cal/cm^2 Flash Hazard at 18 inches**
Class FR	**FR Clothing in excess of available incident energy - layer if necessary**
480 VAC	**Shock Hazard when cover is removed**
00	**Glove Class**
42 inch	**Limited Approach (Fixed Circuit)**
12 inch	**Restricted Approach**
1 inch	**Prohibited Approach**

Bus: Bus #7 Prot: 3A Bus#7

FIGURE 4.39 Typical warning label to be posted on electrical equipment. (*Label generated by SKM Power Systems Analysis Inc., PowerTools for Windows Software.*)

THE ONE-MINUTE SAFETY AUDIT

Inattention is one of the major root causes of all accidents. Employees whose attention is distracted by work-related or personal problems are prone to miss obvious signs of impending hazards. To make certain that employees are constantly aware of potential hazards, the steps listed in Table 4.23 should be performed by each worker who enters an area where any of the electrical hazards may be present.

The following nine simple steps are explanations of each of the items in Table 4.23. Together they should not take more than a minute or two to perform, even in a very large electrical facility. Of course, not every step will apply in each and every situation. For example, some electrical rooms have no transformers; therefore, step 5 would not be performed.

1. Notify responsible personnel of your presence in the area. In the event that a problem does occur, you may not be able to summon the help that you need. If other personnel are aware of your presence in the area, they will know to come to your aid.
2. Listen for any abnormally loud or unusual noises. Sniff for unusual odors. Transformers and other electric equipment do make noise during normal operation. When such noises become too loud or change tone, trouble may be indicated. Burning odors or the smell of ozone can also be signs of incipient faults.
3. Locate all emergency exits. Electrical accidents are often accompanied by smoke, fire, and noise. Such conditions make it difficult or impossible to find the exits. If exit locations are committed to memory before any problems occur, the worker is more likely to remember them after the problem occurs.
4. Locate all fire alarms and telephones. The same reasoning applies here as locating emergency exits. During an emergency, the location of fire alarms and telephones should be a reflex.

TABLE 4.23 The One-Minute Safety Audit

1. Notify responsible personnel of your presence in the area.
2. Listen for any abnormally loud or unusual noises. Sniff for unusual odors.
3. Locate all emergency exits.
4. Locate all fire alarms and telephones.
5. Inspect all transformer liquid level, temperature, and pressure gauges to make certain that they are within acceptable limits.
6. Locate the station one-line diagram and make certain that it is legible and correct.
7. Make certain that the room is neat and tidy. Generally, electrical facilities should not be used for the storage of equipment.
8. Be certain that all required safety equipment is readily available and easily reached.
9. Check to see that all protective relay and other operational flags are properly reset.

5. Inspect all transformer gauges. Transformers will often warn of impending failures by an increase in pressure, temperature, liquid level, or some combination of all these. An inspection of these elements can warn of a problem before it happens.
6. Locate the station one-line diagram. Switching and other such activities must be carried out in a safe manner. To do so requires that the system one-line diagram be accurate. Few things are more hazardous than opening a switch that is believed to be the source to an energized part of the system and then finding out later that, in fact, it was the wrong switch.
7. Make certain that the room is clean and tidy. Storage of some equipment in an electrical facility is a necessary and valuable procedure. For example, what better place to store electrical safety equipment than close to the gear where it may be needed. Ladders, tools, hot sticks, rubber insulating goods, replacement lights, and other such materials are all candidates for storage in the electrical facility. When such storage does take place, however, it should be done in an orderly and safe fashion. Storage cabinets should be procured for small equipment. Ladders should be hung on racks especially mounted for such a purpose. Wire reels and other such materials should not be in the exit paths.
8. Be certain that all safety equipment is readily available. Workers will often use the excuse that "the safety gear is too far away to justify using it for this short procedure." To eliminate such an excuse, the safety equipment should be readily available. Whether it is brought in with the work crew or stored in the electrical facility, it must be readily available.
9. Check all protective relay flags and other indicators. Operation indicators can often be a precursor of a safety problem. A pressure alarm that will not reset, for example, can indicate an impending explosion. Such indicators should be checked. Such an indicator should never be reset without first determining what caused the problem. Also, operations personnel should be contacted before resetting.

These nine simple steps can be quickly accomplished by all personnel . . . and they may save a life.

REFERENCES

1. Stoll, A. M., and Chianta, M. A., "Method and Rating System for Evaluation of Thermal Protection," *Aerospace Medicine*, 40: pp. 1232–1238, 1968.
2. Doughty, Richard L., Neal, Thomas E., and Floyd, H. Landis, *Predicting Incident Energy to Better Manage the Electric Arc Hazard on 600-V Power Distrbution Systems*, IEEE Paper PCIC-98-36, 2000.
3. IEEE Std 1584–2018, *IEEE Guide for Performing Arc-Flash Calculations*, Institute of Electrical and Electronics Engineers (IEEE) Piscataway, NJ.

CHAPTER 5
GROUNDING AND BONDING OF ELECTRICAL SYSTEMS AND EQUIPMENT

INTRODUCTION

Electrical grounding and bonding seems to be one of the most misunderstood subjects and requirements of the National Electrical Code (NEC). The Occupational Safety and Health Administration (OSHA) documented more than 20,000 violations of grounding in the time period of one single year. Most of these violations came from loose, damaged, or missing external grounds on electrical equipment and structures. Just imagine the increased numbers of violations if OSHA inspectors opened electrical equipment to check equipment grounding and bonding connections or if they used test equipment to verify proper grounding!

The real problem is that most electrical equipment *will* operate without a ground connection. For example, an electric drill *will not* work if the hot or neutral wire is open; however, if the ground wire is open, the drill *will* run properly. Even though the drill works properly without a ground connection, it is not safe to use.

This chapter addresses several issues dealing with general grounding and bonding, system grounding, and equipment grounding. Proper grounding is an issue that must be seriously considered for all electrical installations and equipment. Strict adherence to federal and local regulatory standards, such as OSHA 29 CFR 1910.304(g) and NEC article 250, as well as the numerous consensus standards and reference books, reduces the risk of electrical shock from contact with inadvertently energized equipment, enclosures, and structures.

The information in this chapter is not intended to be a substitute for the NEC or OSHA requirements. Nor is it intended to replace or be a substitute for any other standard quoted herein. For proper grounding and bonding of electrical systems and equipment, always comply with the requirements contained in the current editions of standards and regulations.

ELECTRIC SHOCK HAZARD

Energized conductors and circuit components, installed within or on electrical equipment, are insulated from the equipment's metal enclosure to protect personnel who operate or otherwise make contact with the equipment from being exposed to dangerous voltages. When aging or malfunction causes the insulation to break down, the energized conductors within the equipment can make direct contact with the metal enclosure, thereby energizing it. Anyone making contact with an energized equipment enclosure could be injured or killed.

Equipment grounding and bonding are ways of mitigating this shock hazard. An equipment grounding conductor forms a very low-impedance path back to the overcurrent protective device to cause it to trip in the event of a ground fault. Bonding jumpers are installed, where required, to join together all non-current-carrying metal parts, enclosures, and components of electrical equipment to prevent a difference in potential between two or more different surfaces.

Simply stated, proper bonding and grounding of electrical equipment will substantially reduce the risk of electrical shock by effectively eliminating a difference in potential.

GENERAL REQUIREMENTS FOR GROUNDING AND BONDING

Grounding of Electrical Systems

Grounded electrical systems are required to be connected to earth in such a way as to limit any voltages imposed by lightning, line surges, or unintentional contact with higher voltage conductors. Electrical systems are also grounded to stabilize the voltage to earth during normal operation. If, for example, the neutral of a 120/240 V, wye-connected secondary of a transformer were not grounded, instead of being 120 V to ground, the voltage could reach several hundred volts to ground. A wye-connected electrical system becomes very unstable if it is not properly grounded.

The following requirements are taken from OSHA 29 CFR 1910.304(g) and can also be found in the National Electrical Code (NEC) Section 250.20.

(g) Grounding. Paragraphs (g)(1) through (g)(9) of this section contain grounding requirements for systems, circuits, and equipment.

(1) Systems to be grounded. Systems that supply premises wiring shall be grounded as follows:

(i) All 3-wire DC systems shall have their neutral conductor grounded;

(ii) Two-wire DC systems operating at over 50 volts through 300 volts between conductors shall be grounded unless:

(A) They supply only industrial equipment in limited areas and are equipped with a ground detector;

(B) They are rectifier-derived from an AC system complying with paragraphs (g)(1)(iii), (g)(1)(iv), and (g)(1)(v) of this section; or

(C) They are fire-alarm circuits having a maximum current of 0.030 amperes;

(iii) AC circuits of less than 50 volts shall be grounded if they are installed as overhead conductors outside of buildings or if they are supplied by transformers and the transformer primary supply system is ungrounded or exceeds 150 volts to ground;

(iv) AC systems of 50 volts to 1000 volts shall be grounded under any of the following conditions, unless exempted by paragraph (g)(1)(v) of this section:

(A) If the system can be so grounded that the maximum voltage to ground on the ungrounded conductors does not exceed 150 volts; (see Fig. 5.1)

(B) If the system is nominally rated three-phase, four-wire wye connected in which the neutral is used as a circuit conductor; (see Fig. 5.2)

(C) If the system is nominally rated three-phase, four-wire delta connected in which the midpoint of one phase is used as a circuit conductor; (see Fig. 5.3); or

(D) If a service conductor is uninsulated;

(v) AC systems of 50 volts to 1000 volts are not required to be grounded under any of the following conditions:

(A) If the system is used exclusively to supply industrial electric furnaces for melting, refining, tempering, and the like;

(B) If the system is separately derived and is used exclusively for rectifiers supplying only adjustable speed industrial drives;

(C) If the system is separately derived and is supplied by a transformer that has a primary voltage rating less than 1000 volts, provided all of the following conditions are met:

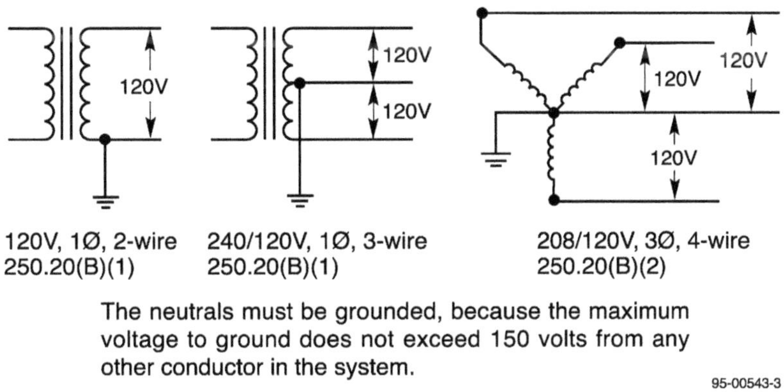

FIGURE 5.1 Voltage-to-ground less than 150 volts. (*Courtesy AVO Training Institute, Inc.*)

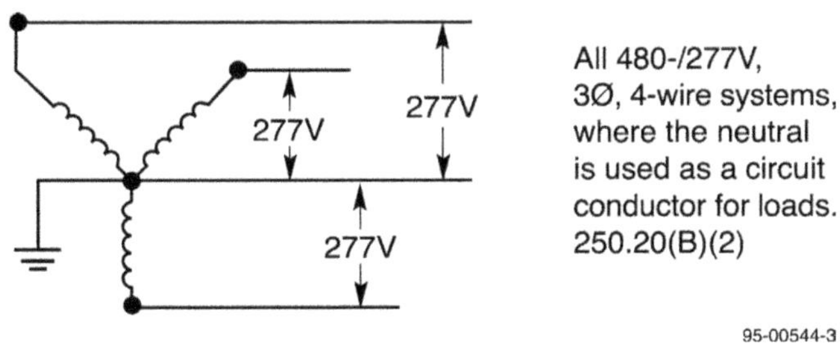

FIGURE 5.2 480/277 volt systems. (*Courtesy AVO Training Institute, Inc.*)

(1) The system is used exclusively for control circuits;

(2) The conditions of maintenance and supervision assure that only qualified persons will service the installation;

(3) Continuity of control power is required; and

(4) Ground detectors are installed on the control system;

(D) If the system is an isolated power system that supplies circuits in health care facilities; or

(E) If the system is a high-impedance grounded neutral system in which a grounding impedance, usually a resistor, limits the ground-fault current to a low value for 3-phase ac systems of 480 volts to 1000 volts provided all of the following conditions are met:

(1) The conditions of maintenance and supervision ensure that only qualified persons will service the installation;

(2) Continuity of power is required;

(3) Ground detectors are installed on the system; and

(4) Line-to-neutral loads are not served.

(2) Conductor to be grounded. The conductor to be grounded for ac premises wiring systems required to be grounded by paragraph (g)(1) of this section shall be as follows:

(i) One conductor of a single-phase, two-wire system shall be grounded;

(ii) The neutral conductor of a single-phase, three-wire system shall be grounded;

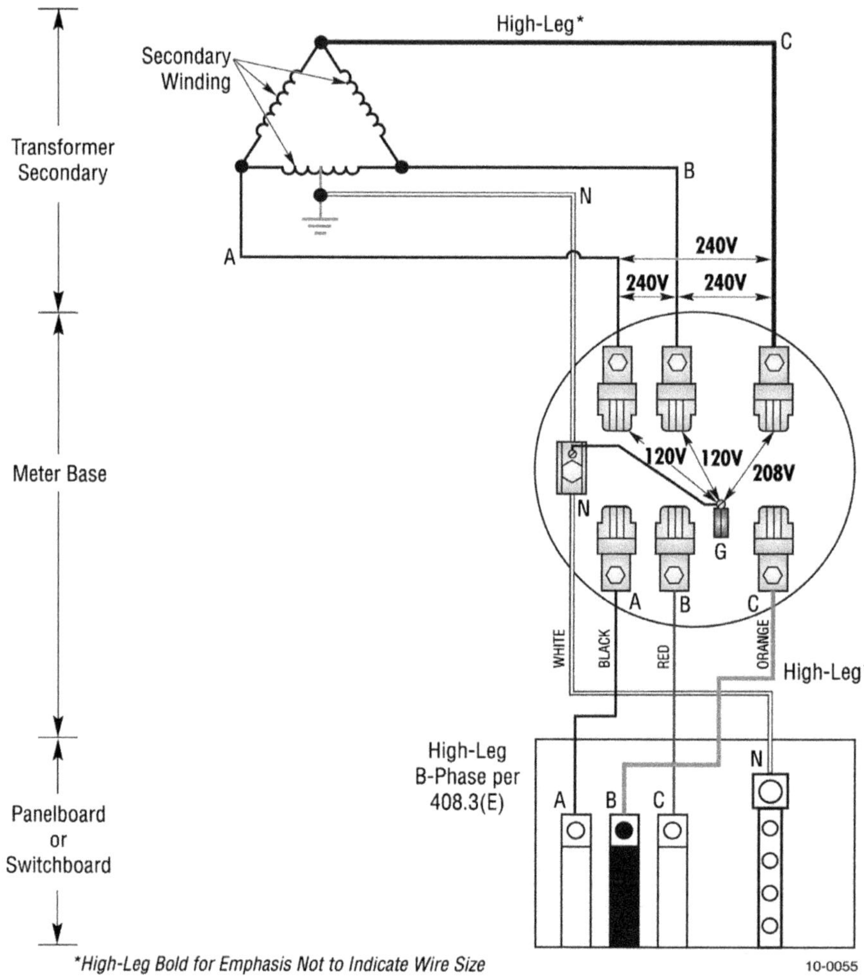

FIGURE 5.3 Four-wire delta system with neutral. (*Courtesy AVO Training Institute, Inc.*)

(iii) The common conductor of a multiphase system having one wire common to all phases shall be grounded;

(iv) One phase conductor of a multiphase system where one phase is grounded shall be grounded; and

(v) The neutral conductor of a multiphase system in which one phase is used as a neutral conductor shall be grounded.

(3) Portable and vehicle-mounted generators. (i) The frame of a portable generator need not be grounded and may serve as the grounding electrode for a system supplied by the generator under the following conditions:

(A) The generator supplies only equipment mounted on the generator or cord- and plug-connected equipment through receptacles mounted on the generator, or both; and

(B) The noncurrent-carrying metal parts of equipment and the equipment grounding conductor terminals of the receptacles are bonded to the generator frame.

(ii) The frame of a vehicle need not be grounded and may serve as the grounding electrode for a system supplied by a generator located on the vehicle under the following conditions:

(A) The frame of the generator is bonded to the vehicle frame;

(B) The generator supplies only equipment located on the vehicle and cord- and plug-connected equipment through receptacles mounted on the vehicle;

(C) The noncurrent-carrying metal parts of equipment and the equipment grounding conductor terminals of the receptacles are bonded to the generator frame; and

(D) The system complies with all other provisions of paragraph (g) of this section.

(iii) A system conductor that is required to be grounded by the provisions of paragraph (g)(2) of this section shall be bonded to the generator frame where the generator is a component of a separately derived system.

(4) Grounding connections. (i) For a grounded system, a grounding electrode conductor shall be used to connect both the equipment grounding conductor and the grounded circuit conductor to the grounding electrode. Both the equipment grounding conductor and the grounding electrode conductor shall be connected to the grounded circuit conductor on the supply side of the service disconnecting means or on the supply side of the system disconnecting means or overcurrent devices if the system is separately derived.

(ii) For an ungrounded service supplied system, the equipment grounding conductor shall be connected to the grounding electrode conductor at the service equipment. For an ungrounded separately derived system, the equipment grounding conductor shall be connected to the grounding electrode conductor at, or ahead of, the system disconnecting means or overcurrent devices.

(iii) On extensions of existing branch circuits that do not have an equipment grounding conductor, grounding-type receptacles may be grounded to a grounded cold water pipe near the equipment if the extension was installed before August 13, 2007. When any element of this branch circuit is replaced, the entire branch circuit shall use an equipment grounding conductor that complies with all other provisions of paragraph (g) of this section.

(5) Grounding path. The path to ground from circuits, equipment, and enclosures shall be permanent, continuous, and effective.

(6) Supports, enclosures, and equipment to be grounded—(i) Metal cable trays, metal raceways, and metal enclosures for conductors shall be grounded, except that:

(A) Metal enclosures such as sleeves that are used to protect cable assemblies from physical damage need not be grounded; and

(B) Metal enclosures for conductors added to existing installations of open wire, knob-and-tube wiring, and nonmetallic-sheathed cable need not be grounded if all of the following conditions are met:

(1) Runs are less than 7.62 m (25.0 ft);

(2) Enclosures are free from probable contact with ground, grounded metal, metal lathes, or other conductive materials; and

(3) Enclosures are guarded against employee contact.

(ii) Metal enclosures for service equipment shall be grounded.

(iii) Frames of electric ranges, wall-mounted ovens, counter-mounted cooking units, clothes dryers, and metal outlet or junction boxes that are part of the circuit for these appliances shall be grounded.

(iv) Exposed non-current-carrying metal parts of fixed equipment that may become energized shall be grounded under any of the following conditions:

(A) If within 2.44 m (8 ft) vertically or 1.52 m (5 ft) horizontally of ground or grounded metal objects and subject to employee contact;

(B) If located in a wet or damp location and not isolated;

(C) If in electrical contact with metal;

(D) If in a hazardous (classified) location;

(E) If supplied by a metal-clad, metal-sheathed, or grounded metal raceway wiring method; or

(F) If equipment operates with any terminal at over 150 volts to ground.

(v) Notwithstanding the provisions of paragraph (g)(6)(iv) of this section, exposed non-current-carrying metal parts of the following types of fixed equipment need not be grounded:

(A) Enclosures for switches or circuit breakers used for other than service equipment and accessible to qualified persons only;

(B) Electrically heated appliances that are permanently and effectively insulated from ground;

(C) Distribution apparatus, such as transformer and capacitor cases, mounted on wooden poles, at a height exceeding 2.44 m (8.0 ft) above ground or grade level; and

(D) Listed equipment protected by a system of double insulation, or its equivalent, and distinctively marked as such.

(vi) Exposed non-current-carrying metal parts of cord- and plug-connected equipment that may become energized shall be grounded under any of the following conditions:

(A) If in hazardous (classified) locations (see § 1910.307);

(B) If operated at over 150 volts to ground, except for guarded motors and metal frames of electrically heated appliances if the appliance frames are permanently and effectively insulated from ground;

(C) If the equipment is of the following types:

(1) Refrigerators, freezers, and air conditioners;

(2) Clothes-washing, clothes-drying, and dishwashing machines, sump pumps, and electric aquarium equipment;

(3) Hand-held motor-operated tools, stationary and fixed motor-operated tools, and light industrial motor-operated tools;

(4) Motor-operated appliances of the following types: hedge clippers, lawn mowers, snow blowers, and wet scrubbers;

(5) Cord- and plug-connected appliances used in damp or wet locations, or by employees standing on the ground or on metal floors or working inside of metal tanks or boilers;

(6) Portable and mobile X-ray and associated equipment;

(7) Tools likely to be used in wet and conductive locations; and

(8) Portable hand lamps.

(vii) Notwithstanding the provisions of paragraph (g)(6)(vi) of this section, the following equipment need not be grounded:

(A) Tools likely to be used in wet and conductive locations if supplied through an isolating transformer with an ungrounded secondary of not over 50 volts; and

(B) Listed or labeled portable tools and appliances if protected by an approved system of double insulation, or its equivalent, and distinctively marked.

(7) Non electrical equipment. The metal parts of the following nonelectrical equipment shall be grounded: frames and tracks of electrically operated cranes and hoists; frames of nonelectrically driven elevator cars to which electric conductors are attached; hand-operated metal shifting ropes or cables of electric elevators; and metal partitions, grill work, and similar metal enclosures around equipment of over 750 volts between conductors.

(8) Methods of grounding fixed equipment. **(i)** Non-current-carrying metal parts of fixed equipment, if required to be grounded by this subpart, shall be grounded by an equipment grounding conductor that is contained within the same raceway, cable, or cord, or runs with or encloses the circuit conductors. For dc circuits only, the equipment grounding conductor may be run separately from the circuit conductors.

(ii) Electric equipment is considered to be effectively grounded if it is secured to, and in electrical contact with, a metal rack or structure that is provided for its support and the metal rack or structure is grounded by the method specified for the non-current-carrying metal parts of fixed equipment in paragraph (g)(8)(i) of this section. Metal car frames supported by metal hoisting cables attached to or running over metal sheaves or drums of grounded elevator machines are also considered to be effectively grounded.

(iii) For installations made before April 16, 1981, electric equipment is also considered to be effectively grounded if it is secured to, and in metallic contact with, the grounded structural metal frame of a building. When any element of this branch circuit is replaced, the entire branch circuit shall use an equipment grounding conductor that complies with all other provisions of paragraph (g) of this section.

(9) Grounding of systems and circuits of 1000 volts and over (high voltage). If high-voltage systems are grounded, they shall comply with all applicable provisions of paragraphs (g)(1) through (g)(8) of this section as supplemented and modified by the following requirements:

(i) Systems supplying portable or mobile high-voltage equipment, other than substations installed on a temporary basis, shall comply with the following:

(A) The system shall have its neutral grounded through an impedance. If a delta-connected high-voltage system is used to supply the equipment, a system neutral shall be derived.

(B) Exposed non-current-carrying metal parts of portable and mobile equipment shall be connected by an equipment grounding conductor to the point at which the system neutral impedance is grounded.

(C) Ground-fault detection and relaying shall be provided to automatically de-energize any high-voltage system component that has developed a ground fault. The continuity of the equipment grounding conductor shall be continuously monitored so as to de-energize automatically the high-voltage feeder to the portable equipment upon loss of continuity of the equipment grounding conductor.

(D) The grounding electrode to which the portable or mobile equipment system neutral impedance is connected shall be isolated from and separated in the ground by at least 6.1 m (20.0 ft) from any other system or equipment grounding electrode, and there shall be no direct connection between the grounding electrodes, such as buried pipe, fence, and so forth.

Grounding of Electrical Equipment

OSHA 1910.304(g)(9)(ii) All non-current-carrying metal parts of portable equipment and fixed equipment including their associated fences, housings, enclosures, and supporting structures shall be grounded. However, equipment that is guarded by location and isolated from ground need not be grounded. Additionally, pole-mounted distribution apparatus at a height exceeding 8 ft above ground or grade level need not be grounded.

In 29 CFR 1910.303, "General," OSHA states under "(b) Examination, installation, and use of equipment (1) Examination" that "Electrical equipment shall be free from recognized hazards that are likely to cause death or serious physical harm to employees." This section continues in paragraph (viii) with "other factors which contribute to the practical safeguarding of persons using or likely to come in contact with the equipment." One of these "other factors" is proper grounding and bonding of the electrical equipment. If the non-current-carrying metal parts of electric equipment are not properly grounded and bonded, and these parts become energized, then any person "using or likely to come in contact with the equipment" is at risk of an electrical shock that may or may not be fatal. This is a risk that must not be taken. Proper grounding and bonding can effectively eliminate this shock hazard by providing a permanent and continuous low-impedance path for ground-fault current to follow in order to clear the circuit protective device(s).

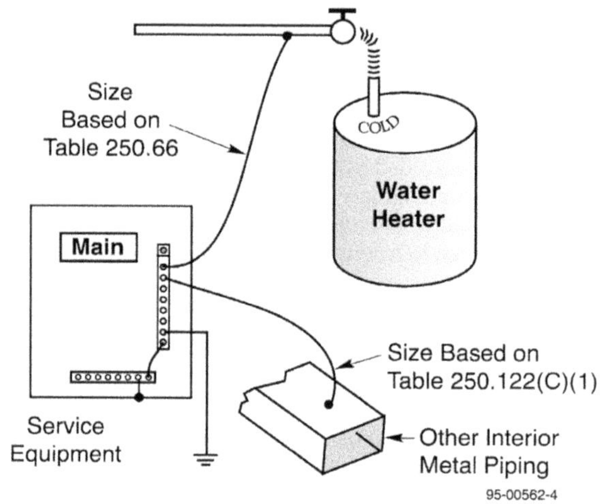

FIGURE 5.4 Bonding of other piping and duct systems. (*Courtesy AVO Training Institute, Inc.*)

Bonding of Electrically Conductive Materials and Other Equipment

Bonding is the permanent joining of metallic parts of materials and equipment. When different metal parts are not bonded together, a difference in potential could exist between the metal parts. This creates an electrically hazardous condition between the parts. Anyone simultaneously coming into contact with the metal parts would be subject to electrical shock, burns, or electrocution. When all conductive materials and parts of equipment are permanently bonded together, there is only one piece of metal and no potential difference exists between the parts. The metal parts must also be grounded to earth in order to be at earth potential. This minimizes the risk of touch-potential and step-potential hazards when working on or around metal enclosures that could become energized.

Other metallic equipment that is in contact with or adjacent to the electrical equipment should also be grounded and bonded to prevent a difference in potential in the event that a ground fault occurs in the electrical equipment. This could include other piping as well as ducts in ventilation systems as shown in Fig. 5.4.

As discussed in the previous section, all non-current-carrying components of electrical equipment must be grounded and bonded. Equipment grounding conductors are installed and connected to the required terminal in the equipment to provide the low-impedance path for fault current to clear the circuit. All other metallic components of the equipment must be bonded to the grounded portion of the equipment to prevent a difference in potential between the components. An example of this would be service enclosures. Figure 5.5 illustrates bonding of service enclosures.

Performance of Fault Path

OSHA makes a very simple statement on the performance of the fault path as found in 1910.304(g)(4), which states: "***Grounding path*** The path to ground from circuits, equipment, and enclosures shall be permanent and continuous." NEC Section 250.4(A)(5) also requires the fault current path to be able to safely carry the maximum ground-fault current, provide a low-impedance path in order to clear the overcurrent devices, and perform in a way that the earth is not used as the sole equipment grounding conductor for fault current.

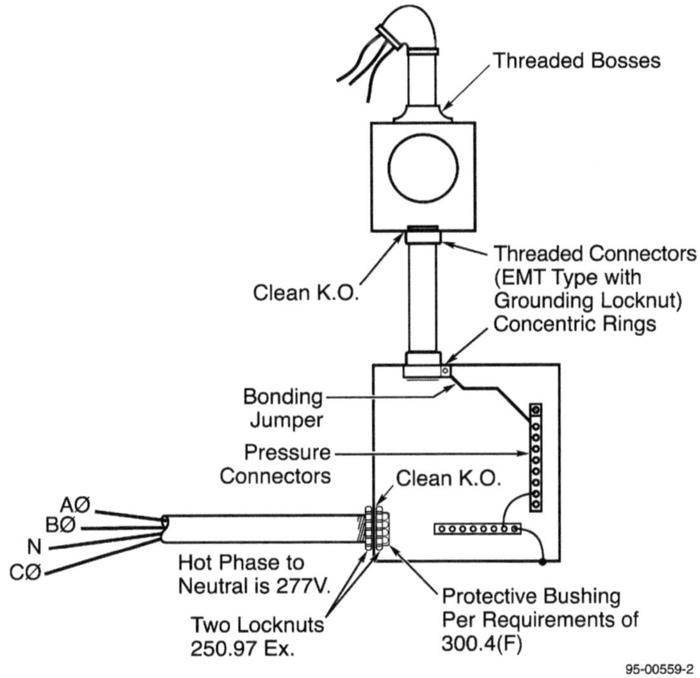

FIGURE 5.5 Bonding service enclosures. (*Courtesy AVO Training Institute, Inc.*)

The fault current path must meet these criteria in order for the overcurrent protection to clear the circuit in the event a ground fault occurs. A low-impedance conductor path must be used because the earth provides an extremely high-impedance path and would not allow a sufficient amount of fault current to flow to clear the overcurrent device. Another factor is that a low-impedance conductor path for fault current to flow through minimizes the possibility of step- and touch-potential hazards by limiting the voltage to ground.

Arrangement to Prevent Objectionable Current

Grounding of electrical equipment and systems must be accomplished in a manner that would prevent objectionable current. The main point here is to install an effective grounding system without creating an objectionable current situation. Circulating current is one form of objectionable current that can occur when multiple grounds are utilized.

Alterations to Stop Objectionable Current

Objectionable current flow can occur when using multiple grounds in electrical systems and equipment. If this occurs, there are several different options to prevent or at least minimize this current flow:

1. One or more of the grounds may be discontinued. Never discontinue all of the grounds.
2. The grounding connection may be changed to another location.
3. The conductive path for the grounding connections may be interrupted.
4. The authority having jurisdiction may grant other remedial action.

Temporary Currents Not Classified as Objectionable Current

One thing to keep in mind when dealing with this subject is that there are times when current will flow in the grounding system or through the non-current-carrying parts of electrical equipment. Ground faults in equipment do occur and that is when the grounding system performs its function. These currents are not classified as objectionable current for the purpose of this discussion.

Connection of Grounding and Bonding Equipment

Conductors for grounding and bonding of electrical equipment must be connected using an approved method such as exothermic welding, listed irreversible compression-type connections, listed pressure connections, listed clamps, and other approved means. Solder must not be used as the sole connection. Solder is too soft and has a very low melting point and, therefore, becomes a fuse in the grounding connection. Also, never use sheet metal screws to make connections between the grounding conductor and the enclosure. The use of sheet metal screws would provide limited metal-to-metal contact between the screw and the enclosure and, therefore, may not provide the required low-impedance connection.

Protection of Ground Clamps and Fittings

All grounding connections must be protected from physical damage either by location or by means of an enclosure made of wood, metal, or equivalent. Damaged grounding conductor connections can result in loss of continuity in the ground path, which will create a potential shock hazard.

Clean Surfaces

If the grounding connection point is contaminated with paint or other such coatings, good continuity may not be accomplished. All surfaces must be cleaned as needed to remove any such coatings or other contaminants that could interfere with the continuity of the grounding connection. As was stated earlier, the grounding system must create a sufficiently low-impedance path for circuit protective devices to clear the circuit in the event of a ground fault.

SYSTEM GROUNDING

Purposes of System Grounding

Power systems are grounded for one or more of the following reasons:

- Control of the voltage to ground to within predictable limits
- To provide for current flow and allow detection and location of ground faults
- To reduce shock hazard to personnel

It is not always possible to accomplish all these goals with a particular method of grounding. In some cases, the chosen method of grounding is a compromise among conflicting goals. Table 5.1 shows the advantages and disadvantages of the various grounding methods.

Grounding Service-Supplied Alternating-Current Systems

The premises wiring system of a grounded electrical service must be grounded. This means that each grounded service must have a grounding electrode conductor connected to the grounding electrode

TABLE 5.1 Comparison of System Grounding Methods

Method of grounding	Advantages	Disadvantages
Ungrounded—no intentional connection to ground	Little or no ground current. Plant does not need to trip for a single ground fault.	Possiblity of large transient overvoltages. Overvoltages on the unfaulted phases. Hard to detect and locate ground faults. Possibility of ferroresonance.
High resistance—ground fault < 10 A	Limit transient overvoltages to 250%. Can usually keep plant running through a single ground fault. Ground fault can be detected.	Overvoltage on the unfaulted phases. Surge arresters need to be rated for the phase-to-phase voltage.
Low resistance—ground fault > 100 A	Limit transient overvoltages to 250%. Immediate and selective fault clearing is possible.	Overvoltage on the unfaulted phases. Surge arresters need to be rated phase-to-phase voltage.
Reactance—ground fault at least 25% and preferably 60% of the three-phase fault $X_o < 10\ X_1$	Prevent serious transient overvoltages. Limit generator phase to ground fault to the magnitude of the three-phase fault.	High ground-fault currents that must be cleared.
Solidly grounded—solid connection between neutral and ground $R_oX_1X_0 < 3X_1$ Note that Z_oO	Limits system overvoltages. Freedom from ferroresonance. Easier detection, location, and selective tripping of ground faults. Can use phase-to-ground-rated surge arresters.	Extreme ground fault magnitudes. Safety risks (arc blast and flash). Stray voltages due to ground faults (shock hazard). Mechanical and thermal stress and damage during ground faults. Ground fault must be detected and cleared, even if the plant must take an outage.

Source: Courtesy AVO Training Institue, Inc.

and to the service equipment. The grounded service conductor is also connected to the grounding electrode conductor. OSHA provides this requirement in 29 CFR 1910.304(g)(4)(i), (ii), and (iii) (NEC Section 250.24 provides a more in-depth requirement) as quoted below:

> ***(4) Grounding connections.*** **(i)** For a grounded system, a grounding electrode conductor shall be used to connect both the equipment grounding conductor and the grounded circuit conductor to the grounding electrode [see Fig. 5.6]. Both the equipment grounding conductor and the grounding electrode conductor shall be connected to the grounded circuit conductor on the supply side of the service disconnecting means, or on the supply side of the system disconnecting means or overcurrent devices if the system is separately derived.
>
> **(ii)** For an ungrounded service-supplied system, the equipment grounding conductor shall be connected to the grounding electrode conductor at the service equipment. For an ungrounded separately derived system, the equipment grounding conductor shall be connected to the grounding electrode conductor at, or ahead of, the system disconnecting means or overcurrent devices.
>
> **(iii)** On extensions of existing branch circuits, which do not have an equipment grounding conductor, grounding-type receptacles may be grounded to a grounded cold water pipe near the equipment if the extension was installed before August 13, 2007. When any element of this branch circuit is replaced, the entire branch circuit shall use an equipment grounding conductor that complies with all other provisions of paragraph (g) of this section.

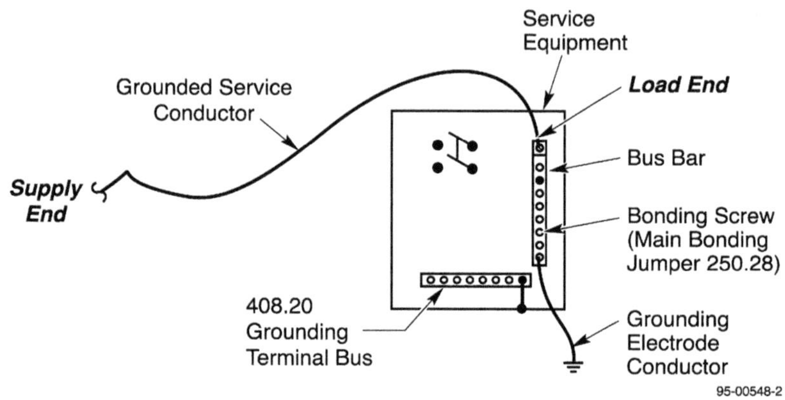

FIGURE 5.6 Grounding service conductor connection to grounding electrode. (*Courtesy AVO Training Institute, Inc.*)

An important point to make here is that the grounding and grounded conductors are only allowed to be connected together on the line side of the service disconnecting means; they are generally not allowed to be connected on the load side. This issue will be addressed in more detail in the section titled "Use of Grounded Circuit Conductor for Grounding Equipment."

Conductors to Be Grounded—Alternating-Current Systems

The following conductors of an ac premises wiring system are required to be grounded:

1. One conductor of a single-phase, two-wire system
2. The neutral conductor of a single-phase, three-wire system
3. The common conductor of a multiphase system where one wire is common to all phases
4. One phase conductor of a multiphase system requiring a grounded phase
5. The neutral conductor of a multiphase system where one phase is used as the neutral

Main Bonding Jumper

In a grounded electrical system, an unspliced main bonding jumper is required to connect all grounding and grounded conductors to the service equipment enclosure. This connection is made within the service equipment enclosure using either a ground bus, a screw, a strap, or a wire. Figure 5.7 further illustrates this requirement.

Grounding Electrode System

A grounding electrode system is made up of a grounding electrode(s), any required bonding jumpers between electrodes, and a grounding electrode conductor. All conductors and jumpers used in the grounding electrode system must be sized in accordance with NEC Section 250.66 and connected as specified in Section 250.70. Several types of grounding electrodes may be available at each building or structure. All available electrodes must be bonded together and used to form a grounding electrode system. Available electrodes include (1) metal underground water pipe, (2) metal in-ground support structure(s), (3) concrete-encased electrodes, (4) ground rings, (5) ground rod, and (6) ground place. Figure 5.8 illustrates several common types of grounding electrodes.

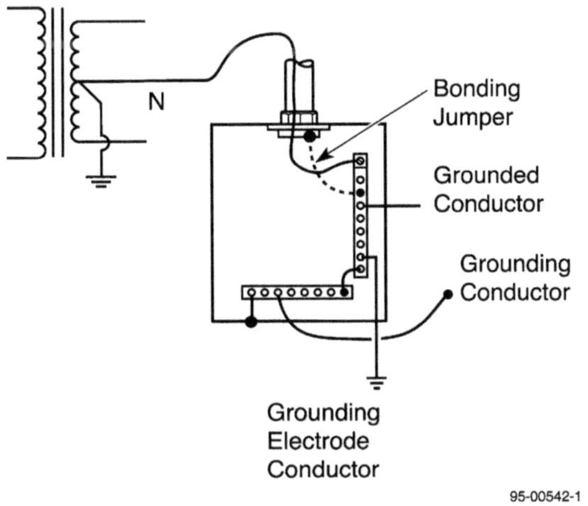

FIGURE 5.7 Location of the main bonding jumper. (*Courtesy AVO Training Institute, Inc.*)

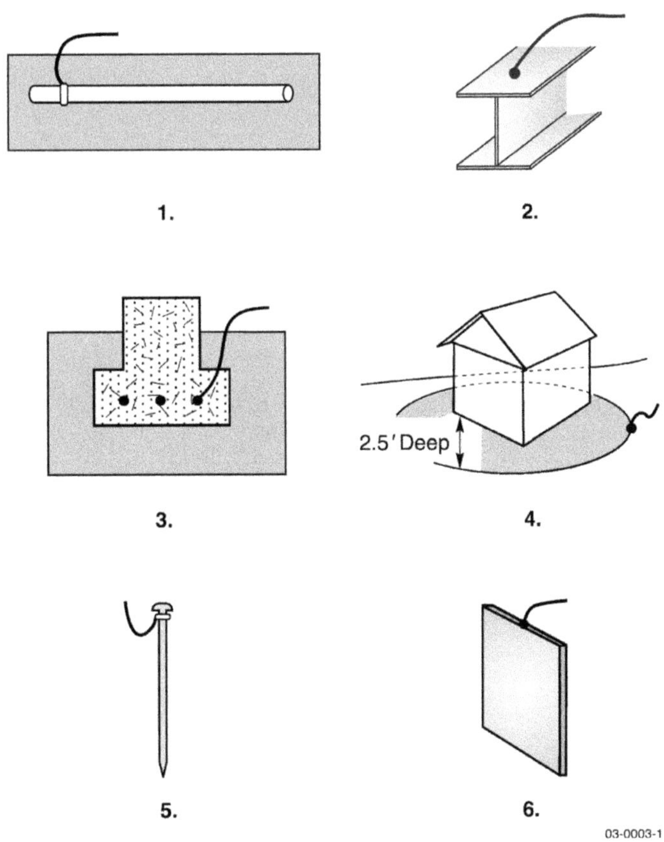

FIGURE 5.8 Types of grounding electrodes. (*Courtesy AVO Training Institute, Inc.*)

If only a water pipe is available for the grounding electrode, it must be supplemented by one of the other electrodes mentioned above or by a made electrode as illustrated in Fig. 5.9.

Made electrodes, such as a ground rod, pipe, or plate, may also be utilized to supplement existing electrodes. Figure 5.10 illustrates the minimum requirements for a ground rod or pipe.

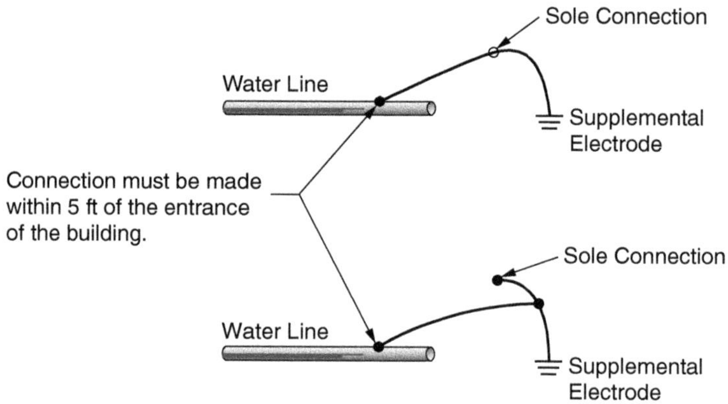

FIGURE 5.9 Sole connections. (*Courtesy AVO Training Institute, Inc.*)

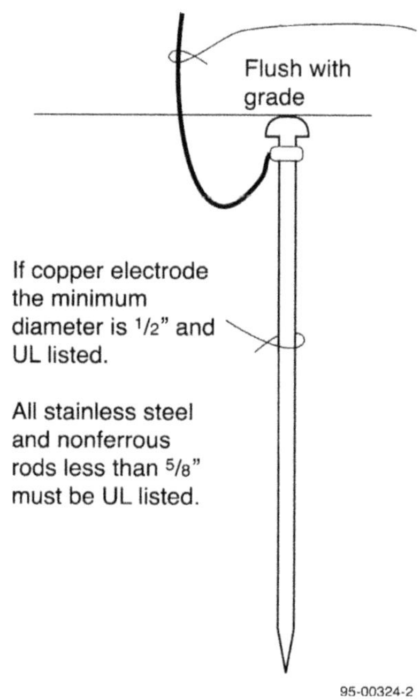

FIGURE 5.10 Minimum requirements for a ground rod or pipe. (*Courtesy AVO Training Institute, Inc.*)

Two alternate installation methods for ground rods or pipes are permitted to be utilized if vertical installation is not possible due to rock or other obstructions. Figure 5.11 illustrates these alternate methods.

Figure 5.12 illustrates the minimum requirements for a ground plate.

Where practicable, these made electrodes are required to be installed below the permanent moisture level and be free from nonconductive coatings. If more than one electrode is used, they must be installed at least 6 ft from each other. A common practice is to place the rods apart at a distance equal to the length of the rod. The best industry practice is to install the rods at least 10 ft apart. This practice will minimize the risk of dissipation overlap in the event of a ground fault.

Grounding Electrode System Resistance

Resistance of all grounding connections must also be addressed briefly. There are three components of resistance to consider: (1) grounding electrode resistance, (2) contact resistance between the electrode and the soil, and (3) the resistance of the soil. These three components are illustrated in Fig. 5.13.

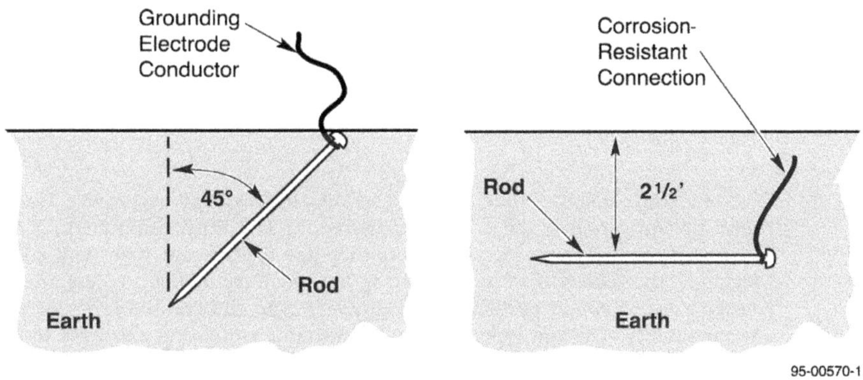

FIGURE 5.11 Alternate burial methods for ground rods. (*Courtesy AVO Training Institute, Inc.*)

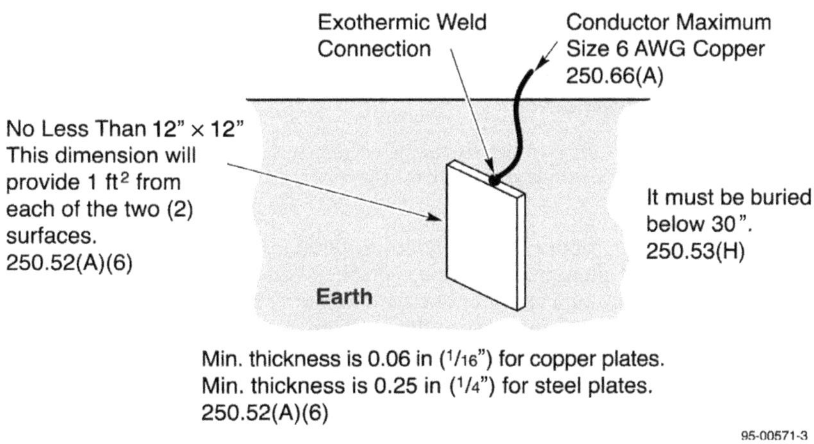

FIGURE 5.12 Plate electrode minimum requirements. (*Courtesy AVO Training Institute, Inc.*)

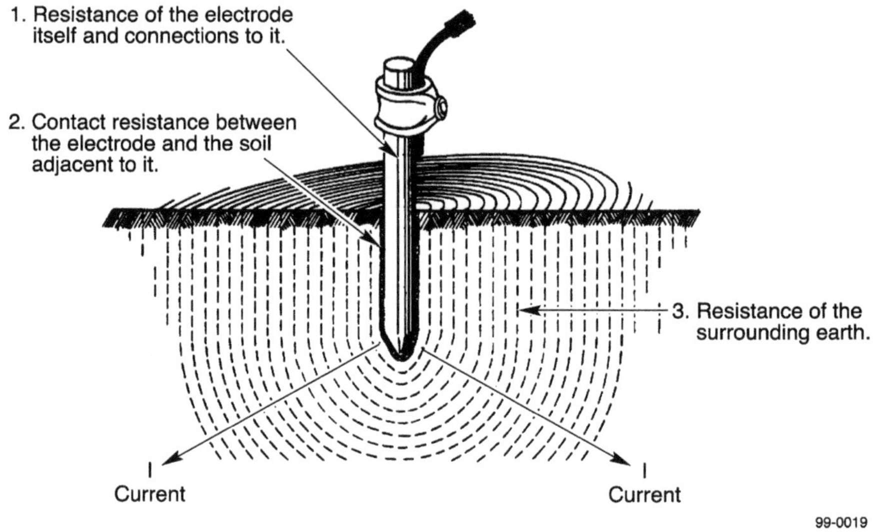

FIGURE 5.13 Components of earth resistance in an earth electrode. (*Courtesy AVO Training Institute, Inc.*)

The grounding electrode, as a general rule, has very low resistance. The contact of the electrode with the surrounding soil is also generally low. The third component, soil, can vary dramatically in resistance from one location to another due to different types and conditions of the soil. For example, the resistivity of inorganic clay can range from 1000 to 5500 ohm/cm whereas gravel can range from 60,000 to 100,000 ohm/cm. Every type and condition of soil will, as can be seen, vary tremendously. The soil type must be known and grounding electrode testing must be done periodically to know whether or not a good grounding system is present. NEC Section 250.53(A)(2) Exception states that if the resistance of a single rod, pipe, or plate electrode is 25 ohm or less, a supplemental electrode is not required. This exception indicates that if the electrode resistance is greater than 25 ohm, an additional electrode(s) would be required to obtain a resistance of 25 ohm or less. As was mentioned earlier, make sure the electrodes are spaced far enough apart to prevent dissipation overlap.

Grounding Electrode Conductor

OSHA 29 CFR 1910.399 defines the grounding electrode conductor as: "The conductor used to connect the grounding electrode to the equipment grounding conductor, to the grounded conductor, or both, of the circuit at the service equipment or at the source of a separately derived system."

Although copper is the preferred material for the grounding electrode conductor, aluminum and copper-clad aluminum are also acceptable. There are, however, restrictions with the latter two materials. The aluminum and copper-clad aluminum conductors are not allowed by code to be installed in direct contact with masonry or the earth, or where corrosive conditions exist. If used, these materials must not be installed within 18 in of earth. Copper, on the other hand, is considered a corrosion-resistant material and, therefore, is permitted to be installed in most locations.

NEC Section 250.64(C) requires the grounding electrode conductor to be installed in one continuous length without splice or joint. There are, however, two deviations from this rule. A splice can be made using a listed irreversible compression-type connector or the exothermic welding process. More on this subject will be addressed in the next section, "Grounding Conductor Connection to Electrode."

The grounding electrode conductor must be sized by the requirements of NEC Section 250.66. The conductor is sized based on the size of the largest service-entrance conductor or equivalent area of parallel conductors. The size can also vary according to the type of connection used, as will be discussed in the next section. Other deviations to the values presented in NEC Table 250.66 are as follows:

1. *Connections to made electrodes.* The sole connection portion of the conductor is not required to be larger than No. 6 copper wire.
2. *Connections to concrete-encased electrodes.* The sole connection portion of the conductor is not required to be larger than No. 4 copper wire.
3. *Connections to ground rings.* The sole connection portion of the conductor is not required to be larger than the conductor used for the ground ring. NEC Section 250.52(A)(4) states "not smaller than No. 2" copper wire.

Grounding Conductor Connection to Electrodes

NEC Section 250.68(B) states that the connection of the grounding electrode conductor must be made in a manner that will ensure a permanent and effective grounding path. Section 250.70 provides the means of connection of the grounding electrode conductor to the grounding electrode. These methods include exothermic welding, listed lugs, listed pressure connectors, listed clamps, and other listed means of connection. Note the term "listed" for various connection methods. These are designed, manufactured, listed, and labeled for the purpose and are the only ones permitted for use in grounding electrode systems. The code goes on to state that connections made with solder must not be used. Solder has a very low melting point and, therefore, would disconnect the grounding connection in the event of a ground fault because the solder would act like a fuse. The National Electrical Safety Code (NESC) states that joints in grounding conductors must be made and maintained so as not to materially increase the resistance of the grounding conductor. The NESC also states that the connection must have appropriate mechanical and corrosion-resistant characteristics.

IEEE Std. 80, "Guide for Safety in AC Substation Grounding," is an excellent resource for determining the minimum size conductor based on the type of connection used. The Onderdonk AC equation is used to calculate fusing current of a conductor based on the type of connection used. Connectors and splice connections must also meet the requirements of IEEE Std. 837 to be acceptable. The following examples will further illustrate this point:

For this example, a shortened version of the Onderdonk equation will be used to determine conductor size and to determine fusing current levels based on the type of connection.

Conductor Size	**Fusing Current**
$A = KI$	$\sqrt{S}I = \dfrac{A}{K\sqrt{S}}$

where A = cable size, circular mil
K = connector factor
S = maximum fault time, seconds
I = maximum fault current, amperes

For copper conductors, these equations use the following K values to represent connection temperature:

- Conductor only—1083°C temperature rating—value of $K = 6.96$
- Welded connections—1083°C temperature rating—value of $K = 6.96$
- Irreversible compression-type connections—1083°C temperature rating— value of $K = 6.96$
- Brazed connections—450°C temperature rating—value of $K = 9.12$
- Pressure-type connections—250°C temperature rating—value of $K = 11.54$

The following three examples, used to determine minimum conductor size, will use a 20,000-A fault current with a protective device clearing time of 5 cycles (0.083 second) using $A = KI\sqrt{S}$:

1. **Pressure-type connection:**

$$A = 11.54 \times 20{,}000 \times \sqrt{0.083}$$

$$A = 66{,}493 \text{ circular mils (minimum size conductor)}$$

According to NEC Table 8, the size that corresponds to the minimum size conductor would be a No. 1 AWG copper conductor (83,690 circular mils).

2. **Brazed connection:**

$$A = 9.12 \times 20{,}000 \times \sqrt{0.083}$$

$$A = 52{,}549 \text{ circular mils (minimum size conductor)}$$

According to NEC Table 8, the size that corresponds to the minimum size conductor would be a No. 3 AWG copper conductor (52,620 circular mils).

3. **Welded or irreversible compression-type connection:**

$$A = 6.96 \times 20{,}000 \times \sqrt{0.083}$$

$$A = 40{,}103 \text{ circular mils (minimum size conductor)}$$

According to NEC Table 8, the size that corresponds to the minimum size conductor would be a No. 4 AWG copper conductor (41,740 circular mils).

As can be seen by the previous examples, if a pressure-type connection were used, it would require a conductor approximately 60 percent larger than a connection made with a welded or irreversible compression-type connection. The brazed connection would require a conductor approximately 31 percent larger than a welded or irreversible compression-type connection. For practical purposes, the welded or irreversible compression-type connection would be the preferred choice.

Another consideration would be the approximate fusing current of the conductor based on the type of connection. This example will use a 4/0 (211.600 circular mils) conductor with a protective device clearing time of 5 cycles (0.083 second) using $I = A/K\sqrt{S}$.

1. **Pressure-type connection:**

$$I = \frac{211{,}600}{11.54\sqrt{0.083}}$$

$$I = 63{,}646 \text{ amperes fusing current}$$

2. **Welded or irreversible compression-type connection:**

$$I = \frac{211{,}600}{6.96\sqrt{0.083}}$$

$$I = 105{,}528 \text{ amperes fusing current}$$

TABLE 5.2 Reference Table for Sizing Bonding Jumpers

Paragraph	Reference for sizing
Supply side of service	Table 250.102(C)(1)
Load side of service	Table 250.122
Interior water pipe	Table 250.102(C)(1)
Multiple occupancies	Table 250.122
Multiple buildings—common service	Section 250.122
Separately derived systems	Table 250.66
Other metal piping	Table 250.122
Structural steel	Table 250.102(C)(1)

As can be seen in the above examples, the conductor utilizing the welded or irreversible compression-type connection will handle much more current before fusing than the pressure-type connection.

When reliability and cost are taken into consideration, it is plain to see that the welded and the irreversible compression-type connections are far superior to any other type of connection.

Bonding

If only one statement were to be made about grounding and bonding, it would be that all non-current-carrying parts of electrical equipment and nonelectrical equipment, likely to become energized, should be effectively grounded and bonded together. By using this general philosophy, there will be a minimum (near zero) risk that the non-current-carrying parts of equipment could become energized. This would greatly reduce the risk of electrical shock or electrocution of any person likely to come into contact with the equipment. As discussed earlier, a general statement in OSHA 29 CFR 1910.303(b)(1)(vii) includes, "other factors which contribute to the practical safeguarding of persons using or likely to come in contact with the equipment." One of these "other factors" refers to grounding and bonding of equipment likely to become energized.

NEC Section 250.90 states that bonding must be provided where necessary to ensure electrical continuity. Bonding is also required to ensure the capacity to safely conduct any fault current that is likely to be imposed on the equipment. This requirement applies to all types of equipment, systems, and structures. NEC Article 250, Part V. Bonding, must be complied with to size the bonding jumper correctly. Table 5.2 illustrates further why Part V must be adhered to.

The key here is to know where the bonding jumper is and what it is being used for; it makes a significant difference. If the wrong section or table is used, the bonding jumper may not be of adequate current-carrying capacity to safely conduct any fault current likely to be imposed on it.

EQUIPMENT GROUNDING

Equipment to Be Grounded

Equipment to be grounded essentially means that all non-current-carrying metal parts of electrical equipment, whether fastened in place or portable, must be grounded. NEC Article 250, Part VI, "Equipment Grounding and Equipment Grounding Conductors," very specifically lays out the requirements for equipment grounding for electrical as well as nonelectrical equipment that could become energized. NEC Section 250.110 lists the requirements for equipment fastened in place (fixed) and has three exceptions. Exception No. 2 itemizes, "distribution apparatus, such as transformer and capacitor cases, mounted on wooden poles, at a height exceeding 8 ft above the ground or grade level." This exception would protect the general public from possible contact with the ungrounded apparatus;

however, it does not protect the person working on the pole. With regard to these issues, OSHA 29 CFR 1910.269(l)(11) states:

> ***Non-current-carrying metal parts.*** Non-current-carrying metal parts of equipment or devices, such as transformer cases and circuit-breaker housings, shall be treated as energized at the highest voltage to which these parts are exposed, unless the employer inspects the installation and determines that these parts are grounded before employees begin performing the work.

The best practice is to always ground and bond every case or enclosure that contains electrical equipment or conductors. When in doubt, ground and bond it.

Grounding Cord- and Plug-Connected Equipment

NEC Section 250.114 states the same philosophy for cord- and plug-connected equipment, as do Sections 250.110 through 250.112 for fixed equipment. It directs that all exposed non-current-carrying metal parts that are likely to become energized must be grounded. A key point here is "likely to become energized." Any metal housing or enclosure that contains electrical components is, at some time or another, likely to become energized. Proper equipment grounding techniques will provide the sufficiently low-impedance path required to cause the overcurrent device to operate and clear the ground-fault condition.

In the material written about electrical safety-related work practices, OSHA has provided specific requirements for the use of portable electrical equipment and extension cords. This requirement is found in OSHA 29 CFR 1910.334 where it states:

> ***Use of equipment. (a) Portable electric equipment. (3) Grounding type equipment.*** **(i)** A flexible cord used with grounding type equipment shall contain an equipment grounding conductor.
>
> **(ii)** Attachment plugs and receptacles may not be connected or altered in a manner which would prevent proper continuity of the equipment grounding conductor at the point where plugs are attached to receptacles. Additionally, these devices may not be altered to allow the grounding pole of a plug to be inserted into slots intended for connection to the current-carrying conductors.
>
> **(iii)** Adapters which interrupt the continuity of the equipment grounding connection may not be used.

OSHA makes it very clear that this type of equipment must be grounded. Note as well the statement in (iii) concerning adapters. The adapters referred to here are used when a grounded plug is required to be used where an ungrounded receptacle exists. These adapters are UL approved and can be used, but only if used properly; that is, the ground connection must be attached to a return ground path. The problem with adapters is that the ground connection device is generally cut off or otherwise not used, thus defeating the ground continuity.

Equipment grounding conductors are required as stated previously; however, grounding alone does not give complete protection when using portable cord- and plug-connected, hand-held equipment and extension cords. The use of a ground-fault circuit interrupter (GFCI) when using portable equipment will provide additional safety for the user. Grounding does provide a path for ground-fault current to flow to cause the overcurrent device to operate; however, it does not provide protection when current leakage occurs due to moisture in a piece of equipment or when there is an undetected cut in the cord or crack in the equipment case. The GFCI is designed to trip at 4 to 6 mA of current. To explain this further, the following example is provided.

Given a 20-ampere rated molded case circuit breaker, the maximum load allowed by code is 16 amperes (80 percent of rating). At 16 amperes, this circuit breaker will, under normal conditions, run indefinitely. Even at 20 amperes, the circuit breaker will remain closed for several minutes to infinity. Taking this into consideration, the amount of current leakage from moisture in the equipment (a hand-held drill, for example) will not trip this circuit breaker; however, a solid ground fault will. Since the circuit breaker will not trip, a person contacting this piece of equipment is exposed to a shock or electrocution hazard. It takes approximately one-tenth of an ampere to cause ventricular fibrillation in most people, which is generally fatal. The point is that the circuit breaker will not protect a person in this situation but a GFCI will.

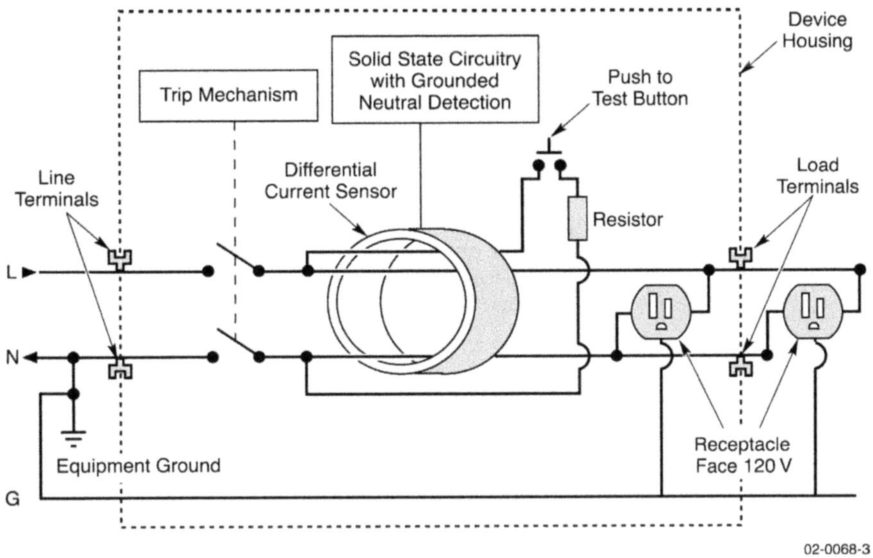

FIGURE 5.14 Internal diagram of a ground-fault circuit interrupter (GFCI). (*Courtesy AVO Training Institute, Inc.*)

NEC Section 590.6 requires ground-fault protection for all temporary wiring installations for construction, remodeling, maintenance, repair, demolition of buildings, structures, equipment, or similar activities. In other words, any time an extension cord is used for one of these activities, a GFCI is required. Although not specifically required here, and because the hazard risk is the same, all hand-held cord- and plug-connected portable electrical equipment should utilize a GFCI for the protection of the worker.

Figure 5.14 further illustrates how a GFCI works.

Use a GFCI; it may save your life. Look at it this way—it is cheap life insurance.

Equipment Grounding Conductors

NEC Section 250.118 identifies several types of equipment grounding conductors. These types vary from an equipment grounding conductor run with circuit conductors or enclosing them. It can consist of one or more of the following:

1. A copper, aluminum, or copper-clad aluminum conductor
2. Rigid metal conduit
3. Intermediate metal conduit
4. Electrical metallic tubing
5. Flexible metal conduit and fittings listed for grounding
6. Liquid-tight flexible metal conduit and fittings listed for grounding
7. Flexible metallic tubing, under conditions
8. Type AC armored cable
9. Copper sheath of type MI cable
10. Type MC cable metal sheath

11. Cable trays meeting the requirements of NEC Section 392.3(C) and 392.7
12. Cablebus framework permitted by NEC Section 370.3
13. Other metal raceways listed for grounding that are electrically continuous
14. Surface metal raceways listed for grounding

The equipment grounding conductor can either be bare, covered, or insulated. When the conductor is insulated, it must be identified by a continuous outer finish of green, or green with one or more yellow stripes, or it must be bare. Equipment grounding conductors larger than No. 6 and multiconductor cable, where one or more conductors are designated for use as an equipment grounding conductor, must be permanently identified as the equipment grounding conductor by green striping, coloring, or taping.

Sizing Equipment Grounding Conductors

Equipment grounding conductors are required to be sized not smaller than is stated in NEC Table 250.122 and are not required to be larger than the supply conductors. Section 250.122(B) states that if conductors are adjusted in size to compensate for voltage drop, the equipment grounding conductor must be adjusted proportionately according to circular mil area. The following example further illustrates this:

A circuit utilizing a No. 1 THW copper conductor that is protected by a 100-ampere circuit breaker would use a No. 8 copper equipment grounding conductor (according to NEC Table 250.122). Due to voltage drop, the ungrounded conductor size must be increased to a 1/0 conductor. To now adjust the size of the equipment grounding conductor proportionately by circular mil area, the following calculation must take place (values of circular mil area are found in NEC Table 8):

Given
No. 1 conductor = 83,690 circular mil area
No. 8 conductor = 16,510 circular mil area
1/0 conductor = 105,600 circular mil area

$$\frac{\text{No. 1 ungrounded conductor}}{\text{No. 8 grounded conductor}} = \frac{83{,}690}{16{,}510} = 5.07$$

$$\frac{\text{No. 1/0 ungrounded conductor}}{\text{Answer from previous equation}} = \frac{105{,}600}{5.07} = 20{,}828$$

This calculation determines that the minimum size of equipment grounding conductor requires a 20,828 circular mil area. Referring again to NEC Table 8, it is noted that this does not correspond to a standard size conductor; therefore, the next larger conductor must be used. In this case, a No. 6 copper conductor must be used, which has a 26,240 circular mil area. This is what the NEC refers to when it says "adjusted proportionately according to circular mil area."

There is an important note just below NEC Table 250.122 that is very often overlooked. It states, "Where necessary to comply with Section 250.4(A)(5) or (B)(4), the equipment grounding conductor shall be sized larger than this table." Section 250.4(A)(5) states that grounding conductors "shall be capable of safely carrying the maximum ground-fault current likely to be imposed on it." Too many times, equipment grounding conductors are sized according to Table 250.122 without considering these other facts. In this case, the conductor may not be able to safely carry the fault current. As was seen earlier, at a given value of current, the conductor will fuse (melt), the ground is now lost, and the equipment case or enclosure would become energized, which would create a shock or electrocution hazard.

Use of Grounded Circuit Conductor for Grounding Equipment

NEC Section 250.24(A)(5) states that the grounded conductor (current-carrying neutral) and the grounding conductor (non-current-carrying) shall not be connected together on the load side of the service disconnecting means (see Fig. 5.15).

Figure 5.16 further illustrates this point.

The hazard of making this connection on the load side is that a parallel return path has now been established. The metallic raceway becomes a parallel return path with the grounded conductor. The metallic raceway will generally have higher impedance than the grounding conductor. As a result, the $I \times R$ values will be different in the parallel return paths. A different voltage rise will occur and thus a difference in potential will exist between the grounded conductor and the metallic raceway. A difference in potential may result in a shock hazard for anyone coming into contact with the metallic raceway. See Fig. 5.17 for more on this issue.

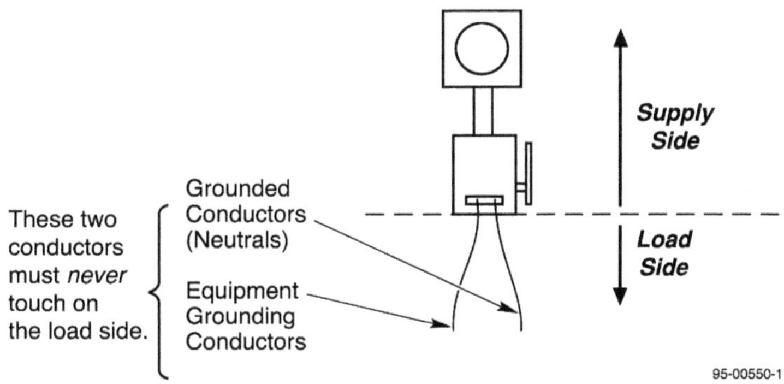

FIGURE 5.15 Neutral and ground to be separated on load side of service. (*Courtesy AVO Training Institute, Inc.*)

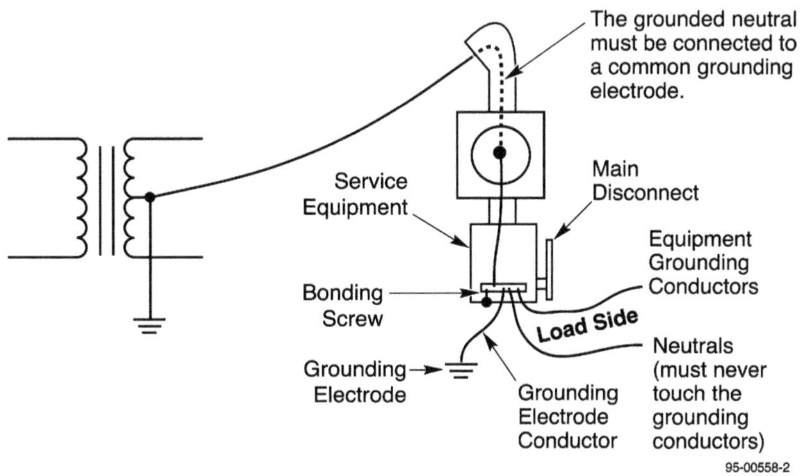

FIGURE 5.16 Grounding requirements at the service (supply side). (*Courtesy AVO Training Institute, Inc.*)

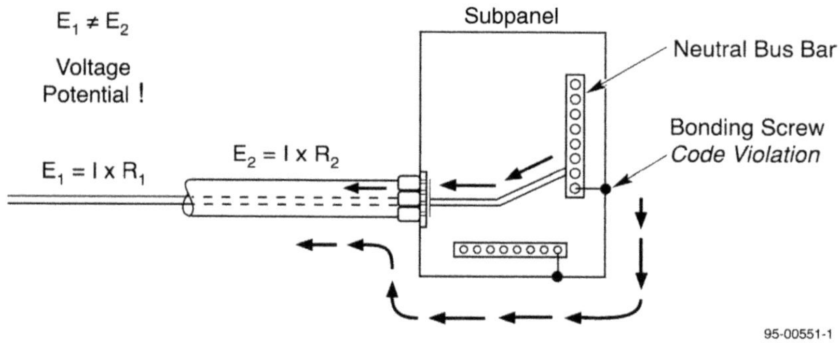

FIGURE 5.17 Ground loop. (*Courtesy AVO Training Institute, Inc.*)

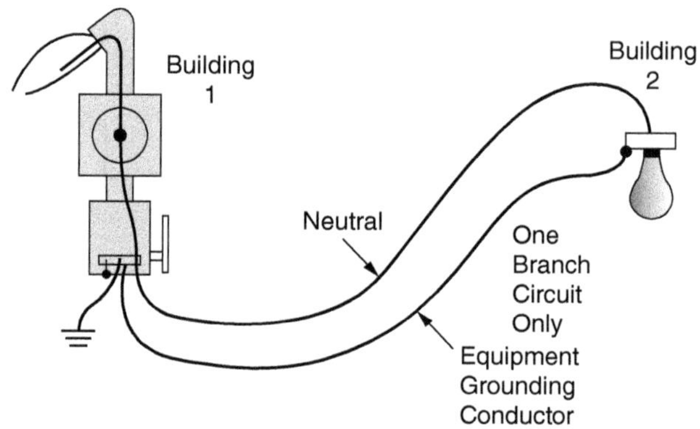

FIGURE 5.18 One branch circuit with equipment grounding conductor run, no electrode is required. (*Courtesy AVO Training Institute, Inc.*)

Another issue that must be addressed is proper grounding for two or more buildings where a common service is used. This issue deals with the grounding electrode, grounded (neutral) conductor, equipment grounding conductor, and bonding. NEC Section 250.50 requires the grounding electrodes at each building to be bonded together to form the grounding electrode system. Where no grounding electrode exists, one must be installed. There are, however, exceptions to this rule. Where the second building has only one branch circuit and an equipment grounding conductor is installed with the supply conductors, the grounding electrode at the main building is all that is required. Figure 5.18 illustrates this in simple form.

Section 250.32(B)(1) requires an equipment grounding conductor to be run to the separate building where there is equipment that is required to be grounded. In this case, the grounding and grounded conductors are prohibited from being connected together. The grounded conductor is considered to "float' in this case because it is not bonded to the equipment enclosure and grounding conductor. See Fig. 5.19 for further information on this.

Section 250.32(B)(1), Exception states another situation that can be used. Where the installation was in compliance with a previous code, and where there is no equipment grounding conductor run with the supply conductors, the buildings are not bonded together through metal raceways or other piping, and there is no ground-fault equipment installed at the common service, the grounded conductor from the main building must be connected to the grounding electrode and bonded to the disconnecting means at the second building. This type of situation is essentially the same as that for individual services at each building. Figure 5.20 illustrates this further.

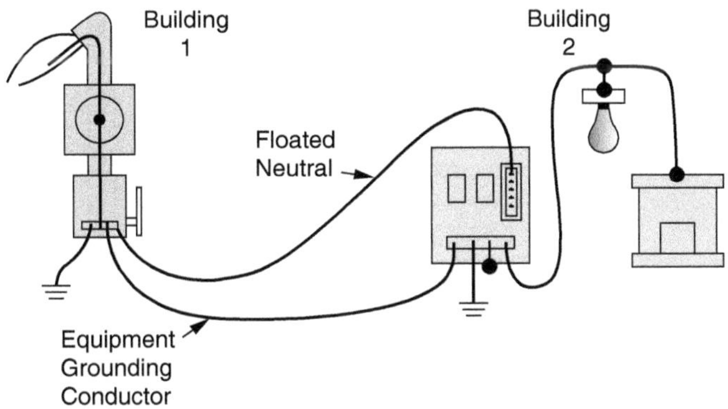

FIGURE 5.19 Equipment grounding conductor run to second building. (*Courtesy AVO Training Institute, Inc.*)

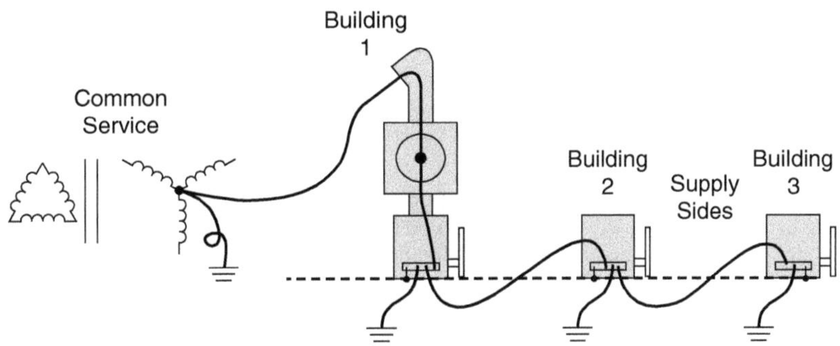

FIGURE 5.20 Common service with no equipment grounding conductor run. (*Courtesy AVO Training Institute, Inc.*)

Ferroresonance

Another very important safety issue associated with proper system grounding is ferroresonance. Ferroresonance is a rare phenomenon in power systems, but one that deserves mention. In response to a voltage transient, phase to ground fault, circuit breaker opening, equipment energization or de-energization, lightning-induced overvoltages, or any number of other sudden changes, the power system can take a sudden nonlinear jump to a sustained condition of severe harmonic distortion and high (several per-unit) overvoltages that can severely damage power system equipment. Ferroresonance is an electrical resonance between the nonlinear inductance of a transformer and system capacitance. Effective grounding decreases the chances of ferroresonance in three ways:

- Placing a very low impedance in parallel with the system capacitance.
- Fixing the electrical neutral of the system. Ferroresonance seems to require one point in the system whose potential is not fixed, like the floating neutral of an ungrounded system.
- Controlling and limiting system overvoltages that can initiate a ferroresonance.

For more information, consult any power system analysis text. An excellent reference is a technical paper by Jefferson Bronfeld, P.E., titled *Ferroresonance*. It is available at http://www.cadickcorp.com/tech004.htm. Another excellent reference by Philippe Ferracci is also titled *Ferroresonance* and can be found at http://www.schneider-electric.com/documents/technical-publications/en/shared/electrical-engineering/electrical-environmental-constraints/general-knowledge/ect190.pdf.

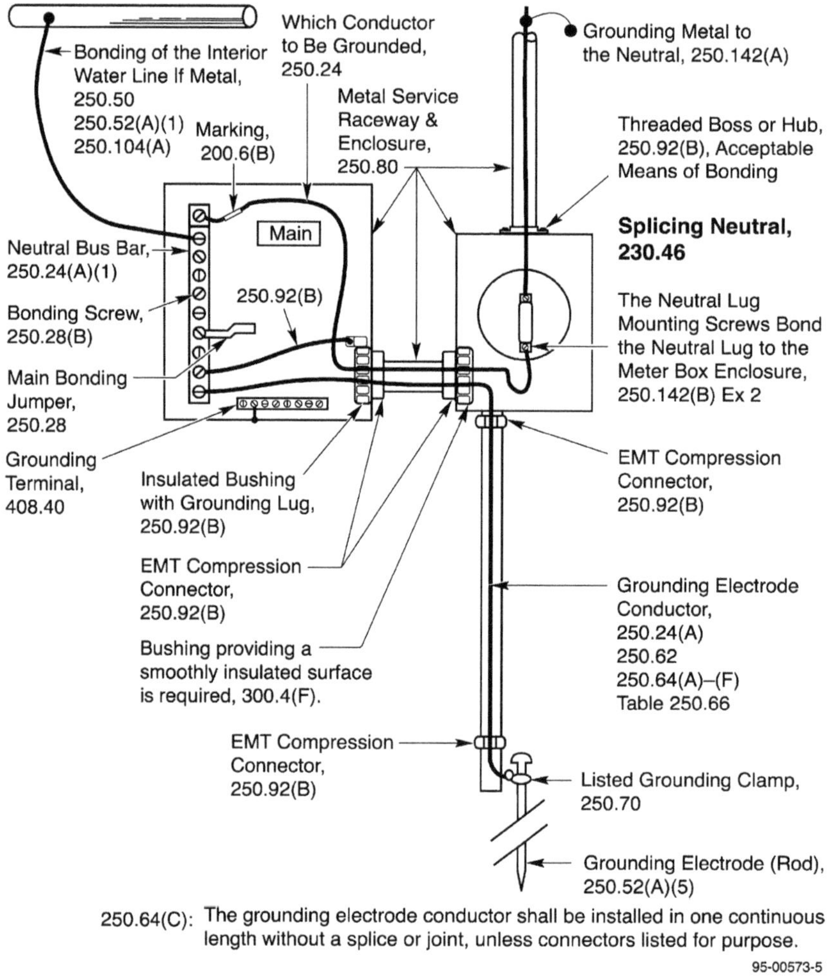

FIGURE 5.21 Service grounding and bonding requirements. (*Courtesy AVO Training Institute, Inc.*)

SUMMARY

Proper grounding is an essential part of electrical safety. Without proper grounding, the non-current-carrying metal components of electrical equipment have the risk of becoming energized. Design and install a grounding system based on the requirements of the National Electrical Code, the IEEE Std. 142 (Green Book), and IEEE Std. 80, as well as any other national standards that are relevant to the application.

Figure 5.21 provides an overall look at electrical system grounding requirements based on the current edition of the National Electrical Code.

Note: Don't take a shortcut *with grounding*; it may cost someone his or her life. That someone might be you.

CHAPTER 6
ELECTRICAL MAINTENANCE AND ITS RELATIONSHIP TO SAFETY

INTRODUCTION

All maintenance and testing of the electrical protective devices addressed in this chapter must be accomplished in accordance with the manufacturer's instructions. The current edition of the ANSI/NETA MTS, *Standard for Maintenance Testing Specifications for Electrical Power Equipment and Systems*[1] is an excellent source of information for performing the required maintenance and testing of these devices. Visit the NETA website (http://www.netaworld.org) for further information.

Because of the electrical hazards to which they may be exposed, all workers involved in electrical testing and maintenance must be qualified personnel as defined in the OSHA and NFPA standards. NFPA 70B, *Recommended Practice for Electrical Equipment Maintenance*, states that equipment deterioration is normal, but equipment failure is not inevitable. Normal deterioration begins as soon as new equipment is installed and if unchecked can eventually cause equipment failure. Deterioration can be accelerated by factors such as environment, overload, or severe duty cycles. A periodic maintenance program can identify these factors and provide measures for coping with them. Other potential causes of equipment failure can also be detected and corrected. Among these are load changes or additions, circuit alterations, improperly set or improperly selected protective devices, and changing voltage conditions.[2]

Without an effective maintenance and testing program, management assumes a greatly increased risk of a serious electrical failure and its consequences. An effective electrical preventive maintenance program can reduce accidents, save lives, and minimize costly breakdowns and unplanned shutdowns of production equipment. Failures can be averted by identifying and correcting impending troubles before they become major problems requiring more expensive, time-consuming solutions.[2]

THE SAFETY-RELATED CASE FOR ELECTRICAL MAINTENANCE

Overview

The relationship between safety and preventive maintenance is not a difficult one to establish. Properly designed equipment that is properly installed is well capable of doing its job when it is new. As equipment ages, however, several factors begin to take their toll on electrical equipment.

- Dust, dirt, and other contaminants collect on equipment, causing the equipment to overheat and bearings and other moving parts to bind.
- Vibration causes hardware to loosen. Subsequent operations of equipment can cause joints and equipment to fail explosively.
- Heat and age can cause insulation to fail, resulting in shock hazards to personnel.

- Increased loads, motor-starting surges, and power quality issues such as harmonics combine to increase the aging process and set the stage for equipment failure.
- Electronic components can drift or fail, causing unstable operation of protective equipment.

Unfortunately, the ultimate failure of unmaintained equipment usually occurs when the equipment is needed the most—during electrical faults. Such failures result in arc flash and blast events that can and do injure or kill workers in the area. They also result in significant downtime, loss of equipment, and construction cost incurred in rebuilding the equipment.

The only way to ensure that electrical equipment continues to operate in an optimal manner is to maintain it so that it stays in factory-new operating condition.

Regulatory

As discussed above and in previous chapters, the catastrophic failure of electrical equipment creates severe hazards for personnel working in the area. Recognizing this, the *Standard for Electrical Safety in the Workplace* (NFPA 70E)[3] requires that electrical equipment be properly maintained to minimize the possibility of failure.

Relationship of Improperly Maintained Electrical Equipment to the Hazards of Electricity

Improperly maintained equipment may expose workers to any of the electrical hazards. For example:

1. Improperly maintained tools or flexible cord sets (extension cords) can have frayed insulation that exposes the energized conductors and allows them to contact the worker or the metallic tool the worker is using. The result is an electric shock.
2. Improperly maintained protective devices, such as circuit breakers, can fail when interrupting an overcurrent. Such a failure could be explosive; consequently, the worker is exposed to electrical arc flash and electrical arc blast.
3. Improperly maintained connections can overheat, resulting in any of the following:
   a. Melted insulation, exposed conductors, and the attendant electrical shock hazard
   b. Fire
   c. Failed connections resulting in electrical arc flash and arc blast
4. Improperly maintained switchgear, motor control centers, or panelboards can fail explosively when an arc occurs internally. This exposes workers to the effects of electrical arc flash and possibly electrical arc blast.

Maintenance and the Potential Impact on an Electrical Arc-Flash

Maintenance of the circuit-protective devices can have a significant impact on the incident energy of an arc-flash. NFPA 70E, Section 130.5(B) states that in estimating the "likelihood of occurrence of injury or damage to health and the potential severity of injury or damage to health shall take into consideration the design of the electrical equipment, including its overcurrent protective device and its operating time, and the electrical equipment operating condition and condition of maintenance." Section 130.5(G) further states that the incident energy analysis method is required to "take into consideration the characteristics of the overcurrent protective device and its fault clearing time, including its condition of maintenance." It should also be noted that improper or inadequate maintenance can result in increased opening time of the overcurrent protective device, thus increasing the incident energy of an arc flash. Maintenance must be a major consideration in an arc flash risk assessment because a major component in determining the likelihood and severity of the occurrence, as well as the arc flash incident energy, is the clearing time of the overcurrent protective device. When the arc flash incident energy analysis is performed, the ratings and settings of the overcurrent protective devices are used, based on what they are supposed to do. The following example is provided to further validate the NFPA 70E requirement for maintenance.

Example:
In this example, we use a low-voltage power circuit breaker that has not been operated or maintained for several years. The lubrication has become sticky or hardened so the circuit breaker could take several additional cycles, seconds, minutes, or longer to clear a fault condition. Two arc-flash hazard analyses will be performed using a 20,000-amp short circuit with the worker 18 in from the arc: (1) Based on what the system is supposed to do, we will use a clearing time of 0.083 second (5 cycles), and (2) due to lack of maintenance, and the resultant sticky mechanism the circuit breaker now has an unintentional time delay and a clearing time of 0.5 second (30 cycles). The following NFPA 70E equation will be used to calculate both conditions: $E_{MB} = 1038.7 D_B^{-1.4738} t_A (0.0093F^2 - 0.3453F + 5.9675)$.

Calculation #1:

E_{MB} = maximum 20-in cubic box incident energy, cal/cm²

D_B = distance from arc electrodes, inches (for working distances 18 in and greater)

t_A = arc duration, seconds

F = short-circuit current, kA (for the range of 16 kA to 50 kA)

(1) $D_B = 18$ in

(2) $t_A = 0.083$ second (5 cycles)

(3) $F = 20$ kA

$$\begin{aligned} E_{MB} &= 1038.7 D_B^{-1.4738} t_A (0.0093F^2 - 0.3453F + 5.9675) \\ &= 1038 \times 0.0141 \times 0.083(0.0093 \times 400 - 0.3453 \times 20 + 5.9675) \\ &= 1.4636 \times 2.7815 \\ &= \mathbf{3.5\ cal/cm^2} \end{aligned}$$

According to NFPA 70E, the required arc-rated clothing and other PPE to be selected is based on this incident energy exposure level. Therefore, the arc-rated clothing and PPE must have an arc rating of at least 3.5 cal/cm².

Calculation #2:

E_{MB} = maximum 20-in cubic box incident energy, cal/cm²

D_B = distance from arc electrodes, inches (for working distances 18 in and greater)

t_A = arc duration, seconds

F = short-circuit current, kA (for the range of 16 kA to 50 kA)

(1) $D_B = 18$ in

(2) $t_A = 0.5$ second (30 cycles)

(3) $F = 20$ kA

$$\begin{aligned} E_{MB} &= 1038.7 D_B^{-1.4738} t_A (0.0093F^2 - 0.3453F + 5.9675) \\ &= 1038 \times 0.0141 \times 0.5(0.0093 \times 400 - 0.3453 \times 20 + 5.9675) \\ &= 7.3179 \times 2.7815 \\ &= \mathbf{20.4\ cal/cm^2} \end{aligned}$$

According to NFPA 70E, the required arc-rated FR clothing and other PPE to be selected is based on this incident energy exposure level. Therefore, the arc-rated FR clothing and PPE must have an arc rating of at least 20.4 cal/cm².

Conclusion:
If the worker is protected based on what the system is supposed to do (0.083 second or 5 cycles) and an unintentional time delay occurs (0.5 second or 30 cycles), the worker could be seriously injured or killed because he or she was under protected.

As can be seen, maintenance of overcurrent protective devices is extremely important to an electrical safety program. Maintenance must be performed according to the manufacturer's instructions, or industry consensus standards, to minimize the risk of having an unintentional time delay in the operation of the overcurrent protective devices.

Hazards Associated with Electrical Maintenance

Any time workers approach electrical equipment, they might face one or more of the electrical hazards. This includes activities associated with electrical maintenance. Both the *National Electrical Code* (NFPA 70)[4] and the *Standard for Electrical Safety in the Workplace* (NFPA 70E)[3] require that:

> Electrical equipment, such as switchboards, switchgear, panelboards, industrial control panels, meter socket enclosures, and motor control centers, that is in other than dwelling units, and is likely to require examination, adjustment, servicing, or maintenance while energized, shall be field or factory marked to warn qualified persons of potential electric arc flash hazards. The marking shall meet the requirements in 110.21(B) and shall be located so as to be clearly visible to qualified persons before examination, adjustment, servicing, or maintenance of the equipment.[3,4]

Figure 6.1 illustrates an example of the minimum labeling requirements, and Fig. 6.2 shows a preferred label for the application. Refer to Chap. 4 for details on the methods for determining the extent of the hazards.

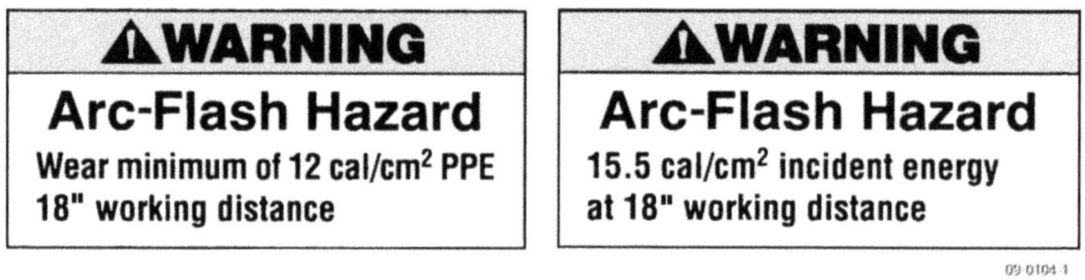

FIGURE 6.1 Minimum required warning label for posting on electrical equipment to warn of hazards.

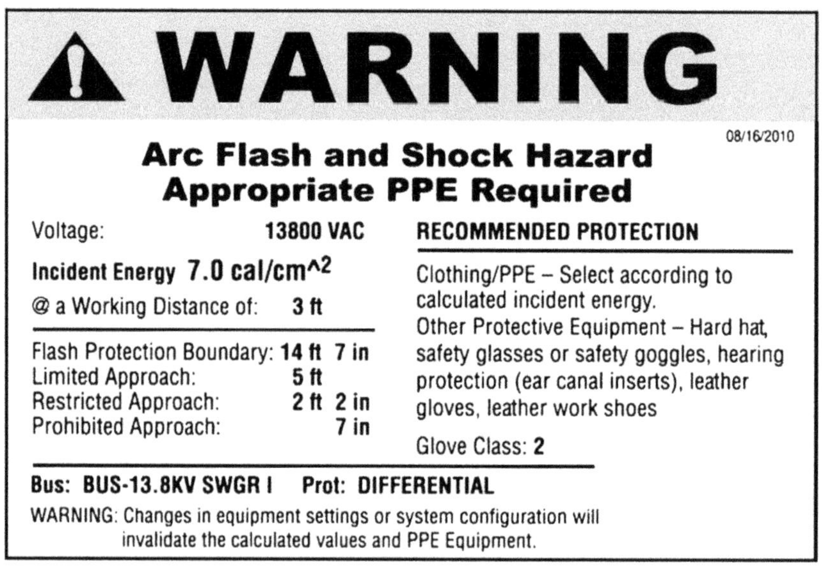

FIGURE 6.2 Preferred required warning label for posting on electrical equipment to warn of hazards.

THE ECONOMIC CASE FOR ELECTRICAL MAINTENANCE

Although a complete discussion of the economics of an electrical maintenance program is beyond the scope of the handbook, Fig. 6.3, taken from NFPA 70B *Recommended Practice for Electrical Equipment Maintenance*,[2] provides a clear definition of the economic importance of such maintenance.

This diagram charts the relationship between costs and maintenance. As can be seen, the cost of electrical preventive maintenance (EPM) decreases as the interval between inspections increases. In other words, the cost of maintenance decreases as less maintenance is done. The cost of equipment repair and replacement, on the other hand, increases as less maintenance is done.

Unfortunately, many managers focus only on the cost of maintenance without taking the costs of equipment repair and replacement into consideration. The correct way to approach the economics of a maintenance program is to look at the total cost as shown in the figure. Little or no maintenance (right-hand side of the figure) results in very high repair and replacement costs. As the electrical maintenance program is intensified (moving to the left on the figure), repair and replacement costs decrease.

At some point, the minimum total cost is realized. This is the point at which the maintenance program should be operated to provide for the best overall efficiency of the system.

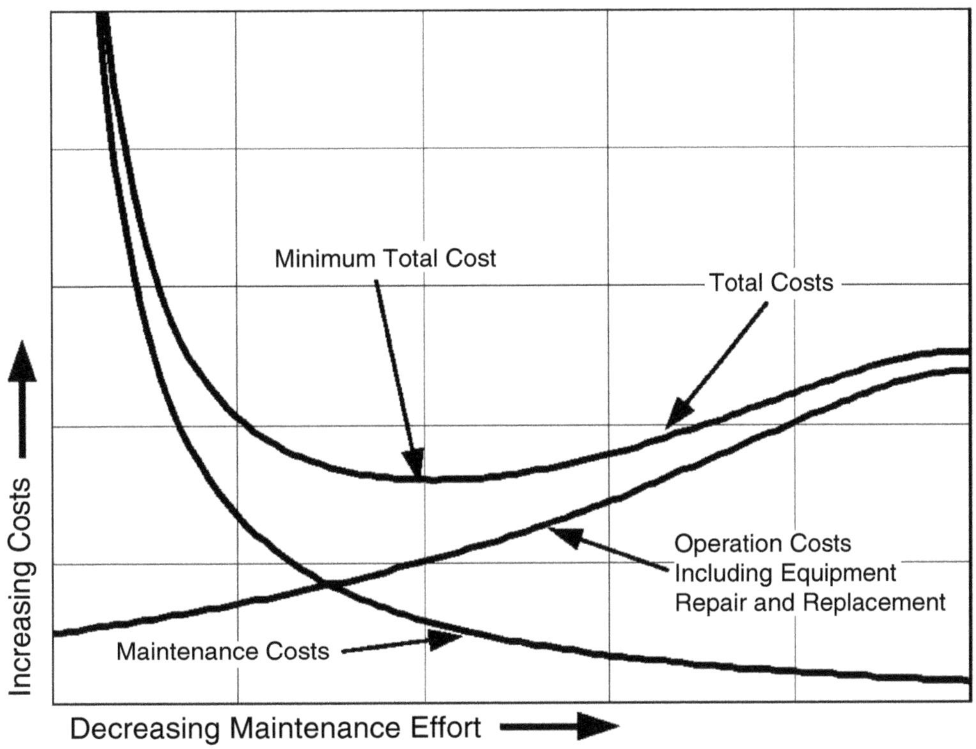

FIGURE 6.3 Effect of electrical maintenance frequency on overall costs. (*Courtesy National Fire Protection Association.*)

RELIABILITY-CENTERED MAINTENANCE (RCM)

(The following paragraphs are taken from *RCM White Paper* by Al Rose, © Cadick Corporation, 2001–2011.)

What Is Reliability-Centered Maintenance?

Reliability-centered maintenance (RCM) is the process of determining the most effective maintenance approach. The RCM philosophy employs preventive maintenance (PM), predictive maintenance (PdM), real-time monitoring (RTM), run-to-failure (RTF; also called reactive maintenance), and proactive maintenance techniques in an integrated manner to increase the probability that a machine or component will function in the required manner over its design life cycle with a minimum of maintenance. The goal of the philosophy is to provide the stated function of the facility, with the required reliability and availability at the lowest cost. RCM requires that maintenance decisions be based on maintenance requirements supported by sound, technical, and economic justification.

A Brief History of RCM

RCM originated in the airline industry in the 1960s. By the late 1950s, the cost of maintenance activities in this industry had become high enough to warrant a special investigation into the effectiveness of those activities. Accordingly, in 1960, a task force was formed consisting of representatives of both the airlines and the Federal Aviation Administration (FAA) to investigate the capabilities of preventive maintenance. The establishment of this task force subsequently led to the development of a series of guidelines for airlines and aircraft manufacturers to use when establishing maintenance schedules for their aircraft.

This led to the 747 Maintenance Steering Group (MSG) document MSG-1, Handbook: Maintenance Evaluation and Program Development from the Air Transport Association in 1968. MSG-1 was used to develop the maintenance program for the Boeing 747 aircraft, the first maintenance program to apply RCM concepts. MSG-2, the next revision, was used to develop the maintenance programs for the Lockheed L-1011 and the Douglas DC-10. The success of this program is demonstrated by comparing maintenance requirements of a DC-8 aircraft, maintained using standard maintenance techniques, and the DC-10 aircraft, maintained using MSG-2 guidelines. The DC-8 aircraft has 339 items that require an overhaul, versus only seven items on a DC-10. Using another example, the original Boeing 747 required 66,000 labor hours on major structural inspections before a major heavy inspection at 20,000 operating hours. In comparison, using standard maintenance programs of the day, the DC-8, a smaller and less sophisticated aircraft, required more than 4 million labor hours before reaching 20,000 operating hours.

In 1974, the U.S. Department of Defense commissioned United Airlines to write a report on the processes used in the civil aviation industry for the development of maintenance programs for aircraft. This report, written by Stan Nowlan and Howard Heap and published in 1978, was titled *Reliability-Centered Maintenance*[5] and has become the report upon which all subsequent reliability-centered maintenance approaches have been based. What Nowlan and Heap found was that many types of failures could not be prevented no matter how intensive the maintenance activities were. Additionally it was discovered that for many items the probability of failure did not increase with age. Consequently, a maintenance program based on age will have little, if any, effect on the failure rate.

RCM in the Industrial and Utility Arena

As with any philosophy, there are many paths or processes that lead to a final goal. This is especially true for RCM where the consequences of failure can vary dramatically.

Rigorous RCM analysis has been used extensively by the aircraft, space, defense, and nuclear industries where functional failures have the potential to result in large losses of life, national security implications, and/or extreme environmental impact. A rigorous RCM analysis is based on a detailed Failure Modes and Effects Analysis (FMEA) and includes probabilities of failure and system reliability calculations. The analysis is used to determine appropriate maintenance tasks to address each of the identified failure modes and their consequences.

While this process is appropriate for these industries, it is not necessarily the most practical or best approach to use for industrial and utility systems maintenance. For these systems, a streamlined or intuitive RCM analysis process may be more appropriate. This is due to the high analysis cost of the rigorous approach, the relative low impact of failure of most industrial systems, the type of systems and components maintained, and the amount of redundant systems in place. The streamlined approach uses the same principles as the rigorous but recognizes that not all failure modes will be analyzed. For most industrial and utility systems, the most economical and efficient approach is to use a combination of rigorous (formal) and intuitive analysis depending on system criticality and failure impact. Failure modes that result in high costs or personnel injury, or where the resultant reliability is still unacceptable in terms of safety or operational impact, still receive the rigorous approach, but all other failure modes will use intuitive analysis.

The Primary RCM Principles

1. **RCM Is Function Oriented**—It seeks to preserve system or equipment function, not just operability for operability's sake. Redundancy of function through multiple equipment improves functional reliability but increases life-cycle cost in terms of procurement and operating costs.
2. **RCM Is System Focused**—It is more concerned with maintaining system function than individual component function.
3. **RCM Is Reliability Centered**—It treats failure statistics in an actuarial manner. The relationship between operating age and the failures experienced is important. RCM is not overly concerned with simple failure rate; it seeks to know the conditional probability of failure at specific ages (the probability that failure will occur in each given operating age bracket).
4. **RCM Acknowledges Design Limitations**—Its objective is to maintain the inherent reliability of the equipment design, recognizing that changes in inherent reliability are the province of design rather than maintenance. Maintenance can, at best, only achieve and maintain the level of reliability for equipment, which is provided for by design. However, RCM recognizes that maintenance feedback can improve on the original design. In addition, RCM recognizes that a difference often exists between the perceived design life and the intrinsic or actual design life, and addresses this through the age exploration (AE) process.
5. **RCM Is Driven by Safety and Economics**—Safety must be ensured at any cost; thereafter, cost-effectiveness becomes the criterion.
6. **RCM Defines Failure as Any Unsatisfactory Condition**—Therefore, failure may be either a loss of function (operation ceases) or a loss of acceptable quality (operation continues).
7. **RCM Uses a Logic Tree to Screen Maintenance Tasks**—This provides a consistent approach to the maintenance of all kinds of equipment. See Fig. 6.4.
8. **RCM Tasks Must Be Applicable**—The tasks must address the failure mode and consider the failure mode characteristics.
9. **RCM Tasks Must Be Effective**—The tasks must reduce the probability of failure and be cost-effective.
10. **RCM Acknowledges Three Types of Maintenance Tasks**
    a. Time-directed (PM)—Scheduled when appropriate.
    b. Condition-directed (PdM and real-time monitoring)—Performed when conditions indicate they are needed.
    c. Failure finding (one of several aspects of proactive maintenance)—Equipment is run-to-failure. This is acceptable for some situations and some types of equipment.
11. **RCM Is a Living System**—It gathers data from the results achieved and feeds this data back to improve design and future maintenance. This feedback is an important part of the proactive maintenance element of the RCM program.

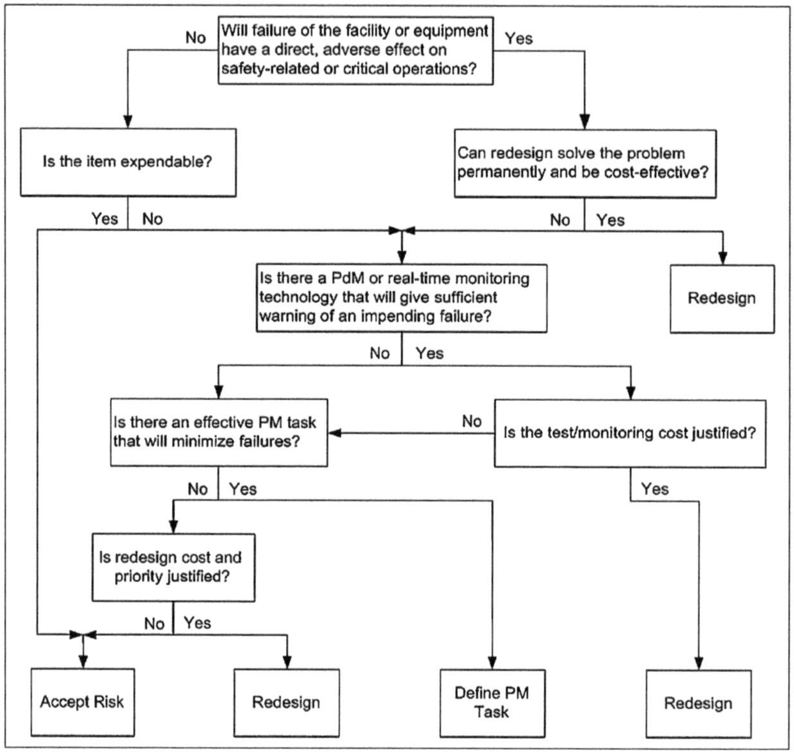

FIGURE 6.4 RCM logic tree.

The RCM analysis carefully considers the following questions:

1. What does the system or equipment do; what is its function?
2. What functional failures are likely to occur?
3. What are the likely consequences of these functional failures?
4. What can be done to reduce the probability of the failure, identify the onset of failure, or reduce the consequences of the failure?

Answers to these four questions determine the actions required to maintain the systems or equipment. Figure 6.4 shows the RCM logic tree used to answer these questions.

Failure

Failure can be defined in many ways. In a broad sense, failure is simply an unsatisfactory condition. RCM, however, requires us to look at failure from not just an equipment standpoint, but from a system standpoint as well. A piece of equipment can be operating (an HVAC unit in a clean room, for example), but if its output is less than required, it would be considered failed. On the other hand, a protective relay in a power system can have failed, but if it has not taken the circuit off-line (the circuit is still energized), the function of the system is unchanged. Essentially, defining failure depends on the function of an item or system and the operating context in which the item or system is used.

Functional Failure. A functional failure is essentially the inability of an item or system to meet its specified performance standard. A complete loss of function is a functional failure; however, in the HVAC example above, if the system's output is less than specified, a functional failure has occurred even if the system is still operating.

Consequences of Failure. The consequences of failure determine the priority of maintenance activities or design improvement required to prevent occurrence. If failure of an item results in little or no consequence, mimimal maintenance activities are generally required. If, however, failure of an item results in a large economic hardship, personnel injury, or environmental damage, maximum maintenance activities or a redesign may be called for.

Maintenance Actions in an RCM Program

RCM goals are to identify the most cost-effective and applicable maintenance techniques to minimize the risk and impact of failure in facility and utility equipment and systems. This allows systems and equipment functionality to be maintained in the most economical manner. Specific RCM objectives as stated by Nowlan and Heap[5] are:

- To ensure realization of the inherent safety and reliability levels of the equipment.
- To restore the equipment to these inherent levels when deterioration occurs.
- To obtain the information necessary for design improvement of those items whose inherent reliability proves to be inadequate.
- To accomplish these goals at a minimum total cost, including maintenance costs, support costs, and economic consequences of operational failures.

To this end, there are four outcomes from an RCM analysis:

1. **Perform No Maintenance**—This is referred to as reactive maintenance, repair, fix-on-fail, or run-to-fail (RTF). This type of maintenance assumes that failure is equally likely to occur in any part, and that a failure is not detrimental to the operation. When this is the sole type of maintenance practiced, high failure rates, large parts inventories, and excessive amounts of overtime become common. A purely RTF maintenance program ignores many of the opportunities to influence equipment survivability.
2. **Perform Preventive Maintenance (PM)**—PM consists of regularly scheduled inspections, adjustments, cleanings, lubrication, and replacement of components and equipment. PM is also referred to as time-driven or interval-based maintenance. It is performed without regard to equipment condition. PM schedules inspection and maintenance at predefined intervals in an attempt to reduce equipment failures. However, as Nowlan and Heap discovered, a PM program can result in a significant increase in inspections and cost without any increase in reliability.
3. **Perform Condition-Based Maintenance (CBM)**—CBM consists of predictive maintenance (PdM) and real-time monitoring. PdM primarily uses nonintrusivee testing techniques to measure and trend equipment performance. Real-time monitoring uses current performance data to asses machinery condition. CBM replaces arbitrarily timed maintenance tasks with maintenance that is scheduled only when warranted by equipment condition. Continuing analysis of equipment condition data allows for planning and scheduling maintenance activities or repairs prior to functional or catastrophic failure.
4. **Redesign**—When failure of a system or piece of equipment is an unacceptable risk and none of the above tasks can help mitigate the failure, a redesign of the equipment or system is in order. In most cases, adding redundancy eliminates the risk and adds very little to overall maintenance costs.

An RCM program will include all four of the above analysis outcomes, as no one is more important than another. Figure 6.5 shows the relationship that maintenance activities have to RCM.

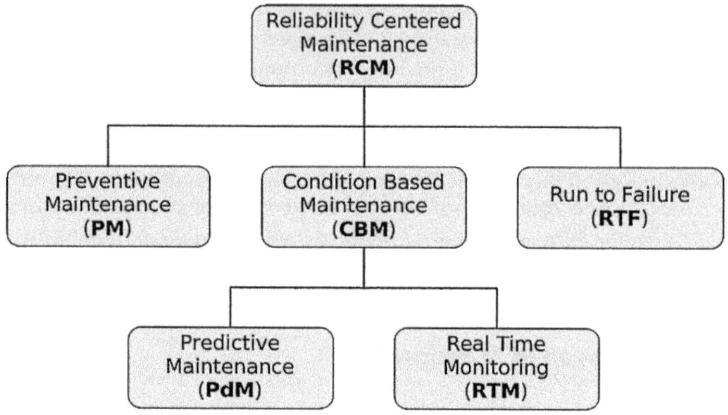

FIGURE 6.5 RCM structure.

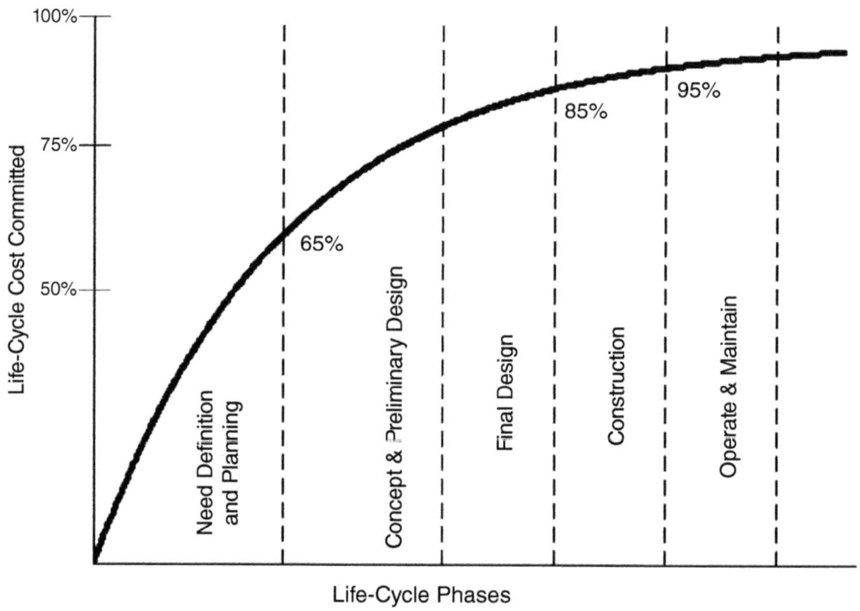

FIGURE 6.6 Life-cycle phases. *Source:* Blanchard, B.S., *Design and Manage to Life Cycle Cost,* Forest Grove, OR, MA Press, 1978.

Impact of RCM on a Facilities Life Cycle

A facilities life cycle is often divided into two broad stages: acquisition (planning, design, and construction) and operations. RCM affects all phases of the acquisition and operations stages to some degree.

Decisions made early in the acquisition cycle profoundly affect the life-cycle cost of a facility. Even though expenditures for plant and equipment may occur later during the acquisition process, their cost is committed at an early stage. As shown conceptually in Fig. 6.6, planning (including conceptual design) fixes two-thirds of the facility's overall life-cycle costs. The subsequent design phases determine an additional 29 percent of the life-cycle cost, leaving only about 5 percent of the life-cycle cost that can be impacted by the later phases.

Thus, the decision to include a facility in an RCM program, including condition monitoring, which will have a major impact on its life-cycle cost, is best made during the planning phase. As RCM decisions are made later in the life cycle, it becomes more difficult to achieve the maximum possible benefit from the RCM program.

Even though maintenance is a relatively small portion of the overall life-cycle cost, typically 3 to 5 percent of a facility's operating cost, RCM is still capable of introducing significant savings during the O&M phase of the facility's life. Savings of 30 to 50 percent in the annual maintenance budget are often obtained through the introduction of a balanced RCM program.

Conclusion

Reliability-centered maintenance is the process of determining the most effective maintenance approach. As the airline industry showed 40 years ago, RCM can not only improve the reliability of a system, it can significantly reduce the required maintenance. In today's competitive world economy, that translates into monies saved, both from reduced failures and reduced work. And a properly implemented RCM program will continue saving money year after year. To remain competitive, most companies can no longer afford to remain on the sidelines doing things like they've always done.

THE EIGHT-STEP MAINTENANCE PROGRAM

Some facilities, especially very small ones, may not want to embark on such an ambitious process as a complete RCM program. For such locations, a minimal program such as the one described below may be more desirable. This information in the following paragraphs is taken from *The Eight Step Maintenance Program* by John Cadick.[6]

Introduction

There are eight simple steps that should be at the heart of an electrical maintenance program. These key steps are plan, inspect, clean, tighten, lubricate, test, record, and evaluate. The following paragraphs briefly describe each step and illustrate the part that they play in a good preventive maintenance program.

Step 1—Plan

Before working on anything, take time to think about what you are going to do and how you are going to do it. A few minutes of thought before the job is started results in a safer and more efficient job. This step is especially important when you are performing emergency repair work.

Start by identifying the specific task or tasks that you intend to perform. In this initial stage, you do not need to go into too much detail; rather, concentrate on the broad goals that you need to accomplish. List each goal, the personnel that you will use, the equipment that you will need, and the amount of time that will be required.

Be certain that you obtain and use the proper instruction manuals and design drawings before you start. The joke of "getting the manual after smoke is present" has too much basis in fact to be very funny for experienced electricians. Troubleshooting without accurate and correct schematics and wiring diagrams is close to impossible in all but the simplest circuits.

Develop detailed plans and procedures from the preliminary plans discussed above. Steps should be detailed as much as required by the complexity of the job and the experience of your personnel.

Always start any procedure with a job briefing as described in Chap. 4.

Step 2—Inspect

Periodic inspections should be set up on a routine basis. Inspection can be done with the senses that nature gave us, called unaided inspection, or with instruments such as infrared viewers, ultrasound transducers, and so on.

Unaided Inspection. Your eyes are your most important inspection tools. They can detect dirt, note current readings, determine levels, discover discolored, overheated insulation, determine the presence of insects and rodents, and find a whole host of other problems. The more experienced you become, the more information your eyes can provide; however, even the most inexperienced electrician can detect spider webs in a circuit breaker operating mechanism. Always observe what you are working on and look for any abnormalities that can give you an idea of anything that can cause these problems.

Your ears can also be useful inspection tools. Overloaded transformers are normally noisier than lightly loaded ones, corona has a distinctive hissing sound, and motors with unbalanced voltages or bad bearings vibrate. Any change in sound should be investigated.

Nothing in electrical systems is more distinctive than the odor of overheated or burning insulation. Your nose can detect such problems long before your other senses. Again, the most useful indicator of problems is a change in odor.

Touch may also be employed to inspect electrical equipment. Feeling for excessive vibration or heating is a time-tested technique. Of course, you should be extremely cautious before touching any piece of equipment. If the equipment is not de-energized, contact should be made only when wearing appropriate insulating PPE, such as rubber gloves.

Inspection with Tools or Instruments. Except for your own senses, infrared viewers are the most cost-effective maintenance tool in existence. Annual infrared scans of your power system invariably uncover problems that, if allowed to continue, could result in severe problems. Whether this service is done "in house" or contracted to an outside service, you *should* perform an annual (minimum) infrared scan of your system.

Ultrasound transducers are normally used on major overhauls to evaluate the condition of metals that are subjected to constant vibration.

Many other such tools and instruments are available to help in the performance of maintenance tasks. Such tools should be used as required by qualified personnel trained in their proper and safe use.

Step 3—Clean

Dirt and electricity don't mix. After you have completed the inspection of your electrical equipment, the next step is to clean all components.

The first step is general cleanliness. Dust, dirt, and foreign material must be cleaned from motors, switchgear, transformers, and other such electrical equipment. For such cleaning, we recommend the use of high suction; insulated hose shop vacuum, clean, lint-free wiping cloth; and general-purpose solvents. A citrus-based general-purpose solvent is effective and biodegradable.

Each insulator or conductor has its own way to be cleaned. Porcelain, for example, can be cleaned with soap, water, and appropriate cleaning brushes or pads. Plastic insulation, on the other hand, cannot take any abrasion. Always refer to the manufacturer's recommendations for specific techniques. Also, always avoid using steel wool or other conductive abrasives in any area where contact with energized conductors is possible.

Conductor connecting surfaces need special attention. Contact surfaces of copper conductors should be cleaned to a bright finish where they are connected together. Aluminum must be cleaned and a protective lubricant must be applied immediately after the cleaning. The lubricant cuts through any aluminum oxide (a good insulator) that may be present and prevents the formation of additional aluminum oxide. (See the section on lubrication.)

Refer to the manufacturer's instruction books, which sometimes specify the cleaning material to be used for its equipment. Be sure to consult this literature before attempting to clean electrical components.

Step 4—Tighten

Tightening fasteners on electrical equipment takes special care and should be done to the torque recommended by the equipment manufacturer. Electrical connections are especially critical. Recommended torque values can be found in the maintenance and acceptance testing specifications published by the InterNational Electrical Testing Association.[1] Be certain to torque connections per this table and/or the manufacturer's recommendations.

Step 5—Lubricate

Lubrication of electrical system components is an often-overlooked or improperly performed procedure. Too much lubricant can be worse than no lubricant at all. This is especially true of devices such as protective relays and circuit breakers, which may go for years without operating. Be especially careful of conductive lubricants to make sure they do not cause a short circuit. Always use the lubricants recommended by the equipment manufacturer or one that has the same characteristics. Lubrication falls into two important areas:

1. Nonconductors

   Nonconductive joints or moving members should be lubricated using an appropriate material. In some cases, grease, oil, or some synthetic lubricant may be recommended. Motor bearings, mechanism pivots, and other such equipment fall into this category. Remember that some pieces of equipment, such as protective relays, do not require any lubricant at all.
2. Conductors

   Most manufacturers make and/or recommend a lubricant to be used on conductors, stabs, and other such equipment. The ubiquitous "black grease" has been and continues to be a good lubricant when used properly. The correct approach to lubrication of electrical conductors is to apply a small amount and then wipe the contact surface clean with a clean, lint-free cloth. Do not *scrub* the surface, however. You wish to leave a fine film of lubricant in place.

   Remember: ***Use the proper lubricants in the proper amounts.***

Step 6—Test

Electrical equipment should be tested periodically. Insulation resistance should be measured, breaker trip times should be checked, relays should be calibrated, and a variety of other procedures should be performed. The only exercise that many protective devices get is during the test interval.

A detailed review of all procedures is beyond the scope of this course; however, the recommendations made by the InterNational Electrical Testing Association (NETA) in its publication titled *Maintenance Testing Specifications*[1] should be reviewed and applied as appropriate.

Step 7—Record

Doing work without keeping records is almost useless. Recording test and other data can help set maintenance intervals, isolate troublesome equipment (or manufacturers), and provide a baseline that helps to indicate when insulation or other components are starting to fail. Besides test results, records should always include the date, equipment identification, and all pertinent data on any problems discovered. System conditions like load current, voltage, temperature, and other such information can help determine how close components are to being overloaded.

Test and record forms should be developed and tailored to your company's specific needs. Examples of such forms may be found at the back of the NFPA 70B *Electrical Equipment Maintenance*. Review that document for ideas and examples.

Step 8—Evaluate

Evaluation of the test results is, possibly, the single most important step in the entire process. Since 1990, significant progress has been made in the statistical analysis of small population data such as that gathered during maintenance intervals. Whether the analysis is done using sophisticated mathematical techniques or simple field rules of thumb, no program is complete without it. Analysis of records allows the company to determine what, if any, additional maintenance needs to be performed and to pinpoint problems that might otherwise go unobserved.

Summary

The organization of an electrical preventive maintenance program may be greatly enhanced by adhering to the simple, eight-step program outlined in this chapter. These steps contain all of the elements required for a comprehensive program. If you *plan, inspect, clean, tighten, lubricate, test, record, and evaluate,* you will be well on your way to a safe, efficient, and profitable preventive maintenance program.

FREQUENCY OF MAINTENANCE

How often should a power system be maintained? If your facility is using a preventive maintenance (PM) program as described in the RCM section and shown in Fig. 6.5, you are basing your intervals on the calendar. For example, you might maintain your large transformers every July, you split your circuit breaker maintenance between May and October, and you maintain one-third of your protective relays once per year.

Referral to Fig. 6.3 shows the problem with that approach. You may be spending way too much money on your maintenance. Moreover, if you do not keep good records, you really do not know if you are maintaining often enough to prevent catastrophic failure.

For many years, maintenance intervals were determined as shown below.

Determining Testing Intervals

1. Start initially with a once-per-year test program. Continue this procedure for the first two intervals.
2. Review the test records from your maintenance intervals.
3. If frequent and/or severe problems are evident, decrease the maintenance intervals.
4. If no major or frequent problems are evident, increase the maintenance intervals.
5. Repeat steps 2 through 5 throughout the life of your maintenance program.

The average maintenance interval in industrial systems is about two years. This means that every two years, most industrial plants completely maintain their electrical system. Electrical utilities have similar intervals, although their average is probably closer to $1^1/_2$ years. Recent advances in statistical analysis have enabled maintenance programs based on technology rather than "estimate." Such *condition-based maintenance* programs will be the way maintenance is done in the 21st century.

CONDITION-BASED MAINTENANCE (CBM)

The condition-based maintenance material is extracted largely from *Condition Based Maintenance: A White Paper Review of CBM Analysis Techniques*, by John Cadick, P.E., and Gabrielle Traugott, © 2010 by Cadick Corporation. A copy of the paper is available at http://www.cadickcorp.com.

Introduction

As discussed earlier in this chapter, condition-based maintenance (CBM) is one of the key building blocks in a reliability-centered maintenance (RCM) program. In a CBM environment, maintenance efforts and expenditures are based on the actual condition of the maintained equipment; that is, equipment that is consistently in good to superior condition does not need to be maintained as frequently as equipment that is deteriorating or has reached an age where deterioration is anticipated. The use of test or running data to statistically model and predict the future condition of the equipment is at the very heart of CBM.

Consider the effect that such an approach would have on the task of changing lubricating oil in an automobile. Car owners know that lubricating oil should be changed every few thousand miles. For example, the car manufacturer may specify that oil should be changed every 3000 miles or every six months, whichever comes first. Such a procedure is purely time (or mileage) based and does not take into consideration the condition of the lubricating oil.

If a CBM approach were employed to this task, chemical tests would be performed on the oil. If the tests predict that the oil has another 1000 miles of life, the owner can extend the oil change interval, thereby reducing operating costs. Conversely, oil that shows rapid deterioration will be replaced ahead of time. Used this way in an electrical power system, CBM is much more cost-effective than PM alone; moreover, CBM will identify incipient failures and allow timely and safe corrective actions when required.

The ability to predict and thereby prevent unplanned downtime is the major strength of CBM. To work properly, data collected either on-line or off-line must be analyzed mathematically. CBM is further broken down into elements or subsets. When implemented properly, CBM offers both technical and/or financial advantages over the more classic methods such as PM or RTF.

The Elements of CBM

Real-Time Monitoring (RTM). Real-time monitoring provides a continuous stream of data from operating equipment. For example, rotating-equipment, RTM systems provide continuous monitoring of speed, vibration, voltage, current, and frequency data. Careful analysis of the collected data allows the user to diagnose incipient faults. An incipient fault is an impending failure that may not be predictable by the normal senses including sight, sound, smell, and touch. The only way to diagnose an incipient fault is through advanced testing techniques.

To be a valid candidate for the use of RTM, electrical equipment must satisfy two criteria:

1. The equipment must be critical or expensive enough to warrant the expenditures for the purchase and installation of the monitoring hardware and software.
2. Analysis of the monitored parameters (voltage, frequency, speed, etc.) must provide meaningful equipment diagnostics and prognostics.

RTM is being used widely on motors, generators, and transformers. However, not all equipment lends itself to RTM; moreover, there are a large variety of off-line tests that yield valuable, trendable results and may actually be less expensive than RTM. This is where PdM fits in.

Predictive Maintenance (PdM). In its simplest form, PdM uses test results taken from PM procedures. These results are statistically evaluated and a prognosis is developed allowing the system operators to increase, decrease, or even eliminate maintenance intervals. In addition, when properly analyzed, the test results can be used to provide a usable estimate of remaining equipment life.

Data Analysis Methods for CBM

The basic methodology for CBM depends upon the collection of either operational data (RTM) and/or maintenance test results (PdM). With either RTM or PdM, the collected data is then mathematically manipulated in one of several ways to determine the condition of the equipment and to develop a prediction of its future behavior. It should be noted that a properly written PdM analysis program is equally effective on data provided by RTM monitors.

Real-Time Monitoring vs. Predictive Maintenance. RTM data is constantly delivered as long as the equipment is on-line and the test equipment connected; conversely, because PdM tests may be performed only very infrequently, the PdM data set is exceptionally sparse. This means that the statistical evaluation of PdM data must use methods that are effective for sparse data sets.

Despite this challenge, PdM methods do offer one very significant advantage over RTM. Consider an insulation resistance test, normally performed on all types of insulation systems during both PM and PdM test intervals. Incipient insulation failures may manifest as only extremely small changes in current and resistance. In fact, changes of only a few milliamperes or milliohms may be the only precursors of an imminent failure. These changes are readily captured by quality PdM test equipment.

On the other hand, since RTM data is collected during the actual operation of the equipment, these very small changes have to be captured from within the normal values of current and voltage. Normal operating values may exceed hundreds of volts or amperes. Although RTM technologies and mathematical models exist that allow this level of discrimination, results are much more readily obtained using the very precise measurements made using specialized, off-line test equipment.

The ideal way to provide accurate diagnosis of impending problems is to combine RTM and PdM. These methodologies are often used together; however, if one of the two methodologies is clearly superior, it may be used alone. The development of these predictions is frequently referred to as *trending*.

What Is Trending and What Do We Expect from It? Trending is the act of using past data values to predict future data values. Consider Fig. 6.7, which shows an example of a simple, linear trending chart. This figure is a graph of insulation resistance readings (in megohms) taken every six months in June and December. As you can see, the insulation resistance tests taken through June were all in the

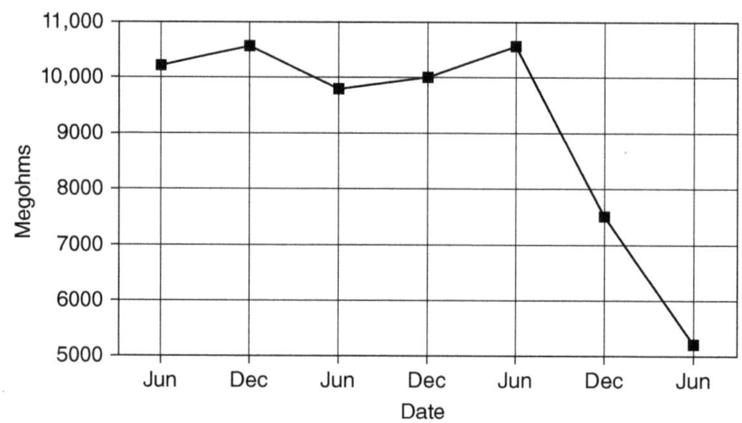

FIGURE 6.7 Linear trending graph.

neighborhood of 10,000 megohms. The last December reading showed a drop to approximately 7500 megohms, and the last June reading dropped again to slightly over 5000 megohms.

It is important to note that every one of the readings falls within the "normal" range for this type of equipment; consequently, a technician might assume that the equipment is in good condition.

Comparison of the last two data points with the previous values clearly shows that the insulation resistance has started a downward slide that points toward complete failure, possibly within one year. Prior to the mid-1980s, records and graphs of this type were rarely kept. Rather, each reading was compared to an industry norm and given a pass or fail accordingly. Both of the last two readings would have passed using industry norms.

Although linear trending represents a major improvement over norm-based systems, even simple linear trending is not the final answer. Figure 6.8 is a trending graph for parts-per-million moisture in transformer insulating oil. The important data on this figure are the Upper Norm Limit, the Actual, and the Predicted lines.

- The Actual curve plots the collected data over a period of years. Prior to the May 13, 2005, results, no trending or prognostication was performed. Note that there are several years when the results peaked above the Upper Norm Limit line.
- The Upper Norm Limit curve is a generally accepted industry value that represents the maximum moisture content that should be in the insulating liquid. Note that a correction was made within a few months each time the test values peaked above the upper limit.
- Perhaps the most important information on the graph is the Predicted curve. Note that when the last result was captured on April 4, 2005, the actual curve trend had dropped a great deal—probably due to a dehydration procedure that was performed after the norm was exceeded on May 13, 2004.

The Predicted line predicts test result values on an annual basis and is extended out many years. Like any statistical analysis, the further out it is extended, the less reliable it is. The method used to develop the trend is a highly sophisticated mathematical algorithm that takes many variables into account, including the following:

- Previous results for this transformer
- Group results from a rich database of other transformers of the same type
- Environmental conditions where the transformer is located

Even though the specimen transformer results are trending lower in the actual, note that the predicted values show a gradual rise over the next several years. Still, the method used for this analysis

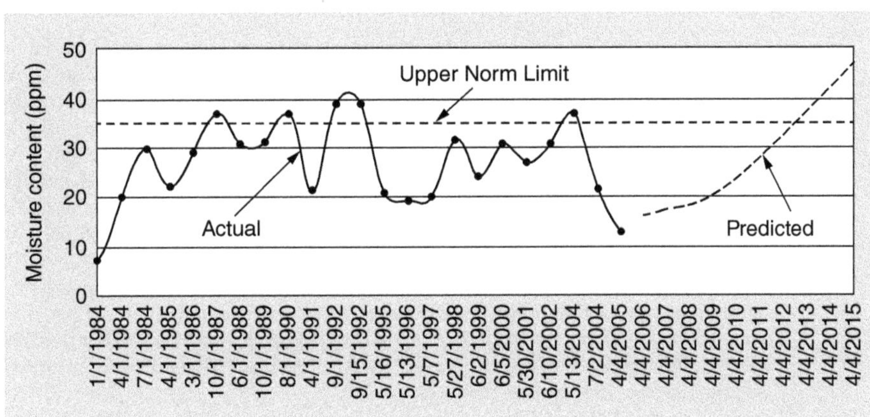

FIGURE 6.8 Parts per million (ppm) moisture graph.

is accurate enough that further testing could be delayed for a period of up to two or even three years. Two points will help to clarify the issue:

1. The trend line is counterintuitive. A visual analysis using simple trending would seem to predict a slow reduction in the moisture content. Although the specific algorithm is not available for this proprietary technique, it is clear that the more sophisticated analysis yields a result that is much more revealing than simple trending. The algorithms used have been vigorously tested and have shown that predictions are well within narrow statistical margins.
2. The last available data point for this measurement is from 2005. Although additional data points might modify the projection somewhat, rigorous testing of the algorithm has shown that over 90 percent of the predictions fall within one standard deviation out to three years beyond the last test.

This example shows that advanced trending methods can reveal useful results even from data that is not intuitively obvious like that shown in Fig. 6.7.

Using such advanced methods, the system owner can reasonably expect ample warning before major problems occur. A trend pattern that shows continued healthy predictions will allow the decision maker to extend the maintenance intervals—cutting the cost of the maintenance effort.

A trend pattern that indicates excessive deterioration or even failure can be used as the basis to schedule physical maintenance that will resolve the problem before it becomes an expensive, unplanned outage and/or failure.

What, then, are the types of mathematical mechanisms that can be and are being used for CBM prognostications?

Fuzzy Logic. Increasing numbers of mathematical trending applications are using *fuzzy logic*. The principles of fuzzy logic were developed by the mathematician Lotfi Zadeh in the mid-1960s. Fuzzy logic is sometimes flippantly referred to as the "life is in shades of gray" theory. That is, any given data point or trend may be something other than true or false, good or bad, 1 or 0. Rather, the data may be some value between 1 and 0. This allows the system to more accurately match the broad spectrum of real-world values.

Fuzzy logic most often uses a set of queries instead of strict parameters for decision making. This makes the process attractive for control systems since it better resembles the way people make decisions. For example, an air-conditioning unit has set parameters for when to turn on and off based on the feedback from a temperature sensor. When the temperature reaches, say 8°F, the A/C unit will turn on and then turn off when the sensor once again reads the 7°F limit. A system implementing fuzzy logic could ask first if the temperature is rising quickly, then turn on the A/C to cool the temperature at a rate to counteract the quick rise. This allows for more control on the environment since it assumes, correctly, that the real world is not static but dynamic.

Artificial Neural Networks. In combination with fuzzy logic, artificial neural networks (ANNs) create an adaptive system for modeling system behaviors. ANNs use a nonlinear approach to statistical data modeling. They are exceptionally useful in identifying patterns in nonsparse, high-volume data sets.

As the name implies, the behavior of an ANN is designed to mimic biological neural networks by creating complex relationships among the components, or nodes, in a system. Present-day ANNs do not self-learn in the quite same way that artificial intelligence (AI) programs can. Rather, ANNs use methodologies that strengthen (or weaken) the multitude of connections among nodes as more data become available. Thus, these networks are always improving to better fit the control system they emulate.

Classic Rule-Based (Algorithmic) Systems. A classic, rule-based approach is most useful when modeling systems that follow a clear pattern that can be quantified. An equation with set parameters, or rules, that can perfectly emulate a real-world process is rare, to say the least; consequently, most classic algorithmic systems are created by compromise between the accuracy of trending results versus the number of variables defined.

The simplest algorithmic system contains only one variable. For trending purposes, this is usually a linear equation over time where the rate of change is a set value. As variables are added, the equation becomes less rigid and can allow for other factors such as age, environment, and usage. There are various approaches to defining these additional variables based on empirical data samples, including—but not limited to—maximum likelihood estimation, inferential statistics, and the method of moments.

Combinations of Fuzzy Logic, ANN Techniques, and Rule-Based Systems. A more recent approach combines the older systems in a method that is especially applicable to sparse data sets. For example, the system used to generate the predictions in Fig. 6.8 applies a combination of a multivariant algorithmic approach and the artificial neural network's dynamic systems with self-training capabilities. Although there are a set number of variables and relationships among them as in a rule-based system, the values of the variables are forever changing as in a self-training system like an ANN. Such an approach works well for sparse data sets in that each individual gains the benefit of the population for predicting future results. The impact of each individual on the population is based on the strength of the relationships so no one "rogue" unit can adversely affect the predictions of the others, a problem faced when trending most small data sets. As these data sets grow, the relationships play a greater part in the trending. When properly developed, the combined approach will work at least as well as ANNs for large data sets.

How About the Safety Aspects. Clearly CBM will improve the financial and technical efficiency of an electrical system, but how about the safety aspects of the system? There are at least two ways that CBM will enhance personnel safety:

1. Every time workers perform maintenance or testing on a piece of equipment, they are in a dangerous situation. An error can cause equipment failure either on the spot or later when some mistake makes itself known. Since using a CBM approach optimizes the testing intervals, workers will have minimal exposure to the electrical hazard.
2. By including the analysis and trending concepts, CBM will pinpoint incipient failures. Using other methods such as a calendar-based approach, workers will not have advance warning of a failure and may be subject to injury or even death.

MAINTENANCE REQUIREMENTS FOR SPECIFIC EQUIPMENT AND LOCATIONS

Although comprehensive maintenance procedures are beyond the scope of this chapter, the following should help to provide a starting point for those new to electrical maintenance or a midcourse correction for those with more experience. For detailed information, refer to the documents in the references of this chapter and other standard texts.

General Maintenance Requirements

Table 6.1 shows the general maintenance requirements for electrical systems that are laid out in NFPA 70E.[3] Some of the requirements have been slightly modified from that document to be consistent with the authors' beliefs about proper maintenance techniques.

Substations, Switchgear, Panelboards, Motor Control Centers, and Disconnect Switches

Enclosures (Electrical Rooms, Buildings, and Other Such Locations). Enclosures must be kept clear of extraneous materials that could cause a tripping, fire, or any other type of hazard.

TABLE 6.1 General Maintenance Requirements for Electrical Systems

Topic	Required maintenance
Qualified persons	Only qualified persons are allowed to perform maintenance on electrical equipment and installations. This means that they must be trained in and familiar with the operation of the equipment, the safety hazards of the equipment, the specific maintenance procedures, and the tests that they will perform.
Single-line diagram	A legible, accurate, and up-to-date single-line diagram should be maintained to assist in day-to-day operations, including maintenance, testing, and switching activities. (Note that NFPA 70E does not require a single-line diagram but does require that it be properly maintained if it exists.)
Spaces about electrical equipment	The working spaces that are required by NFPA 70 and 29CFR 1910.302–1910.308 must be maintained.
Grounding and bonding	All grounding and bonding conductors must be maintained to ensure electrical continuity.
Guarding of live parts	Enclosures around live parts must be maintained to guard against accidental contact.
Safety equipment	Locks, interlocks, and all other types of safety equipment (including personal protective equipment) must be maintained in proper working condition so that they provide the desired protection.
Clear spaces	Access to working space and escape passages must be maintained. (This refers to exits and aisles as opposed to actual working spaces about electrical equipment.)
Identification of components and circuits	All electrical components and circuits should be identified with legible, secure signs, tags, or other such devices. Operating procedures, such as switching and/or lockout-tagout instructions fall under this requirement.
Conductors and cables	All types of conductors and cables, including flexible cord sets (extension cords), should be maintained so that they are not allowed to become damaged, frayed, or strained.

Area Enclosures (Fences and Other Such Enclosures). Fences and other such guards shall be maintained to prevent unauthorized access or accidental exposure to electrical hazards. This requirement is especially important for fences that protect the public from access to electrical facilities.

Conductors. All current-carrying conductors must be maintained so that they can carry rated current without overheating. They must also be maintained so they can carry available short-circuit current without failing and introducing arc and blast hazards.

Electrical Insulation. Electrical insulation must be maintained so that it will continue to support the impressed voltage. The maintenance of insulation is very important. Insulation maintenance typically accounts for over 70 percent of all maintenance procedures.

Protective Devices. Protective devices must be maintained so that they can adequately withstand and/or interrupt the available fault current. They must also be capable of operating as they are designed to do. The next sections provide detail for many of the maintenance procedures for protective devices.

Fuse Maintenance Requirements

Fuses may well be the most widely used electrical protective device. They are certainly the oldest, dating back to the late 1800s when it was noticed that the smallest piece of wire in a series of wires was always the first link to open. Although fuses are usually not electrically tested, the following specific steps should be employed during electrical maintenance:

1. Inspect fuses and fuse blocks closely for any signs of overheating that may be caused by loose connections, overload currents, or environmental conditions.
2. Make certain that the correct size fuse is being used in the application.
3. In a three-phase or two-phase fuse block, the fuses should be of the same size and from the same manufacturer.
4. Make certain that fuses have sufficient interrupting rating for the available fault current at their point of application.
5. Check the tightness of fuse mounting clips and hardware.

Although simple in application, these five easy steps can help to ensure that fuses will operate properly when called upon.

Molded-Case Circuit Breakers

The need for inspection of molded-case breakers will vary depending on operating conditions. Suggested inspection and testing is defined in ANSI/NEMA AB 4, Guidelines for Inspection and Preventive Maintenance of Molded Case Circuit Breakers Used in Commercial and Industrial Applications. As a part of these guidelines, AB 4 also provides some basic procedures for the inspection and maintenance of molded-case circuit breakers, by qualified persons.

Generally, maintenance on molded-case circuit breakers is limited to proper mechanical mounting, electrical connections, and periodic manual operation. Most lighting, appliance, and power panel circuit breakers have riveted frames and are not designed to be opened for internal inspection or maintenance. All other molded-case circuit breakers that are Underwriters Laboratories (UL) approved are factory-sealed to prevent access to the calibrated elements. An unbroken seal indicates that the mechanism has not been tampered with and that it should function as specified by UL. A broken seal voids the UL listing and the manufacturer's warranty of the device. In this case, the integrity of the device would be questionable. The only exception to this would be a seal being broken by a manufacturer's authorized facility.

Molded-case circuit breakers receive initial testing and calibration at the manufacturers' plants. These tests are performed in accordance with UL 489, Standard for Safety, Molded-Case Circuit Breakers, Molded-Case Switches and Circuit Breaker Enclosures. Molded-case circuit breakers, other than the riveted frame types, are permitted to be reconditioned and returned to the manufacturer's original condition. To conform to the manufacturer's original design, circuit breakers must be reconditioned according to recognized standards. An example of a recognized standard is the Professional Electrical Apparatus Recyclers League (PEARL) Reconditioning Standards. To ensure equipment reliability, it is highly recommended that only authorized professionals recondition molded-case circuit breakers.

Circuit breakers installed in a system are often forgotten. Even though the breakers have been sitting in place supplying power to a circuit for years, there are several things that can go wrong. The circuit breaker can fail to open due to a burned-out trip coil or because the mechanism is frozen due to dirt, dried lubricant, or corrosion. The overcurrent device can fail due to inactivity or a burned-out electronic component. Many problems can occur when proper maintenance is not performed and the breaker fails to open under fault conditions. This combination of events can result in fires, damage to equipment, or injuries to personnel.

Very often, a circuit breaker fails because the minimum maintenance (as specified by the manufacturer) was not performed or was performed improperly. Small things, like failing to properly clean and/or lubricate a circuit breaker, can lead to operational failure or complete destruction due to overheating of the internal components. Common sense, as well as manufacturers' literature, must be used when maintaining circuit breakers. Most manufacturers, as well as NFPA 70B,[2] recommend that if a molded-case circuit breaker has not been operated, opened, or closed, either manually or by

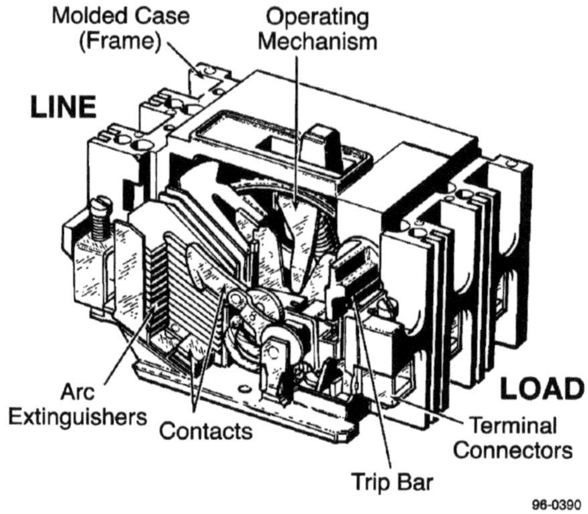

FIGURE 6.9 Principal components of a molded-case circuit breaker.

automatic means, within as little as six months time, it should be removed from service and manually exercised several times. This manual exercise helps to keep the contacts clean, due to their wiping action, and ensures that the operating mechanism moves freely. This exercise, however, does not operate the mechanical linkages in the tripping mechanism, Fig. 6.9. The only way to properly exercise the entire breaker operating and tripping mechanisms is to remove the breaker from service and test the overcurrent and short-circuit tripping capabilities. A stiff or sticky mechanism can cause an unintentional time delay in its operation under fault conditions. This could dramatically increase the arc flash incident energy level to a value in excess of the rating of personal protective equipment.

Another consideration is addressed by OSHA in 29 CFR 1910.334(b)(2), which states:

> Reclosing circuits after protective device operation. After a circuit is deenergized by a circuit protective device, the circuit may NOT be manually reenergized until it has been determined that the equipment and circuit can be safely reenergized. The repetitive manual reclosing of circuit breakers or reenergizing circuits through replaced fuses is prohibited.
>
> NOTE: When it can be determined from the design of the circuit and the overcurrent devices involved that the automatic operation of a device was caused by an overload rather than a fault condition, no examination of the circuit or connected equipment is needed before the circuit is reenergized.

The safety of the worker manually operating the circuit breaker is at risk if the short-circuit condition still exists when reclosing the breaker as stated above. OSHA no longer allows the past practice of resetting a circuit breaker one, two, or three times before investigating the cause of the trip. This previous practice has caused numerous burn injuries that resulted from the explosion of electrical equipment. Before resetting a circuit breaker, it, along with the circuit and equipment, must be tested and inspected by a qualified person, to ensure a short-circuit condition does not exist and that it is safe to reset the breaker.

Any time a circuit breaker has operated and the reason is unknown, the breaker, circuit, and equipment must be inspected for a short-circuit condition. Melted arc chutes will not interrupt fault currents. If the breaker cannot interrupt a second fault, it will fail and may destroy its enclosure and create a hazard for anyone working near the equipment.

To further emphasize this point, the following quote is provided:

> After a high level fault has occurred in equipment that is properly rated and installed, it is not always clear to investigating electricians what damage has occurred inside encased equipment. The circuit breaker may well appear virtually clean while its internal condition is unknown. For such situations, the NEMA AB4 'Guidelines

> for Inspection and Preventive Maintenance of MCCBs Used in Commercial and Industrial Applications' may be of help. Circuit breakers unsuitable for continued service may be identified by simple inspection under these guidelines. Testing outlined in the document is another and more definite step that will help to identify circuit breakers that are not suitable for continued service.
>
> After the occurrence of a short circuit, it is important that the cause be investigated and repaired and that the condition of the installed equipment be investigated. A circuit breaker may require replacement just as any other switching device, wiring or electrical equipment in the circuit that has been exposed to a short circuit. Questionable circuit breakers must be replaced for continued, dependable circuit protection.[7]

The condition of the circuit breaker must be known to ensure that it functions properly and safely before it is put back into service.

Low-Voltage Power Circuit Breakers

Low-voltage power circuit breakers are manufactured under a high degree of quality control, of the best materials available, and with a high degree of tooling for operational accuracy. Manufacturers' tests show these circuit breakers to have durability beyond the minimum standards requirements. All of these factors give these circuit breakers a very high reliability rating when proper maintenance is performed per the manufacturer's instructions. However, because of the varying application conditions and the dependence placed upon them for protection of electrical systems and equipment as well as the assurance of service continuity, inspections and maintenance checks must be made on a regular basis. Several studies have shown that low-voltage power circuit breakers that were not maintained within a five-year period have a 50 percent failure rate.

Maintenance of these breakers will generally consist of keeping them clean and properly lubricated. The frequency of maintenance will depend to some extent on the cleanliness and environmental conditions of the surrounding area. If there were very much dust, lint, moisture, or other foreign matter present, then more frequent maintenance would be required.

Industry standards for and manufacturers of low-voltage power circuit breakers recommend a general inspection and lubrication after a specified number of operations or at least once per year, whichever comes first. Some manufacturers also recommend this same inspection and maintenance be performed after the first six months of service for a new circuit breaker, regardless of the number of operations. If the breaker remains open or closed for a long period of time, it is recommended that arrangements be made to open and close the breaker several times in succession. Environmental conditions would also play a major role in the scheduling of inspections and maintenance. If the initial inspection indicates that maintenance is not required at that time, the period may be extended to a more economical point. However, more frequent inspections and maintenance may be required if severe load conditions exist or if an inspection reveals heavy accumulations of dirt, moisture, or other foreign matter that might cause mechanical, insulation, or electrical failure. Mechanical failure would include an unintentional time delay in the circuit breaker's tripping operation due to dry, dirty, or corroded pivot points, or by hardened or sticky lubricant in the moving parts of the operating mechanism. The manufacturer's instructions must be followed to minimize the risk of any unintentional time delay.

Figure 6.10 provides an illustration of the operating mechanism for such a circuit breaker and the numerous points where lubrication would be required and where dirt, moisture, corrosion, or other foreign matter could accumulate causing a time delay in, or complete failure of, the circuit breaker operation.

Medium-Voltage Circuit Breakers

Most of the inspection and maintenance requirements for low-voltage power circuit breakers also apply to medium-voltage power circuit breakers. Manufacturers recommend that these breakers be removed from service and inspected at least once a year. They also state that the number and severity of interruptions may indicate the need for more frequent maintenance checks. Always follow the

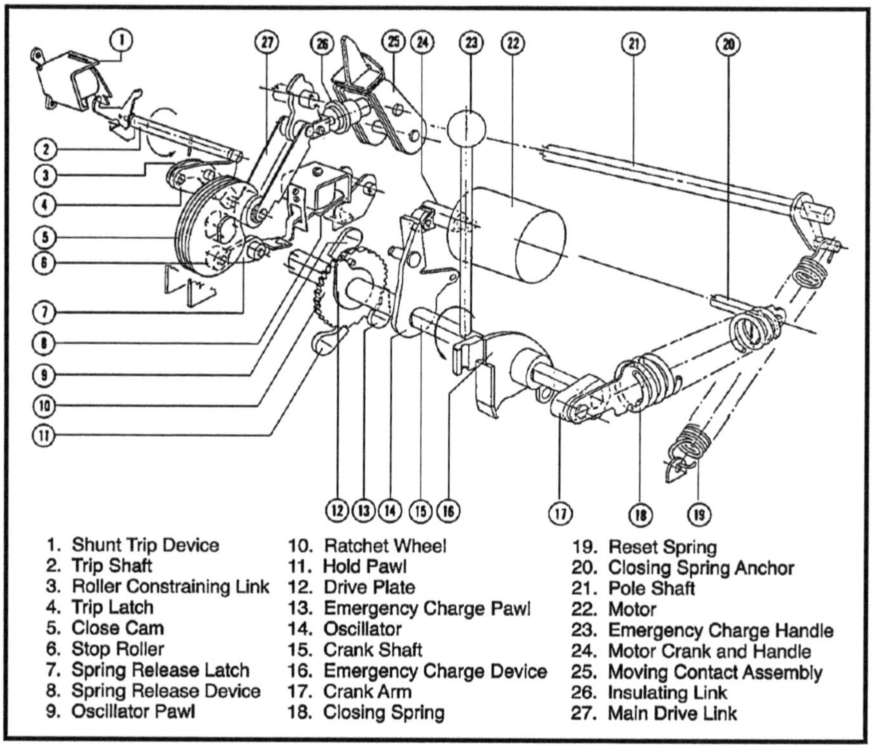

FIGURE 6.10 Power-operated mechanism for a low-voltage power circuit breaker.

manufacturer's instructions because every breaker is different. Figures 6.11 and 6.12 illustrate two types of operating mechanisms for medium-voltage power circuit breakers. These mechanisms are typical of the types used for air, vacuum, oil, and SF-6 (sulfur-hexafluoride) circuit breakers. As can be seen in these figures, there are many points that would require cleaning and lubrication to function properly.

Protective Relays

Relays must continuously monitor complex power circuit conditions, such as current and voltage magnitudes, phase angle relationships, direction of power flow, and frequency. When an intolerable circuit condition, such as a short circuit (or fault) is detected, the relay responds and closes its contacts, and the abnormal portion of the circuit is de-energized via the circuit breaker. The ultimate goal of protective relaying is to disconnect a faulty system element as quickly as possible. Sensitivity and selectivity are essential to ensure that the proper circuit breakers are tripped at the proper speed to clear the fault, minimize damage to equipment, and reduce the hazards to personnel.

A clear understanding of the possible causes of primary relaying failure is necessary for a better appreciation of the practices involved in backup relaying. One of several things may happen to prevent primary relaying from disconnecting a power system fault:

- Current or voltage supplies to the relays are incorrect.
- DC tripping voltage supply is low or absent.
- Protective relay malfunctions.
- Tripping circuit or breaker mechanism hangs up.

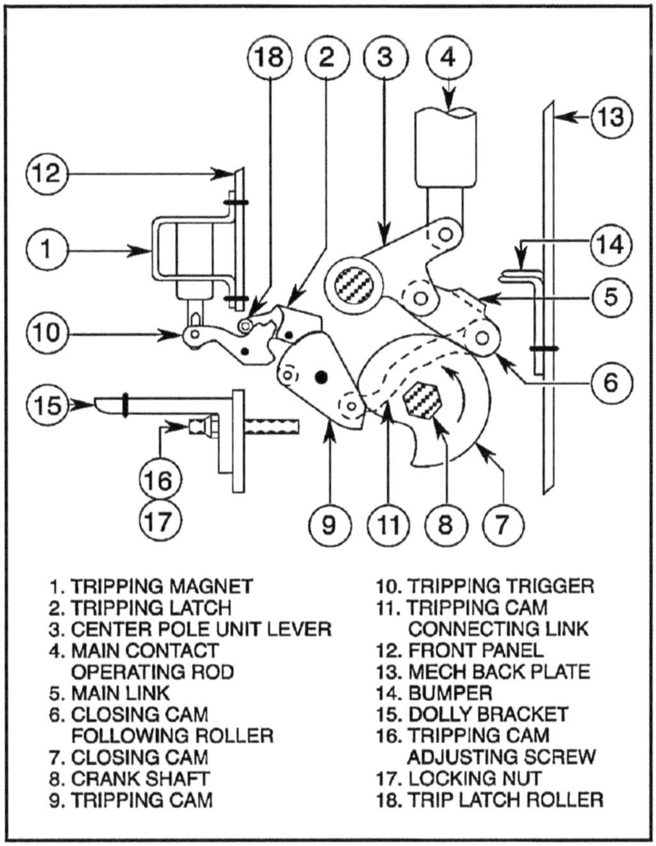

96-0284a

FIGURE 6.11 Operating mechanism for a medium-voltage air circuit breaker.

There are two groups of protective relays: primary and backup. Primary relaying is the so-called first line of defense, and backup relaying is sometimes considered to be a subordinate type of protection. Many companies, however, prefer to supply two lines of relaying and do not think of them as primary and backup. Figure 6.13 shows the zones of protection for a primary relay system. Circuit breakers are found in the connections to each power system element. This provision makes it possible to disconnect only the faulty part of the system. Each element of the system has zones of protection surrounding the element. A fault within the given zone should cause the tripping of all circuit breakers within that zone and no tripping of breakers outside that zone. Adjacent zones of protection can overlap, and in fact, this practice is preferred because for failures anywhere in the zone, except in the overlap region, the minimum number of circuit breakers are tripped.

In addition, if faults occur in the overlap region, several breakers respond and isolate the sections from the power system. Backup relaying is generally used only for protection against short circuits. Since most power system failures are caused by short circuits, short-circuit primary relaying is called on more often than most other types. Therefore, short-circuit primary relaying is more likely to fail.

Voltage and current transformers play a vital role in the power protection scheme. These transformers are used to isolate and protect both people and devices from high voltage and to allow current-carrying devices such as relays, meters, and other instruments to have a reasonable amount of insulation. It should be clearly understood that the performance of a relay is only as good as the

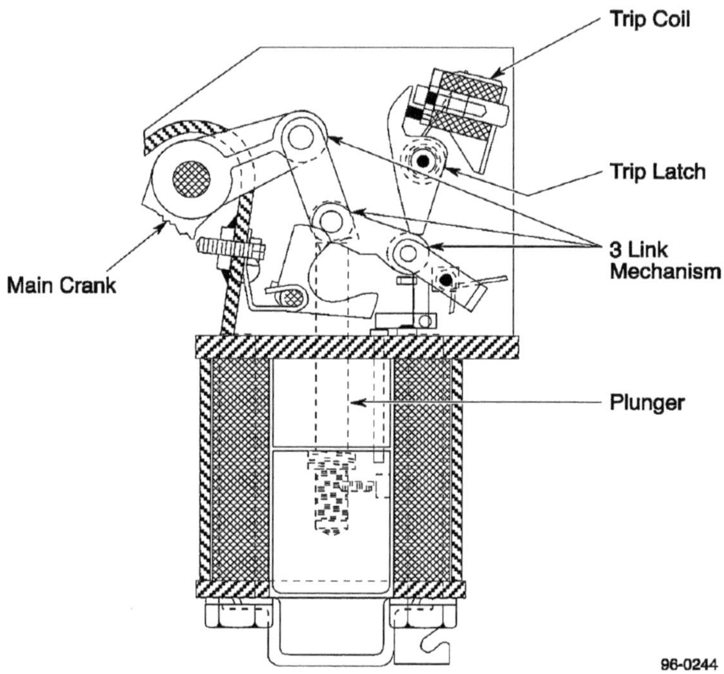

FIGURE 6.12 Solenoid-operated mechanism for a medium-voltage circuit breaker.

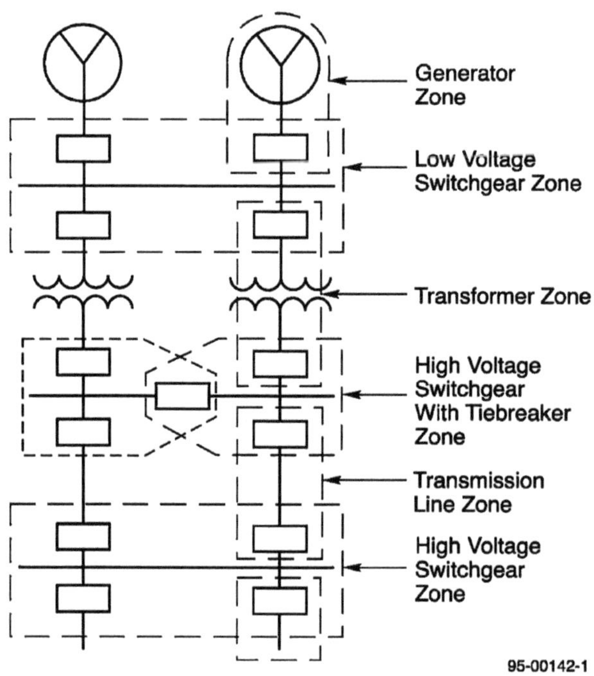

FIGURE 6.13 Zones of protection for a primary protective relaying system.

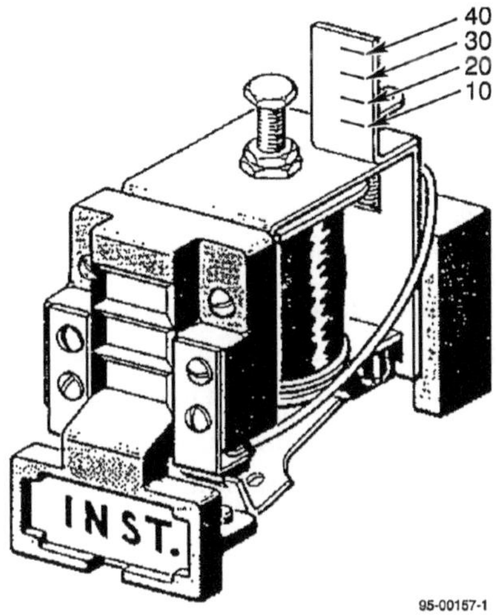

FIGURE 6.14 Electromechanical instantaneous unit.

voltage and current transformers connected to it. A basic understanding of the operating characteristics, application, and function of instrument transformers is essential to the certified relay technician.

Some overcurrent relays are equipped with an instantaneous overcurrent unit, which operates when the current reaches its minimum pickup point (Fig. 6.14). An instantaneous unit is a relay having no intentional time delay. Should an overcurrent of sufficient magnitude be applied to the relay, both the induction disc and the instantaneous unit will operate. However, the instantaneous unit will trip the circuit breaker, since it has no intentional time delay.

The instantaneous trip unit is a small, ac-operated clapper device. A magnetic armature, to which leaf-spring-mounted contacts are attached, is attracted to the magnetic core upon energization. When the instantaneous unit closes, the moving contacts bridge two stationary contacts and complete the trip circuit. The core screw, accessible from the top of the unit, provides the adjustable pickup range. Newer designs also feature tapped coils to allow even greater ranges of adjustment.

The instantaneous unit, like the ICS unit, is equipped with an indicator target. This indication shows that the relay has operated. It is important to know which relay has operated, and no relay target should be reset without the supervisor's knowledge and permission.

As can be seen, several things can go wrong that would prevent the instantaneous unit from operating properly. These things include an open or shunted current transformer, open coil, or dirty contacts. Protective relays, like circuit breakers, require periodic inspection, maintenance, and testing to function properly. Most manufacturers recommend that periodic inspections and maintenance be performed at intervals of one to two years. The intervals between periodic inspection and maintenance will vary depending upon environment, type of relay, and the user's experience with periodic testing.

The periodic inspections, maintenance, and testing are intended to ensure that the protective relays are functioning properly and have not deviated from the design settings. If deviations are found, the relay must be retested and serviced as described in the manufacturer's instructions.

Rotating Equipment

The following maintenance requirements are taken verbatim from NFPA 70E.[3]

Terminal Boxes. Terminal chambers, enclosures, and terminal boxes shall be maintained to guard against accidental contact with live parts and other electrical hazards.

Guards, Barriers, and Access Plates. Guards, barriers, and access plates shall be maintained to prevent employees from contacting moving or energized parts.

Portable Electric Tools and Equipment

The maintenance requirements for these types of devices focus primarily on the attachment plugs, receptacles, cover plates, flexible cords, and cord connectors. These should be maintained to ensure that:

1. There are no breaks, damage, or cracks that may expose energized live parts.
2. All cover plates are in place, undamaged, and secure.
3. All terminations are complete and have no exposed strands or loose terminals.
4. All pins and connectors in the plugs are intact, in place, and firm.
5. The polarity is correct; that is, the hot, neutral, and ground leads are all connected to their correct terminations.

Personal Safety and Protective Equipment

The maintenance, inspection, and testing requirements for this type of equipment are all covered in detail in Chap. 4 of this handbook.

ELECTRICAL SAFETY BY DESIGN

Introduction

Any electrical system design—whether for new construction or refurbishing—is always modified based on a number of competing criteria. Historically, these criteria included such factors as the following:

- Cost
- Available space
- Required types of equipment
- Required size and capacity of equipment
- Proper protective schemes for the power system
- Environmental considerations

If any safety considerations were included in the design, they were almost always *minimum* requirements as found in the National Electrical Code. Clearance distances, size of wire, and grounding requirements are examples.

Consider Fig. 6.15 for example. The classical approach to setting the protective devices shown is as follows:

1. The protective devices must be set in such a way that they will allow full load current to flow continuously. This means that Protective Device 1 (PD1) must carry all the Bus 1 load current without opening, PD2 must carry all the load at Bus 2, and PD3 must carry all the load at Bus 3.
2. All protective devices must operate fast enough so that no thermal or mechanical damage occurs on any equipment that is not already faulted. If, for example, a short circuit were to occur on Bus 1, PD1 must trip fast enough so that T1 will not be damaged by the high levels of current flow.
3. When a short circuit occurs, only the closest upstream protective device to the fault must be the one to trip. This requirement is called *selective tripping*. If a fault occurs on Bus 1, PD1 should

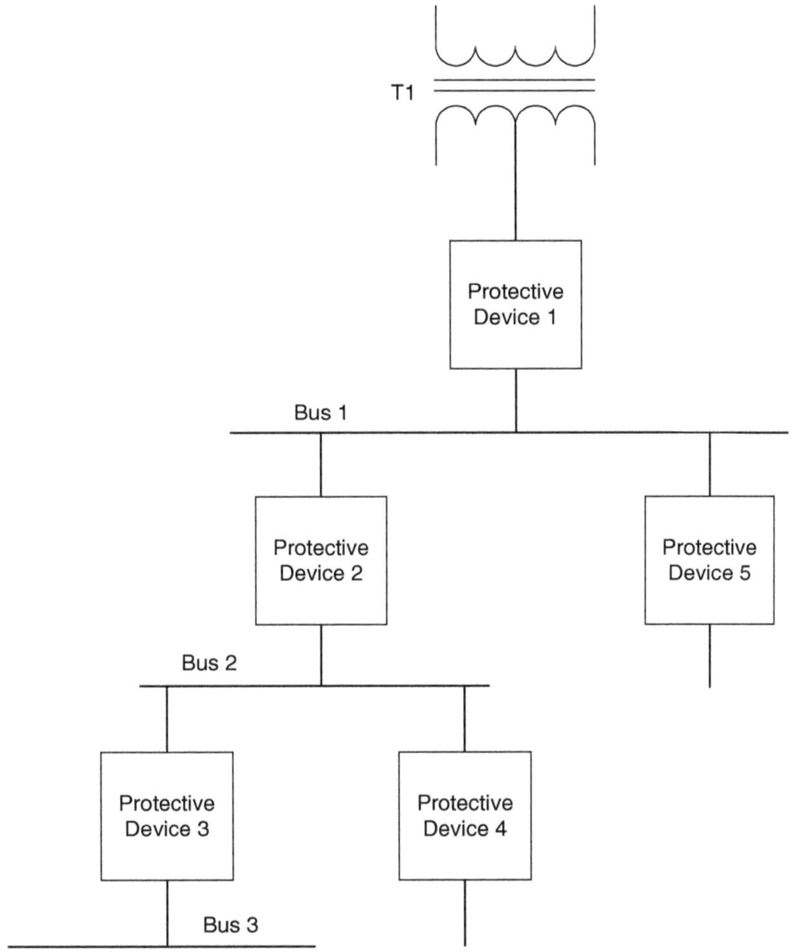

FIGURE 6.15 Simple power distribution system.

open, if the fault occurs on Bus 2, PD2 should open, and so on. Setting the devices in this way allows the load current to continue on other, noninvolved parts of the system. For example, for a fault on Bus 2, PD2 would open and the load connected to PD5 would continue.

4. In the event that the first upstream device fails to trip, the next upstream device should operate after some time delay. If PD3 fails to open for a fault on Bus 3, PD2 should open up after some time delay. This helps to guarantee that the system is adequately and redundantly protected.

Incident energy from an arc is proportional to the length of time that the arc is present. (See Chap. 4 of this handbook.) This means that we would like for every protective device to trip instantaneously. This requirement conflicts with criteria 3 and 4 in the previous list.

Including Safety in Engineering Design Criteria

So how is a system engineered for good safety? Using the example of Fig. 6.15, the engineer may sacrifice selective tripping at the very high values of current flow. Where previously instantaneous tripping would be disabled at PD1 and PD2, now the engineer might enable instantaneous tripping for very high current faults. This will reduce the arc energy that would occur, but it would also eliminate selective tripping for those high current levels.

Another approach might be the inclusion of a *zone-interlocking system*. With this type of system, each of the protective devices includes a communications module. When the device senses an overcurrent condition, the communications module sends a signal to all the other protective devices. For a fault on Bus 3 (Fig. 6.15), PD3, PD2, and PD1 would see the overcurrent and would send signals out. The signal from PD3 will override the others and force them into their normal time-delay operation. PD3 will trip instantaneously. A similar thing happens for a fault on Bus 2, except in this case PD2 will override PD1.

This example is but one specific situation in which an engineer can *design safety* into a power system. There are other broad examples:

- When considering working spaces, such as those required by NFPA 70 (NEC) in Sections 110.26 and 110.34, consider when and why such rules were developed. The working space distances found in such standards are based on two criteria: electric shock hazard and emergency egress. But how about the incident energy from an electrical arc? Even if a worker is wearing appropriate arc-flash protection, being confined in a 1-by-1-meter space is dangerous.
- What about the physical construction of switchgear and its ability to withstand a worst case electrical arc and blast? Even in the early part of the 21st century, there are many engineers who believe that standard switchgear will protect workers from electrical arc and blast. This is simply not true and never was. Unless the switchgear is designed, tested, and built according to the modern *arc-resistance switchgear* standards, a closed door or panel will simply not protect the worker from these hazards.
- Design systems that have adequate backup and bypass so that equipment can be taken out of service for repair and maintenance without having to stop production. This type of approach will greatly limit the number of times that a worker has to approach energized conductors.
- Avoid sizing motors, transformers, cables, and other types of equipment using the minimum requirements of the National Electrical Code. Consider the fact that the minimum requirements by necessity can take into account only the median types of hazards that may be encountered. Short-term expenditures certainly must be taken into account; however, the effect of electrical accidents on long-term costs also must be considered. Remember that one electrical accident can easily cost 10 times, or more, what it would cost to engineer the hazard out.
- There is a great deal of emphasis on utilizing optic sensors and high-speed microprocessor protective relaying to sense an arcing event and clear the circuit before the arc-flash and blast occur. This concept is not entirely true. Granted, the use of these devices will provide high-speed detection of an electrical arc event; however, the circuit can clear the fault only as fast as the electromechanical circuit breaker can operate. Maintenance of the circuit breaker must be considered a vital component for safely clearing the faulted condition, as noted previously in this chapter.
- Another consideration that many are using is current-limiting devices. There are many industries that have installed current-limiting reactors at the secondary of their power transformers to reduce the amount of available short-circuit current that the electrical system and equipment will see. Still another method that many are employing is the use of current-limiting fuses instead of circuit breakers. The advantage of using fast-acting fuses is that they have no moving mechanical parts that are prone to failure and require constant maintenance to help improve their reliability.

Improved Engineering Standards

Starting in the 1990s, there has been a gradual improvement of engineering standards to allow for safety to be "built in." One of the new standards that is in the process of being developed is IEEE Standard Std 1814, titled Recommended Practice for Electrical System Design Techniques to Improve Electrical Safety.

The working group comprises a team of highly skilled engineers, technicians, and safety professionals. The underlying concept of Std 1814 is to develop a reference for the types of procedures that engineers can use to improve the safety of their designs. They will collect, refine, and include information from a number of industry sources including:

- IEEE papers, conference presentations, workshops, and magazines
- Industry end-user best practices
- Manufacturer developments and white papers
- Trade publications

Industry efforts like Std 1814 along with continuing education for design engineers will greatly enhance safety in the workplace with improved, safer electrical equipment and installations.

CONCLUSION

Electrical maintenance is critical to the safety of all personnel who must work in the vicinity of electrical equipment. Whether justified on a safety or an economic basis, all facilities should have a comprehensive electrical maintenance program.

Although the topic is too detailed for complete coverage in this handbook, a number of excellent publications and consultants are available to help with the establishment and continuation of a good electrical maintenance program.

REFERENCES

1. InterNational Electrical Testing Association (NETA), ANSI/NETA *2011 Standard for Maintenance Testing Specification,* Morrison, CO, 2100.
2. NFPA 70B, *Recommended Practice for Electrical Equipment Maintenance*, National Fire Protection Association, Quincy, MA, 2016.
3. NFPA 70E, *Standard for Electrical Safety in the Workplace*, National Fire Protection Association, Quincy, MA, 2018.
4. NFPA 70, *National Electrical Code*, National Fire Protection Association, Quincy, MA, 2017.
5. Nowlan, F. S., and Heap, H. F., *Reliability-Centered Maintenance*, United Airlines, for Office of Assistant Secretary of Defense, Washington, D.C., December 29, 1978.
6. Cadick, John, *The Eight Step Maintenance Program*, Cadick Corporation, Dallas, Texas, 2000–2011.
7. National Equipment Manufacturers Association (NEMA), Vince A. Baclawski, Power Distribution Products, NEMA, *EC&M magazine*, p. 10, January 1995.

CHAPTER 7
REGULATORY AND LEGAL SAFETY REQUIREMENTS AND STANDARDS

INTRODUCTION

The modern trend in electrical safety is toward more and more individual responsibility as employees are being held increasingly responsible for their actions. For individuals to successfully fulfill their responsibilities, they must be aware of the rules that apply to their performance. This means that company safety rules; national and international standards; and local, state, and federal laws must be part of an employee's knowledge base.

At the same time, regulatory bodies such as the Occupational Safety and Health Administration (OSHA) have become increasingly severe in their enforcement of safety standards. Employers must redouble their efforts to interpret and implement the best possible electrical safety program. To do this, employers must become familiar with all of the various standards—both regulatory and consensus—so that they can comply with them.

Electrical safety standards and/or requirements are produced by a variety of organizations. Some of the standards are voluntary and some are local or federal law. Whether voluntary or mandatory, these standards provide guidance for the proper way to work on or around electric energy. Much of the material covered in other parts of this handbook is drawn from the standards that will be covered in this chapter.

Since electrical standards are periodically reviewed and updated, the information included in this chapter is necessarily general. Always refer to the most recent edition of each of these standards when current information is required. Note that not all standards are covered in this chapter. Readers should refer to industry-specific literature for their own industries.

Because of the very critical role that standards organizations play, coverage of them is also included in this chapter. The organizational structure, the standards-making procedure, and the historical development of each standards organization are covered.

THE REGULATORY BODIES

International Electrotechnical Commission (IEC)

The *Electrical Safety Handbook* primarily focuses on U.S. standards and regulations. Other international and country-specific standards and regulations must also be adhered to according to local laws and requirements. The IEC is one such standards body that must be considered.

Founded in 1906, the International Electrotechnical Commission (IEC) is the world's leading organization that prepares and publishes international standards for all electrical, electronic, and related technologies. More than 10,000 experts from industry, commerce, government, test and research labs, academia, and consumer groups participate in IEC standardization work. IEC provides a platform to companies, industries, and governments for meeting, discussing, and developing the international standards they require.

All IEC international standards are fully consensus-based and represent the needs of key stakeholders of every nation participating in IEC work. Every member country, no matter how large or small, has one vote and a say in what goes into an IEC international standard.

Globally Relevant. The IEC is one of three global sister organizations (IEC, ISO, ITU) that develop international standards for the world. When appropriate, IEC cooperates with ISO (International Organization for Standardization) or ITU (International Telecommunication Union) to ensure that international standards fit together seamlessly and complement each other. Joint committees ensure that international standards combine all relevant knowledge of experts working in related areas.

Examples of IEC Requirements

Example 1: IEC 60204 applies to electrical and electronic equipment and systems of machines, including a group of machines working together in a coordinated manner but excluding higher-level system aspects (i.e., communications between systems).

This part of IEC 60204 is applicable to equipment, or parts of equipment, that operate with nominal supply voltages above 1000 V ac or 1500 V dc and not exceeding 36 kV ac or dc with nominal frequencies not exceeding 200 Hz. For higher voltages or frequencies, special requirements may be needed.

In this standard, the term *HV equipment* (high-voltage equipment) also covers the low-voltage equipment (LV equipment) forming an integral part of the equipment operating at high voltage. The requirements in this standard primarily cover the parts operating at high voltage except where explicitly stated otherwise. Reference is made to IEC 60204-1 for those requirements that also apply to HV equipment.

Note 1 Other LV equipment not forming part of the HV equipment and defined as operating at voltages not exceeding 1000 V ac or 1500 V dc is covered by IEC 60204-1.

Note 2 In this standard, the term electrical includes both electrical and electronic matters (i.e., electrical equipment means both the electrical and the electronic equipment).

The electrical equipment covered by this part of IEC 60204 commences at the point of connection of the supply to the *electrical* equipment of the machine.

Note 3 For the requirements for power supply installations, see HD 637.

This part of IEC 60204 is an application standard and is not intended to limit or inhibit technological advancement. It does not cover all the requirements (e.g., guarding, interlocking, or control) that are needed or required by other standards or regulations to safeguard personnel from hazards other than electrical hazards. Each type of machine has unique requirements to be accommodated to provide adequate safety.

Note 4 In the context of this standard, the term *person* refers to any individual; *personnel* are those persons who are assigned and instructed by the user or the user's agent(s) in the use and care of the machine in question.

Example 2: IEC 60204-11 provides requirements and recommendations relating to the HV equipment of machines together with its associated LV equipment so as to promote

safety of persons and property;

consistency of control response;

ease of maintenance.

High performance is not to be obtained at the expense of the essential factors mentioned above.

An example of a possible application of these requirements is a machine or group of machines used for the processing of a material where a failure in such machinery can have serious economic consequences.

Example 3: IEC 60204-33 applies to electrical and electronic equipment associated with semiconductor fabrication equipment for the manufacture, measurement, assembly, and test of semiconductors.

Note 1 In this standard, the term *electrical* includes electrical, electronic, and programmable electronic matters (i.e., electrical equipment means electrical, electronic, and programmable electronic equipment).

Note 2 In the context of this standard, the term *person* refers to any individual and includes those persons who are assigned and instructed by the user or the user's agent(s) in the installation, use, and care of the fabrication equipment in question.

The electrical equipment covered by this standard commences at the point of connection of the supply to the electrical equipment and includes proper instruction for its safe installation.

Note 3 For the requirements for the electrical supply installation in buildings, see IEC 60364 series.

This part is applicable to the electrical equipment or parts of the electrical equipment that operate with nominal supply voltages not exceeding 1000 V for ac and not exceeding 1500 V for dc, and with nominal supply frequencies not exceeding 200 Hz. For higher voltages or frequencies, special requirements may be needed.

Note 4 Electrical equipment within which derived voltages exceed these supply voltage limits is within the scope of this standard.

Included are requirements for protective measures against electrical safety hazards as well as electrical interlock circuits that protect against nonelectrical hazards. However, it does not cover all the requirements that are needed or required by other standards or regulations to safeguard persons from hazards other than electrical hazards (e.g., chemical hazards, mechanical hazards, radiation hazards). Each type of machine has unique requirements to be accommodated to provide adequate safety.

American National Standards Institute (ANSI)

Overview. Founded in 1918, ANSI is a private, nonprofit, membership organization. ANSI coordinates the United States voluntary consensus standards systems and approves American national standards. The ANSI membership includes over 1000 companies; hundreds of professional, technical, trade, labor, and consumer organizations; and dozens of government agencies. ANSI's primary purpose is to ensure that a single set of nonconflicting American national standards is developed by ANSI-accredited standards developers and that all interests concerned have the opportunity to participate in the development process. Since all interested parties are able to participate in the standards-making process, ANSI standards are true consensus standards.

Functions. ANSI serves four extremely important functions for industry:

- *Coordinates voluntary standards activities.* ANSI assists standards developers and standards users from the private sector and government to reach agreement on the need for standards and to establish priorities. It accredits qualified organizations to develop standards, helps them avoid duplication of effort, and offers a neutral forum for resolving differences. ANSI's Executive Standards Council and standards boards spearhead this function.
- *Approves American national standards.* ANSI administers the only recognized U.S. system for establishing American national standards. The ANSI approval procedures ensure that any interested party can participate in a standard's development or comment on its provisions. These requirements for due process guarantee that American national standards earn high levels of confidence and credibility and achieve broad acceptance.
- *Represents U.S. interests in international standardization.* ANSI is the sole U.S. representative to two of the major, nontreaty standards organizations: the International Organization for Standardization (ISO) and the International Electrotechnical Commission (IEC). The ISO, founded in 1946, develops, coordinates, and promotes international standards with national standards organizations from over 70 countries around the world. The IEC, founded in 1906, develops and promotes electrotechnical standards with national committees from over 40 countries. ANSI participates in

almost the entire technical program of both organizations and administers many key committees and subgroups.

- *Provides information on and access to world standards.* ANSI maintains copies of approved and draft American national standards, ISO and IEC standards, proposals of regional groups tied to the European Community, and the specifications of 90 national standards organizations that belong to the ISO. These standards are available to interested parties.

Member Companies. ANSI is a membership organization and draws its principal revenue from the company members. Company members help to govern ANSI by determining policies, procedures, and long-range plans for standards development. They identify commercial and industrial needs for standards, both nationally and internationally. Based on these needs, they encourage ANSI to initiate new activities for standards development and accelerate improvement of existing standards.

Standards Developers. Standards developers are primarily national trade, technical, professional, consumer, and labor organizations. They voluntarily submit standards to ANSI for recognition as national consensus standards. Many of these organizations are also members of ANSI and are represented on the boards, councils, and committees that help govern ANSI and coordinate national and international standardization activities.

Government Agencies. ANSI works with governmental agencies in the same way it works with other organizations. That is, government agencies are ANSI members, and government representatives serve on ANSI boards and councils.

Coordinating committees, formed with OSHA and the Consumer Product Safety Commission, provide forums to discuss activities affecting the voluntary and governmental standards communities.

Institute of Electrical and Electronics Engineers (IEEE)

Overview. IEEE is the world's largest professional technical society, with over 350,000 members from approximately 150 countries around the world. The American Institute of Electrical Engineers (AIEE) was founded in 1884, and the Institute of Radio Engineers (IRE) was founded in 1912. These two organizations merged in 1963 to form IEEE.

IEEE is a membership society composed of electrical engineers and technicians. The interest areas of IEEE members cover the gamut of technical areas including computer engineering, biomedical technology, telecommunications, electric power, aerospace, and consumer products.

Functions. Among IEEE's primary activities are technical publishing, technical conferences, and development of technical standards. IEEE is responsible for approximately 30 percent of the world's published literature in electrotechnology. Ranging from published papers, to textbooks, to technical journals, IEEE literature is typically on the cutting edge of technology.

IEEE is divided into a variety of technical societies with a broad spectrum of technical interests. Primarily through these societies, IEEE holds more than 300 meetings annually, which allow engineers and technicians to exchange information. IEEE also develops and publishes voluntary technical standards that regulate a variety of electrical technical activities.

National Fire Protection Association (NFPA)

Overview. NFPA was founded in 1896 for the express purpose of protecting people, property, and the environment from the effects of fire. To fulfill its purposes, NFPA focuses its efforts on intensive technical activities and extensive educational programs. NFPA is a nonprofit organization and has a membership of approximately 67,000 from 85 nations.

NFPA has an extremely well-recognized and respected set of voluntary consensus codes and standards. These codes and standards, hundreds of them, are used extensively by NFPA's network of fire service, industrial, health care, building code, and other professional members. NFPA technical committees are staffed by thousands of volunteers from a balanced cross section of affected interests.

NFPA is responsible for the National Electrical Code, a document that is one of the most recognized electrical safety standards in the world.

Functions. NFPA is involved in a wide variety of functions and activities.

- The Learn Not to Burn Curriculum is for children from kindergarten through eighth grade. This program, introduced by NFPA in 1979, has a documented record of saving hundreds of lives because of lessons remembered by children.
- Fire Prevention Week is observed each year on October 9—the anniversary of the great Chicago fire of 1871. The first Fire Prevention Week was proclaimed by President Warren G. Harding in 1922. NFPA has sponsored the week ever since.
- NFPA maintains one of the major fire safety literature collections.
- NFPA conducts investigations of major fires. These fires are investigated as a result of their technical or educational interest.

The NFPA Standards-Making System. NFPA and its standards-making efforts grew from a meeting of a small group in Boston in 1895. That group was meeting to discuss sprinkler installation standards in the Boston area. From that initial standard has grown the NFPA standards-making system of today.

The standards-making procedure starts with the submission of a request for a new standard by an interested party. The request is reviewed by the NFPA Standards Council, which publishes an announcement in the membership newsletter, Fire News. The announcement acknowledges the request for a standard and asks for comments on the need for the project, information on organizations that may be active in the subject matter of the proposed project, a listing of resource material that is available, and an indication of who would be willing to participate in the project if it is approved.

If the council determines a need for the proposed project, it assigns the project to either an existing technical committee or a new committee whose membership reflects a fair balance of concerned interests. Members of technical committees are appointed by the standards council and include volunteer experts representing the government, educational institutions, business, insurance companies, industry, and consumers. Each committee member is classified according to his or her interest, and each committee is structured so that no more than one-third of the membership is from a single interest.

Once the technical committee has been established, NFPA issues public notices announcing the committee's next meeting date and calling for specific proposals from interested persons. Anyone may submit a proposal for text to be included in a new document or added to an existing document. The committee then meets to consider all proposals received and drafts a proposed document or amendments to an existing document. Submitters may address the committee on their proposals at the meeting. Letter ballot approval by at least two-thirds of all committee members eligible to vote is required to approve a proposal for inclusion in the document.

After suitable procedures to allow for objections, and a vote of the membership, the document is presented to the standards council for final issuance. If the standards council issues the standard, based on the vote of the membership, it is published in pamphlet form and is also included in the appropriate volume of NFPA's National Fire Codes. If the standard is to be an American national standard, it is submitted to ANSI.

American Society for Testing and Materials (ASTM)

Overview. ASTM was organized in 1898 and has grown into an extremely large, voluntary standards development organization. The organization develops standards in a variety of areas including materials, products, systems, and services. ASTM has over 129 standards-writing committees, which write standards in such diverse areas as metals, paints, plastics, textiles, petroleum, construction, energy, the environment, consumer products, medical services and devices, computerized systems, and electronics.

ASTM headquarters has no technical research or testing facilities because such work is done voluntarily by over 32,000 technically qualified ASTM members located throughout the world. ASTM publishes approximately 10,000 standards each year in the 73 volumes of the *Annual Book of ASTM Standards.*

Function. ASTM develops six principal types of full-consensus standards:

- *Standard test method.* This is a definitive procedure for the identification, measurement, and evaluation of one or more qualities, characteristics, or properties of a material, product, system, or service that produces a test result.
- *Standard specification.* This is a precise statement of a set of requirements to be satisfied by a material, product, system, or service that also indicates the procedures for determining whether each of the requirements is satisfied.
- *Standard practice.* A definitive procedure for performing one or more specific operations or functions that does not produce a test result.
- *Standard terminology.* A document comprising terms, definitions, descriptions of terms, and explanations of symbols, abbreviations, or acronyms.
- *Standard guide.* A series of options or instructions that do not recommend a specific course of action.
- *Standard classification.* A systematic arrangement or division of materials, products, systems, or services into groups based on similar characteristics such as origin, composition, properties, or use.

The ASTM Standards-Making System. Like most consensus standards organizations, ASTM begins its standards-making activities when a need is identified. Task group members prepare a draft standard, which is reviewed by its parent subcommittee through a letter ballot. After the subcommittee approves the document, it is submitted to a main committee letter ballot. Once approved at the main committee level, the document is submitted for balloting to ASTM. All negative votes cast during the balloting process, which must include a written explanation of the voter's objections, must be fully considered before the document can be submitted to the next level in the process. Final approval of a standard depends on concurrence by the ASTM Committee on Standards that proper procedures were followed and due process was achieved.

American Society of Safety Engineers (ASSE)

Overview. ASSE was founded in 1911. It is a membership organization composed of safety professionals throughout the world. The 33,000 members manage, supervise, and consult on safety, health, and environmental issues in industry, government, and education.

ASSE's mission is to enhance the status and promote the advancement of the safety profession and to foster the technical, scientific, managerial, and ethical knowledge, skills, and competency of safety professionals.

ASSE has 32,600 members who manage, supervise, and consult on safety, health, and environmental issues in industry, insurance, government, and education. Safety professionals help prevent accidents, injuries, and occupational diseases; create safe environments for work and leisure; and develop safe products for use in all areas of human activity.

ASSE is guided by a 15-member board of directors, which consists of eight regional vice presidents; three council vice presidents; and the Society president, president-elect, senior vice president, vice president of finance, and executive director. The board is guided in decision making by five standing committees and four councils.

Function. ASSE meets its responsibilities by providing continuing education, membership services, standards-making activities, peer recognition awards, marketing, and safety public relations.

Continuing Education. ASSE sponsors professional development seminars, publications, conference proceedings, training packages, computer resources, and audiovisual aids.

Membership Services. ASSE provides a variety of related membership services including professional development seminars, technical publications, and leadership conferences.

ASSE Standards. ASSE serves as the secretariat for a number of American national standards. Those of special interest to electrical safety include Eye and Face Protection (Z87.1), Safety Requirements for Confined Spaces (Z117.1), and Safety Requirements for Ladders (A14.1 to A14.8).

Occupational Safety and Health Administration (OSHA)

Standards issued by OSHA differ from other standards in one very significant aspect—they are enforceable under federal law. The following sections outline some of the more important facts about OSHA and its standards. You should contact your local OSHA office if more detailed information is required.

The Occupational Safety and Health Act of 1970. Part of the material in the following sections was drawn from the government publication OSHA 2056 titled *All About OSHA*.

In 1970, the United States was faced with the following statistics:

- Job-related accidents accounted for more than 14,000 worker deaths per year.
- Nearly 2.5 million workers were disabled.
- Ten times as many person-days were lost from job-related disabilities than from strikes.
- Estimated new cases of occupational diseases totaled 300,000.

To address these problems, Congress passed the Occupational Safety and Health Act on December 29, 1970 (the Act). The execution of the Act is the responsibility of the Department of Labor under the secretary of labor. The Act created OSHA to continuously create, review, and redefine specific standards and practices.

Coverage under the Act extends to all employers and their employees in the 50 states, the District of Columbia, Puerto Rico, and all other territories under federal government jurisdiction. The Act exempts three important categories:

- Self-employed persons
- Farms at which only immediate members of the farm employer's family are employed
- Working conditions regulated by other federal agencies under other federal statutes

The OSH Act states:

(1) An Act

> "To assure safe and healthful working conditions for working men and women; by authorizing enforcement of the standards developed under the Act; by assisting and encouraging the States in their efforts to assure safe and healthful working conditions; by providing for research, information, education, and training in the field of occupational safety and health; and for other purposes."

The Act has special provisions for some employees and employers. State and local governments and the federal government have special consideration under the Act. Employers and employees should contact their local OSHA office to determine what their obligations are. Section 4. Applicability of This Act, exempts federal agencies as follows:

(b)(1) Nothing in this Act shall apply to working conditions of employees with respect to which other Federal agencies, and State agencies acting under section 274 of the Atomic Energy Act of 1954, as amended (42 U.S.C. 2021), exercise statutory authority to prescribe or enforce standards or regulations affecting occupational safety or health.

However, federal agencies are not exempt from providing safe workplaces for their employees. Section 19 of the OSH Act clarifies this as follows:

19. Federal Agency Safety Programs and Responsibilities
(a) It shall be the responsibility of the head of each Federal agency (not including the United States Postal Service) to establish and maintain an effective and comprehensive occupational safety and health program which is consistent with the standards promulgated under section 6. The head of each agency shall (after consultation with representatives of the employees thereof)—29 USC 668:
(1) Provide safe and healthful places and conditions of employment, consistent with the standards set under section 6;
(2) Acquire, maintain, and require the use of safety equipment, personal protective equipment, and devices reasonably necessary to protect employees;
(3) Keep adequate records of all occupational accidents and illnesses for proper evaluation and necessary corrective action; 84 Stat. 1610 Recordkeeping.
(4) Consult with the Secretary with regard to the adequacy as to form and content of records kept pursuant to subsection (a)(3) of this section; and
(5) Make an annual report to the Secretary with respect to occupational accidents and injuries and the agency's program under this section. Such report shall include any report submitted under section 7902(e)(2) of title 5, United States Code. Annual Report.
(b) The Secretary shall report to the President a summary or digest of reports submitted to him under subsection (a)(5) of this section, together with his evaluations of and recommendations derived from such reports. 80 Stat. 530.

OSHA's Purpose. OSHA is the administration that is given the principal responsibilities for implementation of the Act. Following are a few of OSHA's responsibilities:

- Encourage employers and employees to reduce workplace hazards and to implement new or improved existing safety and health programs.
- Establish separate but dependent responsibilities and rights for employers and employees with respect to achieving safe and healthful working conditions.
- Set and enforce mandatory occupational safety and health standards applicable to businesses affecting interstate commerce.
- Maintain a reporting and record-keeping system to monitor job-related illnesses and injuries.
- Establish training programs to increase the number and competence of occupational safety and health personnel.
- Provide for the development, analysis, evaluation, and approval of state occupational safety and health programs.

OSHA is a branch of the Department of Labor and is responsible to the secretary of labor.

Responsibilities and Rights of Employers. Employers have the principal responsibility to guarantee that all OSHA rules and regulations are properly and promptly administered in the workplace. Employers who do not carry out their responsibilities can be heavily fined. The following list identifies some of the more significant employer responsibilities. Employers must

- Meet their general responsibility to provide a workplace free from recognized hazards that are causing or are likely to cause death or serious physical harm to employees.
- Be familiar with mandatory OSHA standards and make copies available to employees for review upon request.
- Inform all employees about OSHA.
- Examine workplace conditions to make sure they conform to applicable standards.
- Minimize or reduce hazards.

- Make sure employees have, use, and maintain safe tools and equipment. These tools and equipment include appropriate personal protective equipment such as flash suits, rubber gloves, and hard hats.
- Use color codes, posters, labels, or signs when needed to warn employees of potential hazards.
- Establish or update operating procedures and communicate them so that employees follow safety and health requirements.
- Provide medical examinations when required by OSHA standards.
- Provide training required by OSHA standards. For example, the Safety-Related Work Practices rule and the Control of Hazardous Energy Sources rule both require a certain amount of employee training for complete implementation.
- Report to the nearest OSHA office within 48 hours any fatal accident or one that results in the hospitalization of five or more employees.
- Keep OSHA-required records of work-related injuries and illnesses, and post a copy of the totals from the last page of OSHA No. 200 during the entire month of February of each year. *Note:* This applies only to employers with 11 or more employees and employers who are chosen by the Bureau of Labor Statistics.
- Post the OSHA poster (OSHA 2203) informing employees of their rights and responsibilities at a prominent location in the workplace. In states with OSHA-approved state job safety and health programs, the state's equivalent poster and/or the OSHA poster may be required.
- Provide employees, former employees, and their representatives access to the Log and Summary of Occupational Injuries and Illnesses (OSHA No. 200) at a reasonable time and in a reasonable manner.
- Provide access to employee medical records and exposure records to employees or their authorized representatives.
- Cooperate with the OSHA compliance officer by furnishing names of authorized employee representatives who may be asked to accompany the compliance officer during an inspection. (If none, the compliance officer will consult with a reasonable number of employees concerning safety and health in the workplace.)
- Not discriminate against employees who properly exercise their rights under the Act.
- Put OSHA citations at or near the work site involved.
- Abate cited violations within the prescribed period.

Employers also have rights under the OSHA act. The following items list some of the more important employer's rights. Employers have the right to

- Seek advice and off-site consultation as needed by writing, calling, or visiting the nearest OSHA office. OSHA will not inspect merely because an employer requests assistance.
- Be active in their industry association's involvement in safety and health.
- Request and receive proper identification of the OSHA compliance officer prior to inspection.
- Be advised by the compliance officer of the reason for an inspection.
- Have an opening and closing conference with the compliance officer.
- Accompany the compliance officer on the inspection.
- File a Notice of Contest with the OSHA area director within 15 working days of receipt of a notice of citation and proposed penalty.
- Apply to OSHA for a temporary variance from a standard if unable to comply because of the unavailability of materials, equipment, or personnel needed to make necessary changes within the required time.
- Apply to OSHA for a permanent variance from a standard if they can furnish proof that their facilities or method of operation provide employee protection at least as effective as that required by the standard.

- Take an active role in developing safety and health standards through participation in OSHA standards and advisory committees, through nationally recognized standards-setting organizations, and through evidence and views presented in writing or at hearings.
- Be assured of the confidentiality of any trade secrets observed by an OSHA compliance officer during inspection.
- Submit a written request to the National Institute of Occupational Safety and Health (NIOSH) for information on whether any substance in their workplace has potentially toxic effects in the concentrations being used.

Responsibilities and Rights of Employees. The Act was designed to establish workplace rules and regulations that will enhance the health and safety of employees. Because of this, the employers are considered to have principal responsibility for the implementation of OSHA rules. Employees, therefore, are not cited for violations. Despite this, employees still have responsibilities under the Act. Employees must

- Read the OSHA poster at the job site.
- Comply with all applicable OSHA standards.
- Follow all employer safety and health rules and regulations, and wear or use prescribed protective equipment while engaged in work.
- Report hazardous conditions to the supervisor.
- Report any job-related injury or illness to the employer.
- Promptly seek treatment for job-related injuries or illnesses.
- Cooperate with the OSHA compliance officer conducting an inspection if he or she inquires about safety and health conditions in your workplace.
- Exercise their rights under the Act in a responsible manner.

Employees also have significant rights under the Act. Employees have a right to

- Seek safety and health on the job without fear of punishment.
- Complain to an employer, union, OSHA, or any other government agency about job safety and health hazards.
- File safety or health grievances.
- Participate on a workplace safety and health committee or in union activities concerning job safety and health.
- Participate in OSHA inspections, conferences, hearings, or other OSHA-related activities.

OSHA's Electrical Safety Standards. OSHA safety standards are published in the Code of Federal Regulations, Title 29, Subtitle B, Chapter XVII. OSHA has issued or will issue electrical safety standards in at least four categories:

1. Design and installation safety
2. Safety-related work practices
3. Safety-related maintenance requirements
4. Special equipment

In addition, OSHA has issued or will issue standards regulating a variety of related areas such as Control of Hazardous Energy. Electrical safety standards for general industry are found in Part 1910, Subpart S, which includes Paragraphs 1910.301 through 1910.399. Electrical safety standards for employees in the construction industry are found in Part 1926, Subpart K, which includes Paragraphs 1926.400 through 1926.449. Table 7.1 identifies the various locations for the standards.

TABLE 7.1 Partial Listing of Applicable OSHA Safety Standards and the Industries to Which They Apply

Title	Industry	Paragraphs	Comments
Design Safety Standards for Electrical Equipment	General Industry and Maritime	1910.302–1910.308	
Safety-Related Work Practices	General Industry and Maritime	1910.331–1910.335	
Safety-Related Maintenance Requirements	General Industry and Maritime	1910.361–1910.380	Not implemented
Safety Requirements for Special Equipment	General Industry and Maritime	1910.381–1910.398	Not implemented
Installation Safety Requirements	Construction Industry	1926.402–1926.408	
Safety-Related Work Practices	Construction Industry	1926.416–1926.417	
Safety-Related Maintenance and Environmental Considerations	Construction Industry	1926.431–1926.432	
Safety Requirements for Special Equipment	Construction Industry	1926.441–1926.448	Only batteries implemented
Power Generation, Transmission, and Distribution	Construction Industry	1926.950–1926.960	
Power Transmission and Distribution	Power Generation, Transmission, and Distribution	1910.269	
Communications Safety	Telecommunications	1910.268	

TABLE 7.2 Six-Step OSHA Technical Consultation Process

Employer requests assistance	The employer contacts the state agency responsible for OSHA consultation services. Each state has its own agency. Phone numbers may be obtained by contacting the local OSHA office or requesting OSHA bulletins #2056 and #3047. Employer must make initial commitment to correct any observed discrepancies.
Consultant holds on-site opening conference	In this conference, the ground rules for the consultation are thoroughly explained to the employer, the plan for the walk-through visit is determined, and any preliminary questions are answered.
Walk-through inspection	The employer representative and the consultant perform a walk-through inspection of the workplace. Specific hazards are noted and an examination of the aspects of the employer's safety and health program that relate to the scope of the visit are observed.
Consultant holds an on-site closing conference	The closing conference is used to apprise the employer of any extremely serious violations that were observed. Generally only those hazards that represent an immediate and serious hazard to employees are itemized.
Written report	The consultant issues a written report in which the hazards and/or violations of OSHA regulations are delineated.
Corrective actions	The employer corrects those problems that were outlined by the consultant.

Note that by its very nature, Table 7.1 is time-sensitive. Be certain to check the current edition of the OSHA standard for complete information. These standards can be procured from your local OSHA office. The various OSHA standards will be covered in more detail later in this chapter.

OSHA's Other Standards for the Electrical Industry. OSHA electrical standards, which are also published in the Code of Federal Regulations, Title 29, Part 1910, include Subpart J, General Environmental Controls, and Subpart R, Special Industries. Subpart J deals with 1910.146, Permit Required Confined Spaces, and 1910.147, Control of Hazardous Energy Sources (Lockout/Tagout). Subpart R deals with 1910.269, Electric Power Generation, Transmission, and Distribution.

These three additional regulations play a major role in protecting employees from confined space hazards; the release of stored or residual energy or exposure to primary energy sources such as steam, hydraulic, pneumatic, mechanical, or electrical; and the specific hazards and provide safe work practices for electric utilities.

OSHA Technical Consultation. OSHA will provide technical support and consultation at no charge. The consultation services are provided by state government agencies or universities employing professional safety and health consultants. Although intended for smaller employers with more hazardous operations, the consultation service is available to companies of all sizes.

The technical consultation starts when the employer requests the service and makes a commitment to correct any possible violations discovered by the inspector. The complete consultation involves six basic steps: the request, the opening conference, the walk-through inspection, the closing conference, the written report, and the corrective actions. Table 7.2 lists and describes each of these six steps.

Employers are exempted from inspections or fines by OSHA compliance personnel during the consultation process. In some cases, if employers can demonstrate that an effective safety and health program is in operation, they may be exempted from OSHA general schedule enforcement inspections. This exemption does not apply to complaint or accident investigations.

Voluntary Protection Programs. Voluntary protection programs (VPPs) (Table 7.3) are designed to extend worker protection beyond the minimums required by the OSHA standards. These three programs are designed to

1. Recognize outstanding achievements of those who have successfully incorporated comprehensive safety and health programs into their total management system.

TABLE 7.3 OSHA Voluntary Protection Programs

Star program	This is the most demanding and prestigious of the VPPs. It is open to any employer in any industry who has successfully managed a comprehensive safety and health program to reduce injury rates below the national average for the industry. Specific requirements for the program include management commitment and employee participation; a high-quality work site analysis program; hazard prevention and control programs; and comprehensive safety and health training for all employees. These requirements must all be in place and operating effectively.
Merit program	This is a stepping stone to the Star program. An employer with a basic safety and health program built around the Star requirements who is committed to improving the company's program and who has the resources to do so within a specified period of time may work with OSHA to meet Star requirements.
Demonstration program	This VPP is for companies that provide Star-quality worker protection in industries where certain Star requirements may not be appropriate or effective. It allows OSHA both the opportunity to recognize outstanding safety and health programs that would otherwise be unreached by VPP and to determine if general Star requirements can be changed to include these companies as Star participants.

2. Motivate others to achieve excellent safety and health results in the same outstanding way.
3. Establish a relationship between employers, employees, and OSHA that is based on cooperation rather than coercion or confrontation.

Voluntary protection programs are available in all states under federal jurisdiction, and some state OSHA agencies have similar programs. Employers should contact their local OSHA offices for details on the VPPs. OSHA reevaluates VPPs on a regular basis—annually for Merit and Demonstration programs and triennially for Star programs.

Training and Education. OSHA can provide training and education in one of three basic ways. First, OSHA's area offices are full-service centers that provide a wealth of information in the form of pamphlets, brochures, publications, audiovisual aids, technical advice, and availability for speaking engagements.

Second, the OSHA Training Institute in Des Plaines, Illinois, provides both basic and advanced training in areas such as electrical hazards, machine guarding, ventilation, and ergonomics. The institute is a full-service training facility with all necessary classrooms, laboratories, library, and an audiovisual unit. Training courses at the institute are open to federal and state compliance officers; state consultants; federal agency personnel; and private-sector employers, employees, and their representatives.

Finally, OSHA provides funds in the form of grants to nonprofit organizations to conduct workplace training and education in subjects for which OSHA believes there is a lack of current workplace training.

State OSHA Organizations. The Occupational Safety and Health Act encourages states to develop and operate state job safety and health plans. These plans are operated under OSHA guidance. OSHA funds up to 50 percent of the state program's operating costs.

The requirements for development of a state program are beyond the scope of this book. The entire setup process is monitored by OSHA with milestones and performance requirements established from the outset. At the end of the process, if the state program is operating at least as effectively as the federal OSHA program and other administrative requirements are met, OSHA issues final approval and federal authority ends in the areas over which the state has jurisdiction.

TABLE 7.4 Electrical Safety-Related Organizations and Their URLs

Organization	URL
American Burn Association	www.ameriburn.org
American National Standards Institute (ANSI)	www.ansi.org
American Society of Safety Engineers (ASSE)	www.assp.org
American Society for Testing and Materials (ASTM)	www.astm.org
International Electrotechnical Commission (IEC)	www.iec.ch
International Organization for Standardization (ISO)	www.iso.ch
Institute of Electrical and Electronics Engineers (IEEE)	www.ieee.org
National Electrical Safety Foundation (NESF)	www.esfi.org
InterNational Electrical Testing Association (NETA)	www.netaworld.org
National Fire Protection Association (NFPA)	www.nfpa.org
Occupational Safety and Health Administration (OSHA)	www.osha.gov
Petroleum and Chemical Industry Committee (PCIC)	www.ieeepcic.com
IAS Electrical Safety Workshop	www.ewh.ieee.org/cmte/ias-esw/
National Electrical Manufacturers Association	www.nema.org
Professional Electrical Apparatus Reconditioning League (PEARL)	www.pearl1.org
American Academy of Forensic Sciences (AAFS)	www.aafs.org
Canadian Society of Forensic Science (CSFS)	www.csfs.ca
American Society of Training and Development	https://thetranslationcompany.com/translation-directory/translation-associations/american-society-training-development/

Other Electrical Safety Organizations

There are many other organizations that are directly or indirectly involved with electrical safety issues. Table 7.4 identifies a number of groups, including those discussed in this chapter. Note that the World Wide Web is a rapidly changing structure, and that URLs change on a daily basis.

THE NATIONAL ELECTRICAL SAFETY CODE (NESC)—IEEE C-2

General Description

The NESC was first developed in 1913 by the National Bureau of Standards as a consensus standard and remains so today. The NESC is an American national standard published by the IEEE, the secretariat for the NESC. It is intended to provide practical rules for safeguarding personnel during the installation, operation, or maintenance of electric supply and communications lines and associated equipment.

The code has three general rules as follows:

1. All electric supply and communication lines and equipment shall be designed, constructed, operated, and maintained to meet its requirements.
2. The utilities, authorized contractors, or other entities, as applicable, performing design, construction, operation, or maintenance tasks for electric supply or communication lines or equipment covered by the NESC are held responsible for meeting the applicable requirements.
3. For all particulars not specified in the NESC, construction and maintenance should be done in accordance with accepted good practice for the given local conditions.

Industries and Facilities Covered

The NESC rules cover supply and communication lines, equipment, and associated work practices used by both public and private electric supply, communications, railway, or similar utilities. The rules also cover similar systems that are under the control of qualified persons. This means that facilities such as cogeneration plants, industrial complexes, or utility interactive systems are covered by the NESC.

Note that the NESC specifically states it does not cover installations in mines, ships, railway rolling equipment, aircraft, automotive equipment, or utilization wiring except in certain cases. It also specifies that building utilization requirements are covered by the National Electrical Code, ANSI/NFPA 70.

Technical and Safety Items Covered

The NESC covers five major areas in numerous parts. The following sections describe each of the major areas of coverage and list the type of information included in each. The information provided in the following sections is intended to be general information only. Always refer to the most recent edition of the NESC for specific information.

Grounding Methods for Electric Supply and Communications Facilities. This section covers the methods to implement protective grounding for electric supply and communications facilities. The information is technical in nature and includes topics such as the point of connection, the composition of the grounding conductor, the means of connection, grounding electrodes, and many other grounding topics.

Rules for Installation and Maintenance of Electric Supply Stations and Equipment. This section covers those parts of the electrical supply system and associated structural arrangements that are accessible only to qualified personnel. It also covers the conductors and equipment employed primarily for the utilization of electric power when such conductors and equipment are used by the utility in the exercise of its function as a utility. Protective arrangements in electric supply stations, illumination, installation and maintenance of equipment, rotating equipment, storage batteries, transformers and regulators, conductors, circuit breakers, switchgear and enclosed metal bus, and surge arresters are included in this section.

Safety Rules for the Installation and Maintenance of Overhead Electric Supply and Communications Lines. This section applies to overhead lines, associated structural assemblies, and extension of such lines into buildings. Typical examples of topics included in this part are inspection and tests of lines and equipment, grounding of circuits, arrangement of switches, relations between various classes of equipment, joint use of structures, clearances, grades of construction, line insulation, and other such topics.

Safety Rules for the Installation and Maintenance of Underground Electric Supply and Communication Lines. This part is similar to the previous part except the equipment it includes is underground equipment. Topics include general requirements applying to underground lines, underground conduit systems, supply cable, underground structures, direct buried cable, risers, supply cable terminations, and installations in tunnels.

Rules for the Operation of Electric Supply and Communications Lines and Equipment. This final part of the NESC discusses rules for the operation of supply and communications lines and equipment. The first major section provides operational rules for both communications and supply employees. Personal general precautions, general operating routines, overhead line operating procedures, and underground line operating procedures are also included. Included in these work rules are the requirements for performing an arc-flash hazard analysis. The NESC, Section 410.A.3, requires a hazard assessment to determine if an arc-flash hazard exists. It also determines what level of arc-flash personal protective equipment (PPE) would be required. This requirement is consistent with the proposed revision of OSHA 29 CFR 1910.269 and 1926 Subpart V for arc-flash hazard analysis and PPE. See the latest edition of the NESC and OSHA regulations for the specific requirements for performing an electrical arc-flash hazard analysis and for selecting the properly rated PPE to protect employees.

The second major section has additional rules that pertain to communications workers. Approaching energized conductors, joint-use structures, attendant on surface at joint-use manhole, and sheath continuity rules are included.

The final major part lists additional rules that pertain strictly to electrical supply employees. This section has six significant sections including energized conductors or parts, switching control procedures, working on energized lines, de-energizing equipment or lines, protective grounds, and live-line work.

THE NATIONAL ELECTRICAL CODE (NEC)—NFPA 70

General Description

The NEC is arguably the oldest of all the various standards. It is an installation design standard that covers industrial, commercial, and residential electric utilization systems.

The NEC was first developed in 1897 by a variety of insurance, engineering, architectural, electrical, and other such interest groups. In 1911, the sponsorship of the Code was assumed by NFPA. Later, the NEC was approved as an American national standard by ANSI. The NEC is only one of many fire safety codes published by NFPA.

The NEC is unique among the other codes and standards in a very significant way—all or portions of it have been adopted as local law in many states, municipalities, cities, and other such areas. Many countries outside of the United States have adopted the NEC in its entirety, or in part, for their national installation codes. Note that both NFPA and ANSI consider the NEC to be purely an advisory document.

The NEC has become a large document with a wealth of design, installation, and general technical information included. The principal objective of the NEC is to provide standardized, technically sound electrical installation standards to help minimize the possibility of electric fires.

Industries and Facilities Covered

The NEC covers four basic types of installations. Note that all of these installations are so-called utilization installations as opposed to transmission and distribution.

1. Installations of electric conductors and equipment within or on public and private buildings or other structures, including mobile homes, recreational vehicles, and floating buildings. This section also includes premises wiring for such facilities as yards, carnival lots, parking lots, and industrial substations.
2. Installations of conductors and equipment that connect to the supply of electricity.
3. Installations of other outside conductors and equipment on the premises.
4. Installations of optical fiber cable.

The NEC also lists several installations to which it does not apply including ships, railway rolling stock, underground mine installations, railway generation, communications equipment under the control of communications utilities, and certain facilities under the control of electric utilities. The utility facilities that are excluded are the communications, metering, generation, transformation, transmission, and distribution in buildings or substations used specifically for that purpose. The NEC does include utility office buildings and other such public and commercial structures.

Technical and Safety Items Covered

The NEC includes technical specifications and requirements for virtually all electric lines and equipment. The Code has multiple chapters that include rules for wiring and protection, wiring methods and materials, equipment for general use, special occupancies, special equipment, special conditions, and communications systems.

Tables for conductor ampacities are included along with formulas for calculating the required size and number of conductors for special applications, as well as sizing requirements for protective devices. Grounding requirements including what, when, and how are included. Rules for interrupting capacities and other such information are also included. The NEC has become an engineering reference, an installation guide, a design standard, and a fire safety code all in one volume.

ELECTRICAL EQUIPMENT MAINTENANCE—NFPA 70B

General Description

The Electrical Equipment Maintenance document (70B) is a companion publication to the NEC. Many interested parties had been requesting that safety-related maintenance requirements be added to the NEC. A special committee determined that the NEC, which is basically a design and installation standard, is not the correct vehicle for maintenance information.

In studying the need for this document, the committee determined that electrical safety concerns logically fall into four basic categories: product design standards, installation standards (covered by the NEC and the NESC), safety-related maintenance information (covered by NFPA 70E, Chapter 2 and NFPA 70B), and usage instructions. Because of the high number of injuries and deaths related to improper or nonexistent equipment maintenance practices, the committee decided that an electric equipment maintenance document should be developed. To give the document as much prestige as possible, the decision was reached to make this new document a companion piece to the NEC. The final result, NFPA 70B, contains recommended practice information for proper maintenance of electric equipment.

Industries and Facilities Covered

Document 70B covers maintenance of industrial-type electric systems and equipment. The covered systems and equipment are typical of those installed in commercial buildings, industrial plants, and large multifamily dwellings. Document 70B specifically states that it does not cover consumer appliances or home-type equipment. It also states that it is not intended to supersede manufacturer-recommended maintenance procedures.

Technical and Safety Items Covered

Document 70B includes 31 chapters and 16 appendices. One of the most important chapters develops a concise case for the implementation of proper preventive maintenance procedures. Subsequent chapters cover maintenance fundamentals and how to plan a preventive-maintenance program. Then 12 chapters provide specific, detailed maintenance techniques and expected results for switchgear, circuit breakers, cables, motor control centers, rotating equipment, wiring devices, and other such electric equipment.

One chapter details the specifics of electrical testing in areas such as insulation resistance, protective-device testing, infrared, fault-gas analysis, and many other such modern testing procedures. Still other chapters cover the maintenance of equipment that is subject to long intervals between shutdowns and methods to be used for de-energizing equipment in such a way that maintenance personnel are protected.

The appendices include information on walk-through inspections, instruction techniques, symbols, diagrams, and a variety of recommended test sheets and forms to be used for maintenance programs.

STANDARD FOR ELECTRICAL SAFETY IN THE WORKPLACE—NFPA 70E

General Description

Consensus standards are developed by personnel who are intimately familiar with the safety hazards of any given discipline; therefore, they will be accurate and extremely useful to the purpose for which they were intended. Because of this, when OSHA develops a safety regulation, it prefers to work with existing consensus standards. In 1976, the NFPA formed a committee to assist OSHA in the preparation of its electrical safety regulations.

The NFPA 70E is divided into three chapters: Chap. 1, Safety-Related Work Practices; Chap. 2, Safety-Related Maintenance Requirements; and Chap. 3, Safety Requirements for Special Equipment. The 2018 edition also includes 17 Annexes: Annex A, Informative Publications; Informative Annex B, Reserved; Informative Annex C, Limits of Approach; Informative Annex D, Incident Energy and Arc Flash Boundary Calculation Methods; Informative Annex E, Electrical Safety Program; Informative Annex F, Risk Assessment and Risk Control; Informative Annex G, Sample Lockout/Tagout Program; Informative Annex H, Guidance on Selection of Protective Clothing and Other Personal Protective Equipment; Informative Annex J, Energized Electrical Work Permit; Informative Annex K, General Categories of Electrical Hazards; Informative Annex L, Typical Application of Safeguards in the Cell Line Working Zone; Informative Annex M, Layering of Protective Clothing and Total System Arc Rating; Informative Annex N, Example Industrial Procedures and Policies for Working Near Overhead Electrical Lines and Equipment; Informative Annex O, Safety-Related Design Requirements; Informative Annex P, Aligning Implementation of This Standard with Occupational Health and Safety Management Standards; and Informative Annex Q, Human Performance and Workplace Electrical Safety.

Many industries are using the NFPA 70E as the key element in the development of their electrical safety programs and procedures for at least three reasons:

1. OSHA has identified NFPA 70E to be part of, and integral to, its electrical safety rule-making effort.
2. OSHA maintains representation on the NFPA 70E committee, thereby helping to ensure that the NFPA 70E is consistent with the OSHA requirements and intent for electrical safety.
3. NFPA 70E is more detailed and performance-oriented than the equivalent OSHA regulations; consequently, 70E is easier to employ as the basis for a good electrical safety program.

Industries and Facilities Covered

NFPA 70E is a companion document to the NEC; therefore, it covers exactly the same industries as the NEC. Specifically, NFPA 70E addresses electrical safety requirements that are necessary for safeguarding employees in the workplace. It covers three major installations:

1. Building electrical utilization installations including carnival and parking lots, mobile homes and recreational vehicles, and industrial substations
2. Conductors that connect installations to a supply of electricity
3. Other outside premises wiring

Document 70E specifically excludes four basic categories of installation:

1. Ships, watercraft, railway rolling stock, aircraft, and automotive vehicles other than mobile homes and recreational vehicles

2. Installations of railways for generation, transformation, transmission, or distribution of power used exclusively for operation of rolling stock or installations used exclusively for signaling and communication purposes
3. Installation of communication equipment under the exclusive control of communications utilities, located outdoors or in building spaces used exclusively for such installations
4. Installations under the exclusive control of electric utilities used for communication, metering, or for the generation, control transformation, transmission, and distribution of electric energy located in buildings used exclusively by utilities for such purposes or located outdoors on property owned or leased by the utility or on public highways, streets, roads, and so on, or outdoors by established rights on private property

Technical and Safety Items Covered

Chapter 1, Safety-Related Work Practices. Chapter 1 of 70E is intended to give directions to employees in how to perform their work in such a manner that they can avoid electrical hazards. It includes information on safety-related work practices and procedures for employees who are exposed to an electrical hazard, safe de-energizing of electrical equipment and systems, lockout-tagout programs and procedures, safe use of tools and equipment, personal and other protective equipment, safe approach distances for overhead lines, reclosing protective devices after automatic operation, qualified and unqualified employee electrical safety training requirements, and safe voltage-measurement practices.

The majority of the material in Chap. 1 was adopted by OSHA and is included in its safety-related work practices regulation in 29 CFR 1910.331-.335.

Chapter 2, Safety-Related Maintenance Requirements. Chapter 2 includes a variety of information on maintenance; for the most part, it addresses the basic, commonsense procedures for many common types of electric equipment and the requirements for qualified persons only performing electrical equipment maintenance. Included are substation and switchgear assemblies, premises wiring, controller equipment, fuses and molded-case circuit breakers, rotating electric equipment, portable electric tools and equipment, safety and protective equipment, hazardous (classified) locations, and batteries and battery rooms.

Chapter 2 has only minimal broad recommendations. Specific maintenance techniques needed to meet the provisions of Chap. 2 are found in NFPA 70B, *Recommended Practice for Electrical Equipment Maintenance,* ANSI/NETA MTS, *Standard for Maintenance Testing Specifications for Electrical Power Distribution Equipment and Systems,* and IEEE 3007.2, *IEEE Recommended Practice for the Maintenance Industrial and Commercial Power Systems.*

Chapter 3, Safety Requirements for Special Equipment. Chapter 3 covers special safety requirements for employees working on or near electrolytic cells, batteries and battery rooms, lasers, and power electronics equipment. These sections provide critical information for personnel working on or near such equipment.

AMERICAN SOCIETY FOR TESTING AND MATERIALS (ASTM) STANDARDS

ASTM publishes a variety of standards primarily associated with safety equipment design, usage, and testing. Table 7.5 summarizes some of the ASTM standards and briefly describes their coverage. Refer to Chap. 2 for many specific references and uses of the ASTM standards. ASTM Section 10, Volume 10.03 is a volume of ASTM standards that is crucial to the electrical industry, addressing all of the required electrical PPE.

TABLE 7.5 Summary of ASTM Standards for Electrical Protective Equipment

Standard number/name	Description
D 120/Specification for Rubber Insulating Gloves	Acceptance testing of rubber insulating gloves used for worker protection.
D 178/Specification for Rubber Insulating Matting	Acceptance testing of rubber insulating matting used as floor covering.
D 1048/Specification for Rubber Insulating Blankets	Acceptance testing of rubber insulating blankets.
D 1049/Specification for Rubber Insulating Covers	Acceptance testing of rubber insulating covers, insulator hoods, line hose connectors, and so on.
D 1050/Specification for Rubber Insulating Line Hose	Acceptance testing of rubber line hose.
D 1051/Specification for Rubber Insulating Sleeves	Acceptance testing of rubber insulating sleeves.
F 478/In-Service Care of Insulating Line Hose and Covers	Care, inspection, testing, and use voltage of insulating line hose and covers.
F 479/In-Service Care of Insulating Blankets	Care, inspection, testing, and use voltage of insulating blankets for shock protection.
F 496/In-Service Care of Insulating Gloves and Sleeves	Care, inspection, testing, and use voltage of insulating gloves and sleeves for shock protection.
F 696/Specification for Leather Protectors for Rubber Insulating Gloves and Mittens	Specification for the design and manufacture of leather protectors for rubber insulating goods. Note that OSHA requires leather protectors over rubber insulating goods if they may be damaged.
F 711/Specification for Fiberglass Reinforced Plastic Rod and Tube Used in Live Line Tools	Specification for the design and manufacture of insulating foam-filled tubes and rods made from fiberglass-reinforced plastic used for live line tools.
F 712/Test Methods for Electrical Insulating Plastic Guard Equipment for Protection of Workers	Withstand voltage, flashover voltage, and leakage current test methods are defined for line guards, guard connectors, deadend covers, pole guards, and other such equipment.
F 819 of Terms Relating to Electrical Protective Equipment	A concise definition list of key terms.
F 855/Specifications for Temporary Grounding Systems to Be Used on De-Energized Electric Power Lines and Equipment	Construction, testing, and materials used for temporary grounding systems.
F 887/Specifications for Personal Climbing Equipment	Acceptance testing specifications for climbers, climber straps, body belts, and pole straps.
F 914/Test Methods for Acoustic Emission for Insulated Aerial Personnel Devices	Acoustic emission testing methods for insulated aerial personnel devices.

F 968/Specification for Electrically Insulating Plastic Guard Equipment for Protection of Workers	Construction and materials specification for line guards, guard connectors, deadend covers, and other such equipment.
F 1116/Test Method for Determining Dielectric Strength of Overshoe Footwear	Recommended methods for determining the dielectric strength of insulating overshoe footwear.
F 1117/Specification for Dielectric Overshoe Footwear	Acceptance testing of overshoe footwear.
F 1236/Guide for Visual Inspection of Electrical Protective Rubber Products	Recommended methods and techniques for proper in-service inspection of electrical protective rubber products. Chapter 4 describes many of these methods.
F 1505/Specification for Insulated and Insulating Hand Tools	Design, construction, testing, and materials to be used in insulated hand tools below 600 V.
F 1506/Performance Specification for Textile Materials for Wearing Apparel for Use by Electrical Workers Exposed to Momentary Electric Arc and Related Thermal Hazards	Covers the flame-resistance requirements for materials used in electrical workers' clothing. Covers only classic flame testing. Predates the APTV requirement.
F 1564/Specification for Structure-Mounted Insulating Work Platforms for Electrical Workers	Design testing standard for mechanical and electrical characteristics of structure-mounted work platforms. Covers only single-worker platforms.
F 1742/Specification for PVC Insulating Sheeting	Design specs for PVC sheeting material.
F 1796/Specification for High Voltage Detectors—Part 1	Design and construction requirements for capacitive voltage detectors 600 V to 80 kV.
F 1825/Specification for Fixed Length Clampstick T Live Line Tools	Establishes the technical specifications for the design and manufacture of clampsticks.
F 1826/Specification for Telescoping Live Line Tools	Establishes the technical specifications for the design and manufacture of telescoping live line tools.
F 1891/Specification for Arc and Flame-Resistant Rainwear	Similar to F 1506, but covers rainwear materials.
F 1959/Standard Test Method for Determining Arc Thermal Performance Value of Materials for Clothing	Establishes test values and techniques for APTV of flame-resistant clothing.

OCCUPATIONAL SAFETY AND HEALTH ADMINISTRATION (OSHA) STANDARDS

Overview

OSHA standards represent federal law and must be implemented by all covered industries, except in those states that have adopted state safety programs that are approved by OSHA. OSHA has been developing many electrical and related safety standards since its creation in 1970 with the first electrical regulation becoming Final Rule on January 16, 1981.

Because OSHA standards apply so universally, several of the more important ones are reproduced in their entirety, along with explanations from the *Federal Register* preamble as well as the Directorate of Compliance, in this section. *Note*: The standards reproduced in this section are the current standards at the date of publication of this handbook. The user should contact OSHA for the current, up-to-date version of each standard. All OSHA standards are kept updated at its website http://www.osha.gov.

General Industry

OSHA 29 CFR 1910, Subpart S, comprising 1910.302 through 1910.308, Design Safety Standards, and 1910.399, Definitions, was revised and released as Final Rule on February 14, 2007. This revision was based on NFPA 70E-2000 with some additional revisions based on NEC-2002.

Scope. The general industry OSHA standards apply to electric utilization systems installed or used within or on buildings, structures, and other premises. Other premises include:

- Yards
- Carnivals
- Parking and other lots
- Mobile homes
- Recreational vehicles
- Industrial substations
- Conductors that connect the installations to a supply of electricity
- Other outside conductors on the premises

OSHA has issued or will issue standards on design safety; electrical safety-related work practices; lockout-tagout; maintenance-related safety; electric power generation transmission, and distribution; and special equipment safety. All general industry electrical safety standards are published in Subpart S of Part 1910. Some related standards are published in Subpart J and others in Subpart R. Always refer to the most recent edition. They are reproduced here for reference only.

Design Safety Standards. The design safety standards cover all utilization systems that have been installed or undergone major modification since March 15, 1972. These standards provide design and installation requirements. The design safety standards are found in the United States Code of Federal Regulations, Title 29, Part 1910, Subpart S, Paragraphs 1910.302 through 1910.308. They are reproduced in Fig. 7.1. *Note:* Explanations for these paragraphs are found in the *National Electrical Code Handbook.*

Safety-Related Work Practices. This OSHA standard covers the procedures required to ensure that employees use optimum safety-related procedures when working around energized or potentially energized equipment. It covers a variety of topics including lockout-tagout procedures. Electrical lockout-tagout must follow the methods given in Paragraph 1910.333; however, with certain additions, 1910.333 does allow the use of the Control of Hazardous Energy rule (1910.147).

The Electrical Safety-Related Work Practices rule is found in the United States Code of Federal Regulations, Title 29, Part 1910, Subpart S, Paragraphs 1910.331 through 1910.335. It is reproduced in this text, along with explanations extracted from the preamble of the *Federal Register* as well as the Directorate of Compliance, as Fig. 7.2. Additional requirements and explanations for electrical safety-related work practices can be found in the NFPA 70E handbook, *Standard for Electrical Safety in the Workplace.*

Power Generation, Transmission, and Distribution. Employees involved in the operation or maintenance of equipment used for power generation, transmission, or distribution and under the control of an electric utility, as well as equivalent installations of industrial establishments, are covered by this rule. The rule is published in the United States Code of Federal Regulations, Title 29, Part 1910, Subpart R, Paragraph 1910.269, Final Rule April 11, 2014. Figure 7.3 contains the Final Rule.

Control of Hazardous Energy Sources (Lockout/Tagout). This rule is not an electrical standard; in fact, section (a)(1)(ii)(C) specifically excludes electrical hazards. However, the Safety-Related Work Practices rule, Paragraph 1910.333(b)(2), Note 2 allows the use of the Lockout-Tagout rule with certain restrictions. Because of its importance, it is reproduced here, along with explanations from the preamble of the *Federal Register*, as Fig. 7.4. This rule is published in the United States Code of Federal Regulations, Title 29, Part 1910, Subpart J, Paragraph 1910.147.

Definitions. Figure 7.5 lists the definitions that OSHA uses in Subpart S of Part 1910. It is published in the United States Code of Federal Regulations, Part 1910, Subpart S, Paragraph 1910.399.

Construction Industry

Scope. OSHA publishes electrical safety standards for the construction industry in two subparts. Subpart K applies to construction sites and facilities exclusive of those intended exclusively for power generation, transmission, and distribution, which is covered in Subpart V. At the time of this publication, OSHA had not released the Final Rule for revisions of 29 CFR 1926, Subpart V. Contact OSHA for the status of this standard.

Installation Safety Requirements. Figure 7.6 is a reproduction of the Installation Safety Requirements for construction sites. This standard is published in the United States Code of Federal Regulations, Title 29, Part 1926, Subpart K, Paragraphs 1926.402 through 1926.408.

Safety-Related Work Practices. Because of the nature of electrical exposure of construction personnel, this rule is much shorter and less comprehensive than the safety-related work practices rule that applies to the general industry. This standard is published in the United States Code of Federal Regulations, Title 29, Part 1926, Subpart K, Paragraphs 1926.416 and 1926.417. Figure 7.7 is a reproduction of the rule.

Safety-Related Maintenance and Environmental Considerations (Construction). This standard is published in the United States Code of Federal Regulations, Title 29, Part 1926, Subpart K, Paragraphs 431-432. Figure 7.8 is a reproduction of this standard.

Safety Requirements for Special Equipment (Construction). Batteries are the only apparatus included in this rule to date. It is published in the United States Code of Federal Regulations, Part 1926, Subpart K, Paragraph 1926.441. Figure 7.9 reproduces this standard.

Definitions. Figure 7.10 lists the definitions that the OSHA uses in Subpart K of Part 1926. It is published in the United States Code of Federal Regulations, Part 1926, Subpart K, Paragraph 1910.449.

DESIGN SAFETY STANDARDS FOR ELECTRICAL SYSTEMS

Federal Register / Vol. 72, No. 30 / Wednesday, February 14, 2007

§ 1910.302 Electric utilization systems.

Sections 1910.302 through 1910.308 contain design safety standards for electric utilization systems.

(a) *Scope*—(1) *Covered.* The provisions of §§ 1910.302 through 1910.308 cover electrical installations and utilization equipment installed or used within or on buildings, structures, and other premises, including:

(i) Yards;

(ii) Carnivals;

(iii) Parking and other lots;

(iv) Mobile homes;

(v) Recreational vehicles;

(vi) Industrial substations;

(vii) Conductors that connect the installations to a supply of electricity; and

(viii) Other outside conductors on the premises.

(2) *Not covered.* The provisions of §§ 1910.302 through 1910.308 do not cover:

(i) Installations in ships, watercraft, railway rolling stock, aircraft, or automotive vehicles other than mobile homes and recreational vehicles;

(ii) Installations underground in mines;

(iii) Installations of railways for generation, transformation, transmission, or distribution of power used exclusively for operation of rolling stock or installations used exclusively for signaling and communication purposes;

(iv) Installations of communication equipment under the exclusive control of communication utilities, located outdoors or in building spaces used exclusively for such installations; or

(v) Installations under the exclusive control of electric utilities for the purpose of communication or metering; or for the generation, control, transformation, transmission, and distribution of electric energy located in buildings used exclusively by utilities for such purposes or located outdoors on property owned or leased by the utility or on public highways, streets, roads, etc., or outdoors by established rights on private property.

(b) *Extent of application*—(1) *Requirements applicable to all installations.* The following requirements apply to all electrical installations and utilization equipment, regardless of when they were designed or installed:

§ 1910.303(b)—Examination, installation, and use of equipment

§ 1910.303(c)(3)—Electrical connections—Splices

§ 1910.303(d)—Arcing parts

§ 1910.303(e)—Marking

§ 1910.303(f), except (f)(4) and (f)(5)—Disconnecting means and circuits

§ 1910.303(g)(2)—600 volts or less—Guarding of live parts

§ 1910.304(a)(3)—Use of grounding terminals and devices

§ 1910.304(f)(1)(i), (f)(1)(iv), and (f)(1)(v)—Overcurrent protection—600 volts, nominal, or less

§ 1910.304(g)(1)(ii), (g)(1)(iii), (g)(1)(iv), and (g)(1)(v)—Grounding—Systems to be grounded

§ 1910.304(g)(4)—Grounding—Grounding connections

§ 1910.304(g)(5)—Grounding—Grounding path

§ 1910.304(g)(6)(iv)(A) through (g)(6)(iv)(D), and (g)(6)(vi)—Grounding—Supports, enclosures, and equipment to be grounded

§ 1910.304(g)(7)—Grounding—Non-electrical equipment

§ 1910.304(g)(8)(i)—Grounding—Methods of grounding fixed equipment

§ 1910.305(g)(1)—Flexible cords and cables—Use of flexible cords and cables

§ 1910.305(g)(2)(ii) and (g)(2)(iii)—Flexible cords and cables—Identification, splices, and terminations

§ 1910.307, except as specified in

§ 1910.307(b)—Hazardous (classified) locations

(2) *Requirements applicable to installations made after March 15, 1972.* Every electrical installation and all utilization equipment installed or overhauled after March 15, 1972, shall comply with the provisions of §§ 1910.302 through 1910.308, except as noted in paragraphs (b)(3) and (b)(4) of this section.

FIGURE 7.1 Design Safety Standards for Electrical Systems (OSHA-CFR, Title 29, Part 1910, Paragraphs 302–308). (*Courtesy OSHA website.*)

(3) *Requirements applicable only to installations made after April 16, 1981.* The following requirements apply only to electrical installations and utilization equipment installed after April 16, 1981:

§ 1910.303(h)(4)—Over 600 volts, nominal—Entrance and access to work space
§ 1910.304(f)(1)(vii) and (f)(1)(viii)—Overcurrent protection—600 volts, nominal, or less
§ 1910.304(g)(9)(i)—Grounding—Grounding of systems and circuits of 1000 volts and over (high voltage)
§ 1910.305(j)(6)(ii)(D)—Equipment for general use—Capacitors
§ 1910.306(c)(9)—Elevators, dumbwaiters, escalators, moving walks, wheelchair lifts, and stairway chair lifts—Interconnection between multi-car controllers
§ 1910.306(i)—Electrically driven or controlled irrigation machines
§ 1910.306(j)(5)—Swimming pools, fountains, and similar installations—Fountains
§ 1910.308(a)(1)(ii)—Systems over 600 volts, nominal—Aboveground wiring methods
§ 1910.308(c)(2)—Class 1, Class 2, and Class 3 remote control, signaling, and power-limited circuits—Marking
§ 1910.308(d)—Fire alarm systems

(4) *Requirements applicable only to installations made after August 13, 2007.* The following requirements apply only to electrical installations and utilization equipment installed after August 13, 2007:

§ 1910.303(f)(4)—Disconnecting means and circuits—Capable of accepting a lock
§ 1910.303(f)(5)—Disconnecting means and circuits—Marking for series combination ratings
§ 1910.303(g)(1)(iv) and (g)(1)(vii)—600 volts, nominal, or less—Space about electric equipment
§ 1910.303(h)(5)(vi)—Over 600 volts, nominal—Working space and guarding
§ 1910.304(b)(1)—Branch circuits—Identification of multiwire branch circuits
§ 1910.304(b)(3)(i)—Branch circuits—Ground-fault circuit interrupter protection for personnel
§ 1910.304(f)(2)(i)(A), (f)(2)(i)(B) (but not the introductory text to § 1910.304(f)(2)(i)), and (f)(2)(iv)(A)—Overcurrent protection—Feeders and branch circuits over 600 volts, nominal
§ 1910.305(c)(3)(ii)—Switches—Connection of switches
§ 1910.305(c)(5)—Switches—Grounding
§ 1910.306(a)(1)(ii)—Electric signs and outline lighting—Disconnecting means
§ 1910.306(c)(4)—Elevators, dumbwaiters, escalators, moving walks, wheelchair lifts, and stairway chair lifts—Operation
§ 1910.306(c)(5)—Elevators, dumbwaiters, escalators, moving walks, wheelchair lifts, and stairway chair lifts—Location
§ 1910.306(c)(6)—Elevators, dumbwaiters, escalators, moving walks, wheelchair lifts, and stairway chair lifts—Identification and signs
§ 1910.306(c)(7)—Elevators, dumbwaiters, escalators, moving walks, wheelchair lifts, and stairway chair lifts—Single-car and multi-car installations
§ 1910.306(j)(1)(iii)—Swimming pools, fountains, and similar installations—Receptacles
§ 1910.306(k)—Carnivals, circuses, fairs, and similar events
§ 1910.308(a)(5)(v) and (a)(5)(vi)(B)—Systems over 600 volts, nominal—Interrupting and isolating devices
§ 1910.308(a)(7)(vi)—Systems over 600 volts, nominal—Tunnel installations
§ 1910.308(b)(3)—Emergency power systems—Signs
§ 1910.308(c)(3)—Class 1, Class 2, and Class 3 remote control, signaling, and power-limited circuits—Separation from conductors of other circuits
§ 1910.308(f)—Solar photovoltaic systems

(c) *Applicability of requirements for disconnecting means.* The requirement in § 1910.147(c)(2)(iii) that energy isolating devices be capable of accepting a lockout device whenever replacement or major repair, renovation or modification of a machine or equipment is performed, and whenever new machines or equipment are installed after January 2, 1990, applies in addition to any requirements in § 1910.303 through § 1910.308 that disconnecting means be capable of being locked in the open position under certain conditions.

§ 1910.303 General.

(a) *Approval.* The conductors and equipment required or permitted by this subpart shall be acceptable only if approved, as defined in § 1910.399.

FIGURE 7.1 *(Continued)*

(b) *Examination, installation, and use of equipment*—(1) *Examination.* Electric equipment shall be free from recognized hazards that are likely to cause death or serious physical harm to employees. Safety of equipment shall be determined using the following considerations:

(i) Suitability for installation and use in conformity with the provisions of this subpart;

Note to paragraph (b)(1)(i) of this section: Suitability of equipment for an identified purpose may be evidenced by listing or labeling for that identified purpose.

(ii) Mechanical strength and durability, including, for parts designed to enclose and protect other equipment, the adequacy of the protection thus provided;

(iii) Wire-bending and connection space;

(iv) Electrical insulation;

(v) Heating effects under all conditions of use;

(vi) Arcing effects;

(vii) Classification by type, size, voltage, current capacity, and specific use; and

(viii) Other factors that contribute to the practical safeguarding of persons using or likely to come in contact with the equipment.

(2) *Installation and use.* Listed or labeled equipment shall be installed and used in accordance with any instructions included in the listing or labeling.

(3) *Insulation integrity.* Completed wiring installations shall be free from short circuits and from grounds other than those required or permitted by this subpart.

(4) *Interrupting rating.* Equipment intended to interrupt current at fault levels shall have an interrupting rating sufficient for the nominal circuit voltage and the current that is available at the line terminals of the equipment. Equipment intended to interrupt current at other than fault levels shall have an interrupting rating at nominal circuit voltage sufficient for the current that must be interrupted.

(5) *Circuit impedance and other characteristics.* The overcurrent protective devices, the total impedance, the component short-circuit current ratings, and other characteristics of the circuit to be protected shall be selected and coordinated to permit the circuit protective devices used to clear a fault to do so without the occurrence of extensive damage to the electrical components of the circuit. This fault shall be assumed to be either between two or more of the circuit conductors, or between any circuit conductor and the grounding conductor or enclosing metal raceway.

(6) *Deteriorating agents.* Unless identified for use in the operating environment, no conductors or equipment shall be located in damp or wet locations; where exposed to gases, fumes, vapors, liquids, or other agents that have a deteriorating effect on the conductors or equipment; or where exposed to excessive temperatures.

(7) *Mechanical execution of work.* Electric equipment shall be installed in a neat and workmanlike manner.

(i) Unused openings in boxes, raceways, auxiliary gutters, cabinets, equipment cases, or housings shall be effectively closed to afford protection substantially equivalent to the wall of the equipment.

(ii) Conductors shall be racked to provide ready and safe access in underground and subsurface enclosures that persons enter for installation and maintenance.

(iii) Internal parts of electrical equipment, including busbars, wiring terminals, insulators, and other surfaces, may not be damaged or contaminated by foreign materials such as paint, plaster, cleaners, abrasives, or corrosive residues.

(iv) There shall be no damaged parts that may adversely affect safe operation or mechanical strength of the equipment, such as parts that are broken, bent, cut, or deteriorated by corrosion, chemical action, or overheating.

(8) *Mounting and cooling of equipment.* (i) Electric equipment shall be firmly secured to the surface on which it is mounted.

Note to paragraph (b)(8)(i) of this section: Wooden plugs driven into holes in masonry, concrete, plaster, or similar materials are not considered secure means of fastening electric equipment.

(ii) Electric equipment that depends on the natural circulation of air and convection principles for cooling of exposed surfaces shall be installed so that room airflow over such surfaces is not prevented by walls or by adjacent installed equipment. For equipment designed for floor mounting, clearance between top surfaces and adjacent surfaces shall be provided to dissipate rising warm air.

(iii) Electric equipment provided with ventilating openings shall be installed so that walls or other obstructions do not prevent the free circulation of air through the equipment.

FIGURE 7.1 (*Continued*)

(c) *Electrical connections*—(1) *General.* Because of different characteristics of dissimilar metals:

(i) Devices such as pressure terminal or pressure splicing connectors and soldering lugs shall be identified for the material of the conductor and shall be properly installed and used;

(ii) Conductors of dissimilar metals may not be intermixed in a terminal or splicing connector where physical contact occurs between dissimilar conductors (such as copper and aluminum, copper and copper-clad aluminum, or aluminum and copper-clad aluminum) unless the device is identified for the purpose and conditions of use; and

(iii) Materials such as solder, fluxes, inhibitors, and compounds, where employed, shall be suitable for the use and shall be of a type that will not adversely affect the conductors, installation, or equipment.

(2) *Terminals.* (i) Connection of conductors to terminal parts shall ensure a good connection without damaging the conductors and shall be made by means of pressure connectors (including set-screw type), solder lugs, or splices to flexible leads. However, No. 10 or smaller conductors may be connected by means of wire binding screws or studs and nuts having upturned lugs or equivalent.

(ii) Terminals for more than one conductor and terminals used to connect aluminum shall be so identified.

(3) *Splices.* (i) Conductors shall be spliced or joined with splicing devices identified for the use or by brazing, welding, or soldering with a fusible metal or alloy. Soldered splices shall first be spliced or joined to be mechanically and electrically secure without solder and then soldered. All splices and joints and the free ends of conductors shall be covered with an insulation equivalent to that of the conductors or with an insulating device identified for the purpose.

(ii) Wire connectors or splicing means installed on conductors for direct burial shall be listed for such use.

(d) *Arcing parts.* Parts of electric equipment that in ordinary operation produce arcs, sparks, flames, or molten metal shall be enclosed or separated and isolated from all combustible material.

(e) *Marking*—(1) *Identification of manufacturer and ratings.* Electric equipment may not be used unless the following markings have been placed on the equipment:

(i) The manufacturer's name, trademark, or other descriptive marking by which the organization responsible for the product may be identified; and

(ii) Other markings giving voltage, current, wattage, or other ratings as necessary.

(2) *Durability.* The marking shall be of sufficient durability to withstand the environment involved.

(f) *Disconnecting means and circuits*—(1) *Motors and appliances.* Each disconnecting means required by this subpart for motors and appliances shall be legibly marked to indicate its purpose, unless located and arranged so the purpose is evident.

(2) *Services, feeders, and branch circuits.* Each service, feeder, and branch circuit, at its disconnecting means or overcurrent device, shall be legibly marked to indicate its purpose, unless located and arranged so the purpose is evident.

(3) *Durability of markings.* The markings required by paragraphs (f)(1) and (f)(2) of this section shall be of sufficient durability to withstand the environment involved.

(4) *Capable of accepting a lock.* Disconnecting means required by this subpart shall be capable of being locked in the open position.

(5) *Marking for series combination ratings.* (i) Where circuit breakers or fuses are applied in compliance with the series combination ratings marked on the equipment by the manufacturer, the equipment enclosures shall be legibly marked in the field to indicate that the equipment has been applied with a series combination rating.

(ii) The marking required by paragraph (f)(5)(i) of this section shall be readily visible and shall state "Caution—Series Combination System Rated II Amperes. Identified Replacement Component Required."

(g) *600 volts, nominal, or less.* This paragraph applies to electric equipment operating at 600 volts, nominal, or less to ground.

(1) *Space about electric equipment.* Sufficient access and working space shall be provided and maintained about all electric equipment to permit ready and safe operation and maintenance of such equipment.

(i) Working space for equipment likely to require examination, adjustment, servicing, or maintenance while energized shall comply with the following dimensions, except as required or permitted elsewhere in this subpart:

(A) The depth of the working space in the direction of access to live parts may not be less than indicated in Table S-1. Distances shall be measured from the live parts if they are exposed or from the enclosure front or opening if they are enclosed;

(B) The width of working space in front of the electric equipment shall be the width of the equipment or 762 mm (30 in.), whichever is greater. In all cases, the working space shall permit at least a 90-degree opening of equipment doors or hinged panels; and

(C) The work space shall be clear and extend from the grade, floor, or platform to the height required by paragraph (g)(1)(vi) of this section. However, other equipment associated with the electrical installation and located

FIGURE 7.1 *(Continued)*

TABLE S-1 Minimum Depth of Clear Working Space at Electric Equipment, 600 V or Less

	Minimum clear distance for condition[2,3]					
	Condition A		Condition B		Condition C	
Nominal voltage to ground	m	ft	m	ft	m	ft
0–150	[1]0.9	[1]3.0	[1]0.9	[1]3.0	0.9	3.0
151–600	[1]0.9	[1]3.0	1.0	3.5	1.2	4.0

Notes to Table S-1:

1. Minimum clear distances may be 0.7 m (2.5 ft) for installations built before April 16, 1981.

2. Conditions A, B, and C are as follows:

Condition A—Exposed live parts on one side and no live or grounded parts on the other side of the working space, or exposed live parts on both sides effectively guarded by suitable wood or other insulating material. Insulated wire or insulated busbars operating at not over 300 volts are not considered live parts.

Condition B—Exposed live parts on one side and grounded parts on the other side.

Condition C—Exposed live parts on both sides of the work space (not guarded as provided in Condition A) with the operator between.

3. Working space is not required in back of assemblies such as dead-front switchboards or motor control centers where there are no renewable or adjustable parts (such as fuses or switches) on the back and where all connections are accessible from locations other than the back. Where rear access is required to work on deenergized parts on the back of enclosed equipment, a minimum working space of 762 mm (30 in.) horizontally shall be provided.

above or below the electric equipment may extend not more than 153 mm (6 in.) beyond the front of the electric equipment.

(ii) Working space required by this standard may not be used for storage. When normally enclosed live parts are exposed for inspection or servicing, the working space, if in a passageway or general open space, shall be suitably guarded.

(iii) At least one entrance of sufficient area shall be provided to give access to the working space about electric equipment.

(iv) For equipment rated 1200 amperes or more and over 1.83 m (6.0 ft) wide, containing overcurrent devices, switching devices, or control devices, there shall be one entrance not less than 610 mm (24 in.) wide and 1.98 m (6.5 ft) high at each end of the working space, except that:

(A) Where the location permits a continuous and unobstructed way of exit travel, one means of exit is permitted; or

(B) Where the working space required by paragraph (g)(1)(i) of this section is doubled, only one entrance to the working space is required; however, the entrance shall be located so that the edge of the entrance nearest the equipment is the minimum clear distance given in Table S-1 away from such equipment.

(v) Illumination shall be provided for all working spaces about service equipment, switchboards, panelboards, and motor control centers installed indoors. Additional lighting fixtures are not required where the working space is illuminated by an adjacent light source. In electric equipment rooms, the illumination may not be controlled by automatic means only.

(vi) The minimum headroom of working spaces about service equipment, switchboards, panelboards, or motor control centers shall be as follows:

(A) For installations built before August 13, 2007, 1.91 m (6.25 ft); and

(B) For installations built on or after August 13, 2007, 1.98 m (6.5 ft), except that where the electrical equipment exceeds 1.98 m (6.5 ft) in height, the minimum headroom may not be less than the height of the equipment.

(vii) Switchboards, panelboards, and distribution boards installed for the control of light and power circuits, and motor control centers shall be located in dedicated spaces and protected from damage.

(A) For indoor installation, the dedicated space shall comply with the following:

(1) The space equal to the width and depth of the equipment and extending from the floor to a height of 1.83 m (6.0 ft) above the equipment or to the structural ceiling, whichever is lower, shall be dedicated to the electrical installation. Unless isolated from equipment by height or physical enclosures or covers that will afford adequate mechanical protection from vehicular traffic or accidental contact by unauthorized personnel or that complies with paragraph (g)(1)(vii)(A)(2) of this section, piping, ducts, or equipment foreign to the electrical installation may not be located in this area;

FIGURE 7.1 *(Continued)*

(2) The space equal to the width and depth of the equipment shall be kept clear of foreign systems unless protection is provided to avoid damage from condensation, leaks, or breaks in such foreign systems. This area shall extend from the top of the electric equipment to the structural ceiling;

(3) Sprinkler protection is permitted for the dedicated space where the piping complies with this section; and

(4) Control equipment that by its very nature or because of other requirements in this subpart must be adjacent to or within sight of its operating machinery is permitted in the dedicated space.

Note to paragraph (g)(1)(vii)(A) of this section: A dropped, suspended, or similar ceiling that does not add strength to the building structure is not considered a structural ceiling.

(B) Outdoor electric equipment shall be installed in suitable enclosures and shall be protected from accidental contact by unauthorized personnel, or by vehicular traffic, or by accidental spillage or leakage from piping systems. No architectural appurtenance or other equipment may be located in the working space required by paragraph (g)(1)(i) of this section.

(2) *Guarding of live parts.* (i) Except as elsewhere required or permitted by this standard, live parts of electric equipment operating at 50 volts or more shall be guarded against accidental contact by use of approved cabinets or other forms of approved enclosures or by any of the following means:

(A) By location in a room, vault, or similar enclosure that is accessible only to qualified persons;

(B) By suitable permanent, substantial partitions or screens so arranged so that only qualified persons will have access to the space within reach of the live parts. Any openings in such partitions or screens shall be so sized and located that persons are not likely to come into accidental contact with the live parts or to bring conducting objects into contact with them;

(C) By placement on a suitable balcony, gallery, or platform so elevated and otherwise located as to prevent access by unqualified persons; or

(D) By elevation of 2.44 m (8.0 ft) or more above the floor or other working surface.

(ii) In locations where electric equipment is likely to be exposed to physical damage, enclosures or guards shall be so arranged and of such strength as to prevent such damage.

(iii) Entrances to rooms and other guarded locations containing exposed live parts shall be marked with conspicuous warning signs forbidding unqualified persons to enter.

(h) *Over 600 volts, nominal*—(1) *General.* Conductors and equipment used on circuits exceeding 600 volts, nominal, shall comply with all applicable provisions of the paragraphs (a) through (g) of this section and with the following provisions, which supplement or modify the preceding requirements. However, paragraphs (h)(2), (h)(3), and (h)(4) of this section do not apply to the equipment on the supply side of the service point.

(2) *Enclosure for electrical installations.* (i) Electrical installations in a vault, room, or closet or in an area surrounded by a wall, screen, or fence, access to which is controlled by lock and key or other approved means, are considered to be accessible to qualified persons only. The type of enclosure used in a given case shall be designed and constructed according to the hazards associated with the installation.

(ii) For installations other than equipment described in paragraph (h)(2)(v) of this section, a wall, screen, or fence shall be used to enclose an outdoor electrical installation to deter access by persons who are not qualified. A fence may not be less than 2.13 m (7.0 ft) in height or a combination of 1.80 m (6.0 ft) or more of fence fabric and a 305-mm (1-ft) or more extension utilizing three or more strands of barbed wire or equivalent.

(iii) The following requirements apply to indoor installations that are accessible to other than qualified persons:

(A) The installations shall be made with metal-enclosed equipment or shall be enclosed in a vault or in an area to which access is controlled by a lock;

(B) Metal-enclosed switchgear, unit substations, transformers, pull boxes, connection boxes, and other similar associated equipment shall be marked with appropriate caution signs; and

(C) Openings in ventilated dry-type transformers and similar openings in other equipment shall be designed so that foreign objects inserted through these openings will be deflected from energized parts.

(iv) Outdoor electrical installations having exposed live parts shall be accessible to qualified persons only.

(v) The following requirements apply to outdoor enclosed equipment accessible to unqualified employees:

(A) Ventilating or similar openings in equipment shall be so designed that foreign objects inserted through these openings will be deflected from energized parts;

(B) Where exposed to physical damage from vehicular traffic, suitable guards shall be provided;

(C) Nonmetallic or metal-enclosed equipment located outdoors and accessible to the general public shall be designed so that exposed nuts or bolts cannot be readily removed, permitting access to live parts;

FIGURE 7.1 *(Continued)*

(D) Where nonmetallic or metal enclosed equipment is accessible to the general public and the bottom of the enclosure is less than 2.44 m (8.0 ft) above the floor or grade level, the enclosure door or hinged cover shall be kept locked; and

(E) Except for underground box covers that weigh over 45.4 kg (100 lb), doors and covers of enclosures used solely as pull boxes, splice boxes, or junction boxes shall be locked, bolted, or screwed on.

(3) *Work space about equipment.* Sufficient space shall be provided and maintained about electric equipment to permit ready and safe operation and maintenance of such equipment. Where energized parts are exposed, the minimum clear work space may not be less than 1.98 m (6.5 ft) high (measured vertically from the floor or platform) or less than 914 mm (3.0 ft) wide (measured parallel to the equipment). The depth shall be as required in paragraph (h)(5)(i) of this section. In all cases, the work space shall be adequate to permit at least a 90-degree opening of doors or hinged panels.

(4) *Entrance and access to work space.* (i) At least one entrance not less than 610 mm (24 in.) wide and 1.98 m (6.5 ft) high shall be provided to give access to the working space about electric equipment.

(A) On switchboard and control panels exceeding 1.83 m (6.0 ft) in width, there shall be one entrance at each end of such boards unless the location of the switchboards and control panels permits a continuous and unobstructed way of exit travel, or unless the work space required in paragraph (h)(5)(i) of this section is doubled.

(B) Where one entrance to the working space is permitted under the conditions described in paragraph (h)(4)(i)(A) of this section, the entrance shall be located so that the edge of the entrance nearest the switchboards and control panels is at least the minimum clear distance given in Table S-2 away from such equipment.

(C) Where bare energized parts at any voltage or insulated energized parts above 600 volts, nominal, to ground are located adjacent to such entrance, they shall be suitably guarded.

(ii) Permanent ladders or stairways shall be provided to give safe access to the working space around electric equipment installed on platforms, balconies, mezzanine floors, or in attic or roof rooms or spaces.

(5) *Working space and guarding.* (i)(vi) Except as elsewhere required or permitted in this subpart, the minimum clear working space in the direction of access to live parts of electric equipment may not be less than specified in Table S-2. Distances shall be measured from the live parts, if they are exposed, or from the enclosure front or opening, if they are enclosed.

(ii) If switches, cutouts, or other equipment operating at 600 volts, nominal, or less, are installed in a room or enclosure where there are exposed live parts or exposed wiring operating at over 600 volts, nominal, the

TABLE S–2 Minimum Depth of Clear Working Space at Electric Equipment, Over 600 V

	Minimum clear distance for condition[2,3]					
	Condition A		Condition B		Condition C	
Nominal voltage to ground	m	ft	m	ft	m	ft
601–2500 V	0.9	3.0	1.2	4.0	1.5	5.0
2501–9000 V	1.2	4.0	1.5	5.0	1.8	6.0
9001 V–25 kV	1.5	5.0	1.8	6.0	2.8	9.0
Over 25–75 kV[1]	1.8	6.0	2.5	8.0	3.0	10.0
Above 75 kV[1]	2.5	8.0	3.0	10.0	3.7	12.0

Notes to Table S-2:

1. Minimum depth of clear working space in front of electric equipment with a nominal voltage to ground above 25,000 volts may be the same as that for 25,000 volts under Conditions A, B, and C for installations built before April 16, 1981.

2. Conditions A, B, and C are as follows:

Condition A—Exposed live parts on one side and no live or grounded parts on the other side of the working space, or exposed live parts on both sides effectively guarded by suitable wood or other insulating material. Insulated wire or insulated busbars operating at not over 300 volts are not considered live parts.

Condition B—Exposed live parts on one side and grounded parts on the other side. Concrete, brick, and tile walls are considered as grounded surfaces.

Condition C—Exposed live parts on both sides of the work space (not guarded as provided in Condition A) with the operator between.

3. Working space is not required in back of equipment such as dead-front switchboards or control assemblies that has no renewable or adjustable parts (such as fuses or switches) on the back and where all connections are accessible from locations other than the back. Where rear access is required to work on the deenergized parts on the back of enclosed equipment, a minimum working space 762 mm (30 in.) horizontally shall be provided.

FIGURE 7.1 (*Continued*)

high-voltage equipment shall be effectively separated from the space occupied by the low-voltage equipment by a suitable partition, fence, or screen. However, switches or other equipment operating at 600 volts, nominal, or less, and serving only equipment within the high-voltage vault, room, or enclosure may be installed in the high-voltage enclosure, room, or vault if accessible to qualified persons only.

(iii) The following requirements apply to the entrances to all buildings, rooms, or enclosures containing exposed live parts or exposed conductors operating at over 600 volts, nominal: unless they are under the observation of a qualified person at all times; and

(B) Permanent and conspicuous warning signs shall be provided, reading substantially as follows: "DANGER—HIGH VOLTAGE—KEEP OUT."

(iv) Illumination shall be provided for all working spaces about electric equipment.

(A) The lighting outlets shall be arranged so that persons changing lamps or making repairs on the lighting system will not be endangered by live parts or other equipment.

(B) The points of control shall be located so that persons are prevented from contacting any live part or moving part of the equipment while turning on the lights.

(v) Unguarded live parts above working space shall be maintained at elevations not less than specified in Table S-3.

(vi) Pipes or ducts that are foreign to the electrical installation and that require periodic maintenance or whose malfunction would endanger the operation of the electrical system may not be located in the vicinity of service equipment, metal-enclosed power switchgear, or industrial control assemblies. Protection shall be provided where necessary to avoid damage from condensation leaks and breaks in such foreign systems.

Note to paragraph (h)(5)(vi) of this section: Piping and other facilities are not considered foreign if provided for fire protection of the electrical installation.

§ 1910.304 Wiring design and protection.

(a) *Use and identification of grounded and grounding conductors*—(1) *Identification of conductors.* (i) A conductor used as a grounded conductor shall be identifiable and distinguishable from all other conductors.

(ii) A conductor used as an equipment grounding conductor shall be identifiable and distinguishable from all other conductors.

(2) *Polarity of connections.* No grounded conductor may be attached to any terminal or lead so as to reverse designated polarity.

(3) *Use of grounding terminals and devices.* A grounding terminal or grounding-type device on a receptacle, cord connector, or attachment plug may not be used for purposes other than grounding.

(b) *Branch circuits*—(1) *Identification of multiwire branch circuits.* Where more than one nominal voltage system exists in a building containing multiwire branch circuits, each ungrounded conductor of a multiwire branch circuit, where accessible, shall be identified by phase and system. The means of identification shall be permanently posted at each branch circuit panelboard.

(2) *Receptacles and cord connectors.* (i) Receptacles installed on 15- and 20-ampere branch circuits shall be of the grounding type except as permitted for replacement receptacles in paragraph (b)(2)(iv) of this section. Grounding-type receptacles shall be installed only on circuits of the voltage class and current for which they are rated, except as provided in Table S-4 and Table S-5.

TABLE S-3 Elevation of Unguarded Live Parts Above Working Space

Nominal voltage between phases	Elevation	
	m	ft
601–7500 V	[1] 2.8	[1] 9.0
7501 V–35 kV	2.8	9.0
Over 35 kV	2.8 + 9.5 mm/kV over 35 kV	9.0 + 0.37 in./kV over 35 kV

[1]The minimum elevation may be 2.6 m (8.5 ft) for installations built before August 13, 2007. The minimum elevation may be 2.4 m (8.0 ft) for installations built before April 16, 1981, if the nominal voltage between phases is in the range of 601–6600 volts.

FIGURE 7.1 (*Continued*)

TABLE S-4 Maximum Cord- and Plug-Connected Load to Receptacle

Circuit rating (amperes)	Receptacle rating (amperes)	Maximum load (amperes)
15 or 20	15	12
20	20	16
30	30	24

TABLE S-5 Receptacle Ratings for Various Size Circuits

Circuit rating (amperes)	Receptacle rating (amperes)
15	Not over 15
20	15 or 20
30	30
40	40 or 50
50	50

(ii) Receptacles and cord connectors having grounding contacts shall have those contacts effectively grounded except for receptacles mounted on portable and vehicle-mounted generators in accordance with paragraph (g)(3) of this section and replacement receptacles installed in accordance with paragraph (b)(2)(iv) of this section.

(iii) The grounding contacts of receptacles and cord connectors shall be grounded by connection to the equipment grounding conductor of the circuit supplying the receptacle or cord connector. The branch circuit wiring method shall include or provide an equipment grounding conductor to which the grounding contacts of the receptacle or cord connector shall be connected.

(iv) Replacement of receptacles shall comply with the following requirements:

(A) Where a grounding means exists in the receptacle enclosure or a grounding conductor is installed, grounding-type receptacles shall be used and shall be connected to the grounding means or conductor;

(B) Ground-fault circuit-interrupter protected receptacles shall be provided where replacements are made at receptacle outlets that are required to be so protected elsewhere in this subpart; and

(C) Where a grounding means does not exist in the receptacle enclosure, the installation shall comply with one of the following provisions:

(1) A nongrounding-type receptacle may be replaced with another nongrounding-type receptacle; or

(2) A nongrounding-type receptacle may be replaced with a ground-fault circuit-interrupter-type of receptacle that is marked "No Equipment Ground;" an equipment grounding conductor may not be connected from the ground-fault circuit-interrupter-type receptacle to any outlet supplied from the ground-fault circuit-interrupter receptacle; or

(3) A nongrounding-type receptacle may be replaced with a grounding-type receptacle where supplied through a ground-fault circuit-interrupter; the replacement receptacle shall be marked "GFCI Protected" and "No Equipment Ground;" an equipment grounding conductor may not be connected to such grounding-type receptacles.

(v) Receptacles connected to circuits having different voltages, frequencies, or types of current (ac or dc) on the same premises shall be of such design that the attachment plugs used on these circuits are not interchangeable.

(3) *Ground-fault circuit-interrupter protection for personnel.* (i) All 125-volt, single-phase, 15- and 20-ampere receptacles installed in bathrooms or on rooftops shall have ground-fault circuit-interrupter protection for personnel.

(ii) The following requirements apply to temporary wiring installations that are used during maintenance, remodeling, or repair of buildings, structures, or equipment or during similar construction-like activities.

(A) All 125-volt, single-phase, 15-, 20-, and 30-ampere receptacle outlets that are not part of the permanent wiring of the building or structure and that are in use by personnel shall have ground-fault circuit-interrupter protection for personnel.

Note 1 to paragraph (b)(3)(ii)(A) of this section: A cord connector on an extension cord set is considered to be a receptacle outlet if the cord set is used for temporary electric power.

Note 2 to paragraph (b)(3)(ii)(A) of this section: Cord sets and devices incorporating the required ground-fault circuit-interrupter that are connected to the receptacle closest to the source of power are acceptable forms of protection.

FIGURE 7.1 *(Continued)*

(B) Receptacles other than 125 volt, single-phase, 15-, 20-, and 30-ampere receptacles that are not part of the permanent wiring of the building or structure and that are in use by personnel shall have ground-fault circuit-interrupter protection for personnel.

(C) Where the ground-fault circuit-interrupter protection required by paragraph (b)(3)(ii)(B) of this section is not available for receptacles other than 125-volt, single-phase, 15-, 20-, and 30-ampere, the employer shall establish and implement an assured equipment grounding conductor program covering cord sets, receptacles that are not a part of the building or structure, and equipment connected by cord and plug that are available for use or used by employees on those receptacles. This program shall comply with the following requirements:

(1) A written description of the program, including the specific procedures adopted by the employer, shall be available at the jobsite for inspection and copying by the Assistant Secretary of Labor and any affected employee;

(2) The employer shall designate one or more competent persons to implement the program;

(3) Each cord set, attachment cap, plug, and receptacle of cord sets, and any equipment connected by cord and plug, except cord sets and receptacles which are fixed and not exposed to damage, shall be visually inspected before each day's use for external defects, such as deformed or missing pins or insulation damage, and for indications of possible internal damage. Equipment found damaged or defective shall not be used until repaired;

(4) The following tests shall be performed on all cord sets and receptacles which are not a part of the permanent wiring of the building or structure, and cord- and plug-connected equipment required to be grounded:

(i) All equipment grounding conductors shall be tested for continuity and shall be electrically continuous;

(ii) Each receptacle and attachment cap or plug shall be tested for correct attachment of the equipment grounding conductor. The equipment grounding conductor shall be connected to its proper terminal; and

(iii) All required tests shall be performed before first use; before equipment is returned to service following any repairs; before equipment is used after any incident which can be reasonably suspected to have caused damage (for example, when a cord set is run over); and at intervals not to exceed 3 months, except that cord sets and receptacles which are fixed and not exposed to damage shall be tested at intervals not exceeding 6 months;

(5) The employer shall not make available or permit the use by employees of any equipment which has not met the requirements of paragraph (b)(3)(ii)(C) of this section; and

(6) Tests performed as required in paragraph (b)(3)(ii)(C) of this section shall be recorded. This test record shall identify each receptacle, cord set, and cord- and plug-connected equipment that passed the test and shall indicate the last date it was tested or the interval for which it was tested. This record shall be kept by means of logs, color coding, or other effective means and shall be maintained until replaced by a more current record. The record shall be made available on the jobsite for inspection by the Assistant Secretary and any affected employee.

(4) *Outlet devices.* Outlet devices shall have an ampere rating not less than the load to be served and shall comply with the following provisions:

(i) Where connected to a branch circuit having a rating in excess of 20 amperes, lampholders shall be of the heavy-duty type. A heavy-duty lampholder shall have a rating of not less than 660 watts if of the admedium type and not less than 750 watts if of any other type; and

(ii) Receptacle outlets shall comply with the following provisions:

(A) A single receptacle installed on an individual branch circuit shall have an ampere rating of not less than that of the branch circuit;

(B) Where connected to a branch circuit supplying two or more receptacles or outlets, a receptacle may not supply a total cord- and plug-connected load in excess of the maximum specified in Table S-4; and

(C) Where connected to a branch circuit supplying two or more receptacles or outlets, receptacle ratings shall conform to the values listed in Table S-5; or, where larger than 50 amperes, the receptacle rating may not be less than the branch-circuit rating. However, receptacles of cord- and plug-connected arc welders may have ampere ratings not less than the minimum branch-circuit conductor ampacity.

(5) *Cord connections.* A receptacle outlet shall be installed wherever flexible cords with attachment plugs are used. Where flexible cords are permitted to be permanently connected, receptacles may be omitted.

(c) *Outside conductors, 600 volts, nominal, or less.* The following requirements apply to branch-circuit, feeder, and service conductors rated 600 volts, nominal, or less and run outdoors as open conductors.

(1) *Conductors on poles.* Conductors on poles shall have a separation of not less than 305 mm (1.0 ft) where not placed on racks or brackets. Conductors supported on poles shall provide a horizontal climbing space not less than the following:

(i) Power conductors below communication conductors—762 mm (30 in.);

(ii) Power conductors alone or above communication conductors:

(A) 300 volts or less—610 mm (24 in.),

FIGURE 7.1 *(Continued)*

TABLE S-6 Clearances from Ground

	Installations built before August 13, 2007		Installations built on or after August 13, 2007	
Distance	Maximum voltage	Conditions	Voltage to ground	Conditions
3.05 m (10.0 ft)	<600 V	Above finished grade or sidewalks, or from any platform or projection from which they might be reached. (If these areas are accessible to other than pedestrian traffic, then one of the other conditions applies.)	<150 V	Above finished grade or sidewalks, or from any platform or projection from which they might be reached. (If these areas are accessible to other than pedestrian traffic, then one of the other conditions applies.)
3.66 m (12.0 ft)	<600 V	Over areas, other than public streets, alleys, roads, and driveways, subject to vehicular traffic other than truck traffic.	<300 V	Over residential property and driveways. Over commercial areas subject to pedestrian traffic or to vehicular traffic other than truck traffic. (This category includes conditions covered under the 3.05-m (10.0-ft) category where the voltage exceeds 150 V.)
4.57 m (15.0 ft)	<600 V	Over areas, other than public streets, alleys, roads, and driveways, subject to truck traffic.	301 to 600 V	Over residential property and driveways. Over commercial areas subject to pedestrian traffic or to vehicular traffic other than truck traffic. (This category includes conditions covered under the 3.05-m (10.0-ft) category where the voltage exceeds 300 V.)
5.49 m (18.0 ft)	<600 V	Over public streets, alleys, roads, and driveways.	<600 V	Over public streets, alleys, roads, and driveways. Over commercial areas subject to truck traffic. Other land traversed by vehicles, including land used for cultivating or grazing and forests and orchards.

(B) Over 300 volts—762 mm (30 in.);

(iii) Communication conductors below power conductors—same as power conductors; and

(iv) Communications conductors alone—no requirement.

(2) *Clearance from ground.* Open conductors, open multi-conductor cables, and service-drop conductors of not over 600 volts, nominal, shall conform to the minimum clearances specified in Table S-6.

(3) *Clearance from building openings.* (i) Service conductors installed as open conductors or multi-conductor cable without an overall outer jacket shall have a clearance of not less than 914 mm (3.0 ft) from windows that are designed to be opened, doors, porches, balconies, ladders, stairs, fire escapes, and similar locations. However, conductors that run above the top level of a window may be less than 914 mm (3.0 ft) from the window. Vertical clearance of final spans above, or within 914 mm (3.0 ft) measured horizontally of, platforms, projections, or surfaces from which they might be reached shall be maintained in accordance with paragraph (c)(2) of this section.

(ii) Overhead service conductors may not be installed beneath openings through which materials may be moved, such as openings in farm and commercial buildings, and may not be installed where they will obstruct entrance to these building openings.

FIGURE 7.1 *(Continued)*

(4) *Above roofs.* Overhead spans of open conductors and open multi-conductor cables shall have a vertical clearance of not less than 2.44 m (8.0 ft) above the roof surface. The vertical clearance above the roof level shall be maintained for a distance not less than 914 mm (3.0 ft) in all directions from the edge of the roof.

(i) The area above a roof surface subject to pedestrian or vehicular traffic shall have a vertical clearance from the roof surface in accordance with the clearance requirements of paragraph (c)(2) of this section.

(ii) A reduction in clearance to 914 mm (3.0 ft) is permitted where the voltage between conductors does not exceed 300 and the roof has a slope of 102 mm (4 in.) in 305 mm (12 in.) or greater.

(iii) A reduction in clearance above only the overhanging portion of the roof to not less than 457 mm (18 in.) is permitted where the voltage between conductors does not exceed 300 if:

(A) The conductors do not pass above the roof overhang for a distance of more than 1.83 m (6.0 ft), 1.22 m (4.0 ft) horizontally, and

(B) The conductors are terminated at a through-the-roof raceway or approved support.

(iv) The requirement for maintaining a vertical clearance of 914 mm (3.0 ft) from the edge of the roof does not apply to the final conductor span, where the conductors are attached to the side of a building.

(d) *Location of outdoor lamps.* Lamps for outdoor lighting shall be located below all energized conductors, transformers, or other electric equipment, unless such equipment is controlled by a disconnecting means that can be locked in the open position, or unless adequate clearances or other safeguards are provided for relamping operations.

(e) *Services*—(1) *Disconnecting means.* (i) Means shall be provided to disconnect all conductors in a building or other structure from the service entrance conductors. The service disconnecting means shall plainly indicate whether it is in the open or closed position and shall be installed at a readily accessible location nearest the point of entrance of the service-entrance conductors.

(ii) Each service disconnecting means shall simultaneously disconnect all ungrounded conductors.

(iii) Each service disconnecting means shall be suitable for the prevailing conditions.

(2) *Services over 600 volts, nominal.* The following additional requirements apply to services over 600 volts, nominal.

(i) Service-entrance conductors installed as open wires shall be guarded to make them accessible only to qualified persons.

(ii) Signs warning of high voltage shall be posted where unqualified employees might come in contact with live parts.

(f) *Overcurrent protection*—(1) *600 volts, nominal, or less.* The following requirements apply to overcurrent protection of circuits rated 600 volts, nominal, or less.

(i) Conductors and equipment shall be protected from overcurrent in accordance with their ability to safely conduct current.

(ii) Except for motor running overload protection, overcurrent devices may not interrupt the continuity of the grounded conductor unless all conductors of the circuit are opened simultaneously.

(iii) A disconnecting means shall be provided on the supply side of all fuses in circuits over 150 volts to ground and cartridge fuses in circuits of any voltage where accessible to other than qualified persons so that each individual circuit containing fuses can be independently disconnected from the source of power. However, a current-limiting device without a disconnecting means is permitted on the supply side of the service disconnecting means. In addition, a single disconnecting means is permitted on the supply side of more than one set of fuses as permitted by the exception in § 1910.305(j)(4)(vi) for group operation of motors, and a single disconnecting means is permitted for fixed electric space-heating equipment.

(iv) Overcurrent devices shall be readily accessible to each employee or authorized building management personnel. These overcurrent devices may not be located where they will be exposed to physical damage or in the vicinity of easily ignitable material.

(v) Fuses and circuit breakers shall be so located or shielded that employees will not be burned or otherwise injured by their operation. Handles or levers of circuit breakers, and similar parts that may move suddenly in such a way that persons in the vicinity are likely to be injured by being struck by them, shall be guarded or isolated.

(vi) Circuit breakers shall clearly indicate whether they are in the open (off) or closed (on) position.

(vii) Where circuit breaker handles on switchboards are operated vertically rather than horizontally or rotationally, the up position of the handle shall be the closed (on) position.

(viii) Circuit breakers used as switches in 120-volt and 277-volt, fluorescent lighting circuits shall be listed and marked "SWD."

(ix) A circuit breaker with a straight voltage rating, such as 240 V or 480 V, may only be installed in a circuit in which the nominal voltage between any two conductors does not exceed the circuit breaker's voltage rating.

FIGURE 7.1 (*Continued*)

A two-pole circuit breaker may not be used for protecting a 3-phase, corner-grounded delta circuit unless the circuit breaker is marked 1Φ—3Φ to indicate such suitability. A circuit breaker with a slash rating, such as 120/240 V or 480Y/277 V, may only be installed in a circuit where the nominal voltage of any conductor to ground does not exceed the lower of the two values of the circuit breaker's voltage rating and the nominal voltage between any two conductors does not exceed the higher value of the circuit breaker's voltage rating.

(2) *Feeders and branch circuits over 600 volts, nominal.* The following requirements apply to feeders and branch circuits energized at more than 600 volts, nominal:

(i) Feeder and branch-circuit conductors shall have overcurrent protection in each ungrounded conductor located at the point where the conductor receives its supply or at a location in the circuit determined under engineering supervision;

(A) Circuit breakers used for overcurrent protection of three-phase circuits shall have a minimum of three overcurrent relays operated from three current transformers. On three-phase, three-wire circuits, an overcurrent relay in the residual circuit of the current transformers may replace one of the phase relays. An overcurrent relay, operated from a current transformer that links all phases of a three-phase, three-wire circuit, may replace the residual relay and one other phase-conductor current transformer. Where the neutral is not grounded on the load side of the circuit, the current transformer may link all three phase conductors and the grounded circuit conductor (neutral); and

(B) If fuses are used for overcurrent protection, a fuse shall be connected in series with each ungrounded conductor;

(ii) Each protective device shall be capable of detecting and interrupting all values of current that can occur at its location in excess of its trip setting or melting point;

(iii) The operating time of the protective device, the available short-circuit current, and the conductor used shall be coordinated to prevent damaging or dangerous temperatures in conductors or conductor insulation under short-circuit conditions; and

(iv) The following additional requirements apply to feeders only:

(A) The continuous ampere rating of a fuse may not exceed three times the ampacity of the conductors. The long-time trip element setting of a breaker or the minimum trip setting of an electronically actuated fuse may not exceed six times the ampacity of the conductor. For fire pumps, conductors may be protected for short circuit only; and

(B) Conductors tapped to a feeder may be protected by the feeder overcurrent device where that overcurrent device also protects the tap conductor.

(g) *Grounding.* Paragraphs (g)(1) through (g)(9) of this section contain grounding requirements for systems, circuits, and equipment.

(1) *Systems to be grounded.* Systems that supply premises wiring shall be grounded as follows:

(i) All 3-wire dc systems shall have their neutral conductor grounded;

(ii) Two-wire dc systems operating at over 50 volts through 300 volts between conductors shall be grounded unless:

(A) They supply only industrial equipment in limited areas and are equipped with a ground detector;

(B) They are rectifier-derived from an ac system complying with paragraphs (g)(1)(iii), (g)(1)(iv), and (g)(1)(v) of this section; or

(C) They are fire-alarm circuits having a maximum current of 0.030 amperes;

(iii) AC circuits of less than 50 volts shall be grounded if they are installed as overhead conductors outside of buildings or if they are supplied by transformers and the transformer primary supply system is ungrounded or exceeds 150 volts to ground;

(iv) AC systems of 50 volts to 1000 volts shall be grounded under any of the following conditions, unless exempted by paragraph (g)(1)(v) of this section:

(A) If the system can be so grounded that the maximum voltage to ground on the ungrounded conductors does not exceed 150 volts;

(B) If the system is nominally rated three-phase, four-wire wye connected in which the neutral is used as a circuit conductor;

(C) If the system is nominally rated three-phase, four-wire delta connected in which the midpoint of one phase is used as a circuit conductor; or

(D) If a service conductor is uninsulated;

(v) AC systems of 50 volts to 1000 volts are not required to be grounded under any of the following conditions:

(A) If the system is used exclusively to supply industrial electric furnaces for melting, refining, tempering, and the like;

FIGURE 7.1 (*Continued*)

(B) If the system is separately derived and is used exclusively for rectifiers supplying only adjustable speed industrial drives;

(C) If the system is separately derived and is supplied by a transformer that has a primary voltage rating less than 1000 volts, provided all of the following conditions are met:

(1) The system is used exclusively for control circuits;

(2) The conditions of maintenance and supervision ensure that only qualified persons will service the installation;

(3) Continuity of control power is required; and

(4) Ground detectors are installed on the control system;

(D) If the system is an isolated power system that supplies circuits in health care facilities; or

(E) If the system is a high-impedance grounded neutral system in which a grounding impedance, usually a resistor, limits the ground-fault current to a low value for 3-phase ac systems of 480 volts to 1000 volts provided all of the following conditions are met:

(1) The conditions of maintenance and supervision ensure that only qualified persons will service the installation;

(2) Continuity of power is required;

(3) Ground detectors are installed on the system; and

(4) Line-to-neutral loads are not served.

(2) *Conductor to be grounded.* The conductor to be grounded for ac premises wiring systems required to be grounded by paragraph (g)(1) of this section shall be as follows:

(i) One conductor of a single-phase, two-wire system shall be grounded;

(ii) The neutral conductor of a single-phase, three-wire system shall be grounded;

(iii) The common conductor of a multiphase system having one wire common to all phases shall be grounded;

(iv) One phase conductor of a multiphase system where one phase is grounded shall be grounded; and

(v) The neutral conductor of a multiphase system in which one phase is used as a neutral conductor shall be grounded.

(3) *Portable and vehicle-mounted generators.* (i) The frame of a portable generator need not be grounded and may serve as the grounding electrode for a system supplied by the generator under the following conditions:

(A) The generator supplies only equipment mounted on the generator or cord- and plug-connected equipment through receptacles mounted on the generator, or both; and

(B) The non-current-carrying metal parts of equipment and the equipment grounding conductor terminals of the receptacles are bonded to the generator frame.

(ii) The frame of a vehicle need not be grounded and may serve as the grounding electrode for a system supplied by a generator located on the vehicle under the following conditions:

(A) The frame of the generator is bonded to the vehicle frame;

(B) The generator supplies only equipment located on the vehicle and cord- and plug-connected equipment through receptacles mounted on the vehicle;

(C) The non-current-carrying metal parts of equipment and the equipment grounding conductor terminals of the receptacles are bonded to the generator frame; and

(D) The system complies with all other provisions of paragraph (g) of this section.

(iii) A system conductor that is required to be grounded by the provisions of paragraph (g)(2) of this section shall be bonded to the generator frame where the generator is a component of a separately derived system.

(4) *Grounding connections.* (i) For a grounded system, a grounding electrode conductor shall be used to connect both the equipment grounding conductor and the grounded circuit conductor to the grounding electrode. Both the equipment grounding conductor and the grounding electrode conductor shall be connected to the grounded circuit conductor on the supply side of the service disconnecting means or on the supply side of the system disconnecting means or overcurrent devices if the system is separately derived.

(ii) For an ungrounded service supplied system, the equipment grounding conductor shall be connected to the grounding electrode conductor at the service equipment. For an ungrounded separately derived system, the equipment grounding conductor shall be connected to the grounding electrode conductor at, or ahead of, the system disconnecting means or overcurrent devices.

(iii) On extensions of existing branch circuits that do not have an equipment grounding conductor, grounding-type receptacles may be grounded to a grounded cold water pipe near the equipment if the extension was installed before August 13, 2007. When any element of this branch circuit is replaced, the entire branch circuit shall use an equipment grounding conductor that complies with all other provisions of paragraph (g) of this section.

(5) *Grounding path.* The path to ground from circuits, equipment, and enclosures shall be permanent, continuous, and effective.

FIGURE 7.1 *(Continued)*

(6) *Supports, enclosures, and equipment to be grounded.* (i) Metal cable trays, metal raceways, and metal enclosures for conductors shall be grounded, except that:

(A) Metal enclosures such as sleeves that are used to protect cable assemblies from physical damage need not be grounded; and

(B) Metal enclosures for conductors added to existing installations of open wire, knob-and-tube wiring, and nonmetallic-sheathed cable need not be grounded if all of the following conditions are met:

(1) Runs are less than 7.62 meters (25.0 ft);

(2) Enclosures are free from probable contact with ground, grounded metal, metal laths, or other conductive materials; and

(3) Enclosures are guarded against employee contact.

(ii) Metal enclosures for service equipment shall be grounded.

(iii) Frames of electric ranges, wall-mounted ovens, counter-mounted cooking units, clothes dryers, and metal outlet or junction boxes that are part of the circuit for these appliances shall be grounded.

(iv) Exposed non-current-carrying metal parts of fixed equipment that may become energized shall be grounded under any of the following conditions:

(A) If within 2.44 m (8 ft) vertically or 1.52 m (5 ft) horizontally of ground or grounded metal objects and subject to employee contact;

(B) If located in a wet or damp location and not isolated;

(C) If in electrical contact with metal;

(D) If in a hazardous (classified) location;

(E) If supplied by a metal-clad, metal-sheathed, or grounded metal raceway wiring method; or

(F) If equipment operates with any terminal at over 150 volts to ground.

(v) Notwithstanding the provisions of paragraph (g)(6)(iv) of this section, exposed non-current-carrying metal parts of the following types of fixed equipment need not be grounded:

(A) Enclosures for switches or circuit breakers used for other than service equipment and accessible to qualified persons only;

(B) Electrically heated appliances that are permanently and effectively insulated from ground;

(C) Distribution apparatus, such as transformer and capacitor cases, mounted on wooden poles, at a height exceeding 2.44 m (8.0 ft) above ground or grade level; and

(D) Listed equipment protected by a system of double insulation, or its equivalent, and distinctively marked as such.

(vi) Exposed non-current-carrying metal parts of cord- and plug-connected equipment that may become energized shall be grounded under any of the following conditions:

(A) If in hazardous (classified) locations (see § 1910.307);

(B) If operated at over 150 volts to ground, except for guarded motors and metal frames of electrically heated appliances if the appliance frames are permanently and effectively insulated from ground;

(C) If the equipment is of the following types:

(1) Refrigerators, freezers, and air conditioners;

(2) Clothes-washing, clothes-drying, and dishwashing machines, sump pumps, and electric aquarium equipment;

(3) Hand-held motor-operated tools, stationary and fixed motor-operated tools, and light industrial motor-operated tools;

(4) Motor-operated appliances of the following types: hedge clippers, lawn mowers, snow blowers, and wet scrubbers;

(5) Cord- and plug-connected appliances used in damp or wet locations, or by employees standing on the ground or on metal floors or working inside of metal tanks or boilers;

(6) Portable and mobile X-ray and associated equipment;

(7) Tools likely to be used in wet and conductive locations; and

(8) Portable hand lamps.

(vii) Notwithstanding the provisions of paragraph (g)(6)(vi) of this section, the following equipment need not be grounded:

(A) Tools likely to be used in wet and conductive locations if supplied through an isolating transformer with an ungrounded secondary of not over 50 volts; and

(B) Listed or labeled portable tools and appliances if protected by an approved system of double insulation, or its equivalent, and distinctively marked.

FIGURE 7.1 *(Continued)*

(7) *Non-electrical equipment.* The metal parts of the following non-electrical equipment shall be grounded: frames and tracks of electrically operated cranes and hoists; frames of non-electrically driven elevator cars to which electric conductors are attached; hand-operated metal shifting ropes or cables of electric elevators; and metal partitions, grill work, and similar metal enclosures around equipment of over 750 volts between conductors.

(8) *Methods of grounding fixed equipment.* (i) Non-current-carrying metal parts of fixed equipment, if required to be grounded by this subpart, shall be grounded by an equipment grounding conductor that is contained within the same raceway, cable, or cord, or runs with or encloses the circuit conductors. For dc circuits only, the equipment grounding conductor may be run separately from the circuit conductors.

(ii) Electric equipment is considered to be effectively grounded if it is secured to, and in electrical contact with, a metal rack or structure that is provided for its support and the metal rack or structure is grounded by the method specified for the non-current-carrying metal parts of fixed equipment in paragraph (g)(8)(i) of this section. Metal car frames supported by metal hoisting cables attached to or running over metal sheaves or drums of grounded elevator machines are also considered to be effectively grounded.

(iii) For installations made before April 16, 1981, electric equipment is also considered to be effectively grounded if it is secured to, and in metallic contact with, the grounded structural metal frame of a building. When any element of this branch circuit is replaced, the entire branch circuit shall use an equipment grounding conductor that complies with all other provisions of paragraph (g) of this section.

(9) *Grounding of systems and circuits of 1000 volts and over (high voltage).* If high voltage systems are grounded, they shall comply with all applicable provisions of paragraphs (g)(1) through (g)(8) of this section as supplemented and modified by the following requirements:

(i) Systems supplying portable or mobile high voltage equipment, other than substations installed on a temporary basis, shall comply with the following:

(A) The system shall have its neutral grounded through an impedance. If a delta-connected high voltage system is used to supply the equipment, a system neutral shall be derived.

(B) Exposed non-current-carrying metal parts of portable and mobile equipment shall be connected by an equipment grounding conductor to the point at which the system neutral impedance is grounded.

(C) Ground-fault detection and relaying shall be provided to automatically deenergize any high voltage system component that has developed a ground fault. The continuity of the equipment grounding conductor shall be continuously monitored so as to deenergize automatically the high voltage feeder to the portable equipment upon loss of continuity of the equipment grounding conductor.

(D) The grounding electrode to which the portable equipment system neutral impedance is connected shall be isolated from and separated in the ground by at least 6.1 m (20.0 ft) from any other system or equipment grounding electrode, and there shall be no direct connection between the grounding electrodes, such as buried pipe, fence, and so forth.

(ii) All non-current-carrying metal parts of portable equipment and fixed equipment, including their associated fences, housings, enclosures, and supporting structures, shall be grounded. However, equipment that is guarded by location and isolated from ground need not be grounded. Additionally, pole-mounted distribution apparatus at a height exceeding 2.44 m (8.0 ft) above ground or grade level need not be grounded.

§ 1910.305 Wiring methods, components, and equipment for general use.

(a) *Wiring methods.* The provisions of this section do not apply to conductors that are an integral part of factory assembled equipment.

(1) *General requirements.* (i) Metal raceways, cable trays, cable armor, cable sheath, enclosures, frames, fittings, and other metal non-current-carrying parts that are to serve as grounding conductors, with or without the use of supplementary equipment grounding conductors, shall be effectively bonded where necessary to ensure electrical continuity and the capacity to conduct safely any fault current likely to be imposed on them. Any nonconductive paint, enamel, or similar coating shall be removed at threads, contact points, and contact surfaces or be connected by means of fittings designed so as to make such removal unnecessary.

(ii) Where necessary for the reduction of electrical noise (electromagnetic interference) of the grounding circuit, an equipment enclosure supplied by a branch circuit may be isolated from a raceway containing circuits supplying only that equipment by one or more listed nonmetallic raceway fittings located at the point of attachment of the raceway to the equipment enclosure. The metal raceway shall be supplemented by an internal insulated equipment grounding conductor installed to ground the equipment enclosure.

FIGURE 7.1 (*Continued*)

(iii) No wiring systems of any type may be installed in ducts used to transport dust, loose stock, or flammable vapors. No wiring system of any type may be installed in any duct used for vapor removal or for ventilation of commercial-type cooking equipment, or in any shaft containing only such ducts.

(2) *Temporary wiring.* Except as specifically modified in this paragraph, all other requirements of this subpart for permanent wiring shall also apply to temporary wiring installations.

(i) Temporary electrical power and lighting installations of 600 volts, nominal, or less may be used only as follows:

(A) During and for remodeling, maintenance, or repair of buildings, structures, or equipment, and similar activities;

(B) For a period not to exceed 90 days for Christmas decorative lighting, carnivals, and similar purposes; or

(C) For experimental or development work, and during emergencies.

(ii) Temporary wiring shall be removed immediately upon completion of the project or purpose for which the wiring was installed.

(iii) Temporary electrical installations of more than 600 volts may be used only during periods of tests, experiments, emergencies, or construction-like activities.

(iv) The following requirements apply to feeders:

(A) Feeders shall originate in an approved distribution center.

(B) Conductors shall be run as multi-conductor cord or cable assemblies. However, if installed as permitted in paragraph (a)(2)(i)(C) of this section, and if accessible only to qualified persons, feeders may be run as single insulated conductors.

(v) The following requirements apply to branch circuits:

(A) Branch circuits shall originate in an approved power outlet or panelboard.

(B) Conductors shall be multi-conductor cord or cable assemblies or open conductors. If run as open conductors, they shall be fastened at ceiling height every 3.05 m (10.0 ft).

(C) No branch-circuit conductor may be laid on the floor.

(D) Each branch circuit that supplies receptacles or fixed equipment shall contain a separate equipment grounding conductor if run as open conductors.

(vi) Receptacles shall be of the grounding type. Unless installed in a continuous grounded metallic raceway or metallic covered cable, each branch circuit shall contain a separate equipment grounding conductor and all receptacles shall be electrically connected to the grounding conductor.

(vii) No bare conductors nor earth returns may be used for the wiring of any temporary circuit.

(viii) Suitable disconnecting switches or plug connectors shall be installed to permit the disconnection of all ungrounded conductors of each temporary circuit. Multi-wire branch circuits shall be provided with a means to disconnect simultaneously all ungrounded conductors at the power outlet or panelboard where the branch circuit originated.

Note to paragraph (a)(2)(viii) of this section. Circuit breakers with their handles connected by approved handle ties are considered a single disconnecting means for the purpose of this requirement.

(ix) All lamps for general illumination shall be protected from accidental contact or breakage by a suitable fixture or lampholder with a guard. Brass shell, paper-lined sockets, or other metal-cased sockets may not be used unless the shell is grounded.

(x) Flexible cords and cables shall be protected from accidental damage, as might be caused, for example, by sharp corners, projections, and doorways or other pinch points.

(xi) Cable assemblies and flexible cords and cables shall be supported in place at intervals that ensure that they will be protected from physical damage. Support shall be in the form of staples, cables ties, straps, or similar type fittings installed so as not to cause damage.

(3) *Cable trays.* (i) Only the following wiring methods may be installed in cable tray systems: armored cable; electrical metallic tubing; electrical nonmetallic tubing; fire alarm cables; flexible metal conduit; flexible metallic tubing; instrumentation tray cable; intermediate metal conduit; liquidtight flexible metal conduit; liquidtight flexible nonmetallic conduit; metal-clad cable; mineral-insulated, metal-sheathed cable; multi-conductor service-entrance cable; multi-conductor underground feeder and branch-circuit cable; multipurpose and communications cables; nonmetallic-sheathed cable; power and control tray cable; power-limited tray cable; optical fiber cables; and other factory-assembled, multi-conductor control, signal, or power cables that are specifically approved for installation in cable trays, rigid metal conduit, and rigid nonmetallic conduit.

(ii) In industrial establishments where conditions of maintenance and supervision assure that only qualified persons will service the installed cable tray system, the following cables may also be installed in ladder, ventilated-trough, or ventilated-channel cable trays:

FIGURE 7.1 *(Continued)*

(A) Single conductor cable; the cable shall be No. 1/0 or larger and shall be of a type listed and marked on the surface for use in cable trays; where Nos. 1/0 through 4/0 single conductor cables are installed in ladder cable tray, the maximum allowable rung spacing for the ladder cable tray shall be 229 mm (9 in.); where exposed to direct rays of the sun, cables shall be identified as being sunlight resistant; (B) Welding cables installed in dedicated cable trays;

(C) Single conductors used as equipment grounding conductors; these conductors, which may be insulated, covered, or bare, shall be No. 4 or larger; and

(D) Multi-conductor cable, Type MV; where exposed to direct rays of the sun, the cable shall be identified as being sunlight resistant.

(iii) Metallic cable trays may be used as equipment grounding conductors only where continuous maintenance and supervision ensure that qualified persons will service the installed cable tray system.

(iv) Cable trays in hazardous (classified) locations may contain only the cable types permitted in such locations. (See § 1910.307.)

(v) Cable tray systems may not be used in hoistways or where subjected to severe physical damage.

(4) *Open wiring on insulators.* (i) Open wiring on insulators is only permitted on systems of 600 volts, nominal, or less for industrial or agricultural establishments, indoors or outdoors, in wet or dry locations, where subject to corrosive vapors, and for services.

(ii) Conductors smaller than No. 8 shall be rigidly supported on noncombustible, nonabsorbent insulating materials and may not contact any other objects. Supports shall be installed as follows:

(A) Within 152 mm (6 in.) from a tap or splice;

(B) Within 305 mm (12 in.) of a deadend connection to a lampholder or receptacle; and

(C) At intervals not exceeding 1.37 m (4.5 ft), and at closer intervals sufficient to provide adequate support where likely to be disturbed.

(iii) In dry locations, where not exposed to severe physical damage, conductors may be separately enclosed in flexible nonmetallic tubing. The tubing shall be in continuous lengths not exceeding 4.57 m (15.0 ft) and secured to the surface by straps at intervals not exceeding 1.37 m (4.5 ft).

(iv) Open conductors shall be separated from contact with walls, floors, wood cross members, or partitions through which they pass by tubes or bushings of noncombustible, nonabsorbent insulating material. If the bushing is shorter than the hole, a waterproof sleeve of nonconductive material shall be inserted in the hole and an insulating bushing slipped into the sleeve at each end in such a manner as to keep the conductors absolutely out of contact with the sleeve. Each conductor shall be carried through a separate tube or sleeve.

(v) Where open conductors cross ceiling joints and wall studs and are exposed to physical damage (for example, located within 2.13 m (7.0 ft) of the floor), they shall be protected.

(b) *Cabinets, boxes, and fittings*—(1) *Conductors entering boxes, cabinets, or fittings.* (i) Conductors entering cutout boxes, cabinets, or fittings shall be protected from abrasion, and openings through which conductors enter shall be effectively closed.

(ii) Unused openings in cabinets, boxes, and fittings shall be effectively closed.

(iii) Where cable is used, each cable shall be secured to the cabinet, cutout box, or meter socket enclosure. However, where cable with an entirely nonmetallic sheath enters the top of a surface-mounted enclosure through one or more nonflexible raceways not less than 457 mm (18 in.) or more than 3.05 m (10.0 ft) in length, the cable need not be secured to the cabinet, box, or enclosure provided all of the following conditions are met:

(A) Each cable is fastened within 305 mm (12 in.) of the outer end of the raceway, measured along the sheath;

(B) The raceway extends directly above the enclosure and does not penetrate a structural ceiling;

(C) A fitting is provided on each end of the raceway to protect the cable from abrasion, and the fittings remain accessible after installation;

(D) The raceway is sealed or plugged at the outer end using approved means so as to prevent access to the enclosure through the raceway;

(E) The cable sheath is continuous through the raceway and extends into the enclosure not less than 6.35 mm (0.25 in.) beyond the fitting;

(F) The raceway is fastened at its outer end and at other points as necessary; and

(G) Where installed as conduit or tubing, the allowable cable fill does not exceed that permitted for complete conduit or tubing systems.

(2) *Covers and canopies.* (i) All pull boxes, junction boxes, and fittings shall be provided with covers identified for the purpose. If metal covers are used, they shall be grounded. In completed installations, each outlet box shall have a cover, faceplate, or fixture canopy. Covers of outlet boxes having holes through which flexible cord pendants pass

FIGURE 7.1 *(Continued)*

shall be provided with bushings designed for the purpose or shall have smooth, well-rounded surfaces on which the cords may bear.

(ii) Where a fixture canopy or pan is used, any combustible wall or ceiling finish exposed between the edge of the canopy or pan and the outlet box shall be covered with noncombustible material.

(3) *Pull and junction boxes for systems over 600 volts, nominal.* In addition to other requirements in this section, the following requirements apply to pull and junction boxes for systems over 600 volts, nominal:

(i) Boxes shall provide a complete enclosure for the contained conductors or cables.

(ii) Boxes shall be closed by suitable covers securely fastened in place.

Note to paragraph (b)(3)(ii) of this section: Underground box covers that weigh over 45.4 kg (100 lbs) meet this requirement.

(iii) Covers for boxes shall be permanently marked "HIGH VOLTAGE." The marking shall be on the outside of the box cover and shall be readily visible and legible.

(c) *Switches*—(1) *Single-throw knife switches.* Single-throw knife switches shall be so placed that gravity will not tend to close them. Single-throw knife switches approved for use in the inverted position shall be provided with a locking device that will ensure that the blades remain in the open position when so set.

(2) *Double-throw knife switches.* Double-throw knife switches may be mounted so that the throw will be either vertical or horizontal. However, if the throw is vertical, a locking device shall be provided to ensure that the blades remain in the open position when so set.

(3) *Connection of switches.* (i) Single-throw knife switches and switches with butt contacts shall be connected so that the blades are deenergized when the switch is in the open position.

(ii) Single-throw knife switches, molded-case switches, switches with butt contacts, and circuit breakers used as switches shall be connected so that the terminals supplying the load are deenergized when the switch is in the open position. However, blades and terminals supplying the load of a switch may be energized when the switch is in the open position where the switch is connected to circuits or equipment inherently capable of providing a backfeed source of power. For such installations, a permanent sign shall be installed on the switch enclosure or immediately adjacent to open switches that read, "WARNING—LOAD SIDE TERMINALS MAY BE ENERGIZED BY BACKFEED."

(4) *Faceplates for flush-mounted snap switches.* Snap switches mounted in boxes shall have faceplates installed so as to completely cover the opening and seat against the finished surface.

(5) *Grounding.* Snap switches, including dimmer switches, shall be effectively grounded and shall provide a means to ground metal faceplates, whether or not a metal faceplate is installed. However, if no grounding means exists within the snap-switch enclosure, or where the wiring method does not include or provide an equipment ground, a snap switch without a grounding connection is permitted for replacement purposes only. Such snap switches shall be provided with a faceplate of non-conducting, noncombustible material if they are located within reach of conducting floors or other conducting surfaces.

(d) *Switchboards and panelboards*—(1) *Switchboards with exposed live parts.* Switchboards that have any exposed live parts shall be located in permanently dry locations and shall be accessible only to qualified persons.

(2) *Panelboard enclosures.* Panelboards shall be mounted in cabinets, cutout boxes, or enclosures designed for the purpose and shall be dead front. However, panelboards other than the dead front externally-operable type are permitted where accessible only to qualified persons.

(3) *Knife switches mounted in switchboards or panelboards.* Exposed blades of knife switches mounted in switchboards or panelboards shall be dead when open.

(e) *Enclosures for damp or wet locations*—(1) *Cabinets, cutout boxes, fittings, boxes, and panelboard enclosures.* Cabinets, cutout boxes, fittings, boxes, and panelboard enclosures in damp or wet locations shall be installed so as to prevent moisture or water from entering and accumulating within the enclosures and shall be mounted so there is at least 6.35-mm (0.25-in.) airspace between the enclosure and the wall or other supporting surface. However, nonmetallic enclosures may be installed without the airspace on a concrete, masonry, tile, or similar surface. The enclosures shall be weatherproof in wet locations.

(2) *Switches, circuit breakers, and switchboards.* Switches, circuit breakers, and switchboards installed in wet locations shall be enclosed in weatherproof enclosures.

(f) *Conductors for general wiring*—(1) *Insulation.* All conductors used for general wiring shall be insulated unless otherwise permitted in this subpart.

(2) *Type.* The conductor insulation shall be of a type that is approved for the voltage, operating temperature, and location of use.

FIGURE 7.1 (*Continued*)

(3) *Distinguishable.* Insulated conductors shall be distinguishable by appropriate color or other suitable means as being grounded conductors, ungrounded conductors, or equipment grounding conductors.

(g) *Flexible cords and cables—*(1) *Use of flexible cords and cables.* (i) Flexible cords and cables shall be approved for conditions of use and location.

(ii) Flexible cords and cables may be used only for:

(A) Pendants;

(B) Wiring of fixtures;

(C) Connection of portable lamps or appliances;

(D) Portable and mobile signs;

(E) Elevator cables;

(F) Wiring of cranes and hoists;

(G) Connection of stationary equipment to facilitate their frequent interchange;

(H) Prevention of the transmission of noise or vibration;

(I) Appliances where the fastening means and mechanical connections are designed to permit removal for maintenance and repair;

(J) Data processing cables approved as a part of the data processing system;

(K) Connection of moving parts; and

(L) Temporary wiring as permitted in paragraph (a)(2) of this section.

(iii) If used as permitted in paragraphs (g)(1)(ii)(C), (g)(1)(ii)(G), or (g)(1)(ii)(I) of this section, the flexible cord shall be equipped with an attachment plug and shall be energized from an approved receptacle outlet.

(iv) Unless specifically permitted otherwise in paragraph (g)(1)(ii) of this section, flexible cords and cables may not be used:

(A) As a substitute for the fixed wiring of a structure;

(B) Where run through holes in walls, ceilings, or floors;

(C) Where run through doorways, windows, or similar openings;

(D) Where attached to building surfaces;

(E) Where concealed behind building walls, ceilings, or floors; or

(F) Where installed in raceways, except as otherwise permitted in this subpart.

(v) Flexible cords used in show windows and showcases shall be Type S, SE, SEO, SEOO, SJ, SJE, SJEO, SJEOO, SJO, SJOO, SJT, SJTO, SJTOO, SO, SOO, ST, STO, or STOO, except for the wiring of chain-supported lighting fixtures and supply cords for portable lamps and other merchandise being displayed or exhibited.

(2) *Identification, splices, and terminations.* (i) A conductor of a flexible cord or cable that is used as a grounded conductor or an equipment grounding conductor shall be distinguishable from other conductors. Types S, SC, SCE, SCT, SE, SEO, SEOO, SJ, SJE, SJEO, SJEOO, SJO, SJT, SJTO, SJTOO, SO, SOO, ST, STO, and STOO flexible cords and Types G, G–GC, PPE, and W flexible cables shall be durably marked on the surface at intervals not exceeding 610 mm (24 in.) with the type designation, size, and number of conductors.

(ii) Flexible cords may be used only in continuous lengths without splice or tap. Hard-service cord and junior hard-service cord No. 14 and larger may be repaired if spliced so that the splice retains the insulation, outer sheath properties, and usage characteristics of the cord being spliced.

(iii) Flexible cords and cables shall be connected to devices and fittings so that strain relief is provided that will prevent pull from being directly transmitted to joints or terminal screws.

(h) *Portable cables over 600 volts, nominal.* This paragraph applies to portable cables used at more than 600 volts, nominal.

(1) *Conductor construction.* Multi-conductor portable cable for use in supplying power to portable or mobile equipment at over 600 volts, nominal, shall consist of No. 8 or larger conductors employing flexible stranding. However, the minimum size of the insulated ground-check conductor of Type G–GC cables shall be No. 10.

(2) *Shielding.* Cables operated at over 2,000 volts shall be shielded for the purpose of confining the voltage stresses to the insulation.

(3) *Equipment grounding conductors.* Grounding conductors shall be provided.

(4) *Grounding shields.* All shields shall be grounded.

(5) *Minimum bending radii.* The minimum bending radii for portable cables during installation and handling in service shall be adequate to prevent damage to the cable.

(6) *Fittings.* Connectors used to connect lengths of cable in a run shall be of a type that lock firmly together. Provisions shall be made to prevent opening or closing these connectors while energized. Strain relief shall be provided at connections and terminations.

FIGURE 7.1 *(Continued)*

(7) *Splices.* Portable cables may not be operated with splices unless the splices are of the permanent molded, vulcanized, or other approved type.

(8) *Terminations.* Termination enclosures shall be suitably marked with a high voltage hazard warning, and terminations shall be accessible only to authorized and qualified employees.

(i) *Fixture wires*—(1) *General.* Fixture wires shall be approved for the voltage, temperature, and location of use. A fixture wire which is used as a grounded conductor shall be identified.

(2) *Uses permitted.* Fixture wires may be used only:

(i) For installation in lighting fixtures and in similar equipment where enclosed or protected and not subject to bending or twisting in use; or

(ii) For connecting lighting fixtures to the branch-circuit conductors supplying the fixtures.

(3) *Uses not permitted.* Fixture wires may not be used as branch-circuit conductors except as permitted for Class 1 power limited circuits and for fire alarm circuits.

(j) *Equipment for general use*—(1) *Lighting fixtures, lampholders, lamps, and receptacles.* (i) Fixtures, lampholders, lamps, rosettes, and receptacles may have no live parts normally exposed to employee contact. However, rosettes and cleat-type lampholders and receptacles located at least 2.44 m (8.0 ft) above the floor may have exposed terminals.

(ii) Handlamps of the portable type supplied through flexible cords shall be equipped with a handle of molded composition or other material identified for the purpose, and a substantial guard shall be attached to the lampholder or the handle. Metal shell, paper-lined lampholders may not be used.

(iii) Lampholders of the screw-shell type shall be installed for use as lampholders only. Where supplied by a circuit having a grounded conductor, the grounded conductor shall be connected to the screw shell. Lampholders installed in wet or damp locations shall be of the weatherproof type.

(iv) Fixtures installed in wet or damp locations shall be identified for the purpose and shall be so constructed or installed that water cannot enter or accumulate in wireways, lampholders, or other electrical parts.

(2) *Receptacles, cord connectors, and attachment plugs (caps).* (i) All 15- and 20-ampere attachment plugs and connectors shall be constructed so that there are no exposed current-carrying parts except the prongs, blades, or pins. The cover for wire terminations shall be a part that is essential for the operation of an attachment plug or connector (dead-front construction). Attachment plugs shall be installed so that their prongs, blades, or pins are not energized unless inserted into an energized receptacle. No receptacles may be installed so as to require an energized attachment plug as its source of supply.

(ii) Receptacles, cord connectors, and attachment plugs shall be constructed so that no receptacle or cord connector will accept an attachment plug with a different voltage or current rating than that for which the device is intended. However, a 20-ampere T-slot receptacle or cord connector may accept a 15-ampere attachment plug of the same voltage rating.

(iii) Non-grounding-type receptacles and connectors may not be used for grounding-type attachment plugs.

(iv) A receptacle installed in a wet or damp location shall be suitable for the location.

(v) A receptacle installed outdoors in a location protected from the weather or in other damp locations shall have an enclosure for the receptacle that is weatherproof when the receptacle is covered (attachment plug cap not inserted and receptacle covers closed).

Note to paragraph (j)(2)(v) of this section. A receptacle is considered to be in a location protected from the weather when it is located under roofed open porches, canopies, marquees, or the like and where it will not be subjected to a beating rain or water runoff.

(vi) A receptacle installed in a wet location where the product intended to be plugged into it is not attended while in use (for example, sprinkler system controllers, landscape lighting, and holiday lights) shall have an enclosure that is weatherproof with the attachment plug cap inserted or removed.

(vii) A receptacle installed in a wet location where the product intended to be plugged into it will be attended while in use (for example, portable tools) shall have an enclosure that is weatherproof when the attachment plug cap is removed.

(3) *Appliances.* (i) Appliances may have no live parts normally exposed to contact other than parts functioning as open-resistance heating elements, such as the heating elements of a toaster, which are necessarily exposed.

(ii) Each appliance shall have a means to disconnect it from all ungrounded conductors. If an appliance is supplied by more than one source, the disconnecting means shall be grouped and identified.

(iii) Each electric appliance shall be provided with a nameplate giving the identifying name and the rating in volts and amperes, or in volts and watts. If the appliance is to be used on a specific frequency or frequencies, it shall be so marked. Where motor overload protection external to the appliance is required, the appliance shall be so marked.

FIGURE 7.1 *(Continued)*

(iv) Marking shall be located so as to be visible or easily accessible after installation.

(4) *Motors.* This paragraph applies to motors, motor circuits, and controllers.

(i) If specified in paragraph (j)(4) of this section that one piece of equipment shall be "within sight of" another piece of equipment, the piece of equipment shall be visible and not more than 15.24 m (50.0 ft) from the other.

(ii) An individual disconnecting means shall be provided for each controller. A disconnecting means shall be located within sight of the controller location. However, a single disconnecting means may be located adjacent to a group of coordinated controllers mounted adjacent to each other on a multi-motor continuous process machine. The controller disconnecting means for motor branch circuits over 600 volts, nominal, may be out of sight of the controller, if the controller is marked with a warning label giving the location and identification of the disconnecting means that is to be locked in the open position.

(iii) The disconnecting means shall disconnect the motor and the controller from all ungrounded supply conductors and shall be so designed that no pole can be operated independently.

(iv) The disconnecting means shall plainly indicate whether it is in the open (off) or closed (on) position.

(v) The disconnecting means shall be readily accessible. If more than one disconnect is provided for the same equipment, only one need be readily accessible.

(vi) An individual disconnecting means shall be provided for each motor, but a single disconnecting means may be used for a group of motors under any one of the following conditions:

(A) If a number of motors drive several parts of a single machine or piece of apparatus, such as a metal or woodworking machine, crane, or hoist;

(B) If a group of motors is under the protection of one set of branch-circuit protective devices; or

(C) If a group of motors is in a single room within sight of the location of the disconnecting means.

(vii) Motors, motor-control apparatus, and motor branch-circuit conductors shall be protected against overheating due to motor overloads or failure to start, and against short-circuit or ground faults. These provisions do not require overload protection that will stop a motor where a shutdown is likely to introduce additional or increased hazards, as in the case of fire pumps, or where continued operation of a motor is necessary for a safe shutdown of equipment or process and motor overload sensing devices are connected to a supervised alarm.

(viii) Where live parts of motors or controllers operating at over 150 volts to ground are guarded against accidental contact only by location, and where adjustment or other attendance may be necessary during the operation of the apparatus, suitable insulating mats or platforms shall be provided so that the attendant cannot readily touch live parts unless standing on the mats or platforms.

(5) *Transformers.* (i) Paragraph (j)(5) of this section covers the installation of all transformers except the following:

(A) Current transformers;

(B) Dry-type transformers installed as a component part of other apparatus;

(C) Transformers that are an integral part of an X-ray, high frequency, or electrostatic-coating apparatus;

(D) Transformers used with Class 2 and Class 3 circuits, sign and outline lighting, electric discharge lighting, and power-limited fire-alarm circuits; and

(E) Liquid-filled or dry-type transformers used for research, development, or testing, where effective safeguard arrangements are provided.

(ii) The operating voltage of exposed live parts of transformer installations shall be indicated by signs or visible markings on the equipment or structure.

(iii) Dry-type, high fire point liquid-insulated and askarel-insulated transformers installed indoors and rated over 35 kV shall be in a vault.

(iv) Oil-insulated transformers installed indoors shall be installed in a vault.

(v) Combustible material, combustible buildings and parts of buildings, fire escapes, and door and window openings shall be safeguarded from fires that may originate in oil-insulated transformers attached to or adjacent to a building or combustible material.

(vi) Transformer vaults shall be constructed so as to contain fire and combustible liquids within the vault and to prevent unauthorized access. Locks and latches shall be so arranged that a vault door can be readily opened from the inside.

(vii) Any pipe or duct system foreign to the electrical installation may not enter or pass through a transformer vault.

Note to paragraph (j)(5)(vii) of this section. Piping or other facilities provided for vault fire protection, or for transformer cooling, are not considered foreign to the electrical installation.

FIGURE 7.1 (*Continued*)

(viii) Material may not be stored in transformer vaults.

(6) *Capacitors.* (i) All capacitors, except surge capacitors or capacitors included as a component part of other apparatus, shall be provided with an automatic means of draining the stored charge after the capacitor is disconnected from its source of supply.

(ii) The following requirements apply to capacitors installed on circuits operating at more than 600 volts, nominal:

(A) Group-operated switches shall be used for capacitor switching and shall be capable of the following:

(1) Carrying continuously not less than 135 percent of the rated current of the capacitor installation;

(2) Interrupting the maximum continuous load current of each capacitor, capacitor bank, or capacitor installation that will be switched as a unit;

(3) Withstanding the maximum inrush current, including contributions from adjacent capacitor installations; and

(4) Carrying currents due to faults on the capacitor side of the switch;

(B) A means shall be installed to isolate from all sources of voltage each capacitor, capacitor bank, or capacitor installation that will be removed from service as a unit. The isolating means shall provide a visible gap in the electric circuit adequate for the operating voltage;

(C) Isolating or disconnecting switches (with no interrupting rating) shall be interlocked with the load interrupting device or shall be provided with prominently displayed caution signs to prevent switching load current; and

(D) For series capacitors, the proper switching shall be assured by use of at least one of the following:

(1) Mechanically sequenced isolating and bypass switches;

(2) Interlocks; or

(3) Switching procedure prominently displayed at the switching location.

(7) *Storage Batteries.* Provisions shall be made for sufficient diffusion and ventilation of gases from storage batteries to prevent the accumulation of explosive mixtures.

§ 1910.306 Specific purpose equipment and installations.

(a) *Electric signs and outline lighting*—(1) *Disconnecting means.* (i) Each sign and outline lighting system, or feeder circuit or branch circuit supplying a sign or outline lighting system, shall be controlled by an externally operable switch or circuit breaker that will open all ungrounded conductors. However, a disconnecting means is not required for an exit directional sign located within a building or for cord-connected signs with an attachment plug.

(ii) Signs and outline lighting systems located within fountains shall have the disconnect located at least 1.52 m (5.0 ft) from the inside walls of the fountain.

(2) *Location.* (i) The disconnecting means shall be within sight of the sign or outline lighting system that it controls. Where the disconnecting means is out of the line of sight from any section that may be energized, the disconnecting means shall be capable of being locked in the open position.

(ii) Signs or outline lighting systems operated by electronic or electromechanical controllers located external to the sign or outline lighting system may have a disconnecting means located within sight of the controller or in the same enclosure with the controller. The disconnecting means shall disconnect the sign or outline lighting system and the controller from all ungrounded supply conductors. It shall be designed so no pole can be operated independently and shall be capable of being locked in the open position.

(iii) Doors or covers giving access to uninsulated parts of indoor signs or outline lighting exceeding 600 volts and accessible to other than qualified persons shall either be provided with interlock switches to disconnect the primary circuit or shall be so fastened that the use of other than ordinary tools will be necessary to open them.

(b) *Cranes and hoists.* This paragraph applies to the installation of electric equipment and wiring used in connection with cranes, monorail hoists, hoists, and all runways.

(1) *Disconnecting means for runway conductors.* A disconnecting means shall be provided between the runway contact conductors and the power supply. Such disconnecting means shall consist of a motor-circuit switch, circuit breaker, or molded case switch. The disconnecting means shall open all ungrounded conductors simultaneously and shall be:

(i) Readily accessible and operable from the ground or floor level;

(ii) Arranged to be locked in the open position; and

(iii) Placed within view of the runway contact conductors.

(2) *Disconnecting means for cranes and monorail hoists.* (i) Except as provided in paragraph (b)(2)(iv) of this section, a motor-circuit switch, molded case switch, or circuit breaker shall be provided in the leads from the runway contact conductors or other power supply on all cranes and monorail hoists.

(ii) The disconnecting means shall be capable of being locked in the open position.

FIGURE 7.1 *(Continued)*

(iii) Means shall be provided at the operating station to open the power circuit to all motors of the crane or monorail hoist where the disconnecting means is not readily accessible from the crane or monorail hoist operating station.

(iv) The disconnecting means may be omitted where a monorail hoist or hand-propelled crane bridge installation meets all of the following conditions:

(A) The unit is controlled from the ground or floor level;

(B) The unit is within view of the power supply disconnecting means; and

(C) No fixed work platform has been provided for servicing the unit.

(3) *Limit switch.* A limit switch or other device shall be provided to prevent the load block from passing the safe upper limit of travel of any hoisting mechanism.

(4) *Clearance.* The dimension of the working space in the direction of access to live parts that may require examination, adjustment, servicing, or maintenance while alive shall be a minimum of 762 mm (2.5 ft). Where controls are enclosed in cabinets, the doors shall either open at least 90 degrees or be removable.

(c) *Elevators, dumbwaiters, escalators, moving walks, wheelchair lifts, and stairway chair lifts.* The following requirements apply to elevators, dumbwaiters, escalators, moving walks, wheelchair lifts, and stairway chair lifts.

(1) *Disconnecting means.* Elevators, dumbwaiters, escalators, moving walks, wheelchair lifts, and stairway chair lifts shall have a single means for disconnecting all ungrounded main power supply conductors for each unit.

(2) *Control panels.* Control panels not located in the same space as the drive machine shall be located in cabinets with doors or panels capable of being locked closed.

(3) *Type.* The disconnecting means shall be an enclosed externally operable fused motor circuit switch or circuit breaker capable of being locked in the open position. The disconnecting means shall be a listed device.

(4) *Operation.* No provision may be made to open or close this disconnecting means from any other part of the premises. If sprinklers are installed in hoistways, machine rooms, or machinery spaces, the disconnecting means may automatically open the power supply to the affected elevators prior to the application of water. No provision may be made to close this disconnecting means automatically (that is, power may only be restored by manual means).

(5) *Location.* The disconnecting means shall be located where it is readily accessible to qualified persons.

(i) On elevators without generator field control, the disconnecting means shall be located within sight of the motor controller. Driving machines or motion and operation controllers not within sight of the disconnecting means shall be provided with a manually operated switch installed in the control circuit adjacent to the equipment in order to prevent starting. Where the driving machine is located in a remote machinery space, a single disconnecting means for disconnecting all ungrounded main power supply conductors shall be provided and be capable of being locked in the open position.

(ii) On elevators with generator field control, the disconnecting means shall be located within sight of the motor controller for the driving motor of the motor-generator set. Driving machines, motor-generator sets, or motion and operation controllers not within sight of the disconnecting means shall be provided with a manually operated switch installed in the control circuit to prevent starting. The manually operated switch shall be installed adjacent to this equipment. Where the driving machine or the motor-generator set is located in a remote machinery space, a single means for disconnecting all ungrounded main power supply conductors shall be provided and be capable of being locked in the open position.

(iii) On escalators and moving walks, the disconnecting means shall be installed in the space where the controller is located.

(iv) On wheelchair lifts and stairway chair lifts, the disconnecting means shall be located within sight of the motor controller.

(6) *Identification and signs.* (i) Where there is more than one driving machine in a machine room, the disconnecting means shall be numbered to correspond to the identifying number of the driving machine that they control.

(ii) The disconnecting means shall be provided with a sign to identify the location of the supply-side overcurrent protective device.

(7) *Single-car and multi-car installations.* On single-car and multi-car installations, equipment receiving electrical power from more than one source shall be provided with a disconnecting means for each source of electrical power. The disconnecting means shall be within sight of the equipment served.

(8) *Warning sign for multiple disconnecting means.* A warning sign shall be mounted on or next to the disconnecting means where multiple disconnecting means are used and parts of the controllers remain energized from a source other than the one disconnected. The sign shall be clearly legible and shall read "WARNING—PARTS OF THE CONTROLLER ARE NOT DEENERGIZED BY THIS SWITCH."

(9) *Interconnection between multi-car controllers.* A warning sign worded as required in paragraph (c)(8) of this section shall be mounted on or next to the disconnecting means where interconnections between controllers are

FIGURE 7.1 (*Continued*)

necessary for the operation of the system on multi-car installations that remain energized from a source other than the one disconnected.

(10) *Motor controllers.* Motor controllers may be located outside the spaces otherwise required by paragraph (c) of this section, provided they are in enclosures with doors or removable panels capable of being locked closed and the disconnecting means is located adjacent to or is an integral part of the motor controller. Motor controller enclosures for escalators or moving walks may be located in the balustrade on the side located away from the moving steps or moving treadway. If the disconnecting means is an integral part of the motor controller, it shall be operable without opening the enclosure.

(d) *Electric welders—disconnecting means*—(1) *Arc welders.* A disconnecting means shall be provided in the supply circuit for each arc welder that is not equipped with a disconnect mounted as an integral part of the welder. The disconnecting means shall be a switch or circuit breaker, and its rating may not be less than that necessary to accommodate overcurrent protection.

(2) *Resistance welders.* A switch or circuit breaker shall be provided by which each resistance welder and its control equipment can be disconnected from the supply circuit. The ampere rating of this disconnecting means may not be less than the supply conductor ampacity. The supply circuit switch may be used as the welder disconnecting means where the circuit supplies only one welder.

(e) *Information technology equipment*—(1) *Disconnecting means.* A means shall be provided to disconnect power to all electronic equipment in an information technology equipment room. There shall also be a similar means to disconnect the power to all dedicated heating, ventilating, and air conditioning (HVAC) systems serving the room and to cause all required fire/smoke dampers to close.

(2) *Grouping.* The control for these disconnecting means shall be grouped and identified and shall be readily accessible at the principal exit doors. A single means to control both the electronic equipment and HVAC system is permitted.

(3) *Exception.* Integrated electrical systems covered by § 1910.308(g) need not have the disconnecting means required by paragraph (e)(1) of this section.

(f) *X-Ray equipment.* This paragraph applies to X-ray equipment.

(1) *Disconnecting means.* (i) A disconnecting means shall be provided in the supply circuit. The disconnecting means shall be operable from a location readily accessible from the X-ray control. For equipment connected to a 120-volt branch circuit of 30 amperes or less, a grounding-type attachment plug cap and receptacle of proper rating may serve as a disconnecting means.

(ii) If more than one piece of equipment is operated from the same high-voltage circuit, each piece or each group of equipment as a unit shall be provided with a high-voltage switch or equivalent disconnecting means. The disconnecting means shall be constructed, enclosed, or located so as to avoid contact by employees with its live parts.

(2) *Control.* The following requirements apply to industrial and commercial laboratory equipment.

(i) Radiographic and fluoroscopic-type equipment shall be effectively enclosed or shall have interlocks that deenergize the equipment automatically to prevent ready access to live current-carrying parts.

(ii) Diffraction- and irradiation-type equipment shall have a pilot light, readable meter deflection, or equivalent means to indicate when the equipment is energized, unless the equipment or installation is effectively enclosed or is provided with interlocks to prevent access to live current-carrying parts during operation.

(g) *Induction and dielectric heating equipment.* This paragraph applies to induction and dielectric heating equipment and accessories for industrial and scientific applications, but not for medical or dental applications or for appliances.

(1) *Guarding and grounding.* (i) The converting apparatus (including the dc line) and high-frequency electric circuits (excluding the output circuits and remote-control circuits) shall be completely contained within enclosures of noncombustible material.

(ii) All panel controls shall be of dead-front construction.

(iii) Doors or detachable panels shall be employed for internal access. Where doors are used giving access to voltages from 500 to 1000 volts ac or dc, either door locks shall be provided or interlocks shall be installed. Where doors are used giving access to voltages of over 1000 volts ac or dc, either mechanical lockouts with a disconnecting means to prevent access until circuit parts within the cubicle are deenergized, or both door interlocking and mechanical door locks, shall be provided. Detachable panels not normally used for access to such parts shall be fastened in a manner that will make them difficult to remove (for example, by requiring the use of tools).

(iv) Warning labels or signs that read "DANGER—HIGH VOLTAGE—KEEP OUT" shall be attached to the equipment and shall be plainly visible where persons might contact energized parts when doors are opened or closed or when panels are removed from compartments containing over 250 volts ac or dc.

FIGURE 7.1 (*Continued*)

(v) Induction and dielectric heating equipment shall be protected as follows:

(A) Protective cages or adequate shielding shall be used to guard work applicators other than induction heating coils.

(B) Induction heating coils shall be protected by insulation or refractory materials or both.

(C) Interlock switches shall be used on all hinged access doors, sliding panels, or other such means of access to the applicator, unless the applicator is an induction heating coil at dc ground potential or operating at less than 150 volts ac.

(D) Interlock switches shall be connected in such a manner as to remove all power from the applicator when any one of the access doors or panels is open.

(vi) A readily accessible disconnecting means shall be provided by which each heating equipment can be isolated from its supply circuit. The ampere rating of this disconnecting means may not be less than the nameplate current rating of the equipment. The supply circuit disconnecting means is permitted as a heating equipment disconnecting means where the circuit supplies only one piece of equipment.

(2) *Remote control.* (i) If remote controls are used for applying power, a selector switch shall be provided and interlocked to provide power from only one control point at a time.

(ii) Switches operated by foot pressure shall be provided with a shield over the contact button to avoid accidental closing of the switch.

(h) *Electrolytic cells.* This paragraph applies to the installation of the electrical components and accessory equipment of electrolytic cells, electrolytic cell lines, and process power supply for the production of aluminum, cadmium, chlorine, copper, fluorine, hydrogen peroxide, magnesium, sodium, sodium chlorate, and zinc. Cells used as a source of electric energy and for electroplating processes and cells used for production of hydrogen are not covered by this paragraph.

(1) *Application.* Installations covered by paragraph (h) of this section shall comply with all applicable provisions of this subpart, except as follows:

(i) Overcurrent protection of electrolytic cell dc process power circuits need not comply with the requirements of § 1910.304(f);

(ii) Equipment located or used within the cell line working zone or associated with the cell line dc power circuits need not comply with the provisions of § 1910.304(g); and

(iii) Electrolytic cells, cell line conductors, cell line attachments, and the wiring of auxiliary equipment and devices within the cell line working zone need not comply with the provisions of § 1910.303 or § 1910.304(b) and (c).

(2) *Disconnecting means.* If more than one dc cell line process power supply serves the same cell line, a disconnecting means shall be provided on the cell line circuit side of each power supply to disconnect it from the cell line circuit. Removable links or removable conductors may be used as the disconnecting means.

(3) *Portable electric equipment.* (i) The frames and enclosures of portable electric equipment used within the cell line working zone may not be grounded, unless the cell line circuit voltage does not exceed 200 volts DC or the frames are guarded.

(ii) Ungrounded portable electric equipment shall be distinctively marked and shall employ plugs and receptacles of a configuration that prevents connection of this equipment to grounding receptacles and that prevents inadvertent interchange of ungrounded and grounded portable electric equipment.

(4) *Power supply circuits and receptacles for portable electric equipment.* (i) Circuits supplying power to ungrounded receptacles for handheld, cord- and plug-connected equipment shall meet the following requirements:

(A) The circuits shall be electrically isolated from any distribution system supplying areas other than the cell line working zone and shall be ungrounded;

(B) The circuits shall be supplied through isolating transformers with primaries operating at not more than 600 volts between conductors and protected with proper overcurrent protection;

(C) The secondary voltage of the isolating transformers may not exceed 300 volts between conductors; and

(D) All circuits supplied from the secondaries shall be ungrounded and shall have an approved overcurrent device of proper rating in each conductor.

(ii) Receptacles and their mating plugs for ungrounded equipment may not have provision for a grounding conductor and shall be of a configuration that prevents their use for equipment required to be grounded.

(iii) Receptacles on circuits supplied by an isolating transformer with an ungrounded secondary:

(A) Shall have a distinctive configuration;

(B) Shall be distinctively marked; and

(C) May not be used in any other location in the facility.

(5) *Fixed and portable electric equipment.* (i) The following need not be grounded:

FIGURE 7.1 (*Continued*)

(A) AC systems supplying fixed and portable electric equipment within the cell line working zone; and

(B) Exposed conductive surfaces, such as electric equipment housings, cabinets, boxes, motors, raceways and the like that are within the cell line working zone.

(ii) Auxiliary electric equipment, such as motors, transducers, sensors, control devices, and alarms, mounted on an electrolytic cell or other energized surface shall be connected to the premises wiring systems by any of the following means:

(A) Multi-conductor hard usage or extra hard usage flexible cord;

(B) Wire or cable in suitable nonmetallic raceways or cable trays; or

(C) Wire or cable in suitable metal raceways or metal cable trays installed with insulating breaks such that they will not cause a potentially hazardous electrical condition.

(iii) Fixed electric equipment may be bonded to the energized conductive surfaces of the cell line, its attachments, or auxiliaries. If fixed electric equipment is mounted on an energized conductive surface, it shall be bonded to that surface.

(6) *Auxiliary non-electrical connections.* Auxiliary non-electrical connections such as air hoses, water hoses, and the like, to an electrolytic cell, its attachments, or auxiliary equipment may not have continuous conductive reinforcing wire, armor, braids, or the like. Hoses shall be of a nonconductive material.

(7) *Cranes and hoists.* (i) The conductive surfaces of cranes and hoists that enter the cell line working zone need not be grounded. The portion of an overhead crane or hoist that contacts an energized electrolytic cell or energized attachments shall be insulated from ground.

(ii) Remote crane or hoist controls that may introduce hazardous electrical conditions into the cell line working zone shall employ one or more of the following systems:

(A) Isolated and ungrounded control circuit;

(B) Nonconductive rope operator;

(C) Pendant pushbutton with nonconductive supporting means and with nonconductive surfaces or ungrounded exposed conductive surfaces; or

(D) Radio.

(i) *Electrically driven or controlled irrigation machines*—(1) *Lightning protection.* If an irrigation machine has a stationary point, a grounding electrode system shall be connected to the machine at the stationary point for lightning protection.

(2) *Disconnecting means.* (i) The main disconnecting means for a center pivot irrigation machine shall be located at the point of connection of electrical power to the machine or shall be visible and not more than 15.2 m (50 ft) from the machine.

(ii) The disconnecting means shall be readily accessible and capable of being locked in the open position.

(iii) A disconnecting means shall be provided for each motor and controller.

(j) *Swimming pools, fountains, and similar installations.* This paragraph applies to electric wiring for and equipment in or adjacent to all swimming, wading, therapeutic, and decorative pools and fountains; hydromassage bathtubs, whether permanently installed or storable; and metallic auxiliary equipment, such as pumps, filters, and similar equipment. Therapeutic pools in health care facilities are exempt from these provisions.

(1) *Receptacles.* (i) A single receptacle of the locking and grounding type that provides power for a permanently installed swimming pool recirculating pump motor may be located not less than 1.52 m (5 ft) from the inside walls of a pool. All other receptacles on the property shall be located at least 3.05 m (10 ft) from the inside walls of a pool.

(ii) Receptacles that are located within 4.57 m (15 ft), or 6.08 m (20 ft) if the installation was built after August 13, 2007, of the inside walls of the pool shall be protected by ground-fault circuit interrupters.

(iii) Where a pool is installed permanently at a dwelling unit, at least one 125-volt, 15- or 20-ampere receptacle on a general-purpose branch circuit shall be located a minimum of 3.05 m (10 ft) and not more than 6.08 m (20 ft) from the inside wall of the pool. This receptacle shall be located not more than 1.98 m (6.5 ft) above the floor, platform, or grade level serving the pool.

Note to paragraph (j)(1) of this section: In determining these dimensions, the distance to be measured is the shortest path the supply cord of an appliance connected to the receptacle would follow without piercing a floor, wall, or ceiling of a building or other effective permanent barrier.

(2) *Lighting fixtures, lighting outlets, and ceiling suspended (paddle) fans.* (i) In outdoor pool areas, lighting fixtures, lighting outlets, and ceiling-suspended (paddle) fans may not be installed over the pool or over the area extending 1.52 m (5 ft) horizontally from the inside walls of a pool unless no part of the lighting fixture of a

FIGURE 7.1 *(Continued)*

ceiling-suspended (paddle) fan is less than 3.66 m (12 ft) above the maximum water level. However, a lighting fixture or lighting outlet that was installed before April 16, 1981, may be located less than 1.52 m (5 ft) measured horizontally from the inside walls of a pool if it is at least 1.52 m (5 ft) above the surface of the maximum water level and is rigidly attached to the existing structure. It shall also be protected by a ground-fault circuit interrupter installed in the branch circuit supplying the fixture.

(ii) Lighting fixtures and lighting outlets installed in the area extending between 1.52 m (5 ft) and 3.05 m (10 ft) horizontally from the inside walls of a pool shall be protected by a ground-fault circuit interrupter unless installed 1.52 m (5 ft) above the maximum water level and rigidly attached to the structure adjacent to or enclosing the pool.

(3) *Cord- and plug-connected equipment.* Flexible cords used with the following equipment may not exceed 0.9 m (3 ft) in length and shall have a copper equipment grounding conductor with a grounding-type attachment plug:

(i) Cord- and plug-connected lighting fixtures installed within 4.88 m (16 ft) of the water surface of permanently installed pools; and

(ii) Other cord- and plug-connected, fixed or stationary equipment used with permanently installed pools.

(4) *Underwater equipment.* (i) A ground-fault circuit interrupter shall be installed in the branch circuit supplying underwater fixtures operating at more than 15 volts. Equipment installed underwater shall be identified for the purpose.

(ii) No underwater lighting fixtures may be installed for operation at over 150 volts between conductors.

(iii) A lighting fixture facing upward shall have the lens adequately guarded to prevent contact by any person.

(5) *Fountains.* All electric equipment, including power supply cords, operating at more than 15 volts and used with fountains shall be protected by ground-fault circuit interrupters.

(k) *Carnivals, circuses, fairs, and similar events.* This paragraph covers the installation of portable wiring and equipment, including wiring in or on all structures, for carnivals, circuses, exhibitions, fairs, traveling attractions, and similar events.

(1) *Protection of electric equipment.* Electric equipment and wiring methods in or on rides, concessions, or other units shall be provided with mechanical protection where such equipment or wiring methods are subject to physical damage.

(2) *Installation.* (i) Services shall be installed in accordance with applicable requirements of this subpart, and, in addition, shall comply with the following:

(A) Service equipment may not be installed in a location that is accessible to unqualified persons, unless the equipment is lockable; and

(B) Service equipment shall be mounted on solid backing and installed so as to be protected from the weather, unless the equipment is of weatherproof construction.

(ii) Amusement rides and amusement attractions shall be maintained not less than 4.57 m (15 ft) in any direction from overhead conductors operating at 600 volts or less, except for the conductors supplying the amusement ride or attraction. Amusement rides or attractions may not be located under or within 4.57 m (15 ft) horizontally of conductors operating in excess of 600 volts.

(iii) Flexible cords and cables shall be listed for extra-hard usage. When used outdoors, flexible cords and cables shall also be listed for wet locations and shall be sunlight resistant.

(iv) Single conductor cable shall be size No. 2 or larger.

(v) Open conductors are prohibited except as part of a listed assembly or festoon lighting installed in accordance with § 1910.304(c).

(vi) Flexible cords and cables shall be continuous without splice or tap between boxes or fittings. Cord connectors may not be laid on the ground unless listed for wet locations. Connectors and cable connections may not be placed in audience traffic paths or within areas accessible to the public unless guarded.

(vii) Wiring for an amusement ride, attraction, tent, or similar structure may not be supported by another ride or structure unless specifically identified for the purpose.

(viii) Flexible cords and cables run on the ground, where accessible to the public, shall be covered with approved nonconductive mats. Cables and mats shall be arranged so as not to present a tripping hazard.

(ix) A box or fitting shall be installed at each connection point, outlet, switch point, or junction point.

(3) *Inside tents and concessions.* Electrical wiring for temporary lighting, where installed inside of tents and concessions, shall be securely installed, and, where subject to physical damage, shall be provided with mechanical protection. All temporary lamps for general illumination shall be protected from accidental breakage by a suitable fixture or lampholder with a guard.

FIGURE 7.1 *(Continued)*

(4) *Portable distribution and termination boxes.* Employers may only use portable distribution and termination boxes that meet the following requirements:

(i) Boxes shall be designed so that no live parts are exposed to accidental contact. Where installed outdoors, the box shall be of weatherproof construction and mounted so that the bottom of the enclosure is not less than 152 mm (6 in.) above the ground;

(ii) Busbars shall have an ampere rating not less than the overcurrent device supplying the feeder supplying the box. Busbar connectors shall be provided where conductors terminate directly on busbars;

(iii) Receptacles shall have overcurrent protection installed within the box. The overcurrent protection may not exceed the ampere rating of the receptacle, except as permitted in § 1910.305(j)(4) for motor loads;

(iv) Where single-pole connectors are used, they shall comply with the following:

(A) Where ac single-pole portable cable connectors are used, they shall be listed and of the locking type. Where paralleled sets of current-carrying single-pole separable connectors are provided as input devices, they shall be prominently labeled with a warning indicating the presence of internal parallel connections. The use of single-pole separable connectors shall comply with at least one of the following conditions:

(1) Connection and disconnection of connectors are only possible where the supply connectors are interlocked to the source and it is not possible to connect or disconnect connectors when the supply is energized; or

(2) Line connectors are of the listed sequential-interlocking type so that load connectors are connected in the following sequence:

(i) Equipment grounding conductor connection;

(ii) Grounded circuit-conductor connection, if provided; and

(iii) Ungrounded conductor connection; and so that disconnection is in the reverse order; or

(3) A caution notice is provided adjacent to the line connectors indicating that plug connection must be in the following sequence:

(i) Equipment grounding conductor connection;

(ii) Grounded circuit-conductor connection, if provided; and

(iii) Ungrounded conductor connection; and indicating that disconnection is in the reverse order; and

(B) Single-pole separable connectors used in portable professional motion picture and television equipment may be interchangeable for ac or dc use or for different current ratings on the same premises only if they are listed for ac/dc use and marked to identify the system to which they are connected;

(v) Overcurrent protection of equipment and conductors shall be provided; and

(vi) The following equipment connected to the same source shall be bonded:

(A) Metal raceways and metal sheathed cable;

(B) Metal enclosures of electrical equipment; and

(C) Metal frames and metal parts of rides, concessions, trailers, trucks, or other equipment that contain or support electrical equipment.

(5) *Disconnecting means.* (i) Each ride and concession shall be provided with a fused disconnect switch or circuit breaker located within sight and within 1.83 m (6 ft) of the operator's station.

(ii) The disconnecting means shall be readily accessible to the operator, including when the ride is in operation.

(iii) Where accessible to unqualified persons, the enclosure for the switch or circuit breaker shall be of the lockable type.

(iv) A shunt trip device that opens the fused disconnect or circuit breaker when a switch located in the ride operator's console is closed is a permissible method of opening the circuit.

§ 1910.307 Hazardous (classified) locations.

(a) *Scope*—(1) *Applicability.* This section covers the requirements for electric equipment and wiring in locations that are classified depending on the properties of the flammable vapors, liquids or gases, or combustible dusts or fibers that may be present therein and the likelihood that a flammable or combustible concentration or quantity is present. Hazardous (classified) locations may be found in occupancies such as, but not limited to, the following: aircraft hangars, gasoline dispensing and service stations, bulk storage plants for gasoline or other volatile flammable liquids, paint finishing process plants, health care facilities, agricultural or other facilities where excessive combustible dusts may be present, marinas, boat yards, and petroleum and chemical processing plants. Each room, section or area shall be considered individually in determining its classification.

(2) *Classifications.* (i) These hazardous (classified) locations are assigned the following designations:

(A) Class I, Division 1

FIGURE 7.1 *(Continued)*

(B) Class I, Division 2
(C) Class I, Zone 0
(D) Class I, Zone 1
(E) Class I, Zone 2
(F) Class II, Division 1
(G) Class II, Division 2
(H) Class III, Division 1
(I) Class III, Division 2
(ii) For definitions of these locations, see § 1910.399.

(3) *Other sections of this subpart.* All applicable requirements in this subpart apply to hazardous (classified) locations unless modified by provisions of this section.

(4) *Division and zone classification.* In Class I locations, an installation must be classified as using the division classification system meeting paragraphs (c), (d), (e), and (f) of this section or using the zone classification system meeting paragraph (g) of this section. In Class II and Class III locations, an installation must be classified using the division classification system meeting paragraphs (c), (d), (e), and (f) of this section.

(b) *Documentation.* All areas designated as hazardous (classified) locations under the Class and Zone system and areas designated under the Class and Division system established after August 13, 2007 shall be properly documented. This documentation shall be available to those authorized to design, install, inspect, maintain, or operate electric equipment at the location.

(c) *Electrical installations.* Equipment, wiring methods, and installations of equipment in hazardous (classified) locations shall be intrinsically safe, approved for the hazardous (classified) location, or safe for the hazardous (classified) location. Requirements for each of these options are as follows:

(1) *Intrinsically safe.* Equipment and associated wiring approved as intrinsically safe is permitted in any hazardous (classified) location for which it is approved;

(2) *Approved for the hazardous (classified) location.* (i) Equipment shall be approved not only for the class of location, but also for the ignitable or combustible properties of the specific gas, vapor, dust, or fiber that will be present.

Note to paragraph (c)(2)(i) of this section: NFPA 70, the National Electrical Code, lists or defines hazardous gases, vapors, and dusts by "Groups" characterized by their ignitable or combustible properties.

(ii) Equipment shall be marked to show the class, group, and operating temperature or temperature range, based on operation in a 40-degree C ambient, for which it is approved. The temperature marking may not exceed the ignition temperature of the specific gas or vapor to be encountered. However, the following provisions modify this marking requirement for specific equipment:

(A) Equipment of the non-heat-producing type, such as junction boxes, conduit, and fittings, and equipment of the heat-producing type having a maximum temperature not more than 100°C (212°F) need not have a marked operating temperature or temperature range;

(B) Fixed lighting fixtures marked for use in Class I, Division 2 or Class II, Division 2 locations only need not be marked to indicate the group;

(C) Fixed general-purpose equipment in Class I locations, other than lighting fixtures, that is acceptable for use in Class I, Division 2 locations need not be marked with the class, group, division, or operating temperature;

(D) Fixed dust-tight equipment, other than lighting fixtures, that is acceptable for use in Class II, Division 2 and Class III locations need not be marked with the class, group, division, or operating temperature; and

(E) Electric equipment suitable for ambient temperatures exceeding 40°C (104°F) shall be marked with both the maximum ambient temperature and the operating temperature or temperature range at that ambient temperature; and

(3) *Safe for the hazardous (classified) location.* Equipment that is safe for the location shall be of a type and design that the employer demonstrates will provide protection from the hazards arising from the combustibility and flammability of vapors, liquids, gases, dusts, or fibers involved.

Note to paragraph (c)(3) of this section: The National Electrical Code, NFPA 70, contains guidelines for determining the type and design of equipment and installations that will meet this requirement. Those guidelines address electric wiring, equipment, and systems installed in hazardous (classified) locations and contain specific provisions for the following: wiring methods, wiring connections; conductor insulation, flexible cords, sealing and drainage,

FIGURE 7.1 *(Continued)*

transformers, capacitors, switches, circuit breakers, fuses, motor controllers, receptacles, attachment plugs, meters, relays, instruments, resistors, generators, motors, lighting fixtures, storage battery charging equipment, electric cranes, electric hoists and similar equipment, utilization equipment, signaling systems, alarm systems, remote control systems, local loud speaker and communication systems, ventilation piping, live parts, lightning surge protection, and grounding.

(d) *Conduits.* All conduits shall be threaded and shall be made wrenchtight. Where it is impractical to make a threaded joint tight, a bonding jumper shall be utilized.

(e) *Equipment in Division 2 locations.* Equipment that has been approved for a Division 1 location may be installed in a Division 2 location of the same class and group. General-purpose equipment or equipment in general-purpose enclosures may be installed in Division 2 locations if the employer can demonstrate that the equipment does not constitute a source of ignition under normal operating conditions.

(f) *Protection techniques.* The following are acceptable protection techniques for electric and electronic equipment in hazardous (classified) locations.

(1) *Explosionproof apparatus.* This protection technique is permitted for equipment in the Class I, Division 1 and 2 locations for which it is approved.

(2) *Dust ignition proof.* This protection technique is permitted for equipment in the Class II, Division 1 and 2 locations for which it is approved.

(3) *Dust-tight.* This protection technique is permitted for equipment in the Class II, Division 2 and Class III locations for which it is approved.

(4) *Purged and pressurized.* This protection technique is permitted for equipment in any hazardous (classified) location for which it is approved.

(5) *Nonincendive circuit.* This protection technique is permitted for equipment in Class I, Division 2; Class II, Division 2; or Class III, Division 1 or 2 locations.

(6) *Nonincendive equipment.* This protection technique is permitted for equipment in Class I, Division 2; Class II, Division 2; or Class III, Division 1 or 2 locations.

(7) *Nonincendive component.* This protection technique is permitted for equipment in Class I, Division 2; Class II, Division 2; or Class III, Division 1 or 2 locations.

(8) *Oil immersion.* This protection technique is permitted for current-interrupting contacts in Class I, Division 2 locations as described in the Subpart.

(9) *Hermetically sealed.* This protection technique is permitted for equipment in Class I, Division 2; Class II, Division 2; and Class III, Division 1 or 2 locations.

(10) *Other protection techniques.* Any other protection technique that meets paragraph (c) of this section is acceptable in any hazardous (classified) location.

(g) *Class I, Zone 0, 1, and 2 locations*—(1) *Scope.* Employers may use the zone classification system as an alternative to the division classification system for electric and electronic equipment and wiring for all voltage in Class I, Zone 0, Zone 1, and Zone 2 hazardous (classified) locations where fire or explosion hazards may exist due to flammable gases, vapors, or liquids.

(2) *Location and general requirements.* (i) Locations shall be classified depending on the properties of the flammable vapors, liquids, or gases that may be present and the likelihood that a flammable or combustible concentration or quantity is present. Where pyrophoric materials are the only materials used or handled, these locations need not be classified.

(ii) Each room, section, or area shall be considered individually in determining its classification.

(iii) All threaded conduit shall be threaded with an NPT (National (American) Standard Pipe Taper) standard conduit cutting die that provides 3⁄4-in. taper per foot. The conduit shall be made wrench tight to prevent sparking when fault current flows through the conduit system and to ensure the explosionproof or flameproof integrity of the conduit system where applicable.

(iv) Equipment provided with threaded entries for field wiring connection shall be installed in accordance with paragraph (g)(2)(iv)(A) or (g)(2)(iv)(B) of this section.

(A) For equipment provided with threaded entries for NPT threaded conduit or fittings, listed conduit, conduit fittings, or cable fittings shall be used.

(B) For equipment with metric threaded entries, such entries shall be identified as being metric, or listed adaptors to permit connection to conduit of NPT-threaded fittings shall be provided with the equipment. Adapters shall be used for connection to conduit or NPT-threaded fittings.

FIGURE 7.1 *(Continued)*

(3) *Protection techniques.* One or more of the following protection techniques shall be used for electric and electronic equipment in hazardous (classified) locations classified under the zone classification system.

(i) Flameproof "d"—This protection technique is permitted for equipment in the Class I, Zone 1 locations for which it is approved.

(ii) Purged and pressurized—This protection technique is permitted for equipment in the Class I, Zone 1 or Zone 2 locations for which it is approved.

(iii) Intrinsic safety—This protection technique is permitted for equipment in the Class I, Zone 0 or Zone 1 locations for which it is approved.

(iv) Type of protection "n"—This protection technique is permitted for equipment in the Class I, Zone 2 locations for which it is approved. Type of protection "n" is further subdivided into nA, nC, and nR.

(v) Oil Immersion "o"—This protection technique is permitted for equipment in the Class I, Zone 1 locations for which it is approved.

(vi) Increased safety "e"—This protection technique is permitted for equipment in the Class I, Zone 1 locations for which it is approved.

(vii) Encapsulation "m"—This protection technique is permitted for equipment in the Class I, Zone 1 locations for which it is approved.

(viii) Powder Filling "q"—This protection technique is permitted for equipment in the Class I, Zone 1 locations for which it is approved.

(4) *Special precaution.* Paragraph (g) of this section requires equipment construction and installation that will ensure safe performance under conditions of proper use and maintenance.

(i) Classification of areas and selection of equipment and wiring methods shall be under the supervision of a qualified registered professional engineer.

(ii) In instances of areas within the same facility classified separately, Class I, Zone 2 locations may abut, but not overlap, Class I, Division 2 locations. Class I, Zone 0 or Zone 1 locations may not abut Class I, Division 1 or Division 2 locations.

(iii) A Class I, Division 1 or Division 2 location may be reclassified as a Class I, Zone 0, Zone 1, or Zone 2 location only if all of the space that is classified because of a single flammable gas or vapor source is reclassified.

Note to paragraph (g)(4) of this section: Low ambient conditions require special consideration. Electric equipment depending on the protection techniques described by paragraph (g)(3)(i) of this section may not be suitable for use at temperatures lower than −20°C (−4°F) unless they are approved for use at lower temperatures. However, at low ambient temperatures, flammable concentrations of vapors may not exist in a location classified Class I, Zone 0, 1, or 2 at normal ambient temperature.

(5) *Listing and marking.* (i) Equipment that is listed for a Zone 0 location may be installed in a Zone 1 or Zone 2 location of the same gas or vapor. Equipment that is listed for a Zone 1 location may be installed in a Zone 2 location of the same gas or vapor.

(ii) Equipment shall be marked in accordance with paragraph (g)(5)(ii)(A) and (g)(5)(ii)(B) of this section, except as provided in (g)(5)(ii)(C).

Figure S-1—Example Marking for Class I, Zone 0, AEx ia IIC T6

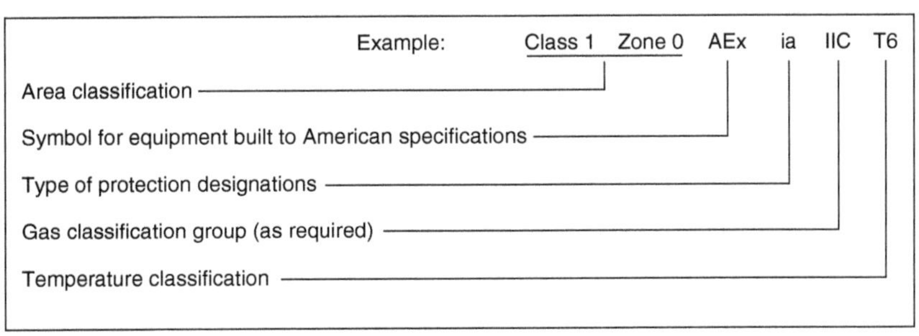

FIGURE 7.1 *(Continued)*

(A) Equipment approved for Class I, Division 1 or Class 1, Division 2 shall, in addition to being marked in accordance with (c)(2)(ii), be marked with the following:

(1) Class I, Zone 1 or Class I, Zone 2 (as applicable);

(2) Applicable gas classification groups; and

(3) Temperature classification; or

(B) Equipment meeting one or more of the protection techniques described in paragraph (g)(3) of this section shall be marked with the following in the order shown:

(1) Class, except for intrinsically safe apparatus;

(2) Zone, except for intrinsically safe apparatus;

(3) Symbol "AEx;"

(4) Protection techniques;

(5) Applicable gas classification groups; and

(6) Temperature classification, except for intrinsically safe apparatus.

Note to paragraph (g)(5)(ii)(B) of this section: An example of such a required marking is "Class I, Zone 0, AEx ia IIC T6." See Figure S-1 for an explanation of this marking.

(C) Equipment that the employer demonstrates will provide protection from the hazards arising from the flammability of the gas or vapor and the zone of location involved and will be recognized as providing such protection by employees need not be marked.

Note to paragraph (g)(5)(ii)(C) of this section: The National Electrical Code, NFPA 70, contains guidelines for determining the type and design of equipment and installations that will meet this provision.

§ 1910.308 Special systems.

(a) *Systems over 600 volts, nominal.* This paragraph covers the general requirements for all circuits and equipment operated at over 600 volts.

(1) *Aboveground wiring methods.* (i) Aboveground conductors shall be installed in rigid metal conduit, in intermediate metal conduit, in electrical metallic tubing, in rigid nonmetallic conduit, in cable trays, as busways, as cablebus, in other identified raceways, or as open runs of metal-clad cable suitable for the use and purpose. In locations accessible to qualified persons only, open runs of Type MV cables, bare conductors, and bare busbars are also permitted. Busbars shall be either copper or aluminum. Open runs of insulated wires and cables having a bare lead sheath or a braided outer covering shall be supported in a manner designed to prevent physical damage to the braid or sheath.

(ii) Conductors emerging from the ground shall be enclosed in approved raceways.

(2) *Braid-covered insulated conductors—open installations.* The braid on open runs of braid-covered insulated conductors shall be flame retardant or shall have a flame-retardant saturant applied after installation. This treated braid covering shall be stripped back a safe distance at conductor terminals, according to the operating voltage.

(3) *Insulation shielding.* (i) Metallic and semiconductor insulation shielding components of shielded cables shall be removed for a distance dependent on the circuit voltage and insulation. Stress reduction means shall be provided at all terminations of factory-applied shielding.

(ii) Metallic shielding components such as tapes, wires, or braids, or combinations thereof, and their associated conducting and semiconducting components shall be grounded.

(4) *Moisture or mechanical protection for metal-sheathed cables.* Where cable conductors emerge from a metal sheath and where protection against moisture or physical damage is necessary, the insulation of the conductors shall be protected by a cable sheath terminating device.

(5) *Interrupting and isolating devices.* (i) Circuit breaker installations located indoors shall consist of metal-enclosed units or fire-resistant cell-mounted units. In locations accessible only to qualified employees, open mounting of circuit breakers is permitted. A means of indicating the open and closed position of circuit breakers shall be provided.

(ii) Where fuses are used to protect conductors and equipment, a fuse shall be placed in each ungrounded conductor. Two power fuses may be used in parallel to protect the same load, if both fuses have identical ratings, and if both fuses are installed in an identified common mounting with electrical connections that will divide the current equally. Power fuses of the vented type may not be used indoors, underground, or in metal enclosures unless identified for the use.

FIGURE 7.1 *(Continued)*

(iii) Fused cutouts installed in buildings or transformer vaults shall be of a type identified for the purpose. Distribution cutouts may not be used indoors, underground, or in metal enclosures. They shall be readily accessible for fuse replacement.

(iv) Where fused cutouts are not suitable to interrupt the circuit manually while carrying full load, an approved means shall be installed to interrupt the entire load. Unless the fused cutouts are interlocked with the switch to prevent opening of the cutouts under load, a conspicuous sign shall be placed at such cutouts reading: "WARNING—DO NOT OPERATE UNDER LOAD."

(v) Suitable barriers or enclosures shall be provided to prevent contact with non-shielded cables or energized parts of oil-filled cutouts.

(vi) Load interrupter switches may be used only if suitable fuses or circuits are used in conjunction with these devices to interrupt fault currents.

(A) Where these devices are used in combination, they shall be coordinated electrically so that they will safely withstand the effects of closing, carrying, or interrupting all possible currents up to the assigned maximum short-circuit rating.

(B) Where more than one switch is installed with interconnected load terminals to provide for alternate connection to different supply conductors, each switch shall be provided with a conspicuous sign reading: "WARNING—SWITCH MAY BE ENERGIZED BY BACKFEED."

(vii) A means (for example, a fuseholder and fuse designed for the purpose) shall be provided to completely isolate equipment for inspection and repairs. Isolating means that are not designed to interrupt the load current of the circuit shall be either interlocked with an approved circuit interrupter or provided with a sign warning against opening them under load.

(6) *Mobile and portable equipment.* (i) A metallic enclosure shall be provided on the mobile machine for enclosing the terminals of the power cable. The enclosure shall include provisions for a solid connection for the grounding terminal to effectively ground the machine frame. The method of cable termination used shall prevent any strain or pull on the cable from stressing the electrical connections. The enclosure shall have provision for locking so only authorized qualified persons may open it and shall be marked with a sign warning of the presence of energized parts.

(ii) All energized switching and control parts shall be enclosed in effectively grounded metal cabinets or enclosures. Circuit breakers and protective equipment shall have the operating means projecting through the metal cabinet or enclosure so these units can be reset without locked doors being opened. Enclosures and metal cabinets shall be locked so that only authorized qualified persons have access and shall be marked with a sign warning of the presence of energized parts. Collector ring assemblies on revolving-type machines (shovels, draglines, etc.) shall be guarded.

(7) *Tunnel installations.* This paragraph applies to installation and use of high-voltage power distribution and utilization equipment that is portable or mobile, such as substations, trailers, cars, mobile shovels, draglines, hoists, drills, dredges, compressors, pumps, conveyors, and underground excavators.

(i) Conductors in tunnels shall be installed in one or more of the following:

(A) Metal conduit or other metal raceway;

(B) Type MC cable; or

(C) Other approved multi-conductor cable.

(ii) Multi-conductor portable cable may supply mobile equipment.

(iii) Conductors and cables shall also be so located or guarded as to protect them from physical damage. An equipment grounding conductor shall be run with circuit conductors inside the metal raceway or inside the multi-conductor cable jacket. The equipment grounding conductor may be insulated or bare.

(iv) Bare terminals of transformers, switches, motor controllers, and other equipment shall be enclosed to prevent accidental contact with energized parts.

(v) Enclosures for use in tunnels shall be drip-proof, weatherproof, or submersible as required by the environmental conditions.

(vi) Switch or contactor enclosures may not be used as junction boxes or raceways for conductors feeding through or tapping off to other switches, unless special designs are used to provide adequate space for this purpose.

(vii) A disconnecting means that simultaneously opens all ungrounded conductors shall be installed at each transformer or motor location.

FIGURE 7.1 *(Continued)*

(viii) All non-energized metal parts of electric equipment and metal raceways and cable sheaths shall be effectively grounded and bonded to all metal pipes and rails at the portal and at intervals not exceeding 305 m (1000 ft) throughout the tunnel.

(b) *Emergency power systems.* This paragraph applies to circuits, systems, and equipment intended to supply power for illumination and special loads in the event of failure of the normal supply.

(1) *Wiring methods.* Emergency circuit wiring shall be kept entirely independent of all other wiring and equipment and may not enter the same raceway, cable, box, or cabinet or other wiring except either where common circuit elements suitable for the purpose are required, or for transferring power from the normal to the emergency source.

(2) *Emergency illumination.* Emergency illumination shall include all required means of egress lighting, illuminated exit signs, and all other lights necessary to provide illumination. Where emergency lighting is necessary, the system shall be so arranged that the failure of any individual lighting element, such as the burning out of a light bulb, cannot leave any space in total darkness.

(3) *Signs.* (i) A sign shall be placed at the service entrance equipment indicating the type and location of onsite emergency power sources. However, a sign is not required for individual unit equipment.

(ii) Where the grounded circuit conductor connected to the emergency source is connected to a grounding electrode conductor at a location remote from the emergency source, there shall be a sign at the grounding location that shall identify all emergency and normal sources connected at that location.

(c) *Class 1, Class 2, and Class 3 remote control, signaling, and power limited circuits*—(1) *Classification.* Class 1, Class 2, and Class 3 remote control, signaling, or power-limited circuits are characterized by their usage and electrical power limitation that differentiates them from light and power circuits. These circuits are classified in accordance with their respective voltage and power limitations as summarized in paragraphs (c)(1)(i) through (c)(1)(iii) of this section.

(i) A Class 1 power-limited circuit shall be supplied from a source having a rated output of not more than 30 volts and 1000 volt-amperes.

(ii) A Class 1 remote control circuit or a Class 1 signaling circuit shall have a voltage not exceeding 600 volts; however, the power output of the source need not be limited.

(iii) The power source for a Class 2 or Class 3 circuit shall be listed equipment marked as a Class 2 or Class 3 power source, except as follows:

(A) Thermocouples do not require listing as a Class 2 power source; and

(B) A dry cell battery is considered an inherently limited Class 2 power source, provided the voltage is 30 volts or less and the capacity is less than or equal to that available from series-connected No. 6 carbon zinc cells.

(2) *Marking.* A Class 2 or Class 3 power supply unit shall be durably marked where plainly visible to indicate the class of supply and its electrical rating.

(3) *Separation from conductors of other circuits.* Cables and conductors of Class 2 and Class 3 circuits may not be placed in any cable, cable tray, compartment, enclosure, manhole, outlet box, device box, raceway, or similar fitting with conductors of electric light, power, Class 1, non-power-limited fire alarm circuits, and medium power network-powered broadband communications cables unless a barrier or other equivalent form of protection against contact is employed.

(d) *Fire alarm systems*—(1) *Classifications.* Fire alarm circuits shall be classified either as non-power-limited or power limited.

(2) *Power sources.* The power sources for use with fire alarm circuits shall be either power limited or non-power-limited as follows:

(i) The power source of non-power-limited fire alarm (NPLFA) circuits shall have an output voltage of not more than 600 volts, nominal; and

(ii) The power source for a power limited fire alarm (PLFA) circuit shall be listed equipment marked as a PLFA power source.

(3) *Separation from conductors of other circuits.* (i) Non-power-limited fire alarm circuits and Class 1 circuits may occupy the same enclosure, cable, or raceway provided all conductors are insulated for maximum voltage of any conductor within the enclosure, cable, or raceway. Power supply and fire alarm circuit conductors are permitted in the same enclosure, cable, or raceway only if connected to the same equipment.

(ii) Power-limited circuit cables and conductors may not be placed in any cable, cable tray, compartment, enclosure, outlet box, raceway, or similar fitting with conductors of electric light, power, Class 1, non-power-limited fire alarm circuit conductors, or medium power network-powered broadband communications circuits.

FIGURE 7.1 *(Continued)*

(iii) Power-limited fire alarm circuit conductors shall be separated at least 50.8 mm (2 in.) from conductors of any electric light, power, Class 1, non-power-limited fire alarm, or medium power network-powered broadband communications circuits unless a special and equally protective method of conductor separation is employed.

(iv) Conductors of one or more Class 2 circuits are permitted within the same cable, enclosure, or raceway with conductors of power-limited fire alarm circuits provided that the insulation of Class 2 circuit conductors in the cable, enclosure, or raceway is at least that needed for the power-limited fire alarm circuits.

(4) *Identification.* Fire alarm circuits shall be identified at terminal and junction locations in a manner that will prevent unintentional interference with the signaling circuit during testing and servicing. Power-limited fire alarm circuits shall be durably marked as such where plainly visible at terminations.

(e) *Communications systems.* This paragraph applies to central-station-connected and non-central-station-connected telephone circuits, radio and television receiving and transmitting equipment, including community antenna television and radio distribution systems, telegraph, district messenger, and outside wiring for fire and burglar alarm, and similar central station systems. These installations need not comply with the provisions of § 1910.303 through § 1910.308(d), except for § 1910.304(c)(1) and § 1910.307.

(1) *Protective devices.* (i) A listed primary protector shall be provided on each circuit run partly or entirely in aerial wire or aerial cable not confined within a block.

(ii) A listed primary protector shall be also provided on each aerial or underground circuit when the location of the circuit within the block containing the building served allows the circuit to be exposed to accidental contact with electric light or power conductors operating at over 300 volts to ground.

(iii) In addition, where there exists a lightning exposure, each inter-building circuit on premises shall be protected by a listed primary protector at each end of the inter-building circuit.

(2) *Conductor location.* (i) Lead-in or aerial-drop cables from a pole or other support, including the point of initial attachment to a building or structure, shall be kept away from electric light, power, Class 1, or non-power-limited fire alarm circuit conductors so as to avoid the possibility of accidental contact.

(ii) A separation of at least 1.83 m (6 ft) shall be maintained between communications wires and cables on buildings and lightning conductors.

(iii) Where communications wires and cables and electric light or power conductors are supported by the same pole or run parallel to each other in span, the following conditions shall be met:

(A) Where practicable, communication wires and cables on poles shall be located below the electric light or power conductors; and

(B) Communications wires and cables may not be attached to a crossarm that carries electric light or power conductors.

(iv) Indoor communications wires and cables shall be separated at least 50.8 mm (2 in.) from conductors of any electric light, power, Class 1, non-power-limited fire alarm, or medium power network-powered broadband communications circuits, unless a special and equally protective method of conductor separation, identified for the purpose, is employed.

(3) *Equipment location.* Outdoor metal structures supporting antennas, as well as self-supporting antennas such as vertical rods or dipole structures, shall be located as far away from overhead conductors of electric light and power circuits of over 150 volts to ground as necessary to prevent the antenna or structure from falling into or making accidental contact with such circuits.

(4) *Grounding.* (i) If exposed to contact with electric light and power conductors, the metal sheath of aerial cables entering buildings shall be grounded or shall be interrupted close to the entrance to the building by an insulating joint or equivalent device. Where protective devices are used, they shall be grounded in an approved manner.

(ii) Masts and metal structures supporting antennas shall be permanently and effectively grounded without splice or connection in the grounding conductor.

(iii) Transmitters shall be enclosed in a metal frame or grill or separated from the operating space by a barrier, all metallic parts of which are effectively connected to ground. All external metal handles and controls accessible to the operating personnel shall be effectively grounded. Unpowered equipment and enclosures are considered to be grounded where connected to an attached coaxial cable with an effectively grounded metallic shield.

(f) *Solar photovoltaic systems.* This paragraph covers solar photovoltaic systems that can be interactive with other electric power production sources or can stand alone with or without electrical energy storage such as batteries. These systems may have ac or dc output for utilization.

(1) *Conductors of different systems.* Photovoltaic source circuits and photovoltaic output circuits may not be contained in the same raceway, cable tray, cable, outlet box, junction box, or similar fitting as feeders or branch circuits of other systems, unless the conductors of the different systems are separated by a partition or are connected together.

FIGURE 7.1 *(Continued)*

(2) *Disconnecting means.* Means shall be provided to disconnect all current-carrying conductors of a photovoltaic power source from all other conductors in a building or other structure. Where a circuit grounding connection is not designed to be automatically interrupted as part of the ground-fault protection system, a switch or circuit breaker used as disconnecting means may not have a pole in the grounded conductor.

(g) *Integrated electrical systems*—(1) *Scope.* Paragraph (g) of this section covers integrated electrical systems, other than unit equipment, in which orderly shutdown is necessary to ensure safe operation. An integrated electrical system as used in this section shall be a unitized segment of an industrial wiring system where all of the following conditions are met:

(i) An orderly shutdown process minimizes employee hazard and equipment damage;

(ii) The conditions of maintenance and supervision ensure that only qualified persons will service the system; and

(iii) Effective safeguards are established and maintained.

(2) *Location of overcurrent devices in or on premises.* Overcurrent devices that are critical to integrated electrical systems need not be readily accessible to employees as required by § 1910.304(f)(1)(iv) if they are located with mounting heights to ensure security from operation by nonqualified persons.

Appendix A—References for Further Information

The references contained in this appendix provide non-mandatory information that can be helpful in understanding and complying with Subpart S of this Part. However, compliance with these standards is not a substitute for compliance with Subpart S of this Part.

ANSI/API RP 500–1998 (2002) *Recommended Practice for Classification of Locations for Electrical Installations at Petroleum Facilities Classified as Class I Division 1 and Division 2.*

ANSI/API RP 505–1997 (2002) *Recommended Practice for Classification of Locations for Electrical Installations at Petroleum Facilities Classified as Class I, Zone 0, Zone 1 and Zone 2.*

ANSI/ASME A17.1–2004 *Safety Code for Elevators and Escalators.*

ANSI/ASME B30.2–2005 *Overhead and Gantry Cranes (Top Running Bridge, Single or Multiple Girder, Top Running Trolley Hoist).*

ANSI/ASME B30.3–2004 *Construction Tower Cranes.*

ANSI/ASME B30.4–2003 *Portal, Tower, and Pedestal Cranes.*

ANSI/ASME B30.5–2004 *Mobile And Locomotive Cranes.*

ANSI/ASME B30.6–2003 *Derricks.*

ANSI/ASME B30.7–2001 *Base Mounted Drum Hoists.*

ANSI/ASME B30.8–2004 *Floating Cranes And Floating Derricks.*

ANSI/ASME B30.11–2004 *Monorails And Underhung Cranes.*

ANSI/ASME B30.12–2001 *Handling Loads Suspended from Rotorcraft.*

ANSI/ASME B30.13–2003 *Storage/Retrieval (S/R) Machines and Associated Equipment.*

ANSI/ASME B30.16–2003 *Overhead Hoists (Underhung).*

ANSI/ASME B30.22–2005 *Articulating Boom Cranes.*

ANSI/ASSE Z244.1–2003 *Control of Hazardous Energy Lockout/Tagout and Alternative Methods.*

ANSI/ASSE Z490.1–2001 *Criteria for Accepted Practices in Safety, Health, and Environmental Training.*

ANSI/IEEE C2–2002 *National Electrical Safety Code.*

ANSI K61.1–1999 *Safety Requirements for the Storage and Handling of Anhydrous Ammonia.*

ANSI/UL 913–2003 *Intrinsically Safe Apparatus and Associated Apparatus for Use in Class I, II, and III, Division 1, Hazardous (Classified) Locations.*

ASTM D3176–1989 (2002) *Standard Practice for Ultimate Analysis of Coal and Coke.*

ASTM D3180–1989 (2002) *Standard Practice for Calculating Coal and Coke Analyses from As-Determined to Different Bases.*

NFPA 20–2003 *Standard for the Installation of Stationary Pumps for Fire Protection.*

NFPA 30–2003 *Flammable and Combustible Liquids Code.*

NFPA 32–2004 *Standard for Drycleaning Plants.*

NFPA 33–2003 *Standard for Spray Application Using Flammable or Combustible Materials.*

NFPA 34–2003 *Standard for Dipping and Coating Processes Using Flammable or Combustible Liquids.*

NFPA 35–2005 *Standard for the Manufacture of Organic Coatings.*

NFPA 36–2004 *Standard for Solvent Extraction Plants.*

FIGURE 7.1 (*Continued*)

NFPA 40–2001 *Standard for the Storage and Handling of Cellulose Nitrate Film.*
NFPA 58–2004 *Liquefied Petroleum Gas Code.*
NFPA 59–2004 *Utility LP-Gas Plant Code.*
NFPA 70–2002 *National Electrical Code. (See also NFPA 70–2005.)*
NFPA 70E–2000 *Standard for Electrical Safety Requirements for Employee Workplaces. (See also NFPA 70E–2004.)*
NFPA 77–2000 *Recommended Practice on Static Electricity.*
NFPA 80–1999 *Standard for Fire Doors and Fire Windows.*
NFPA 88A–2002 *Standard for Parking Structures.*
NFPA 91–2004 *Standard for Exhaust Systems for Air Conveying of Vapors, Gases, Mists, and Noncombustible Particulate Solids.*
NFPA 101–2006 *Life Safety Code.*
NFPA 496–2003 *Standard for Purged and Pressurized Enclosures for Electrical Equipment.*
NFPA 497–2004 *Recommended Practice for the Classification of Flammable Liquids, Gases, or Vapors and of Hazardous (Classified) Locations for Electrical Installations in Chemical Process Areas.*
NFPA 505–2006 *Fire Safety Standard for Powered Industrial Trucks Including Type Designations, Areas of Use, Conversions, Maintenance, and Operation.*
NFPA 820–2003 *Standard for Fire Protection in Wastewater Treatment and Collection Facilities.*
NMAB 353–1–1979 *Matrix of Combustion-Relevant Properties and Classification of Gases, Vapors, and Selected Solids.*
NMAB 353–2–1979 *Test Equipment for Use in Determining Classifications of Combustible Dusts.*
NMAB 353–3–1980 *Classification of Combustible Dust in Accordance with the National Electrical Code.*

[46 FR 4056, Jan. 16, 1981; 46 FR 40185, Aug. 7, 1981; as amended at 53 FR 12123, Apr. 12, 1988; 55 FR 32020, Aug. 6, 1990; 55 FR 46054, Nov. 1, 1990; 72 7215, Feb. 14, 2007]

FIGURE 7.1 *(Continued)*

ELECTRICAL SAFETY-RELATED WORK PRACTICES

(OSHA 29 CFR 1910.331-.335)
(FINAL RULE AUGUST 6, 1990)
1910.331-.335 Electrical Safety-Related Work Practices

Prior to August 6, 1990 the electrical standards in Subpart S of the General Industry Standards covered electrical equipment and installations rather than work practices. The electrical safety-related work practice standards that did exist were distributed in other subparts of 29 CFR 1910. Although unsafe work practices appear to be involved in most workplace electrocutions, OSHA has had very few regulations addressing work practices necessary for electrical safety. Because of this, OSHA determined that standards were needed to minimize these hazards and the risk to employees.

This rule addresses practices and procedures that are necessary to protect employees working on or near exposed energized and deenergized parts of electric equipment. This rule also promotes uniformity and reduces redundancy among the general industry standards. 29 CFR 1910.331-.335 "Electrical Safety-Related Work Practices" is based largely on NFPA 70E "Electrical Safety in the workplace," Chapter I "Safety-Related Work Practices."

On September 1, 1989, OSHA promulgated a generic standard on the control of hazardous energy, 29 CFR 1910.147 (lockout/tagout).

This standard addresses practices and procedures that are necessary to deenergize machinery or equipment and to prevent the release of potentially hazardous energy while maintenance and servicing activities are being performed.

Although this rule is related to electrical energy, it specifically excludes "exposure to electrical hazards from work on, near, or with conductors or equipment in electric utilization installations, which is covered by Subpart S of 29 CFR 1910." Therefore, the lockout/tagout standard does not cover electrical hazards.

The Electrical Safety-Related Work Practice standard has provisions to achieve maximum safety by deenergizing energized parts and, secondly, when lockout/ tagout is used, it is done to ensure that the deenergized state is maintained.

FIGURE 7.2 Electrical Safety-Related Work Practices (OSHA-CFR, Title 29, Part 1910, Paragraphs 331–335). (*Courtesy OSHA website.*)

The effective date for this standard (29 CFR 1910.331-.335) was December 4, 1990, except for 1910.332 (training), which became effective August 6, 1991.

As of August 6, 1991, the training practices of the employer for qualified and unqualified employees will be evaluated by the OSHA inspector to assess whether the training provided is appropriate to the tasks being performed or to be performed.

Summary (from Preamble):

These performance-oriented regulations complement the existing electrical installation standards. The new standard includes requirements for work performed on or near exposed energized and deenergized parts of electric equipment; use of electrical protective equipment; and the safe use of electric equipment. Compliance with these safe work practices will reduce the number of electrical accidents resulting from unsafe work practices by employees.

The following is a table of electrical fatalities that was compiled by the National Institute of Occupational Safety and Health (NIOSH). This is an annual average.

TABLE 4 Electrical Fatalities by Occupation

Job title	Number of fatalities
Laborer	28
Lineman	20
Electrician	16
Painter	13
Truck driver	7
Machine operator	6
Groundman	6
Maintenance man	5
Technician	5
Crew leader/foreman	4
Construction worker	4
Carpenter	3
Iron worker	3
Brick layer	3
Bill poster	3
Welders	2
Restaurant worker	2
Fireman	2

1910.331 Scope

(a) *Covered work by both qualified and unqualified persons.* The provisions of 1910.331 through 1910.335 cover electrical safety-related work practices for both qualified persons (those who have training in avoiding the electrical hazards of working on or near exposed energized parts) . . .

"On" or "near" is mentioned here and will be mentioned throughout this subpart. The definition of "on" is actual contact with the circuit or part (involving either direct contact or contact by means of tools or material); "near" however needs further explanation. "Near" is when a person is near enough to the energized circuits or parts to be exposed to any hazards they present.

More plainly stated; If a person is within reaching, stumbling, or falling distance of an energized circuit or part or close enough to the energized parts that he/she could contact these parts with any part of their body or with any conductive object he/she is handling, that person is doing energized work.

FIGURE 7.2 *(Continued)*

. . . and unqualified persons (those with little or no such training) working on, near, or with the following installations:

(1) ***Premises wiring.*** Installations of electric conductors and equipment within or on buildings or other structures, and on other premises such as yards, carnival, parking, and other lots, and industrial substations;

(2) ***Wiring for connection to supply.*** Installations of conductors that connect to the supply of electricity; and

(3) ***Other wiring.*** Installations of other outside conductors on the premises.

Any wiring system that has been installed in accordance with the NEC and OSHA Subpart S "Electrical Standard").

(4) ***Optical fiber cable.*** Installations of optical fiber cable where such installations are made along with electric conductors.

There are several types of optical fiber cable that contain noncurrent-carrying conductive members such as metallic strength members and metallic vapor barriers.

Note: See 1910.399 for the definition of "qualified person." See 1910.332 for training requirements that apply to qualified and unqualified persons.

According to the National Electrical Code (NEC) and OSHA 29 CFR 1910.399, a qualified person is "one that is familiar with the construction and operation of the equipment and the hazards involved."

OSHA states that qualified persons are intended to be only those who are well acquainted with and thoroughly conversant in the electric equipment and electrical hazards involved with the work being performed.

Definitions applicable to this subpart for a qualified person;

"**Note 1:** Whether an employee is considered to be a qualified person will depend on various circumstances in the workplace. It is possible and, in fact, likely for an individual to be considered qualified with regard to certain equipment in the workplace, but unqualified as to other equipment."

In other words just because a person is a journeyman, master or 1st class electrician does not mean he is qualified on all systems and equipment.

Example: If a new piece of equipment is installed and the electrician is not familiar with the construction, operation and hazards involved, then he/she is unqualified as to that equipment.

Note 2: An employee who is undergoing on-the-job training and who, in the course of such training, has demonstrated an ability to perform duties safely at his or her level of training and who is under the direct supervision of a qualified person is considered to be a qualified person for the performance of those duties."

There are numerous references throughout this subpart that state "qualified persons only." For example: Only qualified persons are allowed to work on or near energized circuits or parts; a qualified person shall operate equipment to verify it cannot restart; a qualified person shall use test equipment to test the circuit to verify it is de-energized; only qualified persons may perform testing work on electric circuits, etc. Therefore, if a person were not qualified he/she could never become qualified because they could never do anything. This is the reason for Note 2, quoted above. A person can become qualified if working under the direct supervision of a qualified person.

(b) ***Other covered work by unqualified persons.*** The provisions of 1910.331 through 1910.335 also cover work performed by unqualified persons on, near, or with the installations listed in paragraphs (c)(1) through (c)(4) of this section.

(c) ***Excluded work by qualified persons.*** The provisions of 1910.331 through 1910.335 do not apply to work performed by qualified persons on or directly associated with the following installations:

(1) ***Generation, transmission, and distribution installations.*** Installations for the generation, control, transformation, transmission, and distribution of electric energy (including communication and metering) located in buildings used for such purposes or located outdoors.

OSHA (D of C) states that "Work on the specified electrical installations is excluded, but work on the other electric equipment in the buildings is not excluded."

Note 1: Work on or directly associated with installations of utilization equipment used for purposes other than generating, transmitting, or distributing electric energy (such as installations which are in office buildings,

FIGURE 7.2 *(Continued)*

warehouses, garages, machine shops, or recreational buildings, or other utilization installations which are not an integral part of a generating installation, substation, or control center) is covered under paragraph (a)(1) of this section.

Note 2: Work on or directly associated with generation, transmission, or distribution installations includes:

1) Work performed directly on such installations, such as repairing overhead or underground distribution lines or repairing a feed-water pump for the boiler in a generating plant.

2) Work directly associated with such installations, such as line-clearance tree trimming and replacing utility poles.

3) Work on electric utilization circuits in a generating plant provided that:

A) Such circuits are commingled with installations of power generation equipment or circuits, and

B) The generation equipment or circuits present greater electrical hazards than those posed by the utilization equipment or circuits (such as exposure to higher voltages or lack of overcurrent protection).

(2) *Communications installations.* Installations of communication equipment to the extent that the work is covered under 1910.268.

(3) *Installations in vehicles.* Installations in ships, watercraft, railway rolling stock, aircraft, or automotive vehicles other than mobile homes and recreational vehicles.

The OSHA Federal Register for ships and watercraft (marine) is 29 CFR 1915.

(4) *Railway installations.* Installations of railways for generation, transformation, transmission, or distribution of power used exclusively for operation of rolling stock or installations of railways used exclusively for signaling and communication purposes.

1910.332 Training

(a) *Scope.* The training requirements contained in this section apply to employees who face a risk of electric shock that is not reduced to a safe level by the electrical installation requirements of 1910.303 through 1910.308.

(NFPA 70E) "Safety Training. Employees shall be trained in the safety-related work practices, safety procedures, and other personnel safety requirements in this standard that pertain to their respective job assignments.

Employees shall not be permitted to work in an area where they are likely to encounter electrical hazards unless they have been trained to recognize and avoid the electrical hazards to which they will be exposed.

For the purpose of this standard, an electrical hazard is a recognizable dangerous electrical condition. Electrical conditions such as exposed energized parts and unguarded electric equipment which may become energized unexpectedly are examples of electrical hazards."

(Preamble) "Employees are required to be trained in the safety-related work practices of this standard, as well as any other practices necessary for safety from electrical hazards."

This would include systems and equipment that are enclosed or otherwise rendered inaccessible to employees or are at voltages less than 50 volts to ground. However, when an enclosure cover has been removed or opened for inspection or maintenance of the energized components, the employee faces a risk of electric shock and the requirements of 1910.331 through 1910.335 would apply.

(D of C) "All employees who face a risk of electric shock, burns, or other related injuries, not reduced to a safe level by the installation safety requirements of Subpart S, must be trained in safety-related work practices required by 29 CFR 1910.331-.335."

(Preamble) "The final rule does not contain provisions addressing the qualifications of trainers or the demonstration of effectiveness of the training."

Note: Employees in occupations listed in Table S-4 face such a risk and are required to be trained …

…Other employees who also may reasonably be expected to face comparable risk of injury due to electric shock or other electrical hazards must also be trained.

This last statement places an additional responsibility on the employer to evaluate the work situations of every employee, not just the ones listed in Table S-4, to determine if they face a risk of electrical shock or other electrical hazards and to provide training for them.

(b) *Content of training.* (1) Practices addressed in this standard. Employees shall be trained in and familiar with the safety-related work practices required by 1910.331 through 1910.335 that pertain to their respective job assignments.

FIGURE 7.2 (*Continued*)

TABLE S-4 Typical Occupational Categories of Employees Facing a Higher Than Normal Risk of Electrical Accident

Occupation
Blue collar supervisors.*
Electrical and electronic engineers.*
Electrical and electronic equipment assemblers.*
Electrical and electronic technicians.*
Electricians.
Industrial machine operators.*
Material handling equipment operators.*
Mechanics and repairers.*
Painters.*
Riggers and roustabouts.*
Stationary engineers.*
Welders.

*Workers in these groups do not need to be trained if their work or the work of those they supervise does not bring them or the employees they supervise close enough to exposed parts of electric circuits operating at 50 volts or more to ground for a hazard to exist.

This statement makes it clear that if an employee is not exposed to all of the hazards or equipment addressed by this requirement, he/she is not required to be trained on it but must be trained on those that do present a hazard.

Example: A painter is not a qualified electrical worker and therefore would not require training in the use of electrical test equipment, the need to defeat an interlock, use electrical protective equipment, or deenergize an electrical system and install a lockout/tagout. However he/she would require training on clearance distances from overhead lines, the purpose of lockout/tagout, requirements for use of portable cord- and plug-connected equipment and the use of flammable liquids near possible spark producing equipment.

(2) *Additional requirements for unqualified persons.* Employees who are covered by paragraph (a) of this section but who are not qualified persons shall also be trained in and familiar with any electrically related safety practices not specifically addressed by 1910.331 through 1910.335 but which are necessary for their safety.

(OSHA's D of C) "In addition to being trained in and familiar with safety-related work practices, unqualified employees must be trained in the inherent hazards of electricity, such as high voltages, electric current, arcing, grounding, and lack of guarding. Any electrically related safety practices not specifically addressed by Sections 1910.331 through 1910.335 but necessary for safety in specific workplace conditions shall be included."

This requirement will cause an employer to evaluate the work hazards of every employee to determine if there are any other hazards that were not addressed in this standard.

(3) *Additional requirements for qualified persons.* Qualified persons (i.e. those permitted to work on or near exposed energized parts)...

(D of C) "The standard defines a qualified person as one familiar with the construction and operation of the equipment and the hazards involved.

Qualified Persons are intended to be only those who are well acquainted with and thoroughly conversant in the electric equipment and electrical hazards involved with the work being performed."

...shall, at a minimum, be trained in and familiar with the following:

(i) The skills and techniques necessary to distinguish exposed live parts from other parts of electric equipment.

(ii) The skills and techniques necessary to determine the nominal voltage of exposed live parts, and

(iii) The clearance distances specified in 1910.333(c) and the corresponding voltages to which the qualified person will be exposed.

Note 1: For the purposes of 1910.331 through 1910.335, a person must have the training required by paragraph (b)(3) of this section in order to be considered a qualified person.

FIGURE 7.2 *(Continued)*

Note 2: Qualified persons whose work on energized equipment involves either direct contact or contact by means of tools or materials must also have the training needed to meet 1910.333(C)(2).

(c) *Type of training.* The training required by this section shall be of the classroom or on-the-job type. The degree of training provided shall be determined by the risk to the employee.

In most cases, both classroom and on-the-job type training would be needed.

(D of C) "The failure to train 'qualified' and 'unqualified' employees, as required for their respective classifications, shall normally be cited as a serious violation."

1910.333 Selection and Use of Work Practices

(Preamble) "The basic intent of 1910.333 is to require employers to take one of three options to protect employees working on electric circuits and equipment: (1) Deenergize the equipment involved and lockout its disconnecting means; or (2) deenergize the equipment and tag the disconnecting means, if the employer can demonstrate that tagging is as safe as locking; or (3) work the equipment energized if the employer can demonstrate that it is not feasible to deenergize it.

Paragraph (a) of 1910.333 sets forth general requirements on the selection and use of work practices. The requirements of this paragraph mainly address accidents which involve the hazards of exposure to live parts of electric equipment. A deenergized part is obviously safer than an energized one. Because the next best method of protecting an employee working on exposed parts of electric equipment (the use of personal protective equipment) would continue to expose that employee to a risk of injury from electric shock, 1910.333(a) makes equipment deenergizing the primary method of protecting employees.

Under certain conditions, however, deenergizing need not be employed. Employees may be allowed to work on or near exposed energized parts as allowed under 1910.333(a)(1)."

(a) *General.* Safety-related work practices shall be employed to prevent electric shock or other injuries resulting from either direct or indirect electrical contacts, when work is performed near or on equipment or circuits which are or may be energized. The specific safety-related work practices shall be consistent with the nature and extent of the associated electrical hazards.

(D of C) "1910.333(a)(1) requires that live parts be deenergized before a potentially exposed employee works on or near them. OSHA believes that this is the preferred method for protecting employees from electrical hazards. The employer is permitted to allow employees to work on or near exposed live parts only:

(1) If the employer can demonstrate that deenergizing introduces additional or increased hazards (see Note 1 below),
(2) If the employer can demonstrate that deenergizing is infeasible due to equipment design or operational limitations (see Note 2 below)."

(1) *Deenergized parts.* Live parts to which an employee may be exposed shall be de-energized before the employee works on or near them, unless the employer can demonstrate that deenergizing introduces additional or increased hazards, or is infeasible due to equipment design or operational limitations. Live parts that operate at less than 50 volts to ground need not be deenergized if there will be no increased exposure to electrical burns or to explosion due to electric arcs.

Note 1: Examples of increased or additional hazards include interruption of life support equipment, deactivation of emergency alarm systems, shutdown of hazardous location ventilation equipment, or removal of illumination for an area.

Note 2: Examples of work that may be performed on or near energized circuit parts because of infeasibility due to equipment design or operational limitations include testing of electric circuits that can only be performed with the circuit energized and work on circuits that form an integral part of a continuous industrial process in a chemical plant that would otherwise need to be completely shut down in order to permit work on one circuit or piece of equipment.

Note 3: Work on or near deenergized parts is covered by paragraph (b) of this section.

(D of C) "Under 1910.333(a)(2) if the employer does not deenergize (under the conditions permitted in 1910.333(a)(1)), then suitable safe work practices for the conditions under which the work is to be performed shall be included in the **written procedures** and **strictly enforced**. These work practices are given in 1910.333(c) and 1910.335.

Only qualified persons shall be allowed to work on energized parts or equipment."

FIGURE 7.2 *(Continued)*

(2) *Energized parts.* If the exposed live parts are not deenergized (i.e., for reasons of increased or additional hazards or infeasibility), other safety-related work practices shall be used to protect employees who may be exposed to the electrical hazards involved. Such work practices shall protect employees against contact with energized circuit parts directly with any part of their body or indirectly through some other conductive object. The work practices that are used shall be suitable for the conditions under which the work is to be performed and for the voltage level of the exposed electric conductors or circuit parts. Specific work practice requirements are detailed in paragraph (c) of this section.

(D of C) "If the employer does not deenergize (under the conditions permitted in 1910.333(a)(1)), then suitable safe work practices for the conditions under which the work is to be performed shall be included in the written procedure and strictly enforced."

(b) *Working on or near exposed deenergized parts.* (1) *Application.* This paragraph applies to work on exposed deenergized parts or near enough to them to expose the employee to any electrical hazard they present. Conductors and parts of electric equipment that have been deenergized but have not been locked out or tagged in accordance with paragraph (b) of this section shall be treated as energized parts, and paragraph (c) of this section applies to work on or near them.

(D of C) "Circuits parts that cannot be deenergized using the procedures outlined in 1910.333(b)(2) must be treated as energized, regardless of whether the parts are turned off and assumed to be deenergized."

The lockout/tagout requirements of this regulation were derived from the existing ANSI and NFPA standards.

(Preamble) "The NFPA 70E requirements were intended to apply any time work is performed on or near deenergized circuit parts or equipment in any situation which presents a danger that the circuit parts or equipment might become unexpectedly energized. Thus, the NFPA provisions not only address the hazard of contact with energized parts, but also cover other hazards which are presented by start-up of equipment during maintenance operations. This is expressed in NFPA 70E, Part II, Section 1.B, second paragraph, which states:

'Where the work to be performed requires employees to work on or near exposed circuit parts or equipment, and there is danger of injury due to electric shock, unexpected movement of equipment, or other electrical hazards, the circuit parts and equipment that endanger the employees shall be deenergized and locked out or tagged out in accordance with the policies and procedures specified.'

In contrast to this, OSHA Subpart S standard is intended to cover employee exposure to electrical hazards which might occur from the unexpected energizing of the circuit parts and does not cover other equipment-related hazards which do not involve exposed live parts. Thus, the Subpart S standard will protect the electrician working on a circuit but does not address a mechanic working on the mechanical parts of an electrically powered machine. The generic lockout standard, 1910.147, covers the hazards of unexpected activation or energization of machinery or equipment during servicing or maintenance activities."

(2) *Lockout and Tagging.* While any employee is exposed to contact with parts of fixed electric equipment or circuits which have been deenergized, the circuits energizing the parts shall be locked out or tagged or both in accordance with the requirements of this paragraph. The requirements shall be followed in the order in which they are presented (i.e., paragraph (b)(2)(i) first, then paragraph (b)(2)(ii), etc.).

Note 1: As used in this section, fixed equipment refers to equipment fastened in place or connected by permanent wiring methods.

Note 2: Lockout and tagging procedures that comply with paragraphs (c) through (f) of 1910.147 will also be deemed to comply with paragraph (b)(2) of this section provided that:

(1) The procedures address the electrical safety hazards covered by this Subpart; and

(2) The procedures also incorporate the requirements of paragraphs (b)(2)(iii)(D) and (b)(2)(iv)(B) of this section.

(Preamble) "In comparing 1910.333(b)(2) with final 1910.147, OSHA determined that the generic lockout standard encompassed all the lockout and tagging requirements contained in the electrical work practices with two exceptions. First, the electrical standard more tightly restricted the use of tags without locks and called for additional protection when tags were permitted. Secondly, the electrical work practices contain specific requirements for testing circuit parts for voltage before they could be considered as deenergized.

OSHA believes that a lockout and tagging program which meets 1910.147 will, with these two exceptions, provide protection for servicing and maintenance involving electrical work and live parts. Accordingly, the final rule on electrical safety-related work practices incorporates this finding."

(D of C) "Only qualified persons may place and remove locks and tags."

FIGURE 7.2 *(Continued)*

(Preamble) "OSHA has decided to accept any lockout and tagging program that conforms to 1910.147 if it also meets paragraphs 1910.333 (b)(2)(iii)(D) and (b)(2)(iv) (B). This will enable employers to use a single lockout and tagout program to cover all hazards addressed by these two standards, as long as that program includes procedures that meet the two additional paragraphs when exposure to electric shock is involved.

OSHA has limited the application of the lockout and tagging provisions to fixed equipment. Employees can safely work on cord- and plug-connected, portable and stationary equipment which is disconnected from the circuit. Generally, such equipment is returned to a maintenance shop for repair, and the danger of accidental energizing of equipment parts is eliminated in most cases. However, in 1910.147 (the generic lockout/ tagout standard) it states: "This standard does not apply to the following:

(A) Work on cord- and plug-connected equipment for which exposure to the hazards of unexpected energization or start-up of the equipment is controlled by the unplugging of the equipment from the energy source and by the plug being under the exclusive control of the employee performing the servicing or maintenance."

This requirement clearly states that if the plug is not under the "exclusive control" of the employee, it must be locked out or tagged out.

"The generic lockout standard (OSHA) and NFPA 70E's lockout provisions contain requirements for lockout procedures to be in writing.

OSHA believes that written procedures are an important part of a successful lockout and tagging program and that they should, therefore, be required. However, the final electrical standard lockout provisions are themselves a step-by-step procedure for safely deenergizing electrical circuits. For this reason, the final electrical work practices standard incorporates a requirement for employers to maintain a copy of the lockout procedure outlined in 1910.333(b)(2) and make it available for inspection by employees and OSHA."

(1910.147) The following is quoted from 1910.147(a)(3) Purpose: "(ii) When other standards in this part (meaning 1910) require the use of lockout or tagout, they shall be used and supplemented by the procedural and training requirements of this section." Therefore, the requirements of the generic standard (1910.147) also apply, along with the additional lockout and tagout requirements of this standard, for all electrical lockout or tagout situations.

(i) *Procedures.* The employer shall maintain a written copy of the procedures outlined in paragraph (b)(2) and shall make it available for inspection by employees and by the Assistant Secretary of Labor and his or her authorized representatives.

(D of C) "During walk around inspections, compliance officers shall evaluate any electrical related work being performed to ascertain conformance with the employer's written procedures as required by 1910.333 (b)(2)(i) and all safety-related work practices in Sections 1910.333 through 1910. 335.

A deficiency in the employer's program that could contribute to a potential exposure capable of producing serious physical harm or death shall be cited as a serious violation."

Note: The written procedures may be in the form of a copy of paragraph (b) of this section.

(ii) *Deenergizing equipment.* (A) Safe procedures for deenergizing circuits and equipment shall be determined before circuits or equipment are deenergized.

1910.147(d)(1) states this requirement much more clearly, as follows:

"Preparation for shutdown. Before an authorized or affected employee turns off a machine or equipment, the authorized employee shall have knowledge of the type and magnitude of the energy, the hazards of the energy to be controlled, and the method or means to control the energy."

(B) The circuits and equipment to be worked on shall be disconnected from all electric energy sources....

There can be several electric energy sources such as the power circuit, control circuit energized from a separate source, and interlocks from other devices. In the case of a power circuit breaker there could be AC and/or DC sources for the trip coil, closing coil, spring charging motor, and any number of circuits that operate through the auxiliary contacts.

...Control circuit devices, such as push buttons, selector switches, and interlocks, may not be used as the sole means for deenergizing circuits or equipment...

It is a good practice to turn a control device off so the load will be removed prior to opening the disconnecting means. This practice will reduce the risk of arcing in the device and a possible flash-over from phase-to-phase and/or phase-to-ground. These control devices should also be locked and/or tagged before the main disconnecting means is opened.

FIGURE 7.2 (*Continued*)

If this is not done, the disconnecting means could be opened at the same time a control is turned to the on position resulting in a current level far beyond its rating. The following example is used to further illustrate:

If the disconnect switch was rated for a load of 30 amperes, and the full load current was 24 amperes, and if the disconnect switch were to be opened at the same time the motor was being started, the disconnect switch contacts would experience approximately locked motor current which is about six times the full load current or, in this case, 144 amperes. In this case a flash-over from phase-to-phase and/or phase-to-ground would be very likely. This could cause the enclosure to explode and injure the person who is switching the device.

...Interlocks for electric equipment may not be used as a substitute for lockout and tagging procedures.

(C) Stored electric energy which might endanger personnel shall be released. Capacitors shall be discharged and high capacitance elements shall be short-circuited and grounded, if the stored electric energy might endanger personnel.

Note: If the capacitors or associated equipment are handled in meeting this requirement, they shall be treated as energized.

(D) Stored non-electrical energy in devices that could reenergize electric circuit parts shall be blocked or relieved to the extent that the circuit parts could not be accidentally energized by the device.

A good example of this would be charged closing springs on a power circuit breaker, or hydraulic or pneumatic pressure in an oil circuit breaker operating mechanism.

(iii) *Application of locks and tags.* **(A)** A lock and a tag shall be placed on each disconnecting means used to deenergize circuits and equipment on which work is to be performed, except as provided in paragraphs (b)(2)(iii)(C) and (b)(2)(iii)(E) of this section. The lock shall be attached so as to prevent persons from operating the disconnecting means unless they resort to undue force or the use of tools.

1910.147 states, "(C) Substantial, (1) Lockout devices. Lockout devices shall be substantial enough to prevent removal without the use of excessive force or unusual techniques, such as with the use of bolt cutters or other metal cutting tools."

(B) Each tag shall contain a statement prohibiting unauthorized operation of the disconnecting means and removal of the tag.

1910.147 states, "Tagout devices shall warn against hazardous conditions if the machine or equipment is energized and shall include a legend such as the following: Do Not Start, Do Not Open, Do Not Close, Do Not Energize, Do Not Operate."

(C) If a lock cannot be applied, or if the employer can demonstrate that tagging procedures will provide a level of safety equivalent to that obtained by the use of a lock, a tag may be used without a lock.

(D) A tag used without a lock, as permitted by paragraph (b)(2)(iii)(C) of this section, shall be supplemented by at least one additional safety measure that provides a level of safety equivalent to that obtained by use of a lock. Examples of additional safety measures include the removal of an isolating circuit element, blocking of a controlling switch, or opening of an extra disconnecting device.

(Preamble) "Because a person could operate the disconnecting means before reading or recognizing the tag, the standard also requires that, where tags are used, one or more additional safety measures be taken to provide added safety. The additional measures used must either: (1) Ensure that the closing of the tagged single switch would not reenergize the circuit on which employees are working or (2) virtually prevent the accidental closing of the disconnecting means. The following examples from the record illustrate protective techniques that are commonly used to supplement tags and protect workers:

(1) The removal of a fuse or fuses for a circuit.
(2) The removal of a draw-out circuit breaker from a switchboard (i.e., racking out the breaker).
(3) The placement of a blocking mechanism over the operating handle of a disconnecting means so that the handle is blocked from being placed in the closed position.
(4) The opening of a switch (other than the disconnecting means) which also opens the circuit between the source of power and the exposed parts on which work is to be performed.
(5) The opening of a switch for a control circuit that operates a disconnect that is itself open and disconnected from the control circuit or is otherwise disabled.
(6) Grounding of the circuit upon which work is to be performed.

FIGURE 7.2 *(Continued)*

The additional safety measure is necessary because, at least for electrical disconnecting means, tagging alone is significantly less safe than locking out. A disconnecting means could be closed by an employee who has failed to recognize the purpose of the tag. The disconnect could also be closed accidentally."

It should be noted that, in addition to the measures noted in the preamble, there are other measures that can be taken, such as:

(1) Removing the control power fuses.
(2) Lifting the wires from the circuit breaker, disconnect, or contactor terminals or from a terminal.

It should also be noted that these additional measures could also be used in the case of a lockout, as an additional safety measure.

(Preamble) "Any measure or combination of measures, that will protect employees as well as the application of a lock, can be used to supplement the tagging of the disconnecting means. In determining whether an individual employer's 'tagging-only' program provides safety equivalent to a lock, OSHA will consider factors such as:

(1) The safety record under the employer's tagging program;
(2) Whether the tagged disconnecting means are accessible to qualified persons only;
(3) Whether the tag and its attachment mechanism clearly identify the disconnecting means that is open and effectively inhibit reenergizing the electric circuit (for example, when a tag is used on an energy isolating device that is capable of being locked out, the tag must be attached at the same location that the lock would have been attached); and
(4) Whether employees are thoroughly familiar with the tagging procedures, especially their duties and responsibilities with respect to the procedures and the meaning and significance of the tags used with the procedure."

It is also a requirement of 1910.147 that if the tag is used in the place of a lock the tag attachment means shall be non-reusable, attachable by hand, self-locking, and non-releasable and have a pull strength of no less than 50 pounds and be at least equivalent to an all environment tolerant nylon cable tie.

It is also a requirement that if the energy isolating device is not capable of being locked out, the tag must be attached to or immediately adjacent to the device in such a manner that it would be obvious to anyone attempting to operate the device, that it is tagged out.

(E) A lock may be placed without a tag only under the following conditions:
{1} Only one circuit or piece of equipment is deenergized, and
{2} The lockout period does not extend beyond the work shift, and
{3} Employees exposed to the hazards associated with reenergizing the circuit or equipment are familiar with this procedure.

(iv) *Verification of deenergized condition.* The requirements of this paragraph shall be met before any circuits or equipment can be considered and worked as deenergized.

(D of C) "Verification of deenergization is mandatory. This verification must be done by a "qualified person." This is done by following the requirements of paragraph (A) and (B) of this section."

(A) A qualified person shall operate the equipment operating controls or otherwise verify that the equipment cannot be restarted.

(B) A qualified person shall use test equipment to test the circuit elements and electrical parts of equipment to which employees will be exposed and shall verify that the circuit elements and equipment parts are deenergized. The test shall also determine if any energized condition exists as a result of inadvertently induced voltage or unrelated voltage backfeed even though specific parts of the circuit have been deenergized and presumed to be safe....

A current one-line diagram can be a very useful tool to verify that the right circuit was deenergized, that all electrical sources were deenergized, and will identify all possible backfeed sources.

The key to this is that the one-line diagram must be current or correct.

...If the circuit to be tested is over 600 volts, nominal, the test equipment shall be checked for proper operation immediately before and immediately after this test.

(Preamble) "Voltages over 600 volts are more likely than lower voltages to cause test equipment itself to fail, leading to false indications of no-voltage conditions. To prevent accidents resulting from such failure of test

FIGURE 7.2 *(Continued)*

equipment, the final standard requires checking operation of the test equipment immediately before and after use, if voltages over 600 volts are involved."

(D of C) "Testing instruments and equipment shall be visually inspected for external defects or damage before being used to determine deenergization according to the requirements of 29 CFR 1910.334(c)(2)."

There is evidence of electrical injuries and fatalities with voltages as low as 50 volts. Because of this hazard, it is a good practice to verify the proper operation of the test equipment on these lower voltages as well as the requirement for over 600 volts.

(v) *Reenergizing equipment.* These requirements shall be met, in the order given, before circuits or equipment are reenergized, even temporarily.

(A) A qualified person shall conduct tests and visual inspections, as necessary, to verify that all tools, electrical jumpers, shorts, grounds, and other such devices have been removed, so that the circuits and equipment can be safely energized.

(B) Employees exposed to the hazards associated with reenergizing the circuit or equipment shall be warned to stay clear of circuits and equipment.

(C) Each lock and tag shall be removed by the employee who applied it or under his or her direct supervision. However, if this employee is absent from the workplace, then the lock or tag may be removed by a qualified person designated to perform this task provided that:

{1} The employer ensures that the employee who applied the lock or tag is not available at the workplace, and

An additional requirement, found in 1910.147 for energy control, is that every reasonable effort be made to contact the employee before locks or tags are removed.

{2} The employer ensures that the employee is aware that the lock or tag has been removed before he or she resumes work at that workplace.

(Preamble) "This requirement ensures that locks and tags are removed with the full knowledge of the employee who applied them. Without such a requirement, it is likely that a lock or a tag could be removed by persons who are not responsible for the lockout, endangering employees working on the 'deenergized parts.' This provision also allows for removal of locks and tags if the employee who applied them is absent from the workplace, under conditions like those in the exception to 1910.147(e)(3)."

(D) There shall be a visual determination that all employees are clear of the circuits and equipment.

Often this would require more than one person, i.e., one at the equipment and one at the disconnecting means, both using two-way radios or other effective communications methods.

(c) *Working on or near exposed energized parts.* (1) *Application.* This paragraph applies to work performed on exposed live parts (involving either direct contact or by means of tools or materials) or near enough to them for employees to be exposed to any hazard they present.

All of 1910.333(c) addresses the required energized work procedures and employee qualification requirements.

(2) *Work on energized equipment.* Only qualified persons may work on electric circuit parts or equipment that have not been deenergized under the procedures of paragraph (b) of this section. Such persons shall be capable of working safely on energized circuits and shall be familiar with the proper use of special precautionary techniques, personal protective equipment, insulating and shielding materials, and insulated tools.

(3) *Overhead lines.* If work is to be performed near overhead lines, the lines shall be deenergized and grounded, or other protective measures shall be provided before work is started. If the lines are to be deenergized, arrangements shall be made with the person or organization that operates or controls the electric circuits involved to deenergize and ground them. If protective measures, such as guarding, isolating, or insulating, are provided, these precautions shall prevent employees from contacting such lines directly with any part of their body or indirectly through conductive materials, tools, or equipment.

(D of C) "This standard for Electrical Safety-Related Work Practices can be applied with respect to electrical hazards related to any size, utilization, or configuration of overhead power lines in general industry; e.g., residential power lines, remotely located overhead power lines, temporarily rigged overhead power lines, and overhead power lines along streets and alleys."

FIGURE 7.2 *(Continued)*

TABLE S-5 Approach Distances for Qualified Employees—Alternating Current

Voltage range (phase to phase)	Minimum approach distance
300V and less	Avoid Contact
Over 300V, not over 750V	1 ft 0 in (30.5 cm).
Over 750V, not over 2kV	1 ft 6 in (46 cm).
Over 2kV, not over 15kV	2 ft 0 in (61 cm).
Over 15kV, not over 37kV	3 ft 0 in (91 cm).
Over 37kV, not over 87.5kV	3 ft 6 in (107 cm).
Over 87.5kV, not over 121kV	4 ft 0 in (122 cm).
Over 121kV, not over 140kV	4 ft 6 in (137 cm).

It should also be noted that overhead power lines and bus structures in substations and switchyards present the same hazard as all other overhead power lines. Therefore the same work practices should be used when work is being performed near these overhead systems.

(D of C) "Working On or Near Overhead Power Lines;

(a) OSHA believes that the preferred method of protecting employees working near overhead power lines is to deenergize and ground the lines when work is to be performed near them.

(b) In addition to other operations, this standard also applies to tree trimming operations performed by tree workers who are not qualified persons. In this respect the exclusion in 1910.331(c)(1) applies only to qualified persons performing line-clearance tree trimming (trimming trees that are closer than 10 feet to overhead power lines).

(c) This standard does not prohibit workers who are not qualified persons from working in a tree that is closer than 10 feet to power lines so long as that person or any object he or she may be using, does not come within 10 feet of a power line. However, it would require qualified persons to perform the work if the worker or any object he or she may be using will come within 10 feet of an exposed energized part or if a branch being cut may be expected to come within 10 feet of an exposed energized part while falling from the tree.

(d) The purpose for the approach distance requirements is to prevent contact with, and/or arcing, from energized overhead power lines. The approach distance applies to tools used by employees as well as the employees themselves. Table S-5 calls for the approach distances for qualified persons only. Unqualified persons are required to adhere to the 10 foot minimum."

Note: The work practices used by qualified persons installing insulating devices on overhead power transmission or distribution lines are not covered by 1910.332 through 1910.335. Under paragraph (c)(2) of this section, unqualified persons are prohibited from performing this type of work.

Qualified persons installing insulating equipment on or deenergizing and grounding the lines and line-clearance tree trimming are covered under 1910.269, "Electric Power Generation, Transmission, and Distribution" Federal Register.

(i) *Unqualified persons.* (A) When an unqualified person is working in an elevated position near overhead lines, the location shall be such that the person and the longest conductive object he or she may contact cannot come closer to any unguarded, energized overhead line than the following distances:

{1} For voltages to ground 50 kV or below - 10 feet (305 cm);

{2} For voltages to ground over 50 kV - 10 feet (305 cm) plus 4 inches (10 cm) for every 10 kV over 50 kV.

(B) When an unqualified person is working on the ground in the vicinity of overhead lines, the person may not bring any conductive object closer to unguarded, energized overhead lines than the distances given in paragraph (c)(3)(i)(A) of this section.

Note: For voltages normally encountered with overhead power line, objects which do not have an insulating rating for the voltage involved are considered to be conductive.

FIGURE 7.2 *(Continued)*

(ii) ***Qualified persons.*** When a qualified person is working in the vicinity of overhead lines, whether in an elevated position or on the ground, the person may not approach or take any conductive object without an approved insulating handle closer to exposed energized parts than shown in Table S-5 unless:

(A) The person is insulated from the energized part (gloves, with sleeves if necessary, rated for the voltage involved are considered to be insulation of the person from the energized part on which work is performed), or

(B) The energized part is insulated both from all other conductive objects at a different potential and from the person, or

(C) The person is insulated from all conductive objects at a potential different from that of the energized part.

(iii) ***Vehicular and mechanical equipment.*** **(A)** Any vehicle or mechanical equipment capable of having parts of its structure elevated near energized overhead lines...

(D of C) "Employees working on or around vehicles and mechanical equipment, such as gin-pole trucks, forklifts, cherry pickers, garbage trucks, cranes, dump trucks, and elevating platforms, who are potentially exposed to hazards related to equipment component contact with overhead lines, shall have been trained by their employers in the inherent hazards of electricity and means of avoiding exposure to such hazards."

...shall be operated so that a clearance of 10 ft (305 cm) is maintained. If the voltage is higher than 50 kV, the clearance shall be increased 4 in (10 cm) for every 10 kV over that voltage. However, under any of the following conditions, the clearance may be reduced:

{1} If the vehicle is in transit with its structure lowered, the clearance may be reduced to 4 ft (122 cm). If the voltage is higher than 50 kV, the clearance shall be increased 4 in (10 cm) for every 10 kV over that voltage.

{2} If insulating barriers are installed to prevent contact with the lines, and if the barriers are rated for the voltage of the line being guarded and are not a part of or an attachment to the vehicle or its raised structure, the clearance may be reduced to a distance within the designed working dimensions of the insulating barrier.

{3} If the equipment is an aerial lift insulated for the voltage involved, and if the work is performed by a qualified person, the clearance (between the uninsulated portion of the aerial lift and the power line) may be reduced to the distance given in Table S-5.

(B) Employees standing on the ground may not contact the vehicle or mechanical equipment or any of its attachments, unless:

(1) The employee is using protective equipment rated for the voltage; or

(2) The equipment is located so that no uninsulated part of its structure (that portion of the structure that provides a conductive path to employees on the ground) can come closer to the line than permitted in paragraph (c)(3)(iii) of this section.

(C) If any vehicle or mechanical equipment capable of having parts of its structure elevated near energized overhead lines is intentionally grounded, employees working on the ground near the point of grounding may not stand at the grounding location whenever there is a possibility of overhead line contact. Additional precautions, such as the use of barricades or insulation, shall be taken to protect employees from hazardous ground potentials, depending on earth resistivity and fault currents, which can develop within the first few feet or more outward from the grounding point.

(4) ***Illumination.*** **(i)** Employees may not enter spaces containing exposed energized parts, unless illumination is provided that enables the employees to perform the work safely.

(ii) Where lack of illumination or an obstruction precludes observation of the work to be performed, employees may not perform tasks near exposed energized parts. Employees may not reach blindly into areas which may contain energized parts.

(5) ***Confined or enclosed work spaces.*** When an employee works in a confined or enclosed space (such as a manhole or vault) that contains exposed energized parts, the employer shall provide, and the employee shall use, protective shields, protective barriers, or insulating materials as necessary to avoid inadvertent contact with these parts.

Doors, hinged panels, and the like shall be secured to prevent their swinging into an employee and causing the employee to contact exposed energized parts.

Confined or enclosed work spaces can present even greater hazards than exposed energized parts. On January 14, 1993, OSHA signed into Final Rule the "Permit Required Confined Spaces for General Industry" Federal Regulation 29 CFR 1910.146. This regulation identifies the major hazards associated with confined spaces. It also requires specific training for anyone entering a confined space. The major hazards of a permit required confined space are:

(1) Oxygen–Less than 19.5% or greater than 23.5%;

(2) Carbon Monoxide–Greater than 50 parts per million (PPM) for an 8-hour exposure;

FIGURE 7.2 *(Continued)*

(3) Hydrogen Sulfide–Greater than 20 PPM "acceptable ceiling concentration" and 50 PPM "acceptable maximum peak above the acceptable ceiling concentration for an 8-hour shift" for a maximum duration of "10 minutes once, only if no other measurable exposure occurs"; and

(4) Greater than 10% by volume of any flammable or explosive substance.

(6) *Conductive materials and equipment.* Conductive materials and equipment that are in contact with any part of an employee's body shall be handled in a manner that will prevent them from contacting exposed energized conductors or circuit parts. If an employee must handle long dimensional conductive objects (such as ducts and pipes) in areas with exposed live parts, the employer shall institute work practices (such as the use of insulation, guarding, and material handling techniques) which will minimize the hazard.

(7) *Portable ladders.* Portable ladders shall have non-conductive siderails if they are used where the employee or the ladder could contact exposed energized parts.

Do not get over confident with wooden or fiberglass ladders. They can become contaminated from moisture, petroleum products, dirt or paint and thus become electrically conductive. A ladder must be treated the same as any other piece of electrical protective equipment, in that it should be inspected and cleaned prior to each use and stored properly when not in use to prevent damage or contamination from occurring.

(8) *Conductive apparel.* Conductive articles of jewelry and clothing (such as watch bands, bracelets, rings, key chains, necklaces, metalized aprons, cloth with conductive thread, or metal headgear) may not be worn if they might contact exposed energized parts. However, such articles may be worn if they are rendered nonconductive by covering, wrapping, or other insulating means.

Additional items have also been noted as presenting a hazard. They are: earrings, metal framed glasses, belt buckles, tool belts and knife scabbards.

It is permissible to render these articles nonconductive so that a person could still wear them. However, it is advisable that these articles be removed. There are numerous accounts of rings being smashed while being on the finger or a person jumping down from a piece of equipment and the ring snagging on the equipment and the person's finger being pulled off. There are other hazards besides the electrical hazard. Be alert to all potentially hazardous situations.

(9) *Housekeeping duties.* Where live parts present an electrical contact hazard, employees may not perform housekeeping duties at such close distances to the parts that there is a possibility of contact, unless adequate safeguards (such as insulating equipment or barriers) are provided. Electrically conductive cleaning materials (including conductive solids such as steel wool, metalized cloth, and silicon carbide, as well as conductive liquid solutions) may not be used in proximity to energized parts unless procedures are followed which will prevent electrical contact.

(10) *Interlocks.* Only a qualified person following the requirements of paragraph (c) of this section may defeat an electrical safety interlock, and then only temporarily while he or she is working on the equipment. The interlock system shall be returned to its operable condition when this work is completed.

(D of C) "Interlocks found on panels, covers and guards are designed to deenergize circuits to prevent electric shock to persons using the equipment or performing minor maintenance or adjustments and shall not be defeated or bypassed by an unqualified person."

A qualified person must be allowed to defeat or bypass a safety interlock in order to perform such tasks as troubleshooting while the circuit is energized. However, as it is stated above, the interlock must be returned to its operable condition when the work is completed so as to prevent access to the energized parts by unqualified persons.

1910.334 Use of Equipment

(a) *Portable electric equipment.* This paragraph applies to the use of cord- and plug-connected equipment, including flexible cord sets (extension cords).

(1) *Handling.* Portable equipment shall be handled in a manner which will not cause damage. Flexible electric cords connected to equipment may not be used for raising or lowering the equipment. Flexible cords may not be fastened with staples or otherwise hung in such a fashion as could damage the outer jacket or insulation.

(2) *Visual inspection.* (i) Portable cord and plug connected equipment and flexible cord sets (extension cords) shall be visually inspected before use on any shift for external defects (such as loose parts, deformed and missing pins,

FIGURE 7.2 (*Continued*)

or damage to outer jacket or insulation) and for evidence of possible internal damage (such as pinched or crushed outer jacket). Cord and plug connected equipment and flexible cord sets (extension cords) which remain connected once they are put in place and are not exposed to damage need not be visually inspected until they are relocated.

(ii) If there is a defect or evidence of damage that might expose an employee to injury, the defective or damaged item shall be removed from service, and no employee may use it until repairs and tests necessary to render the equipment safe have been made.

(iii) When an attachment plug is to be connected to a receptacle (including an on a cord set), the relationship of the plug and receptacle contacts shall first be checked to ensure that they are of proper mating configurations.

(3) *Grounding type equipment.* (i) A flexible cord used with grounding type equipment shall contain an equipment grounding conductor.

(ii) Attachment plugs and receptacles may not be connected or altered in a manner which would prevent proper continuity of the equipment grounding conductor at the point where plugs are attached to receptacles.

Additionally, these devices may not be altered to allow the grounding pole of a plug to be inserted into slots intended for connection to the current-carrying conductors.

(iii) Adapters which interrupt the continuity of the equipment grounding connection may not be used.

It's not that adapters are not safe, they are UL approved. They may not be used because they are used incorrectly most of the time, in that the ground connection is not connected or it is connected to a device that is not grounded. This interrupts the equipment grounding connection and creates a possible shock hazard.

(4) *Conductive work locations.* Portable electric equipment and flexible cords used in highly conductive work locations (such a those inundated with water or other conductive liquids), or in job locations where employees are likely to contact water or conductive liquids, shall be approved for those locations.

(5) *Connecting attachment plugs.* (i) Employees' hands may not be wet when plugging and unplugging flexible cords and cord and plug connected equipment, if energized equipment is involved.

(D of C) "Cord- and plug-connected equipment. Energized equipment here means either the equipment being plugged or the receptacle into which it is being plugged, or both."

(ii) Energized plug and receptacle connections may be handled only with insulating protective equipment if the condition of the connection could provide a conducting path to the employee's hand (if, for example, a cord connector is wet from being immersed in water).

(iii) Locking type connectors shall be properly secured after connection.

Additional safety can be achieved by the use of a ground- fault circuit-interrupter (GFCI) whenever cord- and plug-connected equipment or flexible cord sets are used. This would be for both the grounded type and the double-insulated type of equipment.

(b) *Electric power and lighting circuits.* (1) Routine opening and closing of circuits. Load rated switches, circuit breakers, or other devices specifically designed as disconnecting means shall be used for the opening, reversing, or closing of circuits under load conditions. Cable connectors not of the load break type, fuses, terminal plugs, and cable splice connections may not be used for such purposes, except in an emergency.

It is a good practice to remove the load, if at all possible, before opening any device whether it is load rated or not. This practice will extend the life of the device and reduce the risk of arcing which could cause a flash-over from phase-to-phase and/or from phase-to-ground.

(2) Reclosing circuits after protective device operation. After a circuit is deenergized by a circuit protective device, the circuit may not be manually reenergized until it has been determined that the equipment and circuit can be safely energized. The repetitive manual reclosing of circuit breakers or reenergizing circuits through replaced fuses is prohibited.

Note: When it can be determined from the design of the circuit and the overcurrent devices involved that the automatic operation of a device was caused by an overload rather than a fault condition, no examination of the circuit or connected equipment is needed before the circuit is reenergized.

It has been proven on many occasions that a circuit breaker will explode if it is reclosed on a fault. Sometimes it can be closed several times and on other occasions only once before it explodes. The determining factor being the amount of fault current.

The standard practice in the general industry has been to reset the breaker once or twice before troubleshooting to find the cause. This practice is very dangerous and must be avoided.

FIGURE 7.2 (*Continued*)

(3) *Overcurrent protection modification.* Overcurrent protection of circuits and conductors may not be modified, even on a temporary basis, beyond that allowed by 1910.304(e), the installation safety requirements for overcurrent protection.

(c) *Test instruments and equipment.* (1) *Use.* Only qualified persons may perform testing work on electric circuits or equipment.

(2) *Visual inspection.* Test instruments and equipment and all associated test leads, cables, power cords, probes, and connectors shall be visually inspected for external defects and damage before the equipment is used. If there is a defect or evidence of damage that might expose an employee to injury, the defective or damaged item shall be removed from service, and no employee may use it until repairs and tests necessary to render the equipment safe have been made.

Care should be taken to verify that the setting, scale, and so on are correct.

(3) *Rating of equipment.* Test instruments and equipment and their accessories shall be rated for the circuits and equipment to which they will be connected and shall be designed for the environment in which they will be used.

(d) *Occasional use of flammable or ignitable materials.* Where flammable materials are present only occasionally, electric equipment capable of igniting them shall not be used, unless measures are taken to prevent hazardous conditions from developing. Such materials include, but are not limited to: flammable gases, vapors, or liquids; combustible dust and ignitable fibers or flyings.

Note: Electrical installation requirements for locations where flammable materials are present on a regular basis are contained in 1910.307.

It is very possible to produce a hazardous (classified) location when using solvents and cleaning fluids. This is especially hazardous when it is around electrical equipment that can produce arcing such as circuit breakers, contactors, and relays. If these types of solvents and cleaning fluids are going to be used, adequate ventilation must be utilized to prevent the buildup of an explosive mixture.

1910.335 Safeguards for Personnel Protection

(a) *Use of protective equipment.* (1) *Personal protective equipment.* (i) Employees working in areas where there are potential electrical hazards shall be provided with, and shall use, electrical protective equipment that is appropriate for the specific parts of the body to be protected and for the work to be performed.

Note: Personal protective equipment requirements are contained in subpart I of this part.

(ii) Protective equipment shall be maintained in a safe reliable condition and shall be periodically inspected or tested, as required by 1910.137.

The following is a summary of 1910.137:

OSHA's 1910.137 "Electrical Protective Equipment" provides the requirements for the design, manufacture, and in-service care and use of rubber temporary insulation equipment. (Based on ASTM standards)

The following table outlines six different voltage classifications for rubber protective equipment per ASTM.

OSHA states in 1910.137 that "Electrical protective equipment shall be maintained in a safe, reliable condition." The following specific requirements apply to insulating equipment made of rubber:

1. Maximum use voltage shall conform to those values listed in the above table.
2. Insulating equipment shall be inspected for damage before each day's use and anytime damage is suspected.
3. Gloves shall be given an air test.

Insulating equipment with the following defects may not be used:

1. Holes, tears, punctures, or cuts;
2. Ozone cutting or checking (fine interlacing cracks);
3. Embedded foreign objects;
4. Texture changes such as: swelling, softening, hardening, or becoming sticky or inelastic (swelling, softening and becoming sticky are results of petroleum deterioration and hardening and becoming inelastic are the results of ageing, which can be accelerated by heat); and
5. Any other defects that could damage the insulating properties of the rubber.

FIGURE 7.2 *(Continued)*

Voltage Classifications for Rubber Protective Equipment

ID tag	Class	AC proof test voltage	AC max. use voltage	DC proof test voltage
Beige	00	2,500	500	10,000
Red	0	5,000	1,000	20,000
White	1	10,000	7,500	40,000
Yellow	2	20,000	17,000	50,000
Green	3	30,000	26,500	60,000
Orange	4	40,000	36,000	70,000

Notes: Class 00 does not currently appear in OSHA, 1910.137, but was added by the author based on ASTM F496. (1910.137 is currently being revised by OSHA dated June 15, 2005 and, at the time of this printing, is not yet Final Rule; this revision will include class 00 rubber gloves.)

Protector (leather) gloves should also be inspected for any damage, contamination from petroleum products, any imbedded objects, or anything that could damage the rubber gloves.

Insulating equipment shall be cleaned as needed to remove foreign substances (wash with a mild detergent, no abrasives or bleaching agents, and water, rinse thoroughly and allow the equipment to dry completely before use).

Insulating equipment shall be stored to protect it from the following:

1. Light (primarily sunlight and any light that produces UV; UV produces ozone);
2. Temperature extremes (high temperatures accelerate the ageing process; examples of heat sources are hot forced air, steam lines, near radiators, and so on.);
3. Excessive humidity (because of condensation);
4. Ozone (produced by corona, arcing, and ultra-violet rays); and
5. Other injurious substances and conditions (beware of chemicals and petroleum products of any kind, such as oil, gasoline, fuels, hydraulic fluids, solvents, hand lotions, waterless hand cleaners, ointments, and so on).

Protector gloves shall be worn over insulating gloves, except as follows:

1. Class 0, under limited-use and where high finger dexterity is needed;
2. Other classes, provided there is little chance of physical damage, one class higher is used, and the gloves are tested prior to use on a higher voltage.

Electrical protective equipment shall be subjected to periodic electrical tests, as stated in the following table:

The rubber insulating equipment that fails to pass inspections or electrical tests may not be used, except as follows:

1. Rubber insulating line hose may be used in shorter lengths after the defective portion has been cut off;
2. Rubber insulating blankets may be repaired by using a compatible patch purchased from the manufacturer of the blanket (not a tire patch), or the defective portion may be cut off, provided the undamaged portion is not smaller than 22 inches by 22 inches for Class 1, 2, 3, and 4 blankets (cutting off the defective portion cannot include that part of the blanket that contains the ID tag); and
3. Rubber insulating gloves and sleeves may be repaired if the defect is minor such as small cuts, tears, or punctures. The repairs on gloves are permitted only on the gauntlet area not the hand area. A compatible patch may be used or, if the defect is only a minor surface blemish, a compatible liquid compound may be used.

If any rubber insulating equipment is repaired, it must be retested before it may be used.

The employer shall certify that the equipment has been tested in accordance with 1910.137. This certification must identify the equipment tested and the date it was tested.

Detailed information on rubber protective equipment is available from the American Society for Testing and Materials (ASTM), 1916 Race Street, Philadelphia, Pennsylvania 19103.

FIGURE 7.2 (*Continued*)

Rubber Insulating Equipment Test Intervals

Type of equipment	When to test
Rubber insulating line hose	Upon indication that insulating value is suspect
Rubber insulating covers	Upon indication that insulating value is suspect
Rubber insulating blankets	Before first issue and every 12 months thereafter*
Rubber insulating gloves	Before first issue and every 6 months thereafter*
Rubber insulating sleeves	Before first issue and every 12 months thereafter*

*If the insulating equipment has been electrically tested, but not issued for service, it may not be placed into service unless it has been electrically tested within the previous 12 months.

Standards on Electrical Protective Equipment

Subject	Number and title
Gloves/Sleeves	ASTM F 496, Specification for In-Service Care of Insulating Gloves and Sleeves
Blankets	ASTM F 479, Specification for In-Service Care of Insulating Blankets
Line Hose/Covers	ASTM F 478, Specification for In-Service Care of Insulating Line Hose and Covers

The above listed standards, along with other standards for electrical protective equipment are found in ASTM, Section 10, Volume 10.03.

ASTM Standards for the in-service care of electrical protective equipment are listed in the following table:

(iii) If the insulating capability of protective equipment may be subject to damage during use, the insulating material shall be protected. (For example, an outer covering of leather is sometimes used for the protection of rubber insulating material.)

(iv) Employees shall wear non-conductive head protection wherever there is a danger of head injury from electric shock or burns due to contact with exposed energized parts.

Head protection, for electrical used, must be labeled ANSI Z89.2. All other hard hats are labeled ANSI Z89.1.

(v) Employees shall wear protective equipment for the eyes or face wherever there is danger of injury to the eyes or face from electric arcs or flashes or from flying objects resulting from electrical explosion.

(2) ***General protective equipment and tools.*** **(i)** When working near exposed energized conductors or circuit parts, each employee shall use insulated tools or handling equipment if the tools or handling equipment might make contact with such conductors or parts. If the insulating capability of insulated tools or handling equipment is subject to damage, the insulating material shall be protected.

(OSHA's D of C) "This means a tool incased within material of composition and thickness that is recognized as electrical insulation."

OSHA has stated verbally that tape of any type, heat shrink tubing, or any of the dipping processes do not meet the "recognized as electrical insulation" requirement and that they must not be used.

Insulated tools that are manufactured in accordance with the IEC 900 International Standard as well as the soon to be published ASTM standard will meet this requirement. Additional information is found in IEEE Standard 935.

(A) Fuse handling equipment, insulated for the circuit voltage, shall be used to remove or install fuses when the fuse terminals are energized.

(B) Ropes and handlines used near exposed energized parts shall be nonconductive.

(ii) Protective shields, protective barriers, or insulating materials shall be used to protect each employee from shock, burns, or other electrically related injuries while that employee is working near exposed energized parts which might be accidentally contacted or where dangerous electric heating or arcing might occur. When normally enclosed live parts are exposed for maintenance or repair, they shall be guarded to protect unqualified persons from contact with the live parts.

FIGURE 7.2 *(Continued)*

(b) ***Alerting techniques.*** The following alerting techniques shall be used to warn and protect employees from hazards which could cause injury due to electric shock, burns, or failure of electric equipment parts:

(1) ***Safety signs and tags.*** Safety signs, safety symbols, or accident prevention tags shall be used where necessary to warn employees about electrical hazards which may endanger them, as required by 1910.145.

(2) ***Barricades.*** Barricades shall be used in conjunction with safety signs where it is necessary to prevent or limit employee access to work areas exposing employees to uninsulated energized conductors or circuit parts. Conductive barricades may not be used where they might cause an electrical contact hazard.

(3) ***Attendants.*** If signs and barricades do not provide sufficient warning and protection from electrical hazards, an attendant shall be stationed to warn and protect employees.

These alerting techniques are being used for a reason, a hazard exists. Never attempt to go through or bypass a safety sign or tag, barricade, or an attendant.

[55 FR 32016, Aug. 6, 1990]

FIGURE 7.2 *(Continued)*

Electric Power Generation, Transmission, and Distribution (OSHA 29 CFR 1910.269)
(FINAL RULE APRIL 11, 2014)

1910.269 Electric power generation, transmission, and distribution

(a) ***General.*** **(1)** ***Application.*** **(i)** This section covers the operation and maintenance of electric power generation, control, transformation, transmission, and distribution lines and equipment. These provisions apply to:

(A) Power generation, transmission, and distribution installations, including related equipment for the purpose of communication or metering that are accessible only to qualified employees;

Note to paragraph (a)(1)(i)(A): The types of installations covered by this paragraph include the generation, transmission, and distribution installations of electric utilities, as well as equivalent installations of industrial establishments. Subpart S of this part covers supplementary electric generating equipment that is used to supply a workplace for emergency, standby, or similar purposes only. (See paragraph (a)(1)(i)(B) of this section.)

(B) Other installations at an electric power generating station, as follows:

(1) Fuel and ash handling and processing installations, such as coal conveyors,

(2) Water and steam installations, such as penstocks, pipelines, and tanks, providing a source of energy for electric generators, and

(3) Chlorine and hydrogen systems;

(C) Test sites where employees perform electrical testing involving temporary measurements associated with electric power generation, transmission, and distribution in laboratories, in the field, in substations, and on lines, as opposed to metering, relaying, and routine line work;

(D) Work on, or directly associated with, the installations covered in paragraphs (a)(1)(i)(A) through (a)(1)(i)(C) of this section; and

(E) Line-clearance tree-trimming operations, as follows:

(1) Entire § 1910.269 of this part, except paragraph (r)(1) of this section, applies to line-clearance tree-trimming operations performed by qualified employees (those who are knowledgeable in the construction and operation of the electric power generation, transmission, or distribution equipment involved, along with the associated hazards).

(2) Paragraphs (a)(2), (a)(3), (b), (c), (g), (k), (p), and (r) of this section apply to line-clearance tree-trimming operations performed by line-clearance tree trimmers who are not qualified employees.

(ii) Notwithstanding paragraph (a)(1)(i) of this section, § 1910.269 of this part does not apply:

(A) To construction work, as defined in § 1910.12 of this part, except for line-clearance tree-trimming operations and work involving electric power generation installations as specified in § 1926.950(a)(3) of this chapter; or

(B) To electrical installations, electrical safety-related work practices, or electrical maintenance considerations covered by Subpart S of this part.

Note 1 to paragraph (a)(1)(ii)(B): The Occupational Safety and Health Administration considers work practices conforming to §§ 1910.332 through 1910.335 as complying with the electrical safety-related work-practice requirements of § 1910.269 identified in Table 1 of Appendix A-2 to this section, provided that employers are performing the work on a generation or distribution installation meeting §§ 1910.303 through 1910.308. This table also identifies provisions in § 1910.269 that apply to work by qualified persons directly on, or associated with, installations of

FIGURE 7.3 Electrical Power Generation, Transmission, and Distribution Final Rule (OSHA-CFR, Title 29, part 1910, paragraph 269).

electric power generation, transmission, and distribution lines or equipment, regardless of compliance with §§ 1910.332 through 1910.335.

Note 2 to paragraph (a)(1)(ii)(B): The Occupational Safety and Health Administration considers work practices performed by qualified persons and conforming to § 1910.269 as complying with §§ 1910.333(c) and 1910.335.

(iii) This section applies in addition to all other applicable standards contained in this Part 1910. Employers covered under this section are not exempt from complying with other applicable provisions in Part 1910 by the operation of § 1910.5(c). Specific references in this section to other sections of Part 1910 are for emphasis only.

(2) ***Training.*** **(i)** All employees performing work covered by this section shall be trained as follows:

(A) Each employee shall be trained in, and familiar with, the safety-related work practices, safety procedures, and other safety requirements in this section that pertain to his or her job assignments.

(B) Each employee shall also be trained in and familiar with any other safety practices, including applicable emergency procedures (such as pole-top and manhole rescue), that are not specifically addressed by this section but that are related to his or her work and are necessary for his or her safety.

(C) The degree of training shall be determined by the risk to the employee for the hazard involved.

(ii) Each qualified employee shall also be trained and competent in:

(A) The skills and techniques necessary to distinguish exposed live parts from other parts of electric equipment,

(B) The skills and techniques necessary to determine the nominal voltage of exposed live parts,

(C) The minimum approach distances specified in this section corresponding to the voltages to which the qualified employee will be exposed and the skills and techniques necessary to maintain those distances,

(D) The proper use of the special precautionary techniques, personal protective equipment, insulating and shielding materials, and insulated tools for working on or near exposed energized parts of electric equipment, and

(E) The recognition of electrical hazards to which the employee may be exposed and the skills and techniques necessary to control or avoid these hazards.

Note to paragraph (a)(2)(ii): For the purposes of this section, a person must have the training required by paragraph (a)(2)(ii) of this section to be considered a qualified person.

(iii) Each line-clearance tree trimmer who is not a qualified employee shall also be trained and competent in:

(A) The skills and techniques necessary to distinguish exposed live parts from other parts of electric equipment,

(B) The skills and techniques necessary to determine the nominal voltage of exposed live parts, and

(C) The minimum approach distances specified in this section corresponding to the voltages to which the employee will be exposed and the skills and techniques necessary to maintain those distances.

(iv) The employer shall determine, through regular supervision and through inspections conducted on at least an annual basis, that each employee is complying with the safety-related work practices required by this section.

(v) An employee shall receive additional training (or retraining) under any of the following conditions:

(A) If the supervision or annual inspections required by paragraph (a)(2)(iv) of this section indicate that the employee is not complying with the safety-related work practices required by this section, or

(B) If new technology, new types of equipment, or changes in procedures necessitate the use of safety-related work practices that are different from those which the employee would normally use, or

(C) If he or she must employ safety-related work practices that are not normally used during his or her regular job duties.

Note to paragraph (a)(2)(v)(C): The Occupational Safety and Health Administration considers tasks that are performed less often than once per year to necessitate retraining before the performance of the work practices involved.

(vi) The training required by paragraph (a)(2) of this section shall be of the classroom or on-the-job type.

(vii) The training shall establish employee proficiency in the work practices required by this section and shall introduce the procedures necessary for compliance with this section.

(viii) The employer shall ensure that each employee has demonstrated proficiency in the work practices involved before that employee is considered as having completed the training required by paragraph (a)(2) of this section.

Note 1 to paragraph (a)(2)(viii): Though they are not required by this paragraph, employment records that indicate that an employee has successfully completed the required training are one way of keeping track of when an employee has demonstrated proficiency.

Note 2 to paragraph (a)(2)(viii): For an employee with previous training, an employer may determine that that employee has demonstrated the proficiency required by this paragraph using the following process:

(1) Confirm that the employee has the training required by paragraph (a)(2) of this section,

(2) Use an examination or interview to make an initial determination that the employee understands the relevant safety-related work practices before he or she performs any work covered by this section, and

(3) Supervise the employee closely until that employee has demonstrated proficiency as required by this paragraph.

FIGURE 7.3 *(Continued)*

(3) ***Information transfer.***

(i) Before work begins, the host employer shall inform contract employers of:

(A) The characteristics of the host employer's installation that are related to the safety of the work to be performed and are listed in paragraphs (a)(4)(i) through (a)(4)(v) of this section;

Note to paragraph (a)(3)(i)(A): This paragraph requires the host employer to obtain information listed in paragraphs (a)(4)(i) through (a)(4)(v) of this section if it does not have this information in existing records.

(B) Conditions that are related to the safety of the work to be performed, that are listed in paragraphs (a)(4)(vi) through (a)(4)(viii) of this section, and that are known to the host employer;

Note to paragraph (a)(3)(i)(B): For the purposes of this paragraph, the host employer need only provide information to contract employers that the host employer can obtain from its existing records through the exercise of reasonable diligence. This paragraph does not require the host employer to make inspections of worksite conditions to obtain this information.

(C) Information about the design and operation of the host employer's installation that the contract employer needs to make the assessments required by this section; and

Note to paragraph (a)(3)(i)(C): This paragraph requires the host employer to obtain information about the design and operation of its installation that contract employers need to make required assessments if it does not have this information in existing records.

(D) Any other information about the design and operation of the host employer's installation that is known by the host employer, that the contract employer requests, and that is related to the protection of the contract employer's employees.

Note to paragraph (a)(3)(i)(D): For the purposes of this paragraph, the host employer need only provide information to contract employers that the host employer can obtain from its existing records through the exercise of reasonable diligence. This paragraph does not require the host employer to make inspections of worksite conditions to obtain this information.

(ii) Contract employers shall comply with the following requirements:

(A) The contract employer shall ensure that each of its employees is instructed in the hazardous conditions relevant to the employee's work that the contract employer is aware of as a result of information communicated to the contract employer by the host employer under paragraph (a)(3)(i) of this section.

(B) Before work begins, the contract employer shall advise the host employer of any unique hazardous conditions presented by the contract employer's work.

(C) The contract employer shall advise the host employer of any unanticipated hazardous conditions found during the contract employer's work that the host employer did not mention under paragraph (a)(3)(i) of this section. The contract employer shall provide this information to the host employer within 2 working days after discovering the hazardous condition.

(iii) The contract employer and the host employer shall coordinate their work rules and procedures so that each employee of the contract employer and the host employer is protected as required by this section.

(4) ***Existing characteristics and conditions.*** Existing characteristics and conditions of electric lines and equipment that are related to the safety of the work to be performed shall be determined before work on or near the lines or equipment is started. Such characteristics and conditions include, but are not limited to:

(i) The nominal voltages of lines and equipment,

(ii) The maximum switching-transient voltages,

(iii) The presence of hazardous induced voltages,

(iv) The presence of protective grounds and equipment grounding conductors,

(v) The locations of circuits and equipment, including electric supply lines, communication lines, and fire-protective signaling circuits,

(vi) The condition of protective grounds and equipment grounding conductors,

(vii) The condition of poles, and

(viii) Environmental conditions relating to safety.

(b) ***Medical services and first aid.*** The employer shall provide medical services and first aid as required in § 1910.151. In addition to the requirements of § 1910.151, the following requirements also apply:

(1) ***First-aid training.*** When employees are performing work on, or associated with, exposed lines or equipment energized at 50 volts or more, persons with first-aid training shall be available as follows:

(i) For field work involving two or more employees at a work location, at least two trained persons shall be available. However, for line-clearance tree trimming operations performed by line-clearance tree trimmers who are not

FIGURE 7.3 *(Continued)*

qualified employees, only one trained person need be available if all new employees are trained in first aid within 3 months of their hiring dates.

(ii) For fixed work locations such as substations, the number of trained persons available shall be sufficient to ensure that each employee exposed to electric shock can be reached within 4 minutes by a trained person. However, where the existing number of employees is insufficient to meet this requirement (at a remote substation, for example), each employee at the work location shall be a trained employee.

(2) *First-aid supplies.* First-aid supplies required by § 1910.151(b) shall be placed in weatherproof containers if the supplies could be exposed to the weather.

(3) *First-aid kits.* The employer shall maintain each first-aid kit, shall ensure that it is readily available for use, and shall inspect it frequently enough to ensure that expended items are replaced. The employer also shall inspect each first aid kit at least once per year.

(c) *Job briefing.* (1) *Before each job.* (i) In assigning an employee or a group of employees to perform a job, the employer shall provide the employee in charge of the job with all available information that relates to the determination of existing characteristics and conditions required by paragraph (a)(4) of this section.

(ii) The employer shall ensure that the employee in charge conducts a job briefing that meets paragraphs (c)(2), (c)(3), and (c)(4) of this section with the employees involved before they start each job.

(2) *Subjects to be covered.* The briefing shall cover at least the following subjects: hazards associated with the job, work procedures involved, special precautions, energy-source controls, and personal protective equipment requirements.

(3) *Number of briefings.* (i) If the work or operations to be performed during the work day or shift are repetitive and similar, at least one job briefing shall be conducted before the start of the first job of each day or shift.

(ii) Additional job briefings shall be held if significant changes, which might affect the safety of the employees, occur during the course of the work.

(4) *Extent of briefing.* (i) A brief discussion is satisfactory if the work involved is routine and if the employees, by virtue of training and experience, can reasonably be expected to recognize and avoid the hazards involved in the job.

(ii) A more extensive discussion shall be conducted:

(A) If the work is complicated or particularly hazardous, or

(B) If the employee cannot be expected to recognize and avoid the hazards involved in the job.

Note to paragraph (c)(4): The briefing must address all the subjects listed in paragraph (c)(2) of this section.

(5) *Working alone.* An employee working alone need not conduct a job briefing. However, the employer shall ensure that the tasks to be performed are planned as if a briefing were required.

(d) *Hazardous energy control (lockout/tagout) procedures.* (1) *Application.* The provisions of paragraph (d) of this section apply to the use of lockout/tagout procedures for the control of energy sources in installations for the purpose of electric power generation, including related equipment for communication or metering. Locking and tagging procedures for the deenergizing of electric energy sources which are used exclusively for purposes of transmission and distribution are addressed by paragraph (m) of this section.

Note to paragraph (d)(1): Installations in electric power generation facilities that are not an integral part of, or inextricably commingled with, power generation processes or equipment are covered under § 1910.147 and Subpart S of this part.

(2) *General.* (i) The employer shall establish a program consisting of energy control procedures, employee training, and periodic inspections to ensure that, before any employee performs any servicing or maintenance on a machine or equipment where the unexpected energizing, start up, or release of stored energy could occur and cause injury, the machine or equipment is isolated from the energy source and rendered inoperative.

(ii) The employer's energy control program under paragraph (d)(2) of this section shall meet the following requirements:

(A) If an energy isolating device is not capable of being locked out, the employer's program shall use a tagout system.

(B) If an energy isolating device is capable of being locked out, the employer's program shall use lockout, unless the employer can demonstrate that the use of a tagout system will provide full employee protection as follows:

(1) When a tagout device is used on an energy isolating device which is capable of being locked out, the tagout device shall be attached at the same location that the lockout device would have been attached, and the employer shall demonstrate that the tagout program will provide a level of safety equivalent to that obtained by the use of a lockout program.

(2) In demonstrating that a level of safety is achieved in the tagout program equivalent to the level of safety obtained by the use of a lockout program, the employer shall demonstrate full compliance with all tagout-related

FIGURE 7.3 *(Continued)*

provisions of this standard together with such additional elements as are necessary to provide the equivalent safety available from the use of a lockout device. Additional means to be considered as part of the demonstration of full employee protection shall include the implementation of additional safety measures such as the removal of an isolating circuit element, blocking of a controlling switch, opening of an extra disconnecting device, or the removal of a valve handle to reduce the likelihood of inadvertent energizing.

(C) After November 1, 1994, whenever replacement or major repair, renovation, or modification of a machine or equipment is performed, and whenever new machines or equipment are installed, energy isolating devices for such machines or equipment shall be designed to accept a lockout device.

(iii) Procedures shall be developed, documented, and used for the control of potentially hazardous energy covered by paragraph (d) of this section.

(iv) The procedure shall clearly and specifically outline the scope, purpose, responsibility, authorization, rules, and techniques to be applied to the control of hazardous energy, and the measures to enforce compliance including, but not limited to, the following:

(A) A specific statement of the intended use of this procedure;

(B) Specific procedural steps for shutting down, isolating, blocking and securing machines or equipment to control hazardous energy;

(C) Specific procedural steps for the placement, removal, and transfer of lockout devices or tagout devices and the responsibility for them; and

(D) Specific requirements for testing a machine or equipment to determine and verify the effectiveness of lockout devices, tagout devices, and other energy control measures.

(v) The employer shall conduct a periodic inspection of the energy control procedure at least annually to ensure that the procedure and the provisions of paragraph (d) of this section are being followed.

(A) The periodic inspection shall be performed by an authorized employee who is not using the energy control procedure being inspected.

(B) The periodic inspection shall be designed to identify and correct any deviations or inadequacies.

(C) If lockout is used for energy control, the periodic inspection shall include a review, between the inspector and each authorized employee, of that employee's responsibilities under the energy control procedure being inspected.

(D) Where tagout is used for energy control, the periodic inspection shall include a review, between the inspector and each authorized and affected employee, of that employee's responsibilities under the energy control procedure being inspected, and the elements set forth in paragraph (d)(2)(vii) of this section.

(E) The employer shall certify that the inspections required by paragraph (d)(2)(v) of this section have been accomplished. The certification shall identify the machine or equipment on which the energy control procedure was being used, the date of the inspection, the employees included in the inspection, and the person performing the inspection.

Note to paragraph (d)(2)(v)(E): If normal work schedule and operation records demonstrate adequate inspection activity and contain the required information, no additional certification is required.

(vi) The employer shall provide training to ensure that the purpose and function of the energy control program are understood by employees and that the knowledge and skills required for the safe application, usage, and removal of energy controls are acquired by employees. The training shall include the following:

(A) Each authorized employee shall receive training in the recognition of applicable hazardous energy sources, the type and magnitude of energy available in the workplace, and in the methods and means necessary for energy isolation and control.

(B) Each affected employee shall be instructed in the purpose and use of the energy control procedure.

(C) All other employees whose work operations are or may be in an area where energy control procedures may be used shall be instructed about the procedures and about the prohibition relating to attempts to restart or reenergize machines or equipment that are locked out or tagged out.

(vii) When tagout systems are used, employees shall also be trained in the following limitations of tags:

(A) Tags are essentially warning devices affixed to energy isolating devices and do not provide the physical restraint on those devices that is provided by a lock.

(B) When a tag is attached to an energy isolating means, it is not to be removed without authorization of the authorized person responsible for it, and it is never to be bypassed, ignored, or otherwise defeated.

(C) Tags must be legible and understandable by all authorized employees, affected employees, and all other employees whose work operations are or may be in the area, in order to be effective.

FIGURE 7.3 *(Continued)*

(D) Tags and their means of attachment must be made of materials which will withstand the environmental conditions encountered in the workplace.

(E) Tags may evoke a false sense of security, and their meaning needs to be understood as part of the overall energy control program.

(F) Tags must be securely attached to energy isolating devices so that they cannot be inadvertently or accidentally detached during use.

(viii) Retraining shall be provided by the employer as follows:

(A) Retraining shall be provided for all authorized and affected employees whenever there is a change in their job assignments, a change in machines, equipment, or processes that present a new hazard or whenever there is a change in the energy control procedures.

(B) Retraining shall also be conducted whenever a periodic inspection under paragraph (d)(2)(v) of this section reveals, or whenever the employer has reason to believe, that there are deviations from or inadequacies in an employee's knowledge or use of the energy control procedures.

(C) The retraining shall reestablish employee proficiency and shall introduce new or revised control methods and procedures, as necessary.

(ix) The employer shall certify that employee training has been accomplished and is being kept up to date. The certification shall contain each employee's name and dates of training.

(3) ***Protective materials and hardware.*** **(i)** Locks, tags, chains, wedges, key blocks, adapter pins, self-locking fasteners, or other hardware shall be provided by the employer for isolating, securing, or blocking of machines or equipment from energy sources.

(ii) Lockout devices and tagout devices shall be singularly identified; shall be the only devices used for controlling energy; may not be used for other purposes; and shall meet the following requirements:

(A) Lockout devices and tagout devices shall be capable of withstanding the environment to which they are exposed for the maximum period of time that exposure is expected.

(1) Tagout devices shall be constructed and printed so that exposure to weather conditions or wet and damp locations will not cause the tag to deteriorate or the message on the tag to become illegible.

(2) Tagout devices shall be so constructed as not to deteriorate when used in corrosive environments.

(B) Lockout devices and tagout devices shall be standardized within the facility in at least one of the following criteria: color, shape, size. Additionally, in the case of tagout devices, print and format shall be standardized.

(C) Lockout devices shall be substantial enough to prevent removal without the use of excessive force or unusual techniques, such as with the use of bolt cutters or metal cutting tools.

(D) Tagout devices, including their means of attachment, shall be substantial enough to prevent inadvertent or accidental removal. Tagout device attachment means shall be of a non-reusable type, attachable by hand, self-locking, and nonreleasable with a minimum unlocking strength of no less than 50 pounds and shall have the general design and basic characteristics of being at least equivalent to a one-piece, all environment-tolerant nylon cable tie.

(E) Each lockout device or tagout device shall include provisions for the identification of the employee applying the device.

(F) Tagout devices shall warn against hazardous conditions if the machine or equipment is energized and shall include a legend such as the following: Do Not Start, Do Not Open, Do Not Close, Do Not Energize, Do Not Operate.

Note to paragraph (d)(3)(ii)(F): For specific provisions covering accident prevention tags, see § 1910.145.

(4) ***Energy isolation.*** Lockout and tagout device application and removal may only be performed by the authorized employees who are performing the servicing or maintenance.

(5) ***Notification.*** Affected employees shall be notified by the employer or authorized employee of the application and removal of lockout or tagout devices. Notification shall be given before the controls are applied and after they are removed from the machine or equipment.

Note to paragraph (d)(5): See also paragraph (d)(7) of this section, which requires that the second notification take place before the machine or equipment is reenergized.

(6) ***Lockout/tagout application.*** The established procedures for the application of energy control (the lockout or tagout procedures) shall include the following elements and actions, and these procedures shall be performed in the following sequence:

(i) Before an authorized or affected employee turns off a machine or equipment, the authorized employee shall have knowledge of the type and magnitude of the energy, the hazards of the energy to be controlled, and the method or means to control the energy.

FIGURE 7.3 *(Continued)*

(ii) The machine or equipment shall be turned off or shut down using the procedures established for the machine or equipment. An orderly shutdown shall be used to avoid any additional or increased hazards to employees as a result of the equipment stoppage.

(iii) All energy isolating devices that are needed to control the energy to the machine or equipment shall be physically located and operated in such a manner as to isolate the machine or equipment from energy sources.

(iv) Lockout or tagout devices shall be affixed to each energy isolating device by authorized employees.

(A) Lockout devices shall be attached in a manner that will hold the energy isolating devices in a "safe" or "off" position.

(B) Tagout devices shall be affixed in such a manner as will clearly indicate that the operation or movement of energy isolating devices from the "safe" or "off" position is prohibited.

(1) Where tagout devices are used with energy isolating devices designed with the capability of being locked out, the tag attachment shall be fastened at the same point at which the lock would have been attached.

(2) Where a tag cannot be affixed directly to the energy isolating device, the tag shall be located as close as safely possible to the device, in a position that will be immediately obvious to anyone attempting to operate the device.

(v) Following the application of lockout or tagout devices to energy isolating devices, all potentially hazardous stored or residual energy shall be relieved, disconnected, restrained, or otherwise rendered safe.

(vi) If there is a possibility of reaccumulation of stored energy to a hazardous level, verification of isolation shall be continued until the servicing or maintenance is completed or until the possibility of such accumulation no longer exists.

(vii) Before starting work on machines or equipment that have been locked out or tagged out, the authorized employee shall verify that isolation and deenergizing of the machine or equipment have been accomplished. If normally energized parts will be exposed to contact by an employee while the machine or equipment is deenergized, a test shall be performed to ensure that these parts are deenergized.

(7) *Release from lockout/tagout.* Before lockout or tagout devices are removed and energy is restored to the machine or equipment, procedures shall be followed and actions taken by the authorized employees to ensure the following:

(i) The work area shall be inspected to ensure that nonessential items have been removed and that machine or equipment components are operationally intact.

(ii) The work area shall be checked to ensure that all employees have been safely positioned or removed.

(iii) After lockout or tagout devices have been removed and before a machine or equipment is started, affected employees shall be notified that the lockout or tagout devices have been removed.

(iv) Each lockout or tagout device shall be removed from each energy isolating device by the authorized employee who applied the lockout or tagout device. However, if that employee is not available to remove it, the device may be removed under the direction of the employer, provided that specific procedures and training for such removal have been developed, documented, and incorporated into the employer's energy control program. The employer shall demonstrate that the specific procedure provides a degree of safety equivalent to that provided by the removal of the device by the authorized employee who applied it. The specific procedure shall include at least the following elements:

(A) Verification by the employer that the authorized employee who applied the device is not at the facility;

(B) Making all reasonable efforts to contact the authorized employee to inform him or her that his or her lockout or tagout device has been removed; and

(C) Ensuring that the authorized employee has this knowledge before he or she resumes work at that facility.

(8) *Additional requirements.* (i) If the lockout or tagout devices must be temporarily removed from energy isolating devices and the machine or equipment must be energized to test or position the machine, equipment, or component thereof, the following sequence of actions shall be followed:

(A) Clear the machine or equipment of tools and materials in accordance with paragraph (d)(7)(i) of this section;

(B) Remove employees from the machine or equipment area in accordance with paragraphs (d)(7)(ii) and (d)(7)(iii) of this section;

(C) Remove the lockout or tagout devices as specified in paragraph (d)(7)(iv) of this section;

(D) Energize and proceed with the testing or positioning; and

(E) Deenergize all systems and reapply energy control measures in accordance with paragraph (d)(6) of this section to continue the servicing or maintenance.

FIGURE 7.3 *(Continued)*

(ii) When servicing or maintenance is performed by a crew, craft, department, or other group, they shall use a procedure which affords the employees a level of protection equivalent to that provided by the implementation of a personal lockout or tagout device. Group lockout or tagout devices shall be used in accordance with the procedures required by paragraphs (d)(2)(iii) and (d)(2)(iv) of this section including, but not limited to, the following specific requirements:

(A) Primary responsibility shall be vested in an authorized employee for a set number of employees working under the protection of a group lockout or tagout device (such as an operations lock);

(B) Provision shall be made for the authorized employee to ascertain the exposure status of all individual group members with regard to the lockout or tagout of the machine or equipment;

(C) When more than one crew, craft, department, or other group is involved, assignment of overall job-associated lockout or tagout control responsibility shall be given to an authorized employee designated to coordinate affected work forces and ensure continuity of protection; and

(D) Each authorized employee shall affix a personal lockout or tagout device to the group lockout device, group lockbox, or comparable mechanism when he or she begins work and shall remove those devices when he or she stops working on the machine or equipment being serviced or maintained.

(iii) Procedures shall be used during shift or personnel changes to ensure the continuity of lockout or tagout protection, including provision for the orderly transfer of lockout or tagout device protection between off-going and on-coming employees, to minimize their exposure to hazards from the unexpected energizing or start-up of the machine or equipment or from the release of stored energy.

(iv) Whenever outside servicing personnel are to be engaged in activities covered by paragraph (d) of this section, the on-site employer and the outside employer shall inform each other of their respective lockout or tagout procedures, and each employer shall ensure that his or her personnel understand and comply with restrictions and prohibitions of the energy control procedures being used.

(v) If energy isolating devices are installed in a central location and are under the exclusive control of a system operator, the following requirements apply:

(A) The employer shall use a procedure that affords employees a level of protection equivalent to that provided by the implementation of a personal lockout or tagout device.

(B) The system operator shall place and remove lockout and tagout devices in place of the authorized employee under paragraphs (d)(4), (d)(6)(iv), and (d)(7)(iv) of this section.

(C) Provisions shall be made to identify the authorized employee who is responsible for (that is, being protected by) the lockout or tagout device, to transfer responsibility for lockout and tagout devices, and to ensure that an authorized employee requesting removal or transfer of a lockout or tagout device is the one responsible for it before the device is removed or transferred.

Note to paragraph (d): Lockout and tagging procedures that comply with paragraphs (c) through (f) of § 1910.147 will also be deemed to comply with paragraph (d) of this section if the procedures address the hazards covered by paragraph (d) of this section.

(e) ***Enclosed spaces.*** This paragraph covers enclosed spaces that may be entered by employees. It does not apply to vented vaults if the employer makes a determination that the ventilation system is operating to protect employees before they enter the space. This paragraph applies to routine entry into enclosed spaces in lieu of the permit-space entry requirements contained in paragraphs (d) through (k) of § 1910.146. If, after the employer takes the precautions given in paragraphs (e) and (t) of this section, the hazards remaining in the enclosed space endanger the life of an entrant or could interfere with an entrant's escape from the space, then entry into the enclosed space shall meet the permit-space entry requirements of paragraphs (d) through (k) of § 1910.146.

(1) ***Safe work practices.*** The employer shall ensure the use of safe work practices for entry into, and work in, enclosed spaces and for rescue of employees from such spaces.

(2) ***Training.*** Each employee who enters an enclosed space or who serves as an attendant shall be trained in the hazards of enclosed-space entry, in enclosed-space entry procedures, and in enclosed-space rescue procedures.

(3) ***Rescue equipment.*** Employers shall provide equipment to ensure the prompt and safe rescue of employees from the enclosed space.

(4) ***Evaluating potential hazards.*** Before any entrance cover to an enclosed space is removed, the employer shall determine whether it is safe to do so by checking for the presence of any atmospheric pressure or temperature differences and by evaluating whether there might be a hazardous atmosphere in the space. Any conditions making it unsafe to remove the cover shall be eliminated before the cover is removed.

Note to paragraph (e)(4): The determination called for in this paragraph may consist of a check of the conditions that might foreseeably be in the enclosed space. For example, the cover could be checked to see if it is hot and, if

FIGURE 7.3 *(Continued)*

it is fastened in place, could be loosened gradually to release any residual pressure. An evaluation also needs to be made of whether conditions at the site could cause a hazardous atmosphere, such as an oxygen-deficient or flammable atmosphere, to develop within the space.

(5) ***Removing covers.*** When covers are removed from enclosed spaces, the opening shall be promptly guarded by a railing, temporary cover, or other barrier designed to prevent an accidental fall through the opening and to protect employees working in the space from objects entering the space.

(6) ***Hazardous atmosphere.*** Employees may not enter any enclosed space while it contains a hazardous atmosphere, unless the entry conforms to the permit-required confined spaces standard in § 1910.146.

(7) ***Attendants.*** While work is being performed in the enclosed space, an attendant with first-aid training shall be immediately available outside the enclosed space to provide assistance if a hazard exists because of traffic patterns in the area of the opening used for entry. The attendant is not precluded from performing other duties outside the enclosed space if these duties do not distract the attendant from: monitoring employees within the space or ensuring that it is safe for employees to enter and exit the space.

Note to paragraph (e)(7): See paragraph (t) of this section for additional requirements on attendants for work in manholes and vaults.

(8) ***Calibration of test instruments.*** Test instruments used to monitor atmospheres in enclosed spaces shall be kept in calibration and shall have a minimum accuracy of 10 percent.

(9) ***Testing for oxygen deficiency.*** Before an employee enters an enclosed space, the atmosphere in the enclosed space shall be tested for oxygen deficiency with a direct-reading meter or similar instrument, capable of collection and immediate analysis of data samples without the need for offsite evaluation. If continuous forced-air ventilation is provided, testing is not required provided that the procedures used ensure that employees are not exposed to the hazards posed by oxygen deficiency.

(10) ***Testing for flammable gases and vapors.*** Before an employee enters an enclosed space, the internal atmosphere shall be tested for flammable gases and vapors with a direct-reading meter or similar instrument capable of collection and immediate analysis of data samples without the need for off-site evaluation. This test shall be performed after the oxygen testing and ventilation required by paragraph (e)(9) of this section demonstrate that there is sufficient oxygen to ensure the accuracy of the test for flammability.

(11) ***Ventilation, and monitoring for flammable gases or vapors.*** If flammable gases or vapors are detected or if an oxygen deficiency is found, forced-air ventilation shall be used to maintain oxygen at a safe level and to prevent a hazardous concentration of flammable gases and vapors from accumulating. A continuous monitoring program to ensure that no increase in flammable gas or vapor concentration above safe levels occurs may be followed in lieu of ventilation if flammable gases or vapors are initially detected at safe levels.

Note to paragraph (e)(11): See the definition of "hazardous atmosphere" for guidance in determining whether a specific concentration of a substance is hazardous.

(12) ***Specific ventilation requirements.*** If continuous forced-air ventilation is used, it shall begin before entry is made and shall be maintained long enough for the employer to be able to demonstrate that a safe atmosphere exists before employees are allowed to enter the work area. The forced-air ventilation shall be so directed as to ventilate the immediate area where employees are present within the enclosed space and shall continue until all employees leave the enclosed space.

(13) ***Air supply.*** The air supply for the continuous forced-air ventilation shall be from a clean source and may not increase the hazards in the enclosed space.

(14) ***Open flames.*** If open flames are used in enclosed spaces, a test for flammable gases and vapors shall be made immediately before the open flame device is used and at least once per hour while the device is used in the space. Testing shall be conducted more frequently if conditions present in the enclosed space indicate that once per hour is insufficient to detect hazardous accumulations of flammable gases or vapors.

Note to paragraph (e)(14): See the definition of "hazardous atmosphere" for guidance in determining whether a specific concentration of a substance is hazardous.

Note to paragraph (e): Entries into enclosed spaces conducted in accordance with the permit-space entry requirements of paragraphs (d) through (k) of § 1910.146 are considered as complying with paragraph (e) of this section.

(f) ***Excavations.*** Excavation operations shall comply with Subpart P of Part 1926 of this chapter.

(g) ***Personal protective equipment.*** **(1)** ***General.*** Personal protective equipment shall meet the requirements of Subpart I of this part.

Note to paragraph (g)(1) of this section: Paragraph (h) of § 1910.132 sets employer payment obligations for the personal protective equipment required by this section, including, but not limited to, the fall protection equipment

FIGURE 7.3 *(Continued)*

required by paragraph (g)(2) of this section, the electrical protective equipment required by paragraph (l)(3) of this section, and the flame-resistant and arc-rated clothing and other protective equipment required by paragraph (l)(8) of this section.

(2) *Fall protection.* **(i)** Personal fall arrest systems shall meet the requirements of Subpart M of Part 1926 of this chapter.

(ii) Personal fall arrest equipment used by employees who are exposed to hazards from flames or electric arcs, as determined by the employer under paragraph (l)(8)(i) of this section, shall be capable of passing a drop test equivalent to that required by paragraph (g)(2)(iii)(L) of this section after exposure to an electric arc with a heat energy of 40±5 cal/cm^2.

(iii) Body belts and positioning straps for work-positioning equipment shall meet the following requirements:

(A) Hardware for body belts and positioning straps shall meet the following requirements:

(1) Hardware shall be made of dropforged steel, pressed steel, formed steel, or equivalent material.

(2) Hardware shall have a corrosion-resistant finish.

(3) Hardware surfaces shall be smooth and free of sharp edges.

(B) Buckles shall be capable of withstanding an 8.9-kilonewton (2,000-pound-force) tension test with a maximum permanent deformation no greater than 0.4 millimeters (0.0156 inches).

(C) D rings shall be capable of withstanding a 22-kilonewton (5,000-pound-force) tensile test without cracking or breaking.

(D) Snaphooks shall be capable of withstanding a 22-kilonewton (5,000-pound-force) tension test without failure.

Note to paragraph (g)(2)(iii)(D): Distortion of the snaphook sufficient to release the keeper is considered to be tensile failure of a snaphook.

(E) Top grain leather or leather substitute may be used in the manufacture of body belts and positioning straps; however, leather and leather substitutes may not be used alone as a load-bearing component of the assembly.

(F) Plied fabric used in positioning straps and in load-bearing parts of body belts shall be constructed in such a way that no raw edges are exposed and the plies do not separate.

(G) Positioning straps shall be capable of withstanding the following tests:

(1) A dielectric test of 819.7 volts, AC, per centimeter (25,000 volts per foot) for 3 minutes without visible deterioration;

(2) A leakage test of 98.4 volts, AC, per centimeter (3,000 volts per foot) with a leakage current of no more than 1 mA;

Note to paragraphs (g)(2)(iii)(G)(*1*) and (g)(2)(iii)(G)(*2*): Positioning straps that pass direct-current tests at equivalent voltages are considered as meeting this requirement.

(3) Tension tests of 20 kilonewtons (4,500 pounds-force) for sections free of buckle holes and of 15 kilonewtons (3,500 pounds-force) for sections with buckle holes;

(4) A buckle-tear test with a load of 4.4 kilonewtons (1,000 pounds-force); and

(5) A flammability test in accordance with Table R-2.

(H) The cushion part of the body belt shall contain no exposed rivets on the inside and shall be at least 76 millimeters (3 inches) in width.

(I) Tool loops shall be situated on the body of a body belt so that the 100 millimeters (4 inches) of the body belt that is in the center of the back, measuring from D ring to D ring, is free of tool loops and any other attachments.

TABLE R-2 Flammability Test

Test method	Criteria for passing the test
Vertically suspend a 500-mm (19.7-inch) length of strapping supporting a 100-kg (220.5-lb) weight. Use a butane or propane burner with a 76-mm (3-inch) flame. Direct the flame to an edge of the strapping at a distance of 25 mm (1 inch). Remove the flame after 5 seconds. Wait for any flames on the positioning strap to stop burning.	Any flames on the positioning strap shall self extinguish. The positioning strap shall continue to support the 100-kg (220.5-lb) mass.

FIGURE 7.3 *(Continued)*

(J) Copper, steel, or equivalent liners shall be used around the bars of D rings to prevent wear between these members and the leather or fabric enclosing them.

(K) Snaphooks shall be of the locking type meeting the following requirements:

(1) The locking mechanism shall first be released, or a destructive force shall be placed on the keeper, before the keeper will open.

(2) A force in the range of 6.7 N (1.5 lbf) to 17.8 N (4 lbf) shall be required to release the locking mechanism.

(3) With the locking mechanism released and with a force applied on the keeper against the face of the nose, the keeper may not begin to open with a force of 11.2 N (2.5 lbf) or less and shall begin to open with a maximum force of 17.8 N (4 lbf).

(L) Body belts and positioning straps shall be capable of withstanding a drop test as follows:

(1) The test mass shall be rigidly constructed of steel or equivalent material with a mass of 100 kg (220.5 lbm). For work-positioning equipment used by employees weighing more than 140 kg (310 lbm) fully equipped, the test mass shall be increased proportionately (that is, the test mass must equal the mass of the equipped worker divided by 1.4).

(2) For body belts, the body belt shall be fitted snugly around the test mass and shall be attached to the test structure anchorage point by means of a wire rope.

(3) For positioning straps, the strap shall be adjusted to its shortest length possible to accommodate the test and connected to the test-structure anchorage point at one end and to the test mass on the other end.

(4) The test mass shall be dropped an unobstructed distance of 1 meter (39.4 inches) from a supporting structure that will sustain minimal deflection during the test.

(5) Body belts shall successfully arrest the fall of the test mass and shall be capable of supporting the mass after the test.

(6) Positioning straps shall successfully arrest the fall of the test mass without breaking, and the arrest force may not exceed 17.8 kilonewtons (4,000 pounds-force). Additionally, snaphooks on positioning straps may not distort to such an extent that the keeper would release.

Note to paragraph (g)(2)(iii) of this section: When used by employees weighing no more than 140 kg (310 lbm) fully equipped, body belts and positioning straps that conform to American Society of Testing and Materials *Standard Specifications for Personal Climbing Equipment,* ASTM F887–12e1, are deemed to be in compliance with paragraph (g)(2)(iii) of this section.

(iv) The following requirements apply to the care and use of personal fall protection equipment.

(A) Work-positioning equipment shall be inspected before use each day to determine that the equipment is in safe working condition. Work-positioning equipment that is not in safe working condition may not be used.

Note to paragraph (g)(2)(iv)(A): Appendix F to this section contains guidelines for inspecting work-positioning equipment.

(B) Personal fall arrest systems shall be used in accordance with § 1926.502(d).

Note to paragraph (g)(2)(iv)(B): Fall protection equipment rigged to arrest falls is considered a fall arrest system and must meet the applicable requirements for the design and use of those systems. Fall protection equipment rigged for work positioning is considered work-positioning equipment and must meet the applicable requirements for the design and use of that equipment.

(C) The employer shall ensure that employees use fall protection systems as follows:

(1) Each employee working from an aerial lift shall use a fall restraint system or a personal fall arrest system. Paragraph (c)(2)(v) of § 1910.67 does not apply.

(2) Except as provided in paragraph (g)(2)(iv)(C)(3) of this section, each employee in elevated locations more than 1.2 meters (4 feet) above the ground on poles, towers, or similar structures shall use a personal fall arrest system, work-positioning equipment, or fall restraint system, as appropriate, if the employer has not provided other fall protection meeting Subpart D of this part.

(3) Until March 31, 2015, a qualified employee climbing or changing location on poles, towers, or similar structures need not use fall protection equipment, unless conditions, such as, but not limited to, ice, high winds, the design of the structure (for example, no provision for holding on with hands), or the presence of contaminants on the structure, could cause the employee to lose his or her grip or footing. On and after April 1, 2015, each qualified employee climbing or changing location on poles, towers, or similar structures must use fall protection equipment

FIGURE 7.3 *(Continued)*

unless the employer can demonstrate that climbing or changing location with fall protection is infeasible or creates a greater hazard than climbing or changing location without it.

Note 1 to paragraphs (g)(2)(iv)(C)(2) and (g)(2)(iv)(C)(*3*): These paragraphs apply to structures that support overhead electric power transmission and distribution lines and equipment. They do not apply to portions of buildings, such as loading docks, or to electric equipment, such as transformers and capacitors. Subpart D of this part contains the duty to provide fall protection associated with walking and working surfaces.

Note 2 to paragraphs (g)(2)(iv)(C)(2) and (g)(2)(iv)(C)(*3*): Until the employer ensures that employees are proficient in climbing and the use of fall protection under paragraph (a)(2)(viii) of this section, the employees are not considered "qualified employees" for the purposes of paragraphs (g)(2)(iv)(C)(2) and (g)(2)(iv)(C)(3) of this section. These paragraphs require unqualified employees (including trainees) to use fall protection any time they are more than 1.2 meters (4 feet) above the ground.

(D) On and after April 1, 2015, work-positioning systems shall be rigged so that an employee can free fall no more than 0.6 meters (2 feet).

(E) Anchorages for work-positioning equipment shall be capable of supporting at least twice the potential impact load of an employee's fall, or 13.3 kilonewtons (3,000 pounds-force), whichever is greater.

Note to paragraph (g)(2)(iv)(E): Wood-pole fall-restriction devices meeting American Society of Testing and Materials *Standard Specifications for Personal Climbing Equipment,* ASTM F887–12e1, are deemed to meet the anchorage-strength requirement when they are used in accordance with manufacturers' instructions.

(F) Unless the snaphook is a locking type and designed specifically for the following connections, snaphooks on work-positioning equipment may not be engaged:

(1) Directly to webbing, rope, or wire rope;

(2) To each other;

(3) To a D ring to which another snaphook or other connector is attached;

(4) To a horizontal lifeline; or

(5) To any object that is incompatibly shaped or dimensioned in relation to the snaphook such that accidental disengagement could occur should the connected object sufficiently depress the snaphook keeper to allow release of the object.

(h) *Portable ladders and platforms.* (1) *General.* Requirements for portable ladders contained in Subpart D of this part apply in addition to the requirements of paragraph (h) of this section, except as specifically noted in paragraph (h)(2) of this section.

(2) *Special ladders and platforms.* Portable ladders used on structures or conductors in conjunction with overhead line work need not meet § 1910.25(d)(2)(i) and (d)(2)(iii) or § 1910.26(c)(3)(iii). Portable ladders and platforms used on structures or conductors in conjunction with overhead line work shall meet the following requirements:

(i) In the configurations in which they are used, portable platforms shall be capable of supporting without failure at least 2.5 times the maximum intended load.

(ii) Portable ladders and platforms may not be loaded in excess of the working loads for which they are designed.

(iii) Portable ladders and platforms shall be secured to prevent them from becoming dislodged.

(iv) Portable ladders and platforms may be used only in applications for which they are designed.

(3) *Conductive ladders.* Portable metal ladders and other portable conductive ladders may not be used near exposed energized lines or equipment. However, in specialized high-voltage work, conductive ladders shall be used when the employer demonstrates that nonconductive ladders would present a greater hazard to employees than conductive ladders.

(i) *Hand and portable power equipment.* (1) *General.* Paragraph (i)(2) of this section applies to electric equipment connected by cord and plug. Paragraph (i)(3) of this section applies to portable and vehicle-mounted generators used to supply cord- and plug-connected equipment. Paragraph (i)(4) of this section applies to hydraulic and pneumatic tools.

(2) *Cord- and plug-connected equipment.* Cord- and plug-connected equipment not covered by Subpart S of this part shall comply with one of the following instead of § 1910.243(a)(5):

(i) The equipment shall be equipped with a cord containing an equipment grounding conductor connected to the equipment frame and to a means for grounding the other end of the conductor (however, this option may not be used where the introduction of the ground into the work environment increases the hazard to an employee); or

(ii) The equipment shall be of the double-insulated type conforming to Subpart S of this part; or

(iii) The equipment shall be connected to the power supply through an isolating transformer with an ungrounded secondary of not more than 50 volts.

FIGURE 7.3 (*Continued*)

(3) *Portable and vehicle-mounted generators.* Portable and vehicle-mounted generators used to supply cord- and plug-connected equipment covered by paragraph (i)(2) of this section shall meet the following requirements:

(i) The generator may only supply equipment located on the generator or the vehicle and cord- and plug-connected equipment through receptacles mounted on the generator or the vehicle.

(ii) The non-current-carrying metal parts of equipment and the equipment grounding conductor terminals of the receptacles shall be bonded to the generator frame.

(iii) For vehicle-mounted generators, the frame of the generator shall be bonded to the vehicle frame.

(iv) Any neutral conductor shall be bonded to the generator frame.

(4) *Hydraulic and pneumatic tools.* **(i)** Safe operating pressures for hydraulic and pneumatic tools, hoses, valves, pipes, filters, and fittings may not be exceeded.

Note to paragraph (i)(4)(i): If any hazardous defects are present, no operating pressure is safe, and the hydraulic or pneumatic equipment involved may not be used. In the absence of defects, the maximum rated operating pressure is the maximum safe pressure.

(ii) A hydraulic or pneumatic tool used where it may contact exposed energized parts shall be designed and maintained for such use.

(iii) The hydraulic system supplying a hydraulic tool used where it may contact exposed live parts shall provide protection against loss of insulating value, for the voltage involved, due to the formation of a partial vacuum in the hydraulic line.

Note to paragraph (i)(4)(iii): Use of hydraulic lines that do not have check valves and that have a separation of more than 10.7 meters (35 feet) between the oil reservoir and the upper end of the hydraulic system promotes the formation of a partial vacuum.

(iv) A pneumatic tool used on energized electric lines or equipment, or used where it may contact exposed live parts, shall provide protection against the accumulation of moisture in the air supply.

(v) Pressure shall be released before connections are broken, unless quick-acting, self-closing connectors are used.

(vi) Employers must ensure that employees do not use any part of their bodies to locate, or attempt to stop, a hydraulic leak.

(vii) Hoses may not be kinked.

(j) *Live-line tools.* **(1) *Design of tools.*** Live-line tool rods, tubes, and poles shall be designed and constructed to withstand the following minimum tests:

(i) If the tool is made of fiberglass-reinforced plastic (FRP), it shall withstand 328,100 volts per meter (100,000 volts per foot) of length for 5 minutes, or

Note to paragraph (j)(1)(i): Live-line tools using rod and tube that meet ASTM F711–02 (2007), *Standard Specification for Fiberglass-Reinforced Plastic (FRP) Rod and Tube Used in Live Line Tools,* are deemed to comply with paragraph (j)(1) of this section.

(ii) If the tool is made of wood, it shall withstand 246,100 volts per meter (75,000 volts per foot) of length for 3 minutes, or

(iii) The tool shall withstand other tests that the employer can demonstrate are equivalent.

(2) *Condition of tools.* **(i)** Each live-line tool shall be wiped clean and visually inspected for defects before use each day.

(ii) If any defect or contamination that could adversely affect the insulating qualities or mechanical integrity of the live-line tool is present after wiping, the tool shall be removed from service and examined and tested according to paragraph (j)(2)(iii) of this section before being returned to service.

(iii) Live-line tools used for primary employee protection shall be removed from service every 2 years, and whenever required under paragraph (j)(2)(ii) of this section, for examination, cleaning, repair, and testing as follows:

(A) Each tool shall be thoroughly examined for defects.

(B) If a defect or contamination that could adversely affect the insulating qualities or mechanical integrity of the live-line tool is found, the tool shall be repaired and refinished or shall be permanently removed from service. If no such defect or contamination is found, the tool shall be cleaned and waxed.

(C) The tool shall be tested in accordance with paragraphs (j)(2)(iii)(D) and (j)(2)(iii)(E) of this section under the following conditions:

(1) After the tool has been repaired or refinished; and

(2) After the examination if repair or refinishing is not performed, unless the tool is made of FRP rod or foam-filled FRP tube and the employer can demonstrate that the tool has no defects that could cause it to fail during use.

FIGURE 7.3 *(Continued)*

(D) The test method used shall be designed to verify the tool's integrity along its entire working length and, if the tool is made of fiberglass-reinforced plastic, its integrity under wet conditions.

(E) The voltage applied during the tests shall be as follows:

(1) 246,100 volts per meter (75,000 volts per foot) of length for 1 minute if the tool is made of fiberglass, or

(2) 164,000 volts per meter (50,000 volts per foot) of length for 1 minute if the tool is made of wood, or

(3) Other tests that the employer can demonstrate are equivalent.

Note to paragraph (j)(2): Guidelines for the examination, cleaning, repairing, and in-service testing of live-line tools are specified in the Institute of Electrical and Electronics Engineers' *IEEE Guide for Maintenance Methods on Energized Power Lines,* IEEE Std 516–2009.

(k) *Materials handling and storage.* (1) *General.* Materials handling and storage shall comply with applicable material-handling and material-storage requirements in this part, including those in Subpart N of this part.

(2) *Materials storage near energized lines or equipment.* (i) In areas to which access is not restricted to qualified persons only, materials or equipment may not be stored closer to energized lines or exposed energized parts of equipment than the following distances, plus a distance that provides for the maximum sag and side swing of all conductors and for the height and movement of material-handling equipment:

(A) For lines and equipment energized at 50 kilovolts or less, the distance is 3.05 meters (10 feet).

(B) For lines and equipment energized at more than 50 kilovolts, the distance is 3.05 meters (10 feet) plus 0.10 meter (4 inches) for every 10 kilovolts over 50 kilovolts.

(ii) In areas restricted to qualified employees, materials may not be stored within the working space about energized lines or equipment.

Note to paragraph (k)(2)(ii): Paragraphs (u)(1) and (v)(3) of this section specify the size of the working space.

(l) *Working on or near exposed energized parts.* This paragraph applies to work on exposed live parts, or near enough to them to expose the employee to any hazard they present.

(1) *General.* (i) Only qualified employees may work on or with exposed energized lines or parts of equipment.

(ii) Only qualified employees may work in areas containing unguarded, uninsulated energized lines or parts of equipment operating at 50 volts or more.

(iii) Electric lines and equipment shall be considered and treated as energized unless they have been deenergized in accordance with paragraph (d) or (m) of this section.

(2) *At least two employees.* (i) Except as provided in paragraph (l)(2)(ii) of this section, at least two employees shall be present while any employees perform the following types of work:

(A) Installation, removal, or repair of lines energized at more than 600 volts, (B) Installation, removal, or repair of deenergized lines if an employee is exposed to contact with other parts energized at more than 600 volts,

(C) Installation, removal, or repair of equipment, such as transformers, capacitors, and regulators, if an employee is exposed to contact with parts energized at more than 600 volts,

(D) Work involving the use of mechanical equipment, other than insulated aerial lifts, near parts energized at more than 600 volts, and

(E) Other work that exposes an employee to electrical hazards greater than, or equal to, the electrical hazards posed by operations listed specifically in paragraphs (l)(2)(i)(A) through (l)(2)(i)(D) of this section.

(ii) Paragraph (l)(2)(i) of this section does not apply to the following operations:

(A) Routine circuit switching, when the employer can demonstrate that conditions at the site allow safe performance of this work,

(B) Work performed with live-line tools when the position of the employee is such that he or she is neither within reach of, nor otherwise exposed to contact with, energized parts, and

(C) Emergency repairs to the extent necessary to safeguard the general public.

(3) *Minimum approach distances.* (i) The employer shall establish minimum approach distances no less than the distances computed by Table R-3 for ac systems or Table R-8 for dc systems.

(ii) No later than April 1, 2015, for voltages over 72.5 kilovolts, the employer shall determine the maximum anticipated per-unit transient overvoltage, phase-to-ground, through an engineering analysis or assume a maximum anticipated per-unit transient overvoltage, phase-to-ground, in accordance with Table R-9. When the employer uses portable protective gaps to control the maximum transient overvoltage, the value of the maximum anticipated per-unit transient overvoltage, phase-to-ground, must provide for five standard deviations between the statistical sparkover voltage of the gap and the statistical withstand voltage corresponding to the electrical component of the minimum approach distance. The employer shall make any engineering analysis conducted to determine maximum anticipated per-unit transient overvoltage available upon request to employees and to the Assistant Secretary or designee for examination and copying.

FIGURE 7.3 *(Continued)*

TABLE R-3 AC Live-Line Work Minimum Approach Distance
[The minimum approach distance (MAD; in meters) shall conform to the following equations.]

For phase-to-phase system voltages of 50 V to 300 V:[1] MAD = avoid contact	
For phase-to-phase system voltages of 301 V to 5 kV:[1] *MAD* = *M* + *D*, where *D* = 0.02 m	the electrical component of the minimum approach distance.
M = 0.31 m for voltages up to 750 V and 0.61 m otherwise..........	the inadvertent movement factor.
For phase-to-phase system voltages ot 5.1 kV to 72.5 kV:[1,4] *MAD* = *M* + *AD*, where *M* = 0.61 m..........	the inadvertent movement factor.
A = the applicable value from Table R-5..........	the altitude correction factor.
D = the value from Table R-4 corresponding to the voltage and exposure or the value of the electrical component of the minimum approach distance calculated using the method provided in Appendix B to this section.	the electrical component of the minimum approach distance.

For phase-to-phase system voltages of more than 72.5 kV, nominal:[2,4]

MAD = *0.3048(C +)*$V_{L\text{-}G}$TA + M, where

C = 0.01 for phase-to-ground exposures that the employer can demonstrate consist only of air across the approach distance (gap),
0.01 for phase-to-phase exposures if the employer can demonstrate that no insulated tool spans the gap and that no large conductive object is in the gap, or
0.011 otherwise

$V_{L\text{-}G}$ = phase-to-ground rms voltage, in kV

T = maximum anticipated per-unit transient overvoltage; for phase-to-ground exposures, *T* equals $T_{L\text{-}G}$, the maximum per-unit transient overvoltage, phase-to-ground, determined by the employer under paragraph (l)(3)(ii) of this section; for phase-to-phase exposures, *T* equals $1.35T_{L\text{-}G} + 0.45$

A = altitude correction factor from Table R-5

M = 0.31 m, the inadvertent movement factor

a = saturation factor, as follows:

Phase-to-Ground Exposures				
$V_{Peak} = T_{L\text{-}G}V_{L\text{-}G}\sqrt{2}$ ..... *a*..........	635 kV or less 0	635.1 to 915 kV (V_{Peak}-635)/140,000	915.1 to 1,050 kV (V_{Peak}-645)/135,000	More than 1,050 kV (V_{Peak}-675)/125,000

Phase-to-Phase Exposures[3]					
$V_{Peak} = (1.35T_{L\text{-}G} + 0.45)V_{L\text{-}G}\sqrt{2}$.......... *a*..........	630 kV or less 0	630.1 to 846 kV (V_{Peak}-630)/155,000	848.1 to 1,131 kV (V_{Peak}-633.6)/152,207	1,131.1 to 1,485 kV (V_{Peak}-628)/153,846	More than 1,485 kV (V_{Peak}-350.5)/203,666

[1]Employers may use the minimum approach distances in Table R-6. If the worksite is at an elevation of more than 900 meters [3,000 feet), see footnote 1 to Table R-6.

[2]Employers may use the minimum approach distances in Table R-7, except that the employer may not use the minimun approach distances in Table R-7 for phase-to-phase exposures if an insulated tool spans the gap or if any large conductive object is in the gap. If the worksite is at an elevation of more than 900 meters (3,000 feet), see footnote 1 to Table R-7. Employers may use the minimum approach distances in Table 6 through Table 13 in Appendix B to this section, which calculated MAD for various values of *T*, provided the employer follows the notes to those tables.

[3]Use the equations tor phase-to-ground exposures (with V_{Peak} for phase-to-phase exposures) unless the employer can demonstrate that no insulated tool spans the gap and that no large conductive object is in the gap.

[4]Until March 31, 2015, employers may use the minimum approach distances in Table 6 through Table 13 In Appendix B to this section.

FIGURE 7.3 *(Continued)*

TABLE R-4 Electrical Component of the Minimum Approach Distance at 5.1 to 72.5 kV [D; in meters]

Nominal voltage (kV) phase-to-phase	Phase-to-ground exposure	Phase-to-phase exposure
	D (m)	D (m)
5.1 to 15.0	0.04	0.07
15.1 to 36.0	0.16	0.28
36.1 to 46.0	0.23	0.37
46.1 to 72.5	0.39	0.59

Note to paragraph (l)(3)(ii): See Appendix B to this section for information on how to calculate the maximum anticipated per-unit transient overvoltage, phase-to-ground, when the employer uses portable protective gaps to reduce maximum transient overvoltages.

(iii) The employer shall ensure that no employee approaches or takes any conductive object closer to exposed energized parts than the employer's established minimum approach distance, unless:

(A) The employee is insulated from the energized part (rubber insulating gloves or rubber insulating gloves and sleeves worn in accordance with paragraph (l)(4) of this section constitutes insulation of the employee from the energized part upon which the employee is working provided that the employee has control of the part in a manner sufficient to prevent exposure to uninsulated portions of the employee's body), or

(B) The energized part is insulated from the employee and from any other conductive object at a different potential, or

(C) The employee is insulated from any other exposed conductive object in accordance with the requirements for live-line barehand work in paragraph (q)(3) of this section.

(4) ***Type of insulation.*** **(i)** When an employee uses rubber insulating gloves as insulation from energized parts (under paragraph (l)(3)(iii)(A) of this section), the employer shall ensure that the employee also uses rubber insulating sleeves. However, an employee need not use rubber insulating sleeves if:

(A) Exposed energized parts on which the employee is not working are insulated from the employee; and

(B) When installing insulation for purposes of paragraph (l)(4)(i)(A) of this section, the employee installs the insulation from a position that does not expose his or her upper arm to contact with other energized parts.

(ii) When an employee uses rubber insulating gloves or rubber insulating gloves and sleeves as insulation from energized parts (under paragraph (l)(3)(iii)(A) of this section), the employer shall ensure that the employee:

(A) Puts on the rubber insulating gloves and sleeves in a position where he or she cannot reach into the minimum approach distance, established by the employer under paragraph (l)(3)(i) of this section; and

(B) Does not remove the rubber insulating gloves and sleeves until he or she is in a position where he or she cannot reach into the minimum approach distance, established by the employer under paragraph (l)(3)(i) of this section.

(5) ***Working position.*** **(i)** The employer shall ensure that each employee, to the extent that other safety-related conditions at the worksite permit, works in a position from which a slip or shock will not bring the employee's body into contact with exposed, uninsulated parts energized at a potential different from the employee's.

(ii) When an employee performs work near exposed parts energized at more than 600 volts, but not more than 72.5 kilovolts, and is not wearing rubber insulating gloves, being protected by insulating equipment covering the energized parts, performing work using live-line tools, or performing live-line barehand work under paragraph (q)(3) of this section, the employee shall work from a position where he or she cannot reach into the minimum approach distance, established by the employer under paragraph (l)(3)(i) of this section.

(6) ***Making connections.*** The employer shall ensure that employees make connections as follows:

(i) In connecting deenergized equipment or lines to an energized circuit by means of a conducting wire or device, an employee shall first attach the wire to the deenergized part;

(ii) When disconnecting equipment or lines from an energized circuit by means of a conducting wire or device, an employee shall remove the source end first; and

(iii) When lines or equipment are connected to or disconnected from energized circuits, an employee shall keep loose conductors away from exposed energized parts.

FIGURE 7.3 *(Continued)*

(7) *Conductive articles.* When an employee performs work within reaching distance of exposed energized parts of equipment, the employer shall ensure that the employee removes or renders nonconductive all exposed conductive articles, such as keychains or watch chains, rings, or wrist watches or bands, unless such articles do not increase the hazards associated with contact with the energized parts.

(8) *Protection from flames and electric arcs.* (i) The employer shall assess the workplace to identify employees exposed to hazards from flames or from electric arcs.

(ii) For each employee exposed to hazards from electric arcs, the employer shall make a reasonable estimate of the incident heat energy to which the employee would be exposed.

Note 1 to paragraph (l)(8)(ii): Appendix E to this section provides guidance on estimating available heat energy. The Occupational Safety and Health Administration will deem employers following the guidance in Appendix E to this section to be in compliance with paragraph (l)(8)(ii) of this section. An employer may choose a method of calculating incident heat energy not included in Appendix E to this section if the chosen method reasonably predicts the incident energy to which the employee would be exposed.

Note 2 to paragraph (l)(8)(ii): This paragraph does not require the employer to estimate the incident heat energy exposure for every job task performed by each employee. The employer may make broad estimates that cover multiple system areas provided the employer uses reasonable assumptions about the energy-exposure distribution throughout the system and provided the estimates represent the maximum employee exposure for those areas. For example, the employer could estimate the heat energy just outside a substation feeding a radial distribution system and use that estimate for all jobs performed on that radial system.

(iii) The employer shall ensure that each employee who is exposed to hazards from flames or electric arcs does not wear clothing that could melt onto his or her skin or that could ignite and continue to burn when exposed to flames or the heat energy estimated under paragraph (l)(8)(ii) of this section.

Note to paragraph (l)(8)(iii) of this section: This paragraph prohibits clothing made from acetate, nylon, polyester, rayon and polypropylene, either alone or in blends, unless the employer demonstrates that the fabric has been treated to withstand the conditions that may be encountered by the employee or that the employee wears the clothing in such a manner as to eliminate the hazard involved.

(iv) The employer shall ensure that the outer layer of clothing worn by an employee, except for clothing not required to be arc rated under paragraphs (l)(8)(v)(A) through (l)(8)(v)(E) of this section, is flame resistant under any of the following conditions:

(A) The employee is exposed to contact with energized circuit parts operating at more than 600 volts,

(B) An electric arc could ignite flammable material in the work area that, in turn, could ignite the employee's clothing,

(C) Molten metal or electric arcs from faulted conductors in the work area could ignite the employee's clothing, or

Note to paragraph (l)(8)(iv)(C): This paragraph does not apply to conductors that are capable of carrying, without failure, the maximum available fault current for the time the circuit protective devices take to interrupt the fault.

(D) The incident heat energy estimated under paragraph (l)(8)(ii) of this section exceeds 2.0 cal/cm^2.

(v) The employer shall ensure that each employee exposed to hazards from electric arcs wears protective clothing and other protective equipment with an arc rating greater than or equal to the heat energy estimated under paragraph (l)(8)(ii) of this section whenever that estimate exceeds 2.0 cal/cm^2. This protective equipment shall cover the employee's entire body, except as follows:

(A) Arc-rated protection is not necessary for the employee's hands when the employee is wearing rubber insulating gloves with protectors or, if the estimated incident energy is no more than 14 cal/cm^2, heavy-duty leather work gloves with a weight of at least 407 g/m^2 (12 oz/yd^2),

(B) Arc-rated protection is not necessary for the employee's feet when the employee is wearing heavy-duty work shoes or boots,

(C) Arc-rated protection is not necessary for the employee's head when the employee is wearing head protection meeting § 1910.135 if the estimated incident energy is less than 9 cal/cm^2 for exposures involving single-phase arcs in open air or 5 cal/cm^2 for other exposures,

(D) The protection for the employee's head may consist of head protection meeting § 1910.135 and a face-shield with a minimum arc rating of 8 cal/cm^2 if the estimated incident-energy exposure is less than 13 cal/cm^2 for exposures involving single-phase arcs in open air or 9 cal/cm^2 for other exposures, and

(E) For exposures involving single-phase arcs in open air, the arc rating for the employee's head and face protection may be 4 cal/cm^2 less than the estimated incident energy.

FIGURE 7.3 *(Continued)*

Note to paragraph (l)(8): See Appendix E to this section for further information on the selection of appropriate protection.

(vi) ***Dates.*** **(A)** The obligation in paragraph (l)(8)(ii) of this section for the employer to make reasonable estimates of incident energy commences January 1, 2015.

(B) The obligation in paragraph (l)(8)(iv)(D) of this section for the employer to ensure that the outer layer of clothing worn by an employee is flame-resistant when the estimated incident heat energy exceeds 2.0 cal/cm^2 commences April 1, 2015.

(C) The obligation in paragraph (l)(8)(v) of this section for the employer to ensure that each employee exposed to hazards from electric arcs wears the required arc-rated protective equipment commences April 1, 2015.

(9) ***Fuse handling.*** When an employee must install or remove fuses with one or both terminals energized at more than 300 volts, or with exposed parts energized at more than 50 volts, the employer shall ensure that the employee uses tools or gloves rated for the voltage. When an employee installs or removes expulsion-type fuses with one or both terminals energized at more than 300 volts, the employer shall ensure that the employee wears eye protection meeting the requirements of Subpart I of this part, uses a tool rated for the voltage, and is clear of the exhaust path of the fuse barrel.

(10) ***Covered (noninsulated) conductors.*** The requirements of this section that pertain to the hazards of exposed live parts also apply when an employee performs work in proximity to covered (noninsulated) wires.

(11) ***Non-current-carrying metal parts.*** Non-current-carrying metal parts of equipment or devices, such as transformer cases and circuit-breaker housings, shall be treated as energized at the highest voltage to which these parts are exposed, unless the employer inspects the installation and determines that these parts are grounded before employees begin performing the work.

(12) ***Opening and closing circuits under load.*** **(i)** The employer shall ensure that devices used by employees to open circuits under load conditions are designed to interrupt the current involved.

(ii) The employer shall ensure that devices used by employees to close circuits under load conditions are designed to safely carry the current involved.

(m) ***Deenergizing lines and equipment for employee protection.*** **(1)** ***Application.*** Paragraph (m) of this section applies to the deenergizing of transmission and distribution lines and equipment for the purpose of protecting employees. See paragraph (d) of this section for requirements on the control of hazardous energy sources used in the generation of electric energy. Conductors and parts of electric equipment that have been deenergized under procedures other than those required by paragraph (d) or (m) of this section, as applicable, shall be treated as energized.

(2) ***General.*** **(i)** If a system operator is in charge of the lines or equipment and their means of disconnection, the employer shall designate one employee in the crew to be in charge of the clearance and shall comply with all of the requirements of paragraph (m)(3) of this section in the order specified.

(ii) If no system operator is in charge of the lines or equipment and their means of disconnection, the employer shall designate one employee in the crew to be in charge of the clearance and to perform the functions that the system operator would otherwise perform under paragraph (m) of this section. All of the requirements of paragraph (m)(3) of this section apply, in the order specified, except as provided in paragraph (m)(2)(iii) of this section.

(iii) If only one crew will be working on the lines or equipment and if the means of disconnection is accessible and visible to, and under the sole control of, the employee in charge of the clearance, paragraphs (m)(3)(i), (m)(3)(iii), and (m)(3)(v) of this section do not apply. Additionally, the employer does not need to use the tags required by the remaining provisions of paragraph (m)(3) of this section.

(iv) If two or more crews will be working on the same lines or equipment, then:

(A) The crews shall coordinate their activities under paragraph (m) of this section with a single employee in charge of the clearance for all of the crews and follow the requirements of paragraph (m) of this section as if all of the employees formed a single crew, or

(B) Each crew shall independently comply with paragraph (m) of this section and, if there is no system operator in charge of the lines or equipment, shall have separate tags and coordinate deenergizing and reenergizing the lines and equipment with the other crews.

(v) The employer shall render any disconnecting means that are accessible to individuals outside the employer's control (for example, the general public) inoperable while the disconnecting means are open for the purpose of protecting employees.

(3) ***Deenergizing lines and equipment.*** **(i)** The employee that the employer designates pursuant to paragraph (m)(2) of this section as being in charge of the clearance shall make a request of the system operator to deenergize

FIGURE 7.3 *(Continued)*

the particular section of line or equipment. The designated employee becomes the employee in charge (as this term is used in paragraph (m)(3) of this section) and is responsible for the clearance.

(ii) The employer shall ensure that all switches, disconnectors, jumpers, taps, and other means through which known sources of electric energy may be supplied to the particular lines and equipment to be deenergized are open. The employer shall render such means inoperable, unless its design does not so permit, and then ensure that such means are tagged to indicate that employees are at work.

(iii) The employer shall ensure that automatically and remotely controlled switches that could cause the opened disconnecting means to close are also tagged at the points of control. The employer shall render the automatic or remote control feature inoperable, unless its design does not so permit.

(iv) The employer need not use the tags mentioned in paragraphs (m)(3)(ii) and (m)(3)(iii) of this section on a network protector for work on the primary feeder for the network protector's associated network transformer when the employer can demonstrate all of the following conditions:

(A) Every network protector is maintained so that it will immediately trip open if closed when a primary conductor is deenergized;

(B) Employees cannot manually place any network protector in a closed position without the use of tools, and any manual override position is blocked, locked, or otherwise disabled; and

(C) The employer has procedures for manually overriding any network protector that incorporate provisions for determining, before anyone places a network protector in a closed position, that: The line connected to the network protector is not deenergized for the protection of any employee working on the line; and (if the line connected to the network protector is not deenergized for the protection of any employee working on the line) the primary conductors for the network protector are energized.

(v) Tags shall prohibit operation of the disconnecting means and shall indicate that employees are at work.

(vi) After the applicable requirements in paragraphs (m)(3)(i) through (m)(3)(v) of this section have been followed and the system operator gives a clearance to the employee in charge, the employer shall ensure that the lines and equipment are deenergized by testing the lines and equipment to be worked with a device designed to detect voltage.

(vii) The employer shall ensure the installation of protective grounds as required by paragraph (n) of this section.

(viii) After the applicable requirements of paragraphs (m)(3)(i) through (m)(3)(vii) of this section have been followed, the lines and equipment involved may be considered deenergized.

(ix) To transfer the clearance, the employee in charge (or the employee's supervisor if the employee in charge must leave the worksite due to illness or other emergency) shall inform the system operator and employees in the crew; and the new employee in charge shall be responsible for the clearance.

(x) To release a clearance, the employee in charge shall:

(A) Notify each employee under that clearance of the pending release of the clearance;

(B) Ensure that all employees under that clearance are clear of the lines and equipment;

(C) Ensure that all protective grounds protecting employees under that clearance have been removed; and

(D) Report this information to the system operator and then release the clearance.

(xi) Only the employee in charge who requested the clearance may release the clearance, unless the employer transfers responsibility under paragraph (m)(3)(ix) of this section.

(xii) No one may remove tags without the release of the associated clearance as specified under paragraphs (m)(3)(x) and (m)(3)(xi) of this section.

(xiii) The employer shall ensure that no one initiates action to reenergize the lines or equipment at a point of disconnection until all protective grounds have been removed, all crews working on the lines or equipment release their clearances, all employees are clear of the lines and equipment, and all protective tags are removed from that point of disconnection.

(n) ***Grounding for the protection of employees.*** **(1)** ***Application.*** Paragraph (n) of this section applies to grounding of generation, transmission, and distribution lines and equipment for the purpose of protecting employees. Paragraph (n)(4) of this section also applies to protective grounding of other equipment as required elsewhere in this section.

Note to paragraph (n)(1): This paragraph covers grounding of generation, transmission, and distribution lines and equipment when this section requires protective grounding and whenever the employer chooses to ground such lines and equipment for the protection of employees.

(2) ***General.*** For any employee to work transmission and distribution lines or equipment as deenergized, the employer shall ensure that the lines or equipment are deenergized under the provisions of paragraph (m) of this section and shall ensure proper grounding of the lines or equipment as specified in paragraphs (n)(3) through (n)

FIGURE 7.3 *(Continued)*

(8) of this section. However, if the employer can demonstrate that installation of a ground is impracticable or that the conditions resulting from the installation of a ground would present greater hazards to employees than working without grounds, the lines and equipment may be treated as deenergized provided that the employer establishes that all of the following conditions apply:

(i) The employer ensures that the lines and equipment are deenergized under the provisions of paragraph (m) of this section.

(ii) There is no possibility of contact with another energized source.

(iii) The hazard of induced voltage is not present.

(3) *Equipotential zone.* Temporary protective grounds shall be placed at such locations and arranged in such a manner that the employer can demonstrate will prevent each employee from being exposed to hazardous differences in electric potential.

Note to paragraph (n)(3): Appendix C to this section contains guidelines for establishing the equipotential zone required by this paragraph. The Occupational Safety and Health Administration will deem grounding practices meeting these guidelines as complying with paragraph (n)(3) of this section.

(4) *Protective grounding equipment.* (i) Protective grounding equipment shall be capable of conducting the maximum fault current that could flow at the point of grounding for the time necessary to clear the fault.

(ii) Protective grounding equipment shall have an ampacity greater than or equal to that of No. 2 AWG copper.

(iii) Protective grounds shall have an impedance low enough so that they do not delay the operation of protective devices in case of accidental energizing of the lines or equipment.

Note to paragraph (n)(4): American Society for Testing and Materials *Standard Specifications for Temporary Protective Grounds to Be Used on De-Energized Electric Power Lines and Equipment,* ASTM F855–09, contains guidelines for protective grounding equipment. The Institute of Electrical Engineers *Guide for Protective Grounding of Power Lines,* IEEE Std 1048–2003, contains guidelines for selecting and installing protective grounding equipment.

(5) *Testing.* The employer shall ensure that, unless a previously installed ground is present, employees test lines and equipment and verify the absence of nominal voltage before employees install any ground on those lines or that equipment.

(6) *Connecting and removing grounds.* (i) The employer shall ensure that, when an employee attaches a ground to a line or to equipment, the employee attaches the ground-end connection first and then attaches the other end by means of a live-line tool. For lines or equipment operating at 600 volts or less, the employer may permit the employee to use insulating equipment other than a live-line tool if the employer ensures that the line or equipment is not energized at the time the ground is connected or if the employer can demonstrate that each employee is protected from hazards that may develop if the line or equipment is energized.

(ii) The employer shall ensure that, when an employee removes a ground, the employee removes the grounding device from the line or equipment using a live-line tool before he or she removes the ground-end connection. For lines or equipment operating at 600 volts or less, the employer may permit the employee to use insulating equipment other than a live-line tool if the employer ensures that the line or equipment is not energized at the time the ground is disconnected or if the employer can demonstrate that each employee is protected from hazards that may develop if the line or equipment is energized.

(7) *Additional precautions.* The employer shall ensure that, when an employee performs work on a cable at a location remote from the cable terminal, the cable is not grounded at the cable terminal if there is a possibility of hazardous transfer of potential should a fault occur.

(8) *Removal of grounds for test.* The employer may permit employees to remove grounds temporarily during tests. During the test procedure, the employer shall ensure that each employee uses insulating equipment, shall isolate each employee from any hazards involved, and shall implement any additional measures necessary to protect each exposed employee in case the previously grounded lines and equipment become energized.

(o) *Testing and test facilities.* (1) *Application.* Paragraph (o) of this section provides for safe work practices for high-voltage and high-power testing performed in laboratories, shops, and substations, and in the field and on electric transmission and distribution lines and equipment. It applies only to testing involving interim measurements using high voltage, high power, or combinations of high voltage and high power, and not to testing involving continuous measurements as in routine metering, relaying, and normal line work.

Note to paragraph (o)(1): OSHA considers routine inspection and maintenance measurements made by qualified employees to be routine line work not included in the scope of paragraph (o) of this section, provided that the hazards related to the use of intrinsic high-voltage or high-power sources require only the normal precautions associated with routine work specified in the other paragraphs of this section. Two typical examples of such excluded test work procedures are "phasing-out" testing and testing for a "no-voltage" condition.

FIGURE 7.3 *(Continued)*

(2) *General requirements.* **(i)** The employer shall establish and enforce work practices for the protection of each worker from the hazards of high-voltage or high-power testing at all test areas, temporary and permanent. Such work practices shall include, as a minimum, test area safeguarding, grounding, the safe use of measuring and control circuits, and a means providing for periodic safety checks of field test areas.

(ii) The employer shall ensure that each employee, upon initial assignment to the test area, receives training in safe work practices, with retraining provided as required by paragraph (a)(2) of this section.

(3) *Safeguarding of test areas.* **(i)** The employer shall provide safeguarding within test areas to control access to test equipment or to apparatus under test that could become energized as part of the testing by either direct or inductive coupling and to prevent accidental employee contact with energized parts.

(ii) The employer shall guard permanent test areas with walls, fences, or other barriers designed to keep employees out of the test areas.

(iii) In field testing, or at a temporary test site not guarded by permanent fences and gates, the employer shall ensure the use of one of the following means to prevent employees without authorization from entering:

(A) Distinctively colored safety tape supported approximately waist high with safety signs attached to it,

(B) A barrier or barricade that limits access to the test area to a degree equivalent, physically and visually, to the barricade specified in paragraph (o)(3)(iii)(A) of this section, or (C) One or more test observers stationed so that they can monitor the entire area.

(iv) The employer shall ensure the removal of the safeguards required by paragraph (o)(3)(iii) of this section when employees no longer need the protection afforded by the safeguards.

(4) *Grounding practices.* **(i)** The employer shall establish and implement safe grounding practices for the test facility.

(A) The employer shall maintain at ground potential all conductive parts accessible to the test operator while the equipment is operating at high voltage.

(B) Wherever ungrounded terminals of test equipment or apparatus under test may be present, they shall be treated as energized until tests demonstrate that they are deenergized.

(ii) The employer shall ensure either that visible grounds are applied automatically, or that employees using properly insulated tools manually apply visible grounds, to the high-voltage circuits after they are deenergized and before any employee performs work on the circuit or on the item or apparatus under test. Common ground connections shall be solidly connected to the test equipment and the apparatus under test.

(iii) In high-power testing, the employer shall provide an isolated ground-return conductor system designed to prevent the intentional passage of current, with its attendant voltage rise, from occurring in the ground grid or in the earth. However, the employer need not provide an isolated ground-return conductor if the employer can demonstrate that both of the following conditions exist:

(A) The employer cannot provide an isolated ground-return conductor due to the distance of the test site from the electric energy source, and

(B) The employer protects employees from any hazardous step and touch potentials that may develop during the test.

Note to paragraph (o)(4)(iii)(B): See Appendix C to this section for information on measures that employers can take to protect employees from hazardous step and touch potentials.

(iv) For tests in which using the equipment grounding conductor in the equipment power cord to ground the test equipment would result in greater hazards to test personnel or prevent the taking of satisfactory measurements, the employer may use a ground clearly indicated in the test set-up if the employer can demonstrate that this ground affords protection for employees equivalent to the protection afforded by an equipment grounding conductor in the power supply cord.

(v) The employer shall ensure that, when any employee enters the test area after equipment is deenergized, a ground is placed on the high-voltage terminal and any other exposed terminals.

(A) Before any employee applies a direct ground, the employer shall discharge high capacitance equipment through a resistor rated for the available energy.

(B) A direct ground shall be applied to the exposed terminals after the stored energy drops to a level at which it is safe to do so.

(vi) If the employer uses a test trailer or test vehicle in field testing, its chassis shall be grounded. The employer shall protect each employee against hazardous touch potentials with respect to the vehicle, instrument panels, and other conductive parts accessible to employees with bonding, insulation, or isolation.

FIGURE 7.3 *(Continued)*

(5) *Control and measuring circuits.* **(i)** The employer may not run control wiring, meter connections, test leads, or cables from a test area unless contained in a grounded metallic sheath and terminated in a grounded metallic enclosure or unless the employer takes other precautions that it can demonstrate will provide employees with equivalent safety.

(ii) The employer shall isolate meters and other instruments with accessible terminals or parts from test personnel to protect against hazards that could arise should such terminals and parts become energized during testing. If the employer provides this isolation by locating test equipment in metal compartments with viewing windows, the employer shall provide interlocks to interrupt the power supply when someone opens the compartment cover.

(iii) The employer shall protect temporary wiring and its connections against damage, accidental interruptions, and other hazards. To the maximum extent possible, the employer shall keep signal, control, ground, and power cables separate from each other.

(iv) If any employee will be present in the test area during testing, a test observer shall be present. The test observer shall be capable of implementing the immediate deenergizing of test circuits for safety purposes.

(6) *Safety check.* **(i)** Safety practices governing employee work at temporary or field test areas shall provide, at the beginning of each series of tests, for a routine safety check of such test areas.

(ii) The test operator in charge shall conduct these routine safety checks before each series of tests and shall verify at least the following conditions:

(A) Barriers and safeguards are in workable condition and placed properly to isolate hazardous areas;

(B) System test status signals, if used, are in operable condition;

(C) Clearly marked test-power disconnects are readily available in an emergency;

(D) Ground connections are clearly identifiable;

(E) Personal protective equipment is provided and used as required by Subpart I of this part and by this section; and

(F) Proper separation between signal, ground, and power cables.

(p) *Mechanical equipment.* **(1) *General requirements.*** **(i)** The critical safety components of mechanical elevating and rotating equipment shall receive a thorough visual inspection before use on each shift.

Note to paragraph (p)(1)(i): Critical safety components of mechanical elevating and rotating equipment are components for which failure would result in free fall or free rotation of the boom.

(ii) No motor vehicle or earthmoving or compacting equipment having an obstructed view to the rear may be operated on off-highway jobsites where any employee is exposed to the hazards created by the moving vehicle, unless:

(A) The vehicle has a reverse signal alarm audible above the surrounding noise level, or

(B) The vehicle is backed up only when a designated employee signals that it is safe to do so.

(iii) Rubber-tired self-propelled scrapers, rubber-tired front-end loaders, rubber-tired dozers, wheel-type agricultural and industrial tractors, crawler-type tractors, crawler-type loaders, and motor graders, with or without attachments, shall have rollover protective structures that meet the requirements of Subpart W of Part 1926 of this chapter.

(iv) The operator of an electric line truck may not leave his or her position at the controls while a load is suspended, unless the employer can demonstrate that no employee (including the operator) is endangered.

(2) *Outriggers.* **(i)** Mobile equipment, if provided with outriggers, shall be operated with the outriggers extended and firmly set, except as provided in paragraph (p)(2)(iii) of this section.

(ii) Outriggers may not be extended or retracted outside of the clear view of the operator unless all employees are outside the range of possible equipment motion.

(iii) If the work area or the terrain precludes the use of outriggers, the equipment may be operated only within its maximum load ratings specified by the equipment manufacturer for the particular configuration of the equipment without outriggers.

(3) *Applied loads.* Mechanical equipment used to lift or move lines or other material shall be used within its maximum load rating and other design limitations for the conditions under which the mechanical equipment is being used.

(4) *Operations near energized lines or equipment.* **(i)** Mechanical equipment shall be operated so that the minimum approach distances, established by the employer under paragraph (l)(3)(i) of this section, are maintained from exposed energized lines and equipment. However, the insulated portion of an aerial lift operated by a qualified employee in the lift is exempt from this requirement if the applicable minimum approach distance is maintained between the uninsulated portions of the aerial lift and exposed objects having a different electrical potential.

(ii) A designated employee other than the equipment operator shall observe the approach distance to exposed lines and equipment and provide timely warnings before the minimum approach distance required by paragraph (p)

FIGURE 7.3 *(Continued)*

(4)(i) of this section is reached, unless the employer can demonstrate that the operator can accurately determine that the minimum approach distance is being maintained.

(iii) If, during operation of the mechanical equipment, that equipment could become energized, the operation also shall comply with at least one of paragraphs (p)(4)(iii)(A) through (p)(4)(iii)(C) of this section.

(A) The energized lines or equipment exposed to contact shall be covered with insulating protective material that will withstand the type of contact that could be made during the operation.

(B) The mechanical equipment shall be insulated for the voltage involved. The mechanical equipment shall be positioned so that its uninsulated portions cannot approach the energized lines or equipment any closer than the minimum approach distances, established by the employer under paragraph (l)(3)(i) of this section.

(C) Each employee shall be protected from hazards that could arise from mechanical equipment contact with energized lines or equipment. The measures used shall ensure that employees will not be exposed to hazardous differences in electric potential. Unless the employer can demonstrate that the methods in use protect each employee from the hazards that could arise if the mechanical equipment contacts the energized line or equipment, the measures used shall include all of the following techniques:

(1) Using the best available ground to minimize the time the lines or electric equipment remain energized,

(2) Bonding mechanical equipment together to minimize potential differences,

(3) Providing ground mats to extend areas of equipotential, and

(4) Employing insulating protective equipment or barricades to guard against any remaining hazardous electrical potential differences.

Note to paragraph (p)(4)(iii)(C): Appendix C to this section contains information on hazardous step and touch potentials and on methods of protecting employees from hazards resulting from such potentials.

(q) *Overhead lines and live-line barehand work.* This paragraph provides additional requirements for work performed on or near overhead lines and equipment and for live-line barehand work.

(1) *General.* (i) Before allowing employees to subject elevated structures, such as poles or towers, to such stresses as climbing or the installation or removal of equipment may impose, the employer shall ascertain that the structures are capable of sustaining the additional or unbalanced stresses. If the pole or other structure cannot withstand the expected loads, the employer shall brace or otherwise support the pole or structure so as to prevent failure.

Note to paragraph (q)(1)(i): Appendix D to this section contains test methods that employers can use in ascertaining whether a wood pole is capable of sustaining the forces imposed by an employee climbing the pole. This paragraph also requires the employer to ascertain that the pole can sustain all other forces imposed by the work employees will perform.

(ii) When a pole is set, moved, or removed near an exposed energized overhead conductor, the pole may not contact the conductor.

(iii) When a pole is set, moved, or removed near an exposed energized overhead conductor, the employer shall ensure that each employee wears electrical protective equipment or uses insulated devices when handling the pole and that no employee contacts the pole with uninsulated parts of his or her body.

(iv) To protect employees from falling into holes used for placing poles, the employer shall physically guard the holes, or ensure that employees attend the holes, whenever anyone is working nearby.

(2) *Installing and removing overhead lines.* The following provisions apply to the installation and removal of overhead conductors or cable (overhead lines).

(i) When lines that employees are installing or removing can contact energized parts, the employer shall use the tension-stringing method, barriers, or other equivalent measures to minimize the possibility that conductors and cables the employees are installing or removing will contact energized power lines or equipment.

(ii) For conductors, cables, and pulling and tensioning equipment, the employer shall provide the protective measures required by paragraph (p)(4)(iii) of this section when employees are installing or removing a conductor or cable close enough to energized conductors that any of the following failures could energize the pulling or tensioning equipment or the conductor or cable being installed or removed:

(A) Failure of the pulling or tensioning equipment,

(B) Failure of the conductor or cable being pulled, or

(C) Failure of the previously installed lines or equipment.

(iii) If the conductors that employees are installing or removing cross over energized conductors in excess of 600 volts and if the design of the circuit-interrupting devices protecting the lines so permits, the employer shall render inoperable the automatic-reclosing feature of these devices.

FIGURE 7.3 *(Continued)*

(iv) Before employees install lines parallel to existing energized lines, the employer shall make a determination of the approximate voltage to be induced in the new lines, or work shall proceed on the assumption that the induced voltage is hazardous. Unless the employer can demonstrate that the lines that employees are installing are not subject to the induction of a hazardous voltage or unless the lines are treated as energized, temporary protective grounds shall be placed at such locations and arranged in such a manner that the employer can demonstrate will prevent exposure of each employee to hazardous differences in electric potential.

Note 1 to paragraph (q)(2)(iv): If the employer takes no precautions to protect employees from hazards associated with involuntary reactions from electric shock, a hazard exists if the induced voltage is sufficient to pass a current of 1 milliampere through a 500-ohm resistor. If the employer protects employees from injury due to involuntary reactions from electric shock, a hazard exists if the resultant current would be more than 6 milliamperes.

Note 2 to paragraph (q)(2)(iv): Appendix C to this section contains guidelines for protecting employees from hazardous differences in electric potential as required by this paragraph.

(v) Reel-handling equipment, including pulling and tensioning devices, shall be in safe operating condition and shall be leveled and aligned.

(vi) The employer shall ensure that employees do not exceed load ratings of stringing lines, pulling lines, conductor grips, load-bearing hardware and accessories, rigging, and hoists.

(vii) The employer shall repair or replace defective pulling lines and accessories.

(viii) The employer shall ensure that employees do not use conductor grips on wire rope unless the manufacturer specifically designed the grip for this application.

(ix) The employer shall ensure that employees maintain reliable communications, through two-way radios or other equivalent means, between the reel tender and the pulling rig operator.

(x) Employees may operate the pulling rig only when it is safe to do so.

Note to paragraph (q)(2)(x): Examples of unsafe conditions include: employees in locations prohibited by paragraph (q)(2)(xi) of this section, conductor and pulling line hangups, and slipping of the conductor grip.

(xi) While a power-driven device is pulling the conductor or pulling line and the conductor or pulling line is in motion, the employer shall ensure that employees are not directly under overhead operations or on the crossarm, except as necessary for the employees to guide the stringing sock or board over or through the stringing sheave.

(3) ***Live-line barehand work.*** In addition to other applicable provisions contained in this section, the following requirements apply to live-line barehand work:

(i) Before an employee uses or supervises the use of the live-line barehand technique on energized circuits, the employer shall ensure that the employee completes training conforming to paragraph (a)(2) of this section in the technique and in the safety requirements of paragraph (q)(3) of this section.

(ii) Before any employee uses the live-line barehand technique on energized high-voltage conductors or parts, the employer shall ascertain the following information in addition to information about other existing conditions required by paragraph (a)(4) of this section:

(A) The nominal voltage rating of the circuit on which employees will perform the work,

(B) The clearances to ground of lines and other energized parts on which employees will perform the work, and

(C) The voltage limitations of equipment employees will use.

(iii) The employer shall ensure that the insulated equipment, insulated tools, and aerial devices and platforms used by employees are designed, tested, and made for live-line barehand work.

(iv) The employer shall ensure that employees keep tools and equipment clean and dry while they are in use.

(v) The employer shall render inoperable the automatic-reclosing feature of circuit-interrupting devices protecting the lines if the design of the devices permits.

(vi) The employer shall ensure that employees do not perform work when adverse weather conditions would make the work hazardous even after the employer implements the work practices required by this section. Additionally, employees may not perform work when winds reduce the phase-to-phase or phase-to-ground clearances at the work location below the minimum approach distances specified in paragraph (q)(3)(xiv) of this section, unless insulating guards cover the grounded objects and other lines and equipment.

Note to paragraph (q)(3)(vi): Thunderstorms in the vicinity, high winds, snow storms, and ice storms are examples of adverse weather conditions that make live-line barehand work too hazardous to perform safely even after the employer implements the work practices required by this section.

(vii) The employer shall provide and ensure that employees use a conductive bucket liner or other conductive device for bonding the insulated aerial device to the energized line or equipment.

FIGURE 7.3 *(Continued)*

(A) The employee shall be connected to the bucket liner or other conductive device by the use of conductive shoes, leg clips, or other means.

(B) Where differences in potentials at the worksite pose a hazard to employees, the employer shall provide electrostatic shielding designed for the voltage being worked.

(viii) The employer shall ensure that, before the employee contacts the energized part, the employee bonds the conductive bucket liner or other conductive device to the energized conductor by means of a positive connection. This connection shall remain attached to the energized conductor until the employee completes the work on the energized circuit.

(ix) Aerial lifts used for live-line barehand work shall have dual controls (lower and upper) as follows:

(A) The upper controls shall be within easy reach of the employee in the bucket. On a two-bucket-type lift, access to the controls shall be within easy reach of both buckets.

(B) The lower set of controls shall be near the base of the boom and shall be designed so that they can override operation of the equipment at any time.

(x) Lower (ground-level) lift controls may not be operated with an employee in the lift except in case of emergency.

(xi) The employer shall ensure that, before employees elevate an aerial lift into the work position, the employees check all controls (ground level and bucket) to determine that they are in proper working condition.

(xii) The employer shall ensure that, before employees elevate the boom of an aerial lift, the employees ground the body of the truck or barricade the body of the truck and treat it as energized.

(xiii) The employer shall ensure that employees perform a boom-current test before starting work each day, each time during the day when they encounter a higher voltage, and when changed conditions indicate a need for an additional test.

(A) This test shall consist of placing the bucket in contact with an energized source equal to the voltage to be encountered for a minimum of 3 minutes.

(B) The leakage current may not exceed 1 microampere per kilovolt of nominal phase-to-ground voltage.

(C) The employer shall immediately suspend work from the aerial lift when there is any indication of a malfunction in the equipment.

(xiv) The employer shall ensure that employees maintain the minimum approach distances, established by the employer under paragraph (l)(3)(i) of this section, from all grounded objects and from lines and equipment at a potential different from that to which the live-line barehand equipment is bonded, unless insulating guards cover such grounded objects and other lines and equipment.

(xv) The employer shall ensure that, while an employee is approaching, leaving, or bonding to an energized circuit, the employee maintains the minimum approach distances, established by the employer under paragraph (l)(3)(i) of this section, between the employee and any grounded parts, including the lower boom and portions of the truck and between the employee and conductive objects energized at different potentials.

(xvi) While the bucket is alongside an energized bushing or insulator string, the employer shall ensure that employees maintain the phase-to-ground minimum approach distances, established by the employer under paragraph (l)(3)(i) of this section, between all parts of the bucket and the grounded end of the bushing or insulator string or any other grounded surface.

(xvii) The employer shall ensure that employees do not use handlines between the bucket and the boom or between the bucket and the ground. However, employees may use nonconductive-type handlines from conductor to ground if not supported from the bucket. The employer shall ensure that no one uses ropes used for live-line barehand work for other purposes.

(xviii) The employer shall ensure that employees do not pass uninsulated equipment or material between a pole or structure and an aerial lift while an employee working from the bucket is bonded to an energized part.

(xix) A nonconductive measuring device shall be readily accessible to employees performing live-line barehand work to assist them in maintaining the required minimum approach distance.

(4) ***Towers and structures.*** The following requirements apply to work performed on towers or other structures that support overhead lines.

(i) The employer shall ensure that no employee is under a tower or structure while work is in progress, except when the employer can demonstrate that such a working position is necessary to assist employees working above.

(ii) The employer shall ensure that employees use tag lines or other similar devices to maintain control of tower sections being raised or positioned, unless the employer can demonstrate that the use of such devices would create a greater hazard to employees.

FIGURE 7.3 *(Continued)*

TABLE R-5 Altitude Correction Factor

Altitude above sea level (m)	*A*
0 to 900	1.00
901 to 1,200	1.02
1,201 to 1,500	1.05
1,501 to 1,800	1.08
1,801 to 2,100	1.11
2,101 to 2,400	1.14
2,401 to 2,700	1.17
2,701 to 3,000	1.20
3,001 to 3,600	1.25
3,601 to 4,200	1.30
4,201 to 4,800	1.35
4,801 to 5,400	1.39
5,401 to 6,000	1.44

(iii) The employer shall ensure that employees do not detach the load-line from a member or section until they safely secure the load.

(iv) The employer shall ensure that, except during emergency restoration procedures, employees discontinue work when adverse weather conditions would make the work hazardous in spite of the work practices required by this section.

Note to paragraph (q)(4)(iv): Thunderstorms in the vicinity, high winds, snow storms, and ice storms are examples of adverse weather conditions that make this work too hazardous to perform even after the employer implements the work practices required by this section.

(r) ***Line-clearance tree trimming operations.*** This paragraph provides additional requirements for line-clearance tree-trimming operations and for equipment used in these operations.

(1) ***Electrical hazards.*** This paragraph does not apply to qualified employees.

(i) Before an employee climbs, enters, or works around any tree, a determination shall be made of the nominal voltage of electric power lines posing a hazard to employees. However, a determination of the maximum nominal voltage to which an employee will be exposed may be made instead, if all lines are considered as energized at this maximum voltage.

(ii) There shall be a second line-clearance tree trimmer within normal (that is, unassisted) voice communication under any of the following conditions:

(A) If a line-clearance tree trimmer is to approach more closely than 3.05 meters (10 feet) to any conductor or electric apparatus energized at more than 750 volts or

(B) If branches or limbs being removed are closer to lines energized at more than 750 volts than the distances listed in Table R-5, Table R-6, Table R-7, and Table R-8 or

(C) If roping is necessary to remove branches or limbs from such conductors or apparatus.

(iii) Line-clearance tree trimmers shall maintain the minimum approach distances from energized conductors given in Table R-5, Table R-6, Table R-7, and Table R-8.

(iv) Branches that are contacting exposed energized conductors or equipment or that are within the distances specified in Table R-5, Table R-6, Table R-7, and Table R-8 may be removed only through the use of insulating equipment.

Note to paragraph (r)(1)(iv): A tool constructed of a material that the employer can demonstrate has insulating qualities meeting paragraph (j)(1) of this section is considered as insulated under paragraph (r)(1)(iv) of this section if the tool is clean and dry.

(v) Ladders, platforms, and aerial devices may not be brought closer to an energized part than the distances listed in Table R-5, Table R-6, Table R-7, and Table R-8.

(vi) Line-clearance tree-trimming work may not be performed when adverse weather conditions make the work hazardous in spite of the work practices required by this section. Each employee performing line-clearance tree trim-

FIGURE 7.3 (*Continued*)

TABLE R-6 Alternative Minimum Approach Distances for Voltages of 72.5 kV and Less[1]
[in meters or feet and inches]

Nominal voltage (kV) phase-to-phase	Distance			
	Phase-to-ground exposure		Phase-to-phase exposure	
	m	ft	m	ft
0.50 to 0.300[2]	Avoid contact		Avoid contact	
0.301 to 0.750[2]	0.33	1.09	0.33	1.09
0.751 to 5.0	0.63	2.07	0.63	2.07
5.1 to 15.0	0.65	2.14	0.68	2.24
15.1 to 36.0	0.77	2.53	0.89	2.92
36.1 to 46.0	0.84	2.76	0.98	3.22
46.1 to 72.5	1.00	3.29	1.20	3.94

[1]Employers may use the minimum approach distances in this table provided the worksite is at an elevation of 900 meters (3,000 feet) or less. If employees will be working at elevations greater than 900 meters (3,000 feet) above mean sea level, the employer shall determine minimum approach distances by multiplying the distances in this table by the correction factor in Table R-5 corresponding to the altitude of the work.

[2]For single-phase systems, use voltage-to-ground.

TABLE R-7 Alternative Minimum Approach Distances for Voltages of More Than 72.5 kV[1,2,3]
[in meters or feet and inches]

Voltage range phase to phase (kV)	Phase-to-ground exposure		Phase-to-phase exposure	
	m	ft	m	ft
72.6 to 121.0	1.13	3.71	1.42	4.66
121.1 to 145.0	1.30	4.27	1.64	5.38
145.1 to 169.0	1.46	4.79	1.94	6.36
169.1 to 242.0	2.01	6.59	3.08	10.10
242.1 to 362.0	3.41	11.19	5.52	18.11
362.1 to 420.0	4.25	13.94	6.81	22.34
420.1 to 550.0	5.07	16.63	8.24	27.03
550.1 to 800.0	6.88	22.57	11.38	37.34

[1]Employers may use the minimum approach distances in this table provided the worksite is at an elevation of 900 meters (3,000 feet) or less. If employees will be working at elevations greater than 900 meters (3,000 feet) above mean sea level, the employer shall determine minimun approach distances by multiplying the distances in this table by the correction factor in Table R-5 corresponding to the altitude of the work.

[2]Employers may use the phase-to-phase minimum approach distances in this table provided that no insulated tool spans the gap and no large conductive object is in the gap.

[3]The clear live-line tool distance shall equal or exceed the values for the indicated voltage ranges.

ming work in the aftermath of a storm or under similar emergency conditions shall be trained in the special hazards related to this type of work.

Note to paragraph (r)(1)(vi): Thunderstorms in the immediate vicinity, high winds, snow storms, and ice storms are examples of adverse weather conditions that are presumed to make line-clearance tree trimming work too hazardous to perform safely.

(2) ***Brush chippers.*** **(i)** Brush chippers shall be equipped with a locking device in the ignition system.

(ii) Access panels for maintenance and adjustment of the chipper blades and associated drive train shall be in place and secure during operation of the equipment.

(iii) Brush chippers not equipped with a mechanical infeed system shall be equipped with an infeed hopper of length sufficient to prevent employees from contacting the blades or knives of the machine during operation.

(iv) Trailer chippers detached from trucks shall be chocked or otherwise secured.

(v) Each employee in the immediate area of an operating chipper feed table shall wear personal protective equipment as required by Subpart I of this part.

FIGURE 7.3 *(Continued)*

TABLE R-8 DC Live-Line Minimum Approach Distance with Overvoltage Factor[1] [in meters]

Maximum anticipated per-unit transient overvoltage	Distance (m) Maximum line-to-ground voltage (kV)				
	250	400	500	600	750
1.5 or less	1.12	1.60	2.06	2.62	3.61
1.6	1.17	1.69	2.24	2.86	3.98
1.7	1.23	1.82	2.42	3.12	4.37
1.8	1.28	1.95	2.62	3.39	4.79

[1]The distances specified in this table are for air, bare-hand, and live-line tool conditions. If employees will be working at elevations greater than 900 meters (3,000 feet) above mean sea level, the employer shall determine minimum approach distances by multiplying the distances in this table by the correction factor in Table R-5 corresponding to the altitude of the work.

TABLE R-9 Assumed Maximum Per-Unit Transient Overvoltage

Voltage range (kV)	Type of current (ac or dc)	Assumed maximum per-unit transient overvoltage
72.6 to 420.0	ac	3.5
420.1 to 550.0	ac	3.0
550.1 to 800.0	ac	2.5
250 to 750	dc	1.8

(3) ***Sprayers and related equipment.*** **(i)** Walking and working surfaces of sprayers and related equipment shall be covered with slip-resistant material. If slipping hazards cannot be eliminated, slip-resistant footwear or handrails and stair rails meeting the requirements of Subpart D of this part may be used instead of slip-resistant material.

(ii) Equipment on which employees stand to spray while the vehicle is in motion shall be equipped with guardrails around the working area. The guardrail shall be constructed in accordance with Subpart D of this part.

(4) ***Stump cutters.*** **(i)** Stump cutters shall be equipped with enclosures or guards to protect employees.

(ii) Each employee in the immediate area of stump grinding operations (including the stump cutter operator) shall wear personal protective equipment as required by Subpart I of this part.

(5) ***Gasoline-engine power saws.*** Gasoline-engine power saw operations shall meet the requirements of § 1910.266(e) and the following:

(i) Each power saw weighing more than 6.8 kilograms (15 pounds, service weight) that is used in trees shall be supported by a separate line, except when work is performed from an aerial lift and except during topping or removing operations where no supporting limb will be available.

(ii) Each power saw shall be equipped with a control that will return the saw to idling speed when released.

(iii) Each power saw shall be equipped with a clutch and shall be so adjusted that the clutch will not engage the chain drive at idling speed.

(iv) A power saw shall be started on the ground or where it is otherwise firmly supported. Drop starting of saws over 6.8 kilograms (15 pounds), other than chain saws, is permitted outside of the bucket of an aerial lift only if the area below the lift is clear of personnel.

Note to paragraph (r)(5)(iv): Paragraph (e)(2)(vi) of § 1910.266 prohibits drop starting of chain saws.

(v) A power saw engine may be started and operated only when all employees other than the operator are clear of the saw.

(vi) A power saw may not be running when the saw is being carried up into a tree by an employee.

(vii) Power saw engines shall be stopped for all cleaning, refueling, adjustments, and repairs to the saw or motor, except as the manufacturer's servicing procedures require otherwise.

(6) ***Backpack power units for use in pruning and clearing.*** **(i)** While a backpack power unit is running, no one other than the operator may be within 3.05 meters (10 feet) of the cutting head of a brush saw.

(ii) A backpack power unit shall be equipped with a quick shutoff switch readily accessible to the operator.

(iii) Backpack power unit engines shall be stopped for all cleaning, refueling, adjustments, and repairs to the saw or motor, except as the manufacturer's servicing procedures require otherwise.

FIGURE 7.3 *(Continued)*

(7) *Rope.* **(i)** Climbing ropes shall be used by employees working aloft in trees. These ropes shall have a minimum diameter of 12 millimeters (0.5 inch) with a minimum breaking strength of 10.2 kilonewtons (2,300 pounds). Synthetic rope shall have elasticity of not more than 7 percent.

(ii) Rope shall be inspected before each use and, if unsafe (for example, because of damage or defect), may not be used.

(iii) Rope shall be stored away from cutting edges and sharp tools. Rope contact with corrosive chemicals, gas, and oil shall be avoided.

(iv) When stored, rope shall be coiled and piled, or shall be suspended, so that air can circulate through the coils.

(v) Rope ends shall be secured to prevent their unraveling.

(vi) Climbing rope may not be spliced to effect repair.

(vii) A rope that is wet, that is contaminated to the extent that its insulating capacity is impaired, or that is otherwise not considered to be insulated for the voltage involved may not be used near exposed energized lines.

(8) *Fall protection.* Each employee shall be tied in with a climbing rope and safety saddle when the employee is working above the ground in a tree, unless he or she is ascending into the tree.

(s) *Communication facilities.* **(1) *Microwave transmission.*** **(i)** The employer shall ensure that no employee looks into an open waveguide or antenna connected to an energized microwave source.

(ii) If the electromagnetic-radiation level within an accessible area associated with microwave communications systems exceeds the radiation-protection guide specified by § 1910.97(a)(2), the employer shall post the area with warning signs containing the warning symbol described in § 1910.97(a)(3). The lower half of the warning symbol shall include the following statements, or ones that the employer can demonstrate are equivalent: "Radiation in this area may exceed hazard limitations and special precautions are required. Obtain specific instruction before entering."

(iii) When an employee works in an area where the electromagnetic radiation could exceed the radiation protection guide, the employer shall institute measures that ensure that the employee's exposure is not greater than that permitted by that guide. Such measures may include administrative and engineering controls and personal protective equipment.

(2) *Power-line carrier.* The employer shall ensure that employees perform power-line carrier work, including work on equipment used for coupling carrier current to power line conductors, in accordance with the requirements of this section pertaining to work on energized lines.

(t) *Underground electrical installations.* This paragraph provides additional requirements for work on underground electrical installations.

(1) *Access.* The employer shall ensure that employees use a ladder or other climbing device to enter and exit a manhole or subsurface vault exceeding 1.22 meters (4 feet) in depth. No employee may climb into or out of a manhole or vault by stepping on cables or hangers.

(2) *Lowering equipment into manholes.* **(i)** Equipment used to lower materials and tools into manholes or vaults shall be capable of supporting the weight to be lowered and shall be checked for defects before use.

(ii) Before anyone lowers tools or material into the opening for a manhole or vault, each employee working in the manhole or vault shall be clear of the area directly under the opening.

(3) *Attendants for manholes and vaults.* **(i)** While work is being performed in a manhole or vault containing energized electric equipment, an employee with first-aid training shall be available on the surface in the immediate vicinity of the manhole or vault entrance to render emergency assistance.

(ii) Occasionally, the employee on the surface may briefly enter a manhole or vault to provide nonemergency assistance.

Note 1 to paragraph (t)(3)(ii): Paragraph (e)(7) of this section may also require an attendant and does not permit this attendant to enter the manhole or vault.

Note 2 to paragraph (t)(3)(ii): Paragraph (l)(1)(ii) of this section requires employees entering manholes or vaults containing unguarded, uninsulated energized lines or parts of electric equipment operating at 50 volts or more to be qualified.

(iii) For the purpose of inspection, housekeeping, taking readings, or similar work, an employee working alone may enter, for brief periods of time, a manhole or vault where energized cables or equipment are in service if the employer can demonstrate that the employee will be protected from all electrical hazards.

(iv) The employer shall ensure that employees maintain reliable communications, through two-way radios or other equivalent means, among all employees involved in the job.

FIGURE 7.3 *(Continued)*

(4) *Duct rods.* The employer shall ensure that, if employees use duct rods, the employees install the duct rods in the direction presenting the least hazard to employees. The employer shall station an employee at the far end of the duct line being rodded to ensure that the employees maintain the required minimum approach distances.

(5) *Multiple cables.* When multiple cables are present in a work area, the employer shall identify the cable to be worked by electrical means, unless its identity is obvious by reason of distinctive appearance or location or by other readily apparent means of identification. The employer shall protect cables other than the one being worked from damage.

(6) *Moving cables.* Except when paragraph (t)(7)(ii) of this section permits employees to perform work that could cause a fault in an energized cable in a manhole or vault, the employer shall ensure that employees inspect energized cables to be moved for abnormalities.

(7) *Protection against faults.* **(i)** Where a cable in a manhole or vault has one or more abnormalities that could lead to a fault or be an indication of an impending fault, the employer shall deenergize the cable with the abnormality before any employee may work in the manhole or vault, except when service-load conditions and a lack of feasible alternatives require that the cable remain energized. In that case, employees may enter the manhole or vault provided the employer protects them from the possible effects of a failure using shields or other devices that are capable of containing the adverse effects of a fault. The employer shall treat the following abnormalities as indications of impending faults unless the employer can demonstrate that the conditions could not lead to a fault: Oil or compound leaking from cable or joints, broken cable sheaths or joint sleeves, hot localized surface temperatures of cables or joints, or joints swollen beyond normal tolerance.

(ii) If the work employees will perform in a manhole or vault could cause a fault in a cable, the employer shall deenergize that cable before any employee works in the manhole or vault, except when service-load conditions and a lack of feasible alternatives require that the cable remain energized. In that case, employees may enter the manhole or vault provided the employer protects them from the possible effects of a failure using shields or other devices that are capable of containing the adverse effects of a fault.

(8) *Sheath continuity.* When employees perform work on buried cable or on cable in a manhole or vault, the employer shall maintain metallic-sheath continuity, or the cable sheath shall be treated as energized.

(u) *Substations.* This paragraph provides additional requirements for substations and for work performed in them.

(1) *Access and working space.* The employer shall provide and maintain sufficient access and working space about electric equipment to permit ready and safe operation and maintenance of such equipment by employees.

Note to paragraph (u)(1): American National Standard *National Electrical Safety Code,* ANSI/IEEE C2–2012 contains guidelines for the dimensions of access and working space about electric equipment in substations. Installations meeting the ANSI provisions comply with paragraph (u)(1) of this section. The Occupational Safety and Health Administration will determine whether an installation that does not conform to this ANSI standard complies with paragraph (u)(1) of this section based on the following criteria:

(1) Whether the installation conforms to the edition of ANSI C2 that was in effect when the installation was made,

(2) Whether the configuration of the installation enables employees to maintain the minimum approach distances, established by the employer under paragraph (l)(3)(i) of this section, while the employees are working on exposed, energized parts, and

(3) Whether the precautions taken when employees perform work on the installation provide protection equivalent to the protection provided by access and working space meeting ANSI/IEEE C2–2012.

(2) *Draw-out-type circuit breakers.* The employer shall ensure that, when employees remove or insert draw-out type circuit breakers, the breaker is in the open position. The employer shall also render the control circuit inoperable if the design of the equipment permits.

(3) *Substation fences.* Conductive fences around substations shall be grounded. When a substation fence is expanded or a section is removed, fence sections shall be isolated, grounded, or bonded as necessary to protect employees from hazardous differences in electric potential.

Note to paragraph (u)(3): IEEE Std 80–2000, *IEEE Guide for Safety in AC Substation Grounding,* contains guidelines for protection against hazardous differences in electric potential.

(4) *Guarding of rooms and other spaces containing electric supply equipment.* **(i)** Rooms and other spaces in which electric supply lines or equipment are installed shall meet the requirements of paragraphs (u)(4)(ii) through (u)(4)(v) of this section under the following conditions:

(A) If exposed live parts operating at 50 to 150 volts to ground are within 2.4 meters (8 feet) of the ground or other working surface inside the room or other space,

FIGURE 7.3 *(Continued)*

(B) If live parts operating at 151 to 600 volts to ground and located within 2.4 meters (8 feet) of the ground or other working surface inside the room or other space are guarded only by location, as permitted under paragraph (u)(5)(i) of this section, or

(C) If live parts operating at more than 600 volts to ground are within the room or other space, unless:

(1) The live parts are enclosed within grounded, metal-enclosed equipment whose only openings are designed so that foreign objects inserted in these openings will be deflected from energized parts, or

(2) The live parts are installed at a height, above ground and any other working surface that provides protection at the voltage on the live parts corresponding to the protection provided by a 2.4-meter (8-foot) height at 50 volts.

(ii) Fences, screens, partitions, or walls shall enclose the rooms and other spaces so as to minimize the possibility that unqualified persons will enter.

(iii) Unqualified persons may not enter the rooms or other spaces while the electric supply lines or equipment are energized.

(iv) The employer shall display signs at entrances to the rooms and other spaces warning unqualified persons to keep out.

(v) The employer shall keep each entrance to a room or other space locked, unless the entrance is under the observation of a person who is attending the room or other space for the purpose of preventing unqualified employees from entering.

(5) *Guarding of energized parts.* (i) The employer shall provide guards around all live parts operating at more than 150 volts to ground without an insulating covering unless the location of the live parts gives sufficient clearance (horizontal, vertical, or both) to minimize the possibility of accidental employee contact.

Note to paragraph (u)(5)(i): American National Standard *National Electrical Safety Code,* ANSI/IEEE C2–2002 contains guidelines for the dimensions of clearance distances about electric equipment in substations. Installations meeting the ANSI provisions comply with paragraph (u)(5)(i) of this section. The Occupational Safety and Health Administration will determine whether an installation that does not conform to this ANSI standard complies with paragraph (u)(5)(i) of this section based on the following criteria:

(1) Whether the installation conforms to the edition of ANSI C2 that was in effect when the installation was made,

(2) Whether each employee is isolated from energized parts at the point of closest approach; and

(3) Whether the precautions taken when employees perform work on the installation provide protection equivalent to the protection provided by horizontal and vertical clearances meeting ANSI/IEEE C2–2002.

(ii) Except for fuse replacement and other necessary access by qualified persons, the employer shall maintain guarding of energized parts within a compartment during operation and maintenance functions to prevent accidental contact with energized parts and to prevent dropped tools or other equipment from contacting energized parts.

(iii) Before guards are removed from energized equipment, the employer shall install barriers around the work area to prevent employees who are not working on the equipment, but who are in the area, from contacting the exposed live parts.

(6) *Substation entry.* (i) Upon entering an attended substation, each employee, other than employees regularly working in the station, shall report his or her presence to the employee in charge of substation activities to receive information on special system conditions affecting employee safety.

(ii) The job briefing required by paragraph (c) of this section shall cover information on special system conditions affecting employee safety, including the location of energized equipment in or adjacent to the work area and the limits of any deenergized work area.

(v) *Power generation.* This paragraph provides additional requirements and related work practices for power generating plants.

(1) *Interlocks and other safety devices.* (i) Interlocks and other safety devices shall be maintained in a safe, operable condition.

(ii) No interlock or other safety device may be modified to defeat its function, except for test, repair, or adjustment of the device.

(2) *Changing brushes.* Before exciter or generator brushes are changed while the generator is in service, the exciter or generator field shall be checked to determine whether a ground condition exists. The brushes may not be changed while the generator is energized if a ground condition exists.

(3) *Access and working space.* The employer shall provide and maintain sufficient access and working space about electric equipment to permit ready and safe operation and maintenance of such equipment by employees.

Note to paragraph (v)(3) of this section: American National Standard *National Electrical Safety Code,* ANSI/IEEE C2–2012 contains guidelines for the dimensions of access and working space about electric equipment in

FIGURE 7.3 *(Continued)*

substations. Installations meeting the ANSI provisions comply with paragraph (v)(3) of this section. The Occupational Safety and Health Administration will determine whether an installation that does not conform to this ANSI standard complies with paragraph (v)(3) of this section based on the following criteria:

(1) Whether the installation conforms to the edition of ANSI C2 that was in effect when the installation was made;

(2) Whether the configuration of the installation enables employees to maintain the minimum approach distances, established by the employer under paragraph (l)(3)(i) of this section, while the employees are working on exposed, energized parts, and;

(3) Whether the precautions taken when employees perform work on the installation provide protection equivalent to the protection provided by access and working space meeting ANSI/IEEE C2–2012.

(4) *Guarding of rooms and other spaces containing electric supply equipment.* (i) Rooms and other spaces in which electric supply lines or equipment are installed shall meet the requirements of paragraphs (v)(4)(ii) through (v)(4)(v) of this section under the following conditions:

(A) If exposed live parts operating at 50 to 150 volts to ground are within 2.4 meters (8 feet) of the ground or other working surface inside the room or other space,

(B) If live parts operating at 151 to 600 volts to ground and located within 2.4 meters (8 feet) of the ground or other working surface inside the room or other space are guarded only by location, as permitted under paragraph (v)(5)(i) of this section, or (C) If live parts operating at more than 600 volts to ground are within the room or other space, unless:

(1) The live parts are enclosed within grounded, metal-enclosed equipment whose only openings are designed so that foreign objects inserted in these openings will be deflected from energized parts, or

(2) The live parts are installed at a height, above ground and any other working surface that provides protection at the voltage on the live parts corresponding to the protection provided by a 2.4-meter (8-foot) height at 50 volts.

(ii) Fences, screens, partitions, or walls shall enclose the rooms and other spaces so as to minimize the possibility that unqualified persons will enter.

(iii) Unqualified persons may not enter the rooms or other spaces while the electric supply lines or equipment are energized.

(iv) The employer shall display signs at entrances to the rooms and other spaces warning unqualified persons to keep out.

(v) The employer shall keep each entrance to a room or other space locked, unless the entrance is under the observation of a person who is attending the room or other space for the purpose of preventing unqualified employees from entering.

(5) *Guarding of energized parts.* (i) The employer shall provide guards around all live parts operating at more than 150 volts to ground without an insulating covering unless the location of the live parts gives sufficient clearance (horizontal, vertical, or both) to minimize the possibility of accidental employee contact.

Note to paragraph (v)(5)(i): American National Standard *National Electrical Safety Code,* ANSI/IEEE C2–2002 contains guidelines for the dimensions of clearance distances about electric equipment in substations. Installations meeting the ANSI provisions comply with paragraph (v)(5)(i) of this section. The Occupational Safety and Health Administration will determine whether an installation that does not conform to this ANSI standard complies with paragraph (v)(5)(i) of this section based on the following criteria:

(1) Whether the installation conforms to the edition of ANSI C2 that was in effect when the installation was made;

(2) Whether each employee is isolated from energized parts at the point of closest approach; and

(3) Whether the precautions taken when employees perform work on the installation provide protection equivalent to the protection provided by horizontal and vertical clearances meeting ANSI/IEEE C2–2002.

(ii) Except for fuse replacement and other necessary access by qualified persons, the employer shall maintain guarding of energized parts within a compartment during operation and maintenance functions to prevent accidental contact with energized parts and to prevent dropped tools or other equipment from contacting energized parts.

(iii) Before guards are removed from energized equipment, the employer shall install barriers around the work area to prevent employees who are not working on the equipment, but who are in the area, from contacting the exposed live parts.

(6) *Water or steam spaces.* The following requirements apply to work in water and steam spaces associated with boilers:

(i) A designated employee shall inspect conditions before work is permitted and after its completion. Eye protection, or full-face protection if necessary, shall be worn at all times when condenser, heater, or boiler tubes are being cleaned.

FIGURE 7.3 *(Continued)*

(ii) Where it is necessary for employees to work near tube ends during cleaning, shielding shall be installed at the tube ends.

(7) *Chemical cleaning of boilers and pressure vessels.* The following requirements apply to chemical cleaning of boilers and pressure vessels:

(i) Areas where chemical cleaning is in progress shall be cordoned off to restrict access during cleaning. If flammable liquids, gases, or vapors or combustible materials will be used or might be produced during the cleaning process, the following requirements also apply:

(A) The area shall be posted with signs restricting entry and warning of the hazards of fire and explosion; and

(B) Smoking, welding, and other possible ignition sources are prohibited in these restricted areas.

(ii) The number of personnel in the restricted area shall be limited to those necessary to accomplish the task safely.

(iii) There shall be ready access to water or showers for emergency use.

Note to paragraph (v)(7)(iii): See § 1910.141 for requirements that apply to the water supply and to washing facilities.

(iv) Employees in restricted areas shall wear protective equipment meeting the requirements of Subpart I of this part and including, but not limited to, protective clothing, boots, goggles, and gloves.

(8) *Chlorine systems.* (i) Chlorine system enclosures shall be posted with signs restricting entry and warning of the hazard to health and the hazards of fire and explosion.

Note to paragraph (v)(8)(i): See Subpart Z of this part for requirements necessary to protect the health of employees from the effects of chlorine.

(ii) Only designated employees may enter the restricted area. Additionally, the number of personnel shall be limited to those necessary to accomplish the task safely.

(iii) Emergency repair kits shall be available near the shelter or enclosure to allow for the prompt repair of leaks in chlorine lines, equipment, or containers.

(iv) Before repair procedures are started, chlorine tanks, pipes, and equipment shall be purged with dry air and isolated from other sources of chlorine.

(v) The employer shall ensure that chlorine is not mixed with materials that would react with the chlorine in a dangerously exothermic or other hazardous manner.

(9) *Boilers.* (i) Before internal furnace or ash hopper repair work is started, overhead areas shall be inspected for possible falling objects. If the hazard of falling objects exists, overhead protection such as planking or nets shall be provided.

(ii) When opening an operating boiler door, employees shall stand clear of the opening of the door to avoid the heat blast and gases which may escape from the boiler.

(10) *Turbine generators.* (i) Smoking and other ignition sources are prohibited near hydrogen or hydrogen sealing systems, and signs warning of the danger of explosion and fire shall be posted.

(ii) Excessive hydrogen makeup or abnormal loss of pressure shall be considered as an emergency and shall be corrected immediately.

(iii) A sufficient quantity of inert gas shall be available to purge the hydrogen from the largest generator.

(11) *Coal and ash handling.* (i) Only designated persons may operate railroad equipment.

(ii) Before a locomotive or locomotive crane is moved, a warning shall be given to employees in the area.

(iii) Employees engaged in switching or dumping cars may not use their feet to line up draw heads.

(iv) Draw heads and knuckles may not be shifted while locomotives or cars are in motion.

(v) When a railroad car is stopped for unloading, the car shall be secured from displacement that could endanger employees.

(vi) An emergency means of stopping dump operations shall be provided at railcar dumps.

(vii) The employer shall ensure that employees who work in coal- or ash handling conveyor areas are trained and knowledgeable in conveyor operation and in the requirements of paragraphs (v)(11)(viii) through (v)(11)(xii) of this section.

(viii) Employees may not ride a coal or ash-handling conveyor belt at any time. Employees may not cross over the conveyor belt, except at walkways, unless the conveyor's energy source has been deenergized and has been locked out or tagged in accordance with paragraph (d) of this section.

(ix) A conveyor that could cause injury when started may not be started until personnel in the area are alerted by a signal or by a designated person that the conveyor is about to start.

FIGURE 7.3 (*Continued*)

(x) If a conveyor that could cause injury when started is automatically controlled or is controlled from a remote location, an audible device shall be provided that sounds an alarm that will be recognized by each employee as a warning that the conveyor will start and that can be clearly heard at all points along the conveyor where personnel may be present. The warning device shall be actuated by the device starting the conveyor and shall continue for a period of time before the conveyor starts that is long enough to allow employees to move clear of the conveyor system. A visual warning may be used in place of the audible device if the employer can demonstrate that it will provide an equally effective warning in the particular circumstances involved. However, if the employer can demonstrate that the system's function would be seriously hindered by the required time delay, warning signs may be provided in place of the audible warning device. If the system was installed before January 31, 1995, warning signs may be provided in place of the audible warning device until such time as the conveyor or its control system is rebuilt or rewired. These warning signs shall be clear, concise, and legible and shall indicate that conveyors and allied equipment may be started at any time, that danger exists, and that personnel must keep clear. These warning signs shall be provided along the conveyor at areas not guarded by position or location.

(xi) Remotely and automatically controlled conveyors, and conveyors that have operating stations which are not manned or which are beyond voice and visual contact from drive areas, loading areas, transfer points, and other locations on the conveyor path not guarded by location, position, or guards shall be furnished with emergency stop buttons, pull cords, limit switches, or similar emergency stop devices. However, if the employer can demonstrate that the design, function, and operation of the conveyor do not expose an employee to hazards, an emergency stop device is not required.

(A) Emergency stop devices shall be easily identifiable in the immediate vicinity of such locations.

(B) An emergency stop device shall act directly on the control of the conveyor involved and may not depend on the stopping of any other equipment.

(C) Emergency stop devices shall be installed so that they cannot be overridden from other locations.

(xii) Where coal-handling operations may produce a combustible atmosphere from fuel sources or from flammable gases or dust, sources of ignition shall be eliminated or safely controlled to prevent ignition of the combustible atmosphere.

Note to paragraph (v)(11)(xii): Locations that are hazardous because of the presence of combustible dust are classified as Class II hazardous locations. See § 1910.307.

(xiii) An employee may not work on or beneath overhanging coal in coal bunkers, coal silos, or coal storage areas, unless the employee is protected from all hazards posed by shifting coal.

(xiv) An employee entering a bunker or silo to dislodge the contents shall wear a body harness with lifeline attached. The lifeline shall be secured to a fixed support outside the bunker and shall be attended at all times by an employee located outside the bunker or facility.

(12) ***Hydroplants and equipment.*** Employees working on or close to water gates, valves, intakes, forebays, flumes, or other locations where increased or decreased water flow or levels may pose a significant hazard shall be warned and shall vacate such dangerous areas before water flow changes are made.

(w) ***Special conditions.*** (1) *Capacitors.* The following additional requirements apply to work on capacitors and on lines connected to capacitors.

Note to paragraph (w)(1): See paragraphs (m) and (n) of this section for requirements pertaining to the deenergizing and grounding of capacitor installations.

(i) Before employees work on capacitors, the employer shall disconnect the capacitors from energized sources and short circuit the capacitors. The employer shall ensure that the employee short circuiting the capacitors waits at least 5 minutes from the time of disconnection before applying the short circuit,

(ii) Before employees handle the units, the employer shall short circuit each unit in series-parallel capacitor banks between all terminals and the capacitor case or its rack. If the cases of capacitors are on ungrounded substation racks, the employer shall bond the racks to ground.

(iii) The employer shall short circuit any line connected to capacitors before the line is treated as deenergized.

(2) ***Current transformer secondaries.*** The employer shall ensure that employees do not open the secondary of a current transformer while the transformer is energized. If the employer cannot deenergize the primary of the current transformer before employees perform work on an instrument, a relay, or other section of a current transformer secondary circuit, the employer shall bridge the circuit so that the current transformer secondary does not experience an open-circuit condition.

(3) ***Series streetlighting.*** **(i)** If the open-circuit voltage exceeds 600 volts, the employer shall ensure that employees work on series streetlighting circuits in accordance with paragraph (q) or (t) of this section, as appropriate.

FIGURE 7.3 (*Continued*)

(ii) Before any employee opens a series loop, the employer shall deenergize the streetlighting transformer and isolate it from the source of supply or shall bridge the loop to avoid an open-circuit condition.

(4) ***Illumination.*** The employer shall provide sufficient illumination to enable the employee to perform the work safely.

(5) ***Protection against drowning.*** **(i)** Whenever an employee may be pulled or pushed, or might fall, into water where the danger of drowning exists, the employer shall provide the employee with, and shall ensure that the employee uses, a U.S. Coast Guard approved personal flotation device.

(ii) The employer shall maintain each personal flotation device in safe condition and shall inspect each personal flotation device frequently enough to ensure that it does not have rot, mildew, water saturation, or any other condition that could render the device unsuitable for use.

(iii) An employee may cross streams or other bodies of water only if a safe means of passage, such as a bridge, is available.

(6) *Employee protection in public work areas.* (i) Traffic-control signs and traffic-control devices used for the protection of employees shall meet § 1926.200(g)(2) of this chapter.

(ii) Before employees begin work in the vicinity of vehicular or pedestrian traffic that may endanger them, the employer shall place warning signs or flags and other traffic-control devices in conspicuous locations to alert and channel approaching traffic.

(iii) The employer shall use barricades where additional employee protection is necessary.

(iv) The employer shall protect excavated areas with barricades.

(v) The employer shall display warning lights prominently at night.

(7) ***Backfeed.*** When there is a possibility of voltage backfeed from sources of cogeneration or from the secondary system (for example, backfeed from more than one energized phase feeding a common load), the requirements of paragraph (l) of this section apply if employees will work the lines or equipment as energized, and the requirements of paragraphs (m) and (n) of this section apply if employees will work the lines or equipment as deenergized.

(8) ***Lasers.*** The employer shall install, adjust, and operate laser equipment in accordance with § 1926.54 of this chapter.

(9) ***Hydraulic fluids.*** Hydraulic fluids used for the insulated sections of equipment shall provide insulation for the voltage involved.

(x) ***Definitions.***

Affected employee. An employee whose job requires him or her to operate or use a machine or equipment on which servicing or maintenance is being performed under lockout or tagout, or whose job requires him or her to work in an area in which such servicing or maintenance is being performed.

Attendant. An employee assigned to remain immediately outside the entrance to an enclosed or other space to render assistance as needed to employees inside the space.

Authorized employee. An employee who locks out or tags out machines or equipment in order to perform servicing or maintenance on that machine or equipment. An affected employee becomes an authorized employee when that employee's duties include performing servicing or maintenance covered under this section.

Automatic circuit recloser. A self-controlled device for automatically interrupting and reclosing an alternating-current circuit, with a predetermined sequence of opening and reclosing followed by resetting, hold closed, or lockout.

Barricade. A physical obstruction such as tapes, cones, or A-frame type wood or metal structures that provides a warning about, and limits access to, a hazardous area.

Barrier. A physical obstruction that prevents contact with energized lines or equipment or prevents unauthorized access to a work area.

Bond. The electrical interconnection of conductive parts designed to maintain a common electric potential.

Bus. A conductor or a group of conductors that serve as a common connection for two or more circuits.

Bushing. An insulating structure that includes a through conductor or that provides a passageway for such a conductor, and that, when mounted on a barrier, insulates the conductor from the barrier for the purpose of conducting current from one side of the barrier to the other.

Cable. A conductor with insulation, or a stranded conductor with or without insulation and other coverings (single-conductor cable), or a combination of conductors insulated from one another (multiple-conductor cable).

Cable sheath. A conductive protective covering applied to cables.

Note to the definition of "cable sheath": A cable sheath may consist of multiple layers one or more of which is conductive.

Circuit. A conductor or system of conductors through which an electric current is intended to flow.

FIGURE 7.3 (*Continued*)

Clearance (between objects). The clear distance between two objects measured surface to surface.

Clearance (for work). Authorization to perform specified work or permission to enter a restricted area.

Communication lines. (See *Lines;* (1) *Communication lines.*)

Conductor. A material, usually in the form of a wire, cable, or bus bar, used for carrying an electric current.

Contract employer. An employer, other than a host employer, that performs work covered by this section under contract.

Covered conductor. A conductor covered with a dielectric having no rated insulating strength or having a rated insulating strength less than the voltage of the circuit in which the conductor is used.

Current-carrying part. A conducting part intended to be connected in an electric circuit to a source of voltage. Non-current-carrying parts are those not intended to be so connected.

Deenergized. Free from any electrical connection to a source of potential difference and from electric charge; not having a potential that is different from the potential of the earth.

Note to the definition of "deenergized": The term applies only to current-carrying parts, which are sometimes energized (alive).

Designated employee (designated person). An employee (or person) who is assigned by the employer to perform specific duties under the terms of this section and who has sufficient knowledge of the construction and operation of the equipment, and the hazards involved, to perform his or her duties safely.

Electric line truck. A truck used to transport personnel, tools, and material for electric supply line work.

Electric supply equipment. Equipment that produces, modifies, regulates, controls, or safeguards a supply of electric energy.

Electric supply lines. (See *Lines;* (2) *Electric supply lines.*)

Electric utility. An organization responsible for the installation, operation, or maintenance of an electric supply system.

Enclosed space. A working space, such as a manhole, vault, tunnel, or shaft, that has a limited means of egress or entry, that is designed for periodic employee entry under normal operating conditions, and that, under normal conditions, does not contain a hazardous atmosphere, but may contain a hazardous atmosphere under abnormal conditions.

Note to the definition of "enclosed space": The Occupational Safety and Health Administration does not consider spaces that are enclosed but not designed for employee entry under normal operating conditions to be enclosed spaces for the purposes of this section. Similarly, the Occupational Safety and Health Administration does not consider spaces that are enclosed and that are expected to contain a hazardous atmosphere to be enclosed spaces for the purposes of this section. Such spaces meet the definition of permit spaces in § 1910.146, and entry into them must conform to that standard.

Energized (alive, live). Electrically connected to a source of potential difference, or electrically charged so as to have a potential significantly different from that of earth in the vicinity.

Energy isolating device. A physical device that prevents the transmission or release of energy, including, but not limited to, the following: a manually operated electric circuit breaker, a disconnect switch, a manually operated switch, a slide gate, a slip blind, a line valve, blocks, and any similar device with a visible indication of the position of the device. (Push buttons, selector switches, and other control-circuit-type devices are not energy isolating devices.)

Energy source. Any electrical, mechanical, hydraulic, pneumatic, chemical, nuclear, thermal, or other energy source that could cause injury to employees.

Entry *(as used in paragraph (e) of this section).* The action by which a person passes through an opening into an enclosed space. Entry includes ensuing work activities in that space and is considered to have occurred as soon as any part of the entrant's body breaks the plane of an opening into the space.

Equipment (electric). A general term including material, fittings, devices, appliances, fixtures, apparatus, and the like used as part of or in connection with an electrical installation.

Exposed, Exposed to contact *(as applied to energized parts).* Not isolated or guarded.

Fall restraint system. A fall protection system that prevents the user from falling any distance.

First-aid training. Training in the initial care, including cardiopulmonary resuscitation (which includes chest compressions, rescue breathing, and, as appropriate, other heart and lung resuscitation techniques), performed by a person who is not a medical practitioner, of a sick or injured person until definitive medical treatment can be administered.

Ground. A conducting connection, whether planned or unplanned, between an electric circuit or equipment and the earth, or to some conducting body that serves in place of the earth.

Grounded. Connected to earth or to some conducting body that serves in place of the earth.

FIGURE 7.3 (*Continued*)

Guarded. Covered, fenced, enclosed, or otherwise protected, by means of suitable covers or casings, barrier rails or screens, mats, or platforms, designed to minimize the possibility, under normal conditions, of dangerous approach or inadvertent contact by persons or objects.

Note to the definition of "guarded": Wires that are insulated, but not otherwise protected, are not guarded.

Hazardous atmosphere. An atmosphere that may expose employees to the risk of death, incapacitation, impairment of ability to self-rescue (that is, escape unaided from an enclosed space), injury, or acute illness from one or more of the following causes:

(1) Flammable gas, vapor, or mist in excess of 10 percent of its lower flammable limit (LFL);

(2) Airborne combustible dust at a concentration that meets or exceeds its LFL;

Note to the definition of "hazardous atmosphere" (2): This concentration may be approximated as a condition in which the dust obscures vision at a distance of 1.52 meters (5 feet) or less.

(3) Atmospheric oxygen concentration below 19.5 percent or above 23.5 percent;

(4) Atmospheric concentration of any substance for which a dose or a permissible exposure limit is published in Subpart G, *Occupational Health and Environmental Control,* or in Subpart Z, *Toxic and Hazardous Substances,* of this part and which could result in employee exposure in excess of its dose or permissible exposure limit;

Note to the definition of "hazardous atmosphere" (4): An atmospheric concentration of any substance that is not capable of causing death, incapacitation, impairment of ability to self-rescue, injury, or acute illness due to its health effects is not covered by this provision.

(5) Any other atmospheric condition that is immediately dangerous to life or health.

Note to the definition of "hazardous atmosphere" (5): For air contaminants for which the Occupational Safety and Health Administration has not determined a dose or permissible exposure limit, other sources of information, such as Material Safety Data Sheets that comply with the Hazard Communication Standard, § 1910.1200, published information, and internal documents can provide guidance in establishing acceptable atmospheric conditions.

High-power tests. Tests in which the employer uses fault currents, load currents, magnetizing currents, and line-dropping currents to test equipment, either at the equipment's rated voltage or at lower voltages.

High-voltage tests. Tests in which the employer uses voltages of approximately 1,000 volts as a practical minimum and in which the voltage source has sufficient energy to cause injury.

High wind. A wind of such velocity that one or more of the following hazards would be present:

(1) The wind could blow an employee from an elevated location,

(2) The wind could cause an employee or equipment handling material to lose control of the material, or

(3) The wind would expose an employee to other hazards not controlled by the standard involved.

Note to the definition of "high wind": The Occupational Safety and Health Administration normally considers winds exceeding 64.4 kilometers per hour (40 miles per hour), or 48.3 kilometers per hour (30 miles per hour) if the work involves material handling, as meeting this criteria, unless the employer takes precautions to protect employees from the hazardous effects of the wind.

Host employer. An employer that operates, or that controls the operating procedures for, an electric power generation, transmission, or distribution installation on which a contract employer is performing work covered by this section.

Note to the definition of "host employer": The Occupational Safety and Health Administration will treat the electric utility or the owner of the installation as the host employer if it operates or controls operating procedures for the installation. If the electric utility or installation owner neither operates nor controls operating procedures for the installation, the Occupational Safety and Health Administration will treat the employer that the utility or owner has contracted with to operate or control the operating procedures for the installation as the host employer. In no case will there be more than one host employer.

Immediately dangerous to life or health (IDLH). Any condition that poses an immediate or delayed threat to life or that would cause irreversible adverse health effects or that would interfere with an individual's ability to escape unaided from a permit space.

Note to the definition of "immediately dangerous to life or health": Some materials—hydrogen fluoride gas and cadmium vapor, for example—may produce immediate transient effects that, even if severe, may pass without medical attention, but are followed by sudden, possibly fatal collapse 12–72 hours after exposure. The victim "feels normal" from recovery from transient effects until collapse. Such materials in hazardous quantities are considered to be "immediately" dangerous to life or health.

Insulated. Separated from other conducting surfaces by a dielectric (including air space) offering a high resistance to the passage of current.

FIGURE 7.3 *(Continued)*

Note to the definition of "insulated": When any object is said to be insulated, it is understood to be insulated for the conditions to which it normally is subjected. Otherwise, it is, for the purpose of this section, uninsulated.

Insulation (cable). Material relied upon to insulate the conductor from other conductors or conducting parts or from ground.

Isolated. Not readily accessible to persons unless special means for access are used.

Line-clearance tree trimmer. An employee who, through related training or on-the-job experience or both, is familiar with the special techniques and hazards involved in line-clearance tree trimming.

Note 1 to the definition of "line-clearance tree trimmer": An employee who is regularly assigned to a line-clearance tree-trimming crew and who is undergoing on-the-job training and who, in the course of such training, has demonstrated an ability to perform duties safely at his or her level of training and who is under the direct supervision of a line-clearance tree trimmer is considered to be a line-clearance tree trimmer for the performance of those duties.

Note 2 to the definition of "line-clearance tree trimmer": A line-clearance tree trimmer is not considered to be a "qualified employee" under this section unless he or she has the training required for a qualified employee under paragraph (a)(2)(ii) of this section. However, under the electrical safety-related work practices standard in Subpart S of this part, a line-clearance tree trimmer is considered to be a "qualified employee". Tree trimming performed by such "qualified employees" is not subject to the electrical safety-related work practice requirements contained in §§ 1910.331 through 1910.335 of this part. (See also the note following § 1910.332(b)(3) of this part for information regarding the training an employee must have to be considered a qualified employee under §§ 1910.331 through 1910.335 of this part.)

Line-clearance tree trimming. The pruning, trimming, repairing, maintaining, removing, or clearing of trees, or the cutting of brush, that is within the following distance of electric supply lines and equipment:

(1) For voltages to ground of 50 kilovolts or less—3.05 meters (10 feet);

(2) For voltages to ground of more than 50 kilovolts—3.05 meters (10 feet) plus 0.10 meters (4 inches) for every 10 kilovolts over 50 kilovolts.

Lines.* (1) *Communication lines. The conductors and their supporting or containing structures which are used for public or private signal or communication service, and which operate at potentials not exceeding 400 volts to ground or 750 volts between any two points of the circuit, and the transmitted power of which does not exceed 150 watts. If the lines are operating at less than 150 volts, no limit is placed on the transmitted power of the system. Under certain conditions, communication cables may include communication circuits exceeding these limitations where such circuits are also used to supply power solely to communication equipment.

Note to the definition of "communication lines": Telephone, telegraph, railroad signal, data, clock, fire, police alarm, cable television, and other systems conforming to this definition are included. Lines used for signaling purposes, but not included under this definition, are considered as electric supply lines of the same voltage.

(2) *Electric supply lines.* Conductors used to transmit electric energy and their necessary supporting or containing structures. Signal lines of more than 400 volts are always supply lines within this section, and those of less than 400 volts are considered as supply lines, if so run and operated throughout.

Manhole. A subsurface enclosure that personnel may enter and that is used for installing, operating, and maintaining submersible equipment or cable.

Minimum approach distance. The closest distance an employee may approach an energized or a grounded object.

Note to the definition of "minimum approach distance": Paragraph (l)(3)(i) of this section requires employers to establish minimum approach distances.

Personal fall arrest system. A system used to arrest an employee in a fall from a working level.

Qualified employee (qualified person). An employee (person) knowledgeable in the construction and operation of the electric power generation, transmission, and distribution equipment involved, along with the associated hazards.

Note 1 to the definition of "qualified employee (qualified person)": An employee must have the training required by (a)(2)(ii) of this section to be a qualified employee.

Note 2 to the definition of "qualified employee (qualified person)": Except under (g)(2)(iv)(C)(2) and (g)(2)(iv)(C)(3) of this section, an employee who is undergoing on-the-job training and who has demonstrated, in the course of such training, an ability to perform duties safely at his or her level of training and who is under the direct supervision of a qualified person is a qualified person for the performance of those duties.

Statistical sparkover voltage. A transient overvoltage level that produces a 97.72-percent probability of sparkover (that is, two standard deviations above the voltage at which there is a 50-percent probability of sparkover).

FIGURE 7.3 (*Continued*)

Statistical withstand voltage. A transient overvoltage level that produces a 0.14-percent probability of sparkover (that is, three standard deviations below the voltage at which there is a 50-percent probability of sparkover).

Switch. A device for opening and closing or for changing the connection of a circuit. In this section, a switch is manually operable, unless otherwise stated.

System operator. A qualified person designated to operate the system or its parts.

Vault. An enclosure, above or below ground, that personnel may enter and that is used for installing, operating, or maintaining equipment or cable.

Vented vault. A vault that has provision for air changes using exhaust-flue stacks and low-level air intakes operating on pressure and temperature differentials that provide for airflow that precludes a hazardous atmosphere from developing.

Voltage. The effective (root mean square, or rms) potential difference between any two conductors or between a conductor and ground. This section expresses voltages in nominal values, unless otherwise indicated. The nominal voltage of a system or circuit is the value assigned to a system or circuit of a given voltage class for the purpose of convenient designation. The operating voltage of the system may vary above or below this value.

Work-positioning equipment. A body belt or body harness system rigged to allow an employee to be supported on an elevated vertical surface, such as a utility pole or tower leg, and work with both hands free while leaning.

APPENDIX A TO § 1910.269—FLOW CHARTS

This appendix presents information, in the form of flow charts, that illustrates the scope and application of § 1910.269. This appendix addresses the interface between § 1910.269 and Subpart S of this Part (Electrical), between § 1910.269 and § 1910.146 (Permit-required confined spaces), and between § 1910.269 and § 1910.147 (The control of hazardous energy (lockout/tagout)). These flow charts provide guidance for employers trying to implement the requirements of § 1910.269 in combination with other General Industry Standards contained in Part 1910. Employers should always consult the relevant standards, in conjunction with this appendix, to ensure compliance with all applicable requirements.

Appendix A-l to §1910.269—Application of §1910.269 and Subpart S of this Part to the Design of Electrical Installations

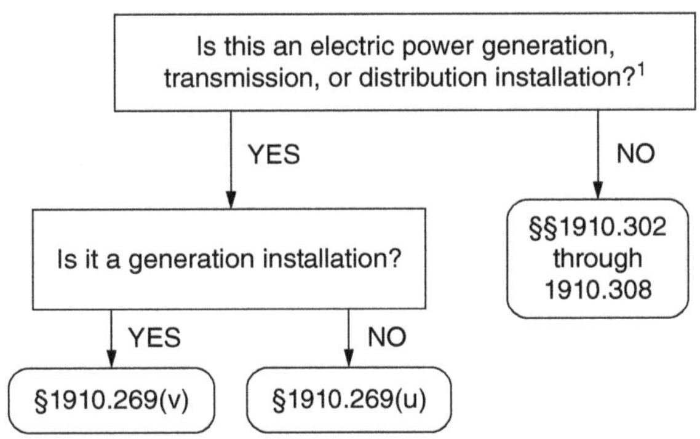

[1]This chart applies to electrical installation design requirements only. See Appendix A-2 for electrical safety-related work practices. Supplementary electric generating epuipment that is used to supply a workplace for emergency, standby, or similar purposes only is not condidered an electric power generation installation.

FIGURE 7.3 (*Continued*)

Appendix A-2 to §1910.269— Application of §1910.269 and Subpart S of this Part to Electrical Safety-Related Work Practices[1]

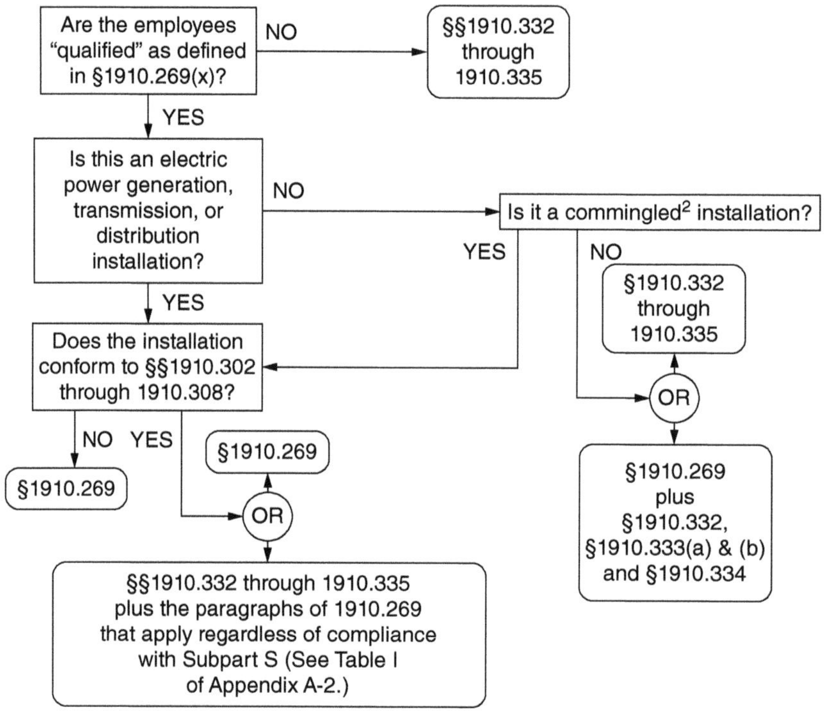

[1]This flowchart applies only to the electrical safety-related work practice and training requirements in §1910.269 and §§1910.332 through 1910.335.

[2]This means commingled to the extent that the electric power generation, transmission, or distribution installation poses the greater hazard.

FIGURE 7.3 (*Continued*)

TABLE 1 Electrical Safety Requirements in §1910.269

Compliance with Subpart S Will Comply with These Paragraphs of §1910.269[1]	Paragraphs that Apply Regardless of Compliance with Subpart S[2]
(d), electric-shock hazards only	(a)(2) and (a)(3)
(h)(3)	(b)
(i)(2)	(c)
(k)	(d), for other than electric-shock hazards
(1)(1) through (l)(4), (1)(6)(i), and (1)(8) through (1)(10)	(e)
(m)	(f)
(p)(4)	(g)
(s)(2)	(h)(1) and (h)(2)
(u)(1) and (u)(3) through (u)(5)	(i)(3) and (i)(4)
(v)(3) through (v)(5)	(j)
(w)(1) and (w)(7)	(1)(5), (1)(6)(ii), (1)(6)(iii), (1)(7), and (1)(11)
	(n)
	(o)
	(p)(1) through (p)(3)
	(q)
	(r)
	(s)(1)
	(t)
	(u)(2) and (u)(6)
	(v)(1), (v)(2), and (v)(6) through (v)(12)
	(w)(2) through (w)(6), (w)(8), and (w)(9)

[1]If the electrical installation meets the requirements of §§1910.302 through 1910.308 of this part, then the electrical installation and any associated electrical safety-related work practices conforming to §§1910.332 through 1910.335 of this part are considered to comply with these provisions of §1910.269 of this part.

[2]These provisions include electrical safety and other requirements that must be met regardless of compliance with Subpart S of this part.

FIGURE 7.3 (*Continued*)

Appendix A-3 to §1910.269—Application of §1910.269 and Subpart S of this Part to Tree-Trimming Operations

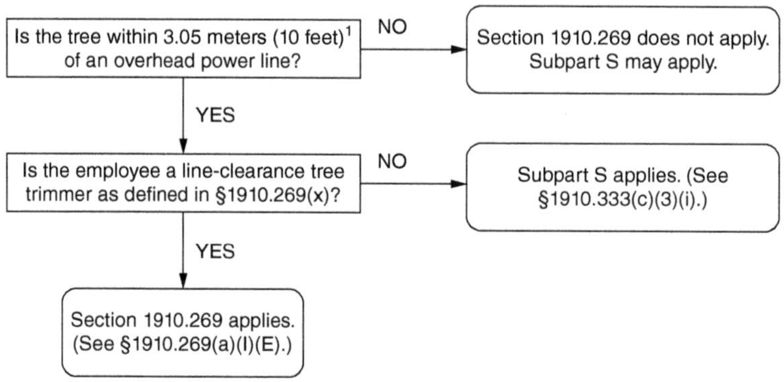

[1]3.05 meters (10 feet) plus 0.10 meters (4 inches) for every 10 kilovolts over 50 kilovolts.

Note: Paragraph (t) of §1910.269 contains additional requirements for work in manholes and underground vaults.

Appendix A-4 to §1910.269—Application of §§1910.147, 1910.269 and 1910.333 to Hazardous Energy Control Procedures (Lockout/Tagout)

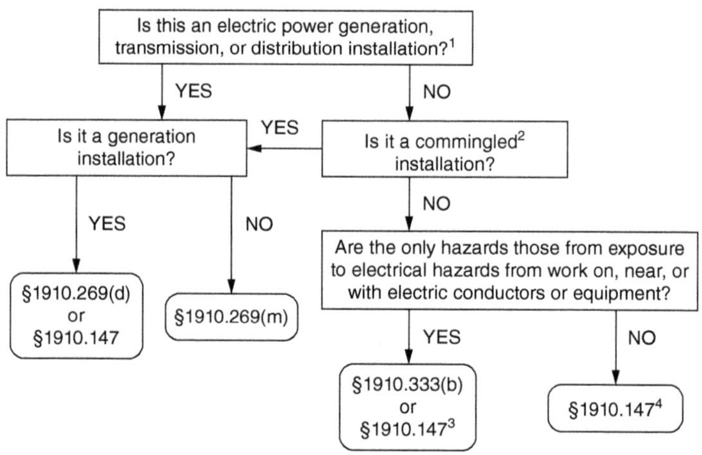

[1]If a generation, transmission, or distribution installation conforms to §§1910.302 through 1910.308, the lockout and tagging procedures of §1910.333(b) may be followed for electric-shock hazards.

[2]This means commingled to the extent that the electric power generation, transmission, or distribution installation poses the greater hazard.

[3]Paragraphs (b)(2)(iii)(D) and (b)(2)(iv)(B) of §1910.333 still apply.

[4]Paragraph (b) of §1910.333 applies to any electrical hazards from work on, near, or with electric conductors and equipment.

FIGURE 7.3 (*Continued*)

Appendix A-5 to §1910.269—Application of §§1910.146 and 1910.269 to Permit-Required Confined Spaces

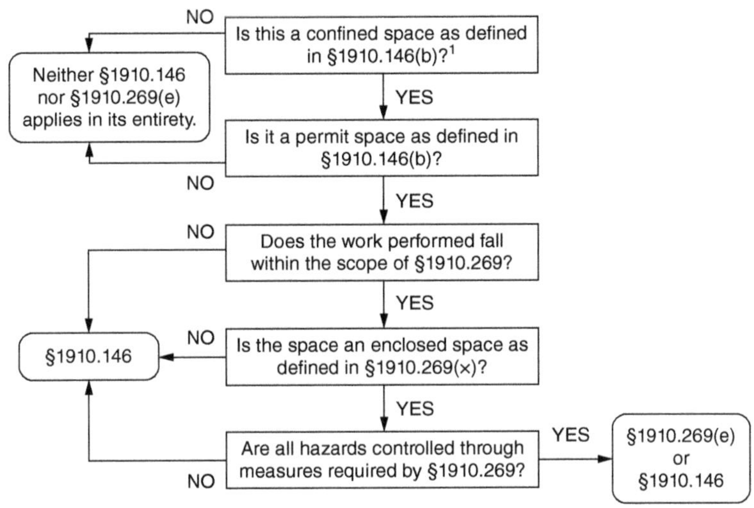

[1]See §1910.146(c) for general nonentry requirements that apply to all confined spaces.

FIGURE 7.3 (*Continued*)

Appendix B to § 1910.269—Working on Exposed Energized Parts

I. Introduction

Electric utilities design electric power generation, transmission, and distribution installations to meet National Electrical Safety Code (NESC), ANSI C2, requirements. Electric utilities also design transmission and distribution lines to limit line outages as required by system reliability criteria 1 and to withstand the maximum overvoltages impressed on the system. Conditions such as switching surges, faults, and lightning can cause overvoltages. Electric utilities generally select insulator design and lengths and the clearances to structural parts so as to prevent outages from contaminated line insulation and during storms. Line insulator lengths and structural clearances have, over the years, come closer to the minimum approach distances used by workers. As minimum approach distances and structural clearances converge, it is increasingly important that system designers and system operating and maintenance personnel understand the concepts underlying minimum approach distances.

The information in this appendix will assist employers in complying with the minimum approach-distance requirements contained in § 1910.269(l)(3) and (q)(3). Employers must use the technical criteria and methodology presented in this appendix in establishing minimum approach distances in accordance with § 1910.269(l)(3)(i) and Table R-3 and Table R-8. This appendix provides essential background information and technical criteria for the calculation of the required minimum approach distances for live-line work on electric power generation, transmission, and distribution installations.

Unless an employer is using the maximum transient overvoltages specified in Table R-9 for voltages over 72.5 kilovolts, the employer must use persons knowledgeable in the techniques discussed in this appendix, and competent in the field of electric transmission and distribution system design, to determine the maximum transient overvoltage.

II. General

A. *Definitions.* The following definitions from § 1910.269(x) relate to work on or near electric power generation, transmission, and distribution lines and equipment and the electrical hazards they present.

Exposed. Not isolated or guarded.

Guarded. Covered, fenced, enclosed, or otherwise protected, by means of suitable covers or casings, barrier rails or screens, mats, or platforms, designed to minimize the possibility, under normal conditions, of dangerous approach or inadvertent contact by persons or objects.

Note to the definition of "guarded": Wires that are insulated, but not otherwise protected, are not guarded.

Insulated. Separated from other conducting surfaces by a dielectric (including air space) offering a high resistance to the passage of current.

Note to the definition of "insulated": When any object is said to be insulated, it is understood to be insulated for the conditions to which it normally is subjected. Otherwise, it is, for the purpose of this section, uninsulated.

Isolated. Not readily accessible to persons unless special means for access are used.

Statistical sparkover voltage. A transient overvoltage level that produces a 97.72-percent probability of sparkover (that is, two standard deviations above the voltage at which there is a 50-percent probability of sparkover).

Statistical withstand voltage. A transient overvoltage level that produces a 0.14-percent probability of sparkover (that is, three standard deviations below the voltage at which there is a 50-percent probability of sparkover).

B. *Installations energized at 50 to 300 volts.* The hazards posed by installations energized at 50 to 300 volts are the same as those found in many other workplaces. That is not to say that there is no hazard, but the complexity of electrical protection required does not compare to that required for high-voltage systems. The employee must avoid contact with the exposed parts, and the protective equipment used (such as rubber insulating gloves) must provide insulation for the voltages involved.

C. *Exposed energized parts over 300 volts AC.* Paragraph (l)(3)(i) of § 1910.269 requires the employer to establish minimum approach distances no less than the distances computed by Table R-3 for ac systems so that employees can work safely without risk of sparkover.[2]

Unless the employee is using electrical protective equipment, air is the insulating medium between the employee and energized parts. The distance between the employee and an energized part must be sufficient for the air to

[1] Federal, State, and local regulatory bodies and electric utilities set reliability requirements that limit the number and duration of system outages.

[2] Sparkover is a disruptive electric discharge in which an electric arc forms and electric current passes through air.

FIGURE 7.3 (*Continued*)

withstand the maximum transient overvoltage that can reach the worksite under the working conditions and practices the employee is using. This distance is the minimum air insulation distance, and it is equal to the electrical component of the minimum approach distance.

Normal system design may provide or include a means (such as lightning arrestors) to control maximum anticipated transient overvoltages, or the employer may use temporary devices (portable protective gaps) or measures (such as preventing automatic circuit breaker reclosing) to achieve the same result. Paragraph (l)(3)(ii) of § 1910.269 requires the employer to determine the maximum anticipated per-unit transient overvoltage, phase-to-ground, through an engineering analysis or assume a maximum anticipated per-unit transient overvoltage, phase-to-ground, in accordance with Table R-9, which specifies the following maximums for ac systems:

- 72.6 to 420.0 kilovolts—3.5 per unit
- 420.1 to 550.0 kilovolts—3.0 per unit
- 550.1 to 800.0 kilovolts—2.5 per unit

See paragraph IV.A.2, later in this appendix, for additional discussion of maximum transient overvoltages.

D. *Types of exposures.* Employees working on or near energized electric power generation, transmission, and distribution systems face two kinds of exposures: Phase-to-ground and phase-to-phase. The exposure is phase-to-ground: (1) With respect to an energized part, when the employee is at ground potential or (2) with respect to ground, when an employee is at the potential of the energized part during live-line barehand work. The exposure is phase-to-phase, with respect to an energized part, when an employee is at the potential of another energized part (at a different potential) during live-line barehand work.

III. Determination of Minimum Approach Distances for AC Voltages Greater Than 300 Volts

A. *Voltages of 301 to 5,000 volts.* Test data generally forms the basis of minimum air insulation distances. The lowest voltage for which sufficient test data exists is 5,000 volts, and these data indicate that the minimum air insulation distance at that voltage is 20 millimeters (1 inch). Because the minimum air insulation distance increases with increasing voltage, and, conversely, decreases with decreasing voltage, an assumed minimum air insulation distance of 20 millimeters will protect against sparkover at voltages of 301 to 5,000 volts. Thus, 20 millimeters is the electrical component of the minimum approach distance for these voltages.

B. *Voltages of 5.1 to 72.5 kilovolts.* For voltages from 5.1 to 72.5 kilovolts, the Occupational Safety and Health Administration bases the methodology for calculating the electrical component of the minimum approach distance on Institute of Electrical and Electronic Engineers (IEEE) Standard 4–1995, *Standard Techniques for High-Voltage Testing.* Table 1 lists the critical sparkover distances from that standard as listed in IEEE Std 516–2009, *IEEE Guide for Maintenance Methods on Energized Power Lines.*

To use this table to determine the electrical component of the minimum approach distance, the employer must determine the peak phase-to-ground transient overvoltage and select a gap from the table that corresponds to that voltage as a withstand voltage rather than a critical sparkover voltage. To calculate the electrical component of the minimum approach distance for voltages between 5 and 72.5 kilovolts, use the following procedure:

1. Divide the phase-to-phase voltage by the square root of 3 to convert it to a phase-to-ground voltage.
2. Multiply the phase-to-ground voltage by the square root of 2 to convert the rms value of the voltage to the peak phase-to-ground voltage.
3. Multiply the peak phase-to-ground voltage by the maximum per-unit transient overvoltage, which, for this voltage range, is 3.0, as discussed later in this appendix. This is the maximum phase-to-ground transient overvoltage, which corresponds to the withstand voltage for the relevant exposure.[3]
4. Divide the maximum phase-to-ground transient overvoltage by 0.85 to determine the corresponding critical sparkover voltage. (The critical sparkover voltage is 3 standard deviations (or 15 percent) greater than the withstand voltage.)
5. Determine the electrical component of the minimum approach distance from Table 1 through interpolation. Table 2 illustrates how to derive the electrical component of the minimum approach distance for voltages from 5.1 to 72.5 kilovolts, before the application of any altitude correction factor, as explained later.

[3] The withstand voltage is the voltage at which sparkover is not likely to occur across a specified distance. It is the voltage taken at the 3s point below the sparkover voltage, assuming that the sparkover curve follows a normal distribution.

FIGURE 7.3 (*Continued*)

TABLE 1 Sparkover Distance for Rod-to-Rod Gap

60 Hz Rod-to-Rod sparkover (kV peak)	Gap spacing from IEEE Std 4–1995 (cm)
25	2
36	3
46	4
53	5
60	6
70	8
79	10
86	12
95	14
104	16
112	18
120	20
143	25
167	30
192	35
218	40
243	45
270	50
322	60

Source: IEEE Std 516–2009.

TABLE 2 Calculating the Electrical Component of MAD 751 V to 72.5 kV

Step	Maximum system phase-to-phase voltage (kV)			
	15	36	46	72.5
1. Divide by √3	8.7	20.8	26.6	41.9
2. Multiply by √2	12.2	29.4	37.6	59.2
3. Multiply by 3.0	36.7	88.2	112.7	177.6
4. Divide by 0.85	43.2	103.7	132.6	208.9
5. Interpolate from Table 1	3+(7.2/10)*1	14+(8.7/9)*2	20+(12.6/23)*5	35+(16.9/26)*5
Electrical component of MAD (cm)	3.72	15.93	22.74	38.25

C. *Voltages of 72.6 to 800 kilovolts.* For voltages of 72.6 kilovolts to 800 kilovolts, this section bases the electrical component of minimum approach distances, before the application of any altitude correction factor, on the following formula:

Equation 1—For Voltages of 72.6 kV to 800 kV

$D = 0.3048(C + a)\ VL\text{-}GT$

Where:
D = Electrical component of the minimum approach distance in air in meters;
C = a correction factor associated with the variation of gap sparkover with voltage;
a = A factor relating to the saturation of air at system voltages of 345 kilovolts or higher;[4]

[4] Test data demonstrates that the saturation factor is greater than 0 at peak voltages of about 630 kilovolts. Systems operating at 345 kilovolts (or maximum system voltages of 362 kilovolts) can have peak maximum transient overvoltages exceeding 630 kilovolts. Table R–3 sets equations for calculating *a* based on peak voltage.

FIGURE 7.3 (*Continued*)

TABLE 3 Equations for Calculating the Surge Factor, *a*

	Phase-to-ground exposures		
$V_{Peak} = T_{L\text{-}G}V_{L\text{-}G}\sqrt{2}$	635 kV or less	635.1 to 915 kV	915.1 to 1,050 kV
a	0	$(V_{Peak}$-635)/140,000	$(V_{Peak}$-645)1/135,0000
$V_{Peak} = T_{L\text{-}G}V_{L\text{-}G}\sqrt{2}$	More than 1,050 kV		
a	$(V_{Peak}$-675)/125,000		
	Phase-to-phase exposures[1]		
$V_{Peak} = (1.35T_{L\text{-}G} + 0.45)V_{L\text{-}G}\sqrt{2}$	630 kV or less	630.1 to 848 kV	848.1 to 1,131 kV
a	0	$(V_{Peak}$-630)/155,000	$(V_{Peak}$-633.6)/152,207
$V_{Peak} = (1.35T_{L\text{-}G} + 0.45)V_{L\text{-}G}\sqrt{2}$	1,131.1 to 1,485 kV		More than 1,485 kV
a	$(V_{Peak}$-628)/153,846		$(V_{Peak}$-350.5)/203,666

[1]Use the equations for phase-to-ground exposures (with V_{Peak} for phase-to-phase exposures) unless the employer can demonstrate that no insulated tool spans the gap and that no large conductive object is in the gap.

VL-G = Maximum system line-to-ground rms voltage in kilovolts—it should be the "actual" maximum, or the normal highest voltage for the range (for example, 10 percent above the nominal voltage); and

T = Maximum transient overvoltage factor in per unit.

In Equation 1, *C* is 0.01: (1) For phase-to-ground exposures that the employer can demonstrate consist only of air across the approach distance (gap) and (2) for phase-to-phase exposures if the employer can demonstrate that no insulated tool spans the gap and that no large conductive object is in the gap. Otherwise, *C* is 0.011.

In Equation 1, the term *a* varies depending on whether the employee's exposure is phase-to-ground or phase-to-phase and on whether objects are in the gap. The employer must use the equations in Table 3 to calculate *a*. Sparkover test data with insulation spanning the gap form the basis for the equations for phase-to-ground exposures, and sparkover test data with only air in the gap form the basis for the equations for phase-to-phase exposures. The phase-to-ground equations result in slightly higher values of *a*, and, consequently, produce larger minimum approach distances, than the phase-to-phase equations for the same value of VPeak.

In Equation 1, *T* is the maximum transient overvoltage factor in per unit. As noted earlier, § 1910.269(l)(3)(ii) requires the employer to determine the maximum anticipated per-unit transient overvoltage, phase-to-ground, through an engineering analysis or assume a maximum anticipated per-unit transient overvoltage, phase-to-ground, in accordance with Table R-9. For phase-to-ground exposures, the employer uses this value, called *TL-G*, as *T* in Equation 1. IEEE Std 516–2009 provides the following formula to calculate the phase-to-phase maximum transient overvoltage, *TL-L*, from *TL-G*:

TL-L = 1.35*TL-G* + 0.45

For phase-to-phase exposures, the employer uses this value as *T* in Equation 1.

D. *Provisions for inadvertent movement.* The minimum approach distance must include an "adder" to compensate for the inadvertent movement of the worker relative to an energized part or the movement of the part relative to the worker. This "adder" must account for this possible inadvertent movement and provide the worker with a comfortable and safe zone in which to work. Employers must add the distance for inadvertent movement (called the "ergonomic component of the minimum approach distance") to the electrical component to determine the total safe minimum approach distances used in live line work.

The Occupational Safety and Health Administration based the ergonomic component of the minimum approach distance on response time-distance analysis. This technique uses an estimate of the total response time to a hazardous incident and converts that time to the distance traveled. For example, the driver of a car takes a given amount of time to respond to a "stimulus" and stop the vehicle. The elapsed time involved results in the car's traveling some distance before coming to a complete stop. This distance depends on the speed of the car at the time the stimulus appears and the reaction time of the driver. In the case of live-line work, the employee must first perceive that he

FIGURE 7.3 (*Continued*)

TABLE 4 Ergonomic Component of Minimum Approach Distance

Voltage range (kV)	Distance	
	m	ft
0.301 to 0.750	0.31	1.0
0.751 to 72.5	0.61	2.0
72.6 to 800	0.31	1.0

Note: The employer must add this distance to the electrical component of the minimum approach distance to obtain the full minimum approach distance.

or she is approaching the danger zone. Then, the worker responds to the danger and must decelerate and stop all motion toward the energized part. During the time it takes to stop, the employee will travel some distance. This is the distance the employer must add to the electrical component of the minimum approach distance to obtain the total safe minimum approach distance. At voltages from 751 volts to 72.5 kilovolts,[5] the electrical component of the minimum approach distance is smaller than the ergonomic component.

At 72.5 kilovolts, the electrical component is only a little more than 0.3 meters (1 foot). An ergonomic component of the minimum approach distance must provide for all the worker's unanticipated movements. At these voltages, workers generally use rubber insulating gloves; however, these gloves protect only a worker's hands and arms. Therefore, the energized object must be at a safe approach distance to protect the worker's face. In this case, 0.61 meters (2 feet) is a sufficient and practical ergonomic component of the minimum approach distance. For voltages between 72.6 and 800 kilovolts, employees must use different work practices during energized line work.

Generally, employees use live-line tools (hot sticks) to perform work on energized equipment. These tools, by design, keep the energized part at a constant distance from the employee and, thus, maintain the appropriate minimum approach distance automatically. The location of the worker and the type of work methods the worker is using also influence the length of the ergonomic component of the minimum approach distance. In this higher voltage range, the employees use work methods that more tightly control their movements than when the workers perform work using rubber insulating gloves. The worker, therefore, is farther from the energized line or equipment and must be more precise in his or her movements just to perform the work. For these reasons, this section adopts an ergonomic component of the minimum approach distance of 0.31 m (1 foot) for voltages between 72.6 and 800 kilovolts. Table 4 summarizes the ergonomic component of the minimum approach distance for various voltage ranges.

The ergonomic component of the minimum approach distance accounts for errors in maintaining the minimum approach distance (which might occur, for example, if an employee misjudges the length of a conductive object he or she is holding), and for errors in judging the minimum approach distance. The ergonomic component also accounts for inadvertent movements by the employee, such as slipping. In contrast, the working position selected to properly maintain the minimum approach distance must account for all of an employee's reasonably likely movements and still permit the employee to adhere to the applicable minimum approach distance. (See Figure 1.) Reasonably likely movements include an employee's adjustments to tools, equipment, and working positions and all movements needed to perform the work. For example, the employee should be able to perform all of the following actions without straying into the minimum approach distance:

- Adjust his or her hardhat,
- maneuver a tool onto an energized part with a reasonable amount of overreaching or underreaching,
- reach for and handle tools, material, and
- equipment passed to him or her, and
- adjust tools, and replace components on them, when necessary during the work procedure.

The training of qualified employees required under § 1910.269(a)(2), and the job planning and briefing required under § 1910.269(c), must address selection of a proper working position.

[5] For voltages of 50 to 300 volts, Table R-3 specifies a minimum approach distance of "avoid contact." The minimum approach distance for this voltage range contains neither an electrical component nor an ergonomic component.

FIGURE 7.3 *(Continued)*

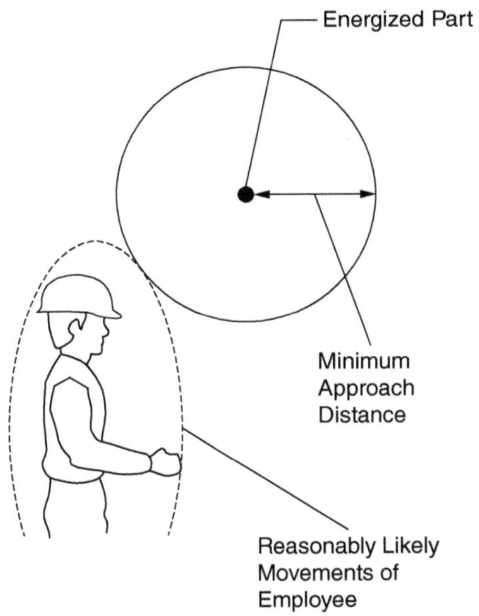

FIGURE 1 Maintaining the Minimum Approach Distance

E. ***Miscellaneous correction factors.*** Changes in the air medium that forms the insulation influences the strength of an air gap. A brief discussion of each factor follows.

1. ***Dielectric strength of air.*** The dielectric strength of air in a uniform electric field at standard atmospheric conditions is approximately 3 kilovolts per millimeter.[6]

The pressure, temperature, and humidity of the air, the shape, dimensions, and separation of the electrodes, and the characteristics of the applied voltage (wave shape) affect the disruptive gradient.

2. ***Atmospheric effect.*** The empirically determined electrical strength of a given gap is normally applicable at standard atmospheric conditions (20°C, 101.3 kilopascals, 11 grams/cubic centimeter humidity). An increase in the density (humidity) of the air inhibits sparkover for a given air gap. The combination of temperature and air pressure that results in the lowest gap sparkover voltage is high temperature and low pressure. This combination of conditions is not likely to occur. Low air pressure, generally associated with high humidity, causes increased electrical strength. An average air pressure generally correlates with low humidity. Hot and dry working conditions normally result in reduced electrical strength. The equations for minimum approach distances in Table R-3 assume standard atmospheric conditions.

3. ***Altitude.*** The reduced air pressure at high altitudes causes a reduction in the electrical strength of an air gap. An employer must increase the minimum approach distance by about 3 percent per 300 meters (1,000 feet) of increased altitude for altitudes above 900 meters (3,000 feet). Table R-5 specifies the altitude correction factor that the employer must use in calculating minimum approach distances.

IV. Determining Minimum Approach Distances

A. Factors Affecting Voltage Stress at the Worksite

1. ***System voltage (nominal).*** The nominal system voltage range determines the voltage for purposes of calculating minimum approach distances. The employer selects the range in which the nominal system voltage falls, as given in the relevant table, and uses the highest value within that range in per-unit calculations.

[6]For the purposes of estimating arc length, § 1910.269 generally assumes a more conservative dielectric strength of 10 kilovolts per 25.4 millimeters, consistent with assumptions made in consensus standards such as the National Electrical Safety Code (IEEE C2–2012). The more conservative value accounts for variables such as electrode shape, wave shape, and a certain amount of overvoltage.

FIGURE 7.3 (*Continued*)

TABLE 5 Magnitude of Typical Transient Overvoltages

Cause	Magnitude (per unit)
Energized 200-mile line without closing resistors	3.5
Energized 200-mile line with one-step closing resistor	2.1
Energized 200-mile line with multistep resistor	2.5
Reclosing with trapped charge one-step resistor	2.2
Opening surge with single restrike	3.0
Fault initiation unfaulted phase	2.1
Fault initiation adjacent circuit	2.5
Fault clearing	1.7 to 1.9

2. *Transient overvoltages.* Operation of switches or circuit breakers, a fault on a line or circuit or on an adjacent circuit, and similar activities may generate transient overvoltages on an electrical system. Each overvoltage has an associated transient voltage wave shape. The wave shape arriving at the site and its magnitude vary considerably. In developing requirements for minimum approach distances, the Occupational Safety and Health Administration considered the most common wave shapes and the magnitude of transient overvoltages found on electric power generation, transmission, and distribution systems. The equations in Table R-3 for minimum approach distances use per-unit maximum transient overvoltages, which are relative to the nominal maximum voltage of the system. For example, a maximum transient overvoltage value of 3.0 per unit indicates that the highest transient overvoltage is 3.0 times the nominal maximum system voltage.

3. *Typical magnitude of overvoltages.* Table 5 lists the magnitude of typical transient overvoltages.

4. *Standard deviation—air-gap withstand.* For each air gap length under the same atmospheric conditions, there is a statistical variation in the breakdown voltage. The probability of breakdown against voltage has a normal (Gaussian) distribution. The standard deviation of this distribution varies with the wave shape, gap geometry, and atmospheric conditions. The withstand voltage of the air gap is three standard deviations (3s) below the critical sparkover voltage. (The critical sparkover voltage is the crest value of the impulse wave that, under specified conditions, causes sparkover 50 percent of the time. An impulse wave of three standard deviations below this value, that is, the withstand voltage, has a probability of sparkover of approximately 1 in 1,000.)

5. *Broken Insulators.* Tests show reductions in the insulation strength of insulator strings with broken skirts. Broken units may lose up to 70 percent of their withstand capacity. Because an employer cannot determine the insulating capability of a broken unit without testing it, the employer must consider damaged units in an insulator to have no insulating value. Additionally, the presence of a live-line tool alongside an insulator string with broken units may further reduce the overall insulating strength. The number of good units that must be present in a string for it to be "insulated" as defined by § 1910.269(x) depends on the maximum overvoltage possible at the worksite.

B. Minimum Approach Distances Based on Known, Maximum-Anticipated Per-Unit Transient Overvoltages

1. *Determining the minimum approach distance for AC systems.* Under §1910.269(l)(3)(ii), the employer must determine the maximum anticipated per-unit transient overvoltage, phase-to-ground, through an engineering analysis or must assume a maximum anticipated per-unit transient overvoltage, phase-to-ground, in accordance with Table R-9. When the employer conducts an engineering analysis of the system and determines that the maximum transient overvoltage is lower than specified by Table R-9, the employer must ensure that any conditions assumed in the analysis, for example, that employees block reclosing on a circuit or install portable protective gaps, are present during energized work. To ensure that these conditions are present, the employer may need to institute new live-work procedures reflecting the conditions and limitations set by the engineering analysis.

2. *Calculation of reduced approach distance values.* An employer may take the following steps to reduce minimum approach distances when the maximum transient overvoltage on the system (that is, the maximum transient overvoltage without additional steps to control overvoltages) produces unacceptably large minimum approach distances:

Step 1. Determine the maximum voltage (with respect to a given nominal voltage range) for the energized part.

Step 2. Determine the technique to use to control the maximum transient overvoltage. (See paragraphs IV.C and IV.D of this appendix.) Determine the maximum transient overvoltage that can exist at the worksite with that form of

FIGURE 7.3 *(Continued)*

control in place and with a confidence level of 3s. This voltage is the withstand voltage for the purpose of calculating the appropriate minimum approach distance.

Step 3. Direct employees to implement procedures to ensure that the control technique is in effect during the course of the work.

Step 4. Using the new value of transient overvoltage in per unit, calculate the required minimum approach distance from Table R-3.

C. Methods of Controlling Possible Transient Overvoltage Stress Found on a System

1. *Introduction.* There are several means of controlling overvoltages that occur on transmission systems. For example, the employer can modify the operation of circuit breakers or other switching devices to reduce switching transient overvoltages. Alternatively, the employer can hold the overvoltage to an acceptable level by installing surge arresters or portable protective gaps on the system. In addition, the employer can change the transmission system to minimize the effect of switching operations. Section 4.8 of IEEE Std 516–2009 describes various ways of controlling, and thereby reducing, maximum transient overvoltages.

2. *Operation of circuit breakers.*[7]

The maximum transient overvoltage that can reach the worksite is often the result of switching on the line on which employees are working. Disabling automatic reclosing during energized line work, so that the line will not be reenergized after being opened for any reason, limits the maximum switching surge overvoltage to the larger of the opening surge or the greatest possible fault-generated surge, provided that the devices (for example, insertion resistors) are operable and will function to limit the transient overvoltage and that circuit breaker restrikes do not occur. The employer must ensure the proper functioning of insertion resistors and other overvoltage-limiting devices when the employer's engineering analysis assumes their proper operation to limit the overvoltage level. If the employer cannot disable the reclosing feature (because of system operating conditions), other methods of controlling the switching surge level may be necessary. Transient surges on an adjacent line, particularly for double circuit construction, may cause a significant overvoltage on the line on which employees are working. The employer's engineering analysis must account for coupling to adjacent lines.

3. *Surge arresters.* The use of modern surge arresters allows a reduction in the basic impulse-insulation levels of much transmission system equipment. The primary function of early arresters was to protect the system insulation from the effects of lightning. Modern arresters not only dissipate lightning-caused transients, but may also control many other system transients caused by switching or faults. The employer may use properly designed arresters to control transient overvoltages along a transmission line and thereby reduce the requisite length of the insulator string and possibly the maximum transient overvoltage on the line.[8]

4. *Switching Restrictions.* Another form of overvoltage control involves establishing switching restrictions, whereby the employer prohibits the operation of circuit breakers until certain system conditions are present. The employer restricts switching by using a tagging system, similar to that used for a permit, except that the common term used for this activity is a "hold-off" or "restriction." These terms indicate that the restriction does not prevent operation, but only modifies the operation during the live-work activity.

D. Minimum Approach Distance Based on Control of Maximum Transient Overvoltage at the Worksite. When the employer institutes control of maximum transient overvoltage at the worksite by installing portable protective gaps, the employer may calculate the minimum approach distance as follows:

Step 1. Select the appropriate withstand voltage for the protective gap based on system requirements and an acceptable probability of gap sparkover.[9]

Step 2. Determine a gap distance that provides a withstand voltage[10] greater than or equal to the one selected in the first step.[11]

[7] The detailed design of a circuit interrupter, such as the design of the contacts, resistor insertion, and breaker timing control, are beyond the scope of this appendix. The design of the system generally accounts for these features. This appendix only discusses features that can limit the maximum switching transient overvoltage on a system.

[8] Surge arrester application is beyond the scope of this appendix. However, if the employer installs the arrester near the work site, the application would be similar to the protective gaps discussed in paragraph IV.D of this appendix.

[9] The employer should check the withstand voltage to ensure that it results in a probability of gap flashover that is acceptable from a system outage perspective. (In other words, a gap sparkover will produce a system outage. The employer should determine whether such an outage will impact overall system performance to an acceptable degree.) In general, the withstand voltage should be at least 1.25 times the maximum crest operating voltage.

[10] The manufacturer of the gap provides, based on test data, the critical sparkover voltage for each gap spacing (for example, a critical sparkover voltage of 665 kilovolts for a gap spacing of 1.2 meters). The withstand voltage for the gap is equal to 85 percent of its critical sparkover voltage.

[11] Switch steps 1 and 2 if the length of the protective gap is known.

FIGURE 7.3 *(Continued)*

Step 3. Use 110 percent of the gap's critical sparkover voltage to determine the phase-to-ground peak voltage at gap sparkover (*VPPG Peak*).

Step 4. Determine the maximum transient overvoltage, phase-to-ground, at the worksite from the following formula:

$$T = \frac{V_{PPG\ Peak}}{V_{L-G}\sqrt{2}}.$$

Step 5. Use this value of *T*[12] in the equation in Table R-3 to obtain the minimum approach distance.

If the worksite is no more than 900 meters (3,000 feet) above sea level, the employer may use this value of *T* to determine the minimum approach distance from Table 7 through Table 14.

Note: All rounding must be to the next higher value (that is, always round up).

Sample protective gap calculations.

Problem: Employees are to perform work on a 500-kilovolt transmission line at sea level that is subject to transient overvoltages of 2.4 p.u. The maximum operating voltage of the line is 550 kilovolts. Determine the length of the protective gap that will provide the minimum practical safe approach distance. Also, determine what that minimum approach distance is.

Step 1. Calculate the smallest practical maximum transient overvoltage (1.25 times the crest phase-to-ground voltage):[13]

$$550\ \text{kV} \times \frac{\sqrt{2}}{\sqrt{3}} \times 1.25 = 561\ \text{kV}.$$

This value equals the withstand voltage of the protective gap.

Step 2. Using test data for a particular protective gap, select a gap that has a critical sparkover voltage greater than or equal to:

561 kV ÷ 0.85 = 660 kV

For example, if a protective gap with a 1.22-m (4.0-foot) spacing tested to a critical sparkover voltage of 665 kilovolts (crest), select this gap spacing.

Step 3. The phase-to-ground peak voltage at gap sparkover (*VPPG Peak*) is 110 percent of the value from the previous step:

665 kV × 1.10 = 732 kV

This value corresponds to the withstand voltage of the electrical component of the minimum approach distance.

Step 4. Use this voltage to determine the worksite value of *T*:

$$T = \frac{732}{564} = 1.7\ \text{p.u.}$$

Step 5. Use this value of *T* in the equation in Table R-3 to obtain the minimum approach distance, or look up the minimum approach distance in Table 7 through Table 14:

MAD = 2.29m (7.6 ft).

E. Location of Protective Gaps

1. ***Adjacent structures.*** The employer may install the protective gap on a structure adjacent to the worksite, as this practice does not significantly reduce the protection afforded by the gap.

2. ***Terminal stations.*** Gaps installed at terminal stations of lines or circuits provide a level of protection; however, that level of protection may not extend throughout the length of the line to the worksite. The use of substation terminal gaps raises the possibility that separate surges could enter the line at opposite ends, each with low enough magnitude to pass the terminal gaps without sparkover. When voltage surges occur simultaneously at each end of a line and travel toward each other, the total voltage on the line at the point where they meet is the arithmetic sum of

[12] IEEE Std 516–2009 states that most employers add 0.2 to the calculated value of *T* as an additional safety factor.

[13] To eliminate sparkovers due to minor system disturbances, the employer should use a withstand voltage no lower than 1.25 p.u. Note that this is a practical, or operational, consideration only. It may be feasible for the employer to use lower values of withstand voltage.

FIGURE 7.3 (*Continued*)

the two surges. A gap installed within 0.8 km (0.5 mile) of the worksite will protect against such intersecting waves. Engineering studies of a particular line or system may indicate that employers can adequately protect employees by installing gaps at even more distant locations. In any event, unless using the default values for *T* from Table R-9, the employer must determine *T* at the worksite.

3. *Worksite.* If the employer installs protective gaps at the worksite, the gap setting establishes the worksite impulse insulation strength. Lightning strikes as far as 6 miles from the worksite can cause a voltage surge greater than the gap withstand voltage, and a gap sparkover can occur. In addition, the gap can sparkover from overvoltages on the line that exceed the withstand voltage of the gap. Consequently, the employer must protect employees from hazards resulting from any sparkover that could occur. In addition, the gap can sparkover from over-voltages on the line that exceed the withstand voltage of the gap. Consequently, the employer must protect employees from hazards resulting from any sparkover that could occur.

F. *Disabling automatic reclosing.* There are two reasons to disable the automatic-reclosing feature of circuit-interrupting devices while employees are performing live-line work:

To prevent reenergization of a circuit faulted during the work, which could create a hazard or result in more serious injuries or damage than the injuries or damage produced by the original fault;

To prevent any transient overvoltage caused by the switching surge that would result if the circuit were reenergized. However, due to system stability considerations, it may not always be feasible to disable the automatic-reclosing feature.

V. Minimum Approach-Distance Tables

A. *Legacy tables.* Employers may use the minimum approach distances in Table 6 through Table 13 until March 31, 2015.

TABLE 6 Minimum Approach Distances Until March 31, 2015

Voltage range phase to phase (kV)	Phase-to-ground exposure		Phase-to-phase exposure	
	m	ft	m	ft
0.05 to 1.0	Avoid Contact		Avoid Contact	
1.1 to 15.0	2.10	0.64	2.20	0.66
15.1 to 36.0	2.30	0.72	2.60	0.77
36.1 to 46.0	2.60	0.77	2.80	0.85
46.1 to 72.5	3.00	0.90	3.50	1.05
72.6 to 121	3.20	0.95	4.30	1.29
138 to 145	3.60	1.09	4.90	1.50
161 to 169	4.00	1.22	5.70	1.71
230 to 242	5.30	1.59	7.50	2.27
345 to 362	8.50	2.59	12.50	3.80
500 to 550	11.30	3.42	18.10	5.50
765 to 800	14.90	4.53	26.00	7.91

Note: The clear live-line tool distance must equal or exceed the values for the indicated voltage ranges.

FIGURE 7.3 (*Continued*)

TABLE 7 Minimum Approach Distances Until March 31, 2015—72.6 to 121.0 kV with Overvoltage Factor

T (p.u.)	Phase-to-ground exposure		Phase-to-phase exposure	
	m	ft	m	ft
2.0	0.74	2.42	1.09	3.58
2.1	0.76	2.50	1.09	3.58
2.2	0.79	2.58	1.12	3.67
2.3	0.81	2.67	1.14	3.75
2.4	0.84	2.75	1.17	3.83
2.5	0.84	2.75	1.19	3.92
2.6	0.86	2.83	1.22	4.00
2.7	0.89	2.92	1.24	4.08
2.8	0.91	3.00	1.24	4.08
2.9	0.94	3.08	1.27	4.17
3.0	0.97	3.17	1.30	4.25

Note 1: The employer may apply the distance specified in this table only where the employer determines the maximum anticipated per-unit transient overvoltage by engineering analysis. (Table 6 applies otherwise.)
Note 2: The distances specified in this table are the air, bare-hand, and live-line tool distances.

TABLE 8 Minimum Approach Distances Until March 31, 2015—121.1 to 145.0 kV with Overvoltage Factor

T (p.u.)	Phase-to-ground exposure		Phase-to-phase exposure	
	m	ft	m	ft
2.0	0.84	2.75	1.24	4.08
2.1	0.86	2.83	1.27	4.17
2.2	0.89	2.92	1.30	4.25
2.3	0.91	3.00	1.32	4.33
2.4	0.94	3.08	1.35	4.42
2.5	0.97	3.17	1.37	4.50
2.6	0.99	3.25	1.40	4.58
2.7	1.02	3.33	1.42	4.67
2.8	1.04	3.42	1.45	4.75
2.9	1.07	3.50	1.47	4.83
3.0	1.09	3.58	1.50	4.92

Note 1: The employer may apply the distance specified in this table only where the employer determines the maximum anticipated per-unit transient overvoltage by engineering analysis. (Table 6 applies otherwise.)
Note 2: The distances specified in this table are the air, bare-hand, and live-line tool distances.

FIGURE 7.3 *(Continued)*

TABLE 9 Minimum Approach Distances Until March 31, 2015—145.1 to 169.0 kV with Overvoltage Factor

T (p.u.)	Phase-to-ground exposure		Phase-to-phase exposure	
	m	ft	m	ft
2.0	0.91	3.00	1.42	4.67
2.1	0.97	3.17	1.45	4.75
2.2	0.99	3.25	1.47	4.83
2.3	1.02	3.33	1.50	4.92
2.4	1.04	3.42	1.52	5.00
2.5	1.07	3.50	1.57	5.17
2.6	1.12	3.67	1.60	5.25
2.7	1.14	3.75	1.63	5.33
2.8	1.17	3.83	1.65	5.42
2.9	1.19	3.92	1.68	5.50
3.0	1.22	4.00	1.73	5.67

Note 1: The employer may apply the distance specified in this table only where the employer determines the maximum anticipated per-unit transient overvoltage by engineering analysis. (Table 6 applies otherwise.)

Note 2: The distances specified in this table are the air, bare-hard, and live-line tool distances.

TABLE 10 Minimum Approach Distances Until March 31, 2015—169.1 to 242.0 kV with Overvoltage Factor

T (p.u.)	Phase-to-ground exposure		Phase-to-phase exposure	
	m	ft	m	ft
2.0	1.17	3.83	1.85	6.08
2.1	1.22	4.00	1.91	6.25
2.2	1.24	4.08	1.93	6.33
2.3	1.30	4.25	1.98	6.50
2.4	1.35	4.42	2.01	6.58
2.5	1.37	4.50	2.06	6.75
2.6	1.42	4.67	2.11	6.92
2.7	1.47	4.83	2.13	7.00
2.8	1.50	4.92	2.18	7.17
2.9	1.55	5.08	2.24	7.33
3.0	1.60	5.25	2.29	7.50

Note 1: The employer may apply the distance specified in this table only where the employer determines the maximum anticipated per-unit transient overvoltage by engineering analysis. (Table 6 applies otherwise.)

Note 2: The distances specified in this table are the air, bare-hand, and live-line tool distances.

FIGURE 7.3 (*Continued*)

TABLE 11 Minimum Approach Distances Until March 31, 2015—242.1 to 362.0 kV with Overvoltage Factor

T (p.u.)	Phase-to-ground exposure		Phase-to-phase exposure	
	m	ft	m	ft
2.0	1.60	5.25	2.62	8.58
2.1	1.65	5.42	2.69	8.83
2.2	1.75	5.75	2.79	9.17
2.3	1.85	6.08	2.90	9.50
2.4	1.93	6.33	3.02	9.92
2.5	2.03	6.67	3.15	10.33
2.6	2.16	7.08	3.28	10.75
2.7	2.26	7.42	3.40	11.17
2.8	2.36	7.75	3.53	11.58
2.9	2.49	8.17	3.68	12.08
3.0	2.59	8.50	3.81	12.50

Note 1: The employer may apply the distance specified in this table only where the employer determines the maximum anticipated per-unit transient overvoltage by engineering analysis. (Table 6 applies otherwise.)

Note 2: The distances specified in this table are the air, bare-hand, and live-line tool distances.

TABLE 12 Minimum Approach Distances Unit March 31, 2015—362.1 to 552.0 kV with Overvoltage Factor

T (p.u.)	Phase-to-ground exposure		Phase-to-phase exposure	
	m	ft	m	ft
1.5	1.83	6.00	2.24	7.33
1.6	1.98	6.50	2.67	8.75
1.7	2.13	7.00	3.10	10.17
1.8	2.31	7.58	3.53	11.58
1.9	2.46	8.08	4.01	13.17
2.0	2.67	8.75	4.52	14.83
2.1	2.84	9.33	4.75	15.58
2.2	3.02	9.92	4.98	16.33
2.3	3.20	10.50	5.23	17.17
2.4	3.43	11.25	5.51	18.08

Note 1: The employer may apply the distance specified in this table only where the employer determines the maximum anticipated per-unit transient overvoltage by engineering analysis. (Table 6 applies otherwise.)

Note 2: The distances specified in this table are the air, bare-hand, and live-line tool distances.

FIGURE 7.3 (*Continued*)

B. *Alternative minimum approach distances.* Employers may use the minimum approach distances in Table 14 through Table 21 provided that the employer follows the notes to those tables.

TABLE 13 Minimum Approach Distances Until March 31, 2015—552.1 to 800.0 kV with Overvoltage Factor

T (p.u.)	Phase-to-ground exposure		Phase-to-phase exposure	
	m	ft	m	ft
1.5	2.95	9.67	3.68	12.08
1.6	3.25	10.67	4.42	14.50
1.7	3.56	11.67	5.23	17.17
1.8	3.86	12.67	6.07	19.92
1.9	4.19	13.75	6.99	22.92
2.0	4.55	14.92	7.92	26.00

Note 1: The employer may apply the distance specified in this table only where the employer determines the maximum anticipated per-unit transient overvoltage by engineering analysis. (Table 6 applies otherwise.)

Note 2: The distances specified in this table are the air, bare-hand, and live-line tool distances.

TABLE 14 AC Minimum Approach Distances—72.6 to 121.0 kV

T (p.u.)	Phase-to-ground exposure		Phase-to-phase exposure	
	m	ft	m	ft
1.5	0.67	2.2	0.84	2.8
1.6	0.69	2.3	0.87	2.9
1.7	0.71	2.3	0.90	3.0
1.8	0.74	2.4	0.93	3.1
1.9	0.76	2.5	0.96	3.1
2.0	0.78	2.6	0.99	3.2
2.1	0.81	2.7	1.01	3.3
2.2	0.33	2.7	1.04	3.4
2.3	0.85	2.8	1.07	3.5
2.4	0.88	2.9	1.10	3.6
2.5	0.90	3.0	1.13	3.7
2.6	0.92	3.0	1.16	3.8
2.7	0.95	3.1	1.19	3.9
2.8	0.97	3.2	1.22	4.0
2.9	0.99	3.2	1.24	4.1
3.0	1.02	3.3	1.27	4.2
3.1	1.04	3.4	1.30	4.3
3.2	1.06	3.5	1.33	4.4
3.3	1.09	3.6	1.36	4.5
3.4	1.11	3.6	1.39	4.6
3.5	1.13	3.7	1.42	4.7

FIGURE 7.3 (*Continued*)

TABLE 15 AC Minimum Approach Distances—121.1 to 145.0 kV

T (p.u.)	Phase-to-ground exposure		Phase-to-phase exposure	
	m	ft	m	ft
1.5	0.74	2.4	0.95	3.1
1.6	0.76	2.5	0.98	3.2
1.7	0.79	2.6	1.02	3.3
1.8	0.82	2.7	1.05	3.4
1.9	0.85	2.8	1.08	3.5
2.0	0.88	2.9	1.12	3.7
2.1	0.90	3.0	1.15	3.8
2.2	0.93	3.1	1.19	3.9
2.3	0.96	3.1	1.22	4.0
2.4	0.99	3.2	1.26	4.1
2.5	1.02	3.3	1.29	4.2
2.6	1.04	3.4	1.33	4.4
2.7	1.07	3.5	1.36	4.5
2.8	1.10	3.6	1.39	4.6
2.9	1.13	3.7	1.43	4.7
3.0	1.16	3.8	1.46	4.8
3.1	1.19	3.9	1.50	4.9
3.2	1.21	4.0	1.53	5.0
3.3	1.24	4.1	1.57	5.2
3.4	1.27	4.2	1.60	5.2
3.5	1.30	4.3	1.64	5.4

TABLE 16 AC Minimum Approach Distances—145.1 to 169.0 kV

T (p.u.)	Phase-to-ground exposure		Phase-to-phase exposure	
	m	ft	m	ft
1.5	0.81	2.7	1.05	3.4
1.6	0.84	2.8	1.09	3.6
1.7	0.87	2.9	1.13	3.7
1.8	0.90	3.0	1.17	3.8
1.9	0.94	3.1	1.21	4.0
2.0	0.97	3.2	1.25	4.1
2.1	1.00	3.3	1.29	4.2
2.2	1.03	3.4	1.33	4.4
2.3	1.07	3.5	1.37	4.5
2.4	1.10	3.6	1.41	4.6
2.5	1.13	3.7	1.45	4.8
2.6	1.17	3.8	1.49	4.9
2.7	1.20	3.9	1.53	5.0
2.8	1.23	4.0	1.57	5.2
2.9	1.26	4.1	1.61	5.3
3.0	1.30	4.3	1.65	5.4
3.1	1.33	4.4	1.70	5.6
3.2	1.36	4.5	1.76	5.8
3.3	1.39	4.6	1.82	6.0
3.4	1.43	4.7	1.88	6.2
3.5	1.46	4.8	1.94	6.4

FIGURE 7.3 (*Continued*)

TABLE 17 AC Minimum Approach Distances—169.1 to 242.kV

T (p.u.)	Phase-to-ground exposure		Phase-to-phase exposure	
	m	ft	m	ft
1.5	1.02	3.3	1.37	4.5
1.6	1.06	3.5	1.43	4.7
1.7	1.11	3.6	1.48	4.9
1.8	1.16	3.8	1.54	5.1
1.9	1.21	4.0	1.60	5.2
2.0	1.25	4.1	1.66	5.4
2.1	1.30	4.3	1.73	5.7
2.2	1.35	4.4	1.81	5.9
2.3	1.39	4.6	1.90	6.2
2.4	1.44	4.7	1.99	6.5
2.5	1.49	4.9	2.08	6.8
2.6	1.53	5.0	2.17	7.1
2.7	1.58	5.2	2.26	7.4
2.8	1.63	5.3	2.36	7.7
2.9	1.67	5.5	2.45	8.0
3.0	1.72	5.6	2.55	8.4
3.1	1.77	5.8	2.65	8.7
3.2	1.81	5.9	2.76	9.1
3.3	1.88	6.2	2.86	9.4
3.4	1.95	6.4	2.97	9.7
3.5	2.01	6.6	3.08	10.1

TABLE 18 AC Minimum Approach Distances—242.1 to 362.0 kV

T (p u.)	Phase-to-ground exposure		Phase-to-phase exposure	
	m	ft	m	ft
1.5	1.37	4.5	1.99	6.5
1.6	1.44	4.7	2.13	7.0
1.7	1.51	5.0	2.27	7.4
1.8	1.58	5.2	2.41	7.9
1.9	1.65	5.4	2.56	8.4
2.0	1.72	5.6	2.71	8.9
2.1	1.79	5.9	2.87	9.4
2.2	1.87	6.1	3.03	9.9
2.3	1.97	6.5	3.20	10.5
2.4	2.08	6.8	3.37	11.1
2.5	2.19	7.2	3.55	11.6
2.6	2.29	7.5	3.73	12.2
2.7	2.41	7.9	3.91	12.8
2.8	2.52	8.3	4.10	13.5
2.9	2.64	8.7	4.29	14.1
3.0	2.76	9.1	4.49	14.7
3.1	2.88	9.4	4.69	15.4
3.2	3.01	9.9	4.90	16.1
3.3	3.14	10.3	5.11	16.8
3.4	3.27	10.7	5.32	17.5
3.5	3.41	11.2	5.52	18.1

FIGURE 7.3 *(Continued)*

TABLE 19 AC Minimum Approach Distances—362.1 to 420.0 kV

T (p u.)	Phase-to-ground exposure		Phase-to-phase exposure	
	m	ft	m	ft
1.5	1.53	5.0	2.40	7.9
1.6	1.62	5.3	2.58	8.5
1.7	1.70	5.6	2.75	9.0
1.8	1.78	5.8	2.94	9.6
1.9	1.88	6.2	3.13	10.3
2.0	1.99	6.5	3.33	10.9
2.1	2.12	7.0	3.53	11.6
2.2	2.24	7.3	3.74	12.3
2.3	2.37	7.8	3.95	13.0
2.4	2.50	8.2	4.17	13.7
2.5	2.64	8.7	4.40	14.4
2.6	2.78	9.1	4.63	15.2
2.7	2.93	9.6	4.87	16.0
2.8	3.07	10.1	5.11	16.8
2.9	3.23	10.6	5.36	17.6
3.0	3.38	11.1	5.59	18.3
3.1	3.55	11.6	5.82	19.1
3.2	3.72	12.2	6.07	19.9
3.3	3.89	12.8	6.31	20.7
3.4	4.07	13.4	6.56	21.5
3.5	4.25	13.9	6.81	22.3

TABLE 20 AC Minimum Approach Distances—420.1 to 550.0 kV

T (p.u.)	Phase-to-ground exposure		Phase-to-phase exposure	
	m	ft	m	ft
1.5	1.95	6.4	3.46	11.4
1.6	2.11	6.9	3.73	12.2
1.7	2.28	7.5	4.02	13.2
1.8	2.45	8.0	4.31	14.1
1.9	2.62	8.6	4.61	15.1
2.0	2.81	9.2	4.92	16.1
2.1	3.00	9.8	5.25	17.2
2.2	3.20	10.5	5.55	18.2
2.3	3.40	11.2	5.86	19.2
2.4	3.62	11.9	6.18	20.3
2.5	3.84	12.6	6.50	21.3
2.6	4.07	13.4	6.83	22.4
2.7	4.31	14.1	7.18	23.6
2.8	4.56	15.0	7.52	24.7
2.9	4.81	15.8	7.88	25.9
3.0	5.07	16.6	8.24	27.0

FIGURE 7.3 *(Continued)*

TABLE 21 Ac Minimum Approach Distances—550.1 to 800.0 kV

T (p.u.)	Phase-to-ground exposure		Phase-to-phase exposure	
	m	ft	m	ft
1.5	3.16	10.4	5.97	19.6
1.6	3.46	11.4	6.43	21.1
1.7	3.78	12.4	6.92	22.7
1.8	4.12	13.5	7.42	24.3
1.9	4.47	14.7	7.93	26.0
2.0	4.83	15.8	8.47	27.8
2.1	5.21	17.1	9.02	29.6
2.2	5.61	18.4	9.58	31.4
2.3	6.02	19.8	10.16	33.3
2.4	6.44	21.1	10.76	35.3
2.5	6.88	22.6	11.38	37.3

Notes to Table 14 through Table 21:

1. The employer must determine the maximum anticipated per-unit transient overvoltage, phase-to-ground, through an engineering analysis, as required by § 1910.269(l)(3)(ii), or assume a maximum anticipated per-unit transient overvoltage, phase-to-ground, in accordance with Table R-9.
2. For phase-to-phase exposures, the employer must demonstrate that no insulated tool spans the gap and that no large conductive object is in the gap.
3. The worksite must be at an elevation of 900 meters (3,000 feet) or less above sea level.

Appendix C to § 1910.269—Protection From Hazardous Differences in Electric Potential

I. Introduction

Current passing through an impedance impresses voltage across that impedance. Even conductors have some, albeit low, value of impedance. Therefore, if a "grounded"[14] object, such as a crane or deenergized and grounded power line, results in a ground fault on a power line, voltage is impressed on that grounded object. The voltage impressed on the grounded object depends largely on the voltage on the line, on the impedance of the faulted conductor, and on the impedance to "true," or "absolute," ground represented by the object. If the impedance of the object causing the fault is relatively large, the voltage impressed on the object is essentially the phase-to-ground system voltage. However, even faults to grounded power lines or to well-grounded transmission towers or substation structures (which have relatively low values of impedance to ground) can result in hazardous voltages.[15] In all cases, the degree of the hazard depends on the magnitude of the current through the employee and the time of exposure. This appendix discusses methods of protecting workers against the possibility that grounded objects, such as cranes and other mechanical equipment, will contact energized power lines and that deenergized and grounded power lines will become accidentally energized.

II. Voltage-Gradient Distribution

A. *Voltage-gradient distribution curve.* Absolute, or true, ground serves as a reference and always has a voltage of 0 volts above ground potential. Because there is an impedance between a grounding electrode and absolute ground, there will be a voltage difference between the grounding electrode and absolute ground underground-fault conditions. Voltage dissipates from the grounding electrode (or from the grounding point) and creates a ground potential gradient. The voltage decreases rapidly with increasing distance from the grounding electrode. A voltage drop associated with this dissipation of voltage is a ground potential. Figure 1 is a typical voltage-gradient distribution curve (assuming a uniform soil texture).

[14] This appendix generally uses the term "grounded" only with respect to grounding that the employer intentionally installs, for example, the grounding an employer installs on a deenergized conductor. However, in this case, the term "grounded" means connected to earth, regardless of whether or not that connection is intentional.

[15] Thus, grounding systems for transmission towers and substation structures should be designed to minimize the step and touch potentials involved.

FIGURE 7.3 (*Continued*)

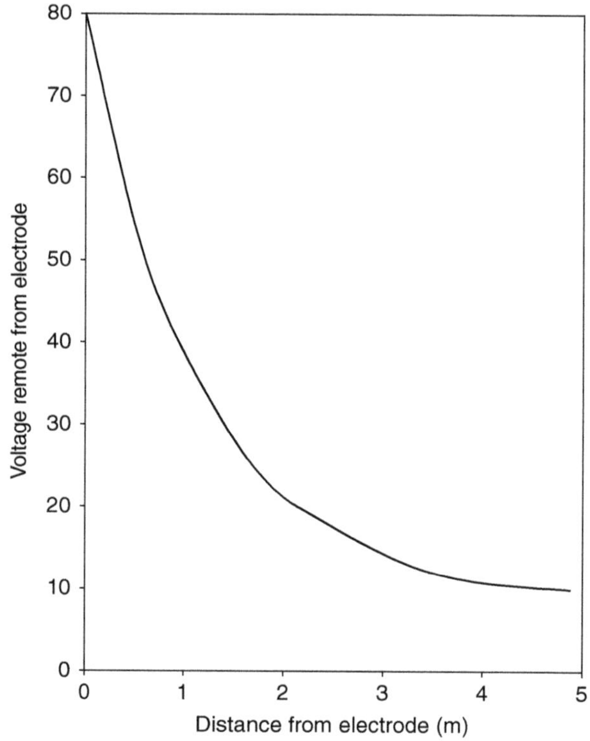

FIGURE 1 Typical Voltage-Gradient Distribution Curve

B. *Step and touch potentials.* Figure 1 also shows that workers are at risk from step and touch potentials. Step potential is the voltage between the feet of a person standing near an energized grounded object (the electrode). In Figure 1, the step potential is equal to the difference in voltage between two points at different distances from the electrode (where the points represent the location of each foot in relation to the electrode). A person could be at risk of injury during a fault simply by standing near the object. Touch potential is the voltage between the energized grounded object (again, the electrode) and the feet of a person in contact with the object. In Figure 1, the touch potential is equal to the difference in voltage between the electrode (which is at a distance of 0 meters) and a point some distance away from the electrode (where the point represents the location of the feet of the person in contact with the object). The touch potential could be nearly the full voltage across the grounded object if that object is grounded at a point remote from the place where the person is in contact with it. For example, a crane grounded to the system neutral and that contacts an energized line would expose any person in contact with the crane or its uninsulated load line to a touch potential nearly equal to the full fault voltage. Figure 2 illustrates step and touch potentials.

III. Protecting Workers From Hazardous Differences in Electrical Potential

A. *Definitions.* The following definitions apply to section III of this appendix:

Bond. The electrical interconnection of conductive parts designed to maintain a common electric potential.

Bonding cable (bonding jumper). A cable connected to two conductive parts to bond the parts together.

Cluster bar. A terminal temporarily attached to a structure that provides a means for the attachment and bonding of grounding and bonding cables to the structure.

Ground. A conducting connection between an electric circuit or equipment and the earth, or to some conducting body that serves in place of the earth.

FIGURE 7.3 *(Continued)*

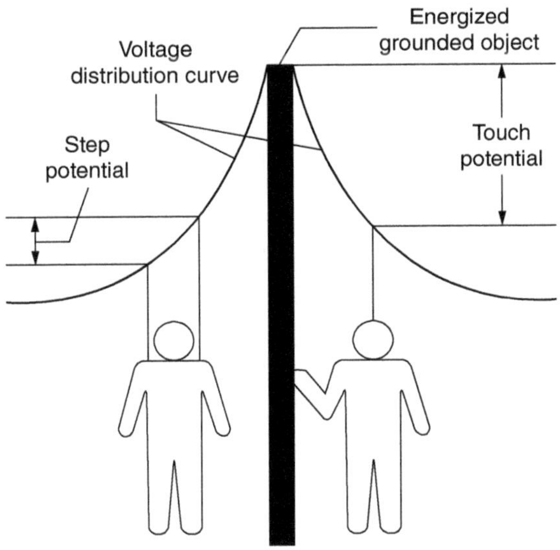

FIGURE 2 Step and Touch Potentials

Grounding cable (grounding jumper). A cable connected between a deenergized part and ground. Note that grounding cables carry fault current and bonding cables generally do not. A cable that bonds two conductive parts but carries substantial fault current (for example, a jumper connected between one phase and a grounded phase) is a grounding cable.

Ground mat (grounding grid). A temporarily or permanently installed metallic mat or grating that establishes an equipotential surface and provides connection points for attaching grounds.

B. ***Analyzing the hazard.*** The employer can use an engineering analysis of the power system under fault conditions to determine whether hazardous step and touch voltages will develop. The analysis should determine the voltage on all conductive objects in the work area and the amount of time the voltage will be present. Based on this analysis, the employer can select appropriate measures and protective equipment, including the measures and protective equipment outlined in Section III of this appendix, to protect each employee from hazardous differences in electric potential. For example, from the analysis, the employer will know the voltage remaining on conductive objects after employees install bonding and grounding equipment and will be able to select insulating equipment with an appropriate rating, as described in paragraph III.C.2 of this appendix.

C. ***Protecting workers on the ground.*** The employer may use several methods, including equipotential zones, insulating equipment, and restricted work areas, to protect employees on the ground from hazardous differences in electrical potential.

1. An equipotential zone will protect workers within it from hazardous step and touch potentials. (See Figure 3.) Equipotential zones will not, however, protect employees located either wholly or partially outside the protected area. The employer can establish an equipotential zone for workers on the ground, with respect to a grounded object, through the use of a metal mat connected to the grounded object. The employer can use a grounding grid to equalize the voltage within the grid or bond conductive objects in the immediate work area to minimize the potential between the objects and between each object and ground. (Bonding an object outside the work area can increase the touch potential to that object, however.) Section III.D of this appendix discusses equipotential zones for employees working on deenergized and grounded power lines.

2. Insulating equipment, such as rubber gloves, can protect employees handling grounded equipment and conductors from hazardous touch potentials. The insulating equipment must be rated for the highest voltage that can be impressed on the grounded objects under fault conditions (rather than for the full system voltage).

3. Restricting employees from areas where hazardous step or touch potentials could arise can protect employees not directly involved in performing the operation. The employer must ensure that employees on the ground in the vicinity of transmission structures are at a distance where step voltages would be insufficient to cause injury.

FIGURE 7.3 *(Continued)*

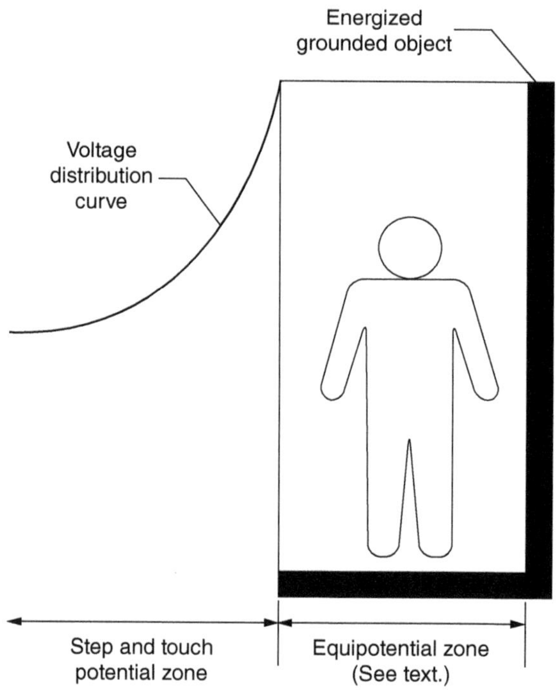

FIGURE 3 Protection and Ground-Potential Gradients

Employees must not handle grounded conductors or equipment likely to become energized to hazardous voltages unless the employees are within an equipotential zone or protected by insulating equipment.

D. *Protecting employees working on deenergized and grounded power lines.* This Section III.D of Appendix C establishes guidelines to help employers comply with requirements in § 1910.269(n) for using protective grounding to protect employees working on deenergized power lines. Paragraph (n) of § 1910.269 applies to grounding of transmission and distribution lines and equipment for the purpose of protecting workers. Paragraph (n)(3) of § 1910.269 requires temporary protective grounds to be placed at such locations and arranged in such a manner that the employer can demonstrate will prevent exposure of each employee to hazardous differences in electric potential.[16] Sections III.D.1 and III.D.2 of this appendix provide guidelines that employers can use in making the demonstration required by § 1910.269(n)(3). Section III.D.1 of this appendix provides guidelines on how the employer can determine whether particular grounding practices expose employees to hazardous differences in electric potential. Section III.D.2 of this appendix describes grounding methods that the employer can use in lieu of an engineering analysis to make the demonstration required by § 1910.269(n)(3). The Occupational Safety and Health Administration will consider employers that comply with the criteria in this appendix as meeting § 1910.269(n)(3).

Finally, Section III.D.3 of this appendix discusses other safety considerations that will help the employer comply with other requirements in § 1910.269(n). Following these guidelines will protect workers from hazards that can occur when a deenergized and grounded line becomes energized.

1. *Determining safe body current limits.* This Section III.D.1 of Appendix C provides guidelines on how an employer can determine whether any differences in electric potential to which workers could be exposed are hazardous as part of the demonstration required by § 1910.269(n)(3).

[16] The protective grounding required by § 1910.269(n) limits to safe values the potential differences between accessible objects in each employee's work environment. Ideally, a protective grounding system would create a true equipotential zone in which every point is at the same electric potential. In practice, current passing through the grounding and bonding elements creates potential differences. If these potential differences are hazardous, the employer may not treat the zone as an equipotential zone.

FIGURE 7.3 *(Continued)*

Institute of Electrical and Electronic Engineers (IEEE) Standard 1048–2003, *IEEE Guide for Protective Grounding of Power Lines,* provides the following equation for determining the threshold of ventricular fibrillation when the duration of the electric shock is limited:

$$I = \frac{116}{\sqrt{t}},$$

where *I* is the current through the worker's body, and *t* is the duration of the current in seconds. This equation represents the ventricular fibrillation threshold for 95.5 percent of the adult population with a mass of 50 kilograms (110 pounds) or more. The equation is valid for current durations between 0.0083 to 3.0 seconds.

To use this equation to set safe voltage limits in an equipotential zone around the worker, the employer will need to assume a value for the resistance of the worker's body. IEEE Std 1048–2003 states that "total body resistance is usually taken as 1000 W for determining . . . body current limits." However, employers should be aware that the impedance of a worker's body can be substantially less than that value. For instance, IEEE Std 1048–2003 reports a minimum hand-to-hand resistance of 610 ohms and an internal body resistance of 500 ohms. The internal resistance of the body better represents the minimum resistance of a worker's body when the skin resistance drops near zero, which occurs, for example, when there are breaks in the worker's skin, for instance, from cuts or from blisters formed as a result of the current from an electric shock, or when the worker is wet at the points of contact.

Employers may use the IEEE Std 1048–2003 equation to determine safe body current limits only if the employer protects workers from hazards associated with involuntary muscle reactions from electric shock (for example, the hazard to a worker from falling as a result of an electric shock). Moreover, the equation applies only when the duration of the electric shock is limited. If the precautions the employer takes, including those required by applicable standards, do not adequately protect employees from hazards associated with involuntary reactions from electric shock, a hazard exists if the induced voltage is sufficient to pass a current of 1 milliampere through a 500-ohm resistor. (The 500-ohm resistor represents the resistance of an employee. The 1-milliampere current is the threshold of perception.) Finally, if the employer protects employees from injury due to involuntary reactions from electric shock, but the duration of the electric shock is unlimited (that is, when the fault current at the work location will be insufficient to trip the devices protecting the circuit), a hazard exists if the resultant current would be more than 6 milliamperes (the recognized let-go threshold for workers[17]).

2. *Acceptable methods of grounding for employers that do not perform an engineering determination.* The grounding methods presented in this section of this appendix ensure that differences in electric potential are as low as possible and, therefore, meet § 1910.269(n)(3) without an engineering determination of the potential differences. These methods follow two principles: (i) The grounding method must ensure that the circuit opens in the fastest available clearing time, and (ii) the grounding method must ensure that the potential differences between conductive objects in the employee's work area are as low as possible.

Paragraph (n)(3) of § 1910.269 does not require grounding methods to meet the criteria embodied in these principles. Instead, the paragraph requires that protective grounds be "placed at such locations and arranged in such a manner that the employer can demonstrate will prevent exposure of each employee to hazardous differences in electric potential." However, when the employer's grounding practices do not follow these two principles, the employer will need to perform an engineering analysis to make the demonstration required by § 1910.269(n)(3).

i. *Ensuring that the circuit opens in the fastest available clearing time.* Generally, the higher the fault current, the shorter the clearing times for the same type of fault. Therefore, to ensure the fastest available clearing time, the grounding method must maximize the fault current with a low impedance connection to ground. The employer accomplishes this objective by grounding the circuit conductors to the best ground available at the worksite. Thus, the employer must ground to a grounded system neutral conductor, if one is present. A grounded system neutral has a direct connection to the system ground at the source, resulting in an extremely low impedance to ground. In a substation, the employer may instead ground to the substation grid, which also has an extremely low impedance to the system ground and, typically, is connected to a grounded system neutral when one is present. Remote system grounds, such as pole and tower grounds, have a higher impedance to the system ground than grounded system neutrals and substation grounding grids; however, the employer may use a remote ground when lower impedance

[17] Electric current passing through the body has varying effects depending on the amount of the current. At the let-go threshold, the current overrides a person's control over his or her muscles. At that level, an employee grasping an object will not be able to let go of the object. The let-go threshold varies from person to person; however, the recognized value for workers is 6 milliamperes.

FIGURE 7.3 (*Continued*)

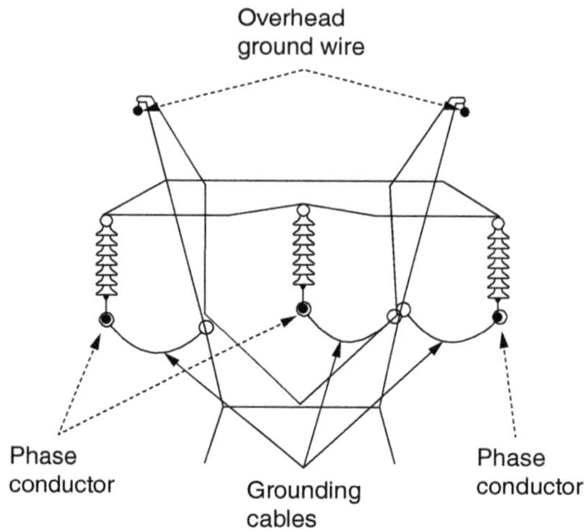

Notes:

1. Employers must ground overhead ground wires that are within reach of the employee.
2. The grounding cable must be as short as practicable; therefore, the attachment points between the grounding cable and the tower may be different from that shown in the figure.

FIGURE 4 Equipotential Zone for Metal Lattice Tower

grounds are not available. In the absence of a grounded system neutral, substation grid, and remote ground, the employer may use a temporary driven ground at the worksite.

In addition, if employees are working on a three-phase system, the grounding method must short circuit all three phases. Short circuiting all phases will ensure faster clearing and lower the current through the grounding cable connecting the deenergized line to ground, thereby lowering the voltage across that cable. The short circuit need not be at the worksite; however, the employer must treat any conductor that is not grounded at the worksite as energized because the ungrounded conductors will be energized at fault voltage during a fault.

ii. *Ensuring that the potential differences between conductive objects in the employee's work area are as low as possible.* To achieve as low a voltage as possible across any two conductive objects in the work area, the employer must bond all conductive objects in the work area. This section of this appendix discusses how to create a zone that minimizes differences in electric potential between conductive objects in the work area.

The employer must use bonding cables to bond conductive objects, except for metallic objects bonded through metal-to-metal contact. The employer must ensure that metal-to-metal contacts are tight and free of contamination, such as oxidation, that can increase the impedance across the connection. For example, a bolted connection between metal lattice tower members is acceptable if the connection is tight and free of corrosion and other contamination. Figure 4 shows how to create an equipotential zone for metal lattice towers.

Wood poles are conductive objects. The poles can absorb moisture and conduct electricity, particularly at distribution and transmission voltages. Consequently, the employer must either: (1) Provide a conductive platform, bonded to a grounding cable, on which the worker stands or (2) use cluster bars to bond wood poles to the grounding cable. The employer must ensure that employees install the cluster bar below, and close to, the worker's feet. The inner portion of the wood pole is more conductive than the outer shell, so it is important that the cluster bar be in conductive contact with a metal spike or nail that penetrates the wood to a depth greater than or equal to the depth the worker's climbing gaffs will penetrate the wood. For example, the employer could mount the cluster bar on a bare pole ground wire fastened to the pole with nails or staples that penetrate to the required depth. Alternatively, the employer may temporarily nail a conductive strap to the pole and connect the strap to the cluster bar. Figure 5 shows how to create an equipotential zone for wood poles.

FIGURE 7.3 (*Continued*)

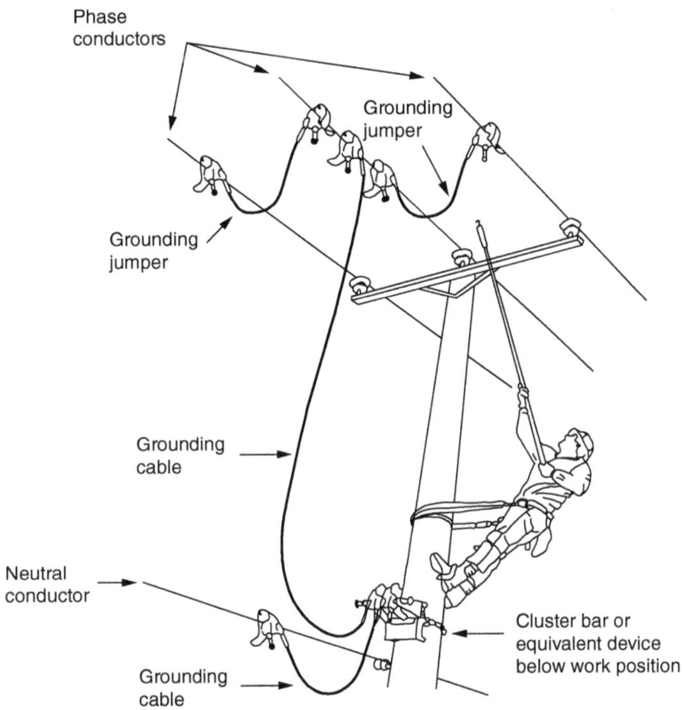

FIGURE 5 Equipotential Grounding for Wood Poles
Figure reprinted with permission from Hubbell Power Systems, Inc. (Hubbell). OSHA revised the figure from Hubbell's original.

For underground systems, employers commonly install grounds at the points of disconnection of the underground cables. These grounding points are typically remote from the manhole or underground vault where employees will be working on the cable. Workers in contact with a cable grounded at a remote location can experience hazardous potential differences if the cable becomes energized or if a fault occurs on a different, but nearby, energized cable. The fault current causes potential gradients in the earth, and a potential difference will exist between the earth where the worker is standing and the earth where the cable is grounded. Consequently, to create an equipotential zone for the worker, the employer must provide a means of connecting the deenergized cable to ground at the worksite by having the worker stand on a conductive mat bonded to the deenergized cable. If the cable is cut, the employer must install a bond across the opening in the cable or install one bond on each side of the opening to ensure that the separate cable ends are at the same potential. The employer must protect the worker from any hazardous differences in potential any time there is no bond between the mat and the cable (for example, before the worker installs the bonds).

3. *Other safety-related considerations.* To ensure that the grounding system is safe and effective, the employer should also consider the following factors:[18]

i. *Maintenance of grounding equipment.* It is essential that the employer properly maintain grounding equipment. Corrosion in the connections between grounding cables and clamps and on the clamp surface can increase the resistance of the cable, thereby increasing potential differences. In addition, the surface to which a clamp attaches, such as a conductor or tower member, must be clean and free of corrosion and oxidation to ensure a low-resistance connection. Cables must be free of damage that could reduce their current-carrying capacity so that they

[18] This appendix only discusses factors that relate to ensuring an equipotential zone for employees. The employer must consider other factors in selecting a grounding system that is capable of conducting the maximum fault current that could flow at the point of grounding for the time necessary to clear the fault, as required by § 1910.269(n)(4)(i). IEEE Std 1048–2003 contains guidelines for selecting and installing grounding equipment that will meet § 1910.269(n)(4)(i).

FIGURE 7.3 (*Continued*)

can carry the full fault current without failure. Each clamp must have a tight connection to the cable to ensure a low resistance and to ensure that the clamp does not separate from the cable during a fault.

ii. *Grounding cable length and movement.* The electromagnetic forces on grounding cables during a fault increase with increasing cable length. These forces can cause the cable to move violently during a fault and can be high enough to damage the cable or clamps and cause the cable to fail. In addition, flying cables can injure workers. Consequently, cable lengths should be as short as possible, and grounding cables that might carry high fault current should be in positions where the cables will not injure workers during a fault.

Appendix D to § 1910.269—Methods of Inspecting and Testing Wood Poles

I. Introduction

When employees are to perform work on a wood pole, it is important to determine the condition of the pole before employees climb it. The weight of the employee, the weight of equipment to be installed, and other working stresses (such as the removal or retensioning of conductors) can lead to the failure of a defective pole or a pole that is not designed to handle the additional stresses.[19] For these reasons, it is essential that, before an employee climbs a wood pole, the employer ascertain that the pole is capable of sustaining the stresses of the work. The determination that the pole is capable of sustaining these stresses includes an inspection of the condition of the pole. If the employer finds the pole to be unsafe to climb or to work from, the employer must secure the pole so that it does not fail while an employee is on it. The employer can secure the pole by a line truck boom, by ropes or guys, or by lashing a new pole alongside it. If a new one is lashed alongside the defective pole, employees should work from the new one.

II. Inspecting Wood Poles

A qualified employee should inspect wood poles for the following conditions:[20]

A. *General condition.* Buckling at the ground line or an unusual angle with respect to the ground may indicate that the pole has rotted or is broken.

B. *Cracks.* Horizontal cracks perpendicular to the grain of the wood may weaken the pole. Vertical cracks, although not normally considered to be a sign of a defective pole, can pose a hazard to the climber, and the employee should keep his or her gaffs away from them while climbing.

C. *Holes.* Hollow spots and woodpecker holes can reduce the strength of a wood pole.

D. *Shell rot and decay.* Rotting and decay are cutout hazards and possible indications of the age and internal condition of the pole.

E. *Knots.* One large knot or several smaller ones at the same height on the pole may be evidence of a weak point on the pole.

F. *Depth of setting.* Evidence of the existence of a former ground line substantially above the existing ground level may be an indication that the pole is no longer buried to a sufficient depth.

G. *Soil conditions.* Soft, wet, or loose soil around the base of the pole may indicate that the pole will not support any change in stress.

H. *Burn marks.* Burning from transformer failures or conductor faults could damage the pole so that it cannot withstand changes in mechanical stress.

III. Testing Wood Poles

The following tests, which are from § 1910.268(n)(3), are acceptable methods of testing wood poles:

A. *Hammer test.* Rap the pole sharply with a hammer weighing about 1.4 kg (3 pounds), starting near the ground line and continuing upwards circumferentially around the pole to a height of approximately 1.8 meters (6 feet). The hammer will produce a clear sound and rebound sharply when striking sound wood. Decay pockets will be indicated by a dull sound or a less pronounced hammer rebound. Also, prod the pole as near the ground line as possible using a pole prod or a screwdriver with a blade at least 127 millimeters (5 inches) long. If substantial decay is present, the pole is unsafe.

B. *Rocking test.* Apply a horizontal force to the pole and attempt to rock it back and forth in a direction perpendicular to the line. Exercise caution to avoid causing power lines to swing together. Apply the force to the pole either by pushing it with a pike pole or pulling the pole with a rope. If the pole cracks during the test, it is unsafe.

[19] A properly guyed pole in good condition should, at a minimum, be able to handle the weight of an employee climbing it.

[20] The presence of any of these conditions is an indication that the pole may not be safe to climb or to work from. The employee performing the inspection must be qualified to make a determination as to whether it is safe to perform the work without taking additional precautions.

FIGURE 7.3 (*Continued*)

Appendix E to § 1910.269—Protection From Flames and Electric Arcs

I. Introduction

Paragraph (l)(8) of § 1910.269 addresses protecting employees from flames and electric arcs. This paragraph requires employers to: (1) Assess the workplace for flame and electric-arc hazards (paragraph (l)(8)(i)); (2) estimate the available heat energy from electric arcs to which employees would be exposed (paragraph (l)(8)(ii)); (3) ensure that employees wear clothing that will not melt, or ignite and continue to burn, when exposed to flames or the estimated heat energy (paragraph (l)(8)(iii)); and (4) ensure that employees wear flame-resistant clothing[21] and protective clothing and other protective equipment that has an arc rating greater than or equal to the available heat energy under certain conditions (paragraphs (l)(8)(iv) and (l)(8)(v)). This appendix contains information to help employers estimate available heat energy as required by § 1910.269(l)(8)(ii), select protective clothing and other protective equipment with an arc rating suitable for the available heat energy as required by § 1910.269(l)(8)(v), and ensure that employees do not wear flammable clothing that could lead to burn injury as addressed by §§ 1910.269(l)(8)(iii) and (l)(8)(iv).

II. Assessing the Workplace for Flame and Electric-Arc Hazards

Paragraph (l)(8)(i) of § 1910.269 requires the employer to assess the workplace to identify employees exposed to hazards from flames or from electric arcs. This provision ensures that the employer evaluates employee exposure to flames and electric arcs so that employees who face such exposures receive the required protection. The employer must conduct an assessment for each employee who performs work on or near exposed, energized parts of electric circuits.

A. Assessment Guidelines Sources electric arcs. Consider possible sources of electric arcs, including:

- o Energized circuit parts not guarded or insulated,
- o Switching devices that produce electric arcs in normal operation,
- o Sliding parts that could fault during operation (for example, rack-mounted circuit breakers), and
- o Energized electric equipment that could fail (for example, electric equipment with damaged insulation or with evidence of arcing or overheating).

Exposure to flames. Identify employees exposed to hazards from flames. Factors to consider include:

- o The proximity of employees to open flames, and
- o For flammable material in the work area, whether there is a reasonable likelihood that an electric arc or an open flame can ignite the material.

Probability that an electric arc will occur. Identify employees exposed to electric-arc hazards. The Occupational Safety and Health Administration will consider an employee exposed to electric-arc hazards if there is a reasonable likelihood that an electric arc will occur in the employee's work area, in other words, if the probability of such an event is higher than it is for the normal operation of enclosed equipment. Factors to consider include:

- o For energized circuit parts not guarded or insulated, whether conductive objects can come too close to or fall onto the energized parts,
- o For exposed, energized circuit parts, whether the employee is closer to the part than the minimum approach distance established by the employer (as permitted by § 1910.269(l)(3)(iii)).
- o Whether the operation of electric equipment with sliding parts that could fault during operation is part of the normal operation of the equipment or occurs during servicing or maintenance, and
- o For energized electric equipment, whether there is evidence of impending failure, such as evidence of arcing or overheating.

B. Examples

Table 1 provides task-based examples of exposure assessments.

III. Protection Against Burn Injury

A. Estimating Available Heat Energy Calculation methods. Paragraph (l)(8)(ii) of § 1910.269 provides that, for each employee exposed to an electric-arc hazard, the employer must make a reasonable estimate of the heat energy to which the employee would be exposed if an arc occurs. Table 2 lists various methods of calculating values of available heat energy from an electric circuit. The Occupational Safety and Health Administration does not endorse any of these specific methods. Each method requires the input of various parameters, such as fault current,

[21] Flame-resistant clothing includes clothing that is inherently flame resistant and clothing chemically treated with a flame retardant. (See ASTM F1506–10a, *Standard Performance Specification for Flame Resistant Textile Materials for Wearing Apparel for Use by Electrical Workers Exposed to Momentary Electric Arc and Related Thermal Hazards,* and ASTM F1891–12 *Standard Specification for Arc and Flame-Resistant Rainwear.*)

FIGURE 7.3 *(Continued)*

TABLE 1 Example Assessments for Various Tasks

Task		Is employee exposed to flame or electric-arc hazard?
Normal operation of enclosed equipment, such as closing or opening a switch.	The employer properly installs and maintains enclosed equipment, and there is no evidence of impending failure.	No.
	There is evidence of arcing or overheating	Yes.
	Parts of the equipment are loose or sticking, or the equipment otherwise exhibits signs of lack of maintenance.	Yes.
Servicing electric equipment, such as racking in a circuit breaker or replacing a switch		Yes.
Inspection of electric equipment with exposed energized parts.	The employee is not holding conductive objects and remains outside the minimum approach distance established by the employer.	No.
	The employee is holding a conductive object, such as a flashlight, that could fall or otherwise contact energized parts (irrespective of whether the employee maintains the minimum approach distance).	Yes.
	The employee is closer than the minimum approach distance established by the employer (for example, when wearing rubber insulating gloves or rubber insulating gloves and sleeves).	Yes.
Using open flames, for example, in wiping cable splice sleeves		Yes.

TABLE 2 Methods of Calculating Incident Heat Energy from An Electric Arc

1. *Standard for Electrical Safety Requirements for Employee Workplaces,* NFPA 70E–2012, Annex D, "Sample Calculation of Flash Protection Boundary."
2. Doughty, T.E., Neal, T.E., and Floyd II, H.L., "Predicting Incident Energy to Better Manage the Electric Arc Hazard on 600 V Power Distribution Systems," *Record of Conference* Papers *IEEE IAS 45th Annual Petroleum and Chemical Industry Conference*, September 28–30, 1998.
3. *Guide for Performing Arc-Flash Hazard Calculations,* IEEE Std 1584–2002, 1584a–2004 (Amendment 1 to IEEE Std 1584–2002), and 15B4b–2011 (Amendment 2: Changes to Clause 4 of IEEE Std 1584–2002).*
4. ARCPRO, a commercially available software program developed by Kinectrics, Toronto, ON, CA.

*This appendix refers to IEEE Std 1584–2002 with both amendments as IEEE Std 1584b–2011.

the expected length of the electric arc, the distance from the arc to the employee, and the clearing time for the fault (that is, the time the circuit protective devices take to open the circuit and clear the fault). The employer can precisely determine some of these parameters, such as the fault current and the clearing time, for a given system. The employer will need to estimate other parameters, such as the length of the arc and the distance between the arc and the employee, because such parameters vary widely.

The amount of heat energy calculated by any of the methods is approximately inversely proportional to the square of the distance between the employee and the arc. In other words, if the employee is very close to the arc, the heat energy is very high; but if the employee is just a few more centimeters away, the heat energy drops substantially. Thus, estimating the distance from the arc to the employee is key to protecting employees.

The employer must select a method of estimating incident heat energy that provides a reasonable estimate of incident heat energy for the exposure involved. Table 3 shows which methods provide reasonable estimates for various exposures.

FIGURE 7.3 *(Continued)*

TABLE 3 Selecting a Reasonable Incident-Energy Calculation Method[1]

Incident-energy calculation method	600 V and Less[2]			601 V to 15 kV[2]			More than 15 kV		
	1Φ	3Φa	3Φb	1Φ	3Φa	3Φb	1Φ	3Φa	3Φb
NFPA 70E–2012 Annex D (Lee equation)	Y–C	Y	N	Y–C	Y–C	N	N[3]	N[3]	N[3]
Doughty. Neal, and Floyd	Y–C	Y	Y	N	N	N	N	N	N
IEEE Std 1594b–2011	Y	Y	Y	Y	Y	Y	N	N	N
ARCPRO	Y	N	N	Y	N	N	Y	Y[4]	Y[4]

Key:
1Φ: Single-phase arc in open air.
3Φa: Three-phase arc in open air.
3Φb: Three-phase arc in an enclosure (box).
Y: Acceptable; produces a reasonable estimate of incident heat energy from this type of electric arc.
N: Not acceptable: does not produce a reasonable estimate of incident heat energy from this type of electric arc.
Y–C: Acceptable; produces a reasonable, but conservative, estimate of incident heat energy from this type of electric arc.

Notes:

[1]Although the Occupational Safety and Health Administration will consider these methods reasonable for enforcement purposes when employers use the methods in accordance with this table, employers should be aware that the listed methods do not necessarily result in estimates that will provide full protection from internal faults in transformers and similar equipment or from arcs in underground manholes or vaults.

[2]At these voltages, the presumption is that the arc is three-phase unless the employer can demonstrate that only one phase is present or that the spacing of the phases is sufficient to prevent a multiphase arc from occurring.

[3]Although the Occupational Safety and Health Administration will consider this method acceptable for purposes of assessing whether incident energy exceeds 2.0 cal/cm2, the results at voltages of more than 15 kilovotts are extremely conservative and unrealist!c.

[4]The Occupational Safety and Health Administraton will deem the results of this method reasonable when the employer adjusts them using the conversion factors for three-phase arcs in Open air or in an enclosure, as indicated in the program's instructions.

TABLE 4 Selecting a Reasonable Distance from the Employee to the Electric Arc

Class of equipment	Single-phase arc mm (inches)	Three-phase arc mm (inches)
Cable	*NA	455 (18)
Low voltage MCCs and panelboards	NA	455 (18)
Low-voltage switchgear	NA	610 (24)
5-kV switchgear	NA	910 (36)
15-kV switchgear	NA	910 (36)
Single conductors in air (up to 46 kilovolts), work with rubber insulating gloves	380 (15)	NA
Single conductors in air, work with live-line tools and live-line barehand work	*MAD* – (2 × kV × 2.54) (*MAD* – (2 × kV /10))†	NA

*NA = not applicable.
†The terms in this equation are:
MAD = The applicable minimum approach distance, and
kV = The system voltage in kilovolts.

Selecting a reasonable distance from the employee to the arc. In estimating available heat energy, the employer must make some reasonable assumptions about how far the employee will be from the electric arc. Table 4 lists reasonable distances from the employee to the electric arc. The distances in Table 4 are consistent with national consensus standards, such as the Institute of Electrical and Electronic Engineers' *National Electrical Safety Code,* ANSI/IEEE C2–2012, and *IEEE Guide for Performing Arc-Flash Hazard Calculations,* IEEE Std 1584b–2011. The employer is free to use other reasonable distances, but must consider equipment enclosure size and the working distance to the employee in selecting a distance from the employee to the arc. The Occupational Safety and Health Administration will consider a distance reasonable when the employer bases it on equipment size and working distance.

FIGURE 7.3 *(Continued)*

TABLE 5 Selecting a Reasonable Arc Gap

Class of equipment	Single-phase arc mm (inches)	Three-phase arc mm[1] (inches)
Cable	NA[2]	13 (0.5).
Low voltage MCCs and panelboards	NA	25 (1.0).
Low-voltage switchgear	NA	32 (1.25).
5-kV switchgear	NA	104 (4.0).
15-kV switchgear	NA	152 (6.0).
Single conductors in air, 15 kV and less.	51 (2.0)	Phase conductor spacing.
Single conductors in air, more than 15 kV	Voltage in kV × 2.54 (Voltage in kV × 0.1), but no less than 51 mm (2 inches).	Phase conductor spacing.

[1]Source: IEEE Std 1584b–2011.
[2]NA = not applicable.

Selecting a reasonable arc gap. For a single-phase arc in air, the electric arc will almost always occur when an energized conductor approaches too close to ground. Thus, an employer can determine the arc gap, or arc length, for these exposures by the dielectric strength of air and the voltage on the line. The dielectric strength of air is approximately 10 kilovolts for every 25.4 millimeters (1 inch). For example, at 50 kilovolts, the arc gap would be 50 ÷ 10 × 25.4 (or 50 × 2.54), which equals 127 millimeters (5 inches).

For three-phase arcs in open air and in enclosures, the arc gap will generally be dependent on the spacing between parts energized at different electrical potentials. Documents such as IEEE Std 1584b–2011 provide information on these distances. Employers may select a reasonable arc gap from Table 5, or they may select any other reasonable arc gap based on sparkover distance or on the spacing between (1) live parts at different potentials or (2) live parts and grounded parts (for example, bus or conductor spacings in equipment). In any event, the employer must use an estimate that reasonably resembles the actual exposures faced by the employee.

Making estimates over multiple system areas. The employer need not estimate the heat-energy exposure for every job task performed by each employee. Paragraph (l)(8)(ii) of § 1910.269 permits the employer to make broad estimates that cover multiple system areas provided that: (1) The employer uses reasonable assumptions about the energy-exposure distribution throughout the system, and (2) the estimates represent the maximum exposure for those areas. For example, the employer can use the maximum fault current and clearing time to cover several system areas at once.

Incident heat energy for single-phase-to-ground exposures. Table 6 and Table 7 provide incident heat energy levels for open-air, phase-to-ground electric-arc exposures typical for overhead systems.[22] Table 6 presents estimates of available energy for employees using rubber insulating gloves to perform work on overhead systems operating at 4 to 46 kilovolts. The table assumes that the employee will be 380 millimeters (15 inches) from the electric arc, which is a reasonable estimate for rubber insulating glove work. Table 6 also assumes that the arc length equals the sparkover distance for the maximum transient overvoltage of each voltage range.[23] To use the table, an employer would use the voltage, maximum fault current, and maximum clearing time for a system area and, using the appropriate voltage range and fault-current and clearing time values corresponding to the next higher values listed in the table, select the appropriate heat energy (4, 5, 8, or 12 cal/cm^2) from the table. For example, an employer might have a 12,470-volt power line supplying a system area. The power line can supply a maximum fault current of 8 kiloamperes with a maximum clearing time of 10 cycles. For rubber glove work, this system falls in the 4.0-to-15.0-kilovolt

[22] The Occupational Safety and Health Administration used metric values to calculate the clearing times in Table 6 and Table 7. An employer may use English units to calculate clearing times instead even though the results will differ slightly.
[23] The Occupational Safety and Health Administration based this assumption, which is more conservative than the arc length specified in Table 5, on Table 410–2 of the 2012 NESC.

FIGURE 7.3 *(Continued)*

TABLE 6 Incident Heat Energy for Various Fault Currents, Clearing Times, and Voltages of 4.0 to 46.0 kV: Rubber Insulating Glove Exposures Involving Phase-to-Ground Arcs in Open Air Only *†‡

Voltage range (kV)**	Fault current (kA)	Maximum clearing time (cycles)			
		4 cal/cm^2	5 cal/cm^2	8 cal/cm2	12 cal/cm^2
4.0 to 15.0	5	46	58	92	138
	10	18	22	36	54
	15	10	12	20	30
	20	6	8	13	19
15.1 to 25.0	5	28	34	55	83
	10	11	14	23	34
	15	7	8	13	20
	20	4	5	9	13
25.1 to 36.0	5	21	26	42	62
	10	9	11	18	26
	15	5	6	10	16
	20	4	4	7	11
36.1 to 46.0	5	16	20	32	48
	10	7	9	14	21
	15	4	5	8	13
	20	3	4	6	9

Notes: *This table is for open-air, phase-to-ground electric-arc exposures. It is not for phase-to-phase arcs or enclosed arcs (arc in a box).

†The table assumes that the employee will be 380 mm (15 in.) from the electric arc. The table also assumes the arc length to be the sparkover distance for the maximum transient overvoltage of each voltage range (see Appendix B to § 1910.269), as follows:

4.0 to 15.0 kV 51 mm (2 in.)
15.1 to 25.0 kV 102 mm (4 in.)
25.1 to 36.0 kV 152 mm (6 in.)
36.1 to 46.0 kV 229 mm (9 in.)

‡The Occupational Safety and Health Administration calculated the values in this table using the ARCPRO method listed in Table 2.

**The voltage range is the phase-to-phase system voltage.

range; the next-higher fault current is 10 kA (the second row in that voltage range); and the clearing time is under 18 cycles (the first column to the right of the fault current column). Thus, the available heat energy for this part of the system will be 4 cal/cm^2 or less (from the column heading), and the employer could select protection with a 5-cal/cm^2 rating to meet § 1910.269(l)(8)(v). Alternatively, an employer could select a base incident-energy value and ensure that the clearing times for each voltage range and fault current listed in the table do not exceed the corresponding clearing time specified in the table. For example, an employer that provides employees with arc-flash protective equipment rated at 8 cal/cm^2 can use the table to determine if any system area exceeds 8 cal/cm^2 by checking the clearing time for the highest fault current for each voltage range and ensuring that the clearing times do not exceed the values specified in the 8-cal/cm^2 column in the table.

Table 7 presents similar estimates for employees using live-line tools to perform work on overhead systems operating at voltages of 4 to 800 kilovolts. The table assumes that the arc length will be equal to the sparkover distance[24] and that the employee will be a distance from the arc equal to the minimum approach distance minus twice the sparkover distance.

The employer will need to use other methods for estimating available heat energy in situations not addressed by Table 6 or Table 7. The calculation methods listed in Table 2 and the guidance provided in Table 3 will help employers do this. For example, employers can use IEEE Std 1584b–2011 to estimate the available heat energy (and to select appropriate protective equipment) for many specific conditions, including lower voltage, phase-to-phase arc, and enclosed arc exposures.

[24] The dielectric strength of air is about 10 kilovolts for every 25.4 millimeters (1 inch). Thus, the employer can estimate the arc length in millimeters to be the phase-to-ground voltage in kilovolts multiplied by 2.54 (or voltage (in kilovolts) × 2.54).

FIGURE 7.3 *(Continued)*

TABLE 7 Incident Heat Energy for Various Fault Currents, Clearing Times, and Voltages: Live-Line Tool Exposures Involving Phase-to-Ground Arcs in Open Air Only*†‡#

Voltage range (kV)**	Fault current (kA)	Maximum clearing time (cycles)			
		4 cal/cm²	5 cal/cm²	8 cal/cm²	12 cal/cm²
4.0 to 15.0	5	197	246	394	591
	10	73	92	147	220
	15	39	49	78	117
	20	24	31	49	73
15.1 to 25.0	5	197	246	394	591
	10	75	94	150	225
	15	41	51	82	122
	20	26	33	52	78
25.1 to 36.0	5	138	172	275	413
	10	53	66	106	159
	15	30	37	59	89
	20	19	24	38	58
36.1 to 46.0	5	129	161	257	386
	10	51	64	102	154
	15	29	36	58	87
	20	19	24	38	57
46.1 to 72.5	20	18	23	36	55
	30	10	13	20	30
	40	6	8	13	19
	50	4	6	9	13
72.6 to 121.0	20	10	12	20	30
	30	6	7	11	17
	40	4	5	7	11
	50	3	3	5	8
121.1 to 145.0	20	12	15	24	35
	30	7	9	15	22
	40	5	6	10	15
	50	4	5	8	11
145.1 to 169.0	20	12	15	24	36
	30	7	9	15	22
	40	5	7	10	16
	50	4	5	8	12
169.1 to 242.0	20	13	17	27	40
	30	8	10	17	25
	40	6	7	12	17
	50	4	5	9	13
242.1 to 362.0	20	25	32	51	76
	30	16	19	31	47
	40	11	14	22	33
	50	8	10	16	25
362.1 to 420.0	20	12	15	25	37
	30	8	10	15	23
	40	5	7	11	16
	50	4	5	8	12

(Continued)

FIGURE 7.3 *(Continued)*

TABLE 7 *(Continued)*

Voltage range (kV)**	Fault current (kA)	Maximum clearing time (cycles)			
		4 cal/cm²	5 cal/cm²	8 cal/cm²	12 cal/cm²
420.1 to 550.0	20	23	29	47	70
	30	14	18	29	43
	40	10	13	20	30
	50	8	9	15	23
550.1 to 900.0	20	25	31	50	75
	30	15	19	31	46
	40	11	13	21	32
	50	8	10	16	24

Notes:

* This table is for open-air, phase-to-ground electric-arc exposures. It is not for phase-to-phase arcs or enclosed arcs (arc in a box).

† The table assumes the arc length to be the sparkover distance for the maximum phase-to-ground voltage of each voltage range (see Appendix B to this section). The table also assumes that the employee will be the minimum approach distance minus twice the arc length from the electric arc.

‡ The Occupational Safety and Health Administration calculated the values in this table using the ARCPRO method listed in Table 2.

For voltages of more than 72.6 kV, employers may use this table only when the minimum approach distance established under §1910.269(1)(3)(1) is greater than or equal to the following values:

72.6 to 121.0 kV 1.02 m.
121.1 to 145.0 kV 1.16 m.
145.1 to 169.0 kV 1.30 m.
169.1 to 242.0 kV 1.72 m.
242.1 to 362.0 kV 2.76 m.
362.1 to 420.0 kV 2.50 m.
420.1 to 550.0 kV 3.62 m.
550.1 to 900.0 kV 4.83 m.

**The voltage range is the phase-to-phase system voltage.

B. Selecting Protective Clothing and Other Protective Equipment

Paragraph (l)(8)(v) of § 1910.269 requires employers, in certain situations, to select protective clothing and other protective equipment with an arc rating that is greater than or equal to the incident heat energy estimated under § 1910.269(l)(8)(ii). Based on laboratory testing required by ASTM F1506–10a, the expectation is that protective clothing with an arc rating equal to the estimated incident heat energy will be capable of preventing second-degree burn injury to an employee exposed to that incident heat energy from an electric arc. Note that actual electric-arc exposures may be more or less severe than the estimated value because of factors such as arc movement, arc length, arcing from reclosing of the system, secondary fires or explosions, and weather conditions. Additionally, for arc rating based on the fabric's arc thermal performance value[25] (ATPV), a worker exposed to incident energy at the arc rating has a 50-percent chance of just barely receiving a second-degree burn. Therefore, it is possible (although not likely) that an employee will sustain a second-degree (or worse) burn wearing clothing conforming to § 1910.269(l)(8)(v) under certain circumstances. However, reasonable employer estimates and maintaining appropriate minimum approach distances for employees should limit burns to relatively small burns that just barely extend beyond the epidermis (that is, just barely a second-degree burn). Consequently, protective clothing and other protective equipment meeting § 1910.269(l)(8)(v) will provide an appropriate degree of protection for an employee exposed to electric-arc hazards.

Paragraph (l)(8)(v) of § 1910.269 does not require arc-rated protection for exposures of 2 cal/cm² or less. Untreated cotton clothing will reduce a 2-cal/cm² exposure below the 1.2- to 1.5-cal/cm² level necessary to cause burn injury, and this material should not ignite at such low heat energy levels. Although § 1910.269(l)(8)(v) does not

[25] ASTM F1506–10a defines "arc thermal performance value" as "the incident energy on a material or a multilayer system of materials that results in a 50% probability that sufficient heat transfer through the tested specimen is predicted to cause the onset of a second-degree skin burn injury based on the Stoll [footnote] curve, cal/cm²." The footnote to this definition reads: "Derived from: Stoll, A. M., and Chianta, M. A., 'Method and Rating System for Evaluations of Thermal Protection,' Aerospace Medicine, Vol 40, 1969, pp. 1232–1238 and Stoll, A. M., and Chianta, M. A., 'Heat Transfer through Fabrics as Related to Thermal Injury,' Transactions—New York Academy of Sciences, Vol 33(7), Nov. 1971, pp. 649–670."

FIGURE 7.3 *(Continued)*

require clothing to have an arc rating when exposures are 2 cal/cm^2 or less, § 1910.269(l)(8)(iv) requires the outer layer of clothing to be flame resistant under certain conditions, even when the estimated incident heat energy is less than 2 cal/cm^2, as discussed later in this appendix. Additionally, it is especially important to ensure that employees do not wear undergarments made from fabrics listed in the note to § 1910.269(l)(8)(iii) even when the outer layer is flame resistant or arc rated. These fabrics can melt or ignite easily when an electric arc occurs. Logos and name tags made from non-flame-resistant material can adversely affect the arc rating or the flame-resistant characteristics of arc-rated or flame-resistant clothing. Such logos and name tags may violate § 1910.269(l)(8)(iii), (l)(8)(iv), or (l)(8)(v).

Paragraph (l)(8)(v) of § 1910.269 requires that arc-rated protection cover the employee's entire body, with limited exceptions for the employee's hands, feet, face, and head. Paragraph (l)(8)(v)(A) of § 1910.269 provides that arc-rated protection is not necessary for the employee's hands under the following conditions:

For any estimated incident heat energy .... When the employee is wearing rubber insulating gloves with protectors.

If the estimated incident heat energy does not exceed 14 cal/cm^2 ..... When the employee is wearing heavy-duty leather work gloves with a weight of at least 407 g/m^2 (12 oz/yd^2).

Paragraph (l)(8)(v)(B) of § 1910.269 provides that arc-rated protection is not necessary for the employee's feet when the employee is wearing heavy-duty work shoes or boots. Finally, § 1910.269(l)(8)(v)(C), (l)(8)(v)(D), and (l)(8)(v)(E) require arc-rated head and face protection as follows:

Exposure	Minimum head and face protection		
	None*	Arc-rated faceshield with a minimum rating of 8 cal/cm^2*	Arc-rated hood or faceshield with balaclava
Single-phase, open air	2–8 cal/cm^2	9-12 cal/cm^2	13 cal/cm^2 or higher †
Three-phase	2–4 cal/cm^2	5–8 cal/cm^2	9 cal/cm^2 or higher ‡

*These ranges assume that employees are wearing hardhats meeting the specifications in § 1910.135 or § 1926.100(b)(2), as applicable.
†The arc rating must be a minimum of 4 cal/cm^2 less than the estimated incident energy. Note that § 1910.269(l)(8)(v)(E) permits this type of head and face protection, with a minimum arc rating of 4 cal/cm^2 less than the estimated incident energy, at any incident energy level.
‡Note that §1910.269(l)(8)(v) permits this type of head and face protection at any incident energy level.

IV. Protection Against Ignition

Paragraph (l)(8)(iii) of § 1910.269 prohibits clothing that could melt onto an employee's skin or that could ignite and continue to burn when exposed to flames or to the available heat energy estimated by the employer under § 1910.269(l)(8)(ii). Meltable fabrics, such as acetate, nylon, polyester, and polypropylene, even in blends, must be avoided. When these fibers melt, they can adhere to the skin, thereby transferring heat rapidly, exacerbating burns, and complicating treatment. These outcomes can result even if the meltable fabric is not directly next to the skin. The remainder of this section focuses on the prevention of ignition.

Paragraph (l)(8)(v) of § 1910.269 generally requires protective clothing and other protective equipment with an arc rating greater than or equal to the employer's estimate of available heat energy. As explained earlier in this appendix, untreated cotton is usually acceptable for exposures of 2.0 cal/cm^2 or less.[26] If the exposure is greater than that, the employee generally must wear flame-resistant clothing with a suitable arc rating in accordance with § 1910.269(l)(8)(iv) and (l)(8)(v). However, even if an employee is wearing a layer of flame-resistant clothing, there are circumstances under which flammable layers of clothing would be uncovered, and an electric arc could ignite them. For example, clothing ignition is possible if the employee is wearing flammable clothing under the flame-resistant clothing and the underlayer is uncovered because of an opening in the flame-resistant clothing. Thus, for purposes of § 1910.269(l)(8)(iii), it is important for the employer to consider the possibility of clothing ignition even when an employee is wearing flame-resistant clothing with a suitable arc rating.

Under § 1910.269(l)(8)(iii), employees may not wear flammable clothing in conjunction with flame-resistant clothing if the flammable clothing poses an ignition hazard.[27] Although outer flame-resistant layers may not have

[26] See § 1910.269(l)(8)(iv)(A), (l)(8)(iv)(B), and (l)(8)(iv)(C) for conditions under which employees must wear flame-resistant clothing as the outer layer of clothing even when the incident heat energy does not exceed 2 cal/cm^2.
[27] Paragraph (l)(8)(iii) of § 1910.269 prohibits clothing that could ignite and continue to burn when exposed to the heat energy estimated under paragraph (l)(8)(ii) of that section.

FIGURE 7.3 *(Continued)*

openings that expose flammable inner layers, when an outer flame-resistant layer would be unable to resist breakopen,[28] the next (inner) layer must be flame-resistant if it could ignite.

Non-flame-resistant clothing can ignite even when the heat energy from an electric arc is insufficient to ignite the clothing. For example, nearby flames can ignite an employee's clothing; and, even in the absence of flames, electric arcs pose ignition hazards beyond the hazard of ignition from incident energy under certain conditions. In addition to requiring flame-resistant clothing when the estimated incident energy exceeds 2.0 cal/cm^2, § 1910.269(l)(8)(iv) requires flame-resistant clothing when: The employee is exposed to contact with energized circuit parts operating at more than 600 volts (§ 1910.269(l)(8)(iv)(A)), an electric arc could ignite flammable material in the work area that, in turn, could ignite the employee's clothing (§ 1910.269(l)(8)(iv)(B)), and molten metal or electric arcs from faulted conductors in the work area could ignite the employee's clothing (§ 1910.269(l)(8)(iv)(C)). For example, grounding conductors can become a source of heat energy if they cannot carry fault current without failure. The employer must consider these possible sources of electric arcs[29] in determining whether the employee's clothing could ignite under § 1910.269(l)(8)(iv)(C).

Appendix F to § 1910.269—Work-Positioning Equipment Inspection Guidelines

I. Body Belts

Inspect body belts to ensure that:

A. The hardware has no cracks, nicks, distortion, or corrosion;
B. No loose or worn rivets are present;
C. The waist strap has no loose grommets;
D. The fastening straps are not 100-percent leather; and
E. No worn materials that could affect the safety of the user are present.

II. Positioning Straps

Inspect positioning straps to ensure that:

A. The warning center of the strap material is not exposed;
B. No cuts, burns, extra holes, or fraying of strap material is present;
C. Rivets are properly secured;
D. Straps are not 100-percent leather; and
E. Snaphooks do not have cracks, burns, or corrosion.

III. Climbers

Inspect pole and tree climbers to ensure that:

A. Gaffs are at least as long as the manufacturer's recommended minimums (generally 32 and 51 millimeters (1.25 and 2.0 inches) for pole and tree climbers, respectively, measured on the underside of the gaff);

Note: Gauges are available to assist in determining whether gaffs are long enough and shaped to easily penetrate poles or trees.

B. Gaffs and leg irons are not fractured or cracked;
C. Stirrups and leg irons are free of excessive wear;
D. Gaffs are not loose;
E. Gaffs are free of deformation that could adversely affect use;
F. Gaffs are properly sharpened; and
G. There are no broken straps or buckles.

Appendix G to § 1910.269—Reference Documents

The references contained in this appendix provide information that can be helpful in understanding and complying with the requirements contained in § 1910.269. The national consensus standards referenced in this appendix contain detailed specifications that employers may follow in complying with the more performance-based requirements of § 1910.269. Except as specifically noted in § 1910.269, however, the Occupational Safety and Health

[28] Breakopen occurs when a hole, tear, or crack develops in the exposed fabric such that the fabric no longer effectively blocks incident heat energy.

[29] Static wires and pole grounds are examples of grounding conductors that might not be capable of carrying fault current without failure. Grounds that can carry the maximum available fault current are not a concern, and employers need not consider such grounds a possible electric arc source.

FIGURE 7.3 (*Continued*)

Administration will not necessarily deem compliance with the national consensus standards to be compliance with the provisions of § 1910.269.

ANSI/SIA A92.2–2009, *American National Standard for Vehicle-Mounted Elevating and Rotating Aerial Devices.*

ANSI Z133–2012, *American National Standard Safety Requirements for Arboricultural Operations—Pruning, Trimming, Repairing, Maintaining, and Removing Trees, and Cutting Brush.*

ANSI/IEEE Std 935–1989, *IEEE Guide on Terminology for Tools and Equipment to Be Used in Live Line Working.*

ASME B20.1–2012, *Safety Standard for Conveyors and Related Equipment.*

ASTM D120–09, *Standard Specification for Rubber Insulating Gloves.*

ASTM D149–09 (2013), *Standard Test Method for Dielectric Breakdown Voltage and Dielectric Strength of Solid Electrical Insulating Materials at Commercial Power Frequencies.*

ASTM D178–01 (2010), *Standard Specification for Rubber Insulating Matting.*

ASTM D1048–12, *Standard Specification for Rubber Insulating Blankets.*

ASTM D1049–98 (2010), *Standard Specification for Rubber Insulating Covers.*

ASTM D1050–05 (2011), *Standard Specification for Rubber Insulating Line Hose.*

ASTM D1051–08, *Standard Specification for Rubber Insulating Sleeves.*

ASTM F478–09, *Standard Specification for In-Service Care of Insulating Line Hose and Covers.*

ASTM F479–06 (2011), *Standard Specification for In-Service Care of Insulating Blankets.*

ASTM F496–08, *Standard Specification for In-Service Care of Insulating Gloves and Sleeves.*

ASTM F711–02 (2007), *Standard Specification for Fiberglass-Reinforced Plastic (FRP) Rod and Tube Used in Live Line Tools.*

ASTM F712–06 (2011), *Standard Test Methods and Specifications for Electrically Insulating Plastic Guard Equipment for Protection of Workers.*

ASTM F819–10, *Standard Terminology Relating to Electrical Protective Equipment for Workers.*

ASTM F855–09, *Standard Specifications for Temporary Protective Grounds to Be Used on De-energized Electric Power Lines and Equipment.*

ASTM F887–12e1, *Standard Specifications for Personal Climbing Equipment.*

ASTM F914/F914M–10, *Standard Test Method for Acoustic Emission for Aerial Personnel Devices Without Supplemental Load Handling Attachments.*

ASTM F1116–03 (2008), *Standard Test Method for Determining Dielectric Strength of Dielectric Footwear.*

ASTM F1117–03 (2008), *Standard Specification for Dielectric Footwear.*

ASTM F1236–96 (2012), *Standard Guide for Visual Inspection of Electrical Protective Rubber Products.*

ASTM F1430/F1430M–10, *Standard Test Method for Acoustic Emission Testing of Insulated and Non-Insulated Aerial Personnel Devices with Supplemental Load Handling Attachments.*

ASTM F1505–10, *Standard Specification for Insulated and Insulating Hand Tools.*

ASTM F1506–10a, *Standard Performance Specification for Flame Resistant and Arc Rated Textile Materials for Wearing Apparel for Use by Electrical Workers Exposed to Momentary Electric Arc and Related Thermal Hazards.*

ASTM F1564–13, *Standard Specification for Structure-Mounted Insulating Work Platforms for Electrical Workers.*

ASTM F1701–12, *Standard Specification for Unused Polypropylene Rope with Special Electrical Properties.*

ASTM F1742–03 (2011), *Standard Specification for PVC Insulating Sheeting.*

ASTM F1796–09, *Standard Specification for High Voltage Detectors—Part 1 Capacitive Type to be Used for Voltages Exceeding 600 Volts AC.*

ASTM F1797–09ε1, *Standard Test Method for Acoustic Emission Testing of Insulated and Non-Insulated Digger Derricks.*

ASTM F1825–03 (2007), *Standard Specification for Clampstick Type Live Line Tools.*

ASTM F1826–00 (2011), *Standard Specification for Live Line and Measuring Telescoping Tools.*

ASTM F1891–12, *Standard Specification for Arc and Flame-Resistant Rainwear.*

ASTM F1958/F1958M–12, *Standard Test Method for Determining the Ignitability of Non-Flame-Resistant Materials for Clothing by Electric Arc Exposure Method Using Mannequins.*

ASTM F1959/F1959M–12, *Standard Test Method for Determining the Arc Rating of Materials for Clothing.*

IEEE Stds 4–1995, 4a-2001 (Amendment to *IEEE Standard Techniques for High-Voltage Testing*), *IEEE Standard Techniques for High-Voltage Testing.*

IEEE Std 62–1995, *IEEE Guide for Diagnostic Field Testing of Electric Power Apparatus—Part 1: Oil Filled Power Transformers, Regulators, and Reactors.*

IEEE Std 80–2000, *Guide for Safety in AC Substation Grounding.*

IEEE Std 100–2000, *The Authoritative Dictionary of IEEE Standards Terms Seventh Edition.*

FIGURE 7.3 (*Continued*)

IEEE Std 516–2009, *IEEE Guide for Maintenance Methods on Energized Power Lines.*
IEEE Std 524–2003, *IEEE Guide to the Installation of Overhead Transmission Line Conductors*
IEEE Std 957–2005, *IEEE Guide for Cleaning Insulators.*
IEEE Std 1048–2003, *IEEE Guide for Protective Grounding of Power Lines.*
IEEE Std 1067–2005, *IEEE Guide for In-Service Use, Care, Maintenance, and Testing of Conductive Clothing for Use on Voltages up to 765 kV AC and ±750 kV DC.*
IEEE Std 1307–2004, *IEEE Standard for Fall Protection for Utility Work.*
IEEE Stds 1584–2002, 1584a-2004 (Amendment 1 to IEEE Std 1584–2002), and 1584b-2011 (Amendment 2: Changes to Clause 4 of IEEE Std 1584–2002), *IEEE Guide for Performing Arc-Flash Hazard Calculations.*
IEEE C2–2012, *National Electrical Safety Code.*
NFPA 70E–2012, *Standard for Electrical Safety in the Workplace.*

FIGURE 7.3 (*Continued*)

THE CONTROL OF HAZARDOUS ENERGY (LOCKOUT/TAGOUT)

(OSHA 29 CFR 1910.147)

(FINAL RULE September 1, 1989)

> 1910.147 The Control of Hazardous Energy (Lockout/Tagout) (Indented statements are quoted from the preamble to this regulation and from the Directorate of Compliance.)

(a) ***Scope, application and purpose***

(1) ***Scope***

(i) This standard covers the servicing and maintenance of machines and equipment in which the unexpected energization or start up of the machines or equipment, or release of stored energy could cause injury to employees. This standard establishes minimum performance requirements for the control of such hazardous energy.

(ii) This standard does not cover the following:

(A) Construction, agriculture, and maritime employment;

(B) Installations under the exclusive control of electric utilities for the purpose of power generation, transmission, and distribution, including related equipment for communication or metering; and

> The lockout/tagout requirements for utilities are contained in 29 CFR 1910.269, "Electric Power Generation, Transmission, and Distribution" Final Rule dated January 31, 1994.

(C) Exposure to electrical hazards from work on, near, or with conductors or equipment in electric utilization installations, which is covered by Subpart S of this part; and

> The lockout/tagout requirements electrical conductors or equipment are contained in 29 CFR 1910.331-.335, "Electrical Safety-Related Work Practices" Final Rule dated August 6, 1990.
>
> OSHA states that the lockout/tagout provisions of this standard are for the protection of general industry workers while performing servicing and maintenance functions and augment the safeguards specified in Subparts O, S, and other applicable portions of 29 CFR 1910.

(D) Oil and gas well drilling and servicing.

> The lockout/tagout requirements oil and gas well drilling and servicing are contained in 29 CFR 1910.290, "Oil and Gas Well Drilling and Servicing."

(2) ***Application.*** **(i)** This standard applies to the control of energy during servicing and/or maintenance of machines and equipment.

(ii) Normal production operations are not covered by this standard (See Subpart O of this Part). Servicing and/or maintenance which takes place during normal production operations is covered by this standard only if:

The standards, in Subpart O, that apply here are 1910.212 (general machine guarding) and 1910.219 (guarding power transmission apparatus).

FIGURE 7.4 Control of Hazardous Energy Source (Lockout/Tagout) (OSHA-CFR, Title 29, part 1910, paragraph 147).

(A) An employee is required to remove or bypass a guard or other safety device; or

(B) An employee is required to place any part of his or her body into an area on a machine or piece of equipment where work is actually performed upon the material being processed (point of operation) or where an associated danger zone exists during a machine operating cycle.

Note: Exception to paragraph (a)(2)(ii): Minor tool changes and adjustments, and other minor servicing activities, which take place during normal production operations, are not covered by this standard if they are routine, repetitive, and integral to the use of the equipment for production, provided that the work is performed using alternative measures which provide effective protection (See Subpart O of this Part).

(iii) This standard does not apply to the following:

(A) Work on cord and plug connected electric equipment for which exposure to the hazards of unexpected energization or start up of the equipment is controlled by the unplugging of the equipment from the energy source and by the plug being under the exclusive control of the employee performing the servicing or maintenance.

This paragraph encompasses the many varieties of portable hand tools that are found in the workplace, as well as cord and plug connected equipment which is intended for use at a fixed location. If the plug is not near the employee doing the maintenance or servicing of the equipment then the plug must be locked out or tagged out to prevent another employee from inadvertently plugging it in. There are plug locking devices available.

OSHA states that "The plug is under the exclusive control of the employee if it is physically in the position of the employee, or in arm's reach and in the line of sight of the employee, or if the employee has affixed a lockout/tagout device on the plug."

OSHA further states that "The company lockout/tagout procedures required by the standard at 29 CFR 1910.147(c)(4)(i) shall specify the acceptable procedure for handling cord and plug connected equipment."

(B) Hot tap operations involving transmission and distribution systems for substances such as gas, steam, water, or petroleum products when they are performed on pressurized pipelines, provided that the employer demonstrates that ***{1}*** continuity of service is essential; ***{2}*** shutdown of the system is impractical; and ***{3}*** documented procedures are followed, and special equipment is used which will provide proven effective protection for employees.

(3) *Purpose.* (i) This section requires employers to establish a program and utilize procedures for affixing appropriate lockout devices or tagout devices to energy isolating devices, and to otherwise disable machines or equipment to prevent unexpected energization, start up or release of stored energy in order to prevent injury to employees.

(ii) When other standards in this part require the use of lockout/tagout, they shall be used and supplemented by the procedural and training requirements of this section.

"This part" means 1910. The requirements of this standard also apply to 1910.331.335 and 1910.269.

(b) *Definitions applicable to this section.*

Affected employee. An employee whose job requires him/her to operate or use a machine or equipment on which servicing or maintenance is being performed under lockout/tagout, or whose job requires him/her to work in an area in which such servicing or maintenance is being performed.

An affected employee is one who does not perform the servicing or implement the energy control procedure, but whose responsibilities are performed in an area in which the energy control procedure is implemented and servicing operations are performed under that procedure. The affected employee does not need to know how to perform lockout/tagout, nor does that employee need to be trained in the detailed implementation of the energy control procedure. Rather, the affected employee need only be able to recognize when the energy control procedure is being implemented, to identify the locks or tags being used, and to understand the purpose of the procedures and the importance of not attempting to start up or use the equipment which has been locked out or tagged out.

The definition of affected employee also recognizes that an affected person and an authorized person may be one and the same person when a machine operator or user must also perform servicing or maintenance on the machine or equipment. In this case, the employee must have the requisite knowledge of an authorized employee.

Authorized employee. A person who locks out or tags out machines or equipment in order to perform servicing or maintenance on that machine or equipment. An affected employee becomes an authorized employee when that employee's duties include performing servicing or maintenance covered under this section.

FIGURE 7.4 *(Continued)*

OSHA agrees that as long as an employee is involved in performing an element of servicing and maintenance which is covered by the energy control procedure, that employee should be considered an authorized employee for the purpose of this standard. This is particularly important in the context of the requirement in paragraph (d)(3) "Machine or equipment isolation" as well as (f)(3)(ii)(D) of this standard, which requires the authorized person to affix a personal lockout or tagout device on the energy isolating device as part of the energy control procedure. This definition assures that when a servicing task is performed by a team or group of employees, each employee who is directly exposed to the hazards of the servicing operation will have the responsibility to affix his/her personal lockout/tagout device before beginning the work and to remove it when he/she completes the work.

Paragraph (c)(5)(ii)(D) of this standard provides additional accountability by requiring such lockout and tagout devices to identify the authorized person responsible for applying them.

"Capable of being locked out." An energy isolating device is capable of being locked out if it has a hasp or other means of attachment to which, or through which, a lock can be affixed, or it has a locking mechanism built into it. Other energy isolating devices are capable of being locked out, if lockout can be achieved without the need to dismantle, rebuild, or replace the energy isolating device or permanently alter its energy control capability.

The capability for lockout does not necessarily mean that the equipment has an actual hasp or other physical attachment point for a lock. For example, the use of chains can be an effective means of facilitating lockout of many types of valves, even if the valve does not have a specific locking point. Many examples of equipment which was lockable with minor modifications have been provided to the record. For equipment of this type, OSHA believes that the lockout capability should be used in order to maximize the protection afforded by this standard.

Energized. Connected to an energy source or containing residual or stored energy.

Energy isolating device. A mechanical device that physically prevents the transmission or release of energy, including but not limited to the following: A manually operated electrical circuit breaker, a disconnect switch, a manually operated switch by which the conductors of a circuit can be disconnected from all ungrounded supply conductors and, in addition, no pole can be operated independently; a line valve; a block; and any similar device used to block or isolate energy. Push buttons, selector switches, and other control circuit type devices are not energy isolating devices.

Energy source. Any source of electrical, mechanical, hydraulic, pneumatic, chemical, thermal, or other energy.

The identification of energy sources, as defined here, is complicated by three very important considerations: (1) Energy is always present in machinery, equipment, or processes; (2) energy is not necessarily dangerous; and (3) danger is only present when energy is released in quantities or at rates that would harm an employee. Generally speaking, however, potentially hazardous energy sources are defined as those that can cause injury to employees working in, on, or around machines or equipment.

Energy as used in this standard means mechanical motion; potential energy due to pressure, gravity, or springs; electrical energy; or thermal energy resulting from high or low temperatures. Some energy sources can be turned on and off, some can be dissipated, some can be eliminated, and some can only be controlled.

The following brief analysis of energy sources may provide the reader with a better understanding of the provisions of this standard.

1. Mechanical motion can be linear translation or rotation, or it can produce work which, in turn, produces changes in temperature. This type of energy can be turned off or left on.
2. Potential energy can be due to pressure (above or below atmospheric) as in hydraulic, pneumatic, or vacuum systems. Potential energy manifested as pressure or in springs can be dissipated or controlled; it cannot be turned off or on.
3. Electrical energy refers to generated electrical power or static electricity. In the case of generated electricity, the electrical power can be turned on or off. Static electricity cannot be turned off; it can only be dissipated or controlled.
4. Thermal energy is manifested by high or low temperatures. This type of energy is the result of mechanical work, radiation, chemical reaction, or electrical resistance. It cannot be turned off or eliminated; however, it can be dissipated or controlled.

FIGURE 7.4 *(Continued)*

Hot tap. A procedure used in the repair maintenance and services activities which involves welding on a piece of equipment (pipelines, vessels, or tanks) under pressure, in order to install connections or appurtenances. It is commonly used to replace or add sections of pipeline without the interruption of service for air, gas, water, steam, and petrochemical distribution systems.

Lockout. The placement of a lockout device on an energy isolating device, in accordance with an established procedure, ensuring that the energy isolating device and the equipment being controlled cannot be operated until the lockout device is removed.

Lockout device. A device that utilizes a positive means such as a lock, either key or combination type, to hold an energy isolating device in the safe position and prevent the energizing of a machine or equipment.

Normal production operations. The utilization of a machine or equipment to perform its intended production function.

Servicing and/or maintenance. Workplace activities such as constructing, installing, setting up, adjusting, inspecting, modifying, and maintaining and/or servicing machines or equipment. These activities include lubrication, cleaning, or unjamming of machines or equipment, and making adjustments or tool changes, where the employee may be exposed to the unexpected energization or startup of the equipment or release of hazardous energy.

Setting up. Any work performed to prepare a machine or equipment to perform its normal production operation.

Tagout. The placement of a tagout device on an energy isolating device, in accordance with an established procedure, to indicate that the energy isolating device and the equipment being controlled may not be operated until the tagout device is removed.

Tagout device. A prominent warning device, such as a tag and a means of attachment, which can be securely fastened to an energy isolating device in accordance with an established procedure, to indicate that the energy isolating device and the equipment being controlled may not be operated until the tagout device is removed.

(c) *General* (1) *Energy control program.* The employer shall establish a program consisting of energy control procedures, employee training, and periodic inspections to ensure that before any employee performs any servicing or maintenance on a machine or equipment where the unexpected energizing, startup or release of stored energy could occur and cause injury, the machine or equipment shall be isolated from the energy source and rendered inoperative.

The OSHA compliance officer is directed to ask the employer for the documentation including:

1. procedures for the control of hazardous energy including:
   a. shutdown,
   b. equipment isolation,
   c. lockout/tagout application,
   d. release of stored energy, and
   e. verification of isolation;
2. certification of periodic inspections; and
3. certification of training.

The documented procedure must identify the specific types of energy to be controlled and, in instances where a common procedure is to be used, the specific equipment covered by the common procedure must be identified at least by type and location. The identification of the energy to be controlled may be by magnitude and type of energy. See the exception to 147(c)(4)(i), "Note". The employer need not document the required procedure for a particular machine or equipment when all eight (8) elements listed in the "Note" exist.

(2) *Lockout/tagout.* (i) If an energy isolating device is not capable of being locked out, the employer's energy control program under paragraph (c)(1) of this section shall utilize a tagout system.

(ii) If an energy isolating device is capable of being locked out, the employer's energy control program under paragraph (c)(1) of this section shall utilize lockout, unless the employer can demonstrate that the utilization of a tagout system will provide full employee protection as set forth in paragraph (c)(3) of this section.

(iii) After January 2, 1990, whenever replacement or major repair, renovation or modification of a machine or equipment is performed, and whenever new machines or equipment are installed, energy isolating devices for such machine or equipment shall be designed to accept a lockout device.

(3) *Full employee protection.* (i) When a tagout device is used on an energy isolating device which is capable of being locked out, the tagout device shall be attached at the same location that the lockout device would have been attached, and the employer shall demonstrate that the tagout program will provide a level of safety equivalent to that obtained by using a lockout program.

FIGURE 7.4 *(Continued)*

(ii) In demonstrating that a level of safety is achieved in the tagout program which is equivalent to the level of safety obtained by using a lockout program, the employer shall demonstrate full compliance with all tagout-related provisions of this standard together with such additional elements as are necessary to provide the equivalent safety available from the use of a lockout device.

Additional means to be considered as part of the demonstration of full employee protection shall include the implementation of additional safety measures such as the removal of an isolating circuit element, blocking of a controlling switch, opening of an extra disconnecting device, or the removal of a valve handle to reduce the likelihood of inadvertent energization.

(4) *Energy control procedure.* **(i)** Procedures shall be developed, documented, and utilized for the control of potentially hazardous energy when employees are engaged in the activities covered by this section.

Note: *Exception:* The employer need not document the required procedure for a particular machine or equipment, when all of the following elements exist: **(1)** The machine or equipment has no potential for stored or residual energy or reaccumulation of stored energy after shut down which could endanger employees: **(2)** the machine or equipment has a single energy source which can be readily identified and isolated: **(3)** the isolation and locking out of that energy source will completely de-energize and deactivate the machine or equipment: **(4)** the machine or equipment is isolated from that energy source and locked out during servicing or maintenance: **(5)** a single lockout device will achieve a locked-out condition: **(6)** the lockout device is under the exclusive control of the authorized employee performing the servicing or maintenance: **(7)** the servicing or maintenance does not create hazards for other employees; and **(8)** the employer, in utilizing this exception, has had no accidents involving the unexpected activation or reenergization of the machine or equipment during servicing or maintenance.

The exception is intended to apply to situations in which the procedure for deenergization, servicing, and reenergization can be carried out without detailed interactions of energy sources, machines, and employees. For example, a motor in a small machine shop is wired into a single electrical disconnect, with no other source, and the motor does not present the hazards of stored or residual energy. When the motor needs repair, the authorized employee can isolate the motor from the single energy source and lock it out, using his/her personal lockout device on the disconnect, in accordance with the procedures set forth in the standard.

Under these conditions, and provided that no other employees are exposed to hazards from the servicing operation, the servicing may be performed without the need to document the energy control procedure.

(ii) The procedures shall clearly and specifically outline the scope, purpose, authorization, rules, and techniques to be utilized for the control of hazardous energy, and the means to enforce compliance including, but not limited to, the following:

(A) A specific statement of the intended use of the procedure;

(B) Specific procedural steps for shutting down, isolating, blocking, and securing machines or equipment to control hazardous energy;

(C) Specific procedural steps for the placement, removal, and transfer of lockout/tagout devices and the responsibility for them; and

(D) Specific requirements for testing a machine or equipment to determine and verify the effectiveness of lockout devices, tagout devices, and other energy control measures.

OSHA uses the word "specific" to emphasize the need to have a detailed procedure, one which clearly and specifically outlines the steps to be followed. Overgeneralization can result in a document which has little or no meaning to the employee who must follow the procedure. However, whereas the procedure is required to be written in detail, this does not mean that a separate procedure must be written for each and every machine or piece of equipment. Similar machines and/or equipment (those using the same type and magnitude energy) which have the same or similar types of controls can be covered with a single procedure.

The written energy control procedure required by this standard need not be overly complicated or detailed, depending on the complexity of the equipment and the control measures to be utilized.

For example, if there is a single machine with a single energy source that must be isolated, and the control means is simple, such as opening an electrical disconnect and locking out that energy source during servicing, the written procedure could be very simple. The steps set forth in the standard can be incorporated into the procedure with very little detail, reflecting the lack of complexity of the control measure. In addition, the employer's procedures may not need to be unique for a single machine or task, but can apply to a group of similar machines, types of energy and tasks if a single procedure can address the hazards and the steps to be taken satisfactorily.

FIGURE 7.4 *(Continued)*

OSHA believes that because of the need to follow the steps in the energy control procedure carefully and specifically, and the number of variables involved in controlling hazardous energy, a documented procedure is necessary for most energy control situations. However, OSHA has determined that in certain limited situations, documentation of the procedure will not add markedly to the projections otherwise provided by the standard.

These situations incorporate several common elements:

1. There is a single source of hazardous energy which can be easily identified and isolated, and there is no potential for stored or residual energy in the equipment. This greatly simplifies the procedure for controlling the energy, since the single energy source is all that needs to be isolated.
2. The isolation and locking out of that single energy source will totally deenergize and deactivate the machine or equipment. There are no collateral sources of energy which need to be addressed.
3. A full lockout of the energy source is achieved by a single lockout device which is under the exclusive control of the authorized employee performing the servicing or maintenance. As used in this provision, exclusive control means that the authorized employee is the only person who can affix or remove the device. The authorized employee follows all steps necessary for deenergizing the equipment, verifying the deenergization, performing the work, and reenergizing the equipment upon completion of servicing. Because the energy control elements are simple, with a single energy source being locked out and no other potential sources of unexpected activation or energization, the authorized employee can perform them without referring to a written document
4. While the equipment is locked out, the servicing or maintenance cannot expose other employees to hazards. For example, shutdown and lockout of a conveyer cannot cause jams or other hazards at other conveyers which feed into the conveyer being serviced.

(5) ***Protective materials and hardware.*** **(i)** Locks, tags, chains, wedges, key blocks, adapter pins, self-locking fasteners, or other hardware shall be provided by the employer for isolating, securing, or blocking of machines or equipment from energy sources.

To meet this requirement, the employer can either issue devices to each employee responsible for implementing energy control measures, or can exercise the option of simply having a sufficient quantity of the devices on hand at any given time and assign or distribute them to the employees as the need arises. All authorized employees will need to have these devices available to attach to energy isolating devices whenever they perform servicing or maintenance using the energy control procedure.

(ii) Lockout/tagout devices shall be singularly identified; shall be the only device(s) used for controlling energy; shall not be used for other purposes; and shall meet the following requirements:

This restriction was imposed to ensure that the sight of a distinctive lock or tag will provide a constant message of the use that the device is being put to and the restrictions which this device is intended to covey. If lockout or tagout devices are used for other purposes, they can lose their significance in the workplace.

For the energy control procedure to be effective, these devices must have a single meaning to employees: "Do not energize the equipment when such a device is affixed to it."

(A) ***Durable.*** ***{1}*** Lockout and tagout devices shall be capable of withstanding the environment to which they are exposed for the maximum period of time that exposure is expected.

{2} Tagout devices shall be constructed and printed so that exposure to weather conditions or wet and damp locations will not cause the tag to deteriorate or the message on the tag to become illegible.

{3} Tags shall not deteriorate when used in corrosive environments such as areas where acid and alkali chemicals are handled and stored.

OSHA requires that tagout devices must be constructed and printed so that exposure to weather or other environmental conditions which exist in the workplace will not cause the tag to become unserviceable and/or the message on the tag to become illegible. For any sign, tag, or other message bearing item, the message must remain legible for the employees to be able to ascertain the meaning and intent of the message. Devices which are not exposed to harsh environments need not be capable of withstanding such exposure.

(B) ***Standardized.*** Lockout and tagout devices shall be standardized within the facility in at least one of the following criteria: Color; shape; or size; and additionally, in the case of tagout devices, print and format shall be standardized.

FIGURE 7.4 *(Continued)*

OSHA requires that lockout or tagout devices be standardized in order to be readily identifiable and distinguished from other similar devices found in the workplace. In addition, there is a requirement for the use of a standardized print and format for tagout devices. This is done to ensure that the tagout devices, which rely exclusively on employee recognition for their effectiveness, will be so unique as to minimize the chances of their being misidentified or their message misinterpreted.

(C) ***Substantial. {1} Lockout devices.*** Lockout devices shall be substantial enough to prevent removal without the use of excessive force or unusual techniques, such as with the use of bolt cutters or other metal cutting tools.

{2} Tagout devices. Tagout devices, including their means of attachment, shall be substantial enough to prevent inadvertent or accidental removal. Tagout device attachment means shall be of a non-reusable type, attachable by hand, self-locking, and non-releasable with a minimum unlocking strength of no less than 50 pounds and having the general design and basic characteristics of being at least equivalent to a one-piece all environment-tolerant nylon cable tie.

OSHA requires that "Tagout devices having reusable, non-locking, easily detachable means of attachment (such as string, cord, or adhesive) are not permitted."

The "tagout device" addressed here is when a tagout device is used instead of a lockout device. If a tag is used in conjunction with a lock then these restrictions do not apply.

OSHA requires that lockout and tagout devices be substantial enough to minimize the possibility of premature removal.

The additional requirements for a tagout device are being imposed to ensure that tags do not become disconnected or lost during use, thereby negating their effectiveness.

(D) ***Identifiable.*** Lockout and tagout devices shall indicate the identity of the employee applying the device(s).

Identification of the user provides an additional degree of accountability to the overall program. It enables the employer to inspect the application of the energy control procedure and determine which employees are properly implementing its requirements. If locks or tags are not being properly attached by an employee, Identification on the locks and tags will enable the employer to locate that employee and correct the problem promptly, including additional training, as necessary. For other employees, this requirement will enable them to determine at a glance which authorized employees are performing a given servicing operation.

It puts them on notice that if questions arise about the servicing or the energy control procedure, the persons listed on the lockout and tagout devices are the appropriate persons to ask. The authorized employee has the additional assurance that other employees know of his/her involvement in the servicing, and that only he/she is allowed to remove the device.

OSHA believes that knowing who applied a lockout device to a machine or equipment can save time and lives. If an employee, upon completing a job, forgets to remove a lockout device, the identity of the employee can be immediately determined and the employee made available to complete the procedure. If that employee cannot be located, it is possible that he/she is still working on the equipment. It would then be possible to check out the area and assure that the employee and others are out of the danger area before the device is removed. Marking a lockout or tagout device is a simple way of identifying the person who applies it, and can prevent the inadvertent reenergization or reactivation of equipment before that employee has been located and has moved clear of the equipment. Thus, marking the identity of the employee who uses a lockout or tagout device is an appropriate safeguard.

Marking of the lockout or tagout devices can also promote a sense of security in employees, in that each device is the individual employee's device, used only for his or her protection. This sense of identity also can be used to encourage willing utilization of the energy control procedure. When an employee can identify with a part of the program he/she controls for his/her own protection, that employee will likely be an active participant in making the program work.

(iii) Tagout devices shall warn against hazardous conditions if the machine or equipment is energized and shall include a legend such as the following: Do Not Start. Do Not Open. Do Not Close. Do Not Energize. Do Not Operate.

OSHA states that the legend (major message) on the tagout devices must warn against hazardous conditions if the equipment is reenergized.

OSHA recognizes, however, that these messages may not be sufficient to cover all conditions involving hazardous energy control. For that reason, the above stated legends are only examples of what must be stated. The

FIGURE 7.4 *(Continued)*

use of graphics, pictographs, or other symbols to convey the message which the tag represents serves the same purpose as the written message and therefore would be acceptable to OSHA.

The significance of this message is imparted through the training of employees and enforcement of the program. The backbone of a tagout system is that when a tagout device is placed on an energy isolating device, it informs employees that the energy isolating device is not to be turned on or otherwise moved to a position which will allow the flow of energy.

The printed message on the tag provides information about what the tag stands for and what it prohibits, and indicates the name of the employee who affixed it to the energy isolating device.

(6) ***Periodic inspection.*** **(i)** The employer shall conduct a periodic inspection of the energy control procedure at least annually to ensure that the procedure and the requirements of this standard are being followed.

(A) The periodic inspection shall be performed by an authorized employee other than the one(s) utilizing the energy control procedure being inspected.

(B) The periodic inspection shall be conducted to correct any deviations or inadequacies identified.

(C) Where lockout is used for energy control, the periodic inspection shall include a review, between the inspector and each authorized employee, of that employee's responsibilities under the energy control procedure being inspected.

(D) Where tagout is used for energy control, the periodic inspection shall include a review, between the inspector and each authorized and affected employee, of that employee's responsibilities under the energy control procedure being inspected, and the elements set forth in paragraph (c)(7)(ii) of this section.

(ii) The employer shall certify that the periodic inspections have been performed. The certification shall identify the machine or equipment on which the energy control procedure was being utilized, the date of the inspection, the employees included in the inspection, and the person performing the inspection.

The periodic inspection is intended to assure that the energy control procedures continue to be implemented properly, and that the employees involved are familiar with their responsibilities under those procedures.

The inspector, who is required to be an authorized person not involved in the energy control procedure being inspected, must be able to determine three things:

1. Whether the steps in the energy control procedure are being followed;
2. Whether the employees involved know their responsibilities under the procedure; and
3. Whether the procedure is adequate to provide the necessary protection, and what changes, if any, are needed.

The inspector will need to observe and talk with the employees in order to make these determinations. This regulation provides some additional guidance as to the inspector's duties in performing periodic inspections, to assure that he or she obtains the necessary information about the energy control procedure and its effectiveness.

Whenever lockout is used. The inspector must review each authorized employee's responsibilities under the procedure with that employee. This does not necessarily require separate one-on-one meetings, but can involve the inspector meeting with the whole servicing crew at one time. Indeed, group meetings can be the most effective way of dealing with this situation, because it reinforces the employee's and that they need to follow the procedure carefully.

Where tagout is used. The inspector's review of responsibilities extends to affected employees as well, because of the increased importance of their role in avoiding accidental or inadvertent activation of the equipment or machinery being serviced.

OSHA believes that periodic inspections by the employer are necessary to ensure continued compliance with the procedure.

The next section addresses the training requirements for the control of hazardous energy.

OSHA states "The standard recognizes three types of employees: (1) 'authorized,' (2) 'affected' and (3) 'other.' Different levels of training are required based upon the respective roles of employees in the control of energy and the knowledge which they must possess to accomplish their tasks safely and to ensure the safety of fellow workers as related to the lockout/tagout procedures."

(7) ***Training and communication.***

OSHA specifies that the employer provide effective initial training, periodic retraining, and certification of such training of employees. OSHA considers these requirements to be of critical importance in helping to ensure that the applicable provisions of the hazardous energy control procedure(s) are known, understood, and strictly adhered to by employees.

FIGURE 7.4 *(Continued)*

In order to provide adequate information, any training program under this standard will need to cover at least three areas; (1) The employer's energy control program; (2) the elements of the energy control procedure which are relevant to the employee's duties; and (3) the requirements of this regulation.

(i) The employer shall provide training to ensure that the purpose and function of the energy control program are understood by employees and that the knowledge and skills required for the safe application, usage, and removal of the energy controls are required by employees. The training shall include the following:

OSHA has determined that the marking or labeling of energy isolating devices is not reasonably necessary for the effectiveness of the energy control program. When employees need to know details on energy sources for protection under the standard, the energy control procedure is required to spell out this information, and the training must incorporate it, as well. For example, authorized employees, in order to perform their servicing or maintenance duties under the energy control procedure, are required to know the type and magnitude of the energy sources which must be controlled. The marking or labeling of the sources themselves will not provide the authorized employees with any additional information. Second, as far as affected or other employees are concerned, their role in the energy control program is essentially to understand what the program is designed to accomplish, and to recognize that when they see an energy isolating device with a tag and/or lock on it, they are not to touch the equipment, regardless of what the type and magnitude of the energy might be. OSHA believes that marking the equipment with this information would not enhance the protection of these employees, because their compliance with the energy control procedure does not depend upon knowledge of these details.

(A) Each authorized employee shall receive training in the recognition of applicable hazardous energy sources, the type and magnitude of the energy available in the workplace, and the methods and means necessary for energy isolation and control.

Because authorized employees are charged with the responsibility for implementing energy control procedures, it is important that they receive training in recognizing and understanding all potentially hazardous energy sources that they might be exposed to during their work assignments, but that they also be trained in the use of adequate methods and means for the control of such energy sources. These employees are the ones authorized to implement the energy control procedure and to perform servicing of the machine or equipment. Therefore, they need extensive training in aspects of the procedure and its proper utilization, together with all relevant information about the equipment being serviced.

OSHA believes that employee knowledge of this information is essential to ensure that the correct energy control devices are used on the proper energy isolating devices and in the proper manner. This provision requires the employee to have that specific information prior to deenergizing the equipment, in order to control the energy and render the machine or equipment safe to work on.

(B) Each affected employee shall be instructed in the purpose and use of the energy control procedure.

Affected employees are important to the overall protection provided in the energy control program, however, because such employees work in areas where the program is being utilized by authorized employees, it is vital to the safety of the authorized employees that the affected employees recognize lockout/tagout devices immediately that they know not to disturb the lockout/tagout devices or the equipment to which the devices are affixed.

The instruction needs to be sufficient to enable the employees to determine if a control measure is in use. The instructions also need to make affected employees aware that disregarding or violating the prohibitions imposed by the energy control program could endanger their own lives or the lives of others.

(C) All other employees whose work operations are or may be in an area where energy control procedures may be utilized, shall be instructed about the procedure, and about the prohibition relating to attempts to restart or reenergize machines or equipment which are locked out or tagged out.

OSHA requires that all other employees shall be instructed about the restrictions imposed upon all employees by the energy control program. This instruction as the employer's lockout/tagout procedure can be conveyed during new employee orientation sessions, by the use of employee handbooks, or through regularly scheduled safety meetings. The training of employees other than authorized and affected employees is considered by OSHA to be essential since other employees working in the plant or facility have been known to have turned on the power to a machine or equipment on which another employee is performing a servicing or maintenance activity. Inadvertent

FIGURE 7.4 *(Continued)*

and intentional activation of machines or equipment by employees other than those working on the machine or equipment is not limited to affected employees. The training requirements for those other employees are minimal, essentially required only that these employees know what the energy control program does and that they are not to touch any locks, tags, or equipment covered by this program.

(ii) When tagout systems are used, employees shall also be trained in the following limitations of tags:

(A) Tags are essentially warning devices affixed to energy isolating devices, and do not provide the physical restraint on those devices that is provided by a lock.

(B) When a tag is attached to an energy isolating means, it is not to be removed without authorization of the authorized person responsible for it, and it is never to be bypassed, ignored, or otherwise defeated.

(C) Tags must be legible and understandable by all authorized employees, affected employees, and all other employees whose work operations are or may be in the area, in order to be effective.

(D) Tags and their means of attachment must be made of materials which will withstand the environmental conditions encountered in the workplace.

(E) Tags may evoke a false sense of security, and their meaning needs to be understood as a part of the overall energy control program.

(F) Tags must be securely attached to energy isolating devices so that they cannot be inadvertently or accidentally detached during use.

(iii) Employee retraining.

OSHA recommends that periodic retraining be provided for authorized employees be conducted at least annually. This retraining may need to be conducted more frequently, that is, whenever an inspection under paragraph (c)(6) reveals or whenever the employer has reason to believe that there are deviations from or inadequacies in the energy control procedure.

OSHA believes that the effectiveness of training diminishes as the time from the last training session increases. OSHA has determined that the provisions of this standard simply rely upon the finding of a problem with the program to trigger the retraining program. Retraining is intended to provide for continued proficiency, and not merely to remedy situations in which such proficiency has been found wanting.

In addition to the periodic retraining, additional retraining is to be conducted whenever a problem is identified during periodic inspections, or whenever the employer has reason to believe that there are problems with the energy control procedure itself or with its implementation.

This retraining should be more concentrated or more encompassing than the routine retraining, based upon the severity of the problem encountered with the use of the energy control program in the workplace.

OSHA is of the opinion that full and uniform utilization of an energy control procedure is necessary in order for that procedure to maintain effectiveness.

Every effort should be made during the periodic inspection to determine whether or not the procedure is being used properly. If deviations are observed, retraining would be required. However, retraining could be triggered by events separate from the findings of the periodic inspection. For example, an employee working with an energy control procedure might be injured in the course of his duties, or there might be a "near miss" where no one is actually injured, but where the energy control program has failed nonetheless. If a subsequent investigation indicates that an employee failed to operate within the guidelines of the control procedure, retraining would be required.

In addition, the investigation might also reveal that the procedure itself was not adequate. Such inadequacies in the procedure could be the result of using a general procedure that does not handle effectively a specific application, or they may arise because changes have been made to the equipment or process that did not take the existing energy control procedure into consideration. In such cases when changes to the energy control procedure must be made, the employer is required to retrain employees in the new or revised procedure.

(A) Retraining shall be provided for all authorized and affected employees whenever there is a change in their job assignments, a change in machines, equipment or processes that presents a new hazard, or when there is a change in the energy control procedures.

(B) Additional retraining shall also be conducted whenever a periodic inspection under paragraph (c)(6) of this section reveals, or whenever the employer has reason to believe that there are deviations from or inadequacies in the employee's knowledge or use of the energy control procedures.

FIGURE 7.4 *(Continued)*

(C) The retraining shall reestablish employee proficiency and introduce new or revised control methods and procedures, as necessary.

In this Final Rule, when lockout is implemented, OSHA is limiting the annual retraining recommendations to "Authorized" employees. These are the employees who must implement the energy control procedure, and their protection is the primary consideration under this standard. Because their safety requires them to follow the steps of the procedure precisely, these employees must be properly trained, and that training must be reinforced to assure their continued proficiency. By contrast, affected employees are not provided with annual retraining under this standard when lockout is used. In these situations, affected employees are initially trained about the energy control procedure and its implementation, and the relevance of that procedure to his/her work. Under lockout conditions, the essential element of the affected employee's training is a simple one: Locks are not to be defeated or bypassed, and locked out equipment must remain deenergized.

This message is reinforced whenever the affected employees work in an area where energy control procedures are being implemented, because this standard requires that such employees be notified before the energy control devices are applied. Further, when a lockout device is attached to a piece of equipment by an authorized employee, an affected employee should not be able to remove the lock, and thus will not have the potential of placing the authorized employee in danger.

OSHA notes that annual retraining should be provided for authorized and affected employees in the use of the tagout system. This additional training is necessary because of the inherent difficulties of tagout systems as opposed to lockout.

The use of tags relies uniquely upon the knowledge and training of the employees involved, and the continued reinforcement of the meaning of the tags. In a lockout system, even if an affected employee has not been adequately trained, the lock will prevent that employee from reenergizing the equipment. Tags, on the other hand, can be inadvertently or intentionally bypassed or ignored by an affected employee, because the tags do not actually prevent the activation of the tagged out equipment. Employees operating under a tagging system must be constantly vigilant, and their awareness of the importance of the tagout device must be frequently reinforced. OSHA believes that when tagout is used, retraining must be provided on at least an annual basis, in order to maximize its effectiveness.

(iv) The employer shall certify that employee training has been accomplished and is being kept up-to-date. The certification shall contain each employee's name and dates of training.

Certifications are intended to cover both the initial training and the periodic retraining. In addition to certifications, the employer must be able to demonstrate that the training includes all elements of the energy control procedure which are directly relevant to the duties of the employee. The adequacy of the training can be evaluated by the employer, employee, and OSHA alike by comparing the elements of the training to the elements of the procedure, which is required to be in written form.

It should be noted that the certification is not intended as a means of evaluating the completeness or efficacy of the training; it only provides an indication that training has been performed. The quality and content of the training are not evaluated through the certification of performance. In evaluating whether an employee has been adequately trained, OSHA will examine the employee's responsibilities under the energy control program in relation to the elements of the standard.

(8) ***Energy isolation.*** Lockout/tagout shall be performed only by the authorized employees who are performing the servicing or maintenance.

These are the only employees who are required to be trained to know in detail about the types of energy available in the workplace and how to control the hazards of that energy. Only properly trained and qualified employees can be relied on to deenergize and to properly lockout/tagout machines or equipment which are being serviced or maintained, in order to ensure that the work will be accomplished safely.

(9) ***Notification of employees.*** Affected employees shall be notified by the employer or authorized employee of the application and removal of lockout devices or tagout devices. Notification shall be given before the controls are applied, and after they are removed from the machine or equipment.

OSHA requires that whenever lockout/tagout control might directly affect another employee's work activities, the employer or authorized employee must notify the affected employee before taking any action to apply or to remove lockout/tagout devices.

FIGURE 7.4 *(Continued)*

OSHA believes that this requirement is an essential component of the total energy control program. Notification of affected employees when lockout/tagout is going to be applied provides the perfect opportunity for the employer or authorized employee who notifies them of the impending interruption of the normal production operation to remind them and reinforce the importance of the restrictions imposed upon them by the energy control program.

OSHA believes that these measures are important to ensure that employees who operate or use machines or equipment do not unknowingly attempt to reenergize those machines or equipment that have been taken out of service and deenergized for the performance of activities covered by this standard. The lack of information regarding the status of the equipment could endanger both the servicing employees and the employees attempting to reenergize or operate the equipment. Such notification is also needed after servicing is completed to assure that employees know when the control measures have been removed. Without such information, employees might mistakenly believe that a system is still deenergized and that it is safe to continue working on or around it.

The main thrust of the standard is to mandate the development, documentation and implementation of control procedures, and this is to be accomplished as outlined in paragraph (d) of this standard. The employer is given considerable flexibility in developing a control program, and such a program will be evaluated by OSHA compliance officers to determine whether it meets all the criteria in this standard.

(d) *Application of control.* The established procedures for the application of energy control (the lockout/tagout procedures) shall cover the following elements and actions, and shall be done in the following sequence:

This section provides that six separate and distinct steps be followed in meeting the procedural requirements of paragraph (c)(1) and the application of energy control (lockout or tagout) measures, and that the actions be taken in the sequence presented.

(1) *Preparation for shutdown.* Before an authorized or affected employee turns off a machine or equipment, the authorized employee shall have knowledge of the type and magnitude of the energy, the hazards of the energy to be controlled, and the method or means to control the energy.

This paragraph requires that in preparation for shutdown of machinery or equipment, the authorized employee must know about the type and magnitude of the energy, the hazards involved and the means of controlling them.

(This would require the authorized person to be qualified on the equipment or system.)

(2) *Machine or equipment shutdown.* The machine or equipment shall be turned off or shut down using the procedures established for the machine or equipment. An orderly shutdown must be utilized to avoid any additional or increased hazard(s) to employees as a result of the equipment stoppage.

This paragraph requires that the machine be turned off or shut down by an authorized employee according to established procedures. This is the starting point for all subsequent actions necessary to put the machine or equipment in a state that will permit employees to work on it safely.

In many operations, activation of an electrical push-button control or the movement of a simple throw switch (electrical, hydraulic, or pneumatic) to the "stop" or "off" mode is sufficient to meet this provision. In other cases, however, such as those found typically in a refining or chemical process, there are control devices that do not necessarily address an "off-on" or "start-stop" condition (i.e., level controls, pressure controllers, etc.). In these instances, a series of predetermined steps may be necessary to achieve a shutdown of the machine or equipment.

OSHA is aware that although an authorized employee would usually have the necessary knowledge and capability to shut down machines or equipment, a machine or equipment operator or user should also be in a position and know how to shut down the machine or equipment he or she is utilizing.

In many cases, allowing a machine or equipment operator or user to shut it down when something goes wrong may save time and money, and may possibly avoid an accident. In many cases, the affected employee may be infinitely more familiar with the shutdown procedure for the machine or equipment, and would be able to accomplish the shutdown more rapidly and safely than an authorized person who does not work with that particular machine or equipment every day.

In the event that a machine or equipment malfunctions, the wise and prudent thing to do in most cases is to require the machine or equipment be immediately shut down. Shutting down a machine or equipment is analogous to stopping the production operation. OSHA believes that stopping the machine's production function is the

FIGURE 7.4 *(Continued)*

necessary and appropriate first step in the procedure. OSHA also believes that the necessary first step is to interrupt the production process to allow non-servicing (affected) personnel to move clear of the machine or equipment. Once this is done and employees are not exposed to a hazard, the machine or equipment can be restarted by the authorized employee under the guidelines of paragraph (f)(1) when necessary to allow positioning of the machine or equipment, or components thereof.

(3) *Machine or equipment isolation.* All energy isolating devices that are needed to control the energy to the machine or equipment shall be physically located and operated in such a manner as to isolate the machine or equipment from the energy source(s).

Following the shutdown of the machine or equipment, paragraph (d)(3), as the next step in the procedure, provides that energy isolation devices be physically located and operated in such a manner as to isolate the machine or equipment from the energy source(s). For example, once an electrical push-button control has been utilized to stop the movement of machine or equipment parts as the first step of the shutdown procedure, isolation can then be accomplished by assuring that the push-button circuitry cannot be supplied with additional electrical energy. For such equipment, the isolation requirement can be accomplished by the employee's actions in tracing the path from the control toward the energy source until he/she locates the energy isolating device, and moving the energy isolating device control lever to the "safe," "off," or "open" position. Performing these actions will prevent the reintroduction of energy to the push-button circuitry and will isolate the operating control and the machine or equipment from the energy source.

(4) *Lockout or tagout device application.* **(i)** Lockout or tagout devices shall be affixed to each energy isolating device by authorized employees.

(ii) Lockout devices, where used, shall be affixed in a manner that will hold the energy isolating devices in a "safe" or "off" position.

(iii) Tagout devices, where used, shall be affixed in such a manner as will clearly indicate that the operation or movement of energy isolating devices from the "safe" or "off" position is prohibited.

(A) Where tagout devices are used with energy isolating devices designed with the capability of being locked, the tag attachment shall be fastened at the same point at which the lock would have been attached.

(B) Where a tag cannot be affixed directly to the energy isolating device, the tag shall be located as close as safely possible to the device, in a position that will be immediately obvious to anyone attempting to operate the device.

As the fourth step in the procedure, paragraph (d)(4) provides that action be taken to secure the energy isolating devices in the "safe" or "off" position. This paragraph requires that appropriate and effective lockout/tagout devices be affixed to each energy isolating device by an authorized employee, and that they be attached so as to prevent reactivation of the machine or equipment.

OSHA is of the opinion that, as a general rule, when it is feasible, the physical protection offered by the use of a lock, when supported by the information provided on a tag used in conjunction with the lock, provides the greatest assurance of employee protection from the release of hazardous energy.

(5) *Stored energy.* **(i)** Following the application of lockout or tagout devices to energy isolating devices, all potentially hazardous stored or residual energy shall be relieved, disconnected, restrained, and otherwise rendered safe.

(ii) If there is a possibility of reaccumulation of stored energy to a hazardous level, verification of isolation shall be continued until the servicing or maintenance is completed, or until the possibility of such accumulation no longer exists.

Paragraph (d)(5) provides that the next step taken in the energy control procedure is to determine the presence of, and relieve, disconnect and/or restrain all potentially hazardous, stored or residual energy in the machine or equipment. Up to this point, the purpose of following all the steps of the procedure has been to enable the employee to isolate and block the source of energy feeding the machine or equipment to be worked on, at a point beyond which it can not be bypassed. However, energy can very easily be trapped in a system down-stream from an energy isolating device, or can be present in the form of potential energy from gravity or from spring action.

Stored or residual energy of this sort cannot be turned on or off; it must be dissipated or controlled.

When energy may still be present in a system that has been isolated from the energy source, this paragraph requires that energy to be controlled before an employee attempts to perform any work covered by the scope of this standard. Compliance with this provision might require, for example, the use of blocks or other physical

FIGURE 7.4 (*Continued*)

restraints to immobilize the machine, machine components, or equipment where necessary for control of the hazard. In the case of electrical circuits, grounding might be necessary to discharge hazardous energy. Hydraulic or pneumatic systems might necessitate the use of bleed valves to relieve the pressure.

OSHA believes that these requirements must be utilized since there is no manner to ensure that some leakage or drainage of energy or energy containing substances, such as super cooled or cryogenic fluids, can occur. In the case of one of those substances being present in a piping, containment, or transport system, a certain amount of leakage may occur without endangering employees. However, if servicing or maintenance must be performed on such a system, the standard requires the employer to continue to verify the isolation of energy sources which may be hazardous, in order to assure that such leakage does not approach a dangerous level. This may involve means, such as continuous monitoring for the displacement of oxygen or the buildup of the concentration of the substance toward the lower explosive limit of the substance, such as could occur with a hydrogen system.

(6) *Verification of isolation.* Prior to starting work on machines or equipment that have been locked/tagged out, the authorized employee shall verify that isolation and deenergization of the machine or equipment have been accomplished.

In this paragraph, as the sixth step in the energy control procedure, the authorized employee must ensure that the previous steps of the procedure have been taken to isolate the machine or equipment effectively. This must be done prior to starting the servicing or maintenance work. The authorized employee needs to verify that the machine or equipment has been turned off or shut down properly; that all energy isolating devices were identified, located and operated; that the lockout or tagout devices have been attached to energy isolating devices; and that stored energy has been rendered safe.

(e) *Release from lockout or tagout.* Before lockout or tagout devices are removed and energy is restored to the machine or equipment, procedures shall be followed and actions taken by the authorized employee(s) to ensure the following:

(1) *The machine or equipment.* The work area shall be inspected to ensure that non-essential items have been removed and to ensure that machine or equipment components are operationally intact.

(2) *Employees.* (i) The work area shall be checked to ensure that all employees have been safely positioned or removed.

(ii) Before lockout or tagout devices are removed and before machines or equipment are energized, affected employees shall be notified that the lockout or tagout devices have been removed.

(iii) After lockout or tagout devices have been removed and before a machine or equipment is started, affected employees shall be notified that the lockout or tagout device(s) have been removed.

(3) *Lockout or tagout devices removal.* Each lockout or tagout device shall be removed from each energy isolating device by the employee who applied the device. ***Exception to paragraph (e)(3):*** When the authorized employee who applied the lockout or tagout device is not available to remove it, that device may be removed under the direction of the employer, provided that specific procedures and training for such removal have been developed, documented and incorporated into the employer's energy control program. The employer shall demonstrate that the specific procedure shall include at least the following elements:

(i) Verification by the employer that the authorized employee who applied the device is not at the facility:

(ii) Making all reasonable efforts to contact the authorized employee to inform him/her that his/her lockout or tagout device has been removed; and

(iii) Ensuring that the authorized employee has this knowledge before he/she resumes work at that facility.

(f) *Additional requirements.* (1) *Testing or positioning of machines, equipment or components thereof.* In situations in which lockout or tagout devices must be temporarily removed from the energy isolating device and the machine or equipment energized to test or position the machine, equipment or component thereof, the following sequence of actions shall be followed:

(i) Clear the machine or equipment of tools and materials in accordance with paragraph (e)(1) of this section;

(ii) Remove employees from the machine or equipment area in accordance with paragraph (e)(2) of this section;

(iii) Remove the lockout or tagout devices as specified in paragraph (e)(3) of this section;

(iv) Energize and proceed with testing or positioning;

(v) De-energize all systems and reapply energy control measures in accordance with paragraph (d) of this section to continue the servicing and/or maintenance.

FIGURE 7.4 *(Continued)*

(2) *Outside personnel (contractors, etc.).* **(i)** Whenever outside servicing personnel are to be engaged in activities covered by the scope and application of this standard, the on-site employer and the outside employer shall inform each other of their respective lockout or tagout procedures.

(ii) The on-site employer shall ensure that his/her employees understand and comply with the restrictions and prohibitions of the outside employer's energy control program.

(3) *Group lockout or tagout.* **(i)** When servicing and/or maintenance is performed by a crew, craft, department, or other group, they shall utilize a procedure which affords the employees a level of protection equivalent to that provided by the implementation of a personal lockout or tagout device.

(ii) Group lockout or tagout devices shall be used in accordance with the procedures required by paragraph (c)(4) of this section including, but not necessarily limited to, the following specific requirements:

(A) Primary responsibility is vested in an authorized employee for a set number of employees working under the protection of a group lockout or tagout device (such as an operations lock);

(B) Provision for the authorized employee to ascertain the exposure status of individual group members with regard to the lockout or tagout of the machine or equipment and

(C) When more than one crew, craft, department, etc. is involved, assignment of overall job-associated lockout or tagout control responsibility to an authorized employee designated to coordinate affected work forces and ensure continuity of protection; and

(D) Each authorized employee shall affix a personal lockout or tagout device to the group lockout device, group lockbox, or comparable mechanism when he or she begins work, and shall remove those devices when he or she stops working on the machine or equipment being serviced or maintained.

(4) *Shift or personnel changes.* Specific procedures shall be utilized during shift or personnel changes to ensure the continuity of lockout or tagout protection, including provision for the orderly transfer of lockout or tagout device protection between off-going and oncoming employees, to minimize exposure to hazards from the unexpected energization or start-up of the machine or equipment, or the release of stored energy.

APPENDIX A—TYPICAL MINIMAL LOCKOUT OR TAGOUT SYSTEM PROCEDURES

General

The following simple lockout procedure is provided to assist employers in developing their procedures so they meet the requirements of this standard. When the energy isolating devices are not lockable, tagout may be used, provided the employer complies with the provisions of the standard which require additional training and more rigorous periodic inspections. When tagout is used and the energy isolating devices are lockable, the employer must provide full employee protection (see paragraph (c)(3)) and additional training and more rigorous periodic inspections are required. For more complex systems, more comprehensive procedures may need to be developed, documented, and utilized.

Lockout/Tagout Procedure

Lockout Procedure for

____

(Name of Company for single procedure or identification of equipment if multiple procedures are used).

Purpose

This procedure establishes the minimum requirements for the lockout of energy isolating devices whenever maintenance or servicing is done on machines or equipment. It shall be used to ensure that the machine or equipment is stopped, isolated from all potentially hazardous energy sources, and locked out before employees perform any servicing or maintenance where the unexpected energization, or start-up of the machine or equipment, or release of stored energy could cause injury.

Compliance With This Program

All employees are required to comply with the restrictions and limitations imposed upon them during the use of lockout. The authorized employees are required to perform the lockout in accordance with this procedure. All employees,

FIGURE 7.4 *(Continued)*

upon observing a machine or piece of equipment which is locked out to perform servicing or maintenance shall not attempt to start, energize, or use that machine or equipment.

Type of compliance enforcement to be taken for violation of the above.

Sequence of Lockout

(1) Notify all affected employees that servicing or maintenance is required on a machine or equipment and that the machine or equipment must be shut down and locked out to perform the servicing or maintenance.

Name(s)/Job Title(s) of affected employees and how to notify.

(2) The authorized employee shall refer to the company procedure to identify the type and magnitude of the energy that the machine or equipment utilizes, shall understand the hazards of the energy, and shall know the methods to control the energy.

Type(s) and magnitude(s) of energy, its hazards and the methods to control the energy.

(3) If the machine or equipment is operating, shut it down by the normal stopping procedure (depress the stop button, open switch, close valve, etc.).

Type(s) and location(s) of machine or equipment operating controls.

(4) De-activate the energy isolating device(s) so that the machine or equipment is isolated from the energy source(s).

Type(s) and location(s) of energy isolating devices.

(5) Lock out the energy isolating device(s) with assigned individual lock(s).
(6) Stored or residual energy (such as that in capacitors, springs, elevated machine members, rotating flywheels, hydraulic systems, and air, gas, steam, or water pressure, etc.) must be dissipated or restrained by methods such as grounding, repositioning, blocking, bleeding down, etc.

Type(s) of stored energy—methods to dissipate or restrain.

(7) Ensure that the equipment is disconnected from the energy source(s) by first checking that no personnel are exposed, then verify the isolation of the equipment by operating the push button or other normal operating control(s) or by testing to make certain the equipment will not operate.
Caution: Return operating control(s) to neutral or "off" position after verifying the isolation of the equipment.

Method of verifying the isolation of the equipment.

(8) The machine or equipment is now locked out.

Restoring Equipment to Service

When the servicing or maintenance is completed and the machine or equipment is ready to return to normal operating condition, the following steps shall be taken.

(1) Check the machine or equipment and the immediate area around the machine to ensure that non-essential items have been removed and that the machine or equipment components are operationally intact.
(2) Check the work area to ensure that all employees have been safely positioned or removed from the area.
(3) Verify that the controls are in neutral.
(4) Remove the lockout devices and reenergize the machine or equipment.
Note: The removal of some forms of blocking may require reenergization of the machine before safe removal.
(5) Notify affected employees that the servicing or maintenance is completed and the machine or equipment is ready for use.

FIGURE 7.4 (*Continued*)

§ 1910.399 Definitions applicable to this subpart.

Acceptable. An installation or equipment is acceptable to the Assistant Secretary of Labor, and approved within the meaning of this Subpart S:

(1) If it is accepted, or certified, or listed, or labeled, or otherwise determined to be safe by a nationally recognized testing laboratory recognized pursuant to § 1910.7; or

(2) With respect to an installation or equipment of a kind that no nationally recognized testing laboratory accepts, certifies, lists, labels, or determines to be safe, if it is inspected or tested by another Federal agency, or by a State, municipal, or other local authority responsible for enforcing occupational safety provisions of the National Electrical Code, and found in compliance with the provisions of the National Electrical Code as applied in this subpart; or

(3) With respect to custom-made equipment or related installations that are designed, fabricated for, and intended for use by a particular customer, if it is determined to be safe for its intended use by its manufacturer on the basis of test data which the employer keeps and makes available for inspection to the Assistant Secretary and his authorized representatives.

Accepted. An installation is "accepted" if it has been inspected and found by a nationally recognized testing laboratory to conform to specified plans or to procedures of applicable codes.

Accessible. (As applied to wiring methods.) Capable of being removed or exposed without damaging the building structure or finish, or not permanently closed in by the structure or finish of the building. (See "concealed" and "exposed.")

Accessible. (As applied to equipment.) Admitting close approach; not guarded by locked doors, elevation, or other effective means. (See "Readily accessible.")

Ampacity. The current, in amperes, that a conductor can carry continuously under the conditions of use without exceeding its temperature rating.

Appliances. Utilization equipment, generally other than industrial, normally built in standardized sizes or types, that is installed or connected as a unit to perform one or more functions.

Approved. Acceptable to the authority enforcing this subpart. The authority enforcing this subpart is the Assistant Secretary of Labor for Occupational Safety and Health. The definition of "acceptable" indicates what is acceptable to the Assistant Secretary of Labor, and therefore approved within the meaning of this subpart.

Armored cable (Type AC). A fabricated assembly of insulated conductors in a flexible metallic enclosure.

Askarel. A generic term for a group of nonflammable synthetic chlorinated hydrocarbons used as electrical insulating media. Askarels of various compositional types are used. Under arcing conditions, the gases produced, while consisting predominantly of noncombustible hydrogen chloride, can include varying amounts of combustible gases depending upon the askarel type.

Attachment plug (Plug cap)(Cap). A device that, by insertion in a receptacle, establishes a connection between the conductors of the attached flexible cord and the conductors connected permanently to the receptacle.

Automatic. Self-acting, operating by its own mechanism when actuated by some impersonal influence, as, for example, a change in current strength, pressure, temperature, or mechanical configuration.

Bare conductor. See Conductor.

Barrier. A physical obstruction that is intended to prevent contact with equipment or live parts or to prevent unauthorized access to a work area.

Bathroom. An area including a basin with one or more of the following: a toilet, a tub, or a shower.

Bonding (Bonded). The permanent joining of metallic parts to form an electrically conductive path that ensures electrical continuity and the capacity to conduct safely any current likely to be imposed.

Bonding jumper. A conductor that assures the necessary electrical conductivity between metal parts required to be electrically connected.

Branch circuit. The circuit conductors between the final overcurrent device protecting the circuit and the outlets.

Building. A structure that stands alone or is cut off from adjoining structures by fire walls with all openings therein protected by approved fire doors.

Cabinet. An enclosure designed either for surface or flush mounting, and provided with a frame, mat, or trim in which a swinging door or doors are or can be hung.

Cable tray system. A unit or assembly of units or sections and associated fittings forming a rigid structural system used to securely fasten or support cables and raceways. Cable tray systems include ladders, troughs, channels, solid bottom trays, and other similar structures.

FIGURE 7.5 Definitions Applying to Part 1910, Subpart S (OSHA-CFR, Title 29, Part 1910, Paragraph 399).

Cablebus. An assembly of insulated conductors with fittings and conductor terminations in a completely enclosed, ventilated, protective metal housing.

Cell line. An assembly of electrically interconnected electrolytic cells supplied by a source of direct current power.

Cell line attachments and auxiliary equipment. Cell line attachments and auxiliary equipment include, but are not limited to, auxiliary tanks, process piping, ductwork, structural supports, exposed cell line conductors, conduits and other raceways, pumps, positioning equipment, and cell cutout or bypass electrical devices. Auxiliary equipment also includes tools, welding machines, crucibles, and other portable equipment used for operation and maintenance within the electrolytic cell line working zone. In the cell line working zone, auxiliary equipment includes the exposed conductive surfaces of ungrounded cranes and crane-mounted cell-servicing equipment.

Center pivot irrigation machine. A multi-motored irrigation machine that revolves around a central pivot and employs alignment switches or similar devices to control individual motors.

Certified. Equipment is "certified" if it bears a label, tag, or other record of certification that the equipment:

(1) Has been tested and found by a nationally recognized testing laboratory to meet nationally recognized standards or to be safe for use in a specified manner; or

(2) Is of a kind whose production is periodically inspected by a nationally recognized testing laboratory and is accepted by the laboratory as safe for its intended use.

Circuit breaker. A device designed to open and close a circuit by non-automatic means and to open the circuit automatically on a predetermined overcurrent without damage to itself when properly applied within its rating.

Class I locations. Class I locations are those in which flammable gases or vapors are or may be present in the air in quantities sufficient to produce explosive or ignitable mixtures. Class I locations include the following:

(1) *Class I, Division 1.* A Class I, Division 1 location is a location:

(i) In which ignitable concentrations of flammable gases or vapors may exist under normal operating conditions; or

(ii) In which ignitable concentrations of such gases or vapors may exist frequently because of repair or maintenance operations or because of leakage; or

(iii) In which breakdown or faulty operation of equipment or processes might release ignitable concentrations of flammable gases or vapors, and might also cause simultaneous failure of electric equipment.

Note to the definition of "Class I, Division 1:" This classification usually includes locations where volatile flammable liquids or liquefied flammable gases are transferred from one container to another; interiors of spray booths and areas in the vicinity of spraying and painting operations where volatile flammable solvents are used; locations containing open tanks or vats of volatile flammable liquids; drying rooms or compartments for the evaporation of flammable solvents; locations containing fat and oil extraction equipment using volatile flammable solvents; portions of cleaning and dyeing plants where flammable liquids are used; gas generator rooms and other portions of gas manufacturing plants where flammable gas may escape; inadequately ventilated pump rooms for flammable gas or for volatile flammable liquids; the interiors of refrigerators and freezers in which volatile flammable materials are stored in open, lightly stoppered, or easily ruptured containers; and all other locations where ignitable concentrations of flammable vapors or gases are likely to occur in the course of normal operations.

(2) *Class I, Division 2.* A Class I, Division 2 location is a location:

(i) In which volatile flammable liquids or flammable gases are handled, processed, or used, but in which the hazardous liquids, vapors, or gases will normally be confined within closed containers or closed systems from which they can escape only in the event of accidental rupture or breakdown of such containers or systems, or as a result of abnormal operation of equipment; or

(ii) In which ignitable concentrations of gases or vapors are normally prevented by positive mechanical ventilation, and which might become hazardous through failure or abnormal operations of the ventilating equipment; or

(iii) That is adjacent to a Class I, Division 1 location, and to which ignitable concentrations of gases or vapors might occasionally be communicated unless such communication is prevented by adequate positive-pressure ventilation from a source of clean air, and effective safeguards against ventilation failure are provided.

Note to the definition of "Class I, Division 2:" This classification usually includes locations where volatile flammable liquids or flammable gases or vapors are used, but which would become hazardous only in case of an accident or of some unusual operating condition. The quantity of flammable material that might escape in case of accident, the adequacy of ventilating equipment, the total area involved, and the record of the industry or

FIGURE 7.5 (*Continued*)

business with respect to explosions or fires are all factors that merit consideration in determining the classification and extent of each location. Piping without valves, checks, meters, and similar devices would not ordinarily introduce a hazardous condition even though used for flammable liquids or gases. Locations used for the storage of flammable liquids or liquefied or compressed gases in sealed containers would not normally be considered hazardous unless also subject to other hazardous conditions. Electrical conduits and their associated enclosures separated from process fluids by single seal or barrier are classed as a Division 2 location if the outside of the conduit and enclosures is a nonhazardous location.

(3) *Class I, Zone 0.* A Class I, Zone 0 location is a location in which one of the following conditions exists:
(i) Ignitable concentrations of flammable gases or vapors are present continuously; or
(ii) Ignitable concentrations of flammable gases or vapors are present for long periods of time.

Note to the definition of "Class I, Zone 0:" As a guide in determining when flammable gases or vapors are present continuously or for long periods of time, refer to *Recommended Practice for Classification of Locations for Electrical Installations of Petroleum Facilities Classified as Class I, Zone 0, Zone 1 or Zone 2,* API RP 505–1997; *Electrical Apparatus for Explosive Gas Atmospheres, Classifications of Hazardous Areas,* IEC 79–10–1995; *Area Classification Code for Petroleum Installations, Model Code—Part 15,* Institute for Petroleum; and *Electrical Apparatus for Explosive Gas Atmospheres, Classifications of Hazardous (Classified) Locations,* ISA S12.24.01–1997.

(4) *Class I, Zone 1.* A Class I, Zone 1 location is a location in which one of the following conditions exists:
(i) Ignitable concentrations of flammable gases or vapors are likely to exist under normal operating conditions; or
(ii) Ignitable concentrations of flammable gases or vapors may exist frequently because of repair or maintenance operations or because of leakage; or
(iii) Equipment is operated or processes are carried on of such a nature that equipment breakdown or faulty operations could result in the release of ignitable concentrations of flammable gases or vapors and also cause simultaneous failure of electric equipment in a manner that would cause the electric equipment to become a source of ignition; or
(iv) A location that is adjacent to a Class I, Zone 0 location from which ignitable concentrations of vapors could be communicated, unless communication is prevented by adequate positive pressure ventilation from a source of clean air and effective safeguards against ventilation failure are provided.
(5) *Class I, Zone 2.* A Class I, Zone 2 location is a location in which one of the following conditions exists:
(i) Ignitable concentrations of flammable gases or vapors are not likely to occur in normal operation and if they do occur will exist only for a short period; or
(ii) Volatile flammable liquids, flammable gases, or flammable vapors are handled, processed, or used, but in which the liquids, gases, or vapors are normally confined within closed containers or closed systems from which they can escape only as a result of accidental rupture or breakdown of the containers or system or as the result of the abnormal operation of the equipment with which the liquids or gases are handled, processed, or used; or
(iii) Ignitable concentrations of flammable gases or vapors normally are prevented by positive mechanical ventilation, but which may become hazardous as the result of failure or abnormal operation of the ventilation equipment; or
(iv) A location that is adjacent to a Class I, Zone 1 location, from which ignitable concentrations of flammable gases or vapors could be communicated, unless such communication is prevented by adequate positive-pressure ventilation from a source of clean air, and effective safeguards against ventilation failure are provided. *Class II locations.* Class II locations are those that are hazardous because of the presence of combustible dust. Class II locations include the following:
(1) *Class II, Division 1.* A Class II, Division 1 location is a location:
(i) In which combustible dust is or may be in suspension in the air under normal operating conditions, in quantities sufficient to produce explosive or ignitable mixtures; or
(ii) Where mechanical failure or abnormal operation of machinery or equipment might cause such explosive or ignitable mixtures to be produced, and might also provide a source of ignition through simultaneous failure of electric equipment, through operation of protection devices, or from other causes; or
(iii) In which combustible dusts of an electrically conductive nature may be present.

Note to the definition of "Class II, Division 1:" This classification may include areas of grain handling and processing plants, starch plants, sugar-pulverizing plants, malting plants, hay-grinding plants, coal pulverizing plants, areas where metal dusts and powders are produced or processed, and other similar locations that contain dust

FIGURE 7.5 *(Continued)*

producing machinery and equipment (except where the equipment is dust-tight or vented to the outside). These areas would have combustible dust in the air, under normal operating conditions, in quantities sufficient to produce explosive or ignitable mixtures. Combustible dusts that are electrically nonconductive include dusts produced in the handling and processing of grain and grain products, pulverized sugar and cocoa, dried egg and milk powders, pulverized spices, starch and pastes, potato and wood flour, oil meal from beans and seed, dried hay, and other organic materials which may produce combustible dusts when processed or handled. Dusts containing magnesium or aluminum are particularly hazardous, and the use of extreme caution is necessary to avoid ignition and explosion.

(2) *Class II, Division 2.* A Class II, Division 2 location is a location where:

(i) Combustible dust will not normally be in suspension in the air in quantities sufficient to produce explosive or ignitable mixtures, and dust accumulations will normally be insufficient to interfere with the normal operation of electric equipment or other apparatus, but combustible dust may be in suspension in the air as a result of infrequent malfunctioning of handling or processing equipment; and

(ii) Resulting combustible dust accumulations on, in, or in the vicinity of the electric equipment may be sufficient to interfere with the safe dissipation of heat from electric equipment or may be ignitable by abnormal operation or failure of electric equipment.

Note to the definition of "Class II, Division 2:" This classification includes locations where dangerous concentrations of suspended dust would not be likely, but where dust accumulations might form on or in the vicinity of electric equipment. These areas may contain equipment from which appreciable quantities of dust would escape under abnormal operating conditions or be adjacent to a Class II Division 1 location, as described above, into which an explosive or ignitable concentration of dust may be put into suspension under abnormal operating conditions.

Class III locations. Class III locations are those that are hazardous because of the presence of easily ignitable fibers or flyings, but in which such fibers or flyings are not likely to be in suspension in the air in quantities sufficient to produce ignitable mixtures. Class III locations include the following:

(1) *Class III, Division 1.* A Class III, Division 1 location is a location in which easily ignitable fibers or materials producing combustible flyings are handled, manufactured, or used.

Note to the definition of "Class III, Division 1:" Such locations usually include some parts of rayon, cotton, and other textile mills; combustible fiber manufacturing and processing plants; cotton gins and cottonseed mills; flax-processing plants; clothing manufacturing plants; woodworking plants, and establishments; and industries involving similar hazardous processes or conditions. Easily ignitable fibers and flyings include rayon, cotton (including cotton linters and cotton waste), sisal or henequen, istle, jute, hemp, tow, cocoa fiber, oakum, baled waste kapok, Spanish moss, excelsior, and other materials of similar nature.

(2) *Class III, Division 2.* A Class III, Division 2 location is a location in which easily ignitable fibers are stored or handled, other than in the process of manufacture.

Collector ring. An assembly of slip rings for transferring electric energy from a stationary to a rotating member.

Competent person. One who is capable of identifying existing and predictable hazards in the surroundings or working conditions that are unsanitary, hazardous, or dangerous to employees and who has authorization to take prompt corrective measures to eliminate them.

Concealed. Rendered inaccessible by the structure or finish of the building. Wires in concealed raceways are considered concealed, even though they may become accessible by withdrawing them. (See Accessible. (*As applied to wiring methods*.))

Conductor—(1) *Bare.* A conductor having no covering or electrical insulation whatsoever.

(2) *Covered.* A conductor encased within material of composition or thickness that is not recognized by this subpart as electrical insulation.

(3) *Insulated.* A conductor encased within material of composition and thickness that is recognized by this subpart as electrical insulation.

Conduit body. A separate portion of a conduit or tubing system that provides access through one or more removable covers to the interior of the system at a junction of two or more sections of the system or at a terminal point of the system. Boxes such as FS and FD or larger cast or sheet metal boxes are not classified as conduit bodies.

FIGURE 7.5 (*Continued*)

Controller. A device or group of devices that serves to govern, in some predetermined manner, the electric power delivered to the apparatus to which it is connected.

Covered conductor. See Conductor.

Cutout. (Over 600 volts, nominal.) An assembly of a fuse support with either a fuseholder, fuse carrier, or disconnecting blade. The fuseholder or fuse carrier may include a conducting element (fuse link), or may act as the disconnecting blade by the inclusion of a non-fusible member.

Cutout box. An enclosure designed for surface mounting and having swinging doors or covers secured directly to and telescoping with the walls of the box proper. (See Cabinet.)

Damp location. See Location.

Dead front. Without live parts exposed to a person on the operating side of the equipment.

Deenergized. Free from any electrical connection to a source of potential difference and from electrical charge; not having a potential different from that of the earth.

Device. A unit of an electrical system that is intended to carry but not utilize electric energy.

Dielectric heating. The heating of a nominally insulating material due to its own dielectric losses when the material is placed in a varying electric field.

Disconnecting means. A device, or group of devices, or other means by which the conductors of a circuit can be disconnected from their source of supply.

Disconnecting (or Isolating) switch. (Over 600 volts, nominal.) A mechanical switching device used for isolating a circuit or equipment from a source of power.

Electrolytic cell line working zone. The cell line working zone is the space envelope wherein operation or maintenance is normally performed on or in the vicinity of exposed energized surfaces of electrolytic cell lines or their attachments.

Electrolytic cells. A tank or vat in which electrochemical reactions are caused by applying energy for the purpose of refining or producing usable materials.

Enclosed. Surrounded by a case, housing, fence, or walls that will prevent persons from accidentally contacting energized parts.

Enclosure. The case or housing of apparatus, or the fence or walls surrounding an installation to prevent personnel from accidentally contacting energized parts, or to protect the equipment from physical damage.

Energized. Electrically connected to a source of potential difference.

Equipment. A general term including material, fittings, devices, appliances, fixtures, apparatus, and the like, used as a part of, or in connection with, an electrical installation.

Equipment grounding conductor. See Grounding conductor, equipment.

Explosion-proof apparatus. Apparatus enclosed in a case that is capable of withstanding an explosion of a specified gas or vapor that may occur within it and of preventing the ignition of a specified gas or vapor surrounding the enclosure by sparks, flashes, or explosion of the gas or vapor within, and that operates at such an external temperature that it will not ignite a surrounding flammable atmosphere.

Exposed. (As applied to live parts.) Capable of being inadvertently touched or approached nearer than a safe distance by a person. It is applied to parts not suitably guarded, isolated, or insulated. (See Accessible and Concealed.)

Exposed. (As applied to wiring methods.) On or attached to the surface, or behind panels designed to allow access. (See Accessible. (*As applied to wiring methods*.))

Exposed. (For the purposes of § 1910.308(e).) Where the circuit is in such a position that in case of failure of supports or insulation, contact with another circuit may result.

Externally operable. Capable of being operated without exposing the operator to contact with live parts.

Feeder. All circuit conductors between the service equipment, the source of a separate derived system, or other power supply source and the final branch-circuit overcurrent device.

Fitting. An accessory such as a locknut, bushing, or other part of a wiring system that is intended primarily to perform a mechanical rather than an electrical function.

Fountain. Fountains, ornamental pools, display pools, and reflection pools.

Note to the definition of "fountain:" This definition does not include drinking fountains.

Fuse. (Over 600 volts, nominal.) An overcurrent protective device with a circuit opening fusible part that is heated and severed by the passage of overcurrent through it. A fuse comprises all the parts that form a unit capable of performing the prescribed functions. It may or may not be the complete device necessary to connect it into an electrical circuit.

FIGURE 7.5 (*Continued*)

Ground. A conducting connection, whether intentional or accidental, between an electric circuit or equipment and the earth, or to some conducting body that serves in place of the earth.

Grounded. Connected to the earth or to some conducting body that serves in place of the earth.

Grounded, effectively. Intentionally connected to earth through a ground connection or connections of sufficiently low impedance and having sufficient current-carrying capacity to prevent the buildup of voltages that may result in undue hazards to connected equipment or to persons.

Grounded conductor. A system or circuit conductor that is intentionally grounded.

Grounding conductor. A conductor used to connect equipment or the grounded circuit of a wiring system to a grounding electrode or electrodes.

Grounding conductor, equipment. The conductor used to connect the noncurrent-carrying metal parts of equipment, raceways, and other enclosures to the system grounded conductor, the grounding electrode conductor, or both, at the service equipment or at the source of a separately derived system.

Grounding electrode conductor. The conductor used to connect the grounding electrode to the equipment grounding conductor, to the grounded conductor, or to both, of the circuits at the service equipment or at the source of a separately derived system.

Ground-fault circuit-interrupter. A device intended for the protection of personnel that functions to deenergize a circuit or a portion of a circuit within an established period of time when a current to ground exceeds some predetermined value that is less than that required to operate the overcurrent protective device of the supply circuit.

Guarded. Covered, shielded, fenced, enclosed, or otherwise protected by means of suitable covers, casings, barriers, rails, screens, mats, or platforms to remove the likelihood of approach to a point of danger or contact by persons or objects.

Health care facilities. Buildings or portions of buildings in which medical, dental, psychiatric, nursing, obstetrical, or surgical care are provided.

Note to the definition of "health care facilities:" Health care facilities include, but are not limited to, hospitals, nursing homes, limited care facilities, clinics, medical and dental offices, and ambulatory care centers, whether permanent or movable.

Heating equipment. For the purposes of § 1910.306(g), the term "heating equipment" includes any equipment used for heating purposes if heat is generated by induction or dielectric methods.

Hoistway. Any shaftway, hatchway, well hole, or other vertical opening or space that is designed for the operation of an elevator or dumbwaiter.

Identified (as applied to equipment). Approved as suitable for the specific purpose, function, use, environment, or application, where described in a particular requirement.

Note to the definition of "identified:" Some examples of ways to determine suitability of equipment for a specific purpose, environment, or application include investigations by a nationally recognized testing laboratory (through listing and labeling), inspection agency, or other organization recognized under the definition of "acceptable."

Induction heating. The heating of a nominally conductive material due to its own I2R losses when the material is placed in a varying electromagnetic field.

Insulated. Separated from other conducting surfaces by a dielectric (including air space) offering a high resistance to the passage of current.

Insulated conductor. See Conductor, Insulated.

Interrupter switch. (Over 600 volts, nominal.) A switch capable of making, carrying, and interrupting specified currents.

Irrigation machine. An electrically driven or controlled machine, with one or more motors, not hand portable, and used primarily to transport and distribute water for agricultural purposes.

Isolated. (As applied to location.) Not readily accessible to persons unless special means for access are used.

Isolated power system. A system comprising an isolating transformer or its equivalent, a line isolation monitor, and its ungrounded circuit conductors.

Labeled. Equipment is "labeled" if there is attached to it a label, symbol, or other identifying mark of a nationally recognized testing laboratory:

(1) That makes periodic inspections of the production of such equipment, and

(2) Whose labeling indicates compliance with nationally recognized standards or tests to determine safe use in a specified manner.

FIGURE 7.5 *(Continued)*

Lighting outlet. An outlet intended for the direct connection of a lampholder, a lighting fixture, or a pendant cord terminating in a lampholder.

Line-clearance tree trimming. The pruning, trimming, repairing, maintaining, removing, or clearing of trees or cutting of brush that is within 305 cm (10 ft) of electric supply lines and equipment.

Listed. Equipment is "listed" if it is of a kind mentioned in a list that:

(1) Is published by a nationally recognized laboratory that makes periodic inspection of the production of such equipment, and

(2) States that such equipment meets nationally recognized standards or has been tested and found safe for use in a specified manner.

Live parts. Energized conductive components.

Location—(1) *Damp location.* Partially protected locations under canopies, marquees, roofed open porches, and like locations, and interior locations subject to moderate degrees of moisture, such as some basements, some barns, and some cold-storage warehouses.

(2) *Dry location.* A location not normally subject to dampness or wetness. A location classified as dry may be temporarily subject to dampness or wetness, as in the case of a building under construction.

(3) *Wet location.* Installations underground or in concrete slabs or masonry in direct contact with the earth, and locations subject to saturation with water or other liquids, such as vehicle-washing areas, and locations unprotected and exposed to weather.

Medium voltage cable (Type MV). A single or multi-conductor solid dielectric insulated cable rated 2001 volts or higher.

Metal-clad cable (Type MC). A factory assembly of one or more insulated circuit conductors with or without optical fiber members enclosed in an armor of interlocking metal tape, or a smooth or corrugated metallic sheath.

Mineral-insulated metal-sheathed cable (Type MI). Type MI, mineral-insulated metal-sheathed, cable is a factory assembly of one or more conductors insulated with a highly compressed refractory mineral insulation and enclosed in a liquidtight and gastight continuous copper or alloy steel sheath.

Mobile X-ray. X-ray equipment mounted on a permanent base with wheels or casters or both for moving while completely assembled.

Motor control center. An assembly of one or more enclosed sections having a common power bus and principally containing motor control units.

Nonmetallic-sheathed cable (Types NM, NMC, and NMS). A factory assembly of two or more insulated conductors having an outer sheath of moisture resistant, flame-retardant, nonmetallic material.

Oil (filled) cutout. (Over 600 volts, nominal.) A cutout in which all or part of the fuse support and its fuse link or disconnecting blade are mounted in oil with complete immersion of the contacts and the fusible portion of the conducting element (fuse link), so that arc interruption by severing of the fuse link or by opening of the contacts will occur under oil.

Open wiring on insulators. Open wiring on insulators is an exposed wiring method using cleats, knobs, tubes, and flexible tubing for the protection and support of single insulated conductors run in or on buildings, and not concealed by the building structure.

Outlet. A point on the wiring system at which current is taken to supply utilization equipment.

Outline lighting. An arrangement of incandescent lamps or electric discharge lighting to outline or call attention to certain features, such as the shape of a building or the decoration of a window.

Overcurrent. Any current in excess of the rated current of equipment or the ampacity of a conductor. It may result from overload, short circuit, or ground fault.

Overhaul means to perform a major replacement, modification, repair, or rehabilitation similar to that involved when a new building or facility is built, a new wing is added, or an entire floor is renovated.

Overload. Operation of equipment in excess of normal, full-load rating, or of a conductor in excess of rated ampacity that, when it persists for a sufficient length of time, would cause damage or dangerous overheating. A fault, such as a short circuit or ground fault, is not an overload. (See Overcurrent.)

Panelboard. A single panel or group of panel units designed for assembly in the form of a single panel; including buses, automatic overcurrent devices, and with or without switches for the control of light, heat, or power circuits; designed to be placed in a cabinet or cutout box placed in or against a wall or partition and accessible only from the front. (See Switchboard.)

Permanently installed decorative fountains and reflection pools. Pools that are constructed in the ground, on the ground, or in a building in such a manner that the fountain or pool cannot be readily disassembled for storage,

FIGURE 7.5 *(Continued)*

whether or not served by electrical circuits of any nature. These units are primarily constructed for their aesthetic value and are not intended for swimming or wading.

Permanently installed swimming, wading, and therapeutic pools. Pools that are constructed in the ground or partially in the ground, and all other capable of holding water in a depth greater than 1.07 m (42 in.). The definition also applies to all pools installed inside of a building, regardless of water depth, whether or not served by electric circuits of any nature.

Portable X-ray. X-ray equipment designed to be hand-carried.

Power and control tray cable (Type TC). A factory assembly of two or more insulated conductors, with or without associated bare or covered grounding conductors under a nonmetallic sheath, approved for installation in cable trays, in raceways, or where supported by a messenger wire.

Power fuse. (Over 600 volts, nominal.) See Fuse.

Power-limited tray cable (Type PLTC). A factory assembly of two or more insulated conductors under a nonmetallic jacket.

Power outlet. An enclosed assembly, which may include receptacles, circuit breakers, fuse-holders, fused switches, buses, and watt-hour meter mounting means, that is intended to supply and control power to mobile homes, recreational vehicles, or boats or to serve as a means for distributing power needed to operate mobile or temporarily installed equipment.

Premises wiring. (Premises wiring system.) The interior and exterior wiring, including power, lighting, control, and signal circuit wiring together with all of their associated hardware, fittings, and wiring devices, both permanently and temporarily installed, that extends from the service point of utility conductors or source of power (such as a battery, a solar photovoltaic system, or a generator, transformer, or converter) to the outlets. Such wiring does not include wiring internal to appliances, fixtures, motors, controllers, motor control centers, and similar equipment.

Qualified person. One who has received training in and has demonstrated skills and knowledge in the construction and operation of electric equipment and installations and the hazards involved.

Note 1 to the definition of "qualified person:" Whether an employee is considered to be a "qualified person" will depend upon various circumstances in the workplace. For example, it is possible and, in fact, likely for an individual to be considered "qualified" with regard to certain equipment in the workplace, but "unqualified" as to other equipment. (See 1910.332(b)(3) for training requirements that specifically apply to qualified persons.)

Note 2 to the definition of "qualified person:" An employee who is undergoing on-the-job training and who, in the course of such training, has demonstrated an ability to perform duties safely at his or her level of training and who is under the direct supervision of a qualified person is considered to be a qualified person for the performance of those duties.

Raceway. An enclosed channel of metal or nonmetallic materials designed expressly for holding wires, cables, or busbars, with additional functions as permitted in this standard. Raceways include, but are not limited to, rigid metal conduit, rigid nonmetallic conduit, intermediate metal conduit, liquidtight flexible conduit, flexible metallic tubing, flexible metal conduit, electrical metallic tubing, electrical nonmetallic tubing, under floor raceways, cellular concrete floor raceways, cellular metal floor raceways, surface raceways, wireways, and busways.

Readily accessible. Capable of being reached quickly for operation, renewal, or inspections, so that those needing ready access do not have to climb over or remove obstacles or to resort to portable ladders, chairs, etc. (See Accessible.)

Receptacle. A receptacle is a contact device installed at the outlet for the connection of an attachment plug. A single receptacle is a single contact device with no other contact device on the same yoke. A multiple receptacle is two or more contact devices on the same yoke.

Receptacle outlet. An outlet where one or more receptacles are installed.

Remote-control circuit. Any electric circuit that controls any other circuit through a relay or an equivalent device.

Sealable equipment. Equipment enclosed in a case or cabinet that is provided with a means of sealing or locking so that live parts cannot be made accessible without opening the enclosure. The equipment may or may not be operable without opening the enclosure.

Separately derived system. A premises wiring system whose power is derived from a battery, a solar photovoltaic system, or from a generator, transformer, or converter windings, and that has no direct electrical connection, including a solidly connected grounded circuit conductor, to supply conductors originating in another system.

FIGURE 7.5 *(Continued)*

Service. The conductors and equipment for delivering electric energy from the serving utility to the wiring system of the premises served.

Service cable. Service conductors made up in the form of a cable.

Service conductors. The conductors from the service point to the service disconnecting means.

Service drop. The overhead service conductors from the last pole or other aerial support to and including the splices, if any, connecting to the service entrance conductors at the building or other structure.

Service-entrance cable. A single conductor or multi-conductor assembly provided with or without an overall covering, primarily used for services, and is of the following types:

(1) *Type SE.* Type SE, having a flame-retardant, moisture resistant covering; and

(2) *Type USE.* Type USE, identified for underground use, having a moisture-resistant covering, but not required to have a flame-retardant covering. Cabled, single-conductor, Type USE constructions recognized for underground use may have a bare copper conductor cabled with the assembly. Type USE single, parallel, or cable conductor assemblies recognized for underground use may have a bare copper concentric conductor applied. These constructions do not require an outer overall covering.

Service-entrance conductors, overhead system. The service conductors between the terminals of the service equipment and a point usually outside the building, clear of building walls, where joined by tap or splice to the service drop.

Service entrance conductors, underground system. The service conductors between the terminals of the service equipment and the point of connection to the service lateral.

Service equipment. The necessary equipment, usually consisting of one or more circuit breakers or switches and fuses, and their accessories, connected to the load end of service conductors to a building or other structure, or an otherwise designated area, and intended to constitute the main control and cutoff of the supply.

Service point. The point of connection between the facilities of the serving utility and the premises wiring.

Shielded nonmetallic-sheathed cable (Type SNM). A factory assembly of two or more insulated conductors in an extruded core of moisture-resistant, flame-resistant nonmetallic material, covered with an overlapping spiral metal tape and wire shield and jacketed with an extruded moisture-, flame-, oil-, corrosion-, fungus-, and sunlight-resistant nonmetallic material.

Show window. Any window used or designed to be used for the display of goods or advertising material, whether it is fully or partly enclosed or entirely open at the rear and whether or not it has a platform raised higher than the street floor level.

Signaling circuit. Any electric circuit that energizes signaling equipment.

Storable swimming or wading pool. A pool that is constructed on or above the ground and is capable of holding water to a maximum depth of 1.07 m (42 in.), or a pool with nonmetallic, molded polymeric walls or inflatable fabric walls regardless of dimension.

Switchboard. A large single panel, frame, or assembly of panels on which are mounted, on the face or back, or both, switches, overcurrent and other protective devices, buses, and (usually) instruments. Switchboards are generally accessible from the rear as well as from the front and are not intended to be installed in cabinets. (See Panelboard.)

Switch—(1) *General-use switch.* A switch intended for use in general distribution and branch circuits. It is rated in amperes, and it is capable of interrupting its rated current at its rated voltage.

(2) *General-use snap switch.* A form of general-use switch constructed so that it can be installed in device boxes or on box covers, or otherwise used in conjunction with wiring systems recognized by this subpart.

(3) *Isolating switch.* A switch intended for isolating an electric circuit from the source of power. It has no interrupting rating, and it is intended to be operated only after the circuit has been opened by some other means.

(4) *Motor-circuit switch.* A switch, rated in horsepower, capable of interrupting the maximum operating overload current of a motor of the same horsepower rating as the switch at the rated voltage.

Switching devices. (Over 600 volts, nominal.) Devices designed to close and open one or more electric circuits. Included in this category are circuit breakers, cutouts, disconnecting (or isolating) switches, disconnecting means, interrupter switches, and oil (filled) cutouts.

Transportable X-ray. X-ray equipment installed in a vehicle or that may readily be disassembled for transport in a vehicle.

Utilization equipment. Equipment that utilizes electric energy for electronic, electromechanical, chemical, heating, lighting, or similar purposes.

FIGURE 7.5 (*Continued*)

Ventilated. Provided with a means to permit circulation of air sufficient to remove an excess of heat, fumes, or vapors.

Volatile flammable liquid. A flammable liquid having a flash point below 38°C (100°F), or a flammable liquid whose temperature is above its flash point, or a Class II combustible liquid having a vapor pressure not exceeding 276 kPa (40 psia) at 38°C (100°F) and whose temperature is above its flash point.

Voltage (of a circuit). The greatest root-mean-square (rms) (effective) difference of potential between any two conductors of the circuit concerned.

Voltage, nominal. A nominal value assigned to a circuit or system for the purpose of conveniently designating its voltage class (as 120/240 volts, 480Y/277 volts, 600 volts). The actual voltage at which a circuit operates can vary from the nominal within a range that permits satisfactory operation of equipment.

Voltage to ground. For grounded circuits, the voltage between the given conductor and that point or conductor of the circuit that is grounded; for ungrounded circuits, the greatest voltage between the given conductor and any other conductor of the circuit.

Watertight. So constructed that moisture will not enter the enclosure.

Weatherproof. So constructed or protected that exposure to the weather will not interfere with successful operation. Rainproof, raintight, or watertight equipment can fulfill the requirements for weatherproof where varying weather conditions other than wetness, such as snow, ice, dust, or temperature extremes, are not a factor.

Wireways. Sheet-metal troughs with hinged or removable covers for housing and protecting electric wires and cable and in which conductors are laid in place after the wireway has been installed as a complete system.

[46 FR 4056, Jan. 16, 1981; 46 FR 40185, Aug. 7, 1981; as amended at 53 FR 12123, Apr. 12, 1988; 55 FR 32020, Aug. 6, 1990; 55 FR 46054, Nov. 1, 1990; 72 7215, Feb. 14, 2007]

FIGURE 7.5 *(Continued)*

§ 1926.402 Applicability.

(a) Covered. Sections 1926.402 through 1926.408 contain installation safety requirements for electrical equipment and installations used to provide electric power and light at the jobsite. These sections apply to installations, both temporary and permanent, used on the jobsite; but these sections do not apply to existing permanent installations that were in place before the construction activity commenced.

Note: If the electrical installation is made in accordance with the National Electrical Code ANSI/NFPA 70-1984, exclusive of Formal Interpretations and Tentative Interim Amendments, it will be deemed to be in compliance with 1926.403 through 1926.408, except for 1926.404(b)(1) and 1926.405(a)(2)(ii)(E), (F), (G), and (J).

(b) Not covered. Sections 1926.402 through 1926.408 do not cover installations used for the generation, transmission, and distribution of electric energy, including related communication, metering, control, and transformation installations. (However, these regulations do cover portable and vehicle-mounted generators used to provide power for equipment used at the jobsite.) See Subpart V of this Part for the construction of power distribution and transmission lines.

§ 1926.403 General requirements.

(a) Approval. All electrical conductors and equipment shall be approved.

(b) Examination, installation, and use of equipment—

(b)(1) Examination. The employer shall ensure that electrical equipment is free from recognized hazards that are likely to cause death or serious physical harm to employees. Safety of equipment shall be determined on the basis of the following considerations:

(b)(1)(i) Suitability for installation and use in conformity with the provisions of this subpart. Suitability of equipment for an identified purpose may be evidenced by listing, labeling, or certification for that identified purpose.

(b)(1)(ii) Mechanical strength and durability, including, for parts designed to enclose and protect other equipment, the adequacy of the protection thus provided.

(b)(1)(iii) Electrical insulation.

(b)(1)(iv) Heating effects under conditions of use.

FIGURE 7.6 Installation Safety Requirements (OSHA-CFR, Title 29, Part 1926, Paragraphs 402–408).

§ 1926.403(b)(1)(v)

(b)(1)(v) Arcing effects.

(b)(1)(vi) Classification by type, size, voltage, current capacity, specific use.

(b)(1)(vii) Other factors which contribute to the practical safeguarding of employees using or likely to come in contact with the equipment.

(b)(2) Installation and use. Listed, labeled, or certified equipment shall be installed and used in accordance with instructions included in the listing, labeling, or certification.

(c) Interrupting rating. Equipment intended to break current shall have an interrupting rating at system voltage sufficient for the current that must be interrupted.

(d) Mounting and cooling of equipment—

(d)(1) Mounting. Electric equipment shall be firmly secured to the surface on which it is mounted. Wooden plugs driven into holes in masonry, concrete, plaster, or similar materials shall not be used.

§ 1926.403(d)(2)

(d)(2) Cooling. Electrical equipment which depends upon the natural circulation of air and convection principles for cooling of exposed surfaces shall be installed so that room air flow over such surfaces is not prevented by walls or by adjacent installed equipment. For equipment designed for floor mounting, clearance between top surfaces and adjacent surfaces shall be provided to dissipate rising warm air. Electrical equipment provided with ventilating openings shall be installed so that walls or other obstructions do not prevent the free circulation of air through the equipment.

(e) Splices. Conductors shall be spliced or joined with splicing devices designed for the use or by brazing, welding, or soldering with a fusible metal or alloy. Soldered splices shall first be so spliced or joined as to be mechanically and electrically secure without solder and then soldered. All splices and joints and the free ends of conductors shall be covered with an insulation equivalent to that of the conductors or with an insulating device designed for the purpose.

(f) Arcing parts. Parts of electric equipment which in ordinary operation produce arcs, sparks, flames, or molten metal shall be enclosed or separated and isolated from all combustible material.

(g) Marking. Electrical equipment shall not be used unless the manufacturer's name, trademark, or other descriptive marking by which the organization responsible for the product may be identified is placed on the equipment and unless other markings are provided giving voltage, current, wattage, or other ratings as necessary. The marking shall be of sufficient durability to withstand the environment involved.

§ 1926.403(h)

(h) Identification of disconnecting means and circuits. Each disconnecting means required by this subpart for motors and appliances shall be legibly marked to indicate its purpose, unless located and arranged so the purpose is evident. Each service, feeder, and branch circuit, at its disconnecting means or overcurrent device, shall be legibly marked to indicate its purpose, unless located and arranged so the purpose is evident. These markings shall be of sufficient durability to withstand the environment involved.

(i) 600 volts, nominal, or less. This paragraph applies to equipment operating at 600 volts, nominal, or less.

(i)(1) Working space about electric equipment. Sufficient access and working space shall be provided and maintained about all electric equipment to permit ready and safe operation and maintenance of such equipment.

(i)(1)(i) Working clearances. Except as required or permitted elsewhere in this subpart, the dimension of the working space in the direction of access to live parts operating at 600 volts or less and likely to require examination, adjustment, servicing, or maintenance while alive shall not be less than indicated in Table K-1. In addition to the dimensions shown in Table K-1, workspace shall not be less than 30 inches (762 mm) wide in front of the electric equipment. Distances shall be measured from the live parts if they are exposed, or from the enclosure front or opening if the live parts are enclosed. Walls constructed of concrete, brick, or tile are considered to be grounded. Working space is not required in back of assemblies such as dead-front switchboards or motor control centers where there are no renewable or adjustable parts such as fuses or switches on the back and where all connections are accessible from locations other than the back.

FIGURE 7.6 *(Continued)*

TABLE K-1 Working Clearances

Nominal voltage to ground	Minimum clear distance for conditions (1)		
	(a) Feet (2)	(b) Feet (2)	(c) Feet (2)
0–150	3	3	3
151–600	3	3½	4

Footnote(1) Conditions (a), (b), and (c) are as follows: [a] Exposed live parts on one side and no live or grounded parts on the other side of the working space, or exposed live parts on both sides effectively guarded by insulating material. Insulated wire or insulated busbars operating at not over 300 volts are not considered live parts. [b] Exposed live parts on one side and grounded parts on the other side. [c] Exposed live parts on both sides of the workplace [not guarded as provided in Condition (a)] with the operator between.

Footnote(2) Note: For International System of Units (SI): one foot = 0.3048 m.

§ 1926.403(i)(1)(ii)

(i)(1)(ii) Clear spaces. Working space required by this subpart shall not be used for storage. When normally enclosed live parts are exposed for inspection or servicing, the working space, if in a passageway or general open space, shall be guarded.

(i)(1)(iii) Access and entrance to working space. At least one entrance shall be provided to give access to the working space about electric equipment.

(i)(1)(iv) Front working space. Where there are live parts normally exposed on the front of switchboards or motor control centers, the working space in front of such equipment shall not be less than 3 feet (914 mm).

(i)(1)(v) Headroom. The minimum headroom of working spaces about service equipment, switchboards, panelboards, or motor control centers shall be 6 feet 3 inches (1.91 m).

(i)(2) Guarding of live parts.

(i)(2)(i) Except as required or permitted elsewhere in this subpart, live parts of electric equipment operating at 50 volts or more shall be guarded against accidental contact by cabinets or other forms of enclosures, or by any of the following means:

(i)(2)(i)(A) By location in a room, vault, or similar enclosure that is accessible only to qualified persons.

§ 1926.403(i)(2)(i)(B)

(i)(2)(i)(B) By partitions or screens so arranged that only qualified persons will have access to the space within reach of the live parts. Any openings in such partitions or screens shall be so sized and located that persons are not likely to come into accidental contact with the live parts or to bring conducting objects into contact with them.

(i)(2)(i)(C) By location on a balcony, gallery, or platform so elevated and arranged as to exclude unqualified persons.

(i)(2)(i)(D) By elevation of 8 feet (2.44 m) or more above the floor or other working surface and so installed as to exclude unqualified persons.

(i)(2)(ii) In locations where electric equipment would be exposed to physical damage, enclosures or guards shall be so arranged and of such strength as to prevent such damage.

(i)(2)(iii) Entrances to rooms and other guarded locations containing exposed live parts shall be marked with conspicuous warning signs forbidding unqualified persons to enter.

(j) Over 600 volts, nominal.

(j)(1) General. Conductors and equipment used on circuits exceeding 600 volts, nominal, shall comply with all applicable provisions of paragraphs (a) through (g) of this section and with the following provisions which supplement or modify those requirements. The provisions of paragraphs (j)(2), (j)(3), and (j)(4) of this section do not apply to equipment on the supply side of the service conductors.

§ 1926.403(j)(2)

(j)(2) Enclosure for electrical installations. Electrical installations in a vault, room, closet or in an area surrounded by a wall, screen, or fence, access to which is controlled by lock and key or other equivalent means, are considered

FIGURE 7.6 *(Continued)*

TABLE K-2 Minimum Depth of Clear Working Space in Front of Electric Equipment

Nominal voltage to ground	Conditions (1)		
	(a) Feet (2)	(b) Feet (2)	(c) Feet (2)
601 to 2,500	3	4	5
2,501 to 9,000	4	5	6
9,001 to 25,000	5	6	9
25,001 to 75 kV	6	8	10
Above 75 kV	8	10	12

Footnote(1) Conditions (a), (b), and (c) are as follows:

(j)(3)(i)(a) Exposed live parts on one side and no live or grounded parts on the other side of the working space, or exposed live parts on both sides effectively guarded by insulating materials. Insulated wire or insulated busbars operating at not over 300 volts are not considered live parts.

(j)(3)(i)(b) Exposed live parts on one side and grounded parts on the other side. Walls constructed of concrete, brick, or tile are considered to be grounded surfaces.

(j)(3)(i)(c) Exposed live parts on both sides of the workspace [not guarded as provided in Condition (a)] with the operator between.

Footnote(2) NOTE: For SI units: one foot = 0.3048 m.

to be accessible to qualified persons only. A wall, screen, or fence less than 8 feet (2.44 m) in height is not considered adequate to prevent access unless it has other features that provide a degree of isolation equivalent to an 8-foot (2.44-m) fence. The entrances to all buildings, rooms or enclosures containing exposed live parts or exposed conductors operating at over 600 volts, nominal, shall be kept locked or shall be under the observation of a qualified person at all times.

(j)(2)(i) Installations accessible to qualified persons only. Electrical installations having exposed live parts shall be accessible to qualified persons only and shall comply with the applicable provisions of paragraph (j)(3) of this section.

(j)(2)(ii) Installations accessible to unqualified persons. Electrical installations that are open to unqualified persons shall be made with metal-enclosed equipment or shall be enclosed in a vault or in an area, access to which is controlled by a lock. Metal-enclosed switchgear, unit substations, transformers, pull boxes, connection boxes, and other similar associated equipment shall be marked with appropriate caution signs. If equipment is exposed to physical damage from vehicular traffic, guards shall be provided to prevent such damage. Ventilating or similar openings in metal-enclosed equipment shall be designed so that foreign objects inserted through these openings will be deflected from energized parts.

§ 1926.403(j)(3)

(j)(3) Workspace about equipment. Sufficient space shall be provided and maintained about electric equipment to permit ready and safe operation and maintenance of such equipment. Where energized parts are exposed, the minimum clear workspace shall not be less than 6 feet 6 inches (1.98 m) high (measured vertically from the floor or platform), or less than 3 feet (914 mm) wide (measured parallel to the equipment). The depth shall be as required in Table K-2. The workspace shall be adequate to permit at least a 90-degree opening of doors or hinged panels.

(j)(3)(i) Working space. The minimum clear working space in front of electric equipment such as switchboards, control panels, switches, circuit breakers, motor controllers, relays, and similar equipment shall not be less than specified in Table K-2 unless otherwise specified in this subpart. Distances shall be measured from the live parts if they are exposed, or from the enclosure front or opening if the live parts are enclosed. However, working space is not required in back of equipment such as deadfront switchboards or control assemblies where there are no renewable or adjustable parts (such as fuses or switches) on the back and where all connections are accessible from locations other than the back. Where rear access is required to work on de-energized parts on the back of enclosed equipment, a minimum working space of 30 inches (762 mm) horizontally shall be provided.

FIGURE 7.6 *(Continued)*

TABLE K-3 Elevation of Unguarded Energized Parts Above Working Space

Nominal voltage between phases	Minimum elevation
601–7,500	8 feet 6 inches.[1]
7,501–35,000	9 feet.
Over 35 kV	9 feet + 0.37 inches per kV above 35 kV.

Footnote(1) NOTE: For SI units: one inch = 25.4 mm; one foot = 0.3048 m

§ 1926.403(j)(3)(ii)

(j)(3)(ii) Lighting outlets and points of control. The lighting outlets shall be so arranged that persons changing lamps or making repairs on the lighting system will not be endangered by live parts or other equipment. The points of control shall be so located that persons are not likely to come in contact with any live part or moving part of the equipment while turning on the lights.

(j)(3)(iii) Elevation of unguarded live parts. Unguarded live parts above working space shall be maintained at elevations not less than specified in Table K-3.

(j)(4) Entrance and access to workspace. At least one entrance not less than 24 inches (610 mm) wide and 6 feet 6 inches (1.98 m) high shall be provided to give access to the working space about electric equipment. On switchboard and control panels exceeding 48 inches (1.22 m) in width, there shall be one entrance at each end of such board where practicable. Where bare energized parts at any voltage or insulated energized parts above 600 volts are located adjacent to such entrance, they shall be guarded.

[61 FR 5507, Feb. 13, 1996]

§ 1926.404 Wiring design and protection.

(a) Use and identification of grounded and grounding conductors—

(a)(1) Identification of conductors. A conductor used as a grounded conductor shall be identifiable and distinguishable from all other conductors. A conductor used as an equipment grounding conductor shall be identifiable and distinguishable from all other conductors.

(a)(2) Polarity of connections. No grounded conductor shall be attached to any terminal or lead so as to reverse designated polarity.

(a)(3) Use of grounding terminals and devices. A grounding terminal or grounding-type device on a receptacle, cord connector, or attachment plug shall not be used for purposes other than grounding.

(b) Branch circuits—

(b)(1) Ground-fault protection—

(b)(1)(i) General. The employer shall use either ground fault circuit interrupters as specified in paragraph (b)(1)(ii) of this section or an assured equipment grounding conductor program as specified in paragraph (b)(1)(iii) of this section to protect employees on construction sites. These requirements are in addition to any other requirements for equipment grounding conductors.

§ 1926.404(b)(1)(ii)

(b)(1)(ii) Ground-fault circuit interrupters. All 120-volt, single-phase 15- and 20-ampere receptacle outlets on construction sites, which are not a part of the permanent wiring of the building or structure and which are in use by employees, shall have approved ground-fault circuit interrupters for personnel protection. Receptacles on a two-wire, single-phase portable or vehicle-mounted generator rated not more than 5kW, where the circuit conductors of the generator are insulated from the generator frame and all other grounded surfaces, need not be protected with ground-fault circuit interrupters.

(b)(1)(iii) Assured equipment grounding conductor program. The employer shall establish and implement an assured equipment grounding conductor program on construction sites covering all cord sets, receptacles which are not a part of the building or structure, and equipment connected by cord and plug which are available for use or used by employees. This program shall comply with the following minimum requirements:

FIGURE 7.6 *(Continued)*

(b)(1)(iii)(A) A written description of the program, including the specific procedures adopted by the employer, shall be available at the jobsite for inspection and copying by the Assistant Secretary and any affected employee.

(b)(1)(iii)(B) The employer shall designate one or more competent persons (as defined in 1926.32(f)) to implement the program.

(b)(1)(iii)(C) Each cord set, attachment cap, plug and receptacle of cord sets, and any equipment connected by cord and plug, except cord sets and receptacles which are fixed and not exposed to damage, shall be visually inspected before each day's use for external defects, such as deformed or missing pins or insulation damage, and for indications of possible internal damage. Equipment found damaged or defective shall not be used until repaired.

§ 1926.404(b)(1)(iii)(D)

(b)(1)(iii)(D) The following tests shall be performed on all cord sets, receptacles which are not a part of the permanent wiring of the building or structure, and cord- and plug-connected equipment required to be grounded:

(b)(1)(iii)(D)(*1*) All equipment grounding conductors shall be tested for continuity and shall be electrically continuous.

(b)(1)(iii)(D)(*2*) Each receptacle and attachment cap or plug shall be tested for correct attachment of the equipment grounding conductor. The equipment grounding conductor shall be connected to its proper terminal.

(b)(1)(iii)(E) All required tests shall be performed:

(b)(1)(iii)(E)(*1*) Before first use;

(b)(1)(iii)(E)(*2*) Before equipment is returned to service following any repairs;

(b)(1)(iii)(E)(*3*) Before equipment is used after any incident which can be reasonably suspected to have caused damage (for example, when a cord set is run over); and

(b)(1)(iii)(E)(*4*) At intervals not to exceed 3 months, except that cord sets and receptacles which are fixed and not exposed to damage shall be tested at intervals not exceeding 6 months.

§ 1926.404(b)(1)(iii)(F)

(b)(1)(iii)(F) The employer shall not make available or permit the use by employees of any equipment which has not met the requirements of this paragraph (b)(1)(iii) of this section.

(b)(1)(iii)(G) Tests performed as required in this paragraph shall be recorded. This test record shall identify each receptacle, cord set, and cord- and plug-connected equipment that passed the test and shall indicate the last date it was tested or the interval for which it was tested. This record shall be kept by means of logs, color coding, or other effective means and shall be maintained until replaced by a more current record. The record shall be made available on the jobsite for inspection by the Assistant Secretary and any affected employee.

(b)(2) Outlet devices. Outlet devices shall have an ampere rating not less than the load to be served and shall comply with the following:

(b)(2)(i) Single receptacles. A single receptacle installed on an individual branch circuit shall have an ampere rating of not less than that of the branch circuit.

(b)(2)(ii) Two or more receptacles. Where connected to a branch circuit supplying two or more receptacles or outlets, receptacle ratings shall conform to the values listed in Table K-4.

(b)(2)(iii) Receptacles used for the connection of motors. The rating of an attachment plug or receptacle used for cord- and plug-connection of a motor to a branch circuit shall not exceed 15 amperes at 125 volts or 10 amperes at 250 volts if individual overload protection is omitted.

§ 1926.404(c)

(c) Outside conductors and lamps—

(c)(1) 600 volts, nominal, or less. Paragraphs (c)(1)(i) through (c)(1)(iv) of this section apply to branch circuit, feeder, and service conductors rated 600 volts, nominal, or less and run outdoors as open conductors.

(c)(1)(i) Conductors on poles. Conductors supported on poles shall provide a horizontal climbing space not less than the following:

(c)(1)(i)(A) Power conductors below communication conductors—30 inches (762 mm).

(c)(1)(i)(B) Power conductors alone or above communication conductors: 300 volts or less—24 inches (610 mm); more than 300 volts—30 inches (762 mm).

(c)(1)(i)(C) Communication conductors below power conductors: with power conductors 300 volts or less—24 inches (610 mm); more than 300 volts—30 inches (762 mm).

FIGURE 7.6 *(Continued)*

TABLE K-4 Receptacle Ratings for Various Size Circuits

Circuit rating amperes	Receptacle rating amperes
15	Not over 15.
20	15 or 20.
30	30.
40	40 or 50.
50	50.

(c)(1)(ii) Clearance from ground. Open conductors shall conform to the following minimum clearances:

(c)(1)(ii)(A) 10 feet (3.05 m)—above finished grade, sidewalks, or from any platform or projection from which they might be reached.

(c)(1)(ii)(B) 12 feet (3.66 m)—over areas subject to vehicular traffic other than truck traffic.

(c)(1)(ii)(C) 15 feet (4.57 m)—over areas other than those specified in paragraph (c)(1)(ii)(D) of this section that are subject to truck traffic.

§ 1926.404(c)(1)(ii)(D)

(c)(1)(ii)(D) 18 feet (5.49 m)—over public streets, alleys, roads, and driveways.

(c)(1)(iii) Clearance from building openings. Conductors shall have a clearance of at least 3 feet (914 mm) from windows, doors, fire escapes, or similar locations. Conductors run above the top level of a window are considered to be out of reach from that window and, therefore, do not have to be 3 feet (914 mm) away.

(c)(1)(iv) Clearance over roofs. Conductors above roof space accessible to employees on foot shall have a clearance from the highest point of the roof surface of not less than 8 feet (2.44 m) vertical clearance for insulated conductors, not less than 10 feet (3.05 m) vertical or diagonal clearance for covered conductors, and not less than 15 feet (4.57 m) for bare conductors, except that:

(c)(1)(iv)(A) Where the roof space is also accessible to vehicular traffic, the vertical clearance shall not be less than 18 feet (5.49 m), or

(c)(1)(iv)(B) Where the roof space is not normally accessible to employees on foot, fully insulated conductors shall have a vertical or diagonal clearance of not less than 3 feet (914 mm), or

(c)(1)(iv)(C) Where the voltage between conductors is 300 volts or less and the roof has a slope of not less than 4 inches (102 mm) in 12 inches (305 mm), the clearance from roofs shall be at least 3 feet (914 mm), or

§ 1926.404(c)(1)(iv)(D)

(c)(1)(iv)(D) Where the voltage between conductors is 300 volts or less and the conductors do not pass over more than 4 feet (1.22 m) of the overhang portion of the roof and they are terminated at a through-the-roof raceway or support, the clearance from roofs shall be at least 18 inches (457 mm).

(c)(2) Location of outdoor lamps. Lamps for outdoor lighting shall be located below all live conductors, transformers, or other electric equipment, unless such equipment is controlled by a disconnecting means that can be locked in the open position or unless adequate clearances or other safeguards are provided for relamping operations.

(d) Services—

(d)(1) Disconnecting means—

(d)(1)(i) General. Means shall be provided to disconnect all conductors in a building or other structure from the service-entrance conductors. The disconnecting means shall plainly indicate whether it is in the open or closed position and shall be installed at a readily accessible location nearest the point of entrance of the service-entrance conductors.

(d)(1)(ii) Simultaneous opening of poles. Each service disconnecting means shall simultaneously disconnect all ungrounded conductors.

(d)(2) Services over 600 volts, nominal. The following additional requirements apply to services over 600 volts, nominal.

(d)(2)(i) Guarding. Service-entrance conductors installed as open wires shall be guarded to make them accessible only to qualified persons.

FIGURE 7.6 *(Continued)*

§ 1926.404(d)(2)(ii)

(d)(2)(ii) Warning signs. Signs warning of high voltage shall be posted where unauthorized employees might come in contact with live parts.

(e) Overcurrent protection—

(e)(1) 600 volts, nominal, or less. The following requirements apply to overcurrent protection of circuits rated 600 volts, nominal, or less.

(e)(1)(i) Protection of conductors and equipment. Conductors and equipment shall be protected from overcurrent in accordance with their ability to safely conduct current. Conductors shall have sufficient ampacity to carry the load.

(e)(1)(ii) Grounded conductors. Except for motor-running overload protection, overcurrent devices shall not interrupt the continuity of the grounded conductor unless all conductors of the circuit are opened simultaneously.

(e)(1)(iii) Disconnection of fuses and thermal cutouts. Except for devices provided for current-limiting on the supply side of the service disconnecting means, all cartridge fuses which are accessible to other than qualified persons and all fuses and thermal cutouts on circuits over 150 volts to ground shall be provided with disconnecting means. This disconnecting means shall be installed so that the fuse or thermal cutout can be disconnected from its supply without disrupting service to equipment and circuits unrelated to those protected by the overcurrent device.

§ 1926.404(e)(1)(iv)

(e)(1)(iv) Location in or on premises. Overcurrent devices shall be readily accessible. Overcurrent devices shall not be located where they could create an employee safety hazard by being exposed to physical damage or located in the vicinity of easily ignitable material.

(e)(1)(v) Arcing or suddenly moving parts. Fuses and circuit breakers shall be so located or shielded that employees will not be burned or otherwise injured by their operation.

(e)(1)(vi) Circuit breakers—

(e)(1)(vi)(A) Circuit breakers shall clearly indicate whether they are in the open (off) or closed (on) position.

(e)(1)(vi)(B) Where circuit breaker handles on switchboards are operated vertically rather than horizontally or rotationally, the up position of the handle shall be the closed (on) position.

(e)(1)(vi)(C) If used as switches in 120-volt, fluorescent lighting circuits, circuit breakers shall be marked "SWD."

(e)(2) Over 600 volts, nominal. Feeders and branch circuits over 600 volts, nominal, shall have short-circuit protection.

(f) Grounding. Paragraphs (f)(1) through (f)(11) of this section contain grounding requirements for systems, circuits, and equipment.

(f)(1) Systems to be grounded. The following systems which supply premises wiring shall be grounded:

§ 1926.404(f)(1)(i)

(f)(1)(i) Three-wire DC systems. All 3-wire DC systems shall have their neutral conductor grounded.

(f)(1)(ii) Two-wire DC systems. Two-wire DC systems operating at over 50 volts through 300 volts between conductors shall be grounded unless they are rectifier-derived from an AC system complying with paragraphs (f)(1)(iii), (f)(1)(iv), and (f)(1)(v) of this section.

(f)(1)(iii) AC circuits, less than 50 volts. AC circuits of less than 50 volts shall be grounded if they are installed as overhead conductors outside of buildings or if they are supplied by transformers and the transformer primary supply system is ungrounded or exceeds 150 volts to ground.

(f)(1)(iv) AC systems, 50 volts to 1000 volts. AC systems of 50 volts to 1000 volts shall be grounded under any of the following conditions, unless exempted by paragraph (f)(1)(v) of this section:

(f)(1)(iv)(A) If the system can be so grounded that the maximum voltage to ground on the ungrounded conductors does not exceed 150 volts;

(f)(1)(iv)(B) If the system is nominally rated 480Y/277 volt, 3-phase, 4-wire in which the neutral is used as a circuit conductor;

(f)(1)(iv)(C) If the system is nominally rated 240/120 volt, 3-phase, 4-wire in which the midpoint of one phase is used as a circuit conductor; or

FIGURE 7.6 *(Continued)*

§ 1926.404(f)(1)(iv)(D)

(f)(1)(iv)(D) If a service conductor is uninsulated.

(f)(1)(v) Exceptions. AC systems of 50 volts to 1000 volts are not required to be grounded if the system is separately derived and is supplied by a transformer that has a primary voltage rating less than 1000 volts, provided all of the following conditions are met:

(f)(1)(v)(A) The system is used exclusively for control circuits,

(f)(1)(v)(B) The conditions of maintenance and supervision assure that only qualified persons will service the installation,

(f)(1)(v)(C) Continuity of control power is required, and

(f)(1)(v)(D) Ground detectors are installed on the control system.

(f)(2) Separately derived systems. Where paragraph (f)(1) of this section requires grounding of wiring systems whose power is derived from generator, transformer, or converter windings and has no direct electrical connection, including a solidly connected grounded circuit conductor, to supply conductors originating in another system, paragraph (f)(5) of this section shall also apply.

§ 1926.404(f)(3)

(f)(3) Portable and vehicle-mounted generators—

(f)(3)(i) Portable generators. Under the following conditions, the frame of a portable generator need not be grounded and may serve as the grounding electrode for a system supplied by the generator:

(f)(3)(i)(A) The generator supplies only equipment mounted on the generator and/or cord- and plug-connected equipment through receptacles mounted on the generator, and

(f)(3)(i)(B) The noncurrent-carrying metal parts of equipment and the equipment grounding conductor terminals of the receptacles are bonded to the generator frame.

(f)(3)(ii) Vehicle-mounted generators. Under the following conditions the frame of a vehicle may serve as the grounding electrode for a system supplied by a generator located on the vehicle:

(f)(3)(ii)(A) The frame of the generator is bonded to the vehicle frame, and

(f)(3)(ii)(B) The generator supplies only equipment located on the vehicle and/or cord- and plug-connected equipment through receptacles mounted on the vehicle or on the generator, and

(f)(3)(ii)(C) The noncurrent-carrying metal parts of equipment and the equipment grounding conductor terminals of the receptacles are bonded to the generator frame, and

(f)(3)(ii)(D) The system complies with all other provisions of this section.

§ 1926.404(f)(3)(iii)

(f)(3)(iii) Neutral conductor bonding. A neutral conductor shall be bonded to the generator frame if the generator is a component of a separately derived system. No other conductor need be bonded to the generator frame.

(f)(4) Conductors to be grounded. For AC premises wiring systems the identified conductor shall be grounded.

(f)(5) Grounding connections—

(f)(5)(i) Grounded system. For a grounded system, a grounding electrode conductor shall be used to connect both the equipment grounding conductor and the grounded circuit conductor to the grounding electrode. Both the equipment grounding conductor and the grounding electrode conductor shall be connected to the grounded circuit conductor on the supply side of the service disconnecting means, or on the supply side of the system disconnecting means or overcurrent devices if the system is separately derived.

(f)(5)(ii) Ungrounded systems. For an ungrounded service-supplied system, the equipment grounding conductor shall be connected to the grounding electrode conductor at the service equipment. For an ungrounded separately derived system, the equipment grounding conductor shall be connected to the grounding electrode conductor at, or ahead of, the system disconnecting means or overcurrent devices.

(f)(6) Grounding path. The path to ground from circuits, equipment, and enclosures shall be permanent and continuous.

§ 1926.404(f)(7)

(f)(7) Supports, enclosures, and equipment to be grounded—

(f)(7)(i) Supports and enclosures for conductors. Metal cable trays, metal raceways, and metal enclosures for conductors shall be grounded, except that:

FIGURE 7.6 *(Continued)*

(f)(7)(i)(A) Metal enclosures such as sleeves that are used to protect cable assemblies from physical damage need not be grounded; and

(f)(7)(i)(B) Metal enclosures for conductors added to existing installations of open wire, knob-and-tube wiring, and nonmetallic-sheathed cable need not be grounded if all of the following conditions are met:

(f)(7)(i)(B)(*1*) Runs are less than 25 feet (7.62 m);

(f)(7)(i)(B)(*2*) Enclosures are free from probable contact with ground, grounded metal, metal laths, or other conductive materials; and

(f)(7)(i)(B)(*3*) Enclosures are guarded against employee contact.

(f)(7)(ii) Service equipment enclosures. Metal enclosures for service equipment shall be grounded.

(f)(7)(iii) Fixed equipment. Exposed noncurrent-carrying metal parts of fixed equipment which may become energized shall be grounded under any of the following conditions:

(f)(7)(iii)(A) If within 8 feet (2.44 m) vertically or 5 feet (1.52 m) horizontally of ground or grounded metal objects and subject to employee contact.

§ 1926.404(f)(7)(iii)(B)

(f)(7)(iii)(B) If located in a wet or damp location and subject to employee contact.

(f)(7)(iii)(C) If in electrical contact with metal.

(f)(7)(iii)(D) If in a hazardous (classified) location.

(f)(7)(iii)(E) If supplied by a metal-clad, metal-sheathed, or grounded metal raceway wiring method.

(f)(7)(iii)(F) If equipment operates with any terminal at over 150 volts to ground; however, the following need not be grounded:

(f)(7)(iii)(F)(*1*) Enclosures for switches or circuit breakers used for other than service equipment and accessible to qualified persons only;

(f)(7)(iii)(F)(*2*) Metal frames of electrically heated appliances which are permanently and effectively insulated from ground; and

(f)(7)(iii)(F)(*3*) The cases of distribution apparatus such as transformers and capacitors mounted on wooden poles at a height exceeding 8 feet (2.44 m) above ground or grade level.

§ 1926.404(f)(7)(iv)

(f)(7)(iv) Equipment connected by cord and plug. Under any of the conditions described in paragraphs (f)(7)(iv)(A) through (f)(7)(iv)(C) of this section, exposed noncurrent-carrying metal parts of cord- and plug-connected equipment which may become energized shall be grounded:

(f)(7)(iv)(A) If in a hazardous (classified) location (see 1926.407).

(f)(7)(iv)(B) If operated at over 150 volts to ground, except for guarded motors and metal frames of electrically heated appliances if the appliance frames are permanently and effectively insulated from ground.

(f)(7)(iv)(C) If the equipment is one of the types listed in paragraphs (f)(7)(iv)(C)(*1*) through (f)(7)(iv)(C)(*5*) of this section. However, even though the equipment may be one of these types, it need not be grounded if it is exempted by paragraph (f)(7)(iv)(C)(6).

(f)(7)(iv)(C)(*1*) Hand held motor-operated tools;

(f)(7)(iv)(C)(*2*) Cord- and plug-connected equipment used in damp or wet locations or by employees standing on the ground or on metal floors or working inside of metal tanks or boilers;

(f)(7)(iv)(C)(*3*) Portable and mobile X-ray and associated equipment;

(f)(7)(iv)(C)(*4*) Tools likely to be used in wet and/or conductive locations;

(f)(7)(iv)(C)(*5*) Portable hand lamps.

§ 1926.404(f)(7)(iv)(C)(6)

(f)(7)(iv)(C)(*6*) Tools likely to be used in wet and/or conductive locations need not be grounded if supplied through an isolating transformer with an ungrounded secondary of not over 50 volts. Listed or labeled portable tools and appliances protected by a system of double insulation, or its equivalent, need not be grounded. If such a system is employed, the equipment shall be distinctively marked to indicate that the tool or appliance utilizes a system of double insulation.

(f)(7)(v) Nonelectrical equipment. The metal parts of the following nonelectrical equipment shall be grounded: Frames and tracks of electrically operated cranes; frames of nonelectrically driven elevator cars to which electric

FIGURE 7.6 (*Continued*)

conductors are attached; hand-operated metal shifting ropes or cables of electric elevators, and metal partitions, grill work, and similar metal enclosures around equipment of over 1 kV between conductors.

(f)(8) Methods of grounding equipment—

(f)(8)(i) With circuit conductors. Noncurrent-carrying metal parts of fixed equipment, if required to be grounded by this subpart, shall be grounded by an equipment grounding conductor which is contained within the same raceway, cable, or cord, or runs with or encloses the circuit conductors. For DC circuits only, the equipment grounding conductor may be run separately from the circuit conductors.

(f)(8)(ii) Grounding conductor. A conductor used for grounding fixed or movable equipment shall have capacity to conduct safely any fault current which may be imposed on it.

§ 1926.404(f)(8)(iii)

(f)(8)(iii) Equipment considered effectively grounded. Electric equipment is considered to be effectively grounded if it is secured to, and in electrical contact with, a metal rack or structure that is provided for its support and the metal rack or structure is grounded by the method specified for the noncurrent-carrying metal parts of fixed equipment in paragraph (f)(8)(i) of this section. Metal car frames supported by metal hoisting cables attached to or running over metal sheaves or drums of grounded elevator machines are also considered to be effectively grounded.

(f)(9) Bonding. If bonding conductors are used to assure electrical continuity, they shall have the capacity to conduct any fault current which may be imposed.

(f)(10) Made electrodes. If made electrodes are used, they shall be free from nonconductive coatings, such as paint or enamel; and, if practicable, they shall be embedded below permanent moisture level. A single electrode consisting of a rod, pipe or plate which has a resistance to ground greater than 25 ohms shall be augmented by one additional electrode installed no closer than 6 feet (1.83 m) to the first electrode.

(f)(11) Grounding of systems and circuits of 1000 volts and over (high voltage)—

(f)(11)(i) General. If high voltage systems are grounded, they shall comply with all applicable provisions of paragraphs (f)(1) through (f)(10) of this section as supplemented and modified by this paragraph (f)(11).

(f)(11)(ii) Grounding of systems supplying portable or mobile equipment. Systems supplying portable or mobile high voltage equipment, other than substations installed on a temporary basis, shall comply with the following:

§ 1926.404(f)(11)(ii)(A)

(f)(11)(ii)(A) Portable and mobile high voltage equipment shall be supplied from a system having its neutral grounded through an impedance. If a delta-connected high voltage system is used to supply the equipment, a system neutral shall be derived.

(f)(11)(ii)(B) Exposed noncurrent-carrying metal parts of portable and mobile equipment shall be connected by an equipment grounding conductor to the point at which the system neutral impedance is grounded.

(f)(11)(ii)(C) Ground-fault detection and relaying shall be provided to automatically de-energize any high voltage system component which has developed a ground fault. The continuity of the equipment grounding conductor shall be continuously monitored so as to de-energize automatically the high voltage feeder to the portable equipment upon loss of continuity of the equipment grounding conductor.

(f)(11)(ii)(D) The grounding electrode to which the portable or mobile equipment system neutral impedance is connected shall be isolated from and separated in the ground by at least 20 feet (6.1 m) from any other system or equipment grounding electrode, and there shall be no direct connection between the grounding electrodes, such as buried pipe, fence or like objects.

§ 1926.404(f)(11)(iii)

(f)(11)(iii) Grounding of equipment. All noncurrent-carrying metal parts of portable equipment and fixed equipment including their associated fences, housings, enclosures, and supporting structures shall be grounded. However, equipment which is guarded by location and isolated from ground need not be grounded. Additionally, pole-mounted distribution apparatus at a height exceeding 8 feet (2.44 m) above ground or grade level need not be grounded.

[54 FR 24334, June 7, 1989; 61 FR 5507, Feb. 13, 1996]

FIGURE 7.6 *(Continued)*

§ 1926.405 Wiring methods, components, and equipment for general use.

(a) Wiring methods. The provisions of this paragraph do not apply to conductors which form an integral part of equipment such as motors, controllers, motor control centers and like equipment.

(a)(1) General requirements—

(a)(1)(i) Electrical continuity of metal raceways and enclosures. Metal raceways, cable armor, and other metal enclosures for conductors shall be metallically joined together into a continuous electric conductor and shall be so connected to all boxes, fittings, and cabinets as to provide effective electrical continuity.

(a)(1)(ii) Wiring in ducts. No wiring systems of any type shall be installed in ducts used to transport dust, loose stock or flammable vapors. No wiring system of any type shall be installed in any duct used for vapor removal or in any shaft containing only such ducts.

§ 1926.405(a)(2)

(a)(2) Temporary wiring—

(a)(2)(i) Scope. The provisions of paragraph (a)(2) of this section apply to temporary electrical power and lighting wiring methods which may be of a class less than would be required for a permanent installation. Except as specifically modified in paragraph (a)(2) of this section, all other requirements of this subpart for permanent wiring shall apply to temporary wiring installations. Temporary wiring shall be removed immediately upon completion of construction or the purpose for which the wiring was installed.

(a)(2)(ii) General requirements for temporary wiring—

(a)(2)(ii)(A) Feeders shall originate in a distribution center. The conductors shall be run as multiconductor cord or cable assemblies or within raceways; or, where not subject to physical damage, they may be run as open conductors on insulators not more than 10 feet (3.05 m) apart.

(a)(2)(ii)(B) Branch circuits shall originate in a power outlet or panelboard. Conductors shall be run as multiconductor cord or cable assemblies or open conductors, or shall be run in raceways. All conductors shall be protected by overcurrent devices at their ampacity. Runs of open conductors shall be located where the conductors will not be subject to physical damage, and the conductors shall be fastened at intervals not exceeding 10 feet (3.05 m). No branch-circuit conductors shall be laid on the floor. Each branch circuit that supplies receptacles or fixed equipment shall contain a separate equipment grounding conductor if the branch circuit is run as open conductors.

§ 1926.405(a)(2)(ii)(C)

(a)(2)(ii)(C) Receptacles shall be of the grounding type. Unless installed in a complete metallic raceway, each branch circuit shall contain a separate equipment grounding conductor, and all receptacles shall be electrically connected to the grounding conductor. Receptacles for uses other than temporary lighting shall not be installed on branch circuits which supply temporary lighting. Receptacles shall not be connected to the same ungrounded conductor of multiwire circuits which supply temporary lighting.

(a)(2)(ii)(D) Disconnecting switches or plug connectors shall be installed to permit the disconnection of all ungrounded conductors of each temporary circuit.

(a)(2)(ii)(E) All lamps for general illumination shall be protected from accidental contact or breakage. Metal-case sockets shall be grounded.

(a)(2)(ii)(F) Temporary lights shall not be suspended by their electric cords unless cords and lights are designed for this means of suspension.

(a)(2)(ii)(G) Portable electric lighting used in wet and/or other conductive locations, as for example, drums, tanks, and vessels, shall be operated at 12 volts or less. However, 120-volt lights may be used if protected by a ground-fault circuit interrupter.

(a)(2)(ii)(H) A box shall be used wherever a change is made to a raceway system or a cable system which is metal clad or metal sheathed.

(a)(2)(ii)(I) Flexible cords and cables shall be protected from damage. Sharp corners and projections shall be avoided. Flexible cords and cables may pass through doorways or other pinch points, if protection is provided to avoid damage.

FIGURE 7.6 *(Continued)*

§ 1926.405(a)(2)(ii)(J)

(a)(2)(ii)(J) Extension cord sets used with portable electric tools and appliances shall be of three-wire type and shall be designed for hard or extra-hard usage. Flexible cords used with temporary and portable lights shall be designed for hard or extra-hard usage.

Note: The National Electrical Code, ANSI/NFPA 70, in Article 400, Table 400-4, lists various types of flexible cords, some of which are noted as being designed for hard or extra-hard usage. Examples of these types of flexible cords include hard service cord (types S, ST, SO, STO) and junior hard service cord (types SJ, SJO, SJT, SJTO).

(a)(2)(iii) Guarding. For temporary wiring over 600 volts, nominal, fencing, barriers, or other effective means shall be provided to prevent access of other than authorized and qualified personnel.

(b) Cabinets, boxes, and fittings.

(b)(1) Conductors entering boxes, cabinets, or fittings. Conductors entering boxes, cabinets, or fittings shall be protected from abrasion, and openings through which conductors enter shall be effectively closed. Unused openings in cabinets, boxes, and fittings shall also be effectively closed.

(b)(2) Covers and canopies. All pull boxes, junction boxes, and fittings shall be provided with covers. If metal covers are used, they shall be grounded. In energized installations each outlet box shall have a cover, faceplate, or fixture canopy. Covers of outlet boxes having holes through which flexible cord pendants pass shall be provided with bushings designed for the purpose or shall have smooth, well-rounded surfaces on which the cords may bear.

§ 1926.405(b)(3)

(b)(3) Pull and junction boxes for systems over 600 volts, nominal. In addition to other requirements in this section for pull and junction boxes, the following shall apply to these boxes for systems over 600 volts, nominal:

(b)(3)(i) Complete enclosure. Boxes shall provide a complete enclosure for the contained conductors or cables.

(b)(3)(ii) Covers. Boxes shall be closed by covers securely fastened in place. Underground box covers that weigh over 100 pounds (43.6 kg) meet this requirement. Covers for boxes shall be permanently marked "HIGH VOLTAGE." The marking shall be on the outside of the box cover and shall be readily visible and legible.

(c) Knife switches. Single-throw knife switches shall be so connected that the blades are dead when the switch is in the open position. Single-throw knife switches shall be so placed that gravity will not tend to close them. Single-throw knife switches approved for use in the inverted position shall be provided with a locking device that will ensure that the blades remain in the open position when so set. Double-throw knife switches may be mounted so that the throw will be either vertical or horizontal. However, if the throw is vertical, a locking device shall be provided to ensure that the blades remain in the open position when so set.

§ 1926.405(d)

(d) Switchboards and panelboards. Switchboards that have any exposed live parts shall be located in permanently dry locations and accessible only to qualified persons. Panelboards shall be mounted in cabinets, cutout boxes, or enclosures designed for the purpose and shall be dead front. However, panelboards other than the dead front externally-operable type are permitted where accessible only to qualified persons. Exposed blades of knife switches shall be dead when open.

(e) Enclosures for damp or wet locations.

(e)(1) Cabinets, fittings, and boxes. Cabinets, cutout boxes, fittings, boxes, and panelboard enclosures in damp or wet locations shall be installed so as to prevent moisture or water from entering and accumulating within the enclosures. In wet locations the enclosures shall be weatherproof.

(e)(2) Switches and circuit breakers. Switches, circuit breakers, and switchboards installed in wet locations shall be enclosed in weatherproof enclosures.

(f) Conductors for general wiring. All conductors used for general wiring shall be insulated unless otherwise permitted in this Subpart. The conductor insulation shall be of a type that is suitable for the voltage, operating temperature, and location of use. Insulated conductors shall be distinguishable by appropriate color or other means as being grounded conductors, ungrounded conductors, or equipment grounding conductors.

(g) Flexible cords and cables—

(g)(1) Use of flexible cords and cables—

(g)(1)(i) Permitted uses. Flexible cords and cables shall be suitable for conditions of use and location. Flexible cords and cables shall be used only for:

FIGURE 7.6 *(Continued)*

§ 1926.405(g)(1)(i)(A)

(g)(1)(i)(A) Pendants;
(g)(1)(i)(B) Wiring of fixtures;
(g)(1)(i)(C) Connection of portable lamps or appliances;
(g)(1)(i)(D) Elevator cables;
(g)(1)(i)(E) Wiring of cranes and hoists;
(g)(1)(i)(F) Connection of stationary equipment to facilitate their frequent interchange;
(g)(1)(i)(G) Prevention of the transmission of noise or vibration; or
(g)(1)(i)(H) Appliances where the fastening means and mechanical connections are designed to permit removal for maintenance and repair.

(g)(1)(ii) Attachment plugs for cords. If used as permitted in paragraphs (g)(1)(i)(C), (g)(1)(i)(F), or (g)(1)(i)(H) of this section, the flexible cord shall be equipped with an attachment plug and shall be energized from a receptacle outlet.

(g)(1)(iii) Prohibited uses. Unless necessary for a use permitted in paragraph (g)(1)(i) of this section, flexible cords and cables shall not be used:

§ 1926.405(g)(1)(iii)(A)

(g)(1)(iii)(A) As a substitute for the fixed wiring of a structure;
(g)(1)(iii)(B) Where run through holes in walls, ceilings, or floors;
(g)(1)(iii)(C) Where run through doorways, windows, or similar openings, except as permitted in paragraph (a)(2)(ii)(1) of this section;
(g)(1)(iii)(D) Where attached to building surfaces; or
(g)(1)(iii)(E) Where concealed behind building walls, ceilings, or floors.

(g)(2) Identification, splices, and terminations—

(g)(2)(i) Identification. A conductor of a flexible cord or cable that is used as a grounded conductor or an equipment grounding conductor shall be distinguishable from other conductors.

(g)(2)(ii) Marking. Type SJ, SJO, SJT, SJTO, S, SO, ST, and STO cords shall not be used unless durably marked on the surface with the type designation, size, and number of conductors.

(g)(2)(iii) Splices. Flexible cords shall be used only in continuous lengths without splice or tap. Hard service flexible cords No. 12 or larger may be repaired if spliced so that the splice retains the insulation, outer sheath properties, and usage characteristics of the cord being spliced.

§ 1926.405(g)(2)(iv)

(g)(2)(iv) Strain relief. Flexible cords shall be connected to devices and fittings so that strain relief is provided which will prevent pull from being directly transmitted to joints or terminal screws.

(g)(2)(v) Cords passing through holes. Flexible cords and cables shall be protected by bushings or fittings where passing through holes in covers, outlet boxes, or similar enclosures.

(h) Portable cables over 600 volts, nominal. Multiconductor portable cable for use in supplying power to portable or mobile equipment at over 600 volts, nominal, shall consist of No. 8 or larger conductors employing flexible stranding. Cables operated at over 2000 volts shall be shielded for the purpose of confining the voltage stresses to the insulation. Grounding conductors shall be provided. Connectors for these cables shall be of a locking type with provisions to prevent their opening or closing while energized. Strain relief shall be provided at connections and terminations. Portable cables shall not be operated with splices unless the splices are of the permanent molded, vulcanized, or other equivalent type. Termination enclosures shall be marked with a high voltage hazard warning, and terminations shall be accessible only to authorized and qualified personnel.

(i) Fixture wires—

(i)(1) General. Fixture wires shall be suitable for the voltage, temperature, and location of use. A fixture wire which is used as a grounded conductor shall be identified.

§ 1926.405(i)(2)

(i)(2) Uses permitted. Fixture wires may be used:

(i)(2)(i) For installation in lighting, fixtures and in similar equipment where enclosed or protected and not subject to bending or twisting in use; or

FIGURE 7.6 *(Continued)*

(i)(2)(ii) For connecting lighting fixtures to the branch-circuit conductors supplying the fixtures.

(i)(3) Uses not permitted. Fixture wires shall not be used as branch-circuit conductors except as permitted for Class 1 power-limited circuits.

(j) Equipment for general use—

(j)(1) Lighting fixtures, lampholders, lamps, and receptacles—

(j)(1)(i) Live parts. Fixtures, lampholders, lamps, rosettes, and receptacles shall have no live parts normally exposed to employee contact. However, rosettes and cleat-type lampholders and receptacles located at least 8 feet (2.44 m) above the floor may have exposed parts.

§ 1926.405(j)(1)(ii)

(j)(1)(ii) Support. Fixtures, lampholders, rosettes, and receptacles shall be securely supported. A fixture that weighs more than 6 pounds (2.72 kg) or exceeds 16 inches (406 mm) in any dimension shall not be supported by the screw shell of a lampholder.

(j)(1)(iii) Portable lamps. Portable lamps shall be wired with flexible cord and an attachment plug of the polarized or grounding type. If the portable lamp uses an Edison-based lampholder, the grounded conductor shall be identified and attached to the screw shell and the identified blade of the attachment plug. In addition, portable handlamps shall comply with the following:

(j)(1)(iii)(A) Metal shell, paperlined lampholders shall not be used;

(j)(1)(iii)(B) Handlamps shall be equipped with a handle of molded composition or other insulating material;

(j)(1)(iii)(C) Handlamps shall be equipped with a substantial guard attached to the lampholder or handle;

(j)(1)(iii)(D) Metallic guards shall be grounded by the means of an equipment grounding conductor run within the power supply cord.

(j)(1)(iv) Lampholders. Lampholders of the screw-shell type shall be installed for use as lampholders only. Lampholders installed in wet or damp locations shall be of the weatherproof type.

§ 1926.405(j)(1)(v)

(j)(1)(v) Fixtures. Fixtures installed in wet or damp locations shall be identified for the purpose and shall be installed so that water cannot enter or accumulate in wireways, lampholders, or other electrical parts.

(j)(2) Receptacles, cord connectors, and attachment plugs (caps)—

(j)(2)(i) Configuration. Receptacles, cord connectors, and attachment plugs shall be constructed so that no receptacle or cord connector will accept an attachment plug with a different voltage or current rating than that for which the device is intended. However, a 20-ampere T-slot receptacle or cord connector may accept a 15-ampere attachment plug of the same voltage rating. Receptacles connected to circuits having different voltages, frequencies, or types of current (ac or dc) on the same premises shall be of such design that the attachment plugs used on these circuits are not interchangeable.

(j)(2)(ii) Damp and wet locations. A receptacle installed in a wet or damp location shall be designed for the location.

(j)(3) Appliances—

(j)(3)(i) Live parts. Appliances, other than those in which the current-carrying parts at high temperatures are necessarily exposed, shall have no live parts normally exposed to employee contact.

(j)(3)(ii) Disconnecting means. A means shall be provided to disconnect each appliance.

§ 1926.405(j)(3)(iii)

(j)(3)(iii) Rating. Each appliance shall be marked with its rating in volts and amperes or volts and watts.

(j)(4) Motors. This paragraph applies to motors, motor circuits, and controllers.

(j)(4)(i) In sight from. If specified that one piece of equipment shall be "in sight from" another piece of equipment, one shall be visible and not more than 50 feet (15.2 m) from the other.

(j)(4)(ii) Disconnecting means—

(j)(4)(ii)(A) A disconnecting means shall be located in sight from the controller location. The controller disconnecting means for motor branch circuits over 600 volts, nominal, may be out of sight of the controller, if the controller is marked with a warning label giving the location and identification of the disconnecting means which is to be locked in the open position.

FIGURE 7.6 *(Continued)*

(j)(4)(ii)(B) The disconnecting means shall disconnect the motor and the controller from all ungrounded supply conductors and shall be so designed that no pole can be operated independently.

(j)(4)(ii)(C) If a motor and the driven machinery are not in sight from the controller location, the installation shall comply with one of the following conditions:

§ 1926.405(j)(4)(ii)(C)(*1*)

(j)(4)(ii)(C)(*1*) The controller disconnecting means shall be capable of being locked in the open position.

(j)(4)(ii)(C)(*2*) A manually operable switch that will disconnect the motor from its source of supply shall be placed in sight from the motor location.

(j)(4)(ii)(D) The disconnecting means shall plainly indicate whether it is in the open (off) or closed (on) position.

(j)(4)(ii)(E) The disconnecting means shall be readily accessible. If more than one disconnect is provided for the same equipment, only one need be readily accessible.

(j)(4)(ii)(F) An individual disconnecting means shall be provided for each motor, but a single disconnecting means may be used for a group of motors under any one of the following conditions:

(j)(4)(ii)(F)(*1*) If a number of motors drive special parts of a single machine or piece of apparatus, such as a metal or woodworking machine, crane, or hoist;

(j)(4)(ii)(F)(*2*) If a group of motors is under the protection of one set of branch-circuit protective devices; or

(j)(4)(ii)(F)(*3*) If a group of motors is in a single room in sight from the location of the disconnecting means.

§ 1926.405(j)(4)(iii)

(j)(4)(iii) Motor overload, short-circuit, and ground-fault protection. Motors, motor-control apparatus, and motor branch-circuit conductors shall be protected against overheating due to motor overloads or failure to start, and against short-circuits or ground faults. These provisions do not require overload protection that will stop a motor where a shutdown is likely to introduce additional or increased hazards, as in the case of fire pumps, or where continued operation of a motor is necessary for a safe shutdown of equipment or process and motor overload sensing devices are connected to a supervised alarm.

(j)(4)(iv) Protection of live parts—all voltages—

(j)(4)(iv)(A) Stationary motors having commutators, collectors, and brush rigging located inside of motor end brackets and not conductively connected to supply circuits operating at more than 150 volts to ground need not have such parts guarded. Exposed live parts of motors and controllers operating at 50 volts or more between terminals shall be guarded against accidental contact by any of the following:

(j)(4)(iv)(A)(*1*) By installation in a room or enclosure that is accessible only to qualified persons;

(j)(4)(iv)(A)(*2*) By installation on a balcony, gallery, or platform, so elevated and arranged as to exclude unqualified persons; or

(j)(4)(iv)(A)(*3*) By elevation 8 feet (2.44 m) or more above the floor.

(j)(4)(iv)(B) Where live parts of motors or controllers operating at over 150 volts to ground are guarded against accidental contact only by location, and where adjustment or other attendance may be necessary during the operation of the apparatus, insulating mats or platforms shall be provided so that the attendant cannot readily touch live parts unless standing on the mats or platforms.

§ 1926.405(j)(5)

(j)(5) Transformers—

(j)(5)(i) Application. The following paragraphs cover the installation of all transformers, except:

(j)(5)(i)(A) Current transformers;

(j)(5)(i)(B) Dry-type transformers installed as a component part of other apparatus;

(j)(5)(i)(C) Transformers which are an integral part of an X-ray, high frequency, or electrostatic-coating apparatus;

(j)(5)(i)(D) Transformers used with Class 2 and Class 3 circuits, sign and outline lighting, electric discharge lighting, and power-limited fire-protective signaling circuits.

(j)(5)(ii) Operating voltage. The operating voltage of exposed live parts of transformer installations shall be indicated by warning signs or visible markings on the equipment or structure.

FIGURE 7.6 (*Continued*)

(j)(5)(iii) Transformers over 35 kV. Dry-type, high fire point liquid-insulated, and askarel-insulated transformers installed indoors and rated over 35 kV shall be in a vault.

(j)(5)(iv) Oil-insulated transformers. If they present a fire hazard to employees, oil-insulated transformers installed indoors shall be in a vault.

§ 1926.405(j)(5)(v)

(j)(5)(v) Fire protection. Combustible material, combustible buildings and parts of buildings, fire escapes, and door and window openings shall be safeguarded from fires which may originate in oil-insulated transformers attached to or adjacent to a building or combustible material.

(j)(5)(vi) Transformer vaults. Transformer vaults shall be constructed so as to contain fire and combustible liquids within the vault and to prevent unauthorized access. Locks and latches shall be so arranged that a vault door can be readily opened from the inside.

(j)(5)(vii) Pipes and ducts. Any pipe or duct system foreign to the vault installation shall not enter or pass through a transformer vault.

(j)(5)(viii) Material storage. Materials shall not be stored in transformer vaults.

(j)(6) Capacitors—

(j)(6)(i) Drainage of stored charge. All capacitors, except surge capacitors or capacitors included as a component part of other apparatus, shall be provided with an automatic means of draining the stored charge and maintaining the discharged state after the capacitor is disconnected from its source of supply.

(j)(6)(ii) Over 600 volts. Capacitors rated over 600 volts, nominal, shall comply with the following additional requirements:

§ 1926.405(j)(6)(ii)(A)

(j)(6)(ii)(A) Isolating or disconnecting switches (with no interrupting rating) shall be interlocked with the load interrupting device or shall be provided with prominently displayed caution signs to prevent switching load current.

(j)(6)(ii)(B) For series capacitors the proper switching shall be assured by use of at least one of the following:

(j)(6)(ii)(B)(*1*) Mechanically sequenced isolating and bypass switches,

(j)(6)(ii)(B)(*2*) Interlocks, or

(j)(6)(ii)(B)(*3*) Switching procedure prominently displayed at the switching location.

[61 FR 5507, Feb. 13, 1996]

§ 1926.406 Specific purpose equipment and installations.

(a) Cranes and hoists. This paragraph applies to the installation of electric equipment and wiring used in connection with cranes, monorail hoists, hoists, and all runways.

(a)(1) Disconnecting means—

(a)(1)(i) Runway conductor disconnecting means. A readily accessible disconnecting means shall be provided between the runway contact conductors and the power supply.

(a)(1)(ii) Disconnecting means for cranes and monorail hoists. A disconnecting means, capable of being locked in the open position, shall be provided in the leads from the runway contact conductors or other power supply on any crane or monorail hoist.

(a)(1)(ii)(A) If this additional disconnecting means is not readily accessible from the crane or monorail hoist operating station, means shall be provided at the operating station to open the power circuit to all motors of the crane or monorail hoist.

(a)(1)(ii)(B) The additional disconnect may be omitted if a monorail hoist or hand-propelled crane bridge installation meets all of the following:

§ 1926.406(a)(1)(ii)(B)(*1*)

(a)(1)(ii)(B)(*1*) The unit is floor controlled;

(a)(1)(ii)(B)(*2*) The unit is within view of the power supply disconnecting means; and

(a)(1)(ii)(B)(*3*) No fixed work platform has been provided for servicing the unit.

FIGURE 7.6 (*Continued*)

(a)(2) Control. A limit switch or other device shall be provided to prevent the load block from passing the safe upper limit of travel of any hoisting mechanism.

(a)(3) Clearance. The dimension of the working space in the direction of access to live parts which may require examination, adjustment, servicing, or maintenance while alive shall be a minimum of 2 feet 6 inches (762 mm). Where controls are enclosed in cabinets, the door(s) shall open at least 90 degrees or be removable, or the installation shall provide equivalent access.

§ 1926.406(a)(4)

(a)(4) Grounding. All exposed metal parts of cranes, monorail hoists, hoists and accessories including pendant controls shall be metallically joined together into a continuous electrical conductor so that the entire crane or hoist will be grounded in accordance with 1926.404(f). Moving parts, other than removable accessories or attachments, having metal-to-metal bearing surfaces shall be considered to be electrically connected to each other through the bearing surfaces for grounding purposes. The trolley frame and bridge frame shall be considered as electrically grounded through the bridge and trolley wheels and its respective tracks unless conditions such as paint or other insulating materials prevent reliable metal-to-metal contact. In this case a separate bonding conductor shall be provided.

(b) Elevators, escalators, and moving walks—

(b)(1) Disconnecting means. Elevators, escalators, and moving walks shall have a single means for disconnecting all ungrounded main power supply conductors for each unit.

(b)(2) Control panels. If control panels are not located in the same space as the drive machine, they shall be located in cabinets with doors or panels capable of being locked closed.

(c) Electric welders—disconnecting means—

(c)(1) Motor-generator, AC transformer, and DC rectifier arc welders. A disconnecting means shall be provided in the supply circuit for each motor-generator arc welder, and for each AC transformer and DC rectifier arc welder which is not equipped with a disconnect mounted as an integral part of the welder.

(c)(2) Resistance welders. A switch or circuit breaker shall be provided by which each resistance welder and its control equipment can be isolated from the supply circuit. The ampere rating of this disconnecting means shall not be less than the supply conductor ampacity.

§ 1926.406(d)

(d) X-Ray equipment—

(d)(1) Disconnecting means—

(d)(1)(i) General. A disconnecting means shall be provided in the supply circuit. The disconnecting means shall be operable from a location readily accessible from the X-ray control. For equipment connected to a 120-volt branch circuit of 30 amperes or less, a grounding-type attachment plug cap and receptacle of proper rating may serve as a disconnecting means.

(d)(1)(ii) More than one piece of equipment. If more than one piece of equipment is operated from the same high-voltage circuit, each piece or each group of equipment as a unit shall be provided with a high-voltage switch or equivalent disconnecting means. This disconnecting means shall be constructed, enclosed, or located so as to avoid contact by employees with its live parts.

(d)(2) Control—Radiographic and fluoroscopic types. Radiographic and fluoroscopic-type equipment shall be effectively enclosed or shall have interlocks that deenergize the equipment automatically to prevent ready access to live current-carrying parts.

§ 1926.407 Hazardous (classified) locations.

(a) Scope. This section sets forth requirements for electric equipment and wiring in locations which are classified depending on the properties of the flammable vapors, liquids or gases, or combustible dusts or fibers which may be present therein and the likelihood that a flammable or combustible concentration or quantity is present. Each room, section or area shall be considered individually in determining its classification. These hazardous (classified) locations are assigned six designations as follows:

Class I, Division 1 Class I, Division 2 Class II, Division 1 Class II, Division 2 Class III, Division 1 Class III, Division 2

FIGURE 7.6 *(Continued)*

For definitions of these locations see 1926.449. All applicable requirements in this subpart apply to all hazardous (classified) locations, unless modified by provisions of this section.

(b) Electrical installations. Equipment, wiring methods, and installations of equipment in hazardous (classified) locations shall be approved as intrinsically safe or approved for the hazardous (classified) location or safe for the hazardous (classified) location. Requirements for each of these options are as follows:

(b)(1) Intrinsically safe. Equipment and associated wiring approved as intrinsically safe is permitted in any hazardous (classified) location included in its listing or labeling.

§ 1926.407(b)(2)

(b)(2) Approved for the hazardous (classified) location—

(b)(2)(i) General. Equipment shall be approved not only for the class of location but also for the ignitable or combustible properties of the specific gas, vapor, dust, or fiber that will be present.

Note: NFPA 70, the National Electrical Code, lists or defines hazardous gases, vapors, and dusts by "Groups" characterized by their ignitable or combustible properties.

(b)(2)(ii) Marking. Equipment shall not be used unless it is marked to show the class, group, and operating temperature or temperature range, based on operation in a 40-degree C ambient, for which it is approved. The temperature marking shall not exceed the ignition temperature of the specific gas, vapor, or dust to be encountered. However, the following provisions modify this marking requirement for specific equipment:

(b)(2)(ii)(A) Equipment of the non-heat-producing type (such as junction boxes, conduit, and fitting) and equipment of the heat-producing type having a maximum temperature of not more than 100 degrees C (212 degrees F) need not have a marked operating temperature or temperature range.

(b)(2)(ii)(B) Fixed lighting fixtures marked for use only in Class I, Division 2 locations need not be marked to indicate the group.

(b)(2)(ii)(C) Fixed general-purpose equipment in Class I locations, other than lighting fixtures, which is acceptable for use in Class I, Division 2 locations need not be marked with the class, group, division, or operating temperature.

§ 1926.407(b)(2)(ii)(D)

(b)(2)(ii)(D) Fixed dust-tight equipment, other than lighting fixtures, which is acceptable for use in Class II, Division 2 and Class III locations need not be marked with the class, group, division, or operating temperature.

(b)(3) Safe for the hazardous (classified) location. Equipment which is safe for the location shall be of a type and design which the employer demonstrates will provide protection from the hazards arising from the combustibility and flammability of vapors, liquids, gases, dusts, or fibers.

Note: The National Electrical Code, NFPA 70, contains guidelines for determining the type and design of equipment and installations which will meet this requirement. The guidelines of this document address electric wiring, equipment, and systems installed in hazardous (classified) locations and contain specific provisions for the following: wiring methods, wiring connections, conductor insulation, flexible cords, sealing and drainage, transformers, capacitors, switches, circuit breakers, fuses, motor controllers, receptacles, attachment plugs, meters, relays, instruments, resistors, generators, motors, lighting fixtures, storage battery charging equipment, electric cranes, electric hoists and similar equipment, utilization equipment, signaling systems, alarm systems, remote control systems, local loud speaker and communication systems, ventilation piping, live parts, lightning surge protection, and grounding. Compliance with these guidelines will constitute one means, but not the only means, of compliance with this paragraph.

(c) Conduits. All conduits shall be threaded and shall be made wrench-tight. Where it is impractical to make a threaded joint tight, a bonding jumper shall be utilized.

[61 FR 5507, Feb. 13, 1996]

FIGURE 7.6 (*Continued*)

§ 1926.408 Special systems

(a) Systems over 600 volts, nominal. Paragraphs (a)(1) through (a)(4) of this section contain general requirements for all circuits and equipment operated at over 600 volts.

(a)(1) Wiring methods for fixed installations—

(a)(1)(i) Above ground. Above-ground conductors shall be installed in rigid metal conduit, in intermediate metal conduit, in cable trays, in cablebus, in other suitable raceways, or as open runs of metal-clad cable designed for the use and purpose. However, open runs of non-metallic-sheathed cable or of bare conductors or busbars may be installed in locations which are accessible only to qualified persons. Metallic shielding components, such as tapes, wires, or braids for conductors, shall be grounded. Open runs of insulated wires and cables having a bare lead sheath or a braided outer covering shall be supported in a manner designed to prevent physical damage to the braid or sheath.

§ 1926.408(a)(1)(ii)

(a)(1)(ii) Installations emerging from the ground. Conductors emerging from the ground shall be enclosed in raceways. Raceways installed on poles shall be of rigid metal conduit, intermediate metal conduit, PVC schedule 80 or equivalent extending from the ground line up to a point 8 feet (2.44 m) above finished grade. Conductors entering a building shall be protected by an enclosure from the ground line to the point of entrance. Metallic enclosures shall be grounded.

(a)(2) Interrupting and isolating devices—

(a)(2)(i) Circuit breakers. Circuit breakers located indoors shall consist of metal-enclosed or fire-resistant, cell-mounted units. In locations accessible only to qualified personnel, open mounting of circuit breakers is permitted. A means of indicating the open and closed position of circuit breakers shall be provided.

(a)(2)(ii) Fused cutouts. Fused cutouts installed in buildings or transformer vaults shall be of a type identified for the purpose. They shall be readily accessible for fuse replacement.

(a)(2)(iii) Equipment isolating means. A means shall be provided to completely isolate equipment for inspection and repairs. Isolating means which are not designed to interrupt the load current of the circuit shall be either interlocked with a circuit interrupter or provided with a sign warning against opening them under load.

§ 1926.408(a)(3)

(a)(3) Mobile and portable equipment—

(a)(3)(i) Power cable connections to mobile machines. A metallic enclosure shall be provided on the mobile machine for enclosing the terminals of the power cable. The enclosure shall include provisions for a solid connection for the ground wire(s) terminal to ground effectively the machine frame. The method of cable termination used shall prevent any strain or pull on the cable from stressing the electrical connections. The enclosure shall have provision for locking so only authorized qualified persons may open it and shall be marked with a sign warning of the presence of energized parts.

(a)(3)(ii) Guarding live parts. All energized switching and control parts shall be enclosed in effectively grounded metal cabinets or enclosures. Circuit breakers and protective equipment shall have the operating means projecting through the metal cabinet or enclosure so these units can be reset without locked doors being opened. Enclosures and metal cabinets shall be locked so that only authorized qualified persons have access and shall be marked with a sign warning of the presence of energized parts. Collector ring assemblies on revolving-type machines (shovels, draglines, etc.) shall be guarded.

(a)(4) Tunnel installations—

(a)(4)(i) Application. The provisions of this paragraph apply to installation and use of high-voltage power distribution and utilization equipment which is associated with tunnels and which is portable and/or mobile, such as substations, trailers, cars, mobile shovels, draglines, hoists, drills, dredges, compressors, pumps, conveyors, and underground excavators.

(a)(4)(ii) Conductors. Conductors in tunnels shall be installed in one or more of the following:

(a)(4)(ii)(A) Metal conduit or other metal raceway,

FIGURE 7.6 *(Continued)*

§ 1926.408(a)(4)(ii)(B)

(a)(4)(ii)(B) Type MC cable, or
(a)(4)(ii)(C) Other suitable multiconductor cable.
Conductors shall also be so located or guarded as to protect them from physical damage. Multiconductor portable cable may supply mobile equipment. An equipment grounding conductor shall be run with circuit conductors inside the metal raceway or inside the multiconductor cable jacket. The equipment grounding conductor may be insulated or bare.
(a)(4)(iii) Guarding live parts. Bare terminals of transformers, switches, motor controllers, and other equipment shall be enclosed to prevent accidental contact with energized parts. Enclosures for use in tunnels shall be drip-proof, weatherproof, or submersible as required by the environmental conditions.
(a)(4)(iv) Disconnecting means. A disconnecting means that simultaneously opens all ungrounded conductors shall be installed at each transformer or motor location.
(a)(4)(v) Grounding and bonding. All nonenergized metal parts of electric equipment and metal raceways and cable sheaths shall be grounded and bonded to all metal pipes and rails at the portal and at intervals not exceeding 1000 feet (305 m) throughout the tunnel.

§ 1926.408(b)

(b) Class 1, Class 2, and Class 3 remote control, signaling, and power-limited circuits—
(b)(1) Classification. Class 1, Class 2, or Class 3 remote control, signaling, or power-limited circuits are characterized by their usage and electrical power limitation which differentiates them from light and power circuits. These circuits are classified in accordance with their respective voltage and power limitations as summarized in paragraphs (b)(1)(i) through (b)(1)(iii) of this section.
(b)(1)(i) Class 1 circuits—
(b)(1)(i)(A) A Class 1 power-limited circuit is supplied from a source having a rated output of not more than 30 volts and 1000 volt-amperes.
(b)(1)(i)(B) A Class 1 remote control circuit or a Class 1 signaling circuit has a voltage which does not exceed 600 volts; however, the power output of the source need not be limited.
(b)(1)(ii) Class 2 and Class 3 circuits—
(b)(1)(ii)(A) Power for Class 2 and Class 3 circuits is limited either inherently (in which no overcurrent protection is required) or by a combination of a power source and overcurrent protection.
(b)(1)(ii)(B) The maximum circuit voltage is 150 volts AC or DC for a Class 2 inherently limited power source, and 100 volts AC or DC for a Class 3 inherently limited power source.
(b)(1)(ii)(C) The maximum circuit voltage is 30 volts AC and 60 volts DC for a Class 2 power source limited by overcurrent protection, and 150 volts AC or DC for a Class 3 power source limited by overcurrent protection.

§ 1926.408(b)(1)(iii)

(b)(1)(iii) Application. The maximum circuit voltages in paragraphs (b)(1)(i) and (b)(1)(ii) of this section apply to sinusoidal AC or continuous DC power sources, and where wet contact occurrence is not likely.
(b)(2) Marking. A Class 2 or Class 3 power supply unit shall not be used unless it is durably marked where plainly visible to indicate the class of supply and its electrical rating.
(c) Communications systems—
(c)(1) Scope. These provisions for communication systems apply to such systems as central-station-connected and non-central-station-connected telephone circuits, radio receiving and transmitting equipment, and outside wiring for fire and burglar alarm, and similar central station systems. These installations need not comply with the provisions of 1926.403 through 1926.408(b), except 1926.404(c)(1)(ii) and 1926.407.
(c)(2) Protective devices—
(c)(2)(i) Circuits exposed to power conductors. Communication circuits so located as to be exposed to accidental contact with light or power conductors operating at over 300 volts shall have each circuit so exposed provided with an approved protector.

FIGURE 7.6 *(Continued)*

§ 1926.408(c)(2)(ii)

(c)(2)(ii) Antenna lead-ins. Each conductor of a lead-in from an outdoor antenna shall be provided with an antenna discharge unit or other means that will drain static charges from the antenna system.

(c)(3) Conductor location—

(c)(3)(i) Outside of buildings—

(c)(3)(i)(A) Receiving distribution lead-in or aerial-drop cables attached to buildings and lead-in conductors to radio transmitters shall be so installed as to avoid the possibility of accidental contact with electric light or power conductors.

(c)(3)(i)(B) The clearance between lead-in conductors and any lightning protection conductors shall not be less than 6 feet (1.83 m).

(c)(3)(ii) On poles. Where practicable, communication conductors on poles shall be located below the light or power conductors. Communications conductors shall not be attached to a crossarm that carries light or power conductors.

(c)(3)(iii) Inside of buildings. Indoor antennas, lead-ins, and other communication conductors attached as open conductors to the inside of buildings shall be located at least 2 inches (50.8 mm) from conductors of any light or power or Class 1 circuits unless a special and equally protective method of conductor separation is employed.

§ 1926.408(c)(4)

(c)(4) Equipment location. Outdoor metal structures supporting antennas, as well as self-supporting antennas such as vertical rods or dipole structures, shall be located as far away from overhead conductors of electric light and power circuits of over 150 volts to ground as necessary to avoid the possibility of the antenna or structure falling into or making accidental contact with such circuits.

(c)(5) Grounding—

(c)(5)(i) Lead-in conductors. If exposed to contact with electric light or power conductors, the metal sheath of aerial cables entering buildings shall be grounded or shall be interrupted close to the entrance to the building by an insulating joint or equivalent device. Where protective devices are used, they shall be grounded.

(c)(5)(ii) Antenna structures. Masts and metal structures supporting antennas shall be permanently and effectively grounded without splice or connection in the grounding conductor.

(c)(5)(iii) Equipment enclosures. Transmitters shall be enclosed in a metal frame or grill or separated from the operating space by a barrier, all metallic parts of which are effectively connected to ground. All external metal handles and controls accessible to the operating personnel shall be effectively grounded. Unpowered equipment and enclosures shall be considered grounded where connected to an attached coaxial cable with an effectively grounded metallic shield.

[61 FR 5507, Feb. 13, 1996]

FIGURE 7.6 *(Continued)*

1926.416 General requirements.

(a) Protection of employees—

(a)(1) No employer shall permit an employee to work in such proximity to any part of an electric power circuit that the employee could contact the electric power circuit in the course of work, unless the employee is protected against electric shock by deenergizing the circuit and grounding it or by guarding it effectively by insulation or other means.

(a)(2) In work areas where the exact location of underground electric powerlines is unknown, employees using jack-hammers, bars, or other hand tools which may contact a line shall be provided with insulated protective gloves.

(a)(3) Before work is begun the employer shall ascertain by inquiry or direct observation, or by instruments, whether any part of an energized electric power circuit, exposed or concealed, is so located that the performance of the work may bring any person, tool, or machine into physical or electrical contact with the electric power circuit. The employer shall post and maintain proper warning signs where such a circuit exists. The employer shall advise employees of the location of such lines, the hazards involved, and the protective measures to be taken.

§ 1926.416(b)

(b) Passageways and open spaces—

(b)(1) Barriers or other means of guarding shall be provided to ensure that workspace for electrical equipment will not be used as a passageway during periods when energized parts of electrical equipment are exposed.

(b)(2) Working spaces, walkways, and similar locations shall be kept clear of cords so as not to create a hazard to employees.

(c) Load ratings. In existing installations, no changes in circuit protection shall be made to increase the load in excess of the load rating of the circuit wiring.

(d) Fuses. When fuses are installed or removed with one or both terminals energized, special tools insulated for the voltage shall be used.

(e) Cords and cables.

(e)(1) Worn or frayed electric cords or cables shall not be used.

(e)(2) Extension cords shall not be fastened with staples, hung from nails, or suspended by wire.

[58 FR 35179, June 30, 1993; 61 FR 9227, March 7, 1996; 61 FR 41738, August 12, 1996]

§ 1926.417 Lockout and tagging of circuits.

(a) Controls. Controls that are to be deactivated during the course of work on energized or deenergized equipment or circuits shall be tagged.

(b) Equipment and circuits. Equipment or circuits that are deenergized shall be rendered inoperative and shall have tags attached at all points where such equipment or circuits can be energized.

(c) Tags. Tags shall be placed to identify plainly the equipment or circuits being worked on.

[58 FR 35181, June 30, 1993; 61 FR 9227, March 7, 1996; 61 FR 41738, August 12, 1996]

FIGURE 7.7 Safety-Related Work Practices for Construction Workers (OSHA-CFR, Title 29, Part 1926, Paragraphs 416–417).

§ 1926.431 Maintenance of equipment.

The employer shall ensure that all wiring components and utilization equipment in hazardous locations are maintained in a dust-tight, dust-ignition-proof, or explosion-proof condition, as appropriate. There shall be no loose or missing screws, gaskets, threaded connections, seals, or other impairments to a tight condition.

§ 1926.432 Environmental deterioration of equipment.

(a) Deteriorating agents—

(a)(1) Unless identified for use in the operating environment, no conductors or equipment shall be located:

(a)(1)(i) In damp or wet locations;

(a)(1)(ii) Where exposed to gases, fumes, vapors, liquids, or other agents having a deteriorating effect on the conductors or equipment; or

(a)(1)(iii) Where exposed to excessive temperatures.

(a)(2) Control equipment, utilization equipment, and busways approved for use in dry locations only shall be protected against damage from the weather during building construction.

(b) Protection against corrosion. Metal raceways, cable armor, boxes, cable sheathing, cabinets, elbows, couplings, fittings, supports, and support hardware shall be of materials appropriate for the environment in which they are to be installed.

FIGURE 7.8 Safety-Related Maintenance and Environmental Considerations (OSHA-CFR, Title 29, Part 1926, Paragraphs 431–432).

§ 1926.441 Batteries and battery charging.

(a) General requirements—

(a)(1) Batteries of the unsealed type shall be located in enclosures with outside vents or in well ventilated rooms and shall be arranged so as to prevent the escape of fumes, gases, or electrolyte spray into other areas.

(a)(2) Ventilation shall be provided to ensure diffusion of the gases from the battery and to prevent the accumulation of an explosive mixture.

(a)(3) Racks and trays shall be substantial and shall be treated to make them resistant to the electrolyte.

(a)(4) Floors shall be of acid resistant construction unless protected from acid accumulations.

(a)(5) Face shields, aprons, and rubber gloves shall be provided for workers handling acids or batteries.

(a)(6) Facilities for quick drenching of the eyes and body shall be provided within 25 feet (7.62 m) of battery handling areas.

§ 1926.441(a)(7)

(a)(7) Facilities shall be provided for flushing and neutralizing spilled electrolyte and for fire protection.

(b) Charging—

(b)(1) Battery charging installations shall be located in areas designated for that purpose.

(b)(2) Charging apparatus shall be protected from damage by trucks.

(b)(3) When batteries are being charged, the vent caps shall be kept in place to avoid electrolyte spray. Vent caps shall be maintained in functioning condition.

FIGURE 7.9 Safety Requirements for Special Equipment (OSHA-CFR, Title 29, Part 1926, Paragraph 441).

§ 1926.449 Definitions applicable to this subpart.

The definitions given in this section apply to the terms used in Subpart K. The definitions given here for "approved" and "qualified person" apply, instead of the definitions given in 1926.32, to the use of these terms in Subpart K.

Acceptable. An installation or equipment is acceptable to the Assistant Secretary of Labor, and approved within the meaning of this Subpart K:

(a) If it is accepted, or certified, or listed, or labeled, or otherwise determined to be safe by a qualified testing laboratory capable of determining the suitability of materials and equipment for installation and use in accordance with this standard; or

(b) With respect to an installation or equipment of a kind which no qualified testing laboratory accepts, certifies, lists, labels, or determines to be safe, if it is inspected or tested by another Federal agency, or by a State, municipal, or other local authority responsible for enforcing occupational safety provisions of the National Electrical Code, and found in compliance with those provisions; or

(c) With respect to custom-made equipment or related installations which are designed, fabricated for, and intended for use by a particular customer, if it is determined to be safe for its intended use by its manufacturer on the basis of test data which the employer keeps and makes available for inspection to the Assistant Secretary and his authorized representatives.

Accepted. An installation is "accepted" if it has been inspected and found to be safe by a qualified testing laboratory.

Accessible. (As applied to wiring methods.) Capable of being removed or exposed without damaging the building structure or finish, or not permanently closed in by the structure or finish of the building. (See "concealed" and "exposed.")

Accessible. (As applied to equipment.) Admitting close approach; not guarded by locked doors, elevation, or other effective means. (See "Readily accessible.")

Ampacity. The current in amperes a conductor can carry continuously under the conditions of use without exceeding its temperature rating.

Appliances. Utilization equipment, generally other than industrial, normally built in standardized sizes or types, which is installed or connected as a unit to perform one or more functions.

Approved. Acceptable to the authority enforcing this Subpart. The authority enforcing this Subpart is the Assistant Secretary of Labor for Occupational Safety and Health. The definition of "acceptable" indicates what is acceptable to the Assistant Secretary of Labor, and therefore approved within the meaning of this Subpart.

Askarel. A generic term for a group of nonflammable synthetic chlorinated hydrocarbons used as electrical insulating media. Askarels of various compositional types are used. Under arcing conditions the gases produced, while consisting predominantly of noncombustible hydrogen chloride, can include varying amounts of combustible gases depending upon the askarel type.

Attachment plug (Plug cap)(Cap). A device which, by insertion in a receptacle, establishes connection between the conductors of the attached flexible cord and the conductors connected permanently to the receptacle.

Automatic. Self-acting, operating by its own mechanism when actuated by some impersonal influence, as for example, a change in current strength, pressure, temperature, or mechanical configuration.

Bare conductor. See *Conductor.*

Bonding. The permanent joining of metallic parts to form an electrically conductive path which will assure electrical continuity and the capacity to conduct safely any current likely to be imposed.

Bonding jumper. A reliable conductor to assure the required electrical conductivity between metal parts required to be electrically connected.

Branch circuit. The circuit conductors between the final overcurrent device protecting the circuit and the outlet(s).

Building. A structure which stands alone or which is cut off from adjoining structures by fire walls with all openings therein protected by approved fire doors.

Cabinet. An enclosure designed either for surface or flush mounting, and provided with a frame, mat, or trim in which a swinging door or doors are or may be hung.

Certified. Equipment is "certified" if it:

(a) Has been tested and found by a qualified testing laboratory to meet applicable test standards or to be safe for use in a specified manner, and

(b) Is of a kind whose production is periodically inspected by a qualified testing laboratory. Certified equipment must bear a label, tag, or other record of certification.

FIGURE 7.10 Definitions Applying to Part 1926, Subpart K (OSHA-CFR, Title 29, Part 1926, Paragraph 449).

Circuit breaker—(a) (600 volts nominal, or less.) A device designed to open and close a circuit by nonautomatic means and to open the circuit automatically on a predetermined overcurrent without injury to itself when properly applied within its rating.

(b) (Over 600 volts, nominal.) A switching device capable of making, carrying, and breaking currents under normal circuit conditions, and also making, carrying for a specified time, and breaking currents under specified abnormal circuit conditions, such as those of short circuit.

Class I locations. Class I locations are those in which flammable gases or vapors are or may be present in the air in quantities sufficient to produce explosive or ignitable mixtures. Class I locations include the following:

(a) Class I, Division 1. A Class I, Division 1 location is a location:

(1) In which ignitable concentrations of flammable gases or vapors may exist under normal operating conditions; or

(2) In which ignitable concentrations of such gases or vapors may exist frequently because of repair or maintenance operations or because of leakage; or

(3) In which breakdown or faulty operation of equipment or processes might release ignitable concentrations of flammable gases or vapors, and might also cause simultaneous failure of electric equipment.

Note: This classification usually includes locations where volatile flammable liquids or liquefied flammable gases are transferred from one container to another; interiors of spray booths and areas in the vicinity of spraying and painting operations where volatile flammable solvents are used; locations containing open tanks or vats of volatile flammable liquids; drying rooms or compartments for the evaporation of flammable solvents; inadequately ventilated pump rooms for flammable gas or for volatile flammable liquids; and all other locations where ignitable concentrations of flammable vapors or gases are likely to occur in the course of normal operations.

(b) Class I, Division 2. A Class I, Division 2 location is a location:

(1) In which volatile flammable liquids or flammable gases are handled, processed, or used, but in which the hazardous liquids, vapors, or gases will normally be confined within closed containers or closed systems from which they can escape only in case of accidental rupture or breakdown of such containers or systems, or in case of abnormal operation of equipment; or

(2) In which ignitable concentrations of gases or vapors are normally prevented by positive mechanical ventilation, and which might become hazardous through failure or abnormal operations of the ventilating equipment; or

(3) That is adjacent to a Class I, Division 1 location, and to which ignitable concentrations of gases or vapors might occasionally be communicated unless such communication is prevented by adequate positive-pressure ventilation from a source of clean air, and effective safeguards against ventilation failure are provided.

Note: This classification usually includes locations where volatile flammable liquids or flammable gases or vapors are used, but which would become hazardous only in case of an accident or of some unusual operating condition. The quantity of flammable material that might escape in case of accident, the adequacy of ventilating equipment, the total area involved, and the record of the industry or business with respect to explosions or fires are all factors that merit consideration in determining the classification and extent of each location.

Piping without valves, checks, meters, and similar devices would not ordinarily introduce a hazardous condition even though used for flammable liquids or gases. Locations used for the storage of flammable liquids or of liquefied or compressed gases in sealed containers would not normally be considered hazardous unless also subject to other hazardous conditions.

Electrical conduits and their associated enclosures separated from process fluids by a single seal or barrier are classed as a Division 2 location if the outside of the conduit and enclosures is a nonhazardous location.

Class II locations. Class II locations are those that are hazardous because of the presence of combustible dust. Class II locations include the following:

(a) Class II, Division 1. A Class II, Division 1 location is a location:

(1) In which combustible dust is or may be in suspension in the air under normal operating conditions, in quantities sufficient to produce explosive or ignitable mixtures; or

(2) Where mechanical failure or abnormal operation of machinery or equipment might cause such explosive or ignitable mixtures to be produced, and might also provide a source of ignition through simultaneous failure of electric equipment, operation of protection devices, or from other causes, or

(3) In which combustible dusts of an electrically conductive nature may be present.

FIGURE 7.10 *(Continued)*

Note: Combustible dusts which are electrically nonconductive include dusts produced in the handling and processing of grain and grain products, pulverized sugar and cocoa, dried egg and milk powders, pulverized spices, starch and pastes, potato and woodflour, oil meal from beans and seed, dried hay, and other organic materials which may produce combustible dusts when processed or handled. Dusts containing magnesium or aluminum are particularly hazardous and the use of extreme caution is necessary to avoid ignition and explosion.

(b) Class II, Division 2. A Class II, Division 2 location is a location in which:

(1) Combustible dust will not normally be in suspension in the air in quantities sufficient to produce explosive or ignitable mixtures, and dust accumulations are normally insufficient to interfere with the normal operation of electrical equipment or other apparatus; or

(2) Dust may be in suspension in the air as a result of infrequent malfunctioning of handling or processing equipment, and dust accumulations resulting therefrom may be ignitable by abnormal operation or failure of electrical equipment or other apparatus.

Note: This classification includes locations where dangerous concentrations of suspended dust would not be likely but where dust accumulations might form on or in the vicinity of electric equipment. These areas may contain equipment from which appreciable quantities of dust would escape under abnormal operating conditions or be adjacent to a Class II Division 1 location, as described above, into which an explosive or ignitable concentration of dust may be put into suspension under abnormal operating conditions.

Class III locations. Class III locations are those that are hazardous because of the presence of easily ignitable fibers or flyings but in which such fibers or flyings are not likely to be in suspension in the air in quantities sufficient to produce ignitable mixtures. Class 111 locations include the following:

(a) Class III, Division 1. A Class III, Division 1 location is a location in which easily ignitable fibers or materials producing combustible flyings are handled, manufactured, or used.

Note: Easily ignitable fibers and flyings include rayon, cotton (including cotton linters and cotton waste), sisal or henequen, istle, jute, hemp, tow, cocoa fiber, oakum, baled waste kapok, Spanish moss, excelsior, sawdust, woodchips, and other material of similar nature.

(b) Class III, Division 2. A Class III, Division 2 location is a location in which easily ignitable fibers are stored or handled, except in process of manufacture.

Collector ring. A collector ring is an assembly of slip rings for transferring electrical energy from a stationary to a rotating member.

Concealed. Rendered inaccessible by the structure or finish of the building. Wires in concealed raceways are considered concealed, even though they may become accessible by withdrawing them. [See "Accessible. (As applied to wiring methods.)"]

Conductor—(a) *Bare.* A conductor having no covering or electrical insulation whatsoever.

(b) *Covered.* A conductor encased within material of composition or thickness that is not recognized as electrical insulation.

(c) *Insulated.* A conductor encased within material of composition and thickness that is recognized as electrical insulation.

Controller. A device or group of devices that serves to govern, in some predetermined manner, the electric power delivered to the apparatus to which it is connected.

Covered conductor. See *Conductor.*

Cutout. (Over 600 volts, nominal.) An assembly of a fuse support with either a fuseholder, fuse carrier, or disconnecting blade. The fuseholder or fuse carrier may include a conducting element (fuse link), or may act as the disconnecting blade by the inclusion of a nonfusible member.

Cutout box. An enclosure designed for surface mounting and having swinging doors or covers secured directly to and telescoping with the walls of the box proper. (See *Cabinet.*)

Damp location. See *Location.*

Dead front. Without live parts exposed to a person on the operating side of the equipment.

Device. A unit of an electrical system which is intended to carry but not utilize electric energy.

Disconnecting means. A device, or group of devices, or other means by which the conductors of a circuit can be disconnected from their source of supply.

FIGURE 7.10 (*Continued*)

Disconnecting (or Isolating) switch. (Over 600 volts, nominal.) A mechanical switching device used for isolating a circuit or equipment from a source of power.

Dry location. See *Location.*

Enclosed. Surrounded by a case, housing, fence or walls which will prevent persons from accidentally contacting energized parts.

Enclosure. The case or housing of apparatus, or the fence or walls surrounding an installation to prevent personnel from accidentally contacting energized parts, or to protect the equipment from physical damage.

Equipment. A general term including material, fittings, devices, appliances, fixtures, apparatus, and the like, used as a part of, or in connection with, an electrical installation.

Equipment grounding conductor. See *Grounding conductor, equipment.*

Explosion-proof apparatus. Apparatus enclosed in a case that is capable of withstanding an explosion of a specified gas or vapor which may occur within it and of preventing the ignition of a specified gas or vapor surrounding the enclosure by sparks, flashes, or explosion of the gas or vapor within, and which operates at such an external temperature that it will not ignite a surrounding flammable atmosphere.

Exposed. (As applied to live parts.) Capable of being inadvertently touched or approached nearer than a safe distance by a person. It is applied to parts not suitably guarded, isolated, or insulated. (See *Accessible* and *Concealed.*)

Exposed. (As applied to wiring methods.) On or attached to the surface or behind panels designed to allow access. [See *Accessible.* (As applied to wiring methods.)"]

Exposed. (For the purposes of 1926.408(d), Communications systems.) Where the circuit is in such a position that in case of failure of supports or insulation, contact with another circuit may result.

Externally operable. Capable of being operated without exposing the operator to contact with live parts.

Feeder. All circuit conductors between the service equipment, or the generator switchboard of an isolated plant, and the final branch-circuit overcurrent device.

Festoon lighting. A string of outdoor lights suspended between two points more than 15 feet (4.57 m) apart.

Fitting. An accessory such as a locknut, bushing, or other part of a wiring system that is intended primarily to perform a mechanical rather than an electrical function.

Fuse. (Over 600 volts, nominal.) An overcurrent protective device with a circuit opening fusible part that is heated and severed by the passage of overcurrent through it. A fuse comprises all the parts that form a unit capable of performing the prescribed functions. It may or may not be the complete device necessary to connect it into an electrical circuit.

Ground. A conducting connection, whether intentional or accidental, between an electrical circuit or equipment and the earth, or to some conducting body that serves in place of the earth.

Grounded. Connected to earth or to some conducting body that serves in place of the earth.

Grounded, effectively (Over 600 volts, nominal.) Permanently connected to earth through a ground connection of sufficiently low impedance and having sufficient ampacity that ground fault current which may occur cannot build up to voltages dangerous to personnel.

Grounded conductor. A system or circuit conductor that is intentionally grounded.

Grounding conductor. A conductor used to connect equipment or the grounded circuit of a wiring system to a grounding electrode or electrodes.

Grounding conductor, equipment. The conductor used to connect the noncurrent-carrying metal parts of equipment, raceways, and other enclosures to the system grounded conductor and/or the grounding electrode conductor at the service equipment or at the source of a separately derived system.

Grounding electrode conductor. The conductor used to connect the grounding electrode to the equipment grounding conductor and/or to the grounded conductor of the circuit at the service equipment or at the source of a separately derived system.

Ground-fault circuit interrupter. A device for the protection of personnel that functions to deenergize a circuit or portion thereof within an established period of time when a current to ground exceeds some predetermined value that is less than that required to operate the overcurrent protective device of the supply circuit.

Guarded. Covered, shielded, fenced, enclosed, or otherwise protected by means of suitable covers, casings, barriers, rails, screens, mats, or platforms to remove the likelihood of approach to a point of danger or contact by persons or objects.

Hoistway. Any shaftway, hatchway, well hole, or other vertical opening or space in which an elevator or dumbwaiter is designed to operate.

FIGURE 7.10 (*Continued*)

Identified (conductors or terminals). Identified, as used in reference to a conductor or its terminal, means that such conductor or terminal can be recognized as grounded.

Identified (for the use). Recognized as suitable for the specific purpose, function, use, environment, application, etc. where described as a requirement in this standard. Suitability of equipment for a specific purpose, environment, or application is determined by a qualified testing laboratory where such identification includes labeling or listing.

Insulated conductor. See *Conductor.*

Interrupter switch. (Over 600 volts, nominal.) A switch capable of making, carrying, and interrupting specified currents.

Intrinsically safe equipment and associated wiring. Equipment and associated wiring in which any spark or thermal effect, produced either normally or in specified fault conditions, is incapable, under certain prescribed test conditions, of causing ignition of a mixture of flammable or combustible material in air in its most easily ignitable concentration.

Isolated. Not readily accessible to persons unless special means for access are used.

Isolated power system. A system comprising an isolating transformer or its equivalent, a line isolation monitor, and its ungrounded circuit conductors.

Labeled. Equipment or materials to which has been attached a label, symbol or other identifying mark of a qualified testing laboratory which indicates compliance with appropriate standards or performance in a specified manner.

Lighting outlet. An outlet intended for the direct connection of a lampholder, a lighting fixture, or a pendant cord terminating in a lampholder.

Listed. Equipment or materials included in a list published by a qualified testing laboratory whose listing states either that the equipment or material meets appropriate standards or has been tested and found suitable for use in a specified manner.

Location—(a) Damp location. Partially protected locations under canopies, marquees, roofed open porches, and like locations, and interior locations subject to moderate degrees of moisture, such as some basements.

(b) *Dry location.* A location not normally subject to dampness or wetness. A location classified as dry may be temporarily subject to dampness or wetness, as in the case of a building under construction.

(c) *Wet location.* Installations underground or in concrete slabs or masonry in direct contact with the earth, and locations subject to saturation with water or other liquids, such as locations exposed to weather and unprotected.

Mobile X-ray. X-ray equipment mounted on a permanent base with wheels and/or casters for moving while completely assembled.

Motor control center. An assembly of one or more enclosed sections having a common power bus and principally containing motor control units.

Outlet. A point on the wiring system at which current is taken to supply utilization equipment.

Overcurrent. Any current in excess of the rated current of equipment or the ampacity of a conductor. It may result from overload (see definition), short circuit, or ground fault. A current in excess of rating may be accommodated by certain equipment and conductors for a given set of conditions. Hence the rules for overcurrent protection are specific for particular situations.

Overload. Operation of equipment in excess of normal, full load rating, or of a conductor in excess of rated ampacity which, when it persists for a sufficient length of time, would cause damage or dangerous overheating. A fault, such as a short circuit or ground fault, is not an overload. (See "Overcurrent.")

Panelboard. A single panel or group of panel units designed for assembly in the form of a single panel; including buses, automatic overcurrent devices, and with or without switches for the control of light, heat, or power circuits; designed to be placed in a cabinet or cutout box placed in or against a wall or partition and accessible only from the front. (See "Switchboard.")

Portable X-ray. X-ray equipment designed to be hand-carried.

Power fuse. (Over 600 volts, nominal.) See "Fuse."

Power outlet. An enclosed assembly which may include receptacles, circuit breakers, fuseholders, fused switches, buses and watt-hour meter mounting means; intended to serve as a means for distributing power required to operate mobile or temporarily installed equipment.

Premises wiring system. That interior and exterior wiring, including power, lighting, control, and signal circuit wiring together with all of its associated hardware, fittings, and wiring devices, both permanently and temporarily installed, which extends from the load end of the service drop, or load end of the service lateral conductors to the outlet(s). Such wiring does not include wiring internal to appliances, fixtures, motors, controllers, motor control centers, and similar equipment.

FIGURE 7.10 *(Continued)*

Qualified person. One familiar with the construction and operation of the equipment and the hazards involved.

Qualified testing laboratory. A properly equipped and staffed testing laboratory which has capabilities for and which provides the following services:

(a) Experimental testing for safety of specified items of equipment and materials referred to in this standard to determine compliance with appropriate test standards or performance in a specified manner;

(b) Inspecting the run of such items of equipment and materials at factories for product evaluation to assure compliance with the test standards;

(c) Service-value determinations through field inspections to monitor the proper use of labels on products and with authority for recall of the label in the event a hazardous product is installed;

(d) Employing a controlled procedure for identifying the listed and/or labeled equipment or materials tested; and

(e) Rendering creditable reports or findings that are objective and without bias of the tests and test methods employed.

Raceway. A channel designed expressly for holding wires, cables, or busbars, with additional functions as permitted in this subpart. Raceways may be of metal or insulating material, and the term includes rigid metal conduit, rigid nonmetallic conduit, intermediate metal conduit, liquidtight flexible metal conduit, flexible metallic tubing, flexible metal conduit, electrical metallic tubing, underfloor raceways, cellular concrete floor raceways, cellular metal floor raceways, surface raceways, wireways, and busways.

Readily accessible. Capable of being reached quickly for operation, renewal, or inspections, without requiring those to whom ready access is requisite to climb over or remove obstacles or to resort to portable ladders, chairs, etc. (See "Accessible.")

Receptacle. A receptacle is a contact device installed at the outlet for the connection of a single attachment plug. A single receptacle is a single contact device with no other contact device on the same yoke. A multiple receptacle is a single device containing two or more receptacles.

Receptacle outlet. An outlet where one or more receptacles are installed.

Remote-control circuit. Any electric circuit that controls any other circuit through a relay or an equivalent device.

Sealable equipment. Equipment enclosed in a case or cabinet that is provided with a means of sealing or locking so that live parts cannot be made accessible without opening the enclosure. The equipment may or may not be operable without opening the enclosure.

Separately derived system. A premises wiring system whose power is derived from generator, transformer, or converter windings and has no direct electrical connection, including a solidly connected grounded circuit conductor, to supply conductors originating in another system.

Service. The conductors and equipment for delivering energy from the electricity supply system to the wiring system of the premises served.

Service conductors. The supply conductors that extend from the street main or from transformers to the service equipment of the premises supplied.

Service drop. The overhead service conductors from the last pole or other aerial support to and including the splices, if any, connecting to the service-entrance conductors at the building or other structure.

Service-entrance conductors, overhead system. The service conductors between the terminals of the service equipment and a point usually outside the building, clear of building walls, where joined by tap or splice to the service drop.

Service-entrance conductors, underground system. The service conductors between the terminals of the service equipment and the point of connection to the service lateral. Where service equipment is located outside the building walls, there may be no service-entrance conductors, or they may be entirely outside the building.

Service equipment. The necessary equipment, usually consisting of a circuit breaker or switch and fuses, and their accessories, located near the point of entrance of supply conductors to a building or other structure, or an otherwise defined area, and intended to constitute the main control and means of cutoff of the supply.

Service raceway. The raceway that encloses the service-entrance conductors.

Signaling circuit. Any electric circuit that energizes signaling equipment.

Switchboard. A large single panel, frame, or assembly of panels which have switches, buses, instruments, overcurrent and other protective devices mounted on the face or back or both. Switchboards are generally accessible from the rear as well as from the front and are not intended to be installed in cabinets. (See *Panelboard.*)

Switches—(a) *General-use switch.* A switch intended for use in general distribution and branch circuits. It is rated in amperes, and it is capable of interrupting its rated current at its rated voltage.

FIGURE 7.10 *(Continued)*

(b) *General-use snap switch.* A form of general-use switch so constructed that it can be installed in flush device boxes or on outlet box covers, or otherwise used in conjunction with wiring systems recognized by this subpart.

(c) *Isolating switch.* A switch intended for isolating an electric circuit from the source of power. It has no interrupting rating, and it is intended to be operated only after the circuit has been opened by some other means.

(d) *Motor-circuit switch.* A switch, rated in horsepower, capable of interrupting the maximum operating overload current of a motor of the same horsepower rating as the switch at the rated voltage.

Switching devices. (Over 600 volts, nominal.) Devices designed to close and/or open one or more electric circuits. Included in this category are circuit breakers, cutouts, disconnecting (or isolating) switches, disconnecting means, and interrupter switches.

Transportable X-ray. X-ray equipment installed in a vehicle or that may readily be disassembled for transport in a vehicle.

Utilization equipment. Utilization equipment means equipment which utilizes electric energy for mechanical, chemical, heating, lighting, or similar useful purpose.

Utilization system. A utilization system is a system which provides electric power and light for employee workplaces, and includes the premises wiring system and utilization equipment.

Ventilated. Provided with a means to permit circulation of air sufficient to remove an excess of heat, fumes, or vapors.

Volatile flammable liquid. A flammable liquid having a flash point below 38 degrees C (100 degrees F) or whose temperature is above its flash point, or a Class II combustible liquid having a vapor pressure not exceeding 40 psia (276 kPa) at 38 deg. C (100 deg. F) whose temperature is above its flash point.

Voltage. (Of a circuit.) The greatest root-mean-square (effective) difference of potential between any two conductors of the circuit concerned.

Voltage, nominal. A nominal value assigned to a circuit or system for the purpose of conveniently designating its voltage class (as 120/240, 480Y/277, 600, etc.). The actual voltage at which a circuit operates can vary from the nominal within a range that permits satisfactory operation of equipment.

Voltage to ground. For grounded circuits, the voltage between the given conductor and that point or conductor of the circuit that is grounded; for ungrounded circuits, the greatest voltage between the given conductor and any other conductor of the circuit.

Watertight. So constructed that moisture will not enter the enclosure.

Weatherproof. So constructed or protected that exposure to the weather will not interfere with successful operation. Rainproof, raintight, or watertight equipment can fulfill the requirements for weatherproof where varying weather conditions other than wetness, such as snow, ice, dust, or temperature extremes, are not a factor.

Wet location. See *Location.*

FIGURE 7.10 (*Continued*)

CHAPTER 8
ACCIDENT PREVENTION, ACCIDENT INVESTIGATION, RESCUE, AND FIRST AID

INTRODUCTION

This chapter is focused on the individual employee's decision making and actions in accident prevention and response. For the purpose of this chapter, the employer's electrical system design, engineering, and management processes to operate and maintain the system are assumed to enable the employee's actions in a safe work environment.

ACCIDENT PREVENTION

No matter how carefully a system is engineered, no matter how carefully employees perform their tasks, and no matter how well trained employees are in the recognition and avoidance of hazards, accidents still happen. This section provides a general approach that may be employed to reduce the number and severity of accidents. Four basic steps—employee responsibility, safe installations, safe work practices, and employee training—combine to create the type of safe work environment that should be the goal of every facility.

Individual Responsibility

The person most responsible for your own personal safety is you. No set of regulations, rules, or procedures can ever replace common sense in the workplace. This statement should not be construed to mean an employer has no responsibility to provide the safest practical work environment, nor does it mean that the injured person is "at fault" in a legal sense. Determining fault for accidents is, in part, a legal problem and is beyond the scope of this handbook.

Time after time, accident investigations reveal that the injured person was the last link in the chain. If the injured person had only been wearing appropriate safety equipment or following proper procedures, or if he or she had only checked one last time, the accident never would have occurred.

Table 8.1 lists five behavioral approaches that will help to improve the safety of all employees. To make certain employees have both the responsibility and the authority to carry out the five steps listed in Table 8.1, employers should adopt a policy similar to the one listed in Table 8.2. Simply putting such a statement in a safety handbook is insufficient. An employer must believe in this principle and must "put teeth into it." Such a credo provides the absolute maximum in individual employee responsibility and authority and will maximize the safety performance of the organization.

TABLE 8.1 Employee Safety Behavior

- Determine the nature and extent of hazards before starting a job.
- Each employee should be satisfied that conditions are safe before beginning work on any job or any part of a job.
- All employees should be thoroughly familiar with and should consistently use the work procedures and the safety equipment required for the performance of the job at hand.
- While working, each employee should consider the effects of each step and do nothing that might endanger themselves or others.
- Each employee should be thoroughly familiar with emergency procedures.

TABLE 8.2 Recommended Safety Credo

If it cannot be done safely, it need not be done!

Installation Safety

Design. Proper design of electrical systems is composed of three parts: selection, installation, and calibration.

- *Selection.* Electric equipment should be selected and applied conservatively. That is, maximum ratings must be well in excess of the quantities to which they will be exposed in the power system. To help with this, manufacturing organizations such as the National Electrical Manufacturers Association (NEMA) have established ratings for equipment that ensure member companies manufacture only the highest-quality equipment. Equipment is tested per manufacturer's procedures by independent laboratories such as the Underwriters Laboratories (UL). Equipment that is rated and labeled by such organizations should be used in electrical systems to help ensure safety. OSHA, NEC,* and NESC requirements should be considered as minimum criteria for safe selection. With increasing cost consciousness, many companies are opting for the installation of recycled rather than brand-new equipment. While the selection and use of such equipment can be a financial advantage, at least three criteria should be considered in the purchase:
  1. Even though used, the equipment should have been originally manufactured by a reputable, professional firm.
  2. The recycled equipment should be thoroughly reconditioned by a professional recycling company such as those represented by the Professional Electrical Apparatus Reconditioning League.
  3. Engineering studies such as short-circuit analysis and coordination studies must be performed to ascertain that the recycled equipment has adequate ratings for the system in which it is being placed. Since the system short-circuit values may change with time, this requirement is true even if the recycled equipment is a direct replacement for the previous equipment.
- *Installation.* Equipment should be installed in a safe and *sensible* manner. Adequate work spaces for safety clearance should be allowed, safety barriers should be provided when necessary, and electrical installations should never be mixed with areas that are used for general public access.
- *Calibration.* Equipment always should be properly calibrated. For example, protective devices should be calibrated so that they will operate for the minimum abnormal system condition. Equipment that is improperly calibrated can result in accidents as though the equipment had been improperly selected to begin with. Calibration is also a two-step process:
  1. Proper engineering should be performed by professional engineers to ensure that the selected calibration settings are suitable for the application. The starting point for such a system is in the performance of an appropriate suite of power system studies, described later in the section.

*National Electrical Code and NEC are registered trademarks of the National Fire Protection Association.

2. Proper testing and physical setting of the devices should be carried out to ensure that the equipment is capable of performing when called upon. Such settings should be executed by professional technicians who are certified to perform this work. Organizations such as the InterNational Electrical Testing Association (NETA) have been formed to ensure quality control on the education and performance of electrical technicians.

Electrical Protective Devices. Protective devices such as circuit breakers, fuses, and switches must be capable of interrupting the currents to which they will be subjected. The National Electrical Code has numerous passages that require proper sizing of protective devices.

Maintenance. Improperly maintained equipment is hazardous. For example, circuit breakers can explode violently if not properly maintained. Equipment should be periodically inspected and tested. If deficiencies are observed, the equipment must be repaired, adjusted, or replaced as required. Properly trained and certified technicians and mechanics should be used for such work. As mentioned earlier, national organizations such as NETA can be used to provide qualified personnel.

Operating Schemes. Many personal injury accidents could be prevented if systems were all designed for safe operating procedures. The following are some of the key design concepts that should be used for any new or refurbished electrical system:

1. Arc-resistant switchgear should be employed. In addition to reducing the possibility of electrical arcs and resulting blasts, arc-resistant switchgear is designed to contain the blast and blast products. Such switchgear also has pressure relief systems that will redirect the release of arc and blast products and energy in directions that are safe for workers.
2. Remote operating controls should be used for all circuit breakers, switches, and other such control devices. This relatively simple and inexpensive design technique allows workers to operate equipment while stationed remotely from the possible results of failure. Whether supervisory control systems or simple remotely placed, hard-wired control switches are used, the distance will allow the workers to operate the equipment in relative safety.
3. Control panels should be designed with protective barriers to prevent shock hazard and contain arcs and arc products. For example, many industrial control panels are fed via 480 V circuit breakers. This breaker, in turn, feeds a step-down transformer. The control voltage output of the transformer feeds the various control circuits throughout the panel. If the 480 V portions of the circuit are designed with appropriate protective covers, workers in the panel need to protect themselves only from the electrical hazards presented by the 120 V circuits.

Power System Studies

Power system studies are engineering procedures that must be performed. There are three distinct times in the life of a power system when such studies should be performed:

1. During the initial design of the system to ensure that all selected equipment will perform properly during the day-to-day operation of the facility.
2. Any time a significant change occurs such as replacement of an existing piece of equipment or addition of new equipment or circuits.
3. At intervals during the life of the system to identify any possible changes that may have occurred externally. For example, electric utility supplies are changing constantly. Such changes can have a profound effect on the short-circuit currents and normal operating voltage levels.

The studies described in this section should be considered safety-related procedures. If they are not performed, equipment may be improperly applied and can fail, putting personnel at risk.

ANSI/IEEE standard Std-399 (*Recommended Practice for Industrial and Commercial Power Systems*) identifies 11 recommended power systems studies. Of these 11, the short-circuit analysis and protective device coordination study are among the most critical with respect to electrical safety. As described later in this section, such studies are required by some industry standards.

The following paragraphs briefly describe these studies and provide basic information about their importance and implementation. The reader is referred to the most recent edition of ANSI/IEEE standard Std-399 for details. Also note that these procedures are safety-related; they should be performed only by engineers and technicians with the education and experience to do them correctly.

Load Flow Analysis. This type of study determines the voltage, current, reactive power, active power, and power factor in an operating power system. It is performed using computer software and can be set up to analyze contingencies of any type. Such studies are used to size and select equipment and will alert system operators to possible hazardous or poor operating conditions.

Stability Analysis. Stability in a power system is defined as either transient or steady state. Steady state stability is the ability of a power system to maintain synchronism among machines within the system following relatively slow load changes. Transient stability is the ability to maintain synchronism after short-term events occur, such as switching and short circuits.

Stability studies are safety-related in that they will allow the power system operator to continue safe operation even when the system is exposed to abnormal or excessive events.

Motor-Starting Analysis. When large motors are started, the high current surges and voltage dips can cause malfunction or failure of other system components. A motor-starting analysis uses a computer program to model the behavior of the system when the motor starts. This can be used to size the power system equipment to prevent outages, and the hazards associated with them will be reduced or eliminated.

Harmonic Analysis. Harmonics and other types of power quality problems can cause premature equipment failure, malfunctions, and other types of hazardous conditions. Such failures can include heating and failure of rotating machinery and power system capacitors and their associated hazards. A harmonic analysis is performed to pinpoint the sources of harmonic distortion and to determine the solutions to such problems.

Switching Transients Analysis. When certain loads are switched and/or when switches are malfunctioning, failures can occur, which can put employees at risk. A switching transients study determines the magnitude of such transients and allows the system engineers to develop solutions.

Reliability Analysis. Reliability is usually expressed as the frequency of interruptions and expected number of interruptions in a year of system operation. When properly applied, the results of a reliability study can be used to make sure the system operates as continuously as possible. This means that workers will not be exposed to the hazards of working on the system to repair it.

Cable Ampacity Analysis. Cable ampacity studies determine the ampacity (current-carrying capacity) of power cables in the power system. Properly selecting power cables and sizing them will help to minimize unexpected failures.

Ground Mat Analysis. The subject of system grounding is covered in detail in Chap. 5. One of the most important pieces of equipment in the grounding system is the ground mat. A properly designed and installed ground mat will reduce step and touch voltages and provide a much safer environment for the worker. According to IEEE standard Std-399, at least five factors need to be considered in the ground mat analysis:

1. Fault current magnitude and duration
2. Geometry of the grounding system
3. Soil resistivity

4. Probability of contact
5. Human's factors such as
   (a) Body resistance
   (b) Standard assumptions on physical conditions of the individual

Short-Circuit Analysis and Protective Device Coordination. A Short Circuit Study determines the magnitude of the currents that flow for faults placed at various buses throughout the power system. This information is used to determine interrupting requirements for fuses and circuit breakers and to set trip points for the overcurrent devices.

A Coordination Study is performed to make certain the overcurrent devices in a system will trip selectively. Selective tripping means that only the nearest upstream device to the short circuit trips to clear the circuit, interrupting only the faulted section.

The two studies, taken together, are used to properly select and calibrate the protective devices used in the power system. The information that they provide is used for the following purposes:

- Fuses and circuit breakers are selected so that they are capable of interrupting the maximum fault current that will flow through them.
- Instantaneous elements are selected so that they will respond (or in some cases not respond) to the short-circuit currents that will flow through them.
- Time curves and instantaneous settings are selected so that the nearest upstream device to the short circuit is the one that operates to clear the fault.
- Protective devices are selected so that the short circuit is cleared in a minimum amount of time with as little collateral damage as possible.
- Protective devices are selected so that fault currents that flow through cables and transformers will not cause thermal or mechanical damage to those pieces of equipment.

Each of these points is critical to the safe and economical operation of a power system. For example, if circuit breakers or fuses are incapable of interrupting fault currents, they may explode violently, injuring personnel in the area. If the wrong protective device operates, a "small" outage may expand to include an entire plant. If a transformer overload relay is too slow, the transformer may be damaged by excessive temperature rise. The only way that such malfunctions can be avoided is to perform a Short-Circuit Analysis and a Coordination Study, and then, to select and set protective devices according to their results.

The NEC is the principal source of regulation in the area of electrical installation and design requirements for industrial and commercial facilities. The NEC has several sections that are pertinent. Table 8.3 reproduces a few of these sections.

In addition, ANSI/NFPA 70B, *Electrical Equipment Maintenance*, also has a section that applies. This is reproduced in Table 8.4. The only way to comply with these requirements is to ensure that a Short-Circuit Analysis and a Coordination Study are performed for the power system.

Few would deny that such studies should be performed during the design phase of an electrical power system. But how about later, as the system ages? Several things happen to require the performance and/or reevaluation of these studies for an existing system:

- Electric utilities constantly add capacity. Your utility may have had a 200,000-kilovoltampere (kVA) fault capacity when the plant was new 20 years ago. Now, however, the utility's capacity may have doubled or even tripled. Such changes can cause fault duties to rise above the ratings of marginal interrupting equipment.
- Many plants are beginning to internally generate electricity. This generation adds to the fault capacity of the system.
- Operating procedures may have changed. A bus tie circuit breaker that was normally open may now be normally closed. Such a change can greatly increase fault capacity.

- Technical standards can change. For example, in 1985 the protection requirements for liquid-filled transformers changed. Studies showed that many transformers were being mechanically damaged by high current through faults. The protection requirements became more stringent for such installations. Although the standards do not require existing systems to be changed, would it not make sense to at least review your system? The protection changes might be minimal.
- New installations or plant modernization may add capacity and other coordination streams to the system. For example, Coordination Studies require that the main breaker coordinate with the largest feeder device. If a larger feeder device is added later, the coordination study must be reviewed.

In general, Short-Circuit Analyses and Coordination Studies should be reviewed at least every five years. These studies should be performed by a registered professional engineer. Many consulting firms have the ability and the experience to perform them; however, since Short-Circuit Analyses and

TABLE 8.3 NEC 70-2011 Requirements for Short-Circuit Analyses and Coordination Studies

Studies Location in NEC	Item
Definitions	Interrupting Rating. The highest current at rated voltage that a device is identified to interrupt under standard test conditions.
Article 110	110-9. Interrupting Rating. Equipment intended to interrupt current at fault levels shall have an interrupting rating not less than the nominal circuit voltage and the current that is available at the line terminals of the equipment. Equipment intended to break current at other than fault levels shall have an interrupting rating at nominal circuit voltage not less than the current that must be interrupted. 110-10. Circuit Impedance, Short-Circuit Current Ratings, and Other Characteristics. The over current protective devices, the total impedance, the equipment short-circuit current ratings, and other characteristics of the circuit to be protected shall be selected and coordinated to permit the circuit-protective devices used to clear a fault to do so without extensive damage to the electrical equipment of the circuit. This fault shall be assumed to be either between two or more circuit conductors, or between any circuit conductor and the equipment grounding conductor(s) permitted in 250.118. Listed equipment applied in accordance with their listing shall be considered to meet the requirements of this section.
Article 240	240-12. Electrical System Coordination. Where an orderly shutdown is required to minimize hazard(s) to personnel and equipment, a system of coordination based on the following two conditions shall be permitted: (1) Coordinated short-circuit protection (2) Overload indication based on monitoring systems or devices. Informational note: The monitoring system may cause the condition to go to alarm, allowing corrective action or an orderly shutdown, thereby minimizing personnel hazard and equipment damage.

TABLE 8.4 ANSI/NFPA 70B 2016 Requirements for Short-Circuit Analyses and Coordination Studies

ANSI/NFPA 70B	Item
Paragraph 8.4.3	An up-to-date Short-Circuit and Coordination Study is essential for the safety of personnel and equipment. As a function of the study, the momentary and interrupting rating requirements of the protective devices should be analyzed and verification made that the circuit breaker or fuse will safely interrupt a fault during fault conditions.
Para 8.4.3.1	Additionally, the study should provide the application of the protective device to realize minimum equipment damage and the least disturbance to the system in the event of a fault by properly clearing downstream devices nearest to the point of a fault.

Coordination Studies are specialized types of engineering services, not all architect and engineering firms have the experience to do them. Closely review the qualifications of the firm that you retain.

ANSI/IEEE Standard Std-399, Power Systems Analysis, is the standard that covers most of the engineering studies that are key to the proper design and performance of an electrical power system. The *Brown Book* is part of the IEEE Color Book Series and should be referenced when you are deciding what studies to perform and how to perform them.

Proper selection, sizing, and calibration of the protective devices in a power system directly affect safety, efficiency, and economics. Common sense and regulatory requirements dictate that a Short-Circuit Analysis and Coordination Study should be performed.

Arc-Flash Study. The Arc-Flash Study is performed to determine the incident energy to which workers will be exposed if an electrical arc occurs. This study is described in some detail in Chap. 4.

FIRST AID

This handbook provides general coverage of the subject with expanded information on handling electrical injuries. Potential first aid givers should remember four very important points. Before an accident happens:

- Obtain hands-on first aid training for yourself and all employees. Such training may be obtained from the American Heart Association, the American Red Cross, or local sources such as fire departments or police departments.

If an accident does happen:

- Act quickly!!! You may be the only person who can prevent a death.
- Do not administer first aid that you are not qualified to administer. Injuries can be aggravated by improperly administered first aid.
- Get qualified medical help quickly. Paramedics and emergency medical technicians are trained to provide emergency first aid and should be summoned as soon as possible.

This handbook is not intended to be used as a first aid training manual. Table 8.5 summarizes the first aid steps that are discussed in the following sections.

General First Aid

Act Quickly. *Remember—you may be the only person between the victim and death.* Whatever you do, do it quickly. This does not imply that you should act impetuously. Your actions should be planned and methodical, but you should not waste any time. Do not attempt to perform procedures for which you have no training or experience. Improperly applied procedures can be deadly.

TABLE 8.5 General First Aid Procedure

- Act quickly.
- Survey the situation.
- Develop a plan.
- Assess the victim's condition.
- Summon help if needed.
- Move the victim only if danger is imminent.
- Establish a de-energized accident scene.
- Administer required first aid:
  Shock
  Electrical burns

TABLE 8.6 First Aid Checklist

- Is the circuit still energized?
- Is the victim contacting the circuit?
- Are noxious gases or materials present that may cause injury?
- Is fire present or possible?

TABLE 8.7 What to Do If The Victim Is Responsive

- Ask the victim what is wrong.
- Assess the victim's condition and treat the injuries as best as possible.
- Treat the worst injuries first.
- When the victim is out of immediate danger, or if you are unable to help because the injuries are beyond your abilities, summon help.
- Attend to the victim(s) and keep them safe until help arrives.
- When help arrives, give the first aid workers your assessment of the situation and stand by to help.

Survey the Situation. Remember that your purpose as a first aid giver is to help resolve the problem, not contribute to it. If you are injured in the process of administering first aid, you cannot help the victim. If your preliminary assessment indicates that you need to wear safety clothing, put it on first, then administer aid. Table 8.6 lists key points that should be checked before you rush in.

Develop a Plan. After the initial survey of the situation, develop the plan of attack. The specifics of any given situation will vary; however, the following guidelines should be used:

- If the victim is in immediate danger, he or she should be moved to a safe position. (See the next section on moving the victim and later sections on rescue techniques.)
- If the victim is nonresponsive, assess his or her condition and respond accordingly. (See the later section on assessing the victim's condition.)
- If the victim is responsive, make him or her as comfortable as possible and summon aid. Do not abandon the victim until aid has arrived.
- Constantly monitor the condition of the victim. Electric shock can cause delayed failures and irregularities of heart rhythm.

Assess the Victim's Condition. The procedures to be used in administering first aid depend on the condition of the victim. If the victim is responsive, no action may be required. Table 8.7 lists the procedures to perform if the victim is awake and responsive.

If the victim is not responsive, you must perform a "hands-on" assessment of his or her condition. Table 8.8 lists the ABCs of First Aid. This memory device can help the first aid giver to remember the proper procedure when examining a nonresponsive victim.

One of the biggest surprises to those who have not worked with accident victims is that the trauma of the accident can induce severe bleeding through the mouth and/or vomiting. Be prepared for these conditions before working with an injured person. When you have prepared yourself for this situation, begin the ABCs.

- **A**—Check the victim's **A**irway. Figure 8.1 illustrates the correct way to clear an injured person's airway. Remember to avoid moving the victim and to keep the victim's spine straight to avoid aggravating an injury. *Caution:* An accident victim may suffer from involuntary muscular reflexes and other such spasms. The strongest muscle in the human body is the jaw. Because of this, rescue workers should put their fingers into the victim's mouth only when absolutely necessary.

TABLE 8.8 The ABCs of First Aid

Airway
Breathing
Circulation
Doctor

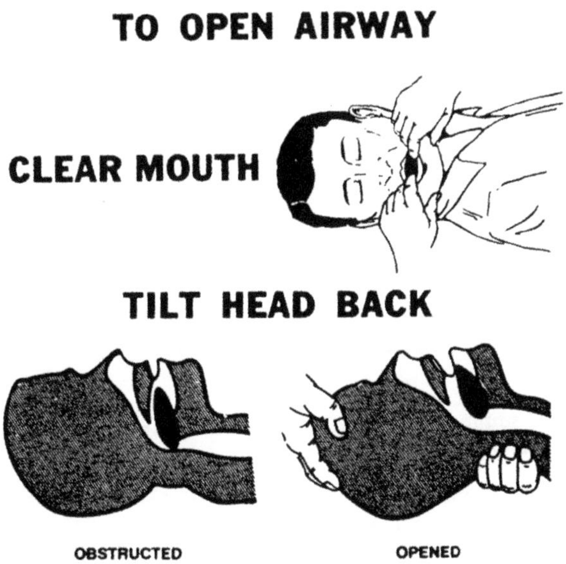

FIGURE 8.1 Clearing the airway of an injured worker.

Start by opening the victim's mouth as shown in Fig. 8.1. Search the mouth for foreign matter or other objects that may be blocking the air passage. Many times the victim's tongue may be blocking the air passage. To fix this problem, put your hand behind the victim's neck, gently pull the jaw forward, and, if required, carefully tilt the head back. If the air passage is clear and the victim is still not breathing, you should perform resuscitation.

- **B**—Check the victim's **B**reathing. First check to see if the victim is breathing. This can be done by observing his or her chest to see if it is moving. Then place your ear close to the victim's mouth and nose and listen carefully. If the victim is breathing but choking or gurgling sounds are heard, clear the victim's airway.
- **C**—Check the victim's **C**irculation. Circulation should be checked by feeling for the victim's pulse at the carotid artery as shown in Fig. 8.2. To find the carotid artery, place your fingertips gently on the victim's larynx. Gently slide the fingers down into the groove between the windpipe and the muscle at the back of the neck. The carotid artery is located in this area. Gently feel for the pulse. Table 8.9 shows the steps to take for the various combinations of problems that may be found.
- **D**—Summon the **D**octor. After the victim's condition has been stabilized, summon help. If the resuscitation efforts are proving unsuccessful, the first aid giver may want to summon more qualified assistance even though the victim is not yet stabilized.

Summon Help If Needed. One of the most difficult decisions is to determine when to summon help. If help is not summoned soon enough, the victim may die. On the other hand, if the first aid giver

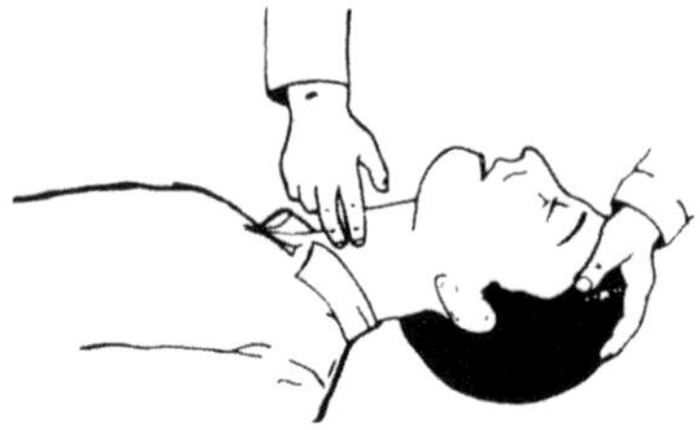

FIGURE 8.2 Checking the circulation.

TABLE 8.9 How to Handle Unresponsive Victims

Breathing—pulse normal	Make victim comfortable. If help has not been summoned, do so and stand by until it arrives.
No breathing—pulse normal	Perform mouth-to-mouth resuscitation until breathing is restored or until help arrives and takes over.
Breathing normal—no pulse	Perform heart-lung resuscitation (CPR) until pulse is restored or until help arrives and takes over.
No breathing—no pulse	Perform heart-lung resuscitation (CPR) until pulse is restored or until help arrives and takes over.

leaves to summon help, the victim may die. No concrete rules can be given here; however, the following guidelines may help:

- Relieve any immediate danger to the victim before summoning help.
- Perform the ABCs before summoning help.
- If the victim is not breathing or has no pulse, perform resuscitation before summoning help.
- If anyone else is in the area, yell or call for help while performing the preliminary accident assessment.

Remember that the first aid giver is in charge of the victim until more qualified help arrives. Do not abandon the victim if immediate aid is required.

Move the Victim If Danger Is Imminent. Unless they are in imminent danger, accident victims should be moved only when necessary and only by personnel who are qualified to move them. A victim of violent injury, such as a fall, may have spinal or other internal injuries. Moving such a victim could cause increased problems including paralysis or even death. Moving an injury victim is discussed in detail in the "Rescue Techniques" section in this chapter.

Establish a De-energized Accident Scene. Before physically approaching a victim, establish that the scene is de-energized. If the scene continues to be supplied electrical energy through an installation failure or downed power line, it is critical to involve utility, police, and emergency medical service personnel. Urgent confirmation is needed that both the electrical energy supply as well as the risk of capacitance are addressed to place the scene in an electrically safe condition for first responders, paramedics, and public safety professionals, including fire and police.

First Aid for Electric Shock. Electric shock is one of the most difficult of all injuries to diagnose. In some cases, even if the injury is fatal, no external signs may be visible. Table 8.10 lists some of the clues and symptoms that may be present when a victim has received an electric shock.

Many prospective first aid givers are themselves injured when they contact an energized wire or a victim who is still in contact with an energized wire. Table 8.11 lists the precautions for working on or

TABLE 8.10 Typical Symptoms of Electric Shock

- Victim may lose consciousness. This may occur at the moment of contact; however, it can also occur later.
- Victim has a weak or irregular pulse.
- Victim has trouble breathing or has stopped breathing.
- Small burns may appear at the entry and exit points of the electric current.

TABLE 8.11 Precautions for Performing First Aid on an Electric Shock Victim

- Do not touch any energized wires with any part of your body or with any conductive tools or equipment.
- Do not touch a victim who is still in contact with an energized wire with any part of your body or with conductive tools or equipment.
- Do not try to move any energized wires unless you are qualified to do so. Qualified in this instance means that you are trained in the performance of such a procedure and are able to avoid electrical hazards.

TABLE 8.12 First Aid Procedures for Conscious Electric Shock Victims Who Exhibit No Symptoms

- Keep the victim still and quiet. Remember that heart and respiratory problems can be delayed in electric shock victims.
- Monitor the victim's condition for at least one hour.
- If the victim continues to show no symptoms, take the victim to a doctor for a thorough examination.

around accident victims who may be in contact with live wires. After cutting the power to the circuits or removing the victim from contact, if the victim is responsive and shows no signs of breathing or heart problems, the procedures listed in Table 8.12 should be followed. After cutting the power to the circuits or removing the victim from contact, if the victim is nonresponsive, the procedures listed in Table 8.13 should be followed.

First Aid for Electrical Burns. Electrical burns may be visible at the skin surface or internal in the victim's body. External burns may be caused by the following:

- Heating from contact with an electrically energized surface
- Flash radiating from an electrical arc
- Fabric or installation ignition and burning due to an electrical arc
- Electrical arc blast

Internal burns are caused by current flow causing resistive heating (Joule heating) to the body's tissues.

Internal burns are virtually impossible to diagnose in the field. The symptoms of internal electrical burns are identical to the symptoms caused by severe electric shock. In addition to the symptoms described in Table 8.10, the victim may also experience significant pain caused by the damaged tissue. External burns are similar to thermally induced burns caused by fire or other heat sources.

For both internal and external burns, the first aid techniques are similar to those given in Tables 8.12 and 8.13.

One extremely important additional procedure for external burn victims is to cool the burns as quickly as possible. A shower or other source of clean, cool water can be used. Since the treatment of burns is a very specialized medical procedure, qualified medical help should be obtained as quickly as possible.

TABLE 8.13 First Aid Procedures for Unconscious Electric Shock Victims with Symptoms

- Check the ABCs. If the victim is not breathing or has heart irregularities, perform resuscitation as described later in this handbook.
- If wounds are evident, cover them with sterile dressings.
- If external burns are evident, they should be cooled using clear, pure water.
- Try to cool burns with sterile compresses.
- Immediately seek medical aid.

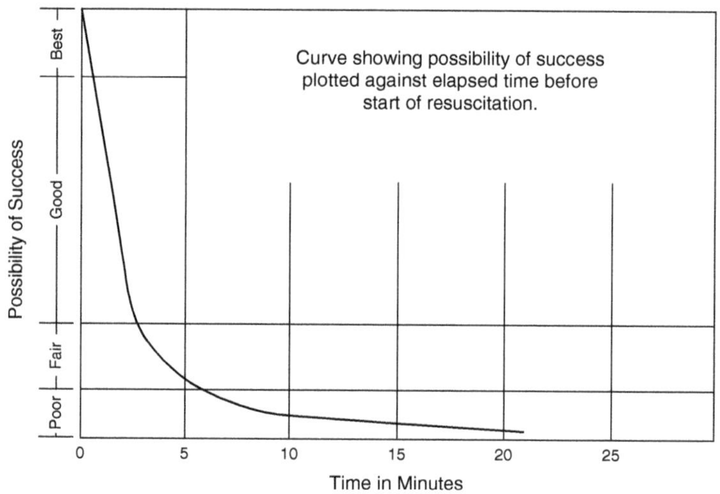

FIGURE 8.3 Elapsed Time Versus Possibility of Success for Artificial Respiration.

Resuscitation (Artificial Respiration)

Breathing trauma is one of the two very serious symptoms that result from severe electric shock or internal burns. When breathing is stopped or made irregular by electricity, it must be restored by resuscitation. Over the years many different types of resuscitation have been developed. Mouth-to-mouth resuscitation is the current preferred technique.

To be most effective, resuscitation must be started as soon as possible after breathing has ceased. Figure 8.3 shows a curve that approximates the possibility of success plotted against elapsed time before the start of resuscitation. Figure 8.4 illustrates a currently accepted procedure for the performance of artificial respiration. *Caution:* With the proliferation of communicable diseases such as acquired immunodeficiency syndrome (AIDS) and hepatitis, the decision of whether to perform life-saving mouth-to-mouth resuscitation has become much more difficult. Instruments such as breathing tubes should be used to protect the victim and the rescuer.

Heart-Lung Resuscitation

Heart-lung resuscitation is also called cardiopulmonary resuscitation (CPR). This technique should be applied when the victim has no pulse and no respiration. The procedure should be started as soon as possible to maximize the probability of successfully restoring full function to the victim.

Remember that while CPR is being performed, the first aid giver is actually pumping blood and breathing for the victim. Do not give up until qualified medical personnel say to. CPR has two separate but equally important parts: rescue breathing and chest compressions. The rescue breathing is performed to force air into an unbreathing victim's lungs. The chest compressions

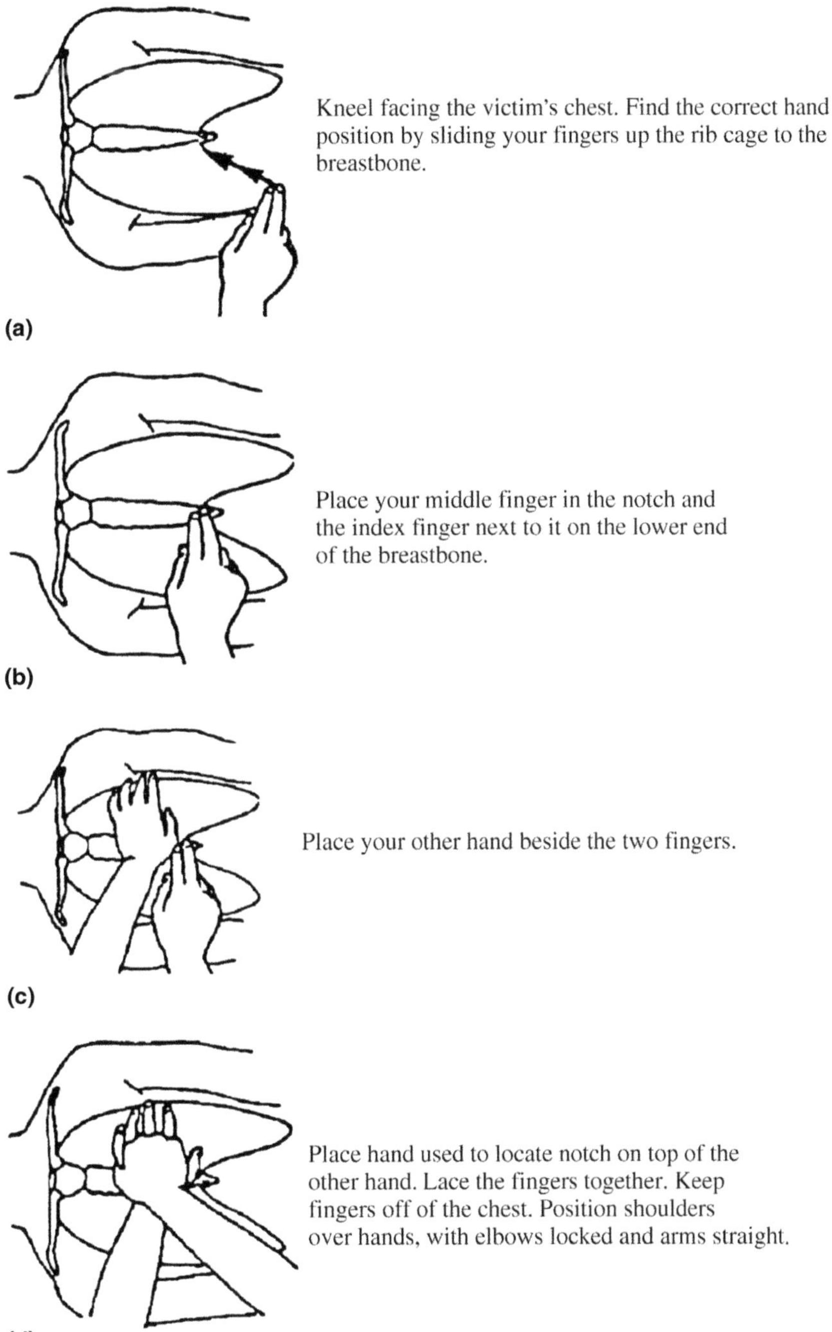

FIGURE 8.4 Steps a, b, c, and d illustrate the procedure to properly position your hands prior to performing CPR.

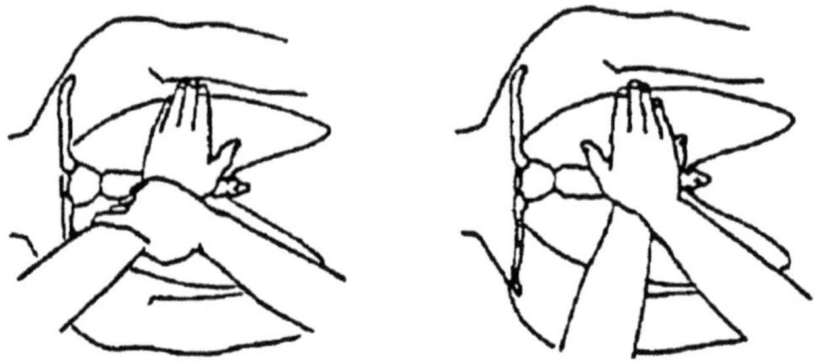

FIGURE 8.5 Alternate positions for hands while applying CPR.

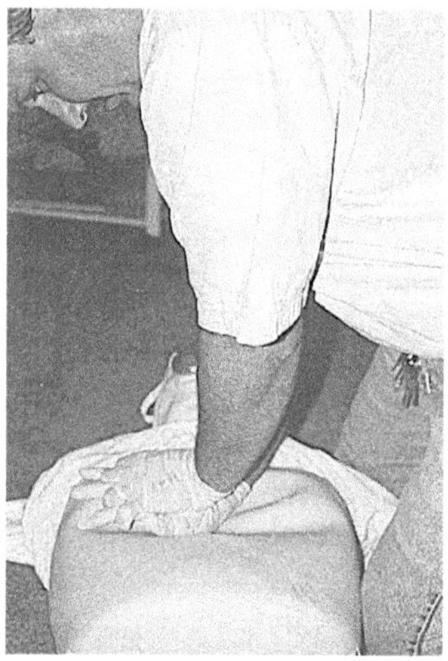

FIGURE 8.6 The weight of the rescuer's upper body is used to compress the chest of the victim.

are used to compress the heart and force blood through the victim's circulatory system. Before CPR is performed, the ABCs described earlier in this chapter should be checked. See Table 8.8 and associated text.

Chest Compressions. If the victim has no pulse, chest compressions should accompany the rescue breathing. The following steps should be used:

1. Properly position your hands on the victim's chest (see Figs. 8.4, 8.5, and 8.6)
   a. Kneel facing the victim's chest. Find the correct hand position by sliding your fingers up the rib cage to the breastbone as shown in Fig. 8.4.
   b. Place your middle finger in the notch and the index finger next to it on the lower end of the breastbone.
   c. Place your other hand beside the two fingers.
   d. Place the hand used to locate the notch on top of the other hand. Lace your fingers together. Be sure to keep your fingers off of the victim's chest. Position your shoulders over your hands with your elbows locked and your arms straight. (See Fig. 8.5 for alternates to this hand position.)

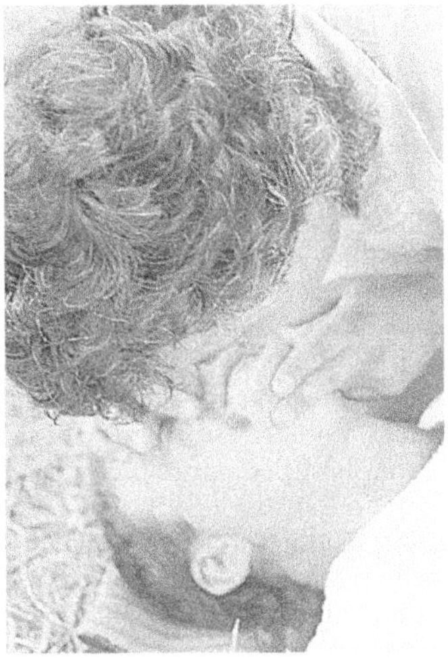

FIGURE 8.7 The rescuer gently pinches the nose closed with the thumb and index finger of the hand on the forehead and breathes into the victim's mouth.

2. At the rate of 100 compressions per minute, compress the breastbone 1½ to 2 inches, as shown in Fig. 8.6.
3. Repeat step 2 thirty times.
4. Stop and give the victim two rescue breaths as described under rescue breathing above. (Note that steps 2, 3, and 4 should take approximately 15 seconds.)
5. Recheck the victim's carotid artery pulse for 5 seconds.
6. If no pulse is found, repeat steps 1 through 5.
7. Continue this process until one of the following occurs
   a. Another trained person takes over CPR from you.
   b. EMS personnel arrive and take over care of the victim.
   c. You are too exhausted and unable to continue.
   d. The scene becomes unsafe.

Rescue Breathing (Mouth-to-Mouth Resuscitation). If the preliminary assessment indicates that the victim is not breathing, mouth-to-mouth resuscitation should be performed as follows:

1. Open the airway, clear the mouth, and tilt the head back, as shown in Fig. 8.1.
2. Gently pinch the victim's nose to seal it closed, as shown in Fig. 8.7.
3. Take a deep breath and seal your lips around the victim's mouth, creating an airtight seal.
4. Gently blow into the victim's mouth for 1½ to 2 seconds. While blowing, watch to be sure the victim's chest is rising.
5. Stop blowing and allow the victim's lungs to exhale.
6. Repeat steps 4 and 5.
7. Stop the procedure and watch the victim for 5 to 10 seconds to determine if the victim is breathing on his/her own.
8. If the victim is not breathing normally, repeat steps 4 and 5 until the victim starts breathing normally or until qualified help arrives to relieve you.

AUTOMATED EXTERNAL DEFIBRILLATOR (AED)

When the heart's electrical system is disturbed, as it is during an electrical shock event that passes electrical shock current through the heart, changes in the heart's rate and rhythm occur. These disruptions cause a large percentage of heart deaths.

The electric impulses in the heart must be coordinated to give a smooth, rhythmic beat. An outside current as small as 60 mA can disturb the nerve impulses and create a situation in which the heartbeat is no longer smooth and rhythmic. The heart goes into fibrillation, defined as rapid and inefficient contraction of muscle fibers of the heart caused by disruption of nerve impulses. When the heart is fibrillating, it does not pump blood through the body. As in sudden cardiac arrest (SCA) in which the heart suddenly stops beating, fibrillation of the heart also creates a situation in which blood stops flowing to the brain and other vital organs. If fibrillation is not treated quickly, each minute of SCA or fibrillation leads to a 10 percent reduction in survival. Any prolonged shock current exceeding 75 mA is almost certain to cause death. Just as with any other muscle, the heart becomes paralyzed if the current flowing through it is of sufficient magnitude. Paralysis of the heart frequently is not fatal if the current is removed quickly enough. Paralysis is used to an advantage in defibrillators. Defibrillators intentionally apply heart-paralyzing current to the patient's chest. Once the heart is paralyzed and the current is removed, the heart is in a relaxed state and ready for the next signal. Frequently, the heart restarts without any other outside intervention.

How an AED Works

To understand how an AED works, you must first understand how the heart works. The heart has an electrical system that controls the rate and rhythm of the heartbeat. Normally, a heartbeat starts in the sinus node, travels at approximately 7 feet/second (2.1 meters/second) through the atrioventricular (AV) node HIS bundle (a collection of heart muscle cells specialized for electrical conduction that transmits the electrical impulses from the AV node, located between the atria and the ventricles, to the point of the apex of the fascicular branches). The fascicular branches then lead to the Purkinje fibers, which provide electrical conduction to the ventricles, causing the cardiac muscle of the ventricles to contract at a paced interval.

Problems with the electrical system can cause abnormal heart rhythms called arrhythmias (ah-RITH-me-ahs). During an arrhythmia, the heart can beat too slowly or too quickly. The most common cause of SCA is an arrhythmia called ventricular fibrillation. In ventricular fibrillation, the ventricles don't beat normally. Instead, they quiver very rapidly and irregularly. In this condition, no blood is being pumped out of the ventricles.

Another arrhythmia that can lead to SCA is ventricular tachycardia (TAK-ih-KAR-de-ah). This is a fast, regular beating of the ventricles that may last for only a few seconds or for longer periods of time.

In people who have either of these arrhythmias, an electric shock from an AED can restore the heart's normal rhythm. Performing CPR—cardiopulmonary resuscitation—on someone having SCA also can improve that person's chance of survival.

AEDs are lightweight, battery-operated, portable devices that are easy to use. Each unit comes with instructions, and AEDs will even give you voice prompts to let you know if and when you should send a shock current to the heart. Learning how to use an AED and taking a CPR course should be mandatory training for any qualified electrical worker. However, if trained personnel are not available, untrained personnel can use an AED to help save a victim's life.

When Should an AED Be Used?

Knowing the signs of SCA can clue you in to when an AED should be used. A person experiencing an SCA may suddenly collapse and lose consciousness. Or you may find the person unconscious and unable to respond when you call or shake him or her. The person may not be breathing or may have an abnormal breathing pattern. If you check, you usually can't find a pulse. The person's skin may become dark or blue from lack of oxygen. Also, the person may not move, or the person's movements may look like a seizure (spasms). An AED can check the victim's heart rhythm and determine whether an electric shock is needed to try to restore normal rhythm.

With modern AEDs, you need not worry about accidentally killing a victim because AEDs have a computer that is able to detect sinus activity and pulse. If either sinus activity (an electric signal being transmitted from the sinus node) or a heartbeat (pulse) is detected, the AED will not deliver a shock to the victim's heart.

The AED is attached to the chest of the victim by use of sticky pads with sensors, called electrodes. The electrodes transmit information about the person's heart rhythm to a computer in the AED. The computer analyzes the heart rhythm to determine whether an electric shock is indicated. If it is, the AED uses voice prompts to tell the operator when to give the shock and the electrodes deliver it.

Facility owners thinking about purchasing and installing AEDs in their facility should provide initial and ongoing training to first responders. It is important to properly maintain an AED and notify local emergency officials where they are located throughout the facility.

How to Use an Automated External Defibrillator

Before applying an AED to a victim, it is important to confirm that the victim can't respond. Shout at and shake the person to make sure he or she isn't sleeping. *Never* shake an infant or young child. Instead, pinch the child to try to wake the child up. Call 9-1-1 or have someone else call 9-1-1. If two rescuers are present, one can provide CPR while the other calls 9-1-1 and gets the AED.

Check the victim's pulse and breathing. If breathing and pulse are irregular, prepare to use the AED as quickly as possible. Death occurs within minutes in cases of SCA. If no one knows how long the person has been unconscious, or if an AED isn't readily available, do two minutes of CPR. Then use the AED, if you have one, to check the victim's condition.

After you use the AED, or if you don't have an AED, give CPR until emergency medical help arrives or until the person begins to move. Try to limit pauses in CPR. After two minutes of CPR, you can use the AED again to check the victim's heart rhythm and give another shock, if needed. If a shock isn't needed, continue CPR.

Before using an AED, check for puddles of water near the person who is unconscious. Move the victim to a dry area and stay away from wetness when delivering shocks because water conducts electricity. Turn on the AED's power. The AED will give you step-by-step instructions. You'll hear voice prompts and see prompts on a screen. Expose the victim's chest. If the victim's chest is wet, dry it because the sticky electrode pads will not adhere to a wet surface. Apply the electrodes to the victim's chest as pictured on the AED's instructions. Apply one pad on the right center of the victim's chest above the nipple. Place the other pad slightly below the other nipple and to the left of the ribcage. Make sure the electrodes have a good connection to the skin. If the victim has a lot of chest hair, you may have to trim it. AEDs generally are equipped with a kit containing scissors and/or a razor. Remove metal necklaces or underwire bras. The metal may conduct electricity and cause burns. You can cut the center of the bra and pull it away from the skin.

Check the person for implanted medical devices, such as a pacemaker or implantable cardioverter defibrillator. The outline of these devices is visible under the skin on the chest or abdomen, and the victim may be wearing a medical alert bracelet. Also check for body piercings. If any of these are found, move the electrode pads at least 1 inch away from the implanted devices or piercings so the electric current can flow freely between the pads. Ensure the wires from the electrodes are connected to the AED. Make sure no one is touching the victim, and then press the AED's "analyze" button. Stay clear while the machine checks the person's heart rhythm.

If a shock is needed, the AED will let you know when to deliver it. Stand clear of the victim and make sure all other personnel in the area are clear before you push the AED's "shock" button to initiate the shock current.

Start or resume CPR until emergency medical help arrives or until the victim starts moving. Stay with the victim until medical help arrives, and report all the information you know about what has happened.

What Risks Are Associated with Using an Automated External Defibrillator?

AEDs are safe to use. There are no reports of AEDs harming bystanders or users. Also, there are no reports of AEDs delivering inappropriate shocks.

Key Points about Automated External Defibrillators

1. An AED is a portable device that analyzes the heart rhythm and can administer an electric shock to the heart to paralyze the heart and try to restore normal heart function. AEDs are used to treat SCA.
2. SCA is a condition in which the heart suddenly and unexpectedly stops beating in normal rhythm. In this condition, no blood is flowing through the body to the brain and other vital organs.
3. In victims having SCA, an electric shock from an AED can paralyze the heart and relax it, so the heart's normal rhythm can be restored. Performing CPR on victims suffering SCA can improve their chances of survival.
4. AEDs are lightweight, battery-operated, portable devices that are easy to use. Each unit comes with instructions, and the device will even give you voice prompts to let you know if and when you should initiate a shock to the victim's heart.
5. Sticky pads with sensors (electrodes) are attached to the chest of the victim having SCA. The electrodes transmit information about the victim's heart rhythm to a computer in the AED.
6. The computer analyzes the heart rhythm to determine whether an electric shock is needed. If a shock is indicated, the AED uses voice prompts to tell you when to initiate the shock and the electrodes deliver it to the chest of the victim.
7. Learning how to use an AED and taking a CPR course should be mandatory for all electrical workers. However, if trained personnel aren't available, untrained personnel can use an AED to help save a victim's life.
8. AEDs are frequently found in locations with large numbers of people such as shopping malls, golf courses, airports, airplanes, casinos, hotels, sports venues, schools, churches, and manufacturing facilities; they can also be purchased for home use.
9. AEDs are very safe to use. There are no reports of AEDs harming bystanders or users, and there are no reports of AEDs delivering inappropriate shocks.

RESCUE TECHNIQUES

General Rescue Procedures

The first priority in any emergency is to remove living victims from the danger area if they cannot escape themselves. This procedure is called *rescue*. In some instances, the rescuer will be risking his or her life to rescue a victim. The decision to risk one's life is a personal one and cannot be regulated by any sort of standard procedures. In any case, good judgment must be exercised by the rescuer. Remember that becoming a second victim does not help anyone.

The American Red Cross method of rescue is shown in Table 8.14. Notice that the steps are virtually identical to those employed in preparing for first aid procedures. Rescue is, in fact, one of the preliminary steps in the performance of first aid.

After the victim has been removed from the hazardous area (in the case of an electrical accident, *do not touch the victim* until electric circuits have been de-energized and/or the victim has been removed from contact of the energized circuits) and after first aid has been administered, the accident area should be secured and made safe for other persons. *Do not disturb* the accident area beyond what

TABLE 8.14 Primary Rescue Steps

1. Survey the scene.
   a. Develop an action plan.
2. Primary survey.
   a. Locate the victim.
   b. Assess the victim's condition.
   c. Try to arouse the victim.
   d. Move the victim to a safe place. *Caution*—do not jostle the victim any more than is necessary when moving. See later sections for a discussion of proper movement techniques.
   e. Apply first aid as soon as the victim is in a safe location.
3. Call for emergency medical assistance.

is necessary to protect other persons. An accident investigation should be carried out by qualified personnel to determine the nature and cause of the accident and what corrective actions need to be taken to prevent the accident from occurring again.

The following outline details the generalized procedures defined by Table 8.14. This outline was taken from the AVO Multi-Amp Institute textbook titled *High Voltage Rescue Techniques.*

1. Survey the scene. A quick survey (1 min or less—time is critical) to enable the rescuer to develop a mental plan. Consider the following:
   a. Anticipating conditions
      (1) Probable life-threatening hazards
      (2) Observed location of victim
      (3) Observed condition of victim
   b. Facts of the situation
      (1) Time and conditions
      (2) Situations—What happened?
      (3) Contributing factors
   c. Weighing facts
      (1) Rescue requirements
      (2) Accomplish rescue alone?
      (3) Equipment and manpower needed/ available
      (4) Additional assistance now or later
      (5) If electrical contact, get de-energized
   d. Determining procedures
      (1) Immediate action
      (2) Course of action
   e. Initial action decision; May be made before a complete survey; May determine a future course of action.
2. Primary survey plan activated. From the initial decision and action until the victim is being transported to medical facilities, the following would be considered:
   a. Victim location
      (1) Accessibility of victim
      (2) Electric circuit contact
      (3) Remove circuit contacts
      (4) Try to arouse victim

(5) Help needed/called

(6) Assess condition of victim (ABCs plus broken bones, burns, twisted torso, etc.)

b. Emergency aid required

(1) What and where to give emergency aid?

(2) How to move victim to safety

(3) Emergency aid given

3. Call emergency medical service.

a. How to move victim to medical aid

b. Continue emergency aid as required

4. Secondary survey.

a. Secure the area

(1) Do not damage or disturb physical evidence

(2) Isolate hazards

(3) Keep untrained personnel away

b. What happened

(1) Ask victim

(2) Ask witnesses

(3) Examine physical evidence

(4) Photographs, sketches

(5) Preserve physical evidence

c. Reports

(1) Written records

(2) Notification

d. Follow up

(1) Reports distributed

(2) Action taken to prevent future similar accidents

Note: The action taken during the actual physical rescue must be started and accomplished very rapidly so that the victim is out of further danger and emergency aid may begin quickly. See Fig. 8.3. Accident investigation techniques are discussed later in this chapter.

Elevated Rescue

The physical techniques that are used for any type of rescue should be learned and practiced in a controlled environment under the direction of experienced, qualified training personnel.

Figure 8.8 graphically depicts a generally accepted method for performing an elevated, pole-top rescue. Note that this rescue technique requires a skilled line person with climbing ability. Do not attempt such a rescue unless you have had the requisite training. All rescue techniques should be accomplished using the basic rescue approach outline, which is illustrated in the previous section.

Confined-Space Rescue

Confined-space rescue is made more complex by two basic problems. First, the confined space may become filled with gases that are not breathable or, worse, are toxic and therefore hazardous to human life. Second, the confined space does not allow for free motion of rescue personnel. This means that rescue workers may not be able to easily get a grip on the victim.

YOU MAY HAVE TO HELP A MAN ON A POLE REACH THE GROUND SAFELY WHEN HE –

- **BECOMES ILL**
- **IS INJURED**
- **LOSES CONSCIOUSNESS**

YOU MUST KNOW –

- **WHEN HE NEEDS HELP**
- **WHEN AND WHY TIME IS CRITICAL**
- **THE APPROVED METHOD OF LOWERING**

Evaluate **THE SITUATION**

CALL TO MAN ON POLE

IF HE DOES NOT ANSWER OR APPEARS STUNNED OR DAZED

- **PREPARE TO GO TO HIS AID**

TIME IS EXTREMELY IMPORTANT

Climb **TO RESCUE POSITION**

- **CLIMB CAREFULLY**

POSITION YOURSELF –

- **TO ENSURE YOUR SAFETY**
- **TO CLEAR THE INJURED FROM HAZARD**
- **TO DETERMINE THE INJURED'S CONDITION**
- **TO RENDER AID AS REQUIRED**
- **TO LOWER INJURED, IF NECESSARY**

THE BEST POSITION WILL USUALLY BE – SLIGHTLY ABOVE THE INJURED

BASIC STEPS IN

POLE TOP RESCUE

Evaluate **THE SITUATION**
Provide **FOR YOUR PROTECTION**
Climb **TO RESCUE POSITION**
Determine **INJURED'S CONDITION**

Then
IF NECESSARY –

- **GIVE FIRST AID**
- **LOWER INJURED**
- **GIVE FOLLOW-UP CARE**
- **CALL FOR HELP**

Provide
FOR YOUR PROTECTION
YOUR SAFETY IS VITAL TO THE RESCUE

– PERSONAL TOOLS AND RUBBER GLOVES (RUBBER SLEEVES, IF REQUIRED)

Check

- ✔ **EXTRA RUBBER GOODS?**
- ✔ **LIVE LINE TOOLS?**
- ✔ **PHYSICAL CONDITION OF POLE?**
  - ✔ **DAMAGED CONDUCTORS, EQUIPMENT?**
  - ✔ **FIRE ON POLE?**
  - ✔ **BROKEN POLE?**
- ✔ **HAND LINE ON POLE AND IN GOOD CONDITION?**

Determine
THE INJURED'S CONDITION

HE MAY BE . . .

- **CONSCIOUS**
- **UNCONSCIOUS BUT BREATHING**
- **UNCONSCIOUS NOT BREATHING**
- **UNCONSCIOUS NOT BREATHING AND HEART STOPPED**

FIGURE 8.8 Illustrated method for elevated, pole-top rescue.

IF THE INJURED IS

CONSCIOUS

- TIME MAY NO LONGER BE CRITICAL
- GIVE NECESSARY FIRST AID ON POLE
- REASSURE THE INJURED
- HELP HIM DESCEND POLE
- GIVE FIRST AID ON GROUND
- CALL FOR HELP, IF NECESSARY

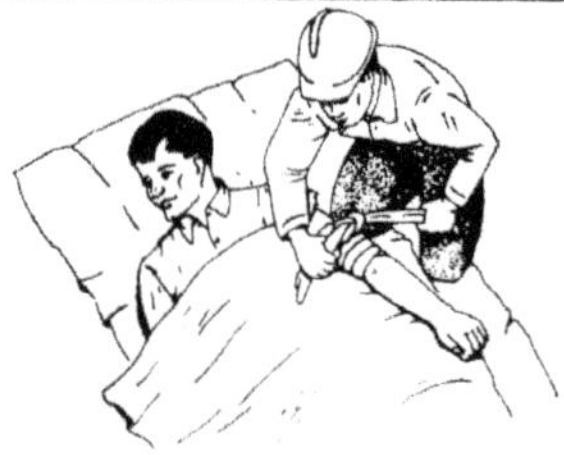

If the injured is not breathing, he/she should be tied off and lowered to the ground for necessary first aid or CPR.

IF THE INJURED IS

UNCONSCIOUS BUT BREATHING

- WATCH HIM CLOSELY IN CASE BREATHING STOPS
- LOWER HIM TO GROUND
- GIVE FIRST AID ON GROUND
- CALL FOR HELP

EQUIPMENT NEEDED

- 1/2 INCH HAND LINE

PROCEDURE . . .

- *Position* HAND LINE
- *Tie* INJURED
- *Remove* SLACK IN HAND LINE
- *Take* FIRM GRIP ON FALL LINE
- *Cut* INJURED'S SAFETY STRAP

Lower INJURED

RESCUER

Position **HAND LINE**

OVER ARM OR OTHER PART OF STRUCTURE

- POSITION LINE FOR CLEAR PATH TO GROUND

Note USUALLY BEST 2 OR 3 FEET FROM POLE

FIGURE 8.8 (*Continued*)

SHORT END OF LINE IS WRAPPED AROUND FALL LINE TWICE (TWO WRAPS AROUND FALL LINE).

RESCUER TIES HAND LINE AROUND VICTIM'S CHEST USING THREE HALF HITCHES.

Tie INJURED–

PASS LINE AROUND INJURED'S CHEST

TIE THREE HALF HITCHES

- **KNOT IN FRONT**
- **NEAR ONE ARM PIT**
- **HIGH ON CHEST**
- **SNUG KNOT**

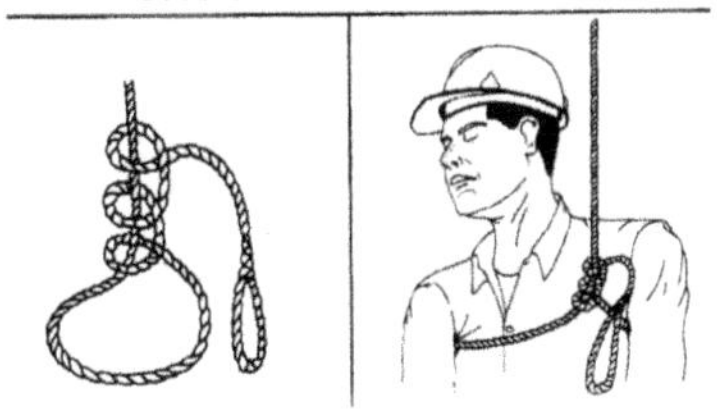

FIGURE 8.8 (*Continued*)

Remove SLACK IN HAND LINE

- **ONE RESCUER–REMOVES SLACK ON POLE**
- **TWO RESCUERS–MAN ON GROUND REMOVES SLACK**

IMPORTANT
GIVE 5 OR 6 QUICK BREATHS . . . IF NECESSARY . . .

THEN

Take FIRM GRIP ON FALL LINE

ONE RESCUER–HOLDS FALL LINE WITH ONE HAND

TWO RESCUERS–MAN ON GROUND HOLDS FALL LINE

Cut INJURED'S SAFETY STRAP

CUT STRAP ON SIDE OPPOSITE DESIRED SWING

Caution

DO NOT CUT YOUR OWN SAFETY STRAP OR THE HAND LINE

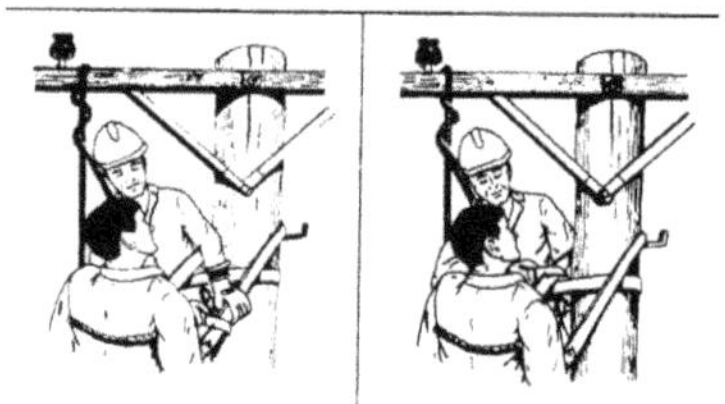

Lower INJURED

ONE RESCUER

- **GUIDE LOAD LINE WITH ONE HAND**
- **CONTROL RATE OF DESCENT WITH THE OTHER HAND**

TWO RESCUERS

- **MAN ON THE POLE GUIDES THE LOAD LINE**
- **MAN ON THE GROUND CONTROLS RATE OF DESCENT**

IF A CONSCIOUS MAN IS BEING ASSISTED IN CLIMBING DOWN . . . THE ONLY DIFFERENCE IS THAT ENOUGH SLACK IS FED INTO THE LINE TO PERMIT HIM CLIMBING FREEDOM

FIGURE 8.8 (*Continued*)

ONE MAN RESCUE | **TWO OR MORE MAN RESCUE**

RESCUES DIFFER ONLY IN CONTROL OF THE FALL LINE

Remember

THE APPROVED METHOD OF LOWERING AN INJURED MAN IS . . .

- *Position* **HAND LINE**
- *Tie* **INJURED**
- *Remove* **SLACK IN HAND LINE**
- *Take* **FIRM GRIP ON FALL LINE**
- *Cut* **INJURED'S SAFETY STRAP**
- *Lower* **INJURED**

IF POLE DOES NOT HAVE CROSSARM, RESCUER PLACES HAND LINE OVER FIBERGLASS BRACKET INSULATOR SUPPORT, OR OTHER SUBSTANTIAL PIECE OF EQUIPMENT SUCH AS A SECONDARY RACK, NEUTRAL BRACKET, OR GUY WIRE ATTACHMENT, STRONG ENOUGH TO SUPPORT THE WEIGHT OF THE INJURED. SHORT END OF LINE IS WRAPPED AROUND FALL LINE TWICE AND TIED AROUND VICTIM'S CHEST USING THREE HALF HITCHES. INJURED'S SAFETY STRAP IS CUT AND RESCUER LOWERS INJURED TO THE GROUND.

LINE MUST BE REMOVED AND VICTIM EASED ON TO GROUND.

FIGURE 8.8 (*Continued*)

LAY VICTIM ON HIS BACK AND OBSERVE IF VICTIM IS CONSCIOUS. IF VICTIM IS CONSCIOUS, TIME MAY NO LONGER BE CRITICAL. GIVE NECESSARY FIRST AID. CALL FOR HELP. IF VICTIM IS UNCONSCIOUS AND NOT BREATHING AND/OR HAS AN ERRATIC OR NO PULSE, PERFORM CPR AS DESCRIBED PREVIOUSLY IN THIS CHAPTER.

BUCKET TRUCK RESCUE TIME IS CRITICAL.

EQUIP PORTION OF INSULATED BOOM OF TRUCK WITH ROPE BLOCKS DESIGNED FOR HOT-LINE WORK.

FIGURE 8.8 (*Continued*)

STRAP IS PLACED AROUND INSULATED BOOM APPROXIMATELY 10 FEET FROM BUCKET TO SUPPORT ROPE BLOCKS. BLOCKS ARE HELD TAUT ON BOOM FROM STRAP TO TOP OF BOOM.

RESCUER ON THE GROUND EVALUATES THE CONDITIONS WHEN AN EMERGENCY ARISES. THE BUCKET IS LOWERED USING THE LOWER CONTROLS. OBSTACLES IN THE PATH OF THE BUCKET MUST BE AVOIDED.

FIGURE 8.8 (*Continued*)

HOOK ON ROPE BLOCKS IS ENGAGED IN A RING ON THE LINEMAN'S SAFETY STRAP. SAFETY STRAP IS RELEASED FROM BOOM OF TRUCK.

ROPE BLOCKS ARE DRAWN TAUT BY RESCUER ON THE GROUND. UNCONSCIOUS VICTIM IS RAISED OUT OF BUCKET WITH ROPE BLOCKS.

FIGURE 8.8 (*Continued*)

RESCUER EASES VICTIM ONTO THE GROUND. CARE SHOULD BE TAKEN TO PROTECT THE INJURED VICTIM FROM FURTHER INJURY.

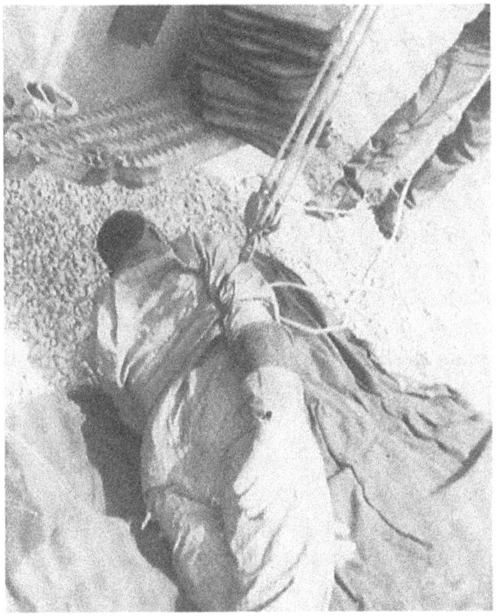

RELEASE ROPE BLOCKS FROM VICTIM.

FIGURE 8.8 (*Continued*)

BASKET OR BUCKET ON AERIAL DEVICE MAY BE CONSTRUCTED TO TILT AFTER BEING RELEASED, ELIMINATING THE NEED FOR SPECIAL RIGGING TO REMOVE AN INJURED PERSON FROM THE BASKET OR BUCKET.

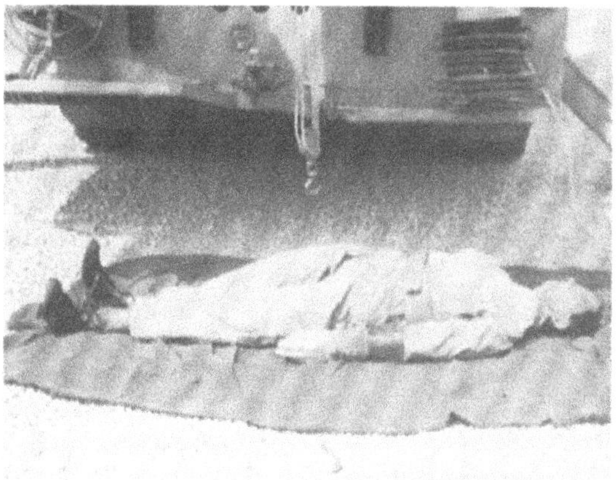

LAY VICTIM ON HIS BACK AND OBSERVE IF VICTIM IS CONSCIOUS. IF VICTIM IS CONSCIOUS, TIME MAY NO LONGER BE CRITICAL. GIVE NECESSARY FIRST AID. CALL FOR HELP. IF VICTIM IS UNCONSCIOUS AND NOT BREATHING AND/OR HAS AN ERRATIC OR NO PULSE, PERFORM CPR AS DESCRIBED PREVIOUSLY IN THIS CHAPTER.

FIGURE 8.8 (*Continued*)

Atmosphere Checking in Confined-Space Rescues. Before the rescue is attempted, the atmosphere must be tested for the presence of adequate supplies of oxygen, excessive quantities of toxic gases, and excessive quantities of combustible gases. Figure 8.9 shows a typical monitoring setup for an underground confined-space area. If the rescuer must descend into a confined space that does not have adequate oxygen concentrations or has excessive quantities of toxic or combustible gases, the rescuer must wear respiration gear.

Oxygen levels must be no less than 19.5 percent and no more than 23.5 percent. Table 8.15 lists the potential effects of insufficient oxygen. The permissible levels of toxic gas vary depending on the type of gas and the individual. Generally, toxic gases should be kept to levels that are less than 50 parts per million (ppm) by volume. Tables 8.16 and 8.17 list the potential effects that carbon monoxide (CO) and hydrogen sulfide (H_2S) will have.

Securing the Victim to the Rescue Line. The method used to secure a victim to the line depends upon the logistics of the given rescue. The nature of the injuries and the space available for the rescue both affect the method used. The simplest way is to tie a hand line around the victim using three half hitches as shown in Fig. 8.10. Figure 8.11 shows a preferred method using a padded harness designed

FIGURE 8.9 Vault atmosphere monitoring.

TABLE 8.15 Potential Effects of an Oxygen-Deficient Atmosphere

Oxygen content (% by volume)	Effects and symptoms* (at atmospheric pressure)
19.5	Minimum permissible oxygen level for normal functioning.
15–19	Decreased ability to work strenuously. May impair coordination and may induce early symptoms in persons with coronary, pulmonary, or circulatory problems.
12–14	Respiration increases in exertion; pulse increases, impaired coordination, perception, judgment.
10–12	Respiration further increases in rate and depth, poor judgment, lips blue.
8–10	Mental failure, fainting, unconsciousness, ashen face, blueness of lips, nausea, and vomiting.
6–8	Eight min, 100% fatal; 6 min, 50% fatal; 4–5 min, recovery with treatment.
4–6	Coma in 40 s, convulsions, respiration ceases, death.

* The effects and symptoms at a given oxygen content are guidelines and will vary with the individual's state of health and physical activity.

specifically for such rescue efforts. If space allows or if the victim's injuries require it, a full body rope harness may be rigged as shown in Fig. 8.12. Figure 8.13 shows a victim being lifted using the body rope harness. Sometimes victims' injuries may be so severe that they must be securely immobilized before being removed from the accident area. Both of the devices shown in Fig. 8.14 may be used.

Retrieving the Victim. After the victim has been secured to the rescue line, he or she must be raised from the confined space to an open area where treatment may be administered. A variety of different methods have been developed to raise victims. Figures 8.15 through 8.19 show various types of rigs that may be used as pulley points to retrieve victims.

The key to a successful victim retrieval is teamwork and adequate planning. Remember that human beings can weigh as much as 300 or 400 lb. Also remember that unconscious victims cannot assist in their own retrieval. Always be certain that the rigging is strong enough.

TABLE 8.16 Potential Effects of Carbon Monoxide Exposure

PPM*	Effects and symptoms†	Time
35	Permissible exposure level limit	8 h
200	Slight headache	3 h
400	Headache, discomfort	2 h
600	Headache, discomfort	1 h
1000–2000	Confusion, headache, nausea	2 h
1000–2000	Tendency to stagger	1¼ h
1000–2000	Slight palpitation of heart	30 min
2000–2500	Unconsciousness	30 min
4000	Fatal	Less than 1 h

* PPM—parts per million; volume measurement of gas concentration.
† The effects and symptoms at a given exposure are guidelines and will vary with the individual's state of health and physical activity.

TABLE 8.17 Potential Effects of Hydrogen Sulfide Exposure

PPM	Effects and symptoms	Time
10	Permissible exposure level limit	8 h
50–100	Mild eye irritation, mild respiratory irritation	1 h
200–300	Marked eye irritation, marked respiratory irritation	1 h
500–700	Unconsciousness, death	¼–1 h
1000 or more	Unconsciousness, death	Minutes

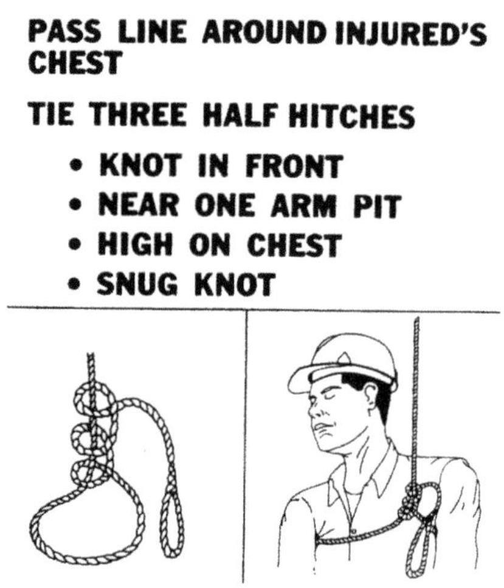

FIGURE 8.10 Securing the victim using a hand line.

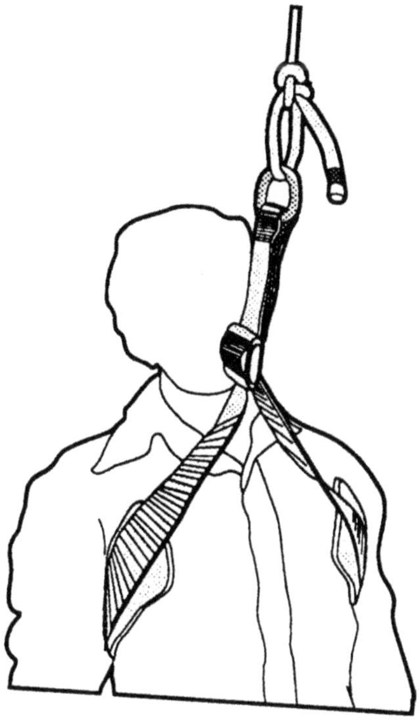

FIGURE 8.11 Using a padded safety harness to secure the rescue line to the victim. (*Courtesy AVO Training Institute, Inc.*)

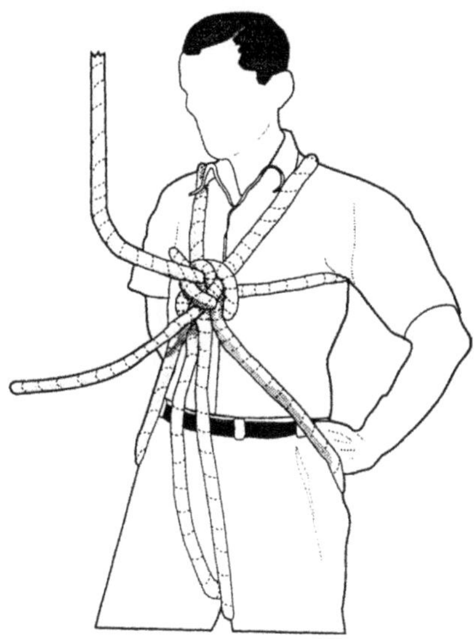

FIGURE 8.12 Full body rope harness. (*Courtesy AVO Training Institute, Inc.*)

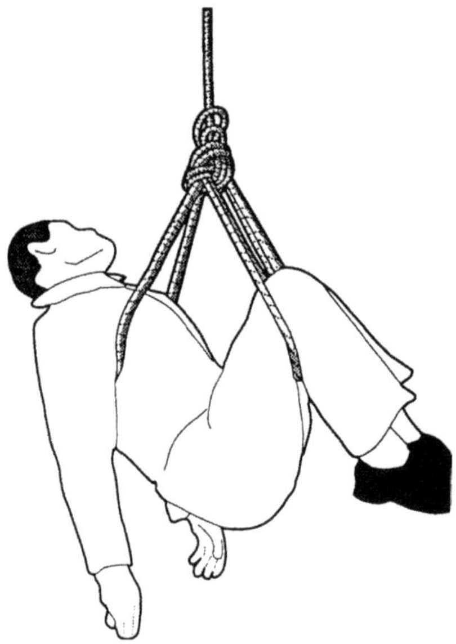

FIGURE 8.13 Full body rope harness being used to lift the victim. (*Courtesy AVO Training Institute, Inc.*)

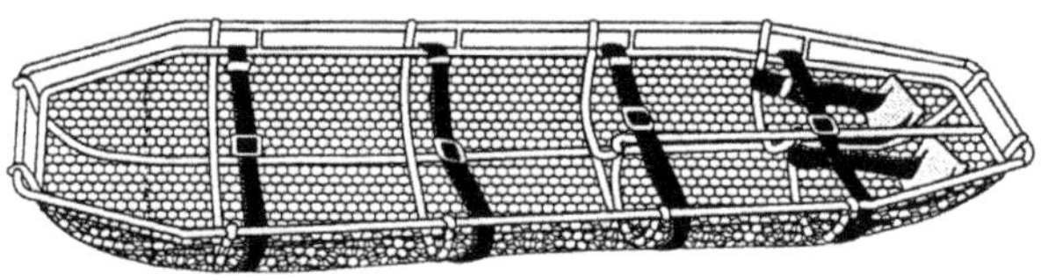

STOKES NAVY STRETCHER

(a)

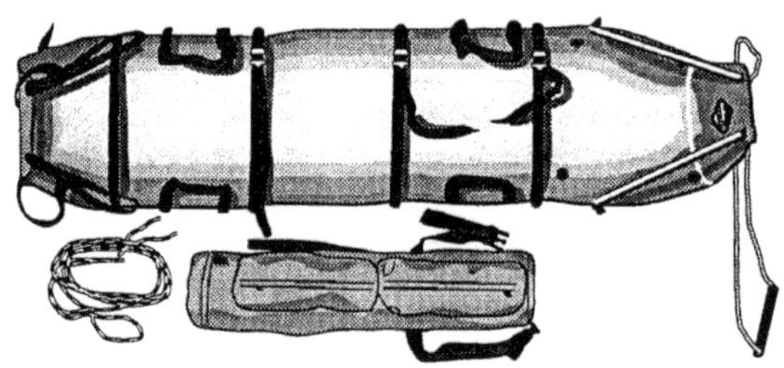

SKED SYSTEM MODEL SK-200

(b)

FIGURE 8.14 (*a*) Stokes navy stretcher. (*b*) SKED system model SK-200. (*Courtesy AVO Training Institute, Inc.*)

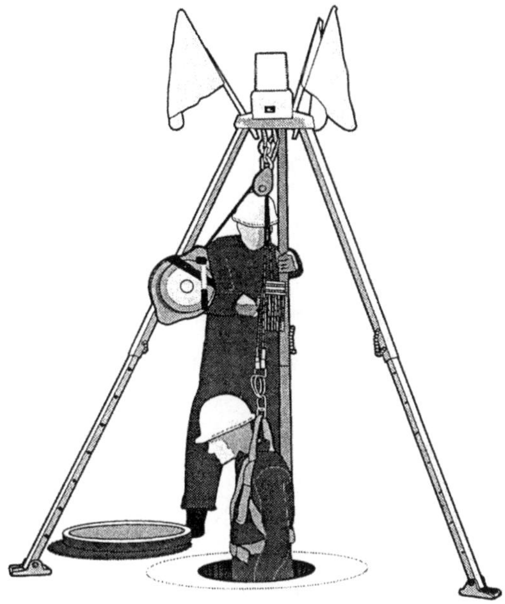

FIGURE 8.15 Tripod hoist operation. (*Courtesy AVO Training Institute, Inc.*)

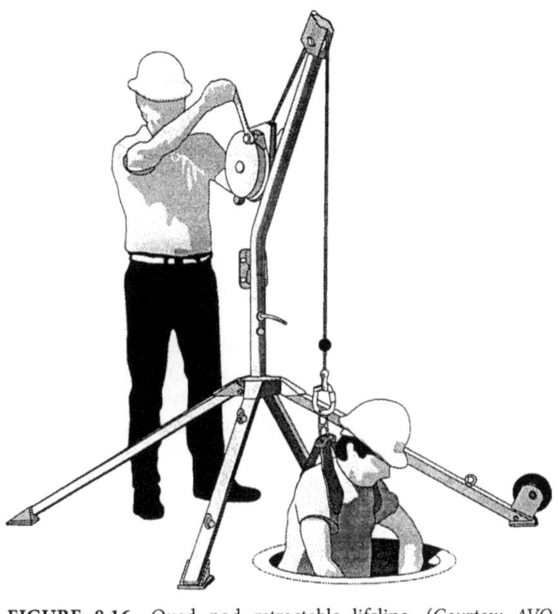

FIGURE 8.16 Quad pod retractable lifeline. (*Courtesy AVO Training Institute, Inc.*)

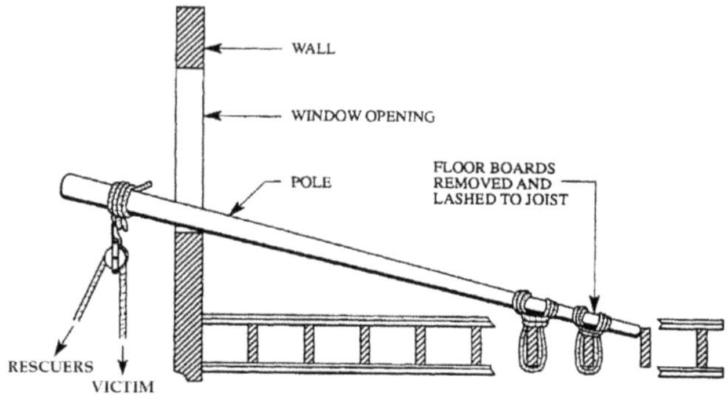

FIGURE 8.17 Gin pole hoist. (*Courtesy AVO Training Institute, Inc.*)

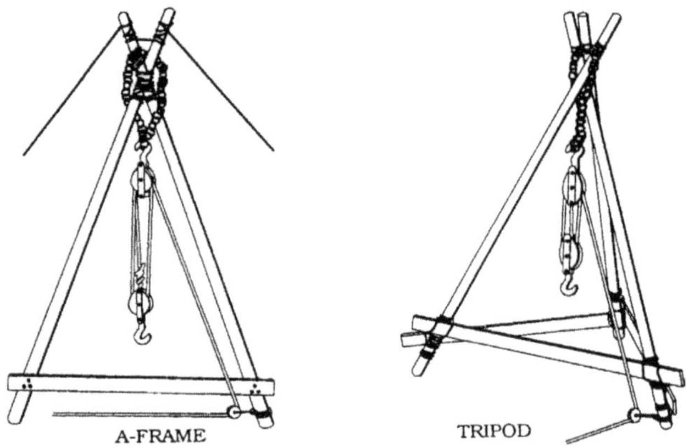

FIGURE 8.18 Rope-tied a-frame and tripod hoists. (*Courtesy AVO Training Institute, Inc.*)

Ground-Level Rescue

The Problem. Ground-level victims do not require rescue as often as elevated or confined-space victims. Occasionally, however, victims do need retrieval, especially from metal-clad switchgear assemblies. Walk-in metal-clad switchgear has circuit breaker cubicles that may represent a limited space problem. They also have doorways that must be traversed. In addition to metal-clad gear problems, victims may be endangered by fire, chemical release, or any number of other similar hazards. Regardless of where the rescue is taking place, the general rescue procedures should be the starting point for all rescues.

Retrieving the Victim. Since moving a victim can induce additional injuries or aggravate existing injuries, always make certain that moving the victim is absolutely necessary. If the victim must be moved, immobilizing him or her using a stretcher or other technique is preferred. However, if time is critical, one of the following carries may be employed.

The method used to move the victim depends on the nature and severity of the hazard to which he or she is exposed. For example, Figs. 8.20 and 8.21 illustrate two methods that may be used to remove a victim from a fire area. In these types of situations, smoke and heat may require that the rescuer stay low to the ground.

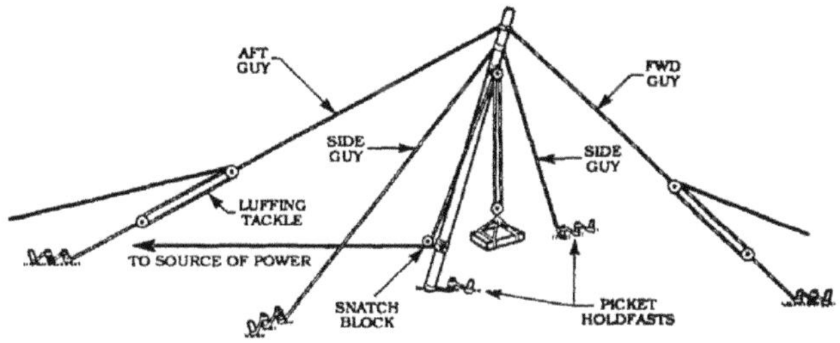

GIN POLE

LADDER GIN POLE

FIGURE 8.19 Gin pole hoists. (*Courtesy AVO Training Institute, Inc.*)

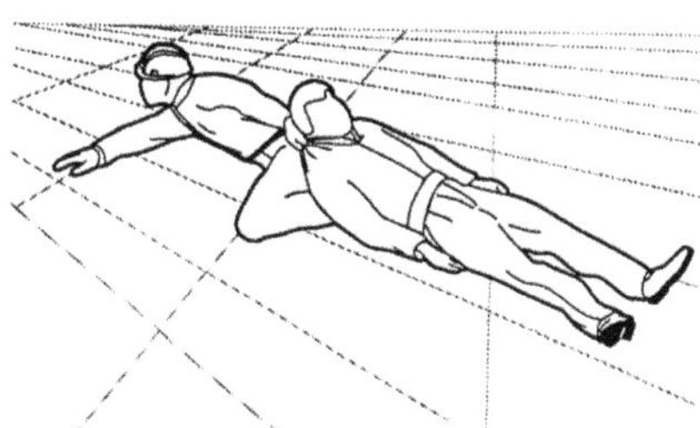

FIGURE 8.20 Clothes drag. (*Courtesy AVO Training Institute, Inc.*)

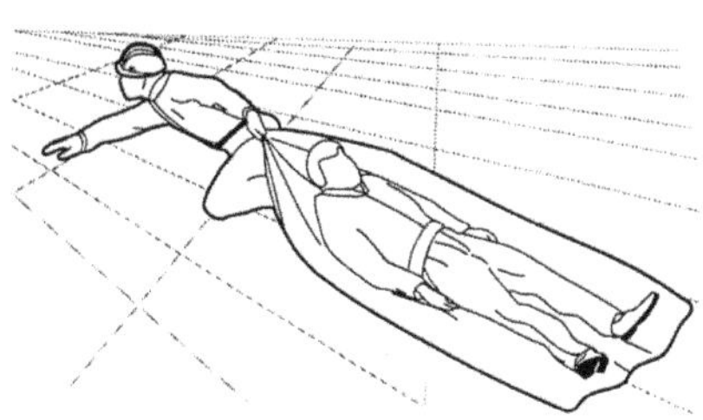

FIGURE 8.21 Blanket drag. (*Courtesy AVO Training Institute, Inc.*)

FIGURE 8.22 Pack-strap carry (conscious victim). (*Courtesy AVO Training Institute, Inc.*)

FIGURE 8.23 Firefighter's carry. (*Courtesy AVO Training Institute, Inc.*)

If only one person is available for the rescue, and the victim is conscious and mobile, the pack-strap carry (Fig. 8.22) may be employed. The firefighter's carry (Fig. 8.23) may be used if the victim is not conscious.

If two people are available for the rescue, any of the carries shown in Figs. 8.24 through 8.26 may be used. The carry shown in Fig. 8.25 generally requires that the victim be conscious.

FIGURE 8.24 Carry by extremities. *(Courtesy AVO Training Institute, Inc.)*

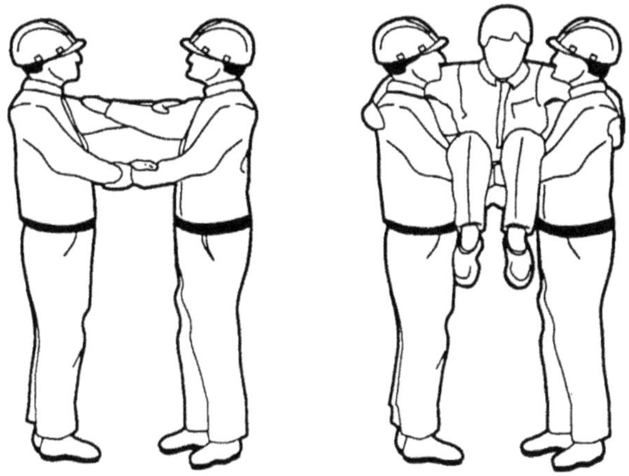

FIGURE 8.25 Two-person seat carry. *(Courtesy AVO Training Institute, Inc.)*

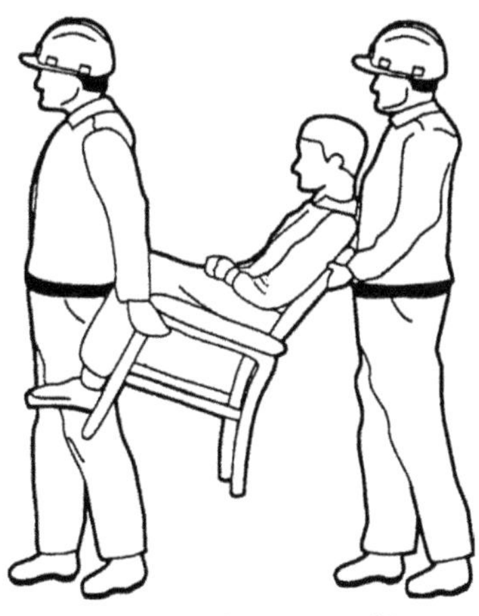

FIGURE 8.26 Chair carry. *(Courtesy AVO Training Institute, Inc.)*

ACCIDENT INVESTIGATION

Purpose

When accidents occur, they must be investigated thoroughly. An investigation serves two major purposes as shown in Table 8.18. Note that accident investigations should be performed only by qualified professionals who are experienced in such matters. Litigation can often ensue and result in corporate, personal, or product liability; errors and omissions; insurance subrogation; or other claims.

Accident Prevention. Accidents occur for a reason, and sometimes the reason is not obvious. A thorough investigation can reveal underlying causes. Employers must then act to eliminate those causes, so such accidents will not occur again.

Litigation. Modern society is becoming increasingly litigious. Evidence gathered during an accident investigation can be used in court to either defend or prosecute a lawsuit. Another aspect of our litigious society is the tendency of companies and individuals to "keep everything secret until the attorney approves it." Sadly, this approach may seem necessary to prevent frivolous lawsuits and/or to protect assets. It should be remembered by all parties involved in an accident investigation that the understanding of the accident and prevention of a recurrence should be the most important issues. Ultimately, the value of a single human life exceeds any corporate asset.

General Rules

Several general rules for carrying out accident investigations are listed in Table 8.19. Many investigations have been hampered by well-meaning, inexperienced individuals who remove or alter evidence in the interests of cleaning up the area. Remember that only absolutely necessary modifications should be made at the accident scene, and even these should be documented, preferably with photographs and/or videotapes.

One of the major mistakes made by employers is to wait for the lawsuit before an accident investigation begins. Employers owe it to themselves and their employees to investigate accidents completely and thoroughly, as soon after the accident as possible. This is especially true of witness interviews. Time quickly alters or destroys memories of traumatic events; therefore, people must be interviewed immediately.

Whenever possible, *retain an expert.* Forensic engineering firms specialize in the investigation and analysis of accidents. Such organizations have a wealth of factual information available to them.

TABLE 8.18 Reasons for Performing Accident Investigations

- Future accidents may be prevented by elimination of accident causes.
- Evidence gathered during the investigation may be required in litigation.

TABLE 8.19 General Rules for Performance of an Accident Investigation

- Investigations should be performed only by qualified, experienced personnel.
- Until the investigation is complete, the accident scene should be modified only to the extent necessary to eliminate immediate hazards to personnel.
- No evidence should be destroyed or thrown away for at least 5 to 10 years after the accident.
- Photograph everything from many different angles.
- Document everything that is observed or taken in interviews.
- Perform the investigation as soon as possible. Remember that human memory is volatile.

TABLE 8.20 Steps for an Accident Site Survey

- Visually inspect the area, making notes about general observations.
- Photograph or videotape the scene.
- Make a detailed inspection and develop a sketch of the entire scene.

They can determine causes in cases where inexperienced investigators can only guess. For information on forensic consultants, contact the American Academy of Forensic Sciences or the Canadian Society of Forensic Science.

Data Gathering

Site Survey. The gathering of data should start with a survey of the accident scene. If the basic rules have been followed, the scene will be virtually unchanged. The investigator should follow the steps outlined in Table 8.20.

The first visual inspection can produce amazing insights into the nature and causes of the accident. Following are some things to look for:

- Scorch and burn marks
- Damaged equipment
- Evidence of safety equipment
- Position of victim and or objects

The photographs or videotape, preferably both, can permanently store vital evidence. Because of the two-dimensional nature of photography, shots should be taken from all angles including from above to get a complete perspective. A scale reference can be included in the scene. An audio or written record should be made to clearly define the nature of each photograph, where it was taken, and any specific information that should be noted.

The final detailed inspection and sketch can be handmade, clearly labeled, and dimensioned. Dimensions are particularly important since they may prove or disprove the possibility of any given event during the accident. The sketch should also reference photographs or videotape evidence that was taken.

Witness Interviews. Witness interviews help to establish what, how, where, and when. However, the investigator must remember that of all the evidence he or she will gather, eyewitness accounts are the least reliable. Human beings tend to be influenced by their own personal backgrounds and prejudices. Within a few hours, or even minutes, of the accident, eyewitnesses' memories start to modify. Witnesses should be interviewed as soon as possible and on audiotape if possible.

The cautions in the previous paragraph are especially true of the accident victim. The human mind will literally shut off to prevent the agony associated with an accident. In case after case, victims remember nothing from several hours or even days before and after the accident. Information supplied by a victim should always be viewed with skepticism. The more traumatic the injury, the less credible will be the victim's statement.

When interviewing witnesses, the following points should be followed:

- When possible, conduct the interview at the accident site. The familiar surroundings may help to arouse memories that would be lost.
- Prepare a series of questions to be asked before the interview, and start with these questions. If the witness seems more comfortable with a freewheeling statement, allow it; however, avoid rambling and nonrelevant statements since these will only cloud the results.
- Try to keep the witness focused on the facts of the case. Blame, fault, and other such subjective conclusions should be left to the analysis stage of the investigation.

- When the interview is completed, the interview notes should be carefully stored and dated. If you make a summary of the interview, it should be stored with the notes; however, it should not replace the interview notes.
- Arrange for a follow-up interview to deal with questionable or conflicting accounts.

Physical Evidence. One of the most frequent mistakes made by inexperienced investigators is the alteration, destruction, or loss of physical evidence. A scorched or melted wire, a bit of flaked insulation, a piece of torn cloth, or a damaged circuit breaker can all be extremely important evidence.

Physical evidence should be documented in the preliminary stages of the investigation by the use of photographs, sketches, and written notes. After it has been documented, it should be carefully marked and stored. Physical evidence should be kept for a minimum of 5 years and possibly as long as 10 years depending upon the circumstances of the accident.

Accident Analysis

After the evidence has been gathered and the interviews have been taken, the analysis phase begins. The actual steps taken in this portion depend to a large extent on the specifics of the accident itself. However, at least two major steps may be involved.

Experimentation. Actual experimentation includes tests that are performed to evaluate or simulate certain conditions during an accident. For example, pork can be subjected to electric currents to evaluate the degree of burning that occurs for various currents, electric arcs can be staged with manikins to determine the degree of thermal burning that may occur, and air gaps can be checked to see how far an arc will jump in given circumstances. Experimentation is a detailed and sophisticated procedure and should be performed only by organizations with the necessary expertise.

Report. A report is developed based on a comprehensive analysis of all the evidence and experimental data that have been gathered. The report should contain, at a minimum, the sections listed in Table 8.21.

- *Overview.* This section should include an executive summary type of description of the events leading up to the accident, the accident itself, and the accident aftermath. Only very general information should be included, such as the number of injured persons, the basic cause of the injuries, and the generalized suggestions for preventing such a problem in the future.
- *Step-by-step description of the accident.* Here the investigator takes the reader through each step of the accident from beginning to end. All the physical evidence is discussed, and eyewitness accounts are integrated into this portion of the report. All relevant information should be included, such as the activities of the victim prior to the accident that may have affected his or her behavior, the specific actions of the victim during the accident, the safety equipment that was in use at the time, the actions of witnesses, the duration of the electrical contact, the points on the body that were contacted, and other such details.
- *Conclusion.* The investigator should provide an evaluation of the accident including such information as the steps that could be taken to prevent similar accidents in the future, changes in safety equipment or rules that may be indicated, changes in employee training that may be required,

TABLE 8.21 Major Elements of the Final Report of an Accident Investigation

• Overview
• Step-by-step description of the accident
• Conclusion
• Appendices

and/or changes in work procedures that may be indicated. Since this is an engineering report, the investigator should avoid comments about who may be responsible for the accident. Such issues are better left to the legal system.

- *Appendices.* All the collected data can be included for reference. Remember that others may draw different conclusions than that of the investigator and should, therefore, be given the advantage of being able to examine all the evidence that has been collected. In addition, the evidence must be present to support the conclusions drawn. *Do not* include the originals of photographs and videotapes. Only copies should be included. The originals should be kept in a secure location such as a locked file cabinet or a safe-deposit box. Of course, physical evidence should not be included except at a formal presentation of the report. However, photographs of physical evidence can be included.

Please note that any and all content of such a report will be subject to insurance contract directives; insurance adjuster's instructions; attorney-client privilege; powers granted to workers' compensation investigators; and powers granted to federal, state, and provincial agencies that have jurisdiction in the investigations.

CHAPTER 9
MEDICAL ASPECTS OF ELECTRICAL TRAUMA

INTRODUCTION

This handbook serves as a reference on how to be safe around electrical energy. In this chapter, the consequences of an electrical safety failure related to generation, transmission, distribution, or operation with commercial power are presented from a medical point of view.

As noted in the Preface, each year many people are killed or injured by electrical events. Individuals in these situations often are not aware of how electrical hazards might create injury or loss.

Statistical Survey

In 2016, 5,190 public and private industry workers were killed on the job, with more than 14 deaths occurring every day, according to the U.S. Occupational Safety Health Administration (Fig. 9.1).[1] In this same year, 154 fatalities were caused by exposure to electricity. Improvement in this number has not continued as earlier in the decade. This is shown on the dual-axis graph for the period 2011–2016 with the trend line of electrical fatalities compared to all causes of worker fatalities (Fig. 9.2).[2]

Private industry construction workers accounted for one out of every five worker fatalities in 2016. Not including highway collisions, electrocution was the third most frequent cause of private industry construction worker deaths, and accounted for 82 fatal injuries, or 8.2 percent (Fig. 9.3).[1] Notably relevant to safety management, based on 2012 experience, construction workers are typically employed in small business or enterprises with less than 500 employees (Fig. 9.4).[3]

Nonoccupational Electrical Trauma

Compared to workplace electrical trauma, nonoccupational electrical injury and fatality statistics are not collected routinely, limiting injury and fatality estimates availability. The U.S. Consumer Product Safety Commission (CPSC) has published reports that offer a perspective.

For the period 2013–2015, the CPSC report on residential fire loss estimates considers U.S. national estimates of fires, deaths, injuries, and property losses from unintentional fires. An estimated annual average of 9,600 fires was attributable to electrical equipment, or about 2.7 percent of the estimated

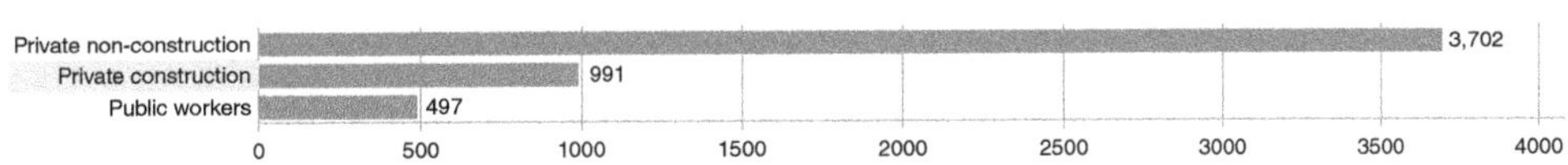

FIGURE 9.1 U.S. occupational fatalities, 2016. In 2016, a total of 5,190 workers were killed on the job. Of the 4,693 worker fatalities in private industry, 991 or 21% were in construction.[1]

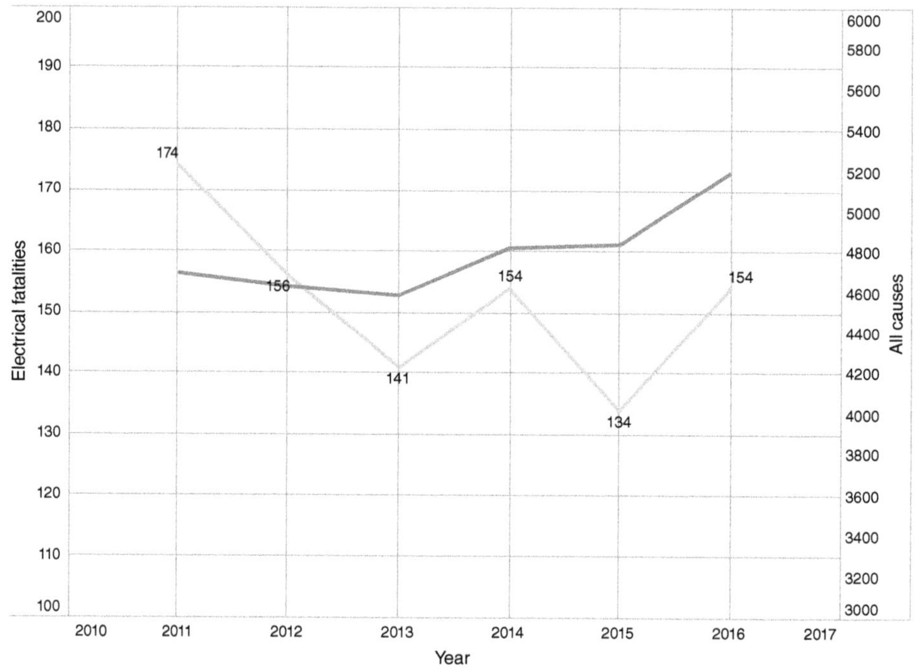

FIGURE 9.2 Occupational electrical and all causes U.S. occupational fatalities, 2011–2016. Progress in reduction of occupational electrical fatalities has slowed mid-decade, as illustrated on this dual-axis graph illustrating occupational electrical fatalities compared to all causes of occupational fatalities for the period 2011–2016.[2] (*Data adapted from Table 2 of the "National Census of Fatal Occupational Injuries in 2016" News Release.*)

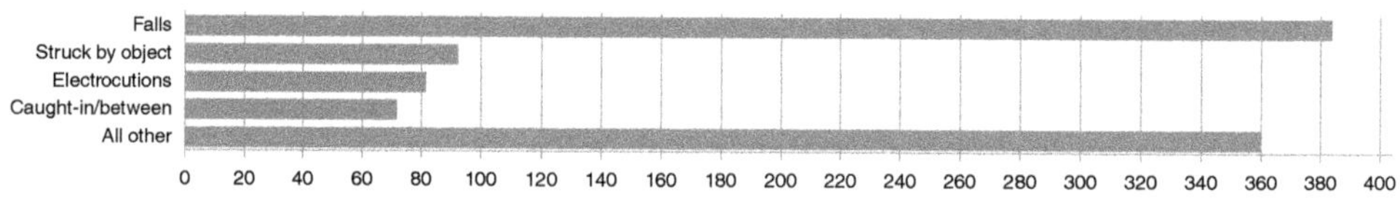

FIGURE 9.3 Leading causes of construction worker fatalities, 2016. Excluding highway collisions, the leading causes of private industry worker deaths in construction were as follows: falls—38.7%; —9.4%; electrocutions—8.2%; and caught-in/between–7.3%. Together these "top 4" accounted for 631 private construction fatalities. All other causes combined accounted for 36.3% or 360 private construction fatalities.[1]

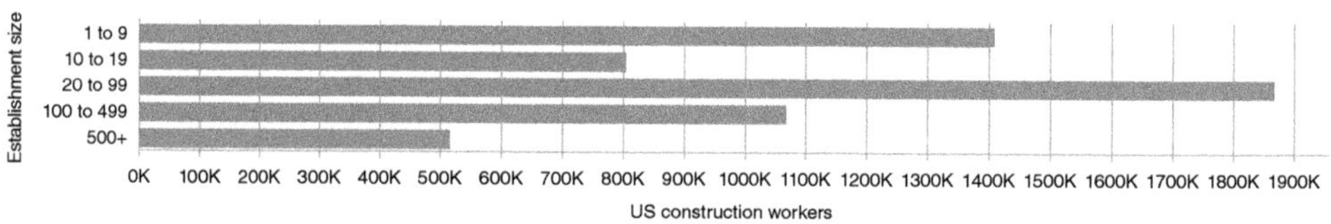

FIGURE 9.4 Establishment size and U.S. construction workers, 2012. Construction workers are typically employed in enterprises with less than 500 employees.[3]

annual average number of residential fires for this period (Fig. 9.5).[4] The associated annual average death estimate was 130 deaths. The annual average injury estimate was 450.

For the period 2004–2013, the CPSC report on electrocutions associated with consumer products summarized reported causes of fatal electrical injury (Fig. 9.6).[5] Large appliances were the most frequently identified, followed by small appliances, power tools, lighting equipment, and ladders.

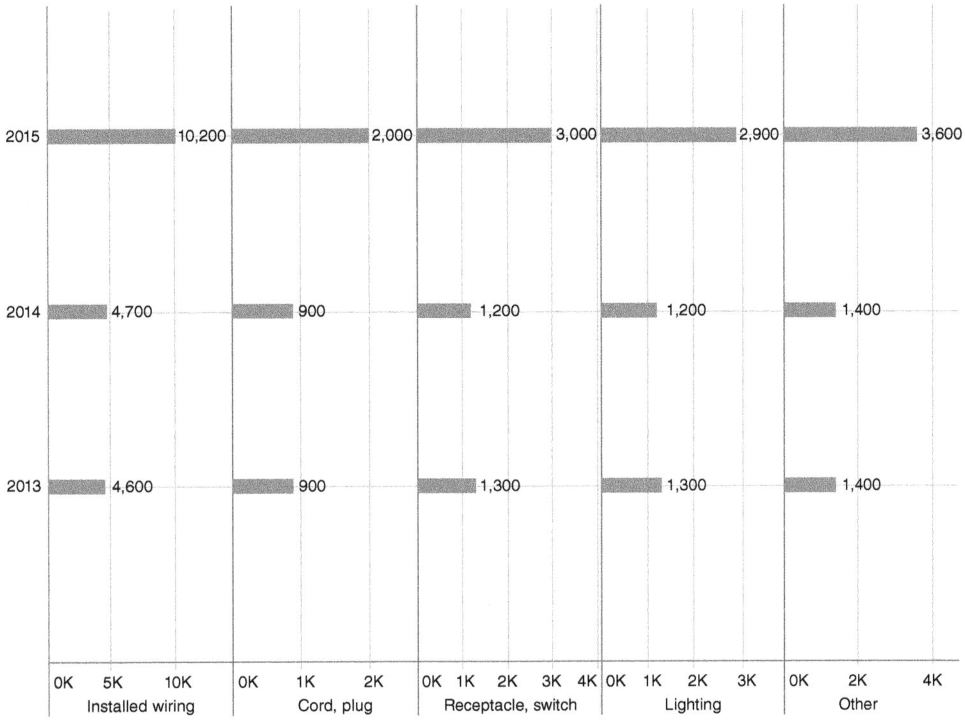

FIGURE 9.5 U.S. residential fires and electrical equipment causes, 2013–2015. An estimated annual average of 9,600 fires were attributable to electrical equipment, or about 2.7% of the estimated annual average number of residential fires for this period. The associated annual average death estimate is 130 deaths and the injury estimates averaged 450.[4] (*Adapted from Table 1a of Miller [2018].*)

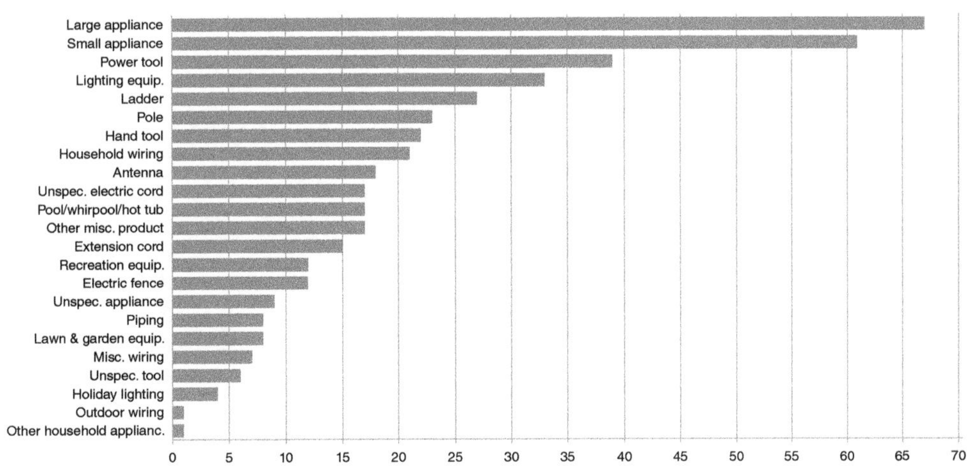

FIGURE 9.6 U.S. consumer electrocutions, 2004–2013. For the period 2004–2013, causes of consumer electrocutions are shown by product category.[5]

ELECTRICAL EVENTS

An electrical event may scale from an almost unnoticeable release of energy, such as localized heating in a conductor or a very brief light flash in a low voltage installation, to a massive explosion from an escalating arcing fault in a multi-phase high-voltage system. The scale and time factors lead to differences in the amount of energy in an electrical event, and by extension, differences in the destructive effects (Fig. 9.7).

Consider when the electrical energy release in an event is relatively small and slow, as for example, with ohmic heating in fluorescent fixture ballast. This heating effect may only be noticeable during a preventive maintenance check of the fixture's performance. However, the electrical hazard can lead to a potentially fatal consequence if it serves as an ignition source in the presence of combustible dust, vapor, or other ignitable materials. Fires and explosions from combustion of these hazardous materials can follow, placing workers and other occupants and neighbors of the facility at risk.

Another aspect of the hazardous electrical energy relates to the physical response it may produce. For example, if an employee brings a metal ladder under the light to repair the fixture while electrical current is flowing, the employee may be startled by a shock that causes the loss of balance on the ladder. Or worse, the employee may be electrocuted and die. For this reason, use of nonconductive equipment and tools as well as local disconnects at luminaires are potentially lifesaving.

In another scenario an employee may come in unprotected contact with an electrical shock hazard. Then electrical energy can flow from the hazard to the employee, essentially as if the employee has become part of the source circuit. The employee's connection to the circuit can be "in series," as when current flows "into" or "through" the body, or the connection to the circuit can be "in parallel," as when current flows over the employee without creating a potential electrical gradient across specific tissues in the body. Complex connection in both series and parallel paths may also occur.

Finally, the scenario when an employee or equipment the employee is using enters an air gap between conductors may happen initially intentionally as, for example, in the high-voltage scenario of live line work. Another possibility is that the employee or his or her equipment may enter the air gap between conductors initially accidentally, as with the operation of equipment too near to a power line.

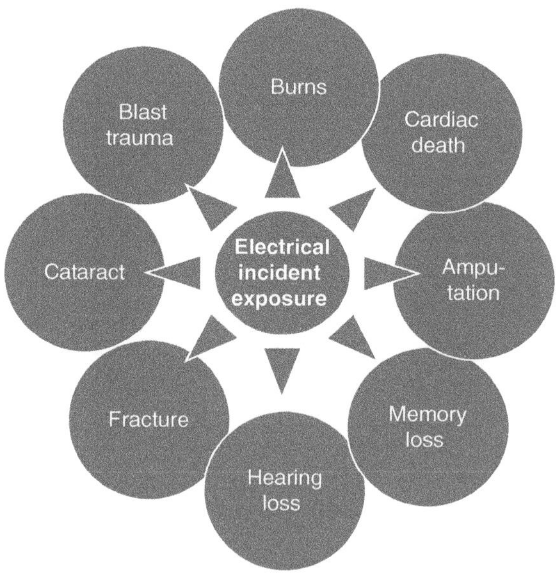

FIGURE 9.7 Possible trauma following an electrical event. This bubble diagram illustrates the kinds of medical problems that may follow the exposure to an electrical event. The body's response to the transferred energy from the event in part depends on the form of energy, the dose, and the duration of the exposure.

As a result of entry into an air gap between conductors, a conductive channel can open between the source and the employee or their tool, resulting in an electrical arc fault. With an electrical arc fault, a large amount of electrical energy can flow in an ultrashort time, creating an event that transforms electrical energy into multiple forms of energy flowing from the source into the electrical event space (Fig. 9.8).

As suggested in Chap. 1, an electrical event can be formally described using the following:

- Equipment voltage and involved current
- Occurrence of an outage resulting in disruption of service, manufacturing, or industrial process
- Destructive effects of the energy released in the incident

As earlier discussions about the National Electrical Code have indicated, hazard rating of electrical installations is variable, depending on the location of the installation as well as the surrounding operations and conditions. While there is consensus on how to rate electrical installations, there is no widely agreed on scheme to rate an event's electrical failure severity.

Electrocution and Electrical Fatalities

The term *electrocution* refers to an electrical event with electrical current exposure resulting in death. The implication is that the current flow has caused an electrical shock with subsequent physiologic

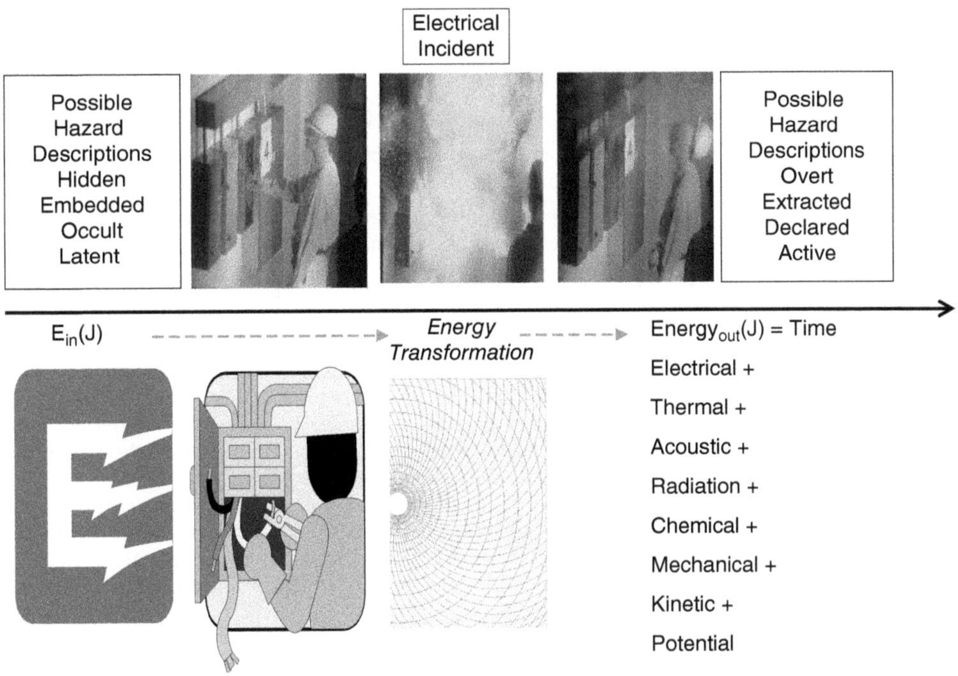

FIGURE 9.8 Schematic of electrical event energy transformation and transfer with photographs of experimental simulation of a "real world" scenario.[10,11] Industrial scenario with electrical energy input to a wall panel box while worker is standing nearby. At the moment of the electrical incident event, electrical energy transformation occurs. Uncontrolled energy release follows, resulting in the transfer of electrical and other energy forms to the space and nearby worker. Simulation photographs parallel schematic illustration. A staged test with a mannequin industrial worker standing nearby a panel box is shown under high-voltage laboratory conditions with 480 V and approximately 22,000 A available. An electrical failure is staged to show a typical workplace scenario. (*Schematic is adapted from Bowen et al., 2004. Photographs by the author are from source documentation reported in Jones et al., 2000.*)

effects interfering with life. While an informed speaker may use the term precisely, sometimes survivors or their caregivers use the term *electrocution* to refer to a "near death" electrical exposure rather than a more commonly experienced "shock" that might be experienced as "routine" or "innocent."

Electrical accident fatality is a general-use phrase seen in news reports meaning either electrocution or death resulting at the time of the electrical accident. This phrase may include fatalities, associated shock, or other forms of energy released at the time of the electrical accident, particularly those causing physical changes including burns, blast effects, or radiation damage.

Electrical injury mortality is a medical statistics phrase that suggests that a person who was injured in an electrical accident lived long enough to receive medical care for their injuries, but the medical care was not followed by survival.

It's important to appreciate that an electrical event can produce a fatality or injury even when there is no electrical current flow to the victim or electrical shock. This might be the situation, for example, when a victim is caught in an electrical ignition fire, explosion, or blast. In this type of scenario, the "root cause" of the accident is electrical, but the mechanism of death or injury is from thermal, acoustic, radiation, or blast exposure related to electro-thermal-chemical (ETC) combustion.

Employees can also be killed or injured after an electrical event in that they are surprised by an energized source, either through a spark, such as a static "zap" to exposed skin, or through a noise, such as a sharp "gunshot" type sound close to the head. The surprise can lead to an unintended body movement that might be characterized as a "startle response." If the startle occurs at the top of a ladder or scaffold, the direct mechanism of death or injury can be through a fall. If the startle occurs in proximity to other energized equipment that is moving, death or injury can be with a body part being caught in or by the moving equipment. If the startle results in a tool or small metal piece such as a screw falling into operating equipment, the metal object can serve to initiate a fault condition that disrupts the equipment's performance, possibly leading to a cascade of events that is more damaging than the spark or static discharge causing the startle.

Fatal and nonfatal electrical incidents share three characteristics:

1. *Unintentional* exposure of employees to electrical energy
2. Compliance *failure* in at least one aspect of electrical design, installation, policies, procedures, practices, or personal protection
3. Energy *transfer* to exposed employees in some combination of electrical, thermal, radiation, acoustic (pressure), mechanical, light, kinetic, or potential energy

What is the difference between fatal and nonfatal electrical incidents? The answer depends in part on whether the question is asked hypothetically, as in a "what if" planning scenario, or whether the question is asked retrospectively after a traumatic accident has occurred.

Hypothetically, based on human physical and biological characteristics, we know that a fatal electrical event transfers a greater amount of energy to its victim than a nonfatal situation. This knowledge about the fatal risk of energy transfer underlies the use of equipment designs (e.g., required doors, specified space clearances, venting systems on equipment to discharge combustion products, "umbilical corded" controls, and infrared monitoring ports for "doors closed" heat monitoring) and barrier protection (such as PPE, including leather gloves, flash suits, safety glasses, face shields, long sticks, extended handles, and flame-resistant clothing) discussed in this handbook.

By reducing the amount of possible energy transfer during an unintentional electrical exposure, strategies including equipment design and barrier protection can increase the likelihood of survival after an electrical incident.

Retrospectively, if two people are present in an electrical incident when one dies and the other survives, the difference in survival may come down to nuances in the victims' innate individual differences and their spatial and temporal relationship to the electrical hazard at the time of the energy release, transformation, and transfer. Medical and legal privacy protections tend to reduce accessibility to accident details, so systematic information is lacking about how various scenarios unfold.

Generally, different forms of energy vary in their lethal doses. When multiple forms of energy are involved in an electrical event, lethal and sublethal energy exposures may result. Injuries can vary depending on the energy transferred to the body's tissues.

Medical Aspects

The medical consequences of an electrical incident cannot be readily predicted from an event description. This is a frustrating aspect to understanding how survivors look or feel after an electrical exposure and complicates the medical evaluation and management of injured patients.

Three questions are central to understanding the fatalities or injuries seen as health effects following an electrical failure:

1. How much energy was released during the electrical event?
2. What forms of energy were released during the electrical event?
3. What parts of the released energy transferred to the affected worker?

A schematic of an electrical incident with photographs of an experimental simulation of a "real-world" scenario is shown in Fig. 9.8. As the illustration suggests, an electrical incident unfolds over time. Again, when an electrical fault occurs, electrical energy is transformed and transferred leading to multiple possible exposures. Tissue damage is related to the amount of energy that builds up or is absorbed into the tissue as a result of the exposure.

Voltage ratings, available fault current estimates, and incident energy calculations permit a guess at the amount of electrical energy that may be involved in an electrical incident. Practically, when more than one person is present at an electrical incident, more than one health effect from the incident is possible, ranging from no effect to fatality. Such varying effects result from individual differences in size, shape, position, and body composition.[6]

Although it may seem unreasonable to employers and coworkers, the same exposure can and often does produce differing levels of injury in individuals involved in the same event. Figure 9.7 illustrates the possible traumatic outcomes.

People can vary in their susceptibility to any single exposure (i.e., electricity, heat, radiation, noise, and pressure) because of their individual differences. Moreover, an exposure itself can vary, depending, first, on the event timing and, second, on the positions in space where the hazard is released relative to the affected employees. Since two or more people cannot occupy the same point and time in space, even during the same accident their positions cannot be exactly the same.

Much scientific knowledge exists about how the body responds to direct electromechanical contact with 50–60 Hz electricity generated for electrical power, as discussed in Chap. 1. For fatality and injury scenarios in which electrical shock as the dominant exposure, the tissues' "vulnerability" to electrical forces, is key because biologic tissues vary in their electrical conductivity. This variation explains uneven patterns of damage on a finger or limb, where bone, muscle, tendon, blood vessels, and nerves may demonstrate differing degrees of loss. In electrical exposures to children and young adults, individual differences as relates to growth, development, and body composition are also important.[7,8]

Nonelectrical Effects in Electrical Events

As power density increases across more compact spaces, there is more fault energy available in electrical installations. With installations designed to be compact, there are smaller spatial cushions and consequently less physical or spatial barrier protection emerging in innovative designs. As a result, there may be less time and less physical distance through which an unintentional release of energy can be dissipated during an electrical event. With higher power density and smaller, more compact spaces, there is an increased risk of collateral damage from the nonelectrical forms of energy.

High power density in a confined space can have bomb-like potential. Consequently, an electrical event can behave like a closed blast explosion. For example, the potential ignition of 1 MW is roughly equivalent to the potential ignition of one stick of dynamite. One stick of dynamite is roughly 1/3 lb of TNT equivalent. So, a 100-MW scenario can be represented as loaded with 100 sticks of dynamite. Or, a 300-MW scenario can be represented as loaded with 100 lbs of TNT.

A closed space with an electrical event creates the possibility of closed blast health effects. At a pole top in a vacant field, the space is more like an open area suggesting an electrical scenario at a pole top is more likely to cause injuries similar to those seen with a free blast. Blast trauma may include destruction from shrapnel, confined space air contamination, acoustic or pressure effects, and acceleration-deceleration brain effects. In sum, when an electrical event involves more than a direct electromechanical current exposure, the traumatic consequences can vary, as the bubble diagram suggests (Fig. 9.7). In fatal injuries, determination of electrical and nonelectrical traumatic effects may require confirmation by forensic pathology studies.[9]

Apart from the electrical sources of thermal energy (ohmic or Joule heating, radiation with arcing), fire and molten metals at the scene can lead to heat transfer from the combustion of nearby installation, building, and equipment materials. Inhalational injury associated with fire can complicate the survivor's chances for recovery, especially in an event that has also included a blast (Fig. 9.8).[10,11]

Survivor Experience

As suggested in Table 9.1, an electrical event may unfold in fractions of a second. In electrical scenarios, a full second is a long time, suggesting a 60-cycle event. Compare this to the investigation of an electrical incident that may take weeks to months. The litigation accompanying such an event typically takes years. Note that the logistics of work do not permit a site to go on hold while these activities are underway.

Obviously, a survivor experiences an electrical event on a different time frame than his or her workplace. Practically, within a few work shifts, an electrical accident site may be operating routinely. In contrast, the survivor experience may unfold from the moment of the electrical event through minutes of triage to years of rehabilitation.

Worker Reflexes

At the scene of an electrical event, a survivor relies on his or her reflexes and senses. As discussed in Chap. 12, to see, think, and then respond to a scenario demanding a choice require at least 1.5 seconds. Physical responses to an unfolding electrical event in a shorter amount of time occur with reflexes, such as eye blinks in response to flash, or quick hand withdrawal from a hot surface.

Sometimes the injured survivor can "walk away" from the electrical event. However, this immediate post-incident behavior may be deceptive, given the possible medical consequences of electrical trauma.

TABLE 9.1 Timing of Workplace and Survivor Experiences After Electrical Events

Experience	Timing
Event:	Fractions of milliseconds
Investigation:	Weeks to months
Litigation:	Years
Survivor	
Worker reflexes	Seconds
Triage and medical evacuation	Event to hour 1
Stabilization and initial evaluation	Hour 1 to hour 8
Medical and surgical intervention	Hour 8 to hour 24
Hospitalization	Day 2 to day 9
Discharge to outpatient services	Day 10 to day 30
Rehabilitation	Day 31 to day 89
Return to work planning	Day 90 to day 119
Re-entry to employment	Month 4 to month 60
Plateau in recovery	Beyond 5 years

Instead of walking away, a person may collapse with the loss of regular heart function if a nonfatal electrical contact is made. As the heart's pumping action becomes inefficient, the blood supply to the lungs and brain can be severely limited, reducing the delivery of oxygen and glucose (the body's fuel). When oxygen and glucose are not available because of circulation failure, a survivor cannot breathe or remain upright or awake. This experience may be later recalled by a survivor as "near death" or "out of body." In this time, successful cardiopulmonary resuscitation (CPR) is essential to rescue the affected individual, as discussed in Chap. 8. Anyone who needs CPR should then be sent to the hospital without exception for further evaluation and treatment.

Once help arrives to the scene, an alert electrical incident survivor is routinely questioned. As first responders, emergency medical technicians, or paramedics become involved, the survivor is expected to tell these clinicians a history of their difficulties.

Even while a survivor needs to explain how they were injured to first responder and emergency medical providers, it may be difficult for the employee to respond if their thinking ability is clouded by the event or if the electrical accident has occurred quickly. For example, a 30-cycle event in a 60-Hz electrical power system occurs in 0.5 seconds. A scenario that takes a half second *to complete* is faster than most people can even recognize, much less remember in detail.

Triage and Medical Evacuation

Medical triage, evacuation, and treatment require minutes, hours, and longer, as discussed below. In the emergency or first-response period, the American Burn Association (ABA) Advanced Burn Life Support guidelines outline triage, initial resuscitation, and emergency transport protocols. On admission to the emergency medical services, physical examination, laboratory studies, and diagnostic radiology evaluation are prioritized, depending on the patient's clinical situation.

When the survivor cannot remember accident details, there is little information available to clinicians for rigorous decision-making about the exposure circumstances, possible mechanisms of injury, and need to obtain extensive laboratory studies or diagnostic imaging.

If in coma or heavily medicated, the survivor may not be able to coherently answer questions such as the following:

- What happened in terms of the energy such as volts, amps, cycles?
- Was there a blast or an explosion?
- Was the space small or confined?
- Did you fall or lose consciousness?
- Were you wearing protection?
- Did you need CPR?
- Was anyone else hurt or killed at the scene?

Further confusion between a survivor who becomes a patient and clinicians may arise when the circumstances around the injury event simply cannot be immediately known. For example, when a work situation is believed to be safely deenergized but leads to an electrical accident, engineers may investigate over weeks to months to establish the root causes and available energy to the fault. During that investigation time, the electrical incident survivor may be repeatedly asked by clinicians, "What happened? Why didn't you turn off the power?" From the survivor's perspective, each episode of questioning by clinicians is an opportunity for self-doubt about their accident recollection.

Commonly, survivors report they were doing a manual task or handling equipment when caught in an electrical event. Upper extremities, the body from the fingertips to the shoulders, are involved most often in electrical trauma. However, depending on the employee's body position at the time of the electrical fault, contact wounds from electrical shock may be found in more than one body location. Current flow in the body may follow a path as short at the distance between two fingers. Or current may travel along the length from the employee's arm span or his or her height in a standing position.

From a clinician's viewpoint, knowledge of the electrical event's voltage conditions alone is of limited helpfulness because the extent of the survivor's electrical injury is a function of the power frequency energy transfer due to a combination of the current density in the tissue and the contact duration. The geometry of the body influences the current density.[8,12,13]

Similarly, from a medical perspective, the heat, radiation, and pressure damage to the body are predicated on the efficiency of the electromechanical and electrochemical coupling between the hazardous source, event space, and the body. This information is not usually available to the medical team and may only come to light after months of investigation.

Physically, when an extremity is exposed to electrical current, the skeletal muscle in the current path through the extremity can be significantly injured from the electrical effect, extreme heat, and damage to proteins. At the electrical accident site, first responders may see minimal skin wounds, or they might notice profound charring and vaporization of tissue. Even in the absence of impressive wounds, the survivor's risk of amputation can be high.

After a severe electrical contact injury, a survivor's injured muscles can swell massively, resulting in a situation referred to as "compartment syndrome." In a compartment syndrome, the blood supply to local muscle is literally squeezed by swelling. With this swelling, oxygen supply to the muscle can be "choked." Without oxygen from the blood supply, muscle tissue dies. If muscle tissue dies, proteins are lost as waste and filtered by the kidneys. These proteins can block the kidneys and result in kidney or renal failure.

Other significant features of severe electrical exposure include disturbances in the survivor's heart electrical pattern, called refractory cardiac arrhythmias discussed in Chaps. 1 and 8. Further serious problems may include the following:

- Neck and back fractures from the electrical current exposure, falls, or blast
- Disruption in the blood serum's balance of sodium, potassium, and chloride
- Transient lifeless signs, including a condition referred to as keraunoparalysis
- Shock lung, perforated ear drums, and ruptured gastrointestinal organs subsequent to blast trauma
- Traumatic brain injury, including prolonged post-concussive symptoms

In high-voltage electrical incidents, current passing through a limb across opposite sides of a joint can set up large skin or "transdermal" potentials, resulting in burns called "kissing" wounds. The most frequent sites of this trauma include the following:

- The armpit or axilla, with injury between the limb and skin of the chest
- The leg by the space in back of the knee, where the skin of the upper and lower leg can meet when the survivor is caught in a crouched position

Other skin burns may result in the following ways:

- With ohmic or joule heating at the electrical contact points between the survivor and the electrical hazard
- From the ignition of clothing burning in place against the survivor's skin
- At unprotected skin exposed in the extremes of heat generated by and radiating from electrical arcing

Stabilization and Initial Evaluation. Severely electrically injured victims are difficult to resuscitate, and the early phase of medical management can be very complex. Survivors are best served by admission to a burn trauma center where staffing is experienced in the management of patients with massive loss of tissue from electrical current damage and burns.

From the survivor's perspective, after a severely traumatic accident, a breathing tube, intravenous fluids, and massive pain control medicines obliterate awareness of all that happens in the early treatment period.

Because nerves are very sensitive to electrical forces, even minor electrical exposures may cause temporary nerve dysfunction in the affected limbs or body areas. Pain may dominate. Nerve symptoms called by the medical terms *anesthesia* (no feeling), *paresthesia* (some feelings), or *dysesthesia* (prickly, pin-prick, or knife-like feelings) may occur. With a less severe hand electrical contact, the survivor may experience loss of feeling or "funny" feelings in the contact areas referred to as "stunning," which can hamper how effective hand grip is in terms of strength, endurance, and reliability. Stunning usually completely resolves within hours or days.

Although less common, electrical contacts may also lead to a nerve problem called temporary autonomic nervous system dysfunction, which may show up in the patient as specifically reflex sympathetic dystrophy (RSD), and hypertension or high blood pressure.

Less quickly resolved for a survivor are the classic complaints of burning pain and exquisite sensitivity, which may develop within hours after an electrical injury. Swelling (called edema) and excessive sweats may then manifest during the next 3–6 months. This may be followed in 6–12 months with loss of muscle mass, nails, and hair. Osteoporosis may develop in the limb. To evaluate this problem, clinicians conduct diagnostic nerve tests and treat with injections of medication near the nerves to produce a "nerve blockade."

Medical and Surgical Intervention

For a survivor with a severe electrical exposure, polytrauma can require more intensive medical and surgical care. Polytrauma refers to the injury of two or more body parts or systems that lead to physical, cognitive, psychological, or psychosocial impairments.[14] An initial goal of medical and surgical intervention is to completely diagnose the extent of injury. To distinguish central nervous system damage from peripheral nerve injury, testing includes key steps summarized in Table 9.2. The diagnostic evaluation of muscle and nerve injury unfolds while the patient's clinical condition is being stabilized and treated.

Even when an employee injury at the scene of the electrical event may appear less severe to coworkers or witnesses, diagnostic testing in a hospital is still necessary because the health changes resulting from electrical exposure may not be obvious for hours. An employee who walks away from an electrical event may not want to draw more attention to their accident by seeking or attending a medical evaluation. Yet because the body damage may not be detectable by the eye, an exposed employee should always be required to have a medical checkup.

Hospitalization Experience

Employees surviving a serious occupational electrical event are often hospitalized. During days or weeks of hospitalization, the uncertainty in answering a clinicians' questions (e.g., "What happened?" "Why did this happen?" "How were you hurt?") can impact their medical treatment and amplify their distress at their injury.

Clinicians may not be aware of the extended time required to complete engineering root cause analyses of the factors contributing to an electrical event. Employer representatives may be cautious, limiting their statements to the employee about their investigation findings due to medical concerns for the employee or risk management considerations. Therefore, minimal information may flow to the affected employee and the employee's family or caregiver regarding the electrical accident.

Consequently, an experienced, knowledgeable, and competent employee can be made to feel ignorant, uninformative, manipulative, or uncooperative—as if he or she doesn't have the "right answer." After surviving a workplace electrical accident, the hospitalized employee essentially moves from the occupational environment where he may be considered "a good worker" or an "expert at his job," to the health care environment and the medical care delivery system, where the survivor may only be known by reference to a medical record number, a date of admission, and a treating physician's name. Concurrently, for family or coworkers who have not been hurt, the hospitalization experience can be alienating and frightening.

TABLE 9.2 Determination of Muscle and Peripheral Nerve Injury

Diagnostic	Description
Determination of the exposure circumstances	Determination of the electrical current exposure circumstances is fundamental to understanding the survivor's pattern of injury. Exposures may involve direct mechanical contact with an energized surface by an unprotected body part, or may occur through electrical arc current flow. Personal protection equipment such as industrial garments and gloves, face shields, safety glasses, and arc enclosures may moderate the intensity of the incidental exposure. For example, safety gloves may minimize skin wounds, or safety glasses may protect the eyes from the ultraviolet radiation released by an electrical arc. The duration of the electrical current flow, release of multiple forms of energy at time of the current flow, and geographic location of the event may increase the potential severity of the patient's trauma.
Documentation of the physical findings	Physical findings may include open wounds, traumatic amputation, fractue dislocations, impaired neuromuscular function, or edema. Physical examination supported by laboratory studies for the serum markers of muscle damage (e.g., serum enzymes, urine myoglobin) provides correlative information for radiology and electro-physiology diagnostic studies. When external wounds are not observed, a high degree of clinical suspicion of occult neurologic injury is advised.
Radiology studies	Radiographic and nuclear medicine examination can evaluate nervous system structural damage and confirm fractures, local tissue edema, and focal areas of inflammation. For example, in the acute period, the localization of tissue edema in an upper extremity is an early sign of muscle injury.
Electro-physiology studies	Electro-physiology studies are guided by the patient's history and physical examination. Muscle weakness, easy fatigue, loss of endurance, paralysis, and pain are indications for electromyography, and peripheral nerve conduction studies (sensory and motor). Further evaluation may be needed with peripheral nerve refractory period spectroscopy and magnetic resonance spectroscopy. Neuro-physiologic evaluation may also be indicated by additional complaints. For example, history of loss of consciousness or seizures, central nervous system complaints such as headache, impaired memory, attentional problems, personality changes, and cognitive changes suggest electroencephalography (EEG) and electrocardiography (EKG). A history of temporary or persistent loss of hearing, ringing in the ears, or change in memory to verbal communication are indications for auditory evoked potentials (AEP). A change in visual acuity or visual disturbance is an indication for visual evoked potentials (VEP). Cardiac rhythm disturbances, syncope, chest pain, and easy fatigue are indications for electrocardiography (EKG).

Outpatient Care

Even without a visible physical injury like burns or electrical contact wounds requiring hospitalization, electrical injury survivors may report medical complaints. These complaints may seem unrelated to the electrical event sometimes because they show up hours, days, or weeks after the occurrence, or because they seem more or less severe than the event itself.

This situation can go from bad to worse when the facts of the injury scenario do emerge and survivors learn that they have experienced a near-fatal situation. Hearing comments such as "You are lucky to be alive!" can disintegrate the workers' belief in their professional competence. Employees who work around electricity don't survive on luck. Moreover, having a near-death accident doesn't feel lucky to most.

In particular, electrical professionals hold to an occupational identity and set of beliefs that suggest workplace policies, procedures, planning, and personal protection can provide for safe occupational activity. When an employee's identity or beliefs are threatened, anxiety, adjustment reactions, or more serious psychiatric issues should be expected. If the facts of the injury scenario emerge as attributing blame for electrical incident on the survivor, then personal and professional guilt are added to the list of issues the worker may need to confront.

Besides difficulty in remembering event details because of how quickly the event occurred, the employee who does not have an easily noticed physical injury may still have problems which suggest a brain or nerve injury. These problems directly affect employability, and may be described as weakness, pain, headache, memory changes, disorientation, slowing of mental processes, agitation, or confusion. Personality changes such as irritability, moodiness, nightmares, difficulty sleeping, or feelings of depression or post-traumatic stress may also occur. These changes can interfere with personal and workplace relationships, straining a worker's support network.

Rehabilitation Focus and Return to Work Planning

Rehabilitation and return to work (RTW) planning starts at the time of hospital admission and may extend months after hospital discharge. With massive trauma, years may be necessary to return a survivor to active employment. Re-training to allow for change of professions may be required.

Successful rehabilitation often does not lead to the return of the patient to their preinjury job. The disability for electrical incident survivors is disproportionate to the incidence of this preventable condition. This means that for relatively few injuries, there is a relatively high frequency of permanent disability. With the typical youth of those injured in electrical accidents and the loss of potential productivity in economic terms, the health effects of electrical incidents carry significant costs for victims, their families, and their employers.

Reentry to Employment Settings

When workers have difficulty in the use of their limbs after an electrical incident, their security in completing tasks (such as exerting a forceful grip, climbing a ladder, using hands to lift a load, and assisting a coworker in a hazardous activity) may be unacceptably compromised. This difficulty can drastically reduce employment options. Workers may wish to return to their job but face resistance in this step from their family because of their concerns for reinjury. When the electrical incident is felt to be the victim's "fault," coworkers may harbor undisclosed resentments, acting as a roadblock to reentry to employment.

Plateau in Recovery

The complexities of electrical injury rehabilitation are often under appreciated by the medical community. As a general guide, three elements are essential to successful recovery following an electrical incident: *team, time,* and *talk.* The involvement of an experienced occupational rehabilitation team is needed to avoid the catastrophe of repeated failure to return to work, depression, and loss of self-respect that may evolve when an injured worker fears for his or her livelihood. Time is necessary for the survivor to go through the healing stages following repeated surgeries, possible amputation, and long hospitalizations. Talk, or excellent communication, is critical to maintaining the employee's relationships with family, friends, coworkers, and caregivers.

Research suggests electrical trauma survival is associated with significant functional impairment. While in the late 20th and early 21st centuries the scope of impairments was detailed in a limited number of scientific reports,[12,15] continued studies and research publications now point to an evolving consensus on post-injury neuropsychological challenges.[16,17]

In a landmark retrospective study of employees of a national electrical energy company, Gourbiere et al.[18] reviewed the electrical trauma survival experience during 1970–1989 for a workforce between

100,000 and120,000 during this period. Electrical burns affected 2,080 workers. Of these, 515 patients or 25 percent were noted to haven post-injury problems, including the following:

- 63 percent burn related, with amputations in 5 percent
- 18 percent neuropsychiatric
- 12 percent sensory
- 5 percent orthopedic
- 1 percent cardiovascular

Sense organ problems were also noted, including the following:

- Vision-related changes due to conjunctivitis, keratitis, and cataracts
- Auditory late effects, with conductive or sensorineural hearing loss, tinnitus, and vertigo
- Anosmia, or the loss of sense of smell

In 59 of the 515 patients, disability was considered serious, with impairment rating 31–100 percent.[18]

Neuropsychological changes after an electrical injury may be categorized as post-traumatic stress disorder or be assigned diagnoses such as depression or anxiety[19] subsequent to electrical contacts that do not appear to have an electrical path through the brain. In one study, all patients had peripheral electrical contacts, with no evidence on history or examination of direct mechanical electrical contact with the head. Forty-four patients were injured on the job and 9 were injured during nonvocational activities. At the time of follow-up study interviews, 30 (56.6 percent) patients were working again, 18 (44.0 percent) patients were unemployed or retired, 1 patient was deceased, and 4 patients could not be contacted.[19]

Electrical trauma patients meet with varying degrees of success in their return to work. While some electrical trauma patients are able to return to their preinjury activities without function impairment, others are not. Both survivors who are owner operators of their business as well as employees of larger enterprises see postinjury job loss. Recent studies suggest that patients with low-voltage and high-voltage injuries may have similar neuropsychiatric and functional outcomes,[16] and may require continued evaluation and management by mental health professionals.[17]

While postinjury medical care may not interfere with job placement, survivors may find that their job may not exist due to competition from other vendors, contract completion, restaffing, or workforce reductions. If the survivor wishes to return to work, and their preinjury job is not available, the clinician can be influential in assisting placement into new work. To limit the potentially destructive effect of financial uncertainty on the survivor and their personal or family situation, when a clinician is advising on placement in new work, a vocational evaluation is recommended as soon as possible to establish the local employment area's market for the survivor's skills set, occupational abilities, and training.

For example, driving is a common requirement in service jobs where travel is necessary to deliver tasks to client sites, such as construction locations, factories, or customer businesses. Under U.S. Department of Transportation (DOT) rules, a person cannot be medically certified for a commercial driver's license if there is a recent history of loss of consciousness, vision less than 40/40 in each eye and both eyes, certain cardiac conditions, and certain medications.[20] Individuals with amputations must demonstrate performance in driving in a vehicle that may fall under DOT rules. When the patient cannot drive a DOT-regulated vehicle, designed work support is necessary to allow access to job sites and equipment.

Another aspect of return to commercial and industrial work is the need for bimanual secure grip. To climb a ladder, lift a load, or handle a heavy tool, secure, forceful bimanual grip that can be reliably and repeatedly used is an essential function of the job. When finger or hand amputation is a consequence of electrical trauma, supported work to offset the limitation of secure bimanual grip is needed.

At a certain point after an electrical trauma, a survivor reaches a plateau in his or her recovery. With consideration for numerous surgeries, retraining for employment, and mental health care, plateaus are assessed on an individual basis. Nevertheless, a common characteristic is that the survivor rarely goes "back" to the life he or she had before his or her incident. A survivor typically goes "forward" by living differently. The differences can be small, as with an attitude change, or dramatic, as with a job change. The bottom line for the survivor is that after electrical trauma life may never be the same.

REFERENCES

1. Occupational Safety and Health Administration (OSHA), *OSHA Commonly Used Statistics*, November 2018. Retrieved from https://www.osha.gov/oshstats/commonstats.html.
2. Bureau of Labor Statistics (BLS), *National Census of Fatal Occupational Injuries in 2016*. U.S. Department of Labor, 2017. USDL-17-1667. Retrieved from www.bls.gov/charts/census-of-fatal-occupational-injuries/.
3. Center for Construction Research and Training (CPWR), *The Construction Chart Book: The US Construction Industry and Its Workers*,6th ed. eChart Book, 2018. Retrieved from https://www.cpwr.com/publications/research-findings-articles/construction-chart-book.
4. Miller, D., *2013–2015 Residential Fire Loss Estimates: US National Estimates of Fires, Deaths, Injuries and Property Losses from Unintentional Fires*. U.S. Consumer Product Safety Commission, Bethesda, MD, 2018. Retrieved from https://cpsc.gov/s3fs-public/2011to2013ResidentialFireLossEstimates-Final.pdf.
5. Hanaway, S., and Siddiqui, S., *Electrocutions Associated with Consumer Products: 2004–2013*. U.S. Consumer Product Safety Commission, Bethesda, MD, May 2017. Retrieved from https://cpsc.gov/s3fs-public/Electrocution-Report-2004-to-2013.pdf?V_9Zl10pv4Wz03uBPRx78IctKRABjYDv.
6. Lee, D. H., Desai, M. J., and Gauger, E. M., Electrical injuries of the hand and upper extremity, *Journal of the American Academy of Orthopaedic Surgeons*. pp. 1–8, 2018.
7. Roberts, S., and Meltzer, J. A. An evidence-based approach to electrical injuries in children, *Pediatric Emergency Medicine Practice*. 10(9): pp. 1–20, 2013. Retrieved from www.ebmedicine.net.
8. Gentges, J., Schieche, C., Nusbaum, J., and Gupta, N., Electrical injuries in the emergency department: An evidence-based review, *Emergency Medicine Practice*, November 2018. Retrieved from www.ebmedicine.net.
9. Aquila, I., Gratteri, S., Amirante, C., Fineschi, V., Frati, P., and Ricci, P., Electric or traumatic injury? The role of histopathological investigations, *Medico-Legal Journal*. 86(2): pp. 85–88, 2018.
10. Jones R, Liggett DP, Capelli-Schellpfeffer M, Downey RE, Jamil S, Macalady T, McClung LB, et al., Staged tests increase awareness of arc-flash hazards in electrical equipment, *IEEE Transactions of Industry Applications Society*. 36(2): pp. 659–667, 2000.
11. Bowen, J. E, Wactor, M. W., Miller, G. W., and Capelli-Schellpfeffer, M., Catch the wave, IEEE *Industry Applications Magazine*. 10(4): pp. 59–67, July/August 2004.
12. McGowen, J. C., Shkolnikov, Y. P., Sala, J., Ray, R. M., Diffuse electrical injury. *Proceedings of the Canadian Conference on Electrical and Computer Engineering*, Niagara Falls, Canada, CCECE 2008, pp. 1977–1980, May 5–7, 2008.
13. Lee, R. C., Zhang, D., Hannig, J., Biophysical injury mechanism in electrical shock trauma, *Annual Review of Biomedical Engineering*. 2: pp. 477–509, 2000.
14. Dobscha, S. K., Campbell, R., Morasco, B. J., Freeman, M., and Helfand, M., *Pain in Patients with Polytrauma: A Systematic Review*. Department of Veterans Affairs, Washington, DC, September 2008. Retrieved from http://www.ncbi.nlm.nih.gov/books/NBK49088/. Accessed July 31, 2011.
15. Pliskin, N. H., Meyer, G. J., Dolske, M. C., Heilbronner, R. L., Kelley, K. M., and Lee, R. C., Neuropsychiatric aspects of electrical injury: A review of neuropsychological research, *Annals of the New York Academy of Sciences*. 720(1): pp. 219–223, 1994.
16. Chudasama, S., Goverman, J., Donaldson, J. H., Van Aalst, J., Cairns, B. A., and Hultman, C. S., Does voltage predict return to work and neuropsychiatric sequelae following electrical burn injury? *Annals of Plastic Surgery*. 64(5): pp. 522–525, 2010.
17. Andrews, C. J., Reisner, A. D., and Cooper, M. A., Post electrical or lightning injury syndrome: a proposal for an American Psychiatric Association's Diagnostic and Statistical Manual formulation with implications for treatment, *Neural Regeneration Research*. 12(9): pp. 1405–1412, 2017.

18. Gourbiere, E., Corbut, J-P., Bazin, Y., Functional consequence of electrical injury, In *Electrical Injury: A Multi-Disciplinary Approach to Therapy, Prevention, and Rehabilitation.* Lee, R. C., Capelli-Schellpfeffer, M., Kelley, K. M. NYAS, 720: pp. 259–271, 1994.

19. Pliskin, N., Capelli-Schellpfeffer, M., Malina, A., Law, R., Kelley, K. M., and Lee, R. C., Neuropsychological sequelae of electrical shock, *Journal of Trauma.* 44(4): pp. 709–715, 1998.

20. Gerbo, R. M., What criteria are disqualifying for a driver undergoing a Department of Transportation medical evaluation, Schwerha, J. J., ed. *JOEM.* 46(7): pp. 755–756, 2004.

CHAPTER 10
LOW-VOLTAGE SAFETY SYNOPSIS

INTRODUCTION

Each year 120-V circuits cause more deaths and injuries than circuits of all other voltage levels combined. Such low-voltage circuits are extremely hazardous for two reasons. First, low-voltage circuits are the most common. Because they are the final distribution voltage, 240-V and 120-V circuits are used throughout residential, commercial, industrial, and utility systems.

The second reason for the extreme danger of low-voltage circuits is user apathy. Comments such as "It can't hurt you; it's only 120 volts" are heard all too often. Reference to Table 10.1 shows that 120-V circuits can produce currents through the human body that can easily reach fibrillation levels. Consider a perspiring worker using a metal electric drill with one foot immersed in water. Table 10.1 clearly shows that under such conditions, a worker can be subjected to a lethal shock. Furthermore, if sustained for a sufficient period, 120-V contact can create severe burns.

A 480-V circuit is more than four times as lethal as a 120-V circuit. A 480-V circuit has sufficient energy to sustain arcing faults and to create severe blast conditions. This chapter summarizes some of the safety-related concerns that apply to low-voltage circuits—that is, circuits of 1000 V ac and less and 250 V dc and less. Generally, workers should treat low-voltage circuits with the same degree of respect afforded medium- and high-voltage circuits. Refer to Chaps. 3 and 4 for detailed information. Note that some of the safety-related information covered in this chapter also applies to medium-voltage and high-voltage systems. Where necessary, the information is repeated in Chap. 11.

Electrical safety requirements when working on or near electronic circuits can be a problem. For example, the dictates of electrical safety would seem to require that circuit parts and workers be insulated from one another. In some electronic circuits, however, the prevention of static electricity damage requires that the worker be grounded. Also, many workers develop a sense of false security believing that 12,000 V in an electronic circuit is somehow less hazardous than 12,000 V in a power system. This chapter will also present and explain the electrical safety procedures to be employed when working on or near electronic circuits.

LOW-VOLTAGE EQUIPMENT

Hand tools and extension cords are the most commonly used pieces of low-voltage equipment. Each of these items is responsible for hundreds of injuries and deaths each year. The following sections summarize the types of usage procedures that should be employed with such equipment.

Extension Cords

Flexible cord sets (extension cords) are used to extend the reach of the power cord for low-voltage-operated tools and equipment. They typically have a male plug at one end and a female receptacle

TABLE 10.1 Possible Current Flow in 120-V Circuit

Body part	Resistance, Ω
Wet hand around drill handle	500
Foot immersed in water	100
Internal resistance of body	200
Total resistance	800

Total current flow possible in 120-V circuit = $I = (120/800) = 150$ mA

TABLE 10.2 Visual Inspection Points for Extension Cords and Cord-Connected Tools

- Missing, corroded, or damaged prongs on connecting plugs
- Frayed, worn, or missing insulation
- Improperly exposed conductors
- Loose screws or other poorly made electrical connections
- Missing or incorrectly sized fuses
- Damaged or cracked cases
- Burns or scorch marks

at the other. The tool's power plug is inserted into the female receptacle, and the extension cord's male plug is inserted into the power source or into yet another extension cord's female receptacle. Extension cords can be used to supply power to tools that are located many meters away from the source of power.

Extension cords can be extremely hazardous if not used properly. The following precautions should always be observed when using extension cords:

- Closely inspect extension cords before each use. Table 10.2 lists the types of items that should be looked for during the inspection. (Note that Table 10.2 applies to both extension cords and portable tools.)
- Never use an extension cord or power cord to lift or support a tool.
- Make certain the ground connection is complete from one end of the cord to the other.
- Never alter the plug or receptacle on an extension cord. This applies especially to altering or removing the ground connection.
- If the extension cord is of the locking or twist-locking type, the plugs should be securely locked before using the cord.
- Do not use an extension cord in wet or hazardous environments unless it is rated for such service by the manufacturer. If extension cords are used in wet or hazardous environments, insulating safety equipment such as rubber gloves with appropriate leather protectors should be used.
- Only personnel who are authorized and trained in the use of extension cords should be allowed to use them.
- Extension cords should be subjected to the electrical tests outlined in Table 10.3. If the cord fails any of the listed tests, it should be replaced or repaired.
- When not in use, extension cords should be carefully rolled and stored in such a way that they cannot be damaged.
- Extension cords should never be used for permanent power installations. If power is required for more than the short duration, permanent electrical installation should be used. The maximum

TABLE 10.3 Recommended Periodic Tests and Test Results for Extension Cords

Test	Description	Pass/fail criteria
Ground continuity test	25 A minimum is passed through the cord's ground circuit.	Voltage drop across the cord should not exceed 2.5 V.
Insulation breakdown	High voltage is applied to the cord's insulation system and the leakage current or insulation resistance is measured (3000 V maximum direct current applied).	Leakage current no more than 6 μA @ 3000 V (500 MΩ).

allowed is generally 90 days for seasonal types of installations, such as Christmas lights and other types of decorations.

- All temporary power connections in industrial and commercial facilities should be supplied from a circuit protected by a ground-fault circuit interrupter (GFCI). Alternatively, the cord itself should be equipped with a GFCI. (See Chap. 3 for more information on GFCIs.)

An extension cord that exhibits any of the visual inspection problems listed in Table 10.2 or that does not pass the tests listed in Table 10.3 should be removed from service until it can be repaired or replaced.

Electric Hand Tools

Electric hand tools generally fall into one of three categories:

1. Metal tools with ground
2. Double-insulated types of tools
3. Battery-powered tools

Metal Tools with Ground. Metal tools are equipped with a three-conductor power cord. Figure 10.1 is a pictorial diagram of such an assembly. The three conductors are the hot (or line), the neutral, and the safety ground. The hot wire and the neutral wire form the actual power circuit of the tool. The safety ground is connected to the metal frame of the tool. If the hot wire is short-circuited to the case of the tool, the safety ground wire forms a continuous, low-impedance path back to the service box. In this situation, the worker will form a parallel path to ground, causing current to flow through the worker's body. Metal tools should always be connected to the power through a GFCI.

Double-Insulated Tools. Even the relatively low-impedance path provided by the safety ground in a metal tool can create lethal voltages from the hands to the feet. The double-insulated tool does not employ a metallic case. Instead, its case is made of a high-strength, nonconductive plastic or composite material. The power connection for such tools is a two-conductor power cord with no safety ground. Since the tool case is nonconductive, the user is protected by both the case and the normal insulation of the electric circuit. The double-insulated tool can still present a shock hazard if it is dropped in water or if water enters the case. Because of this fact, many companies and workers have opted for the battery-powered tools described below.

Battery-Powered Tools. Figure 10.2 is a photo of a typical battery-powered tool. This battery-powered drill (also a driver and hammer) is powered by an 18-V rechargeable battery. Heavier-duty battery-powered tools for use in industrial and construction environments will normally have

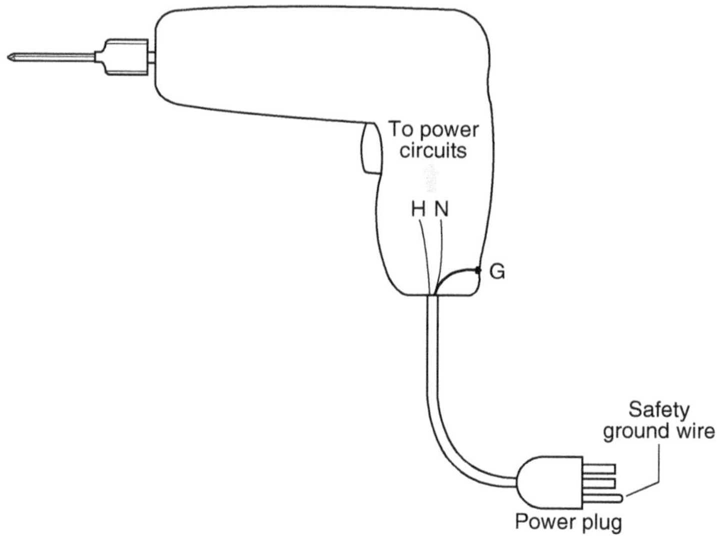

FIGURE 10.1 Typical metallic case hand tool.

FIGURE 10.2 Battery-powered drill.

higher-voltage batteries for reduced current and longer life. Most workers have two or three batteries that are kept fully charged for backup. When the tool battery runs down, it takes only seconds to replace it with a fully charged backup.

Because the battery-powered tool is not connected to an ac power circuit, and because it has an insulated case assembly, it is considered to be intrinsically safe from an electrical hazard standpoint. However, battery-powered tools are physically dangerous and must be handled properly to prevent injury.

TABLE 10.4 Recommended Periodic Tests and Recommended Results for Cord-Powered Tools

Test	Description	Pass/fail criteria
Ground continuity test	25 A minimum is passed through the cord's ground circuit.	Voltage drop across the cord should not exceed 2.5 V.
Insulation breakdown	High voltage is applied to the cord's insulation system and the leakage current or insulation resistance is measured (3000 V maximum direct current applied)*.	Leakage current not more than 6 mA @ 3000 V (500 MΩ).
Leakage test	Measures the current (0 to 10 mA) that would flow through the operator if he or she were to provide a path to ground at normal operating voltage.	
Operational check	Operates the tool to verify proper operation and to indicate operating current.	Operating current should be within nominal nameplate values.

* Refer to tool manufacturer's directions for allowed maximum test voltages.

Precautions for AC-Powered Tools. Both insulated case and metal case should be subject to the same types of precautions as those outlined for extension cords in the previous section. Specifically, the following should be observed:

- Closely inspect tools before each use. Table 10.2 lists the types of items that should be looked for during the inspection. (Note that Table 10.2 applies to both extension cords and portable tools.)
- Never use the tool power cord to lift the tool. If the tool must be lifted, tie a hand line or rope to the tool.
- Make certain the tool's ground connection is complete. For metal-cased tools, the safety ground connection is the difference between life and death when an internal short circuit occurs.
- Never alter the plug or receptacle on a tool. This applies especially to altering the ground connection.
- If a tool employs a twist-lock or locking type of plug, it should be securely fastened before the tool is used.
- Do not use a tool in wet or hazardous environments unless it is rated for such service by the manufacturer.
- If tools are used in wet or hazardous environments, insulating safety equipment such as rubber gloves with appropriate leather protectors should be worn.
- Only personnel who are authorized and trained in the use of power tools should be allowed to use them.
- Cord-connected tools should be subjected to the electrical tests outlined in Table 10.4. If the tool fails any of the listed tests, it should be replaced or repaired.
- When not in use, tools should be carefully stored in such a way that they cannot be damaged.
- All temporary power connections in industrial and commercial facilities should be supplied from a circuit protected by a GFCI. (See Chap. 3 for more information on GFCIs.)

Any tool that does not meet the pass/fail criteria listed in Table 10.4 should be removed from service until it can be repaired or replaced.

Current Transformers

The safety hazards of current transformers are identical for low-, medium-, and high-voltage circuits. Refer to Chap. 11 for a detailed coverage of the nature of current transformer hazards and methods for protecting workers from those hazards.

GROUNDING LOW-VOLTAGE SYSTEMS

The subject of electrical grounding is a complex one. The following sections focus on the grounding concepts and requirements of low-voltage systems as they relate to safety. Chapter 5 of this handbook provides a much more in-depth coverage of electrical grounding principles and requirements. Also, the reader may refer to the various ANSI/IEEE standards including IEEE Guide for Safety in AC Substation Grounding (ANSI/IEEE standard 80) and IEEE Recommended Practice for Grounding of Industrial and Commercial Power Systems (ANSI/IEEE standard 142).

What Is a Ground?

A *ground* is an electrically conducting connection between equipment or an electric circuit and the earth, or to some conducting body that serves in place of the earth. If a ground is properly made, the earth or conducting body and the circuit or system will all maintain the same relative voltage.

Figure 10.3*a* shows an electrical system that is not grounded. In such a situation, a voltage will exist between the ground and some of the metallic components of the power system. In Fig. 10.3*b*,

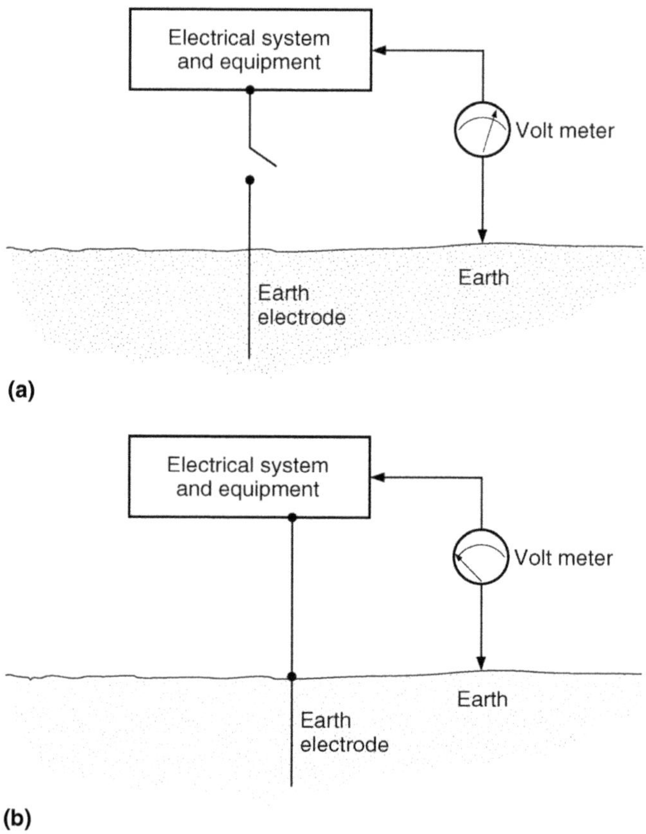

FIGURE 10.3 Grounding. (*a*) Before system is grounded, a voltage exists between the system and the earth; (*b*) when ground connection is made, no voltage exists.

the earth connection has been made. When the system is grounded, the voltage is reduced to zero between the previously energized sections of the system.

Note that Fig. 10.3 assumes that the earth is a perfect conductor. The earth is not a perfect conductor; consequently, a voltage drop may exist between the metal and the measurement point. This is covered in more detail in the Voltage Hazards sections below.

Bonding versus Grounding

Bonding is the permanent joining of metallic parts to form a continuous, conductive path. Since the earth is generally not a good conductor, bonding is used to provide a low-impedance metallic path between all metallic parts. This, in essence, bypasses the earth and overcomes its relatively high impedance. A good grounding system is a combination of solid connections between metallic parts and the earth as well as between all metallic parts.

Voltage Hazards

Figure 10.4 illustrates the four standard voltage hazards that are associated with and relieved by proper system grounding. If a system is ungrounded and the nonenergized metallic parts become energized, the metallic parts will have a measurable voltage between themselves and ground. If the system is grounded by driving ground electrodes into the earth, current will flow from the rods into the earth. At each ground electrode, the voltage will rise relative to the remote reference point. The

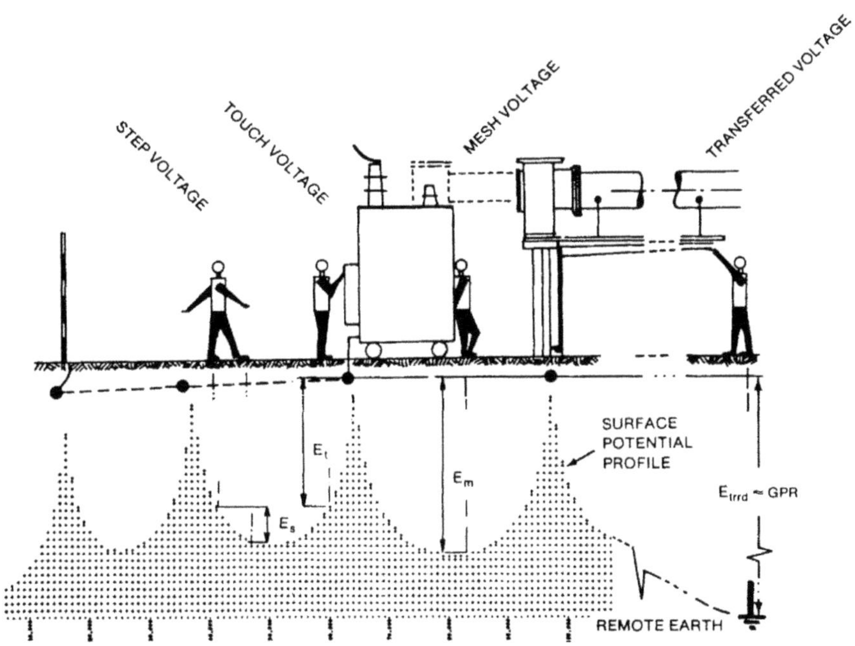

FIGURE 10.4 Four voltage hazards related to system grounding. (*Courtesy Institute of Electrical and Electronics Engineers.*)

voltage will drop off away from the electrodes and peak at each electrode. This situation creates the four types of voltage hazards as follows:

- *Step voltage.* As a worker steps across the ground, the front foot will be at a different potential than the rear foot. This effect is caused by the voltage gradient created by the ground electrodes (Fig. 10.4). Step voltage can easily reach lethal levels. It can be mitigated by increasing the number of grounding electrodes, increasing bonding between the metal parts, and employing a grounding grid.
- *Touch voltage.* Because of the voltage gradient, the voltage of the earth only a short distance from the grounded metallic equipment will be different from the voltage of the equipment. Thus, if a worker touches the grounded equipment, his or her feet will be at a different potential than his or her hands. This voltage can be lethal. Touch voltage is mitigated by increasing the number of grounding electrodes, increasing bonding between metal parts, and employing a grounding grid.
- *Mesh voltage.* Mesh voltage is the worst case of touch voltage. Cause, effect, and mitigation methods are identical to those described for touch voltage.
- *Transferred voltage.* Because metal parts have a much lower impedance than the earth, the voltage drop between remote locations is lower on the metallic connections than the earth. This means that the earth may be at a significantly different voltage than the metallic connections. Transferred voltages are particularly noticeable on neutral wires that are grounded at the service point and nowhere else. Transferred potential can be mitigated by providing the entire area with a ground grid; however, this solution is infeasible in any but the smallest systems. Transferred voltage must be mitigated by avoiding contact with conductors from remote locations and/or using rubber insulating gloves.

System Grounds

What Is a System Ground? A *system ground* is the connection of one of the conductors to the earth. Such a connection is accomplished by connecting an electric wire to the selected system conductor and the grounding electrode.

Why Are Systems Grounded? Power systems have conductors grounded for a variety of safety and operational reasons, including the following:

- Grounded systems provide sufficient short-circuit current for efficient operation of protective equipment.
- Grounded systems are less prone to transient overvoltages, which can cause insulation failures.
- Grounded systems are generally more easily protected from lightning.
- Solidly grounded systems are less prone to resonant conditions, which can cause equipment and insulation failures.

What Systems Must Be Grounded? The NEC requires that both ac and dc systems be grounded. Table 10.5 summarizes low-voltage dc grounding requirements, and Table 10.6 summarizes low-voltage ac grounding requirements.

How Are Systems Grounded? Electrical systems are grounded by connecting one of the electric conductors to earth. The conductor chosen and the location of the ground are determined as part of the engineering design for the system. Figures 10.5 through 10.8 illustrate four different low-voltage circuits and how they are grounded. Note that in some cases, the point of ground is determined by regulatory requirements such as the NEC.

The wire(s) that must be grounded depend on the voltage level and the system application. For dc systems, one of the conductors or the neutral wire is to be grounded depending on the type of system (Table 10.5). For ac systems, the NEC specifies five different locations for the selection of the grounding point. Table 10.7 identifies each of the five locations.

Systems with voltages between 480 and 1000 V phase to phase may be grounded through an impedance to limit the amount of fault current. In the circuit shown in Fig. 10.9, the resistor will

TABLE 10.5 DC Circuits That Require Grounding

Type of circuit	Circuits to be grounded	Exceptions/comments
Two-wire dc	All	Systems less than 50 V or greater than 300 V between conductors need not be grounded. Limited-area industrial systems with ground detectors need not be grounded. Certain rectifier-derived dc systems do not need to be grounded. DC fire-protective signaling circuits with no more than 0.030 A may not need to be grounded.
Three-wire dc	All	The neutral wire is grounded.

Note: See the current edition of the National Electrical Code for details.

TABLE 10.6 AC Circuits That Require Grounding

Type of circuit	Circuits to be grounded	Exceptions/comments
AC systems 50 to 1000 V	If maximum voltage to ground is less than 50 V Three-phase, 4-wire wye, if neutral is a circuit conductor Three-phase, 4-wire delta with midpoint grounded on one leg In some cases when grounded service conductor is uninsulated	The National Electrical Code has many exceptions to these grounding requirements.

Note: See the current edition of the National Electrical Code for details.

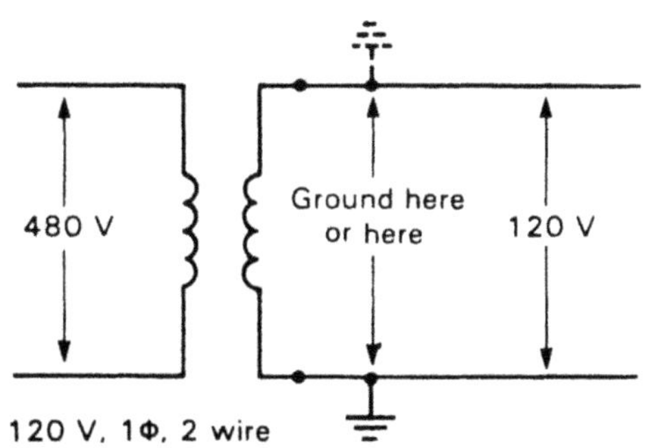

FIGURE 10.5 Grounding a 120-V single-phase circuit.

oppose current flow between the earth and the phase wires. For example, assume that the phase A wire falls to the earth. The circuit that is formed will be composed of the power system's phase-to-neutral voltage (277 V) impressed across the series combination of the power system's impedances plus the grounding resistor. If the grounding resistor is properly sized, it will limit the fault current to any maximum value that is chosen during the design.

By limiting the amount of fault current, resistance-grounded systems provide a somewhat higher level of safety. However, the engineer making the design decision to ground through a resistor must take into consideration more variables than just safety. Protective systems, ground-fault current, voltage transients, and many other such concerns must be considered before the engineer decides to use a resistance-grounded system.

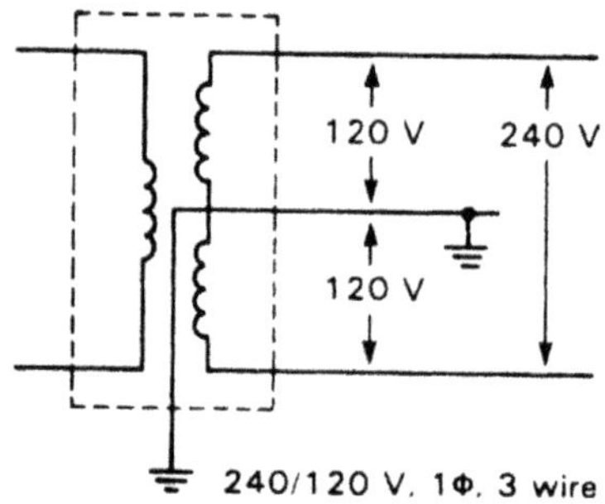

FIGURE 10.6 Grounding a 240/120-V single-phase, three-wire circuit.

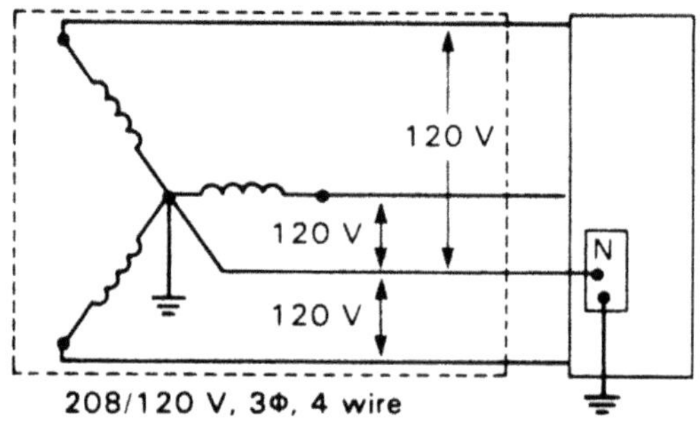

FIGURE 10.7 Grounding a 208/120-V three-phase, four-wire system.

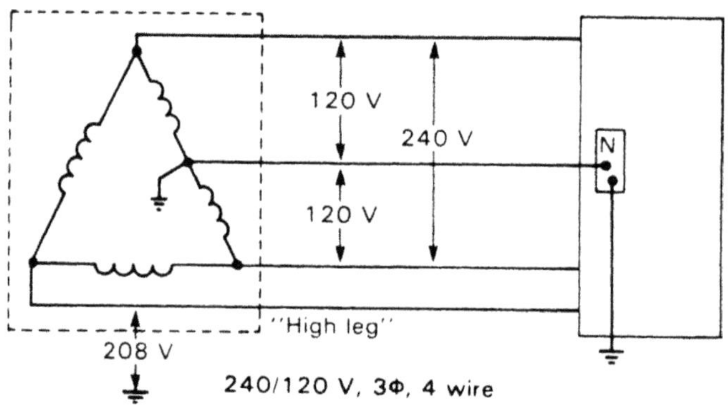

FIGURE 10.8 Grounding a 240/120-V, three-phase, four-wire system.

Equipment Grounds

What Is an Equipment Ground? An *equipment ground* is an electrically conductive connection between the metallic parts of equipment and the earth. For example, transformer cases and cores are connected to the earth—this connection is called an equipment ground. Note that the non-current-carrying metallic parts of equipment are grounded and bonded together. This bonding serves to reduce the voltage potential between all metallic parts and the earth.

TABLE 10.7 Typical Grounding Requirements for Low-Voltage Systems

Type of premises wiring circuit	Location of ground
One-phase, 2-wire	Either conductor (Fig. 10.5)
One-phase, 3-wire	The neutral conductor (Fig. 10.6)
Multiphase with one wire common to all phases	The common conductor (not illustrated)
Multiphase systems required on grounded phase	One of the phase conductors (not illustrated)
Three-phase, 4-wire circuits	The neutral conductor (Fig. 10.7)
240/120, 3-phase, 4-wire	The center point of the grounded leg (Fig. 10.8)

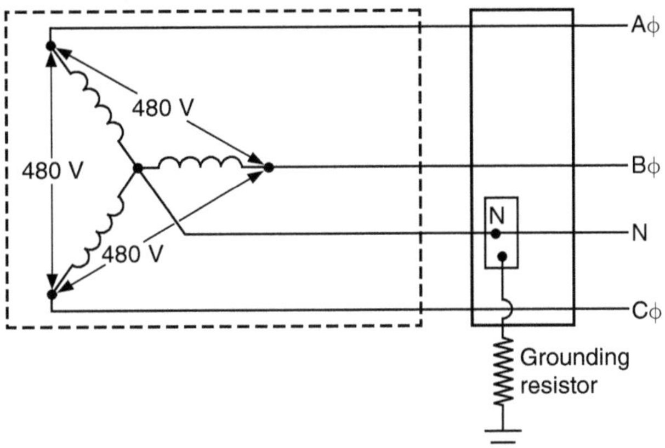

FIGURE 10.9 Resistance grounded 480/277-V, three-phase, four-wire system.

TABLE 10.8 Equipment Grounding Requirements

- Must be permanent and continuous
- Must have the capacity to conduct any fault current likely to be imposed on it
- Must have sufficiently low impedance to limit the voltage to ground and to facilitate the operation of the circuit protective devices

Why Is Equipment Grounded? The equipment ground is one of the most important safety aspects of grounding. Workers are constantly in contact with transformer shells, raceways, conduits, switchgear frames, and all the other conductive, non-current-carrying parts. Proper equipment grounding and bonding ensures that the voltage to which workers will be subjected is kept to a minimum. Proper bonding and grounding mitigates touch and step voltages.

How Is Equipment Grounded? The NEC requires that the path to ground from circuits, equipment, and metal enclosures must meet the conditions listed in Table 10.8. The NEC does not allow the earth to be the sole equipment grounding conductor. Simply setting metal equipment on the earth is insufficient. Equipment must be connected to the earth via metal electrodes and conductors.

Equipment used in grounded systems is grounded by bonding the equipment grounding conductor to the grounded service conductor and the grounding electrode conductor. That is, the equipment ground is connected to the system ground. Equipment used in ungrounded systems is grounded by bonding the equipment grounding conductor to the grounding electrode conductor.

What Equipment Must Be Grounded? Tables 10.9 through 10.12 list the types of equipment that must be grounded according to the NEC. Always refer to the current edition of the NEC for up-to-date information.

TABLE 10.9 Grounding Requirements for Equipment Fastened in Place or Connected by Permanent Wiring (by Location)

Must ground	Exceptions
• Equipment within 8 ft (2.44 m) horizontally or 5 ft (1.52 m) vertically of ground or grounded metal objects subject to human contact • Equipment located in wet or damp locations • Equipment in electrical contact with other metallic objects • Equipment located in classified hazardous locations • Equipment that is supplied by metal-clad, metal-sheathed, metal-raceway, or other wiring method that provides an equipment ground • Equipment that operates with any terminal in excess of 150 V to ground	• Enclosures for switches or circuit breakers used for other than service equipment and accessible to qualified persons only. • Metal frames of electrically heated appliances; exempted by special permission, in which case the frames shall be permanently and effectively insulated from ground. • Distribution apparatus, such as transformer and capacitor cases, mounted on wooden poles, at a height exceeding 8 ft (2.44 m) above ground or grade level. • Listed equipment protected by a system of double insulation, or its equivalent, shall not be required to be grounded. Such equipment must be distinctively marked.

TABLE 10.10 Grounding Requirements for Equipment Fastened in Place or Connected by Permanent Wiring (by Type)

Must ground	Exceptions
Switchboard frames and structures	Frames of 2-wire dc switchboards where effectively insulated
Pipe organs (generator and motor frames)	Where the generator is effectively insulated from ground and from the motor driving it
Motor frames	
Enclosures for motor controllers	Enclosures attached to ungrounded portable equipment
	Lined covers of snap switches
Elevators and cranes	
Garages, theaters, and motion picture studios	Pendant lampholders supplied by circuits less than 150 V to ground
Electric signs	
Motion picture projection equipment	
Remote-control, signaling, and fire-protective signaling circuits	
Lighting fixtures	
Motor-operated water pumps	

TABLE 10.11 Nonelectric Equipment Grounding Requirements

Must ground
Cranes
Elevator cars
Electric elevators
Metal partitions
Mobile homes and recreational vehicles

TABLE 10.12 Grounding Requirements for Equipment Connected by Cord and Plug

Must ground	Exceptions
In classified hazardous locations	
In systems that are operated in excess of 150 V to ground	Motors, where guarded. Metal frames of electrically heated appliances, exempted by special permission, in which case the frames shall be permanently and effectively insulated from ground. Listed equipment protected by a system of double insulation, or its equivalent, shall not be required to be grounded. Where such a system is employed, the equipment shall be distinctively marked.
In residential occupancies • Refrigerators, freezers, and air conditioners • Clothes washing/drying machines, sump pumps, aquariums • Handheld motor-operated tools including snow blowers, hedge clippers, etc. • Portable handlamps	Listed equipment protected by a system of double insulation, or its equivalent, shall not be required to be grounded. Where such a system is employed, the equipment shall be distinctively marked.
In other than residential occupancies • Refrigerators, freezers, air conditioners • Clothes washing/drying machines, electronic computers/data-processing equipment, sump pumps, aquariums • Handheld, motor-operated tools including hedge clippers, lawn mowers, snow blowers, etc. • Cord- and plug-connected appliances used in damp or wet locations • Tools likely to be used in wet or conductive locations • Portable handlamps	Tools and handlamps in wet locations when supplied through an isolating transformer with an ungrounded secondary of not over 50 V. Listed equipment protected by a system of double insulation, or its equivalent, shall not be required to be grounded. Where such a system is employed, the equipment shall be distinctively marked.

Ground-Fault Circuit Interrupters

GFCIs are described in detail in Chap. 3. Although few mandatory standards require universal applications of GFCI devices, prudence and common sense suggest that they should be applied in all industrial/commercial environments. Their sensitivity and operating speed (5 mA, 25 m/s) make them the only type of protective device that is capable of being used to protect human lives.

Arc-Fault Circuit Interrupters

Even in low-voltage systems, most faults will have some level of arcing. The NEC (2011) requires that arc-fault circuit interrupters (AFCIs) be installed in residential family rooms, dining rooms, living rooms, parlors, libraries, dens, bedrooms, sunrooms, recreation rooms, closets, hallways, or similar rooms. For those locations where the NEC is the baseline document for the authority having jurisdiction, AFCIs must be installed in all new locations. See Chap. 3 for more information on AFCIs.

SAFETY EQUIPMENT

Overview

Because of the delusion that "low voltage can't hurt you," many workers do not use safety equipment when working on or near energized, low-voltage conductors. In fact, low-voltage systems are extremely lethal and should be treated with the respect they deserve. The following sections describe the types of safety equipment that should be worn when working on or near energized, low-voltage conductors. The recommendations given are minimum recommendations. If additional or more stringent protection is desired or required, it should be worn. Refer to the tables in Chap. 3 for specific recommendations.

Hard Hats

Protective headgear for persons working on or near energized, low-voltage circuits should provide both mechanical and electrical protection. Since the ANSI Z89.1 class C helmet provides no electrical protection, class C helmets should not be worn. Workers should be supplied with and should wear either ANSI Z89.1 class G or class E helmets. If workers are never required to work around high-voltage circuits, the ANSI Z89.1 class G helmet may be used. If, however, workers are required to work around both high- and low-voltage circuits, they should be supplied with and should wear ANSI Z89.1 class E helmets. Table 10.13 summarizes the characteristics of the three ANSI Z89.1 classes. Note that class G and E were formerly class A and B, respectively.

Eye Protection

Even low-voltage systems are capable of producing extremely powerful and hazardous electric arcs and blasts. This is especially true of 480-V and 575-V systems. Because of this, eye protection for electrical workers should provide protection against heat and optical radiation. The most recent edition of ANSI standard Z87.1 provides a selection chart as well as a chart that illustrates the various protection options available. The selection chart is reproduced in this handbook as Table 10.14, and eye protection options are shown in Fig. 10.10.

Electrical workers should consider using eye protection that combines both heat and optical radiation protection. Table 10.14 and Fig. 10.10 show several different types of equipment that will provide such protection including types B, C, D, E, and F. If arcing and molten splashing is possible, as with open-door switching or racking of circuit breakers, the type N eye protection should be employed.

TABLE 10.13 Summary of the Characteristics for ANSI Z89.1 Class C, E, and G Hard Hats

Class	Description	Comments
G	Reduce the impact of falling objects and reduce danger of contact with exposed, low-voltage conductors. Representative sample shells are proof-tested at 2200 V phase to ground.	Recommended to be worn by personnel working around only low-voltage circuits
E	Reduce the impact of falling objects and reduce danger of contact with exposed high-voltage conductors. Representative sample shells are proof-tested at 20,000 V phase to ground.	Recommended to be worn by personnel working around high- and low-voltage circuits
C	Intended to reduce the force of impact of falling objects. This class offers no electrical protection.	Should not be worn by personnel working on or around energized conductors of any voltage

TABLE 10.14 Eye Protection Selection Chart

		Assessment SEE NOTE (1)	Protector type[a]	Protectors	Limitations	Not recommended
IMPACT	Chipping, grinding, machining, masonry work, riveting, and sanding	Flying fragments, objects, large chips, particles, sand, dirt, etc.	B, C, D, E, F, G, H, I, J, K, L, N	Spectacles, goggles, face shields SEE NOTES (1) (3) (5) (6) (10). For severe exposure, add N.	Protective devices do not provide unlimited protection. SEE NOTE (7).	Protectors that do not provide protection from side exposure. Filter or tinted lenses that restrict light transmittance, unless it is determined that a glare hazard exists. Refer to Optical Radiation.
HEAT	Furnace operations, pouring, casting, hot dipping, gas cutting, and welding	Hot sparks	B, C, D, E, F, G, H, I, J, K, L, *N	Face shields, goggles, spectacles *For severe exposure, add N. SEE NOTES (2) (3).	Spectacles, cup and cover type goggles do not provide unlimited facial protection. SEE NOTE (2).	Protectors that do not provide protection from side exposure.
		Splash from molten metals	*N	*Face shields worn over goggles H, K SEE NOTES (2) (3).		
		High temperature exposure	N	Screen face shields, reflective face shields SEE NOTES (2) (3).	SEE NOTE (3).	
CHEMICAL	Acid and chemicals handling, degreasing, and plating	Splash	G, H, K *N	Goggles, eyecup and cover types *For severe exposure, add N.	Ventilation should be adequate but well protected from splash entry.	Spectacles, welding helmets, hand shields.
		Irritating mists	G	Special-purpose goggles	SEE NOTE (3).	
DUST	Woodworking, buffing, general dusty conditions	Nuisance dust	G, H, K	Goggles, eyecup and cover types	Atmospheric conditions and the restricted ventilation of the protector can cause lenses to fog. Frequent cleaning may be required.	

(Continued)

TABLE 10.14 (Continued)

		Assessment SEE NOTE (1)	Protector type[a]	Protectors: Typical filter lens shade	Protectors: Protectors	Limitations	Not recommended
OPTICAL RADIATION							
	Welding:			SEE NOTE (9).			
	Electric arc		O, P, Q	10-14	Welding helmets or welding shields	Protection from optical radiation is directly related to filter lens density. SEE NOTE (4). Select the darkest shade that allows adequate task performance.	Protectors that do not provide protection from optical radiation. SEE NOTE (4).
	Welding:			SEE NOTE (9).			
	Gas		J, K, L, M, N, O, P, Q	4-8	Welding goggles or welding face shield		
	Cutting			3-6			
	Torch brazing			3-4		SEE NOTE (3).	
	Torch soldering		B, C, D, E, F, N	1.5-3	Spectacles or welding face shield		
	Glare		A, B	Spectacle SEE NOTES (9) (10).		Shaded or special-purpose lenses, as suitable. SEE NOTE (8).	

[a] Refer to Fig. 10.10 for protector types.

NOTE: For NOTES referred to in this table, see text accompanying Fig. 10.10.

Source: Courtesy American National Standards Institute.

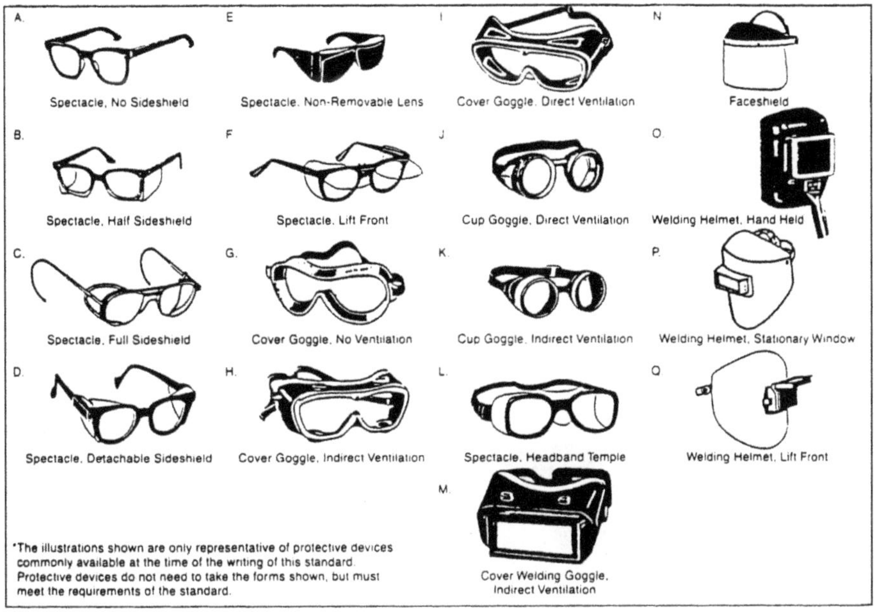

NOTES:

(1) Care shall be taken to recognize the possibility of multiple and simultaneous exposure to a variety of hazards. Adequate protection against the highest level of each of the hazards must be provided.

(2) Operations involving heat may also involve optical radiation. Protection from both hazards shall be provided.

(3) Faceshields shall only be worn over primary eye protection.

(4) Filter lenses shall meet the requirements for shade designations in Table 1.

(5) Persons whose vision requires the use of prescription (Rx) lenses shall wear either protective devices fitted with prescription (Rx) lenses or protective devices designed to be worn over regular prescription (Rx) eyewear.

(6) Wearers of contact lenses shall also be required to wear appropriate covering eye and face protection devices in a hazardous environment. It should be recognized that dusty and/or chemical environments may represent an additional hazard to contact lens wearers.

(7) Caution should be exercised in the use of metal frame protective devices in electrical hazard areas.

(8) Refer to Section 6.5, Special Purpose Lenses.

(9) Welding helmets or handshields shall be used only over primary eye protection.

(10) Non-sideshield spectacles are available for frontal protection only.

FIGURE 10.10 Eye protection devices. (Refer to Table 10.14 for selection criteria.) (*Courtesy American National Standards Institute.*)

Arc Protection

Low-voltage systems can create and sustain significant electric arcs accompanied by electric blast. Employees performing work in 575-V, 480-V, or 208-V phase-to-phase systems should wear the type of clothing listed in Table 10.15. Refer to Chap. 4 for methods that can be used to calculate flash clothing weights.

Rubber Insulating Equipment

The introduction of the so-called low-voltage rubber glove has made the use of insulating protection much more convenient than in the past. Personnel working on or near energized low-voltage conductors should wear low-voltage rubber gloves with appropriate leather protectors. Such gloves will have either beige (class 00) or red (class 0) ANSI labels and are rated for use in circuits of up to and including 500 V rms (class 00) or 1000 V rms (class 0). Many workers find the class 00 to be preferable because of the higher flexibility. However, improvements in insulation materials and designs have allowed manufacturers to greatly improve the flexibility of the class 0 gloves.

TABLE 10.15 Recommended Arc Protection and Clothing for Persons Working on or Near Energized Low-Voltage Conductors

Description of work	Type of clothing
Routine work on or close to energized conductors (circuits above 50 V to ground)	Flame-resistant work clothing
Open box switching and/or fuse removal (circuits 208 V phase to phase and higher)	Flame-resistant work clothing and/or flash suits
Installation or removal of low-voltage circuit breakers and/or motor starters with energized bus (circuits 208 V phase to phase or higher)	Flame-resistant work clothing and/or flash suits

* Head, eye, and insulating protection should also be worn. See Chap. 3 for additional information on the use of various types of protective clothing.

† See Chap. 4 for calculating the required weight of protective clothing.

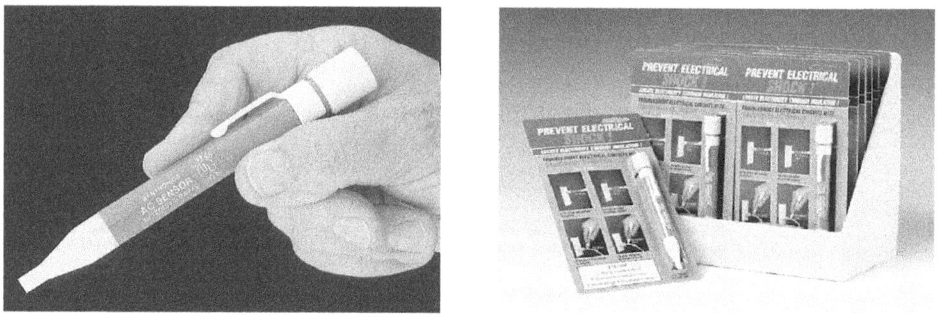

FIGURE 10.11 Volt-stick proximity voltage sensor for use on circuits up to 600 V. (*Courtesy Santronics, Inc.*)

FIGURE 10.12 Safety voltage/continuity tester for circuits of 600 V ac or dc and less. (*Courtesy Ideal Industries, Inc.*)

Voltage-Testing Devices

Proximity or contact testers intended for use in low-voltage circuits should be used to test circuits and to verify that they are de-energized and safe to work on. Voltage-testing devices are described in Chap. 3. Examples of low-voltage–measuring instruments are shown in Figs. 10.11 through 10.13. Workers also should use receptacle and GFCI testers such as that shown in Fig. 10.14. These testers are especially important for the verification of the ground path in a duplex receptacle.

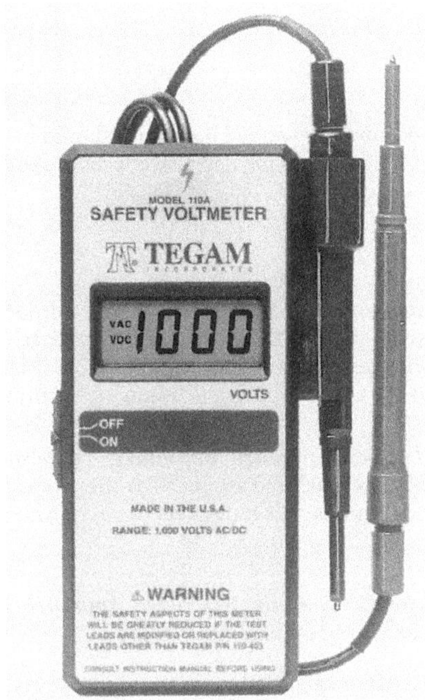

FIGURE 10.13 Digital readout contact-type safety voltmeter for circuits of 1000 V ac and dc or less. (*Courtesy Tegam, Inc.*)

FIGURE 10.14 Receptacle and GFCI tester. (*Courtesy Direct Safety Company, Phoenix, Arizona.*)

SAFETY PROCEDURES

General

The general procedures described in Chap. 4 should be used on all circuits in excess of 50 V to ground. The following sections describe key procedures that apply to low-voltage circuits.

Approach Distances

Although approach distances are laid out in a number of OSHA references, NFPA 70E provides a much more useful and practical concept for approach distances. The following paragraphs identify methods that may be used to determine approach distances. Refer to Chap. 4, Fig. 4.35 and Table 4.15 for reference to the terms used in the following section.

Crossing the Limited Approach Boundary. Qualified workers may cross the limited approach boundary if they are qualified to perform the work. Unqualified workers are allowed to cross the limited approach boundary as long as they are continuously escorted by and under the constant direct supervision of a qualified person.

Requirements for Crossing the Restricted Approach Boundary. To cross the restricted approach boundary, the following criteria must be met:

- The worker must be qualified to do the work.
- There must be a plan in place that is documented and approved by the employer.
- The worker must be certain that no part of the body crosses the prohibited approach boundary.
- The worker must work to minimize the risk that may be caused by inadvertent movement by keeping as much of the body out of the restricted space as possible. Allow only protected body parts to enter the restricted space as necessary to complete the work.
- Personal protective equipment must be used appropriate for the hazards of the exposed energized conductor.

Requirements for Crossing the Prohibited Approach Boundary. NFPA 70E considers crossing the prohibited approach boundary to be the same as working on or contacting an energized conductor. To cross into the prohibited space, the following requirements must be met:

- The worker must have specified training required to work on energized conductors or circuit parts.
- There must be a plan in place that is documented and approved by the employer.
- A complete risk analysis must be performed.
- Authorized management must review and approve the plan and the risk analysis.
- Personal protective equipment must be used appropriate for the hazards of the exposed energized conductor.

Voltage Measurement

The voltage-measurement techniques defined in Chap. 4 should be employed on circuits of all voltages, including 1000 V and below. Note that although OSHA standards do not require the three-step measurement process for circuits of below 600 V, best safety practice does call for instrument checks both before and after the actual circuit measurement.

Locking and Tagging

Lockout-tagout and energy control procedures apply to circuits of all voltage levels. Refer to Chap. 4 for detailed coverage of lockout-tagout procedures.

Closing Protective Devices After Operation

Workers should never reclose any protective device after it has operated until it has been determined that it is safe to do so. Several criteria may be used to determine whether it is safe to reclose the protective device (Table 10.16). Other conditions may or may not indicate a safe reclosing situation. Under no circumstances should a device be closed multiple times to search for the source of a fault.

ELECTRICAL SAFETY AROUND ELECTRONIC CIRCUITS

Modern technology requires many persons to work on or near electronic circuitry. Such circuits can present special or unusual hazards. The following sections provide information about the nature of the hazard and some specific procedures that may be used by workers to enhance their personal safety.

The Nature of the Hazard

Frequencies. The relationship of frequency to electrical hazards is discussed in Chap. 1. Generally, the following points apply:

- DC currents and ac currents up to approximately 100 Hz seem to affect the body in a very similar manner. For all practical purposes, when working around a dc circuit, the worker should use the same types of procedures as when working around power system frequencies.
- Above 100 Hz, the threshold of perception increases. Between 10 and 100 kHz, the threshold increases from 10 to 100 mA.

Capacitive Discharges. According to the NFPA 70E, the following are true with respect to capacitor discharges:

- A current caused by the discharge of a 1-μF, 10,000-V capacitor may cause ventricular fibrillation.
- A current caused by the discharge of a 20-μF, 10,000-V capacitor will probably cause ventricular fibrillation.

Specific Hazards of Electronic Equipment. Although seemingly harmless, electronic circuits present a number of hazards, including the following:

1. Electrical shock from 120-, 240-, or 480-V ac power supplies
2. High power supply voltages

TABLE 10.16 Typical Situations That May Allow Reclosing of a Protective Device That Has Operated

• The faulted section of the system is found and repaired.
• The nature of the protective device makes it clear that no hazard is present. For example, if the device that operated is an overload type of device, it may be safe to reclose.
• The reclosing operation can be made in such a way that the workers are not exposed to additional hazard. For example, if the reclosing operation can be made by remote control, and if all personnel are kept away from all parts of the circuit, it may be reclosed.

3. Possible shock and burn hazards caused by radio frequency (RF) fields on or around antennas and antenna transmission lines
4. RF energy–induced voltages
5. Ionizing (x-radiation) hazards from magnetrons, klystrons, thyratrons, cathode ray tubes (CRTs), and other such devices
6. Nonionizing RF radiation hazards from
   (a) Radar equipment
   (b) Radio communication equipment
   (c) Satellite earth-transmitters
   (d) Industrial scientific and medical equipment
   (e) RF induction heaters and dielectric heaters
   (f) Industrial microwave heaters and diathermy radiators

Special Safety Precautions

The following methods are offered in addition to the other safety equipment and procedures that are discussed throughout this handbook.

AC and DC Power Supplies. The nature of these hazards is similar to the hazards that are discussed throughout Chap. 4 and this chapter. One piece of equipment that is finding increasing use in protecting workers from these types of hazards is the PVC sheeting that can be placed over the exposed circuit parts. This PVC material provides an insulating blanket for voltages up to 1000 V and will allow the worker to perform the necessary tasks in the equipment.

Protection from Shock and Burn Caused by RF Energy on Antennas and/or Transmission Lines. Avoidance of contact is the best possible protection for this type of hazard. Transmitting equipment should always be disabled and made electrically safe before workers are allowed to approach antennas or transmission lines.

Electrical Shock Caused by RF-Induced Voltage. Electric shocks from contacting metallic objects that have induced RF voltages on them can be dangerous in at least two ways:

1. The surprise effect of the shock can cause the victim to fall from a ladder or other elevated location.
2. RF discharge can cause ventricular fibrillation under the right circumstances.

Three methods can be used to protect personnel from induced RF voltages:

1. De-energize the RF circuits to eliminate the energy.
2. Use insulating barriers to isolate the metal objects from the worker.
3. Ground and bond all non-current-carrying metal parts such as chassis, cabinets, covers, and so on. Proper RF ground wires must be very short compared to the wavelength of the RF. If a solid ground cannot be reached because of distance, a counterpoise type of ground can be employed. The design of such a ground is beyond the scope of this handbook. The reader should refer to one of the many engineering texts available.

Radiation (Ionizing and Nonionizing Hazards). The best methods for protecting workers from this type of hazard are:

1. De-energize the circuit so that the worker is not exposed to the radiation.
2. Protect the worker from the radiation by using appropriate shielding.

STATIONARY BATTERY SAFETY

Introduction

Stationary batteries (Fig. 10.15) are used for various types of standby and emergency power requirements throughout electrical power systems. Batteries are usually connected to the power system as shown in Fig. 10.16. Because of their construction and energy capacity, batteries offer a special type of safety hazard that includes chemical, electrical, and explosive hazards.

Basic Battery Construction

Stationary batteries operate on the basic principle of galvanic action; that is, two dissimilar materials will produce a voltage when they are put close together. While the basic chemistry of stationary batteries may vary and change, the two most common types in use today are the lead-acid and the nickel-cadmium types.

FIGURE 10.15 Partial view of a typical stationary battery installation.

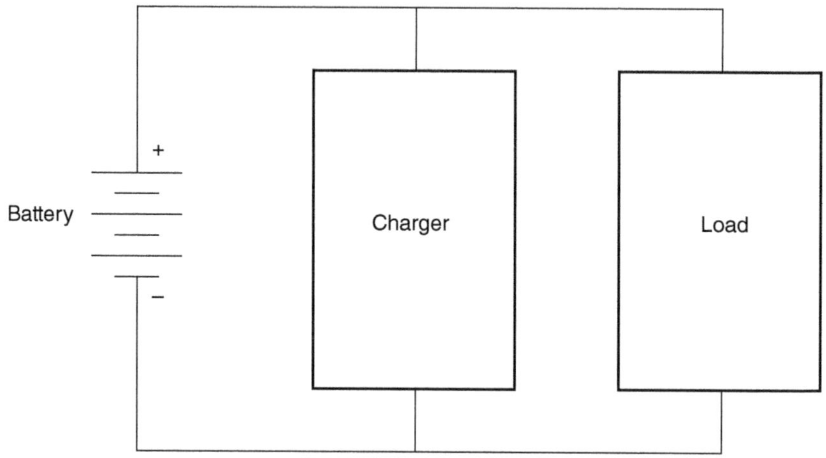

FIGURE 10.16 Connection diagram of a typical stationary battery installation.

Lead-Acid Batteries. The lead-acid battery uses lead (Pb) and lead peroxide (PbO_2) for its negative and positive plates, respectively. The electrolyte in which the plates are immersed is a solution of sulfuric acid (H_2SO_4) and water (H_2O). The basic chemical reaction is shown in Eq. 10.1.

$$PbO_2 + Pb + 2H_2SO_4 \underset{\text{Discharging}}{\overset{\text{Charging}}{\rightleftarrows}} 2PbSO_4 + 2H_2O \tag{10.1}$$

Modern lead-acid batteries are constructed in one of two general formats:

1. Vented cell batteries (also known as flooded cells) are a mature technology in which the plates are completely immersed in the electrolyte. The containers are generally open to the atmosphere with flame arresters used to minimize the chance of explosion or fire.
2. Valve-regulated cell (VRLA) is also called the starved electrolyte cell. Such a cell is essentially sealed except for the presence of a relief valve. The electrolyte is restrained internally either by gelling the material or by insertion of a fiber mat. The VRLA is, in its essence, lead-acid technology.

NiCad (Nickel-Cadmium) Batteries. The NiCad batteries used for stationary applications use nickel hydrate [$Ni(OH)_3$] for the positive plate and cadmium (Cd) for the negative plate. The electrolyte commonly used in the NiCad battery is potassium hydroxide (KOH). The basic chemical action of the NiCad battery is shown in Eq. 10.2. Notice that the electrolyte does not take part in the chemical reaction.

$$2Ni(OH)_3 + Cd \underset{\text{Discharging}}{\overset{\text{Charging}}{\rightleftarrows}} 2Ni(OH)_2 + Cd(OH)_2 \tag{10.2}$$

Safety Hazards of Stationary Batteries

Electrical Hazards. Stationary batteries have sufficient stored energy to represent both shock and arcing hazards. Additionally, the high current capacity of stationary batteries can cause extremely dangerous heat. Severe burns have been caused by the high battery currents through personal jewelry (such as wedding rings) and tools.

Chemical Hazards. The electrolytic solutions from both of the major types of batteries are destructive to human tissues. Although not normally in strong concentrations, the sulfuric acid and potassium hydroxide solutions can destroy eye tissue and cause serious burns on more hardy locations.

Explosion Hazards. Explosions of stationary batteries result from two different sources:

1. Excessive heat from ambient conditions, excessive charging, or excessive discharging can cause a cell to pressurize and explode if it cannot properly vent. This hazard exists to some degree for all batteries; however, it tends to be more of an issue with VRLAs and NiCads.
2. The chemical reaction during charging of a lead-acid battery is not 100 percent efficient. In fact, if the battery is charged too quickly, not all of the hydrogen will find a sulfate radical with which it can combine. This can cause the release of hydrogen to the air. Concentrations of hydrogen in air of more than 4 percent or 5 percent by volume can explode violently.

Battery Safety Procedures

Electrical Safety. Table 10.17 lists the minimum safety procedures and equipment that should be available to personnel working on or near stationary batteries. Refer to manufacturer's instructions for more specific recommendations.

ELECTRICAL HAZARDS OF THE HOME-BASED BUSINESS

For many of us, the daily commute to and from our place of business is a thing of the past. More and more workers are being required or opting to work at home. The reasons for this change include the following:

- The Internet and other modern communication methods have greatly reduced the need for face-to-face meetings.
- Many entrepreneurs are starting and running businesses that are technology based.
- Even those workers who must meet with or visit clients can keep their records and scheduling at home and visit their clients' locations as they need to.

TABLE 10.17 Recommended Safety Procedures/Equipment for Stationary Batteries

Hazard	Protection procedure/equipment
Electrical	• Low-voltage rubber gloves (class 00 or class 0). • Insulated tools. • Arc protection (face shield and flame-retardant clothing, minimum).

Chemical	• Chemical protective apron.
	• Chemically protective face shield and goggles.
	• Chemically resistant gloves.
	• Safety shoes.
	• Ample supply of pure water.
	• Eye and body wash station.
	• Neutralizing solution. (Use with caution and only with the approval of the battery manufacturer.)
	  • NiCad—7 oz boric acid/gal H_2O.
	  • Lead-acid—1 lb baking soda/gal H_2O.

Explosion	• Be sure that battery room is adequately ventilated. Typically, hydrogen concentrations should be kept to less than 1%. • Use nonsparking insulated tools. • A class C fire extinguisher should be immediately available.

- Even if customer records, job records, and scheduling information must be maintained on a remote office computer, the proliferation of smartphones, tablet computers, and the Internet makes it simple to connect and access the needed information.
- Hotel meeting rooms and business centers that will rent meeting space on a short-term basis are much more economical than sustaining a brick-and-mortar storefront.
- Established businesses can save substantial amounts of money and resources by reducing their on-site workforce. Knowledgeable workers are especially suitable for working from home.
- The costs and time required for commutes are virtually eliminated.
- The additional cost of working at home is substantially less than the savings that can be realized.
- Some workers are able to take income tax deductions when they dedicate a work space to their business.

In many cases, the only equipment needed for a work-at-home business is a dedicated work space or office, a computer, an Internet connection, and a telephone. Even traditional file cabinets, which used to occupy many square meters of space, are not required since most files are now stored digitally on a hard drive.

Working at home presents one significant drawback—safety, in particular electrical safety. When we work at home, we are no longer under the umbrella of an employer-regulated safety program. Further, governmental safety regulations are generally not applicable when you are working in your own home. The only safety regulation commonly applied at home is the National Electrical Code (NFPA 70). But the NEC is an installation-based standard and has very little to say about behavioral or performance safety issues in the home.

This section of the chapter provides information for those who work at home. The reader should also refer to the other sections of this chapter and the appropriate sections of Chaps. 3 and 4.

Electrical Hazards in the Home

Practicing electrical safety is an easy task if you have an understanding of the electrical hazards. Always remember that electricity will flow through all available paths between two locations of different voltage. It will go through wire, metal, wet objects ... or you. It's invisible, but very real, so treat it with respect. Electricity is often referred to as a "silent killer" because it cannot be seen, smelled, heard, or tasted.

Common electrical hazards in the home and home-based business include electrical shock or electrocution and fires caused by overloading circuits. Each year hundreds of people are electrocuted in their homes, and thousands are injured in electricity-related accidents, accidents that can be prevented with a little foresight and some common sense.

We need to consider not only those who are working from a home office or business but also their family members, especially children. Many areas of a home-based business, such as bathrooms, garages, and/or shops, and sometimes even the office area, are shared by family members.

Here are some basic home business safety tips:

- Don't overload electrical outlets. It is very easy to overload a circuit in a home office. This is generally because we have a lot of computer-related equipment to plug in and few receptacles to plug into. The additional need for receptacles often leads to the use of plug strips. The use of a single plug strip is generally not an issue; it is when we start to "daisy-chain" them that we run into trouble. Sometimes we overload the receptacle and not the circuit and end up with overheating and fires. If fuses blow or circuit breakers trip frequently, you should have your circuits and wiring checked by a qualified/licensed electrician.
- Never unplug or carry anything by its cord. And don't run cords under carpets or furniture; the cords can overheat or become frayed and cause a fire.

- Use only equipment and appliances approved by Underwriters Laboratories (look for the UL listing on the label) or other recognized testing laboratories.
- Make sure you comply with the NEC requirements for GFCI-protected receptacles. Additionally, make sure you test the GFCI monthly, as required by the listing and labeling of the device, to ensure its proper operation. See Chaps. 3 and 11 of this handbook for more information on GFCIs.
- Make sure that the NEC requirements for AFCI-protected receptacles are complied with. See Chaps. 3 and 11 of this handbook for more information on AFCIs.
- Tamper-resistant receptacles should be used where needed to prevent children from inserting metal objects into the slots of the receptacle.
- Keep all radios, hair dryers, and other appliances secured or out of bathrooms. Appliances like hair dryers should never be used near water-filled tubs and sinks. Teach your children that electricity and water don't mix.
- Unplug equipment and appliances before you clean them and when they are not in use. Make sure you use all three prongs of your electric plugs, and replace worn or frayed cords immediately. Never force a plug into an outlet if it doesn't fit, and never nail or tack cords to walls or floors.
- Keep electrical cords out of the reach of children.
- Teach your children not to poke items into electrical outlets. Use plug covers or inserts in all your outlets.

The Electrical Safety Foundation International (ESFI) reports that "home electrical problems account for an estimated 51,000 fires each year, resulting in almost 500 deaths, more than 1,400 injuries and $1.3 billion in property damage. Electrical distribution systems are the third leading cause of home structure fires."

The ESFI *Electrical Safety Workbook* (free download online) provides an introduction to the basic home electrical system, as well as information to help you answer questions about your home.

ESFI also provides some information concerning older homes, as follows:

> **Know the Dangers in Your Older Home**
> Many of these home fires occur in aging homes. Our dependence on electricity is increasing every day, and we are expecting more out of our home's electrical system. According to the U.S. Census Bureau, the average home in the U.S. is 37 years old. These homes were built before many of the electronics and appliances we use today were even invented. Unfortunately, our increased demand for energy can overburden an older home's electrical system, resulting in fires or electrocutions.
>
> By educating yourself on the dangers commonly found in older homes, you can take an active role in protecting yourself. ESFI's *Know the Dangers in Your Older Home* booklet provides you with a simple and easy checklist that can help you identify electrical hazards in your home. It also introduces new, safer technologies that can protect your family from injury and your home from fire.

Contact the Electrical Safety Foundation International at www.esfi.org for more information on home and workplace electrical safety.

Working Alone

When working alone, you are the one responsible for electrical safety. The preceding information is vital to your electrical safety success in a home-based business. You should also be aware of warning signs at your home office. If you smell a bad odor coming from electrical outlets, hear high-pitched noises from areas, or notice electricity going on and off, consider hiring an electrician to take care of the problem. Another important thing to always remember is not to overload outlets as noted previously. Always make sure you unplug unused equipment and appliances to reduce the risk of an electrical fire. Keep an electrical accident from happening by following the suggested electrical safety work practices identified in this chapter.

Working with Employees

If you have employees in your home-based business, then you have at least two additional responsibilities to ensure their safety.

- Having workers in your home exposes you to the liability of any type of accident—electrical or not.
- If you have employees in your home, your home may be subject to OSHA safety rules. You should obtain the advice of an attorney to be made aware of your liabilities.

Again, the information provided here is basic and vital to your success with regard to electrical safety.

The work environment can be worrying and often overwhelming. Employers may sometimes be too caught up in projects to notice the kinds of hazards that need attention. So employers should hire maintenance personnel to take care of problems that may result in accidents or injuries for employees in order to maintain electrical safety at their home business.

EVALUATING ELECTRICAL SAFETY

Electrical safety does not just happen; it requires knowledge of the electrical hazards and potentially hazardous conditions and situations, and it takes constant effort and vigilance to maintain an electrical safety program, whether you are working alone or with employees.

There are several good reasons for consistently evaluating the electrical safety program. We could refer to this as an electrical safety checkup. Consider the following:

1. Maintain safety in the operation of your home-based business.
2. Insurance agencies may require a risk assessment inspection, and if you maintain one, this is accomplished without any additional effort.
3. Find and correct all safety hazards.
4. Save energy and cut costs.
5. Peace of mind.

Electrical Safety Checklists

The use of an electrical safety checklist or inspection procedure can be very helpful in maintaining a consistent electrical safety program for all who may be exposed to the hazards. Maintaining electrical equipment and systems in a safe, reliable condition is essential to personnel safety as well as reducing the risk of an electrical fire.

The following are common areas found on many electrical safety checklists and should be considered.

- Lightbulb wattage: If the fixture calls for a maximum of 60 watts, *do not* put in a 100-watt bulb just because you want more light.
- Switch and wall outlet operation and condition: If the light switch does not always turn on the light and you have to wiggle it to get the light to come on, you have a problem that could cause a fire. If, when you plug in equipment, you have to wiggle it around to get a connection, a problem exists with either the cord and plug or the receptacle that could cause a fire.
- Shock or electrocution: *Any* exposed electrical conductor or circuit part presents a shock or electrocution hazard. Fix the problem, *now*.
- Check arc-fault circuit interrupters (AFCIs): Checking AFCIs is very beneficial to prevent a fire from an arc-fault in equipment, circuits, light switches, and receptacles.

- Ground-fault circuit interrupters (GFCIs): Their use will prevent electrocution. Whether you have a newer home with three-wire grounded circuits or an older home with two-wire ungrounded circuits, a GFCI will work on either system and will provide you with needed protection. GFCIs must be tested monthly, as specified by the listing and labeling of the device. How do you know it works if you don't test it? Think of the GFCI as cheap life insurance.
- Check for safety and security lighting: This is especially a good idea when working at night or in other low-light situations.
- Check grounding systems: The problem with grounding is that it is out of sight and out of mind. Electrical equipment will work just fine without grounding, but it is not safe because there is not a ground-fault path to cause a circuit breaker to trip or a fuse to blow in the event of a ground fault.
- Check for appropriate surge protection: Surge protection is a good practice if you want to save your valuable electronic equipment, especially your computer equipment.
- Portable heater safety: It may be cold and you want that extra heat under your desk to warm up your feet and legs, but using a portable heater is a very bad idea. Portable space heaters generally draw quite a bit of current, and when plugged in under the desk with all of your computer equipment, the receptacle is generally overloaded, especially when everything is plugged into a power strip or "daisy-chained" power strips. This is a formula for disaster, usually a fire.
- Check heating and air-conditioning systems: If you want to stay warm in the winter and cool in the summer, this is vital. Sensitive electronic equipment works more efficiently when at consistent, moderate temperatures.
- Check for proper placement and presence of smoke detectors: If there is a fire, these devices could very well save your life.
- Test smoke detectors: Make sure they work so they can provide the early warning that is needed to get out of the building if there is a fire.
- Test carbon monoxide detectors, if present: These are a good idea if you have an attached garage and need to warm up a vehicle in cold weather. They are also useful if you are using a portable generator during power outages. The biggest hazard with using portable generators is that they are generally placed too close to doors and windows where the exhaust (carbon monoxide) enters the building.
- Check the electrical panel for appropriate labels and operation: The National Electrical Code specifies a minimum clearance around electrical panels. The NEC also specifies that each disconnecting device (circuit breaker) be labeled to identify its purpose—accurately fill out the panel directory. If you need to turn off (de-energize) a circuit, it is always good to know which circuit.

Electrical Inspections by Professionals

If you don't have the electrical knowledge (qualification), then you need to consider hiring a qualified electrical professional (contractor) to perform the inspections noted previously. Electricity is not something you want to deal with if you are not qualified to do so. Remember the statistics noted earlier by the Electrical Safety Foundation International—you don't want to become one of those statistics.

CHAPTER 11
MEDIUM- AND HIGH-VOLTAGE SAFETY SYNOPSIS

INTRODUCTION

High-voltage systems* have three uniquely hazardous characteristics. First, the likelihood of an arc occurring increases as the system voltage increases. Although arc energies in high-voltage systems are often smaller than low-voltage circuits, the chance of an arc occurring is greater. Second, most workers do not have much experience working on or around such systems. Personnel inexperience can lead to very serious accidents and injuries. And finally, high voltages can puncture through the keratin skin layer, virtually eliminating the only significant skin resistance. Lack of knowledge can be corrected by personnel training. Chapter 14 covers the topic of personnel training in great detail.

The energy content and skin-puncturing characteristics of high-voltage systems are inherent to their nature. Safety precautions must be strictly observed to reduce the possibility of accidents, and safety equipment must be worn to reduce the severity of accidents.

This chapter highlights some of the key or critical safety concepts related to high-voltage systems. Many of the types of safety precautions used in high-voltage systems are identical to those used in low-voltage systems. Such precautions have been repeated in this chapter.

HIGH-VOLTAGE EQUIPMENT

Current Transformers

Note: The material in this section applies to current transformers (CTs) used in low-, medium-, and high-voltage circuits. It is included in this chapter because CTs are much more common in medium- and high-voltage systems.

CTs (Fig. 11.1) are used to reduce primary current levels to lower values that are usable by instruments such as meters and protective relays. In doing this, the primary winding of the CT is connected in series with the system load current and the secondary winding is connected to the instruments (Fig. 11.2).

If the secondary of an energized transformer is open-circuited, an extremely high voltage will appear at the secondary (Fig. 11.3). Depending on the type of transformer and the conditions at the time of the open circuit, this high voltage can create shock, arc, and blast hazards. CTs can explode violently when their secondary circuits are opened.

Caution: Never open-circuit the secondary of an energized CT. A CT secondary circuit should always be terminated with rated load or a short circuit.

* ANSI/IEEE standard 141 defines low-voltage systems as those that are less than 1000 V, medium-voltage systems as those equal to or greater than 1000 V and less than 100,000 V, and high-voltage systems as those equal to or greater than 100,000 V. To simplify notation, this chapter will refer to all systems of 1000 V or higher as *high-voltage* systems.

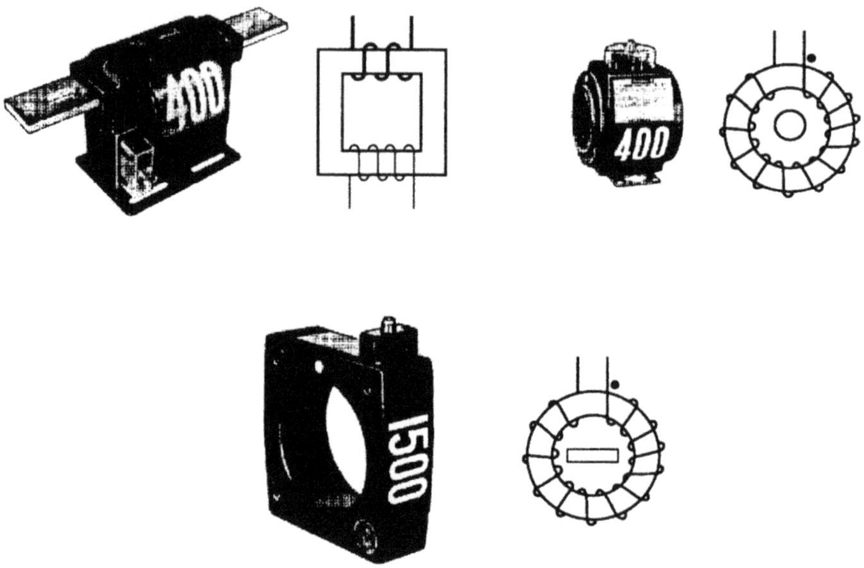

FIGURE 11.1 Typical current transformers. (*Courtesy AVO Training Institute.*)

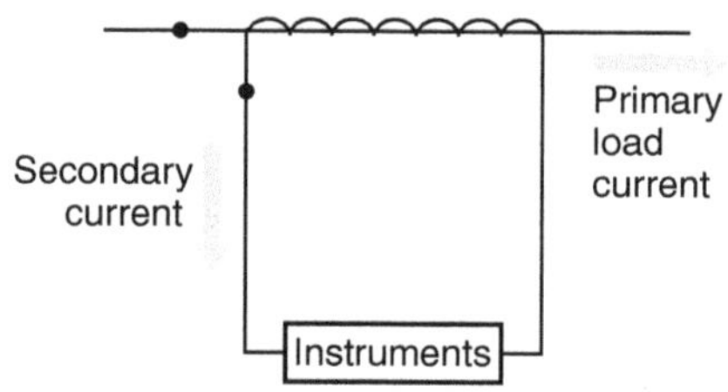

FIGURE 11.2 Schematic diagram of a current transformer connection.

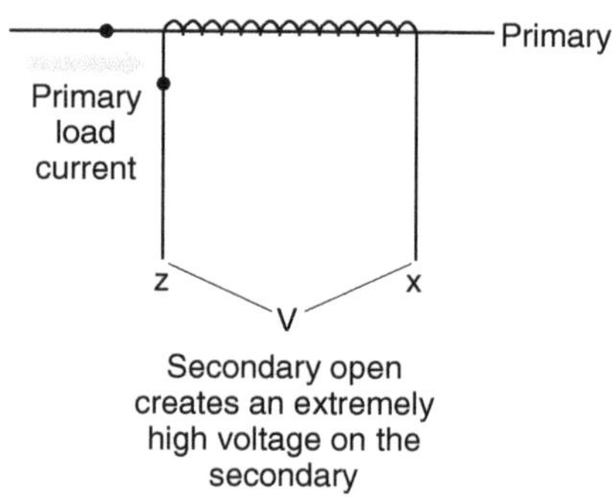

FIGURE 11.3 Open circuit on secondary of current transformer.

When working on or near CT circuits, personnel should always be aware of the hazard and should wear rubber gloves, arc protective clothing, and face shields if working on the wiring for the circuit. The primary circuit should be de-energized before the secondary wiring is opened.

GROUNDING SYSTEMS OF OVER 1000 V

The subject of electrical grounding is a complex one. The following sections focus on the grounding concepts and requirements of medium- and high-voltage systems as they relate to safety. Chapter 5 of this handbook provides a much more in-depth coverage of electrical grounding principles and requirements. Also, the reader may refer to the various ANSI/IEEE standards including IEEE Guide for Safety in AC Substation Grounding.

What Is a Ground?

A ground is an electrically conducting connection between equipment or an electric circuit and the earth, or to some conducting body that serves in place of the earth. If the ground is properly made, the earth or conducting body and the circuit or system will all maintain the same relative voltage.

Figure 11.4*a* shows an electrical system that is not grounded. In such a situation, a voltage will exist between the ground and some of the metallic components of the power system. In Fig. 11.4*b*,

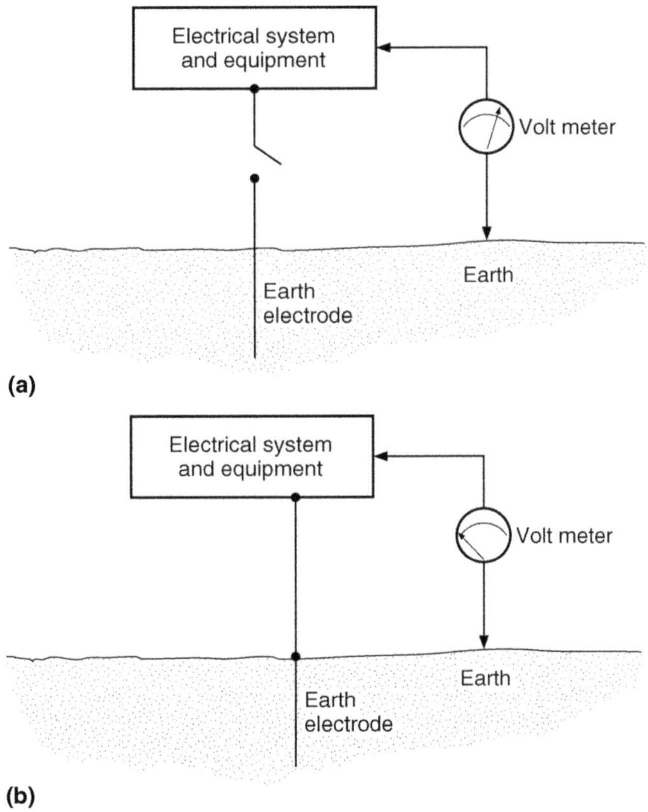

FIGURE 11.4 Grounding. (*a*) Before system is grounded, a voltage exists between the system and the earth; (*b*) when ground connection is made, no voltage exists.

the earth connection has been made. When the system is grounded, the voltage is reduced to zero between the previously energized sections of the system.

Bonding versus Grounding

Bonding is the permanent joining of metallic parts to form a continuous, conductive path. Since the earth is generally not a good conductor, bonding is used to provide a low-impedance metallic path between all metallic parts. This, in essence, bypasses the earth and overcomes its relatively high impedance. A good grounding system is a combination of solid connections between metallic parts and the earth as well as between all metallic parts.

Voltage Hazards

Figure 11.5 illustrates the four standard voltage hazards that are associated with and relieved by proper system grounding. If a system is ungrounded and the nonenergized metallic parts become energized, the metallic parts will have a measurable voltage between themselves and ground. If the system is grounded by driving ground electrodes into the earth, current will flow from the rods into the earth. At each ground electrode, the voltage will rise relative to the remote reference point. The voltage will drop off away from the electrodes and peak at each electrode. This situation creates the four types of voltage hazards as follows:

- *Step voltage.* As a worker steps across the ground, the front foot will be at a different potential than the rear foot. This effect is caused by the voltage gradient created by the ground electrodes (Fig. 11.5). Step voltage can easily reach lethal levels. Step voltage can be mitigated by increasing the number of grounding electrodes, increasing bonding between the metal parts, and employing a grounding grid.

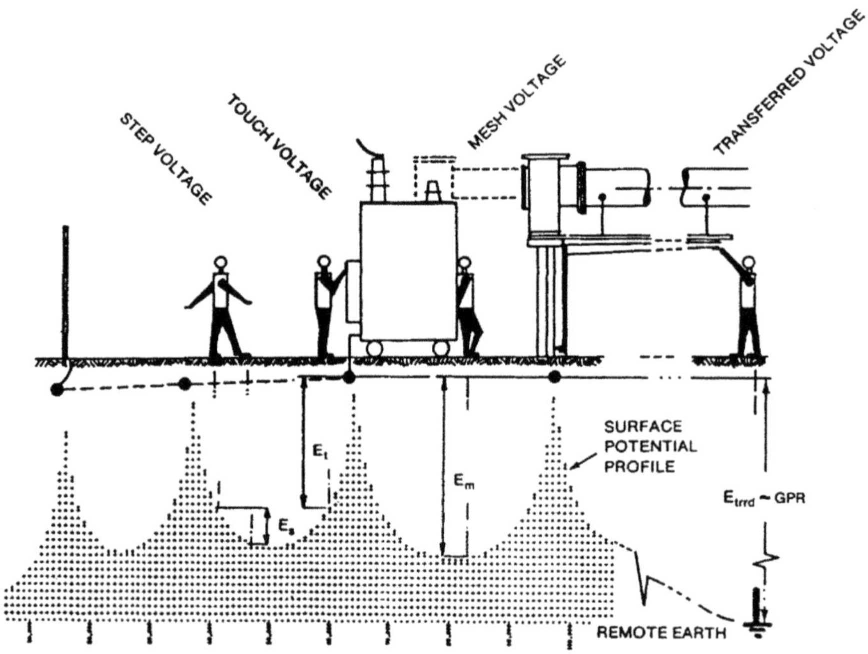

FIGURE 11.5 Four voltage hazards related to system grounding. (*Courtesy Institute of Electrical and Electronics Engineers.*)

- *Touch voltage.* Because of the voltage gradient, the voltage of the earth, only a short distance from the grounded metallic equipment, will be different from the voltage of the equipment. Thus, if a worker touches the grounded equipment, his or her feet will be at a different potential than his or her hands. This voltage can be lethal. Touch voltage is mitigated by increasing the number of grounding electrodes, increasing bonding between metal parts, and employing a grounding grid.
- *Mesh voltage.* Mesh voltage is the worst case of touch voltage. Cause, effect, and mitigation methods are identical to those described previously under touch voltage.
- *Transferred voltage.* Because metal parts have a much lower impedance than the earth, the voltage drop between remote locations is lower on the metallic connections than the earth. This means that the earth may be at a significantly different voltage than the metallic connections. Transferred voltages are particularly noticeable on neutral wires, which are grounded at the service point and nowhere else. Transferred potential can be mitigated by providing the entire area with a ground grid; however, this solution is infeasible in any but the smallest systems. Transferred voltage must be mitigated by avoiding contact with conductors from remote locations and/or using rubber insulating gloves.

System Grounds

What Is a System Ground? A system ground is the connection of one of the conductors to the earth. Such a connection is accomplished by connecting an electric wire to the selected system conductor and the grounding electrode.

Why Are Systems Grounded? Power systems have conductors grounded for a variety of safety and operational reasons, including the following:

- Grounded systems provide sufficient short-circuit current for efficient operation of protective equipment.
- Grounded systems are less prone to transient overvoltages that can cause insulation failures.
- Grounded systems are generally more easily protected from lightning.
- Solidly grounded systems are less prone to resonant conditions that can cause equipment and insulation failures.

What Systems Must Be Grounded? The NEC allows the grounding of permanently installed electrical utilization systems and requires the grounding of mobile electrical supply systems. If the mobile supply system is supplied by a high-voltage delta system, the NEC requires that a ground be derived using a grounding transformer. The National Electrical Safety Code requires that electric utility supply systems be grounded.

How Are Systems Grounded? Electrical systems are grounded by connecting one of the electrical conductors to earth or some conductive object that replaces the earth. Ship hulls are examples of grounds that do not involve the earth. The conductor chosen and the location of the ground are determined as part of the engineering design for the system. With systems that have delta transformer or generator windings, the neutral, and therefore the ground point, can be derived by using a grounding transformer. Figure 11.6 shows a grounded wye-delta transformer used to derive a system neutral and ground. Figure 11.7 shows a zigzag transformer used for the same purpose.

The zigzag transformer (Fig. 11.7) is the most commonly used method for deriving grounds on high-voltage systems. The windings consist of six equal parts, each designed for one-third of the line-to-line voltage; two of these parts are placed on each leg and connected as shown in Fig. 11.7. In the case of a ground fault on any line, the ground current flows equally in the three legs of the autotransformer, and the interconnection offers the minimum impedance to the flow of the single-phase fault current.

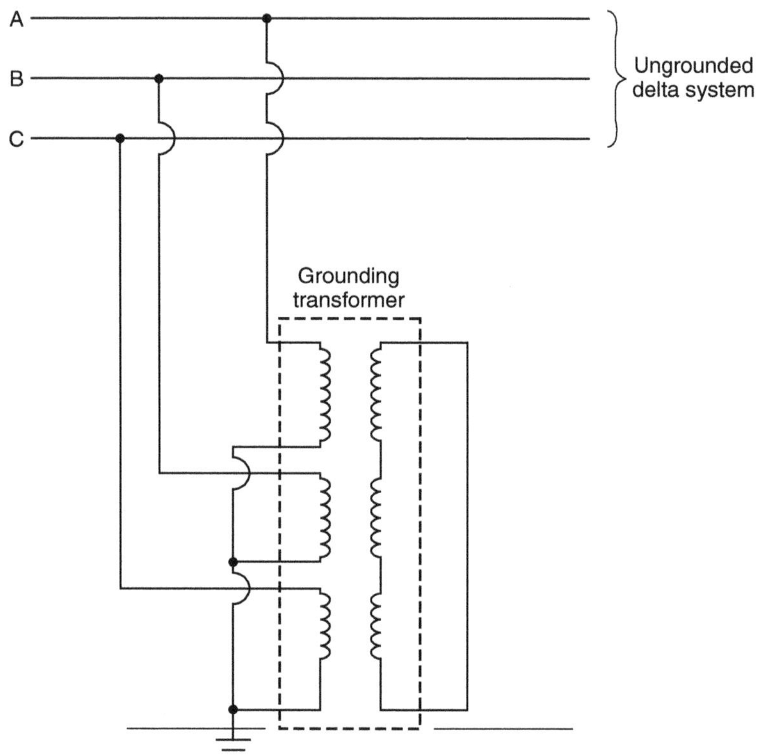

FIGURE 11.6 Grounded wye-delta transformer used to derive a system ground.

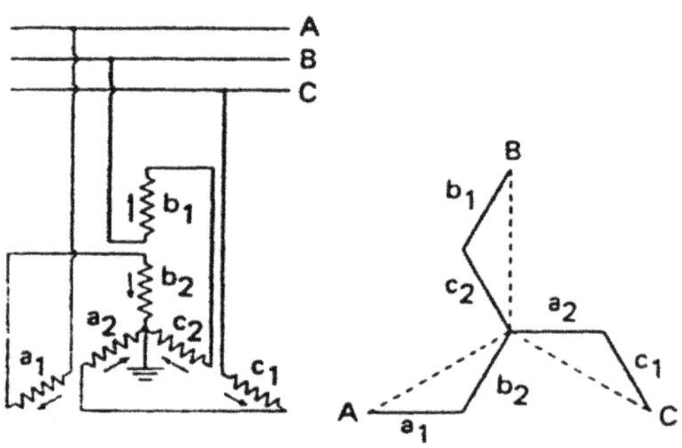

FIGURE 11.7 Grounding autotransformer with interconnected Y (also called zigzag windings).

Equipment Grounds

What Is an Equipment Ground? An *equipment ground* is an electrically conductive connection between the metallic parts of equipment and the earth. For example, transformer cases and cores are connected to the earth—this connection is called an equipment ground. Note that the metallic parts of equipment are grounded and bonded together. This bonding serves to reduce the voltage potential between all metallic parts and the earth.

TABLE 11.1 Equipment Grounding Requirements

- Must be permanent and continuous
- Must have the capacity to conduct any fault current likely to be imposed on it
- Must have sufficiently low impedance to limit the voltage to ground and to facilitate the operation of the circuit-protective devices

Why Is Equipment Grounded? The equipment ground is one of the most important aspects of grounding as far as safety is concerned. Workers are constantly in contact with transformer shells, raceways, conduits, switchgear frames, and all the other conductive, non-current-carrying parts. Proper equipment grounding and bonding ensures that the voltage to which workers will be subjected is kept to a minimum. Proper bonding and grounding mitigates touch and step voltages.

How Is Equipment Grounded? The NEC requires that the path to ground from circuits, equipment, and metal enclosures must meet the conditions listed in Table 11.1. Equipment used in grounded systems is grounded by bonding the equipment grounding conductor to the grounded service conductor and the grounding electrode conductor. That is, the equipment ground is connected to the system ground. Equipment used in ungrounded systems is grounded by bonding the equipment grounding conductor to the grounding electrode conductor.

What Equipment Must Be Grounded? All non-current-carrying metal parts of fixed, portable, and mobile high-voltage equipment and associated fences, housings, enclosures, and supporting structures shall be grounded.

The NEC does allow two exceptions to this rule: High-voltage equipment does not have to be grounded if

1. it is isolated from ground and located so as to prevent any person who can make contact with ground from contacting the metal parts when the equipment is energized.
2. it is certain pole-mounted distribution equipment that is exempted.

Refer to the current edition of the NEC for detailed information.

SAFETY EQUIPMENT

Overview

Since the amount of available energy in the system determines its lethal effects, and since high-voltage systems are generally very high-energy systems, high-voltage systems are extremely hazardous and should be treated with the utmost respect.

The following sections describe the types of safety equipment that should be worn by persons working on or near energized, high-voltage conductors. The recommendations given are minimum recommendations. If additional or more stringent protection is desired or required, it should be worn. Refer to the tables in Chap. 4 for specific recommendations.

Hard Hats

Protective headgear for persons working on or near energized, high-voltage circuits should provide both mechanical and electrical protection. Personnel working on or near energized high-voltage circuits should be supplied with and should wear ANSI class E helmets. ANSI class G and class C helmets are not acceptable. The ANSI class G helmet is not rated or tested for use in high-voltage

TABLE 11.2 Summary of the Characteristics for ANSI Class C, G, and E Hard Hats

Class	Description	Comments
C	Intended to reduce the force of impact of falling objects. This class offers no electrical protection.	Should not be worn by personnel working on or around energized conductors of any voltage
E	Reduce the impact of falling objects and reduce danger of contact with exposed high-voltage conductors. Representative sample shells are proof-tested at 20,000 V phase to ground.	Recommended to be worn by personnel working around high- and low-voltage circuits
G	Reduce the impact of falling objects and reduce danger of contact with exposed, low-voltage conductors. Representative sample shells are proof-tested at 2200 V phase to ground.	Recommended to be worn by personnel working around only low-voltage circuits

circuits and the ANSI class G helmet provides no electrical insulation at all. Table 11.2 summarizes the characteristics of the three ANSI classes. Note that class G and E were formerly class A and B, respectively.

Eye Protection

High-voltage systems are capable of producing extremely powerful and hazardous electric arcs and blasts; therefore, eye protection for electrical workers should provide protection against heat and optical radiation. ANSI standard Z87.1 provides a selection chart as well as a chart that illustrates the various protection options available. The selection chart is reproduced in this chapter as Table 11.3, and eye protection options are shown in Fig. 11.8.

Electrical workers should consider using eye protection that combines both heat and optical radiation protection. Table 11.3 and Fig. 11.8 show several different types of equipment that will provide such protection including types B, C, D, E, and F. If arcing and molten splashing are possible, as with open-door switching or racking of circuit breakers, the type N eye protection should be employed.

Arc Protection

High-voltage systems create and sustain significant electric arcs accompanied by electric blast. Workers performing routine work on or around energized, high-voltage conductors should wear flame-retardant work clothing or its equivalent with a rating determined by the incident energy to which they will be exposed. An arc-flash study is the best approach to determine the required rating. Workers operating open-air switches or performing open-door switching procedures should wear flame-retardant flash suits or their equivalent with an arc rating also determined by the incident energy to which they will be exposed. Refer to Chap. 4 for methods that can be used to calculate required flash clothing weights.

Rubber Insulating Equipment

Rubber insulating equipment and appropriate leather protectors should be used by personnel who are working on or around energized, high-voltage conductors. Rubber gloves, rubber sleeves, rubber blankets, line hose, and other such protective devices should be employed any time workers must approach such conductors. Rubber insulating equipment used around high-voltage systems must

TABLE 11.3 Eye Protection Selection Chart

		Assessment SEE NOTE (1)	Protector type†	Protectors	Limitations	Not recommended
IMPACT	Chipping, grinding, machining, masonry work, riveting, and sanding	Flying fragments, objects, large chips, particles, sand, dirt, etc.	B, C, D, E, F, G, H, I, J, K, L, N	Spectacles, goggles, face shields SEE NOTES (1) (3) (5) (6) (10). For severe exposure, add N.	Protective devices do not provide unlimited protection. SEE NOTE (7).	Protectors that do not provide protection from side exposure. Filter or tinted lenses that restrict light transmittance, unless it is determined that a glare hazard exists. Refer to Optical Radiation.
HEAT	Furnace operations, pouring, casting, hot dipping, gas cutting, and welding	Hot sparks	B, C, D, E, F, G, H, I, J, K, L, *N	Face shields, goggles, spectacles *For severe exposure, add N. SEE NOTES (2) (3).	Spectacles, cup and cover type goggles do not provide unlimited facial protection. SEE NOTE (2).	Protectors that do not provide protection from side exposure.
		Splash from molten metals	*N	*Face shields worn over goggles H, K SEE NOTES (2) (3).		
		High temperature exposure	N	Screen face shields, reflective face shields SEE NOTES (2) (3).	SEE NOTE (3).	
CHEMICAL	Acid and chemicals handling, degreasing, plating	Splash	G, H, K *N	Goggles, eyecup and cover types *For severe exposure, add N.	Ventilation should be adequate but well protected from splash entry.	Spectacles, welding helmets, hand shields.
		Irritating mists	G	Special-purpose goggles	SEE NOTE (3).	
DUST	Woodworking, buffing, general dusty conditions	Nuisance dust	G, H, K	Goggles, eyecup and cover types	Atmospheric conditions and the restricted ventilation of the protector can cause lenses to fog. Frequent cleaning may be required.	

(*Continued*)

TABLE 11.3 (Continued)

	Assessment SEE NOTE (1)	Protector type†	Protectors: Typical filter lens shade	Protectors: Protectors	Limitations	Not recommended
OPTICAL RADIATION						
Welding:			SEE NOTE (9).			
Electric arc		O, P, Q	10-14	Welding helmets or welding shields	Protection from optical radiation is directly related to filter lens density. SEE NOTE (4). Select the darkest shade that allows adequate task performance.	Protectors that do not provide protection from optical radiation. SEE NOTE (4).
Welding:			SEE NOTE (9).			
Gas		J, K, L, M, N, O, P, Q	4-8	Welding goggles or welding face shield		
Cutting			3-6			
Torch brazing			3-4		SEE NOTE (3).	
Torch soldering		B, C, D, E, F, N	1.5-3	Spectacles or welding face shield		
Glare		A, B	Spectacle SEE NOTES (9) (10).		Shaded or special-purpose lenses, as suitable. SEE NOTE (8).	

† Refer to Fig. 11.8 for protector types.
NOTE: For NOTES referred to in this table, see text accompanying Fig. 10.10.
Source: Courtesy American National Standards Institute.

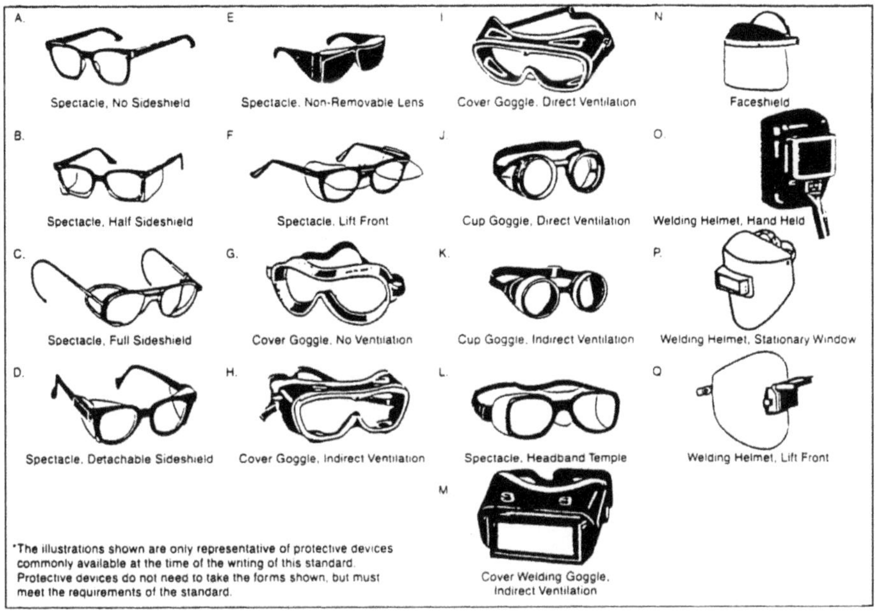

FIGURE 11.8 Eye protection devices. Refer to Table 11.3 for selection criteria.

have a rating sufficient for the voltage of the system. ANSI class 1, 2, 3, or 4 rubber insulating goods should be employed as required.

Voltage-Testing Devices

Proximity or contact testers intended for use in high-voltage circuits should be used to test circuits and to verify that they are de-energized and safe to work on. Voltage-testing devices are described in Chap. 3. An example of a high-voltage–measuring instrument is shown in Fig. 11.9. ASTM standard F 1796 is the standard defining proximity testers.

SAFETY PROCEDURES

General

The general procedures described in Chap. 4 should be used on all high-voltage circuits. The following sections describe key procedures that apply to high-voltage circuits.

Approach Distances

Although approach distances are laid out in a number of OSHA references, the NFPA 70E provides a much more useful and practical concept for approach distances. The following paragraphs identify methods that may be used to determine approach distances. Refer to Chap. 4—Fig. 4.35 and Table 4.15—for reference to the terms used in the following section.

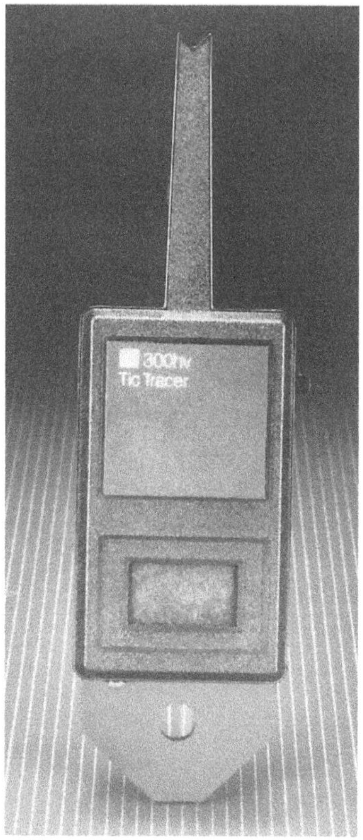

FIGURE 11.9 "TIC" tracer proximity voltage sensor for use on circuits up to 35,000 V. (*Courtesy TIF Electronics.*)

Crossing the Limited Approach Boundary. Qualified workers may cross the limited approach boundary if they are qualified to perform the work. Unqualified workers are allowed to cross the limited approach boundary as long as they are continuously escorted by and under the constant direct supervision of a qualified person.

Requirements for Crossing the Restricted Approach Boundary. To cross the restricted approach boundary, the following criteria must be met:

- The worker must be qualified to do the work.
- There must be a plan in place that is documented and approved by the employer.
- The worker must be certain that no part of the body crosses the restricted approach boundary.
- The worker must work to minimize the risk that may be caused by inadvertent movement by keeping as much of the body out of the restricted space as possible. Allow only protected body parts to enter the restricted space as necessary to complete the work.
- Personal protective equipment must be used appropriate for the hazards of the exposed energized conductor.

Requirements for Crossing the Prohibited Approach Boundary. NFPA 70E considers crossing the prohibited approach boundary to be the same as working on or contacting an energized conductor.

TABLE 11.4 Typical Situations That May Allow Reclosing of a Protective Device That Has Operated

• The faulted section of the system is found and repaired. • The nature of the protective device makes it clear that no hazard is present. For example, if the device that operated is an overload type of device, it may be safe to reclose. • The reclosing operation can be made in such a way that the workers are not exposed to additional hazard. For example, if the reclosing operation can be made by remote control, and if all personnel are kept away from all parts of the circuit, it may be reclosed.

To cross into the prohibited space, the following requirements must be met:

- The worker must have specified training required to work on energized conductors or circuit parts.
- There must be a plan in place that is documented and approved by the employer.
- A complete risk analysis must be performed.
- Authorized management must review and approve the plan and the risk analysis.
- Personal protective equipment must be used appropriate for the hazards of the exposed energized conductor.

Voltage Measurement

The voltage-measurement techniques defined in Chap. 4 should be employed on circuits of all voltages.

Locking and Tagging

Lockout-tagout and energy control procedures apply to circuits of all voltage levels. Refer to Chap. 4 for detailed discussions of lockout-tagout procedures.

Closing Protective Devices after Operation

Workers should never reclose any protective device after it has operated, until it has been determined that it is safe to do so. Several criteria may be used to determine whether it is safe to reclose the protective device (Table 11.4). Other conditions may or may not indicate a safe reclosing situation.

CHAPTER 12
HUMAN FACTORS IN ELECTRICAL SAFETY

INTRODUCTION

The energy sector has significantly contributed to our understanding about the influence of human factors in complex and technically challenging work. Notably, as data collection and analyses have accumulated over the past two decades, the sizable role of human factors in safety failures and accidents has become increasingly clear.

Historically, a 1995–1999 Institute of Nuclear Power Operations (INPO) analysis of significant events in the commercial power industry indicated that three out of every four events were attributed to human error.[1] Another report, an Idaho National Energy and Environment Laboratory (INEEL) 2001 study[2] commissioned to analyze the human error contribution to risk in operating events found for 37 qualitatively analyzed events the following:

- Each event had four or more human errors in combination with hardware failures.
- Engineering contributed to 81 percent of latent error.
- Maintenance contributed to 76 percent of latent error.
- Management and supervision contributed to 30 percent of latent error.

Latent errors (i.e., errors committed prior to an accident but not discovered until after the accident) were present four times more often than active errors (i.e., errors occurring during the event response). Root causes of errors were considered by human performance observations (Table 12.1) of employee assignments categorized as the following:

- Task based
- Rule based
- Knowledge based

The generation, transmission, distribution, design, installation, operation, and maintenance of electrical power occur in diverse situations. Additional examples of human factors considerations in the electrical industry are presented in Table 12.2.

Given the diversity in workplaces, the applicable human factors depend in part on where in the power system the employee may be working. Employees assigned to a control room at a nuclear power station can reasonably be expected to confront demands in their work that differ from those of employees wiring a house under construction or doing maintenance in a cogeneration facility. So it follows then that the human factors in a control room can be expected to differ from those at the construction site.

TABLE 12.1 Common Work Characteristics

	Characteristics	Examples	Error mode
Skill based	Highly practiced, physical actions Habitual movements with little conscious monitoring Behavior driven by training or instructions Relies on self-check	Using a hammer Operating manual controls Hanging a tag	Inattention to execution requirements
Rule based	Modified work due to environmental change detection covered by procedures or past practices	Responding to control alarm Fitting a pipe	Misinterpretation
Knowledge based	Work being done when significant knowledge or familiarity is missing Response to unfamiliar situation	Performing an engineering calculation Resolving conflicting control board indicators	Inaccurate information Inaccurate mental models

Adapted from *DOE Human Performance Handbook*, U.S. Department of Energy, Chap. 2, pp. 22–29, Washington, DC, 2009.

TABLE 12.2 Power Systems Environments and Human Factors Considerations

Workplace	Human factors considerations
Generation	Dust, fumes, noise in fossil fuel environments Common features to control rooms,[3] including Compact workstations using visual displays Large overview displays Increased cognitive workloads as staffing changes Information multiplicity Virtual workspaces[17] with Serial access to information and controls More time spent on secondary tasks
Transmission	Live line work[4] with Helicopter approaches at high elevations Moving parts (helicopter rotars) Required calculations: minimum approach distance Placing person with tools in air gap[5]
Distribution	Voltage protection personal equipment, including Rubber goods Extended tool handles Recognition of minimal power line approach distances At or above shoulder work requiring stressful postures Work in vaults or confined space requiring respirators Repetitive motion and lifting heavy loads[6]
Construction	Crouching, kneeling or reaching in various spaces Vibration in powered equipment Environmental temperature extremes Heavy equipment operation, with noise, moving parts Potential contact with power lines above and below

Industry data confirms the expected variation.[7] With 1.5 million person-years of work analyzed from 17 utilities during the period from 1995 to 2008, review of 52,000 recordable and lost-time injures[8] showed the following:

- Percent total injuries were highest for line workers (18.2 percent), mechanics (14 percent), meter readers (13.4 percent), maintenance workers (8.7 percent), and plant and equipment operators (7.9 percent).
- Percent total injuries were lowest for custodians/cooks (0.5 percent), managers (0.9 percent), coordinators (0.9 percent), security (1.1 percent), drivers/delivery drivers/inspectors (1.3 percent), engineers (1.4 percent), machinists (1.4 percent), foremen (1.4 percent), and welders (1.8 percent).
- Percent total injuries fell in the mid-range for technical/professional support (5.9 percent), supervisors (3.7 percent), electricians (3.7 percent), administrative support (3.3 percent), and material handlers (2.4 percent).

Overall reported injury rates per 100 employee-years were highest in transmission and distribution (9.94) and shop (9.11) locations, and lowest in office (2.15), customer service (2.08), and training (1.08) locations. Mid-range rates were reported for field (4.77), substation (4.23), and generation station (4.18) locations.

Across all years of available data among 16 utilities, 1,409 full-time equivalents were lost due to injury or illness, and 988 of these losses were due to ergonomics injuries.[7]

Total medical claims amounts were dominated by sprains and strains cases (45.8 percent), fracture/dislocation (14.6 percent), and cumulative trauma/repetitive strain injuries (12.8 percent), leading to further research and interventions to reduce injury risk, considering in particular human factors in ergonomic tool and task design.[8]

Electrical safety risks were also recognized in the construction, operation, and maintenance of wind turbines and solar arrays. The risks included electric shock and arcing, grounding, lightning, and electric fires related to insulation breakdown. Other hazards include work at heights, heavy exertional forces, and moving machinery with transporting, lifting, and field assembly of large equipment components.[8,9]

Data surveys by the Electrical Safety Foundation International (ESFI) using the U.S. Bureau of Labor Statistics' (BLS) Census of Fatal Occupational Injuries (CFOI) and the Survey of Occupational Injuries (SOII) have tracked fatal and nonfatal occupational electrical injuries across the 23-year period from 1992 to 2016, and deeper analysis of 2003–2016 data. In 2016, ESFI found that 3 percent of reported U.S. occupational fatalities from all causes involved electricity. Between 2015 and 2016, electrical fatalities increased 15 percent, with 53 percent of these fatalities occurring in the construction industry, affecting younger workers 2.3 times more often than experienced workers. During the same period, nonfatal electrical injuries decreased 33 percent, with 1,640 nonfatal electrical injuries resulting in days away from work reported in 2016.[10]

OVERVIEW

The following section includes a brief review of the safety strategy called Defense in Depth as well as the evolution of human factors. The topics of visualization and cognitive ergonomics are also presented.

Defense in Depth

As demand for energy has escalated, security and reliability considerations along with safety priorities have given rise to the Defense in Depth strategy. Originally used to defeat enemies during war, Defense in Depth is a layering tactic to achieve comprehensive protective coverage.[11] First applied to power generation, transmission, and distribution as an outcome of analyses of possible terrorist

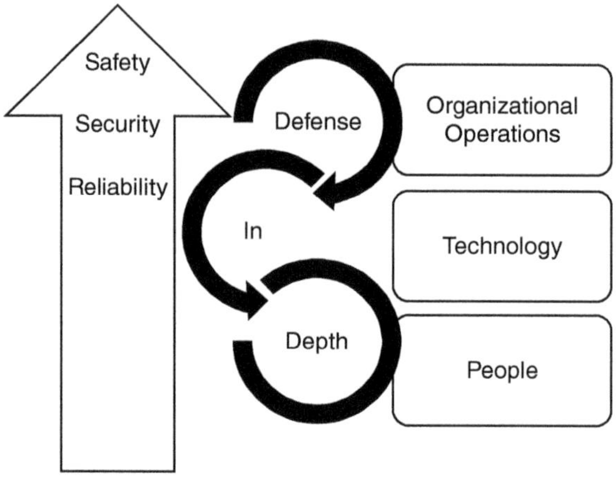

FIGURE 12.1 Defense in Depth strategy. Increasing safety, security, and reliability in the energy sector depends on complex, layered interactions among people, technology, and organizational operations. (*Translated to energy sector based on NSA illustration.*[11])

TABLE 12.3 Examples of Error Precursors

Precursor type	Examples
Task	Time pressures High workload Multistep or multiaction requirements Boring or monotonous action requirements Ambiguous goals, roles, or assignments Absence of standard work
Work environment	Multiple distractions Multiple interruptions Unexpected conditions Unresolved conflicts

Adapted from *DOE Human Performance Handbook*, U.S. Department of Energy, Chap. 2, p. 32, Washington, DC, 2009.

threats against utilities in the Global War on Terror, the strategy has as its foundation the creation of multiple independent and redundant defensive layers to compensate for potential unintentional or intentional human as well as mechanical failures.[12]

The Defense in Depth strategy allows electrical safety to be viewed as an interdependence between individuals, engineered systems, and operational environment, including the organizational culture. The premise is that continuous interactions—between people, technology, and organizational operations—drive safety, security, and reliability performance (Fig. 12.1). With this illustration providing a framework, the contributing factors or "error precursors" to safety failures can be identified (Table 12.3). Advances in how people work with technology and improvements in an organization's operations are basic approaches used in the Defense in Depth strategy.

Evolution of Human Factors

The study of human factors evolves from the obvious statement that "People are not machines." In the early 19th century, during the early periods of industrialization, a persistent question was how to optimize the placement of people in "machine systems."[13] Pre-dating electrical power, when jobs involved assembly-line operations with repetitive tasks requiring uniformity and efficiency of physical movements between operations, materials, and employees, management used studies of "time-and-motion" and human "anthropomorphics" or "average body part sizes" to figure out problems like "how far can an average worker reach to push or pull this piece" on the assembly line.

In the early 20th century, as electrification enabled industrialization to move beyond manufacturing across all economic sectors, the complexity of machine systems increased. More than one process could be pulled into a work environment, leading to multiple operations situated in geographic proximity, running concurrently, and producing output at high speeds.

With electrification, increasing industrial complexity, and faster production rates, the demands on people working with the machine systems grew. As suggested by Rasmussen,[13] at this stage, human factors more commonly studied measures of human physical abilities along with measures of physical reliability. In this context, "knob-and-dial" ergonomics was a research focus—for example, identifying the correct size of knob for an employee to easily grasp doing a turning task, and measuring how often the knob would be incorrectly versus correctly turned.

Then, in the 1930s, as automation was introduced into the chemical process industries, the link between machines and people often became indirect.[13] Automatic control engineering made it possible for individual employees to have a wide reach over large-scale and significantly hazardous production.

As production failures occurred, interest in human factors extended from human physical reliability to include the reliability people might show in understanding process information and executing successful decisions. This specialty later became known as "cognitive ergonomics."

Research on human error developed as an aspect of studies into human factors, with the goal being to "break down" or "untangle" causes of human physical and/or cognitive malfunction from machine or process malfunction in post-accident studies. Engineers planning installation design and development also considered human error probability—or human reliability assessment (HRA)—as an aspect of human factors during operational engineering analysis.

In the mid-20th century, military experience in World War II aviation and the post–World War II growth in civilian aviation further emphasized how interactions between humans and engineered systems could lead to potentially tragic results, as demonstrated by news reports of plane crashes. By the early 1970s, a generally accepted professional approach within the engineering community led to practices regarding how to evaluate the fit between work and workers.

Concurrently, within the electrical power industry, the post–World War II development of the civilian nuclear power industry resulted in an active application of ideas regarding human factors to the management of civilian nuclear power operations. This transfer of ideas developed further following two major tragedies: the Unit 2 scram at the Three Mile Island Power Station (TMI-2) in Pennsylvania on March 28, 1979, and the explosion of Unit 4 of the Chernobyl Nuclear Power Station in the Ukrainian former Soviet Union on April 26, 1986.[14]

With TMI-2 and Chernobyl, nations learned that human error in complex operational environments producing electrical power could potentially result in global disaster. Investigations identified human factors in the causal chain of events leading to these tragedies. This knowledge drew additional focus on basic concepts of human performance.

Continuing to the present, the topics of human factors have moved beyond "human machine interactions"[15] to "human-systems interfaces" or HSIs.[16,17] The term "human factors" may have two meanings. Human factors may refer to various traits or "elements of the human"[14] as individuals which should be priority considerations for safe and effective results from engineered systems. Or, the term may mean "the applied science technology" relating fundamental human sciences (such as anatomy, physiology, neuropsychology) to industrial systems.[14] The discussion below regarding vision highlights the two meanings.

Researchers turned their focus to environment influences on human factors as failures continued to occur. Relating Hawkins's[18] "SHEL" illustration of this concept, Kawano[14] described the acronym as follows:

- S: for software, including, for example, computer software, paper documents or instructions, permits, or procedures
- H: for hardware, or the engineered or constructed aspects of the environment
- E: for the environment, or ambient conditions, such as geography, meteorology, humidity, or altitude
- L: for "liveware," or the people involved directly or indirectly in the work situation

Kawano added an "M" to Hawkins's SHEL acronym to include "managing systems" as a dynamic part of the model. As suggested by Hawkins and amended by Kawano, the M-SHEL model provided a step toward an integrated systems approach to human factors. The model offered another strategy for thinking about electrical safety so that the programmatic focus could go beyond the individual to include the technologies, engineering, interpersonal, and environmental contributions to the individual's experience in the system of production where he or she works.

Today, human factors, errors, risk, and performance are addressed with Defense in Depth strategies and included in integrated safety management, probabilistic risk analyses, reliability engineering, and inherently safe design programs, policies, and practices. For example, U.S. Department of Energy (DOE) publications focus on design for the ease of maintenance[1] and performance improvement.[19]

Developing from research on the utility industry, ergonomics handbooks are available for electrical workers, plant operators, and mechanics in fossil-fueled power plants and generating stations.[3,20–22]

Another resource is a guide for developing an ergonomics program titled *Elements of Ergonomics Programs: A Primer Based on Workplace Evaluations of Musculoskeletal Disorders*, released by the National Institute of Occupational Safety and Health (NIOSH). The document includes a seven-step approach to reducing musculoskeletal injuries, including guidance for proactive human factors design (Table 12.4).[23]

Examples noted in Table 12.5 are adapted from another NIOSH resource published as a series of ergonomics tip sheets for construction enterprises and their employees for use during tasks such as the following:

- Floor and ground-level work
- Overhead work
- Lifting, holding, and handling materials
- Hand-intensive work

Visualization

Anything that inhibits an employee's vision creates risk for an electrical event. Consider the two definitions of human factors applied to vision. As a human factor relating to the individual, vision may be understood as limited to the measurement of visual acuity (or accuracy of how well an employee sees a wall chart). A test of visual acuity ensures that an employee has adequate vision ability to read a computer screen or see a stoplight. On the other hand, vision can be understood as a target of the engineering advancement of vision assistance devices, such as these examples:

- Infrared cameras installed in vehicle dashboards for night driving
- Infrared intrusion detection systems
- Infrared monitors used in heat tracing as a safety monitoring strategy to identify excessive ohmic heating of installed conductors

Vision is a critically important human factor. About 80 percent of information about machines and systems comes to an employee via their vision.[24] Note that visual acuity decreases as the complexity of

TABLE 12.4 Proactive Design Suggestions to Optimize Human Factors

Design focus	Examples
Workstation	Adjust heights to fit for work, seating, and reading surfaces Position equipment to reduce twisting/reaching/pulling Position work above elbow height for fine visual tasks Position workload below elbow height for heavy physical effort Eliminate excessive noise, heat, humidity, cold, or poor lighting Minimize work placement causing arm movements that are Straight-lined or jerking Extended repeatedly through more than 15 inches Pivoted around the shoulder
Repetitive hand/wrist	Optimize neutral "handshake" position at wrist/hand Substitute partial or full automation to reduce repetitions Limit force required for hand gripping, carrying, or pinch grasp Limit trigger finger in favor of full finger or automation controls Control or limit vibration to hands/wrists through Power tool selection Protective gloves and vibration absorbing materials

Adapted from *Elements of Ergonomics Programs*, by A. L. Cohen, C. C. Gjessing, L. Fine, B. P. Bernard, and J. D. McGlothin, 1997, DHHS (NIOSH) Publication No. 97–117, Washington DC, Tray 9, pp. 124–127.

TABLE 12.5 Examples of Tips to Reduce Ergonomic Injuries during Construction

Ergonomic hazard	Common work activity	Solution
Repetitive stooping with task completed at floor or ground	Bending over to attach screw to floor level fixture	Auto-feed screw gun with an attachment
Screeding concrete bent over with heavy gripping to pull board over concrete	Concrete finishing	Motorized screed
Repetitive bending and reaching with hand tools	Tying rebar by hand	Rebar-tying tool with extension
Frequent kneeling, squatting, or stooping	Kneeling to lay floor tile	Using kneeling creepers with chest support
Bending and twisting while lifting	Loading masonry onto a scaffold	Split-level adjustable scaffolding
Repetitive strain at shoulder	Overhead screwing	Bit extension shaft on screw gun
Forceful pushing at shoulder/upper back	Finishing drywall	Spring-assisted mud box Pneumatic drywall finishing system
Heavy repetitive lifting	Laying standard concrete block	Lightweight block materials

Adapted from *Simple Solutions: Ergonomics for Construction Workers*, by J. T. Albers and C. Estill, 2007, DHHS (NIOSH) Publication No. 2007–122, Cincinnati, OH.

the visual target increases. In other words, when there is more to look at, it is harder to see. Multiple physical factors influence visual acuity, including the following:

- Illumination
- Contrast
- Time of exposure to the visual target
- Color of the visual target and the target's background

TABLE 12.6 Cognitive Aids with Computer Visualization Technologies

Cognitive challenge	Cognitive challenge
Multiple permits per job step	Reducing search via data grouping or structure with multiple steps in a job
Monitoring for breaker trips, power surges, or faults	Perceptual monitoring using pre-attentive visual characteristics
Switching procedures in power outage or routine maintenance	Enhancing data recognition, abstraction, and aggregation

Note: Visualization methods are based on Tory and Mőller's report[24] in Table 1, which is adapted in part from work by Card et al., *Information Visualization:* Using Vision to Think. San Francisco: Morgan Kaufmann Publishers, 1999.

The smallest detectable threshold for vision is 10^{-6} mL. Vision ability is decreased with vibration, hypoxia or low oxygen, and motion of the visual target.

Rasmussen[13] suggested that automation relieved employees of repetitive manual tasks; consequently, work with more decision-making or trouble-shooting content became common in the industrial setting. Today, information technology, including the use of computer-based approaches that require employees to do intense visual work with multiple screens, has placed volumes of data at the center of employee tasks.

Tory and Mőller[25] describe the methods available with computer visualization and graphics which can support perception and cognition, or the human abilities to sense (via eyes, ears, nose, taste, touch) and think. Table 12.6 provides electrical safety examples where these methods can serve as cognitive aids.

Cognitive Ergonomics

The early 21st century is notable for the promise of new horizons in research advancing cognitive ergonomics to fully benefit the industrial workplace. As brain science maps the terrain of thought and mind–body interactions using scanning technologies such as magnetic resonance imaging (MRI), positron emission tomography (PET), and single photon emission computed tomography (SPECT), data are being accumulated to explain how people think in different scenarios.

A key insight from this research is that context is hugely influential over thought. Using the metaphor of the brain as a computer, one way to appreciate the role of context is to suggest that no person comes to their work with a "blank hard drive or screen." Rather people do their jobs with the advantage, and disadvantage, of what they see, hear, smell, and touch in their surroundings; what they believe; and how they physically feel. This information loads the "drive and screen" occupying space alongside of where "the programs run" or thought happens.

Cognitive ergonomics specifically addresses how to modify what occupies an employee's "hard drive and screen" when work is being done to improve the way thought unfolds, and thereby, improve decisions.

Context is not simply the employee's physical background. Context may take the form of detrimental attitudes and background beliefs. Attitudes highlighted in the *Human Performance Improvement Handbook* published by DOE[26] and illustrated in Table 12.7 shape employee actions as do background beliefs or "work myths." The dictionary defines a myth as "an old traditional story or legend, especially one concerning fabulous or supernatural beings, giving expression to early beliefs, aspirations, and perceptions of a people and serving to explain natural phenomena or the origins of a people."[27]

Following are examples of myths which can affect electrical safety. Exploring these myths offers promise to improve electrical safety by helping employees better understand how their detrimental attitudes and inaccurate background beliefs can influence their safety.

"I'm experienced, so I won't get injured." This belief assumes that experiences *protect against* injury and death. However, as the statistics shown in Chap. 9 suggest, the employee who is *most at risk* of an

TABLE 12.7 Employee Attitudes Detrimental to Safety

	Attitude	Attitude expression
Pride	High opinion of self without recognizing the importance of team	"I'm doing it my way!"
Heroic	Fearlessness and boldness emphasized	"I'm always the fireman here!"
Fatalistic	Defeated or worse feeling	"Nothing to do about it … it is what it is."
Invulnerability	Immunity to the threats, forces, limits, or risks	"That's never going to happen to me."
Pollyanna	Finding the good in everything	"This is routine and never goes wrong."
Complacency	Comfortable and accepting of deviations	"Always happens … so what?"

Adapted from *DOE Human Performance Handbook*, U.S. Department of Energy, Chap. 2, pp. 5–6, Washington, DC, 2009.

electrical event is between the ages of 25 and 45 years and with accident-free years on the job. To the extent that employees' experience dulls their awareness of the distinctive features present in the job they are performing, their experience may create a false context for thinking about situational facts such as the following:

- Task electrical configuration
- Personal protection and barrier needs
- Required resources (such as people, equipment, and time) to complete the job

"Electrical accidents happen when an employee isn't paying attention." This mythic belief is based on a logic that goes as follows: As long as an employee does pay attention, no accidents will occur; therefore, if an accident happens, an employee must have been inattentive.

Comments such as the following presume inattention as an explanation:

- "He wandered off in his thinking."
- "She was daydreaming on the job."
- "He was worried about something else."

However, through many debriefing of electrical accidents, engineers have come to appreciate modern electrical work unfolds in highly complicated situations. If an electrical safety failure occurs, as the earlier data presented about latent and active errors suggest, typically more than one thing goes wrong. Multiple system faults or personnel errors contribute to the event. This understanding suggests that attention alone is not *sufficient* to prevent an electrical accident.

Given the embedded complexity in the power system, multiple actions are required to preserve safety. How to sharpen or focus the actions of human attention is a research target of cognitive ergonomics. Of special interest is the question of how multitasking may impair attention when tasks require cognitive control.[28]

Related to attention is the ability to hear. If employees can't hear instructions they are being given verbally, they can be expected to have difficulty in their focus on the instructions or attention to the verbal guidance given.

Hearing as a limited human factor depends on sound. Sound pressures needed for hearing are a function of the material or media (e.g., air) through which the sound or acoustic waves are propagating.[24] The threshold for hearing in the frequency range of the spoken word, from about 1,000 to 5,000 Hz, is about 20 μPascals (2.9×10^{-9} psi).

According to Chapanis, a listener's ability to detect sound depends on multiple factors, including the following:

- The listener's age
- The listener's history of past or ongoing noise exposure
- Whether the listener is using one ear or two to hear, with the use on one ear (monoaural listening) requiring 3 dB more sound pressure
- The sound's acoustic frequency, expressed in hertz
- The presence of competing sounds, or masking
- The complexity of the bandwidth tones
- The sound's duration, with durations less than 200 ms requiring increased intensity (i.e., as acoustic signal duration is halved, the intensity of the signal must double to be audible)

Returning to the M-SHEL model of human factors, the interaction of "liveware" or people depends on successful communication. To the extent that employees cannot adequately communicate because they cannot hear each other talk, safety is jeopardized. So the physical ability or individual human factor of hearing can affect the ability to communicate, which can affect attention, which is influenced in part by background belief, or context.

"As long as I don't touch an electrical source, I won't get shocked." This belief correctly identifies the need to avoid exposure to shock hazard. However, mechanical contact is *not necessary* for an employee to be shocked. Returning to the M-SHEL model, the mythic belief here underplays the role of environment in its influence over the individual human factors of body size and positioning in relationship to electrical conductivity, resistivity, and impedance.

Electricity is conducted along copper wires in power generation, transmission, and distribution. Depending on the current, when an employee's body comes sufficiently near to an electrical source, the charge that is carried by electrons in copper or aluminum wire may be converted through an electrochemical reaction to charge conducted by the ions in the human body. Employee size and positioning as well as meteorological conditions and geography play a role in whether electricity can cross an air gap by arcing and flow through or around the employee. The resulting shock can be destructive, even fatal, if adequate personal protection is not being used.

Studies suggest that employees have little opportunity to react during an electrical hazard exposure. As a human factor, reaction times for responses to stimuli (such as buzzing or light shocks), "word" information, or other prompts have been studied. These times are studied in laboratory and real world scenarios. Generally, reaction times vary from person to person, and sometimes between successive trials by the same person.

Environmental stress, such as heat exhaustion, altitude sickness, or hyperbaric conditions (such as work in mining, undersea, or in certain medical facilities), can change mental efficiency and lengthen reactions. Responses have relatively slower reaction time when language is involved. For example:

- A printed word is registered in the reader's brain in about an eighth of a second (0.125 s).
- A spoken word is accessed by a listener's brain in about a fifth of a second (0.200 s), before the speaker has finished pronouncing the word.
- The brain takes about a quarter second (0.250 s) to find a word to name an object and another quarter of a second (0.250 s) to program the mouth and tongue to pronounce the name (total: 0.500 s).[30]

Psychologists have studied attention and voluntary action in responses, finding the following:

- An average physical movement (motor) response time of >0.600 s in healthy people tested
- An average nonmotor response time of >1.050 s in adults asked to *say* verbs for printouts of words shown as *visually presented* nouns

In Chap. 1, the thresholds for neuromuscular responses in response to electrical current were reviewed. In comparison to electrical responses, motor and nonmotor responses to language stimuli generally take much longer.

The implication of this human performance limitation can be shown with this example: If a coworker needs 250 ms to process a spoken word such as "STOP" or "HELP" while an electrical incident such as an arcing fault is unfolding, say in less than 6 cycles, there is going to be a mismatch between the following:

- The amount of time the coworker needs to detect, process, and respond to the fault and
- The amount of time wherein maximum risk is present from the fault.

This time mismatch is even more pronounced when there is a need to sequence perception, thought, and response in a rapid amount of time, as in the following:

- Sense physical stimuli
- Perceive and process information
- Act

Age influences reaction time, with times slower for those older than 60 and younger than 15 years. Certain situational conditions generally slow reaction time and increase reaction time variability for a person, including the following:

- Sleep deprivation
- Fatigue
- Time of day
- Environmental extremes
- Alcohol or other drug use
- Medical problems
- Nutritional status

Notably, sleep research studies have found that on-the-job reaction times may be many times longer than what is found in a "lab" set-up, especially for specific kinds of tasks requiring physical exertion and mental concentration,[29] such as those found in electrical work.

SUMMARY

Innovations resulting from the study of human factors have led to increased safety, security, and reliability of electrical energy generation, transmission, and distribution. To maximize success, layered Defense in Depth strategies must extend beyond the individual across technology platforms and organizations. Employees have physical and mental limits when they perform complex and potentially hazardous tasks. Reducing precursor and latent errors requires management, engineering, and maintenance approaches.

REFERENCES

1. U.S. Department of Energy, *Human Factors/Ergonomics Handbook for the Design for Ease of Maintenance*, DOE-HDBK-1140-2001, Washington, DC, February 2001. Chapter 1, p. 10.
2. Gertman, D. I., Hallbert, B. P., Parrish, M. W., Sattision, M. B., Brownson, D., and Tortorelli, J. P., *Review of Findings for Human Error Contribution to Risk in Operating Events*. Publication reference: NUREG INEEL/

EXT-01-01166 prepared for Office of Nuclear Regulatory Research, Division of Systems Analysis and Regulatory Effectiveness, U.S. Nuclear Regulatory Commission, Washington DC, August 2001.

3. Lee, J. W, Oh, I. S., Lee, H. C., Lee, Y. H., and Sim, B.S., Human factors researches in KAERI for nuclear power plants. In *Global Perspectives of Human Factors in Power Generation.* IEEE Sixth Conference on Human Factors and Power Plants, Orlando, Conference Proceedings, 1997, pp. 13/11–13/16.
4. IEEE Task Force 15.07.05.05, Recommended practices for helicopter bonding procedures for live line work, *IEEE Transactions on Power Delivery*, 15(1): pp. 333–349, January 2000.
5. ESMOL Subcommittee 15.07, "Safety Considerations When Placing a Person with Tools in an Air Gap to Change Porcelain and Glass Insulators on Transmission Systems of 345 kV and Above, Using Ladder and Aerial Lift Methods," *IEEE Transactions on Power Delivery,* 17 (3): pp. 805–808, July 2002.
6. OSHA, Ergonomics solutions for electrical contractors. Retrieved from www.osha.gov/SLTC/etools/electricalcontractors/index.html.
7. Electric Power Research Institute (EPRI), *Ergonomics Handbook for the Electric Power Industry: Ergonomic Interventions for Electrical Workers in Fossil-fueled Power Plants, 2008.* EPRI Electric Power Research Institute Publication No. 1014042, Palo Alto, CA, 2008.

   Electric Power Research Institute (EPRI), *Ergonomics Handbook for the Electric Power Industry: Ergonomic Interventions for Plant Operators and Mechanics in Fossil-Fueled Generating Stations, 2008.* EPRI Publication No. 1015631, Palo Alto, CA, 2008.

   Electric Power Research Institute (EPRI), *Success Story: Public Service Electric & Gas Uses EPRI Handbooks to Build a Case for Ergonomics.* EPRI Publication No. 1018282, Palo Alto, CA, October 2008.

   Electric Power Research Institute (EPRI), *Ergonomics Handbook for the Electric Power Industry: Ergonomic Design Handbook for Fossil-Fueled Electrical Generating Stations, 2008.* EPRI Publication No. 1014942, Palo Alto, CA, 2008.
8. Miller, M. J., EPRI occupational health and safety research program, EEI Occupational Safety and Health Conference, Orlando, FL, September 27, 2010. Retrieved from http://www.esafetyline.com/eei/conference%20pdf%20files/EEI%20Fall%202010/EPRI_Miller.pdf. Accessed April 8, 2011.
9. Mustafa, A. M., and Al-Mahadin, A., Risk assessment due to the installation and maintenance of onshore wind turbines. In *2018 Advances in Science and Engineering International Conferences (ASET) Proceedings.* Dubai, UAE. March 2018, pp. 1–7.
10. Electrical Safety Foundation International (ESFI), Workplace fatalities and injuries, 2003–2016. Last accessed April 22, 2019 @ http://www.esfi.org/resource/workplace-fatalities-and-injuries-2003-2016-644.
11. NSA, 2011.
12. Diaz, N. J., *The Very Best-Laid Plans (the NRC's Defense-in-Depth Philosophy)*, Publication No. S-04-009, U.S. Nuclear Regulatory Commission, Washington DC, June 3, 2004.
13. Rasmussen, J., Human factors in high risk systems, IEEE Fourth Conference on Human Factors and Power Plants, Conference Record, 1988, pp. 43-48.
14. Kawano, R., "Steps toward the Realization of 'Human-Centered Systems'—An Overview of the Human Factors Activities at TEPCO." In *Global Perspectives of Human Factors in Power Generation,* Proceedings of IEEE Sixth Conference on Human Factors and Power Plants, Orlando, FL, pp. 13/27–13/32, 1997.
15. Hawkins, W. H., "Where Does Human Factors Fit in R&D Organizations?" *IEEE Aerospace and Electronic Systems Magazine,* pp. 31–33, September 1990.
16. O'Hara, J., and Higgins, J., Human-system interfaces for automatic systems. Presented at the Seventh American Nuclear Society International Topical Meeting on Nuclear Plant Instrumentation, Control and Human-Machine Interface Technologies (NPIC & HMIT). Las Vegas, NV. 2010. Brookhaven National Laboratory Publication BNL-93950-2010-CP, 2010, pp. 1–11.
17. O'Hara, J., Stubler, W. F., and Kramer, J., Human factors considerations in control room modernization: Trends and personnel performance issues. In *Global Perspectives of Human Factors in Power Generation.* IEEE Sixth Conference on Human Factors and Power Plants, Orlando, Conference Proceedings, 1997, pp. 4/7–4/10.
18. Hawkins, F. H., *Human Factors in Flight.* Gower Technical Press, Hants, England, 1987.
19. Department of Energy (DOE), *DOE Human Performance Handbook*, U.S. Department of Energy, Washington, DC, 2009.
20. Electric Power Research Institute (EPRI), *Ergonomics Handbook for the Electric Power Industry: Overhead Distribution Line Workers Interventions, 2001,* EPRI Publication No. 1005199, Palo Alto, CA, 2001.

21. Electric Power Research Institute (EPRI), *Ergonomics Handbook for the Electric Power Industry: Ergonomic Interventions for Manhole, Vault and Conduit Applications, 2004,* EPRI Publication No. 1005430, Palo Alto, CA, 2004.
22. Electric Power Research Institute (EPRI), *Ergonomics Handbook for the Electric Power Industry: Ergonomic Interventions for Direct-Buried Cable Applications, 2005,* EPRI Publication No. 1005574, Palo Alto, CA, 2005.
23. Cohen, A. L., Gjessing, C. C., Fine, L., Bernard, B. P., and McGlothin, J. D., *Elements of Ergonomics Programs,* DHHS (NIOSH) Publication No. 97-117, Washington DC, Tray 9, *1997,* pp. 124–127.
24. Chapanis, A., *Human Factors in Systems Engineering,* John Wiley & Sons, New York, pp. 186, 218–220, 228, 1996.
25. Tory, M., and Mőller, T., Human factors in visualization research, *IEEE Transactions on Visualization and Computer Graphics,* 10(1): pp. 72–84, 2004.
26. Department of Energy (DOE), *DOE Human Performance Handbook,* U.S. Department of Energy, Chap. 2, pp. 22–29, Washington, DC, 2009.
27. *New Lexicon Webster's Dictionary of the English Language.* New York: Lexicon Publications, 1989, p. 660.
28. Floyd, A. H. L., Multitasking and the illusion of safety, *IEEE Industry Applications Magazine.* 18(3): pp. 18–22, 2012.
29. Beare, A. N., Dorris, R. E., and Kozinsky, E. J., Response times of nuclear plant operations: Comparison of field and simulator data." In *Proceedings of the Human Factors Society 26th Annual Meeting,* pp. 669–673, Human Factors Society, Santa Monica, CA, 1982.
30. Pinker, S., *Words and Rules,* New York: HarperCollins, 1999, p. 335.

RECOMMENDED READINGS

Albers, J. T., and Estill, C., *Simple Solutions: Ergonomics for Construction Workers,* DHHS (NIOSH) Publication No. 2007-122, Cincinnati, OH, August 2007.

Rigot, B., Implementation of human performance (HP) for engineers at DOE's Savannah River Site. Retrieved from http://www.hss.gov/HealthSafety/ism/Workshop/ 2007Nov/TrackD/Wednesday/400_530/communication

CHAPTER 13
SAFETY MANAGEMENT AND ORGANIZATIONAL STRUCTURE

INTRODUCTION

An electrical safety program is effective only if management makes a strong commitment to support it. This chapter develops some of the key management concepts and procedures that must be present for a safety program to work. Of course, electrical safety is only part of an overall safety program; consequently, much of the material in this chapter is applicable to the entire safety effort.

The procedures introduced in this chapter should be applicable to all types of electrical installations. However, the effectiveness of any specific program must be determined by ongoing evaluations.

Safety organizations should be responsible to the very highest management levels and generally should not report to operations. Safety-related decisions should not be made by personnel with direct, bottom-line responsibility.

Additionally, the decisions made by legal counsel should be closely evaluated in terms of their effect on personnel safety. Many well-meaning attorneys and/or senior-level managers are required to make decisions that will maximize shareholder returns and/or limit corporate liabilities. Such decisions are, unfortunately, not always consistent with long-term worker safety.

Problems also may be introduced by labor organizations in their attempt to secure the best overall package for their members. Care should be exercised to avoid using safety as a "bargaining chip." This caution applies to labor and management.

CHANGING THE SAFETY CULTURE

One of the biggest roadblocks to improved electrical safety is culture. Worker behavior is often controlled by the local culture in terms of attitudes, procedures, and the use of safety equipment. Since most of us rely on anecdotal experiences in our lives, we often make inaccurate assumptions and do things that put us in harm's way.

Electrical safety regulatory and standards groups have tried especially hard over the past 20 years to change wrong-headed thinking and inaccurate and dangerous assumptions such as the following:

- *It's only 120 volts. It can't hurt you.*
- *These leather gloves provide me with plenty of insulation to work on the 240-volt circuit.*
- *I'll just ground this by wrapping this chain around the 480-volt bus.* (This practice was once so common that even in the 21st century safety grounds are often referred to as *grounding chains.*)
- *As long as I keep one hand in my pocket, I can't get shocked.*
- *I won't ever be exposed to an electrical blast—I keep the doors to the switchgear firmly latched.*

Although there is still a long way to go and a lot of work to do, we have made great strides in changing the electrical safety culture. The following paragraphs explain.

An article titled "Occupational Electrical Injury and Fatality Trends and Statistics: 1992–2007," written by Brent C. Brenner in May 2009 and published in the *IAEI* (International Association of Electrical Inspectors) *Magazine*, stated that since the creation of the U.S. Department of Labor Occupational Safety and Health Administration (OSHA), overall workplace fatalities have been cut by more than 60 percent and occupational injury and illness rates have declined by 40 percent. This decline occurred during an era of growth in which the total number of workers more than doubled. By focusing on the industrywide trends in electrically related fatalities and injuries, we can confirm that efforts to create a "culture of safety" are, in fact, beginning to have an impact. The numbers studied demonstrated substantially fewer deaths and injuries in 2007 than were experienced 15 years earlier in 1992.

In 1994, there were 348 electrically related fatalities and 6018 injuries. In contrast, in 2007, there were a total of 212 electrically related fatalities and 2540 injuries reported. It is important to note, however, even with the overall decline, fatality rates in certain industries remained static or had risen over the five years between 2003 and 2007. The Electrical Safety Foundation International (ESFI) has found that worker contact with electric current in some shape or form was responsible for 1213 fatal workplace accidents during this period and 13,150 workers were so severely injured from electrical contacts that their injuries required time off from work.

Brenner stated in his article that contact with overhead conductors accounted for 43 percent of all occupational electrical fatalities between 2003 and 2007, and the second leading category of electrical fatalities involved workers coming in contact with wiring, transformers, or other electrical components. This is the type of accident that occurs more often to employees whose jobs routinely involve working with electrical components such as an electrician or contractor. This category accounted for 28 percent of electrical fatalities and 37 percent of nonfatal electrical accidents. The third leading category of electrical fatalities involved workers coming in contact with electric current from machines, tools, appliances, or light fixtures. This type of accident occurs more often to workers whose job duties include mechanical and electrical maintenance. Accidental electrocution due to contact with tools and apparatus whose grounding conductors were faulty or missing is included in this category. These accidents accounted for 18 percent of all electrical fatalities and 35 percent of nonfatal electrical accidents.

Historically, the construction industry has had the greatest number of electrical fatalities. Between 2003 and 2007, the construction industry, with a workforce numbering 9.5 million workers in 2007, accounted for 52 percent of the occupational electrical fatalities. Electricians sustained 47 percent of the electrical fatalities in construction followed by laborers at 23 percent. Painters, roofers, and carpenters suffered 6 percent in each occupation. During the study period from 2003 to 2007, the electrical burn rate per 100,000 workers in the construction industry hovered between 0.5 and 1.0. Construction was one of only two industries, along with the utility industry, in which the number of nonfatal electrical burn injuries exceeded the number of electrical shock injuries. There were 2390 electrical burn injuries in construction and 1710 electrical shock injuries.

These statistics demonstrate that we have made progress, but there is still a lot of room for improvement. The only way to improve these grim statistics is by creating a change in the electrical safety culture. This can only happen by changing the way employees think about their personal safety. Personal safety is exactly what the term implies: the workers exposed to electrical hazards must take personal responsibility for their own health and welfare. No matter how many rules are in place, how much personal protective equipment is supplied, or how much information is available for an individual worker's use, individual workers are the only ones in a position to make good judgments that will keep them safe. Common sense coupled with a thorough understanding of basic fundamentals of electricity, safety rules, federal and state regulations, safety practices, and up-to-date procedures will increase the worker's level of safety awareness.

Working together, owners, managers, and electrical workers can create an environment in each individual workplace that will incubate and nurture positive changes in the electrical safety culture.

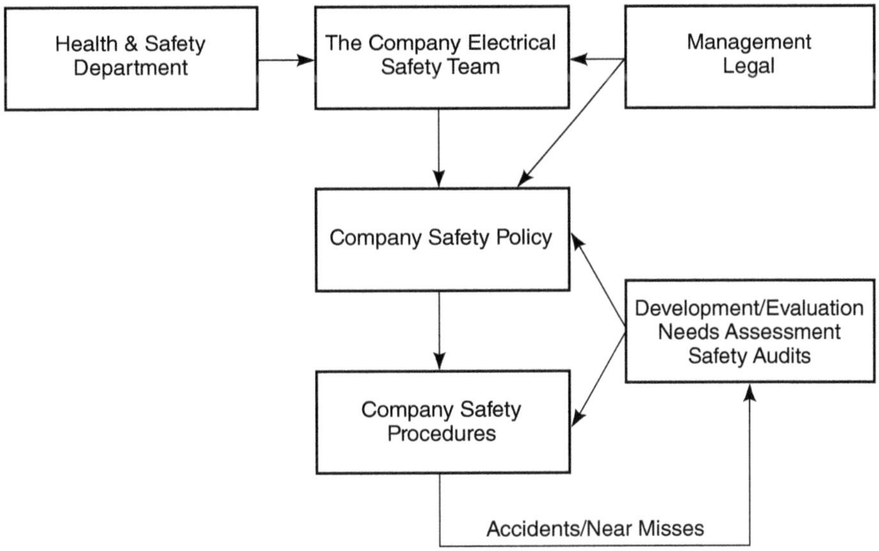

FIGURE 13.1 The electrical safety program structure.

ELECTRICAL SAFETY PROGRAM STRUCTURE

Figure 13.1 is a suggested design for the overall structure of a company electrical safety program. Of course, such a structure must be integrated into the overall safety program; however, the unique needs of the electrical safety program should be included in the design in a manner similar to the one shown.

Each of the various elements of the design is described in the following sections.

ELECTRICAL SAFETY PROGRAM DEVELOPMENT

An electrical safety program is composed of a variety of procedures, techniques, rules, and methods. Each of these individual items must be developed independently; the sum of them then adds up to the overall safety program. The development of an entire program becomes the accumulation of all the individual procedures.

This does not mean that the development of a safety program is a hodgepodge effort. On the contrary, the whole procedure should be viewed as an engineering design problem. The program should be put together by a corporate electrical safety team composed of management, safety, and technical personnel.

Some companies take regulatory and consensus standards such as the OSHA Electrical Safety-Related Work Practices rule and the NFPA 70E and apply them almost verbatim for their safety policies, procedures, and/or rules. This is a very poor practice. Even the NFPA 70E standard should be looked at only as a set of minimum requirements. In addition to regulatory requirements, company policies and procedures must include specific local requirements and must be developed in a way that is consistent with the facility culture.

Company Electrical Safety Team

Introduction. One of the best overall safety concepts of the last part of the 20th century was that of the worker *team*. Nowhere is the team of more use than in the safety arena. The following paragraphs describe the concept of the electrical safety team and provide enough information to serve as a template for the setup of such a body.

Note that the company health and safety department (HSD) has the ultimate responsibility for the safety program. The HSD should work closely with the team and review all procedures and policies developed by the electrical safety team. As described in the following, the HSD should have a representative on the team.

Responsibilities. The company electrical safety team (CEST) should have the following responsibilities and authorities:

- The CEST should have overall responsibility for the development, implementation, evaluation, and modification of the company electrical safety procedures. Either directly or through delegation, the CEST should develop the entire program, working closely with the health and safety professionals.
- The degree of authority vested in the CEST must be a matter of individual company policy. Generally, the CEST should have the maximum authority allowable under existing company structure.
- If the company has multiple locations, the CEST should appoint or sponsor elections for employee electrical safety teams (EESTs) at each site. The local teams will participate in local accident reviews, evaluate procedures and determine the best way to apply them locally, counsel employees, and implement and plan safety meetings.
- The CEST should have representation at the management level for the purpose of participating in the development of the company safety policy.

Employee Electrical Safety Teams. Employee electrical safety teams should be put in place to perform the actual fieldwork and legwork required by the CEST. The EEST will participate in accident investigation, program development, and any other activities deemed necessary. Some companies appoint permanent employee electrical safety teams as part of their ongoing corporate structure.

Structure. Generally, the safety team should include the following representatives/members (Table 13.1):

TABLE 13.1 Recommended Electrical Safety Team Members

Electrical workers	An electrical worker should serve as the chairperson of the team. In addition to the chairperson, sufficient members should be included from the workforce to ensure that the workers have a significant say in their overall safety program. Working supervisors should be eligible for these positions; however, supervision should never dominate the team.
Health and safety professionals	At least one health and safety professional should be included on the team to advise and assist in the areas in which his or her expertise applies.
Management	A member of management should be present as an adviser. This person can direct the team with respect to company policies to avoid the team conflicting with company directives. The management representative should be as high a level as possible. The management representative should always be willing to consider modification of conflicting company policies in the interest of improved safety.
Legal representation	Company counsel should be represented on the team; however, they should be there strictly in an advisory capacity.

Company Safety Policy

All safety programs should be underwritten by a company and/or departmental safety policy. Although specific policy statements must vary from industry to industry, all policies should contain the following key statements:

1. The company is committed to safe work practices.
2. At a minimum, all company safety policies and procedures shall comply with applicable federal, state, and local standards as well as recognized consensus standards.
3. Safety is the premier consideration in performing work.
4. Employees will be required to follow all company safety procedures.
5. If a job cannot be safely done, it need not be done.
6. Each individual employee is uniquely responsible for his or her own personal safety.
7. The cooperation of all personnel will be required to sustain the safety program.

Assessing the Need

The development or revision of an electrical safety program should begin with an evaluation of any existing programs. This initial survey should closely examine and catalog the number and types of electrical accidents. Investigators should also be creative in their analysis. That is, they should identify potential hazards as well as demonstrated ones. One of the most effective ways to catalog safety hazards is to perform a safety audit. Safety audits are discussed later in this chapter.

Problems and Solutions

Based on the results of the needs assessment, accidents and potential problem areas should be cataloged into *cause* categories. The specific categories selected should be chosen to fit the industry. Good starting points are the seven categories used by the Occupational Safety and Health Administration in the development of the Safety-Related Work Practices rule. These categories, illustrated in Table 13.2, can be used for most electrical accidents. The specific installation should develop additional categories or subcategories as required so that very few accidents fall into category 7.

Identification of the safety problems will inevitably lead to solution concepts. These solutions should take the form of specific plans and programs that can be incorporated in the new or existing electrical safety program.

TABLE 13.2 Categories to Classify Accidents or Potential Accidents

1. Use of equipment or material too close to exposed energized lines
a. Vehicles (e.g., cranes and dump trucks)
b. Other mechanical equipment (e.g., augers and derricks)
c. Tools and materials (e.g., ladders and tree limbs)
2. Failure to use electrical protective equipment
3. Assuming an unsafe position
4. Failure to de-energize (and lockout-tagout) equipment
5. Use of visibly defective electric equipment
6. Blind reaching, drilling, digging, etc.
7. No unsafe work practice or not enough information to classify

Program Implementation

After the analysis phase, the solutions can be integrated into the facility's safety program. The method used here will depend upon the facility; however, a reasonable starting point would be the development of an energy control program similar to that described in Chap. 4.

Examples

The following examples (Tables 13.3 through 13.6) illustrate the steps just outlined. These examples are drawn from actual industry experiences and illustrate the specific kinds of problems that workers may face every day.

Electric Shock While Replacing Low-Voltage Fuses. Assume that your electrical safety audit has identified several instances in which employees were shocked during the removal and/or installation of 120-V fuses. Table 13.3 shows the problem, solution, and implementation required to provide a safer work environment.

Protection for Arc Injury during Switching Operations. Not all safety procedures are developed based on local facility experience. In this example, we assume that the local safety personnel have

TABLE 13.3 Example of Fuse Removal Safety Procedure Development

Need assessment	In the past five years, 15 employees have been shocked while removing and/or replacing low-voltage cartridge fuses. Five of the injuries were extremely serious. No deaths have occurred.
Problem(s)	Review of the problems indicates that employees must, from time to time, work on such circuits when they are energized. Further investigation shows that employees are using channel lock type pliers to remove the fuses. Employees are not using rubber insulating gloves during such "hot" work.
Solutions	To resolve this problem, employees will be supplied with insulated tools that are specifically designed for the removal and replacement of fuses. Employees are already supplied with rubber insulating gloves.
Implementation	A procedure is added to the employee safety procedures manual, which specifies the following: When removing or replacing fuses, the bus that feeds them shall first be de-energized unless doing so introduces additional hazard or unless de-energizing requires a major plant outage.* Removing energized fuses is classified as hot work. Only qualified, trained personnel shall be allowed to remove and replace energized fuses. Low-voltage, cartridge type fuses shall be removed using only approved, insulated fuse pullers. Employees shall wear at least the minimum required safety equipment when removing or replacing energized low-voltage, cartridge type fuses. Minimum personnel protective equipment required to remove and replace energized low-voltage, cartridge safety type fuses shall include hard hat, safety glasses or goggles, long pants and long-sleeve shirt, leather safety shoes with rubber soles, and rubber electrical insulating gloves with leather protectors approved for the voltage level being worked on. The new procedure should be introduced at an employee safety training meeting. The new procedure must be demonstrated and employees must demonstrate competency to a management representative before being considered qualified to perform this energized work procedure.

*See Chap. 3 for a specific procedure to assess additional hazards and/or outages.

reviewed other industrial safety programs with respect to flash protection. They have discovered that other companies have experienced severe injuries during routine switching and racking operations. Furthermore, they discover that other companies have implemented successful programs that involve the use of flash suits. The number of flash-related injuries has dropped at these other companies. Table 13.4 was developed based on this analysis.

TABLE 13.4 Example of Flash Suit Safety Procedure Development

Need assessment	Companies of similar size and business, with comparably sized and arranged electrical systems, have had problems with severe flash injuries. These same companies have implemented flash suit use during certain phases of their operations and have exhibited a marked decrease in the number of injuries.
Problem(s)	Employees must, from time to time, perform open-door circuit breaker switching, remove motor starters from cubicles, and operate open-air disconnect switches. This exposes them to more than a normal electrical arc hazard.
Solutions	To resolve this problem, the company will supply and require the use of flame-retardant flash suits.
Implementation	A procedure is added to the employee safety procedures manual that specifies that employees shall be required to wear flame-retardant flash suits when performing open-door switching, motor starter removal or installation, or open-air switch operation. This requirement applies to all medium-voltage circuits and all low-voltage circuits with ampacities in excess of 100 A. This requirement is in addition to other safety equipment including hard hat, eye protection, and rubber insulating gloves with leather protectors. The new procedure should be introduced at an employee safety training meeting. This new procedure must be demonstrated and employees must demonstrate competency to a management representative before being considered qualified to perform this energized work procedure.

TABLE 13.5 Example of Recognition of Exposed Live Parts Procedure Development

Need assessment	An employee was working in a 480-V cabinet. His supervisor had previously told him that the cabinet was de-energized. The employee had looked at the cabinet and had measured the *obvious* energized locations. They were dead.
Problem(s)	There were exposed fasteners in the front of the cabinet that were not de-energized. They were not obvious and the employee missed them. The employee believed his supervisor and did not inspect the cabinet closely enough. The employee did not perform a thorough prework inspection to check for energized parts. He made several assumptions.
Solutions	Employees must be retrained to closely inspect for exposed energized parts.
Implementation	At a safety meeting, the entire incident was discussed and the problem brought to the forefront. All employees were reminded to inspect and check for energized parts no matter how certain they are that the system is safe. A training program was implemented in which the various specific pieces of equipment are identified and employees learn where all of the possible exposed parts are located. A formal procedure was added to require that employees carefully inspect equipment before they work on it.

TABLE 13.6 Example of Voltage Measurement Procedure Development

Need assessment	In addition to the problem discussed in the previous example (Table 13.5), other employees have been observed performing voltage measurements incorrectly.
Problem(s)	OSHA and other regulations require that voltage measurements must be made prior to working on a piece of de-energized equipment. All qualified electrical workers must be thoroughly trained in the proper and safe methods of voltage measurement.
Solutions	Employees are to be retrained in the proper methods of voltage measurement.
Implementation	The correct methods of voltage measurement were discussed and reinforced in employee safety meetings. Specific, hands-on training programs were implemented and all employees were tested to make certain that they were thoroughly familiar with the procedures. Additionally, the methods of selecting and inspecting measuring instruments were reinforced.

Recognizing Exposed Energized Parts. Unfortunately, many procedures are developed and modified based on an accident or injury. The example shown in Table 13.5 is one such instance. In this example, the employee was qualified and trained; he simply made a mistake.

Despite this, a significant amount of effort must go into the continuous retraining of personnel.

Voltage Measurement. OSHA states that one of the basic requirements for workers to be considered electrically qualified is that they shall possess the ability to discern what is energized and/or de-energized. This requirement means electrical workers must be able to examine the construction of the electrical system and determine the maximum voltage that could be present by looking at the construction, insulation, and conductor spacing. Once this is determined, workers can then know what type and class of measuring device they must use to measure for voltage. OSHA mandates that all voltage measurements be a three-step process—test the tester to make sure it is operable, measure the circuit or equipment being verified as de-energized, and retest the tester to make sure the instrument is still operable and a valid measurement has been obtained. On three-phase systems, measuring the equipment being verified requires three phase-to-ground measurements and three phase-to-phase measurements. Table 13.6 is an example of voltage measurement procedure development.

Company Safety Procedures

The actual development and structure of company electrical safety procedures are specific to each company. The methods and formats described in Chaps. 3 and 4 of this handbook should be referred to as a basis.

Also, the OSHA procedures have suggestions and formats for such plans.

Results Assessment

Few, if any, procedures will endure forever without modification. Personnel replacements, equipment changes, enhanced operating experience, and new safety innovations will require that any safety program be periodically reviewed and updated. The following criteria should be used to evaluate the need to revise an existing safety procedure:

1. Accidents and near misses
2. Employee suggestions
3. Employee electrical safety committee recommendations (see next section)
4. Changes in regulatory or consensus standards

EMPLOYEE ELECTRICAL SAFETY TEAMS

Reason

Electrical safety programs that are developed and/or administered without employee input will be ineffective. Employees know their jobs better than anyone. They understand those procedures with which they are comfortable and those that seem wrong. An electrical safety program that is developed without complete employee involvement has very little chance of success.

Method

Employee safety teams provide the best method for ensuring employee participation in a safety program. Such teams should be composed of a minimum of three employees chosen by their peers. The team should have regularly scheduled meetings on a monthly basis—more often if required by accidents or emergency conditions. The meetings should be held on company time.

Employee safety teams should be intimately involved with the entire safety program from initial design to ongoing results evaluations. They should be officially sanctioned by management and should have direct access to executive management. A limited, but useful, travel budget should be made available to the team to allow them to attend seminars and visit other facilities to study other safety programs. Although the specific responsibilities of such a team may vary from one company to another, the following may be used as guidelines.

Safety Program Development. Because of their familiarity with their jobs, employees are in a unique position to evaluate safety needs. What appears to be a perfectly safe procedure to a layperson may be obviously unsafe to a skilled, qualified worker. Furthermore, employees will often talk to their peers more freely than to supervisors or management. An employee safety team will be able to identify problems and solutions that outsiders might miss. Because of this, the safety team should be directly involved with the safety program development and/or modification described earlier in this chapter.

Safety Meetings. Employees should control all or at least part of their safety meetings. Briefings, permits, certifications, audits, and other activities required by legislation or regulations should be scheduled by management and safety personnel; however, other safety-related presentations should be generated by employees and administrated through the employee safety team.

Accident Investigation. The team should be involved in the investigation of accidents. The best way to ensure this involvement is to have the employee safety team appoint a member to serve on the accident investigation team. Accident investigation is a very demanding science and should not be performed by those unfamiliar with the information presented in Chap. 4.

Employee Training. The safety team should be allowed to review and recommend employee safety-related training. Employees who have been to outside safety training courses should be interviewed by the safety team. The team should then issue an annual recommendation for additional or modified training. This report should be considered a primary source of information when training budgets and schedules are produced.

SAFETY MEETINGS

The venerable safety meeting is one of the most universal of all safety vehicles. This one brief period combines training, program review, new concept presentation, and hazard notification all in one. The specific design and implementation of a safety meeting is necessarily dependent on the organizational arrangement of the company. The following items highlight some of the key imperatives in the structure of a safety meeting.

Who Attends

Safety meetings should be attended by all personnel affected by the safety topics that will be discussed. Those who should attend electrical safety meetings include the electricians, electrical technicians, electronics technicians, electrical supervisors, electrical department management, safety personnel with electrical responsibility, and anyone else who may be exposed to any of the electrical hazards.

Safety meetings should be chaired by employees, preferably members of the employee safety team. Remember that the company's attitude toward proper safety can make an enormous difference in employee attitudes.

What Material Should Be Covered

The safety meeting should have a standard, but flexible, schedule. Table 13.7 shows the type of schedule that should be included. Flexibility is the key word. One of the most frequent complaints about safety meetings is that they always contain the same old messages, are presented the same old way, and use the same old films or videos. The program should be flexible and new. Employees should be consulted as to topics that they feel may be relevant.

Avoid using the safety meeting time for non-safety-related information. General company announcements should be made at general employee meetings, not at safety meetings. Also remember that employees who are injured at home cannot report to work. Safety meetings should also address home safety issues. Topics such as safe use of appliances, house wiring safety, home lightning protection, and first aid can be useful to employees both on and off the job.

When Meetings Should Be Held

To accommodate work schedules, many companies hold their safety meetings at the beginning of the work day. Others may wait until the end of the day. Try moving the time of the meetings around. When the results of the meeting are evaluated, as described later, the most appropriate time will be that in which the greatest amount of information is imparted. For real variety, some companies have found that continuously moving the meeting times is effective. Some companies even hold meetings at night and invite spouses to attend. This approach allows the whole family to become involved in safety and allows the employer to present home safety topics more effectively.

Safety meetings should be canceled only when absolutely no other option is available. Major operational emergencies such as fires and major outages are the only acceptable reasons for canceling a safety meeting.

TABLE 13.7 Typical Safety Meeting Schedule

- Welcome and Review of Topics to Be Covered
- Accident Report
  - Summary of Accidents
  - Analysis and Lessons Learned
  - Procedure/Policy Modifications If Any
- Training Schedules and/or Safety Awards
- Meeting Topic Presentation
- Employee Safety Suggestions
- Questions/Answers
- Announcement of Next Meeting and Adjournment

Where Meetings Should Be Held

Variety should also be exploited when choosing meeting locations. Most meetings will be held at the work site for economic reasons; however, some meeting topics might call for different locations. For example, a demonstration of the proper use of hot sticks might best be held at a substation area, while tool safety training might be most effective if presented in the shop. *Remember that when demonstrations are given, even more extreme safety precautions should be taken.*

Evening safety meetings can be held at private banquet rooms or restaurants with meeting facilities. The meeting can be integrated with a social occasion that will allow management personnel, safety personnel, employees, and families to discuss problems and share solutions on a more informal basis. When such meetings are held, they should remain serious, with safety as the premier topic.

How Long Meetings Should Be

Meetings held at the workplace should be kept to a maximum of one hour unless some special topic or presentation requires more time. Remember that a safety meeting is a training presentation. Adult training sessions are best kept to short segments. Thus, if a meeting requires more than the one-hour time period, be certain that breaks are given every 45 minutes or so. Evening meetings may be somewhat longer if a dinner or other social event is integrated with the meeting.

Evaluation of Safety Meetings

Like all parts of a safety program, meetings should be constantly evaluated. Employee questionnaires and quizzes should be used to evaluate the quality of the meetings and the amount of information retained. These evaluations can be done on an anonymous basis so that employees feel free to share negative as well as positive comments. When tests or questionnaires identify problem areas, the meeting structure should be modified to correct the problem.

Follow-through. The safety program will not work if the company is not truly committed. When safety suggestions are made, the company should follow through and report back to the employee in writing. The employee safety committee should monitor suggestions and intervene if the company is not following through properly.

OUTAGE REPORTS

Safety hazards and electrical outages are frequently related to each other. Knowledge of the details that surround an electrical outage can be invaluable in pinpointing safety hazards. Outage reports can be used to

1. target potential or existing safety hazards.
2. justify maintenance or additions to a power system to increase reliability.
3. gather data for designing better power systems or additions to other plants or facilities.

Outage reports should include the following:

- Time and length of outage
- Cause of outage if known
- Results of outage (costs, injuries, etc.)
- Suggestions to prevent such an outage in the future

SAFETY AUDITS

Description

Depending on the intended extent, audits vary from short, simple inspections of specific problems or areas to large, companywide reviews that explore every facet of a safety program. Whatever their size or complexity, a safety audit is performed to review and assess the ongoing safety elements of a business. Audits are designed to identify the weaknesses and strengths of a safety program.

Process. A comprehensive audit starts with a review of upper-management safety awareness and attitudes and continues through the entire organization. During the procedure, every facet of the safety program is reviewed, cataloged, critiqued, and—if necessary—modified.

Attitudes. The attitudes of company personnel can make or break the effectiveness of a safety audit. Many employees tend to take safety audits very personally. A properly run safety audit should not be performed as an inquisition. Honest errors should be noted and corrected without hostile indictments or defensive recriminations. All personnel should be aware that a safety audit is only one tool used to make the workplace safer for everyone.

Audits versus Inspections. Some references differentiate between an audit and an inspection. In such references, audits are portrayed as major, comprehensive efforts that include all facets of the safety program. Inspections, on the other hand, are smaller, more concise efforts that may assess only one small area or procedure.

While there may be some psychological advantage in separating the two terms, remember that both are intended to identify, and subsequently eliminate, existing or potential safety hazards.

Purposes

The single most important purpose of a safety audit is to identify and eliminate safety hazards. There are, however, additional side benefits and purposes that are realized, including the following:

1. Employee morale is improved as safety problems are eliminated.
2. Audits provide a dynamic record of safety performance.
3. A positive cycle of safety improvement is created. That is, departments that get good scores tend to work harder to maintain their high scores. Departments that do not do as well tend to work harder to improve for the next audit.
4. Managers at all levels are made aware of many safety problems and procedures that they might not otherwise discover. For example, an audit that includes analysis of OSHA compliance may make managers aware of OSHA rules that were previously unknown.
5. When performed as a "family" procedure, a safety audit can actually enhance employer-employee relations. Employees respect management teams that truly put employee safety first.

Procedure

The safety audit should review and evaluate each element of the safety procedure. Site inspection trips, personnel interviews, task observation, and documentation review are among the methods that may be used. Each portion of the company that is reviewed will have its own specific concerns. Note that the items presented in the following sections are intended for example only. Specific facilities may require more or different audit criteria.

Facilities. Checkpoints for the electrical physical plant include the following:

1. Is the electric equipment kept clean?
2. Are safety exits clearly marked and unblocked by wire reels, ladders, and other such equipment?
3. Are relay flags and other such indicators kept reset?
4. Is explosion-proof equipment properly maintained and sealed?
5. Are all electrical grounds in place and secure?

Employees. Properly trained, alert employees are at the heart of any safety program. An individual employee is always the person most responsible for his or her own personal safety. The following points are typical of the types of checks that should be made to audit the safety awareness of individual employees:

1. Are employees adequately trained?
2. Are employees familiar with their safety handbook?
3. Do employees know where all electrical safety equipment is located?
4. Can employees successfully perform safety-related procedures such as safety grounding and voltage measurement?
5. Are personnel familiar with the electrical safety one-line diagram?
6. Are employees familiar with the safety techniques that are unique to the various voltage levels to which they will be exposed?

Management. Good safety practice starts at the top. If management includes itself in the audit process, employees will accept that they too must be reviewed. Management involvement is more than just window dressing, however. Other management audit points include the following:

1. Does the facility have a formal, written safety policy?
2. Are safety rules enforced uniformly?
3. Is there an employee safety committee to which management defers when technical and day-to-day decisions must be made?
4. Does management promote both on-the-job and off-the-job safety?
5. Does management provide, at company expense, CPR and other such training?

Safety Equipment. Various rules and regulations require the use of approved electrical safety equipment. The following points are typical of safety equipment checks during an audit:

1. Is safety equipment readily available for all personnel?
2. Have all rubber goods been tested within the required period?
3. Are meters and instruments mechanically sound and electrically operable?
4. Are test equipment fuses properly sized?
5. Are safety interlocks operational and not bypassed?
6. Are safety signs, tags, warning tapes, and other such warning devices readily available to all employees?

Safety Procedures. Energy control and other such safety procedures must be properly developed and implemented. A few of the key points for procedures include the following:

1. Are written switching orders required for planned outages?
2. Is a written copy of the lockout-tagout procedure available to all personnel?

3. Are all employees trained in electrical safety grounding?
4. Are safety procedures followed for each and every operation or are they bypassed in the interest of production?

Documentation. Safety electrical one-line diagrams and other such documents are too frequently ignored and allowed to fall behind or remain inaccurate. The following key points should be checked for safety-related documentation:

1. Are the electrical portions of the company safety handbook up to date and consistent with currently accepted safety practice?
2. Are copies of all safety standards and practices readily available to all personnel?
3. Is the safety electrical one-line diagram up to date, accurate, and legible?
4. Are accident reporting forms, test sheets, procedures, and other such operational aids readily available?

The Audit Team

Audit teams should be composed of safety personnel, management personnel, and employees. Because of their familiarity with safety auditing techniques, safety personnel should usually be the *team leaders*. The audit and assessment tasks should be shared equally by all team members.

The number of team members will, of course, depend on the size of the facility or facilities being audited. A minimum of three members is recommended even for very small facilities. This will allow representations for safety, management, and employees.

Some companies use the employee safety team to spearhead safety audits. The safety team assigns audit personnel, reviews audit results, and continues to follow up on changes.

Audit Tools

The principal tool of the safety audit is the audit form (Figs. 13.2 to 13.7). A form of this type should be developed for every area involved in the audit. In addition to such forms, audit team members should be supplied with copies of safety procedures, standards, and other such documents. Team members can thus evaluate whether the procedures and standards are being followed. Note that the management and employee questionnaires can be used for statistical analysis.

Follow-Up

Emergency or life-threatening problems should be corrected immediately. Problems that are left uncorrected continue to put employees and equipment at risk. Moreover, major audits that find problems that are not corrected send very negative signals to employees.

Follow-up to correct a problem should occur within a very short time after the results of the audit. Major safety problems should be corrected immediately. Action on less serious problems should begin within one month of the audit.

When audit results call for additional employee training, major system modification, or other such long-term expenditures, management should develop the plan within one month. No one expects major expenditures to be performed on a short-term basis; however, safety considerations must be met in a timely and effective manner. The audit report should be written as soon as the audit is

(Text continues on page 612.)

Outdoor Substations

Company ______________________________

Location ______________________________

Substation ______________________________

Date ______________ Auditors ______________________________

General Conditions

Item	Rating	Comments
Properly fenced		
Restricted access		
Fence grounding		
Rock and gravel (no vegetation)		
Corrosion-free		
Proper clearances		
"Danger High Voltage" signs		
Control room properly ventilated		
Additional observations and comments:		

(a)

FIGURE 13.2 Facilities audit form. (*a*) Outdoor substations; (*b*) indoor substation/electrical room; (*c*) miscellaneous outdoor equipment; (*d*) power transformers; (*e*) metal-clad switchgear; (*f*) motor control centers; (*g*) battery stations; (*h*) panelboards. (*Courtesy Cadick Corporation.*)

Indoor Substation/Electrical Room

Company ______________________________

Location ______________________________

Substation/Area ______________________________

Date ______________ Auditors ______________________________

General Conditions

Item	Rating	Comments
Restricted access		
"Danger High Voltage" signs		
Improper material storage		
Room properly ventilated		
Proper clearances		
Voltage level markings		
General housekeeping		
Corrosion-free		
Fire extinguishers		
Clear egress		
Additional observations and comments:		

(b)

FIGURE 13.2 (*Continued*)

Miscellaneous Outdoor Equipment

Company ______________________________

Location ______________________________

Substation ______________________________

Date ______________ Auditors ______________________________

Item	Rating	Comments
Lightning arresters		
Air disconnect switches		
Capacitor banks		
Oil circuit breakers		
Reclosers		
Fused disconnects		
CTs and PTs		
Control transformers		
Additional observations and comments:		

(c)

FIGURE 13.2 (*Continued*)

Power Transformers

Company ______________________________

Location ______________________________

Substation/Area ______________ Designation(s) ______________

Date ____________ Auditors ______________________

Item	Rating	Comments
Insulators		
Gauges operable		
Proper clearances		
No signs of leaks		
Equipment grounding		
Grounding plates		
Neutral ground resistors		
Proper ventilation		
Type of maintenance program		
PCB stickers		
Additional observations and comments:		

(d)

FIGURE 13.2 *(Continued)*

Metal-Clad Switchgear

Company ______________________________

Location ______________________________

Substation/Area ______________ Designation(s) ______________

Date ____________ Auditors ______________________

Item	Rating	Comments
Door and cover properly secured		
Proper ventilation		
Proper working space & clearances		
Panel identification markings		
Circuit identification markings		
Voltage level markings		
Protective device indicators		
Type of maintenance program		
Clear egress		
Equipment grounding		
Panel meters operative		
Additional observations and comments:		

(e)

FIGURE 13.2 (*Continued*)

Motor Control Centers

Company ____

Location ____

Substation/Area ________________ Designation(s) ________________

Date __________ Auditors ________________________

Item	Rating	Comments
Door and cover properly secured		
Proper ventilation		
Proper working space & clearances		
Panel identification markings		
Circuit identification markings		
Voltage level markings		
Indicators		
Type of maintenance program		
Clear egress		
Equipment grounding		
Panel meters operative		
Additional observations and comments:		

(f)

FIGURE 13.2 (*Continued*)

Battery Stations

Company ____________________

Location ____________________

Substation/Area ____________ Designation(s) ____________

Date __________ Auditors ____________________

Item	Rating	Comments
Restricted access		
Proper ventilation		
Eyewash station		
Neutralizing solution		
Proper working space & clearances		
Electrolyte level		
Signs of outgassing		
Voltage level markings		
Corrosion-free		
Type of maintenance program		
Clear egress		
Equipment grounding		
Availability of battery PPE		
Battery charger		
Additional observations and comments:		

(g)

FIGURE 13.2 (*Continued*)

Panelboards

Company ______________________________

Location ______________________________

Substation/Area ______________ Designation(s) ______________

Date ____________ Auditors ______________________

Item	Rating	Comments
Door and cover properly secured		
Proper ventilation		
Proper working space & clearances		
Panel identification markings		
Circuit identification markings		
Voltage level markings		
Type of maintenance program		
Equipment grounding		
Additional observations and comments:		

(h)

FIGURE 13.2 (*Continued*)

Company/Location:			Date:			Auditor:										Page - 1
Item	Accessibility/Availability	Properly Rated	Tested or inspected		Functional	Physical Condition 1 = OK 2 = Marginal 3 = Bad					Proper Storage (Yes or No)				Comments
			Interval (Months)	Yes or No		Insulation	Connectors/Cables	Cleanliness	Case	Proper Fusing	Dry	UV Protected	Proper Containers	Physical Protection	
Rubber Insulating Good															
Rubber Gloves			6												
Leather Protectors															
Rubber Blankets			12												
Rubber Sleeves			12												
Line Hose															
Rubber Covers															
Rubber Mats															

(a)

FIGURE 13.3 Safety equipment audit form. (*a*) Page 1; (*b*) page 2; (*c*) page 3. (*Courtesy Cadick Corporation.*)

Company/Location: Date: Auditor: Page -2

Item	Accessibility/Availability	Properly Rated	Tested or inspected		Functional	Physical Condition 1 = OK 2 = Marginal 3 = Bad					Proper Storage (Yes or No)				Comments
			Interval (Months)	Yes or No		Insulation	Connectors/Cables	Cleanliness	Case	Proper Fusing	Dry	UV Protected	Proper Containers	Physical Protection	
Eye Protection															
Safety Glasses															
Face Shields															
Thermal Protection															
Thermal Work Clothes															
Flash Suits															
Fire Blankets															
Warning Devices															
Barricades/Barrier Tape															
Signs															

(b)

FIGURE 13.3 (*Continued*)

Company/Location:					Date:				Auditor:						Page - 3
Item	Accessibility/Availability	Properly Rated	Tested or inspected		Functional	Physical Condition 1 = OK 2 = Marginal 3 = Bad					Proper Storage (Yes or No)				Comments
			Interval (Months)	Yes or No		Insulation	Connectors/Cables	Cleanliness	Case	Proper Fusing	Dry	UV Protected	Proper Containers	Physical Protection	
Measuring Instruments															
Voltage Testers															
Ground Measurement Equip.															
Current Measuring Devices															
Thermographic Devices															
Other Safety Equipment															
Safety Grounds															
Hard Hats															
Hot Sticks			24												
Insulated Tools															
Ladders															
Foot Protection															

(c)

FIGURE 13.3 (*Continued*)

Cadick Corporation

Electrical Safety Audit Form
Safety Policies & Procedures

Rev 1.2

Company/Location:					Date:	Auditor:	
Item	(Y or N)		Review		Utilization	Accessibility	Comments
	Written	Oral	Interval	Last Reviewed	1 = Always 2 = Sometimes 3 = Never		
First Aid and CPR							
Accident Investigation							
Protective Clothing							
Electrical Firefighting							
Emergency Comm.							
Energy Control Procedures							
Switching Procedures							
Lockout/Tagout							
Voltage Measurement							
Safety Grounding							

FIGURE 13.4 Safety procedures audit form. (*Courtesy Cadick Corporation.*)

Cadick Corporation

Electrical Safety Audit Form
Documentation

Rev 1.2

Company/Location: Date: Auditors:

	Documentation	Document Exists 1 = Yes 2 = Some 3 = No	Current Revision 1 = Yes 2 = Some 3 = No	Accessibility 1 = Yes 2 = Some 3 = No	Legibility 1 = Good 2 = Marginal 3 = Unacceptable	Comments
	Electrical Safety Manual					
	Safety Procedures					
	Safety Equipment Records					
	Safety Equipment Manuals					
	Accident Investigation					
Engineering	One-lines					
	Three-lines					
	Schematics					
	Wiring diagrams					
	SCA and Coordination					
	Flash Hazard Analysis					
Maintenance Records	PM Records					
	Protective Device Tests					
	Oil & Gas Analysis					
	Grounding Tests					
	Battery Tests					

FIGURE 13.5 Documentation audit form. (*Courtesy Cadick Corporation.*)

Electrical Safety Program

Employee questionnaire—page 1	Company/location:
This interview form is anonymous. Please feel free to answer honestly.	

The answers that you provide on this form are critical to the successful implementation and maintenance of a Zero Incidents Safety Program. Please provide your reaction to each statement by checking the appropriate box. You may be asked for an interview to further expand on your answers.

		Strongly Agree	Agree	Disagree	Strongly Disagree
1.	I am not able to perform my job safely because I have not been provided with the right safety equipment.	☐	☐	☐	☐
2.	It is not appropriate to point out my coworkers' at-risk behavior.	☐	☐	☐	☐
3.	Many of the safety procedures we are supposed to follow are unnecessary.	☐	☐	☐	☐
4.	I follow the correct safety procedures and use the appropriate equipment.	☐	☐	☐	☐
5.	Management at my company supports a company-sponsored electrical safety training program.	☐	☐	☐	☐
6.	Management at my company supports the electrical safety program.	☐	☐	☐	☐
7.	I would not hesitate to report one of my coworkers who was exhibiting at-risk behavior.	☐	☐	☐	☐
8.	I am provided with all the necessary safety procedures to allow me to perform my job safely.	☐	☐	☐	☐
9.	My supervisor encourages productivity at the expense of safety.	☐	☐	☐	☐
10.	My company says it supports the electrical safety program, but in reality it discourages anything that detracts from production.	☐	☐	☐	☐
11.	My supervisor supports the electrical safety program.	☐	☐	☐	☐
12.	My coworkers do not follow the correct safety procedures.	☐	☐	☐	☐
13.	I enjoy my job.	☐	☐	☐	☐
14.	I know what to do in an electrical emergency.	☐	☐	☐	☐
15.	My team has adequate skills to perform its job safely and efficiently.	☐	☐	☐	☐
16.	My company does not provide adequate skills training.	☐	☐	☐	☐

(a)

FIGURE 13.6 Employee audit questionnaire. (*a*) Page 1; (*b*) page 2; (*c*) page 3; (*d*) page 4. (*Courtesy Cadick Corporation.*)

Electrical Safety Program

Employee questionnaire—page 2	Company/location:
This interview form is anonymous. Please feel free to answer honestly.	

The answers that you provide on this form are critical to the successful implementation and maintenance of a Zero Incidents Safety Program. Please provide your reaction to each statement by checking the appropriate box. You may be asked for an interview to further expand on your answers.

	Statement	Strongly Agree	Agree	Disagree	Strongly Disagree
1.	I am skeptical about the ultimate value of our electrical safety program.	☐	☐	☐	☐
2.	The concept of "zero incidents" is unrealistic.	☐	☐	☐	☐
3.	On occasion, electrical safety must be compromised for production.	☐	☐	☐	☐
4.	The concept of "zero incidents" is fine in theory but not achievable at a reasonable cost.	☐	☐	☐	☐
5.	I am aware of incidents where safety has been compromised for production.	☐	☐	☐	☐
6.	I believe the company is committed to correcting, in a timely manner, deficiencies that are identified by this audit.	☐	☐	☐	☐
7.	All my coworkers have adequate electrical safety and skills training.	☐	☐	☐	☐
8.	I endorse a "zero incidents" electrical safety program.	☐	☐	☐	☐
9.	Electrical safety should never be compromised for production.	☐	☐	☐	☐
10.	We have adequate electrical personal protective equipment.	☐	☐	☐	☐

(b)

FIGURE 13.6 (*Continued*)

Electrical Safety Program

Employee questionnaire—page 3	Company/location:
This interview form is anonymous. Please feel free to answer honestly.	

The following questions are designed to determine your knowledge of the various safety-related equipment and procedures. Please check the box that best describes your understanding of the subject matter.

		In-depth Understanding	Competent	Need Additional Training	Totally Unfamiliar
11.	Company safety handbook (electrical section)	☐	☐	☐	☐
12.	Switching procedures for plant distribution system	☐	☐	☐	☐
13.	Voltage measurement procedures	☐	☐	☐	☐
14.	Use of drawings (one-lines, schematics, etc.)	☐	☐	☐	☐
15.	Lockout/tagout procedures	☐	☐	☐	☐
16.	Use of temporary safety grounds	☐	☐	☐	☐
17.	First aid and CPR	☐	☐	☐	☐
18.	Personal protective equipment usage (tools and clothing)	☐	☐	☐	☐
19.	Electrical firefighting	☐	☐	☐	☐
20.	Hazard analysis/recognition	☐	☐	☐	☐
21.	Tool usage	☐	☐	☐	☐
22.	Familiarity with job safety analysis	☐	☐	☐	☐

- Please fill out the form on the next page -

(c)

FIGURE 13.6 *(Continued)*

Electrical Safety Program

Employee questionnaire—page 4	Company/location:
This interview form is anonymous. Please feel free to answer honestly.	

Type of training		Training received	Training current	How trained (check all appropriate)		Comments
				Classroom	OJT	
Print reading						
Troubleshooting						
National Electrical Code						
Safety procedures and use of equipment	CPR & first aid					
	Electrical firefighting					
	Protective clothing					
	Hazard risk analysis					
	Confined space					
	Fall restraint					
	Switching procedures					
	Lockout/tagout					
	Voltage measurement					
	Safety grounding					
Equipment-specific training						

Directions: Please enter a (Y)es or (N)o in each box.
Training is considered current if it is done within 10 years except CPR, which is 1 year.

(d)

FIGURE 13.6 *(Continued)*

Electrical Safety Program

Management questionnaire—page 1	Company/location:
This interview form is anonymous. Please feel free to answer honestly.	

The answers that you provide on this form are critical to the successful implementation and maintenance of a Zero Incidents Safety Program. Please provide your reaction to each statement by checking the appropriate box. You may be asked for an interview to further expand on your answers.

	Statement	Strongly Agree	Agree	Disagree	Strongly Disagree
1.	Our electrical employees are not able to perform their job safely because they have not been provided with the right safety equipment.	☐	☐	☐	☐
2.	It is not appropriate for an employee to point out coworkers' at-risk behavior.	☐	☐	☐	☐
3.	Many of the safety procedures we are supposed to follow are unnecessary.	☐	☐	☐	☐
4.	Our employees follow the correct safety procedures and use the appropriate equipment.	☐	☐	☐	☐
5.	I support a company-sponsored electrical safety training program.	☐	☐	☐	☐
6.	I support the electrical safety program.	☐	☐	☐	☐
7.	I would not hesitate to counsel an employee who was exhibiting at-risk behavior.	☐	☐	☐	☐
8.	The company provides all the necessary safety procedures to allow our workers to perform their jobs safely.	☐	☐	☐	☐
9.	Our supervisors have encouraged productivity at the expense of safety.	☐	☐	☐	☐
10.	Our company says it supports the electrical safety program, but in reality it discourages anything that detracts from production.	☐	☐	☐	☐
11.	All supervisors support the electrical safety program.	☐	☐	☐	☐
12.	Our employees know what to do in an electrical emergency.	☐	☐	☐	☐
13.	Our electrical team has adequate skills to perform its job safely and efficiently.	☐	☐	☐	☐
14.	Our company does not provide adequate skills training.	☐	☐	☐	☐
15.	I am skeptical about the ultimate value of our electrical safety program.	☐	☐	☐	☐

(a)

FIGURE 13.7 Management audit questionnaire. (*a*) Page 1; (*b*) page 2. (*Courtesy Cadick Corporation.*)

Electrical Safety Program

Management questionnaire—page 2	Company/location:
This interview form is anonymous. Please feel free to answer honestly.	

The answers that you provide on this form are critical to the successful implementation and maintenance of a Zero Incidents Safety Program. Please provide your reaction to each statement by checking the appropriate box. You may be asked for an interview to further expand on your answers.

	Statement	Strongly Agree	Agree	Disagree	Strongly Disagree
16.	The concept of "zero incidents" is unrealistic.	☐	☐	☐	☐
17.	On occasion, electrical safety must be compromised for production.	☐	☐	☐	☐
18.	The concept of "zero incidents" is fine in theory but not achievable at a reasonable cost.	☐	☐	☐	☐
19.	I am aware of incidents where safety has been compromised for production.	☐	☐	☐	☐
20.	I am committed to correcting, in a timely manner, deficiencies that are identified by this audit.	☐	☐	☐	☐
21.	All of my employees have adequate electrical safety and skills training.	☐	☐	☐	☐
22.	I endorse a "zero incidents" electrical safety program.	☐	☐	☐	☐
23.	Electrical safety should never be compromised for production.	☐	☐	☐	☐
24.	All of my employees have adequate electrical personal protective equipment.	☐	☐	☐	☐

(b)

FIGURE 13.7 (*Continued*)

complete. The timetable for the correction of observed problems should be based on three criteria as follows:

1. How serious is the problem? If an accident caused by this problem would result in very severe injuries or high levels of damage, the problem should be corrected immediately. This criterion is the most important of the three, and it should outweigh any other considerations.
2. What is the likelihood that an accident will occur as a result of the problem noted in the audit? If there is little possibility of an accident, correction of the problem may be relegated to a lower status.
3. How much will it cost to fix the problem?

Internal versus External Audits

Although companies usually employ their own personnel in the performance of a safety audit, sometimes an external firm may be used. Each approach has its own advantages and disadvantages. Three principal reasons may call for the use of an external consultant in the performance of a safety audit:

1. Consultants perform such audits on a routine basis and, therefore, tend to be more efficient than inexperienced employees.
2. Consultants tend to be dispassionate and therefore more objective in their analysis.
3. In-house personnel may not be available because of work schedules.

Of course, the use of consultants has some disadvantages as well, such as the following:

1. Consultants lack familiarity with their clients' in-house systems and conditions.
2. Employees may not speak as freely with an outsider as they do with their own coworkers.

CHAPTER 14
SAFETY TRAINING METHODS AND SYSTEMS

INTRODUCTION

"Here's your rubber gloves. Go follow Joe." For much of the second half of the 20th century, this was the ultimate safety training method for many facilities. Fortunately, safety training professionals have managed to get two critical messages across:

1. Bad training is worse than no training at all.
2. Good training is both an art and a science.

This chapter defines "good" training and shows how it can augment a safety program. Employers and consultants alike can use the information. Consultants can use the information to help set up training programs for their clients. Employers can use it to evaluate training consultants and/or take preliminary steps to set up their own training programs. *Note:* This information is applicable to all types of technical training. Refer to the American Society for Training & Development website, https://thetranslationcompany.com/translation-directory/translation-associations/american-society-training-development/, for training details.

Safety Training Definitions

For the purposes of this chapter, training is defined as a formal process used to generate a desired positive change in the behavior of an individual or group of individuals. Notice the use of the word *formal*. Training must be "on purpose." Never allow training to just happen. Organize it; develop clear, measurable, and achievable objectives; administer the training, and evaluate the results. The training program and the administration of the training must then be documented.

Training is not the same thing as education, although training may well include education. Education is used to add to the cognitive skills and abilities of an individual or individuals. In other words, education teaches you how to think, while training modifies your behavior and teaches you how to do.

Behavioral modification is the core of training; however, there are many different types of behavioral modification. Three examples follow.

1. Many training companies offer training courses in which they attempt to teach general electrical safety procedures. While these courses are generally excellent, they do not directly address the specific procedures of each student's company. Rather, the courses emphasize the importance of good safety and show clients what should be done. The behavioral change that occurs is an increase in the safety effort. Such an increase will lead to many benefits; however, students will need additional safety training in their own specific job requirements.
2. Many in-house training programs teach the safety aspects of the maintenance or operation of a range of specific types of equipment. For example, a four-day program that teaches the maintenance of low-voltage power circuit breakers will spend great amounts of time discussing and

showing the safety aspects of the particular equipment. This type of behavior modification is relatively specific in that it teaches detailed procedures for a class of equipment—low-voltage power circuit breakers. The desired behavior is the use of proper safety techniques to ensure that students will safely maintain their breakers.

3. A manufacturer's training program on a specific piece of equipment is a third example. When the manufacturer of an uninterruptible power supply (UPS) offers a course on the operation and maintenance of its UPS system, very specific safety procedures are taught. In this instance, the manufacturer knows that good maintenance and operation will result in fewer problems. However, the manufacturer also knows that trained personnel must perform the maintenance safely. Here, the desired modification is reduced to a relatively few, very specific procedures.

In each of the examples just given, the training programs should be prefaced with educational objectives. An educational objective is a specific, measurable behavioral change required of the student. For example, in a lockout-tagout course, an objective might be as follows: "The student is expected to determine that a given system has been de-energized by correctly performing the three-step voltage-measurement process." A complete list of objectives should precede every module in a training program. The use of objectives allows the trainer and the trainee to be measured accurately and objectively. All the training discussed in this chapter will be objective-based training.

Training Myths

The electrical safety industry has at least two major training myths. These myths hurt or in some cases eliminate the implementation of good training programs.

Myth #1: My People Can Learn by Just Doing Their Job. This is the ultimate in "go follow Joe" thinking. The fact is that today's electrical systems are too complex to permit self-training unless it is very well organized and documented. The best result from such an approach will be slow, very expensive learning curves. The more likely result will be employees who learn improperly and make potentially grave errors.

Misconceptions are among the biggest problems for professional trainers. Adult learners have substantially more trouble unlearning a concept than they do learning one. While experience is definitely the best teacher, experience should be prefaced with a properly organized and presented training experience, with well-defined objectives.

Of course, well-designed self-training programs are an excellent source of training. The key word here, however, is *designed*. Simply turning an employee loose with an instruction book is not effective and can be very dangerous for the employee.

Myth #2: My Supervisors Can Train My People. Training is a part of every supervisor's job, but it should not be allowed to become the only part for three basic reasons. First, in today's highly regulated workplace, supervisors have their hands full with the day-to-day requirements of supervising. They simply do not have the time to put in the detailed effort required for good training. Second, good technical or supervision skills do not necessarily include good training skills. The ability to train well is an art. The third consideration is the "person with a briefcase" syndrome. Employees tend to listen closely and learn from a trainer from outside the company. The feeling is that this person brings a fresh viewpoint to the job. Frankly, this belief has some merit. The outside viewpoint can often instill skills and ideas that are not available from inside.

Conclusion

1. Insurance companies are increasing safety requirements for their insured companies.
2. The technology of power systems continues to spiral upward.
3. Systems grow at an ever-increasing rate.
4. Regulatory requirements continue to increase in number and complexity.

All these facts demand that employees be trained. Good training is available from a variety of sources.

COMPARISON OF THE FOUR MOST COMMONLY USED METHODS OF ADULT TRAINING

Introduction

Four types of training methods are in common use today: classroom, computer-based training (CBT), Internet (Web-based) training (WBT), and simple video training. CBT and WBT will be called *self-training* in later parts of this chapter. Caution must be used with the CBT, WBT, and video-based training programs. OSHA, in two interpretation letters dated October 11, 1994, and November 22, 1994, states the following:

> In OSHA's view, self-paced, interactive computer-based training can serve as a valuable training tool in the context of an overall training program. However, use of computer-based training by itself would not be sufficient to meet the intent of most of OSHA's training requirements. Our position on this matter is essentially the same as our policy on the use of training videos, since the two approaches have similar shortcomings. OSHA urges employers to be wary of relying solely on generic "packaged" training programs in meeting their training requirements.

Safety and health training involves the presentation of technical material to audiences that typically have not had formal education in technical or scientific disciplines. In an effective training program, it is critical that trainees have the opportunity to ask questions where material is unfamiliar to them. In a computer-based program, these requirements may be by providing a telephone hotline so that trainees will have direct access to a qualified trainer.

Equally important is the use of hands-on training and exercises to provide trainees with an opportunity to become familiar with equipment and safe practices in a nonhazardous setting. Many industrial operations can involve many complex and hazardous tasks. It is imperative that employees be able to perform such tasks safely. Traditional, hands-on training is the preferred method to ensure that workers are prepared to safely perform these tasks. The purpose of hands-on training, for example in the donning and doffing of personal protective equipment, is twofold: first, to ensure that workers have an opportunity to learn by experience, and second, to assess whether workers have mastered the necessary skills. It is unlikely that sole reliance on a computer-based training program is likely to achieve these objectives.

Thus, OSHA believes that computer-based training programs can be used as part of an effective safety and health training program to satisfy OSHA training requirements, provided that the program is supplemented by the opportunity for trainees to ask questions of a qualified trainer, and provides trainees with sufficient hands-on experience. In order for the training to be effective, trainees must have the opportunity to ask questions. The trainees' mastery of covered knowledge and skills must also be assessed.

In conclusion, it is possible in some cases to use computer-based training in meeting refresher training requirements, provided that the computer-based training covers topics relevant to workers' assigned duties and is supplemented by the opportunity to ask questions of a qualified trainer, as well as an assessment of worker skill degradation through auditing of hands-on performance of work tasks.

This section will compare and contrast each of these methods of training as they apply to adult learning.

John Mihall and Helen Belletti's "Adult Learning Styles and Training Methods" presentation handouts, dated February 16, 1999, state:

> The term "pedagogy" was derived from the Greek words "paid" (meaning child) and "agogus" (meaning leading). It is defined as the art and science of teaching children. The term "Andragogy" was coined by researchers of adult learning in order to contrast their beliefs about learning to the pedagogical mode. Malcolm Knowles first introduced the concept in the United States in 1968. The concept of andragogy implies self-directedness and an active student role, as well as solution-centered activities. It was derived from the Greek word "aner" (with the stem andr-) meaning man, not boy.

Mihall and Belletti's handouts also point out the following differences between children and adults as learners.

1. Adults decide for themselves what is important to learn. Children rely on others to decide what is important.
2. Adults need to validate the information based on their beliefs and life experiences. Children accept the presentation at face value.
3. Adults expect what they are learning to be immediately useful. Children expect what they are learning to be useful in their long-term future.
4. Adults have experience upon which to draw and may have fixed viewpoints. Children have little or no experience upon which to draw and are relatively clean slates.
5. Adults have significant ability to serve as a knowledgeable resource to the trainer and fellow learners. Children do not have the ability to serve as a knowledgeable resource to teachers or classmates.

Basic Assumptions for Adult Learners. Adult learners expect and enjoy independence; they like control and like to take control. Learning, for them, is a process of sharing with one another. They need to know why they need to learn something before undertaking the process of learning it. Therefore, the subject matter must be applicable to their life, presented so adult learners understand why they need to know the information; they can then "buy into" and absorb what is being presented. To achieve this, the training must have credibility with the audience.

Adult learners are pragmatic and want to be able to apply new knowledge immediately. Generally speaking, most adults have trouble tolerating the idea of studying anything that can't be applied to tasks they plan and expect to perform. They are problem or task centered and the adult learning experience is a process in which the adult increases competence to achieve maximum potential in life. Most have many experiences; therefore, the training must draw on adult-learner experiences and anyone in the class can also share their experiences.

Most adult learners come to the classroom highly motivated and eager to learn, especially if they have arrived there voluntarily. They learn in order to cope with real-life tasks and generally do not group together by age or sex but by their experiences.

Adult Learning Principles. Again, referencing Mihall and Belletti's handouts, several adult learning principles must be considered in the development of any training program. The training must

1. focus on real-world situations and problems.
2. emphasize how the learning can be immediately applied.
3. relate the learning to the goals of the learner.
4. relate the materials to the learners' past experiences.
5. allow debate and challenge of ideas.
6. listen to and respect the opinions of the learners.
7. encourage learners to be resources to each other (and the trainer) in a classroom setting.
8. treat learners as adults.
9. give learners orderly control.

Rate of Retention When Adults Learn. Some of these statistics may surprise many people, but over a three-day period of time, using the different methods of learning, adults retain different percentages of what they learn. Adults retain 20 percent of what they hear only, 30 percent of what they see only, 50 percent of what they see and hear, and 90 percent of what they "say as they do."

It becomes obvious from these statistics that the more interactive the training is and the more that learners draw on their own life experiences, the more knowledge is retained.

Classroom Presentation

Following are the main advantages of classroom training:

- The learners are kept together and on the same point.
- It is easier to control the training time.
- It is very useful and cost-effective for large groups (20 or more people).
- If properly facilitated by the trainer, incorporated interaction draws on the learners' experiences and knowledge.

When classroom training is adequately interactive, large volumes of knowledge can be disseminated to all class participants, making the training extremely beneficial to both the learners and the facilitator.

Following are some disadvantages of classroom training:

- The training can be extremely dull and boring if it is not broken up into 45- or 50-minute increments.
- If proper interaction between the learners and facilitator isn't incorporated, it is difficult to gauge if the participants are actually learning anything. This has to be controlled through the use of proper, ongoing evaluation of training as previously described in this chapter.

Classroom presentation that is structured with exercises and/or role play not only has the advantages listed above but also aids retention. Remember, if adults only hear things, the average retention is only 20 percent. Structured presentation keeps learners actively involved and allows them to practice new skills. Adult learners who combine saying and doing the tasks retain 90 percent of the information disseminated in the classroom. Structured classroom training requires more preparation time, may be more difficult to tailor to every learner's situation, and needs sufficient time built into the training time for exercise completion and feedback.

The authors of this handbook agree that structured classroom training is the type of training that consistently provides the desired results and develops a more highly skilled and accident-free workforce.

Computer-Based Training (CBT) and Web-Based Training (WBT)

Self-training programs such as computer-based training (CBT) and Web-based training (WBT) have increased in popularity in recent years. Both CBT and WBT are excellent for some training tasks and, if used properly, can be an extremely efficient use of training budgets. The biggest advantage of CBT and WBT is that either or both can be worked into an employee's schedule very easily, and they are very cost-effective ways of administering and documenting employee training.

In an article by Gary James titled "Advantages and Disadvantages of Online Learning," he states: "There are several distinct advantages and disadvantages of designing, developing, and delivering web-based training (WBT). By carefully weighing your audience and training content against this list of advantages and disadvantages, you should be able to better judge if what you have in mind is right for Web dissemination."

Following are some of the advantages of Web-based training:

1. Turnaround of finished product is quicker.
2. Training delivery is easy and affordable.
3. Cross platform—you can deliver your training course to any machine over the Internet or company intranet without having to develop a different course for each unique platform.
4. No separate and distinct distribution mechanism is needed—WBT can be accessed from any computer anywhere in the world while at the same time keeping delivery costs down.
5. WBT requires less technical support.

6. Content can be easily updated.
7. Installation on private networks is an option for security or greater bandwidth.
8. WBT saves on travel costs and time.
9. Internet connections are widely available.
10. WBT-based development is easier to learn than computer-based training development.
11. Access is controllable.
12. Direct access is available to many other training resources.

Following are some of the disadvantages of Web-based training:

1. Limited formatting of content in browsers: WBT may not resemble CBT because of bandwidth constraints. If content relies on media bells and whistles, or particular formatting, the Internet may not be the best delivery system.
2. Bandwidth/browser limitations may restrict instructional methodologies. If the content relies on much use of video, audio, or intense graphics, and your audience isn't on the proper equipment, Internet delivery frustrates the learners.
3. Limited bandwidth means slower performance for sound, video, and large graphics.
4. Someone must provide Web server access and control usage.
5. Downloading applications requires time.
6. Student assessment and feedback is limited.
7. Many of today's Web-based training programs are too static with very little interactivity.

CBT and WBT training programs are very similar. CBT has many of the same advantages and disadvantages as WBT except for the Internet limitations of WBT and the ease of updating WBT.

With humans, memory fades quickly. If you don't use a skill or a thought process for a long period of time, you tend to lose the information. OSHA recognized this long ago. OSHA 29 CFR 1910.269(2)(iv) states: "An employee shall receive additional training (or retraining) under any of the following conditions: [one of the conditions listed in 1910.269 states the following] NOTE: OSHA would consider tasks that are performed less often than once per year to necessitate retraining before the performance of the work practices involved." This statement means that if employees do not perform a task for a period of 12 months, they must be retrained on that task for OSHA to consider them "qualified" to perform the task safely. OSHA 29 CFR 1910.332(c) states: "Type of training. The training required by this section shall be of the classroom or on-the-job type. The degree of training provided shall be determined by the risk to the employee." CBT and WBT are considered "classroom training" and are excellent sources of refresher training for qualified electrical workers to remind them of the rules or proper techniques.

CBT can be very cost-effective if "off-the-shelf" packages can be implemented for generic training. If the training must be customized to suit the customer's needs, CBT can be very expensive at first. After the initial investment, you own a 24-hour-a-day trainer that can handle any number of learners simultaneously.

WBT is also very cost-effective. Both WBT and CBT generally enable the owner to reduce or eliminate travel costs associated with training. When adult learners are utilizing WBT or CBT, they feel in control of their training and receive immediate feedback as to how they are progressing. The interaction with the computer screen actively engages the adult learner and provides more trainee satisfaction. Adults enjoy working at their own pace and on their own schedule.

For the owner, both WBT and CBT provide computerized systems for tracking the progress of trainees and provide inexpensive standardization of training when training is being facilitated in several different departments, plants, and locations.

Computer-based training and Web-based training are used widely by colleges and universities to reach a greater audience than is available by use of traditional classroom instruction. Another advantage of CBT or WBT type of training is that learners can use the computer and/or computer software during free time and pace the learning to suit their schedule and perceived needs. Some of these characteristics of CBT and WBT have built-in disadvantages. One big disadvantage is that training at a very slow pace allows learners to lose interest and drop out of the training altogether.

CBT can be an efficient method for upgrading employees' skills, but this is not always the result. If CBT is your choice of training, it is imperative that you select the best product, product developer, and presentation technique. CBT, like all other training, must be supervised. This type of training is a complete failure if the learner is simply placed in a room with a computer and left alone. If no one is available to answer questions and make sure that the proper level of understanding is developed, no quality learning is taking place. Adults can watch a video presentation several times, but if they leave the training with more questions than answers, the experience may not be worthwhile. To maximize the CBT training experience, a supervisor or mentor must be available to answer questions and develop understanding on information that is not immediately understandable to the trainee.

The only way computer-based training is truly an efficient and effective training tool is if the sessions are monitored and supervised by a qualified person who can ensure that training schedules are followed and questions are answered and who is available to develop complete understanding of the training materials presented in the training program.

Video Training

Video training, when properly used, is an extremely effective tool. When overused or used improperly, video training loses effectiveness very rapidly.

Short, high-impact training videos are an excellent tool for use in safety meetings or for training sessions to either teach new skills or simply elevate awareness on any particular topic. They are also an excellent tool for educating employees on certain processes such as the use of specialty tools like exothermic welding and a myriad of other topics.

The insertion of short (5 to 20 minutes) videos, dealing with the topic being taught, into a training presentation serves several purposes. Adding the visual aspect aids in knowledge retention, breaks up the monotony of a lecture, clarifies the material being presented in the lecture, and serves as a refreshing break for the learners. It has been said millions of times that a picture is worth a thousand words. This is a very true statement and one that needs to be considered when a trainer is selecting video materials to include in any training program.

Video producers tend to overdo the dramatics in videos. For example, in electrical safety training videos dealing with arc-flash, the video producers put some type of fire agent on the back side of the electrical worker. The worker then performs some act that produces an electric arc to simulate igniting the electrician's clothes as an electrical arc would do in the event of an arcing failure. The worker then flails around with his back side on fire and eventually falls on his face and continues to burn. This is not real life and these episodes water down the video and diminish its credibility. Typically, electrical workers are facing the arc source and therefore are burned on the front of their body. We teach very small children to stop, drop, and roll in the event their clothing catches fire. So the safety video has just depicted a situation that adults with life experiences do not relate to very well.

Another common mistake witnessed in safety videos is to picture an electrical worker wearing class 00 rubber gloves with a beige label and rated for 500 volts ac, or class 0 gloves with a red label rated for 1000 volts ac with class 2 leather protectors over them. OSHA mandates a minimum distance between the rolled bead of the rubber glove and the end of the gauntlet on the leather protectors for class 00 and class 0 rubber gloves of at least ½ inch, 1 inch on class 1 gloves, 2 inches on class 2 gloves, 3 inches on class 3 gloves, and 4 inches on class 4 gloves. Common sense must prevail throughout the training session, including during video presentation, to maintain credibility with the folks being trained in the industrial workforce. But in the case of electrical safety videos, the number of producers is very limited and the market is so small, there is not much choice but to use the videos that are available and on the market.

When using safety videos in a training program, the training facilitator needs to mention and refer to them in a positive way during the presentation and build interest among the trainees in seeing the videos. Then when it is time to introduce the video, the trainer must use some technique to grab the interest of the learners prior to starting the video. One great way to do this is to tell the class something like this: "This is one of the finest electrical safety videos on the market; it includes actual accidents that have been reenacted for the video" or "There are five flagrant violations of OSHA and NFPA 70E rules in this video. See how many you can identify" or "This is an excellent video but it contains violations concerning the wearing and inspection of rubber insulating gloves; see if you can identify them." When trainers use this technique, learners are tuned into the video presentation because they have been offered a challenge and expect to have to respond to questioning once the video presentation has ended. Training facilitators need to follow up on their challenge to the students and ask a few open-ended questions about the video that require the response of several learners.

Conclusion

In summary, CBT and WBT are excellent for some training, but both of these types of training need to be monitored by someone who can act as a mentor to ensure that an acceptable amount of knowledge is transferred to the learner.

The highest-quality training is achieved by use of structured lecture with quality handouts and limited use of video training incorporated into the program. The use of written quizzes and final exams forms the basis of an excellent record-keeping system and provides documentation of exactly what topics were covered in the formal training session. Also keep in mind that 100 percent is the only acceptable score on any test that relates to any type of workplace safety. Electrons are not very smart, but they move at the speed of light, or 186,000 miles per second. Electrical workers can't move their hands or run faster than the speed of light. Therefore, if they can't outrun electrons, they must possess enough knowledge about basic current flow and electrical safety to allow them to outthink electricity. They deserve the highest-quality training they can be provided.

The next section details some of the requirements of training delivery and explores the pros and cons of training delivery methods. In addition, on-the-job training is discussed in some depth.

ELEMENTS OF A GOOD TRAINING PROGRAM

If you do not know what it is made of, it is hard to build. Because of this fact, many companies, even those with strong training commitment, have poor training programs. From a safety training standpoint, three types of training can be defined—classroom, on-the-job, and self-training. A complete adult training package must include all three of these methods assembled in a formal, planned program. Each method has certain critical elements, some of them shared.

Element 1: Classroom Training

Classrooms. When most of us think of classrooms, we think of the rooms where we went to grade school—the small, cramped desks with books underneath the seat, chalk dust, and the spinsterly schoolmarm. Adult classrooms need to be designed for adults.

1. The classroom should be spacious but not huge. Students should not be cramped together and should have sufficient table or desk space to spread out their texts and other materials.
2. The classroom should be well lit and quiet. Dim lighting is more conducive to sleeping than learning, and noise from a nearby production facility can completely destroy the learning process.
3. The classroom should be as far from the students' work area as possible. This serves two purposes.
    a. A remote location tends to focus the student on learning.
    b. Interruptions by well-meaning supervisors are minimized.
4. All required audiovisual equipment should be readily available. Screens should be positioned for easy viewing and controls should be readily accessible to the instructor. Avoid using a whiteboard for a screen due to the glare.

Laboratories. The adult learner must learn by doing. For safety training, learning by doing means laboratory sessions and/or on-the-job training. Many of the design features mentioned for classrooms also apply to laboratories. However, there are a few special considerations.

1. Avoid using the same room for both laboratory and classroom. A room big enough for both is usually too large for effective classroom presentation. Too small a room, on the other hand, crowds the students during lecture and laboratory. Equipment needed to train students in procedures should be placed in positions where it does not interfere with the lecture presentation.
2. Initial training on safety procedures should be done on de-energized equipment.
3. Students should work in groups to help support each other.

Materials. Student materials can easily make or break a training program. Although a complete material design treatise is beyond the scope of this book, a few critical considerations follow:

1. All student text chapters should be prefaced by concise, measurable training objectives. See the "Safety Training Definitions" section earlier in this chapter.
2. Illustrations should be legible and clear. Avoid photocopies when possible. Also, put illustrations on separate pages that may be folded out and reviewed while the student reads the text. Nothing detracts more than flipping back and forth from text to illustration.
3. Students should have copies of all visual aids used by the instructor.
4. Laboratories should have lab booklets that clearly define the lab objectives and have all required information at the students' fingertips.
5. Avoid using manufacturer's instruction literature for texts. While these types of documents make excellent reference books, they are not designed for training purposes.

Instructors. No one has really defined what makes the difference between a good instructor and a mediocre one. Some of the most technically qualified individuals make poor instructors while others are excellent. Some excellent public speakers are very poor instructors, while others are quite good. A few points are common to all good instructors.

1. Good instructors earn the respect of their students by being honest with them. No student expects an instructor to know everything there is to know about a subject. An instructor who admits that he or she does not know has taken a giant step toward earning respect.
2. Instructors must have field experience in the area they are teaching. Learning a subject from a textbook or an instruction manual does not provide sufficient depth. (Unless, of course, the material is theoretical.)
3. Instructors must be given adequate time to prepare for a presentation. Telling an instructor on Friday that a course needs to be taught on the following Monday will lead to poor training. Even a course an instructor has taught many times before should be thoroughly reviewed before each session.
4. Avoid using a student's supervisor for the instructor. Supervisors will often expect the student to learn because they are the supervisor. In addition, the student may feel uncomfortable asking questions.

Frequency of Training. How often to train is an extremely complex problem. Management must balance the often-conflicting requirements of production schedules, budgetary constraints, insurance regulations, union contracts, and common sense with the needs of employees to know what they are doing.

Overshadowing all this, however, is one basic fact—the employee who is not learning is moving backward. Technology, standards, and regulations are changing constantly; therefore, employees must be trained constantly. In general, the following points can be made:

1. Employees in technical positions should be constantly engaged in a training program. Two or even three short courses per year are not unreasonable or unwarranted. The courses should be intensive and related directly to the job requirements of the employee.

2. Employee training should be scheduled strategically to occur before job requirements. In other words, an employee charged with cable testing during an outage should go to a cable testing school a few weeks prior to the outage. Few of us can remember a complex technical procedure for two or three years.
3. Training should be provided at least as often as required by regulatory requirements. Remember that standards and regulations only call out minimum requirements.
4. Regulatory requirements must be met. For example, OSHA rules require annual review of lockout-tagout. Training is required if problems are noted.

Element 2: On-the-Job Training (OJT)

On-the-job training is discussed in great detail in a later section. Regarding the elements of laboratories, instructors, materials, and frequency of training, virtually all the previous discussion applies to on-the-job training, with the following additions:

1. The laboratory for OJT is the whole plant or facility. The student learns while actually on the job.
2. In OJT, the supervisor (if technically qualified) is often the most logical choice for an instructor. In this role, however, the instructor does not usually get involved in a direct one-on-one training session. Rather, the instructor gives the student the objectives for each OJT session and then evaluates progress. *Note:* Throughout industry we are seeing more and more multicraft shops that are supervised by nonelectrical people, such as mechanics or pipe fitters; therefore, the supervisor should not be the instructor for electrical classroom instruction or on-the-job-training.
3. Good training is done on purpose. A complete training program will have OJT materials that include objectives, reading materials, and examples of what is to be done. In addition, the student should have a record card (see Fig. 14.1) (sometimes called a *qual* card in the military) that identifies each area of responsibility. When the student has performed satisfactorily, the instructor checks off on the card to indicate successful completion.

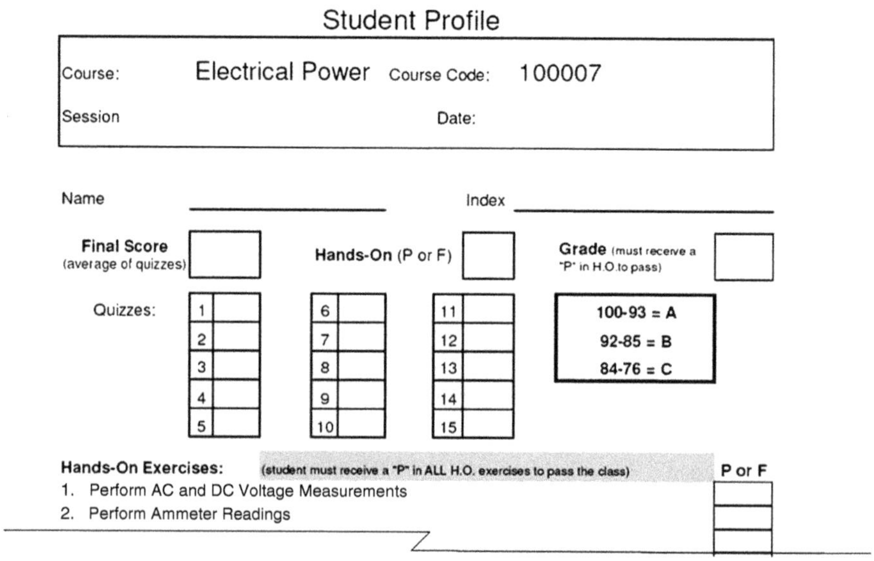

FIGURE 14.1 A student qualification sheet.

4. Since employees are on the job every day, they are (theoretically) learning every day. This is not true, however, of OJT. Remember that all good training is done on purpose. Typically a student should be given one specific OJT assignment at least once per week. An instructor (qualified person) should observe the successful completion of the assignment, and the employee should be given credit for the job.

Element 3: Self-Training

Most well-intentioned self-training fails simply because it is not organized. Adults must know what is expected before they can perform. In this sense, then, self-training has much in common with OJT, and virtually all the comments that apply to OJT also apply to self-training.

The most important element of self-training is materials. The materials for self-training must be written very carefully. Small, self-supporting modules that are prefaced with clear objectives should serve as the core for self-training materials. Each module should end with a self-progress quiz. After the student completes each module or related group of modules, the instructor should administer an examination.

Conclusion

When confronted with requirements for training, such as those previously outlined, many companies are staggered by what appears to be a very expensive investment. The investment made in training is returned many times over in improved safety, morale, and productivity. Furthermore, a training program that does not have the proper elements will probably not be good training, and poor training is worse than no training at all.

ON-THE-JOB TRAINING

The major difference between 25 years of experience and 1 year of experience repeated 25 times is OJT. Good, well-planned, on-purpose OJT is the heart and soul of an adult learning program. Adults must learn by doing, and the workplace is the best laboratory. The three major segments of OJT include setup, implementation, and evaluation. A well-defined lesson plan is as important with OJT as it is for classroom instruction.

You must involve the employees in the development of this program. Their input and feedback will be critical every step of the way. The employee safety committee can also be used as a training committee and should oversee every aspect of the training program development and presentation.

Note that safety topics are especially suitable for presentation in an OJT format. The best way to memorize critical safety procedures is while actually performing them under field conditions. Note also, however, that special supervision is required when personnel are learning safety procedures with energized equipment.

Setup

As noted previously, all training must be properly planned. Good training implies that it is being done on purpose. The setup portion is the most critical step of all and must be properly executed because it serves as the foundation.

Step 1. Define the employees' job functions. This step, often called a job analysis, is critical to all training but especially to OJT. Since OJT takes place while actually on the job, you must know what your employees do before you can train them. This is a simple concept, but it is surprising how many organizations miss this step. The performance of the job analysis can be as simple as asking your employees to make a list of what they do. The analysis of time sheets is another good approach. A final source of information is to ask your supervisors to list job functions for employees.

Notice that interviewing supervisors is the last of the three approaches. Unfortunately, company management and supervision are not always as closely in tune with day-to-day employee activities as they should be. This is not always true, but too often it is.

At the conclusion of this step, you should have a complete listing of the major jobs performed by your employees; for example, one major job is to remove motor starters from motor control centers with energized buses.

To simplify this step, concentrate on major job functions. Detail can be added as the OJT program develops over the years.

Step 2. Isolate the job functions that are practical and measurable. For example, you may have one task such as troubleshooting the south plant UPS system. While troubleshooting is a trainable skill, it is very difficult to measure troubleshooting ability in an OJT environment. On the other hand removal of starters from motor control centers is measurable and practical. Initially, only the practical and measurable skills should be included in the OJT program. As your program's sophistication increases, you may wish to add more cognitive skills.

As each job function is determined, it should be identified with a simple numbering system. For example, the starter removal might be job function number 1; cable testing, job function number 2; and so on. This makes it possible to keep records for which job functions employees have received training.

Step 3. Break each job function into tasks, subtasks, and detailed procedures. Continuing with the starter removal as an example, we might develop several tasks including the de-energizing of the cubicle, proper use of safety equipment, mechanically removing the starter, and so on. The safety subtasks would include, but not necessarily be limited to, determining the proper personal protective equipment required to perform the job, the inspection of rubber insulating gloves and leather protectors, inspecting and testing the proper voltage-measurement device, performing proper voltage measurements, application of proper lockout-tagout procedures, proper technique for removal of the starter, proper inspection and application of safety grounds, and so on. The detailed procedures can be developed either in-house or from manufacturers' literature.

Step 4. Develop training objectives for the tasks and subtasks. The objectives should have the measurement criteria contained within them. The following examples use the motor starter removal as their subject.

Task

1. The employee shall independently remove a low-voltage motor starter from its cubicle. The job shall be performed using all applicable company safety standards.

Subtasks

a. The employee shall perform a safety inspection of a pair of class 00 low-voltage rubber insulating gloves and the leather protectors.

b. The employee shall wear a flash suit, previously inspected rubber insulating gloves, hard hat, and face shield.

c. The employee shall stop the motor and de-energize the starter module. He or she shall demonstrate proper procedure and stand in a safe work location while performing these steps.

Here we have shown only one of the objectives; there would be several more, one for each safety task and one for the mechanical procedures.

Step 5. Develop the written materials necessary to support the program. The following should be developed as a minimum:

1. Complete set of written objectives.
2. Written, detailed description of each task. To accomplish this portion, manufacturers' literature may augment materials developed in-house.
3. Diagrams, illustrations, or examples.
4. Employee record cards. The cards should have space for the employee's name, classification, and other demographic data. The majority of the card should contain space for completed job functions (by identification code), time of completion, and instructor's initials indicating successful completion.

Implementation

The implementation phase of an OJT program should be a simple extension of the setup phase.

Step 1. Select several of the job functions to be used as trial runs of the OJT program. At the same time, select an experienced group of employees to receive the initial training. Develop the record cards for each employee.

Step 2. Thoroughly indoctrinate each of the test employees in the program and give them advance knowledge of what is expected. Of course, if you have been properly including the employees in the development phase, this step will not be necessary.

Step 3. Introduce the training, one job function at a time. Scheduling should be made by working the program into the normal work schedule. That is, when plant scheduling requires relay testing, use the opportunity to provide the OJT.

Step 4. Have the employee's supervisor go over the task with the employee and make certain the employee thoroughly understands what is required of him or her. Remember, this is training and the supervisor may be the instructor. The employee should perform the work, with the instructor giving guidance. The instructor should watch the employee perform each task and make certain all tasks are performed within measurement criteria.

Step 5. After each training session is completed, thoroughly evaluate the results. See the next section for a more detailed description of evaluation.

Step 6. Gradually involve each existing and new employee in the OJT program. Any time an employee is called upon to perform a new job function or task, the OJT program should be implemented and the employee's record card should be updated accordingly.

Evaluation

Step 1. The results each employee obtains should be measured against the criteria established during the setup phase. Do not make any assumptions based on just one or two employees. Poor performance by several employees could mean the program is bad, or it could mean that more classroom training is required.

Step 2. Interview the employees in a confidential manner. Give them a chance to critique the entire training program. All aspects—materials, implementation, and objectives—should be reviewed.

Step 3. Interview the instructor to get his or her feedback.

Step 4. Based on the first three steps, modify the program as required.

Step 5. Every two years, review the program from top to bottom.

Conclusion

One major misstep taken by many organizations when they enter the world of OJT is to forget the key word—*training*. On-the-job *training* is not an on-the-job test. Rather it is on-the-job *training*. Do not forget to provide the necessary field instruction as the program progresses.

In general, OJT will not stand alone. It must be accompanied by good classroom instruction and/or self-training properly implemented and monitored. As stated in the previous section, OJT can be the most cost-effective of all the parts of a training program.

TRAINING CONSULTANTS AND VENDORS

Only very large companies can justify having their own, completely self-supporting, in-house training organizations. Even when in-house organizations exist, they frequently use outside vendors for support and special programs. In other words, you should use training consultants and vendors for the same reasons you use any consultant or vendor—when time, personnel, expertise, or economics makes it impossible to perform the task yourself.

Training consultants fall roughly into three categories—those who provide canned programs, those who develop and present tailored programs, and those who provide training analysis. Of course, some companies fit into more than one category.

Canned Programs and Materials

Programs. Companies that provide the short one-week, manufacturer's type of seminars have proliferated since the early 1960s. Training programs are available in electrical safety, circuit breaker maintenance, relay testing, substation maintenance, transformer testing, uninterruptible power supplies, programmable logic controllers, and a host of other such topics. Usually these courses are from three days to two weeks in duration. They are normally presented either in a training center or at a hotel with adequate meeting facilities. Some of them are equipment-specific, while others are generic in nature. Several of these programs consist of classroom as well as hands-on training, which provides a more thorough training program, essentially consisting of classroom instruction and OJT in one package. Pricing varies, but these seminars generally offer a very economical approach when you have only one or two employees to train.

Learning-by-mail programs, called correspondence schools, also fall into this general category. Some companies have been providing programmed mail-order instruction for decades. The correspondence schools are especially useful for vocational-type training at a more generic level.

The vocational technical schools (votechs), both private and public, fill yet another need in this arena. Many companies are requiring two-year associate degrees or the equivalent for employment in certain technical areas. The votech schools provide such programs. Most of the votechs have evening courses, allowing students to attend in their free time.

Each of these three types of programs has its own advantages. The one-week seminar is very intensive and provides training in very specialized areas in a hurry. The correspondence schools are very effective for students who have schedules that make attending classes difficult. The votech types of programs are excellent for those students who wish to start at the beginning and learn the entire discipline.

Materials. Some organizations sell training materials such as programmed courses, videotapes, interactive video, and/or computer- or Web-based programs. The principal difference between these materials and the correspondence school is that the materials may be used to present training to many people in one organization. These packages may also cover very specific equipment maintenance, safety, and/or operations.

Tailored Programs

When a company is truly committed to highly cost-effective training, it should opt for the tailored program. Often surprisingly affordable, tailoring may be as simple as taking a canned one-week course and teaching only about the equipment that is in use at the particular plant or company. Most of the large vendors are glad to do this sort of tailoring. Other vendors will develop entire tailored presentations to a client's specification.

This type of service is most useful when your organization has a large number of people to train in a given area or areas. For example, immediately prior to a major maintenance shutdown, you may wish to have five or 10 of your maintenance personnel trained in cable testing and maintenance at your own facility. Such training is usually much more affordable than sending individual employees to single presentations.

Some training consultants will develop completely tailored programs based on existing industry standard texts, for example, safety programs based on NFPA 70E or substation maintenance based on an electric equipment text. Such programs are extremely effective and are generally no more costly than those described previously.

Training Analysis

Sometimes, a company may have no idea what its training needs are. This is less common than in years past, but it does still happen. Training consultants can be retained to provide an entire training systems analysis. The major areas in such a development include *needs analysis*, *job and task analysis*, and *curriculum development*.

Needs Analysis. Also called a front-end analysis, the needs analysis is the starting point for any training program. The overall method is covered in many industry standards such as those by ASTD. The general principle is quite simple. Desired performance is compared to actual performance for the workers under consideration. If the desired performance is below nominal, training is indicated.

Job and Task Analysis. This portion is usually a part of the needs or front-end analysis. The procedure involves defining each employee's job functions in measurable, objective terms; breaking each job function into tasks and subtasks; and finally writing these job functions, tasks, and subtasks as measurable learning objectives. These learning objectives are then used as the basis for the training programs.

Curriculum Development. The actual curriculum is written and/or selected to fulfill the learning objectives developed during the job and task analysis. However, how this is accomplished may be quite flexible. For example, if a given group of employees has the responsibility to perform routine maintenance on the uninterruptible power supply, the learning objectives may be met by sending the group to one of the canned courses previously discussed.

Evaluating Training Vendors and Consultants

Evaluation of any training must be based on the learning objectives. To know how well you are doing, you must know what we intended to do. Training must be objective-based and done on purpose.

Proper evaluation demands that such an approach be used. With this in mind, vendor training can be evaluated using four simple criteria.

Does the Vendor Provide Properly Documented, Objective-Based Training? If the answer is no, do not use that vendor. Of course, this does not apply directly to some of the situations described in this book. However, even training analysis consultants can be checked for competency in and knowledge of objective-based learning.

Does the Vendor Listen Closely to Your Desires and Counsel You If He or She Believes That You Are Wrong? One of the most difficult things for any consultant to do is to argue with his or her client. Remember, the client pays the bills. Therefore, if your consultant listens, and then counsels (or argues) with you when he or she believes you are wrong, that consultant has your best interests at heart.

Does the Vendor Provide Only Qualified and Competent Personnel? Consultants are only as good as the people they employ. Take the time to quiz the individuals that your consultant provides. Ask for resumes, and, where applicable, interview the individuals. Even qualified people will not be effective if they have a personality conflict with you.

Did the Program Meet the Objectives That Were Set Out at the Beginning? Do not depend on course critiques and/or personal evaluations of attendees. Give quizzes where applicable. Sit down with employees and have them review the material they learned. Develop your own short quiz based on the course materials, and administer it to your employees. The quiz can be administered orally, in written form, or as a field-type performance quiz. (For example, one task might be to "test this relay like they taught you at school.")

As a corollary to this last question, always insist that a vendor administer performance quizzes as part of a training program. Quizzes accomplish three objectives:

1. They measure the student's performance.
2. They measure the trainer's performance.
3. They measure the course's success.

In some cases, union contracts may make quizzing difficult or impossible; however, since the quizzes also measure the quality of the training that occurs, many unions are eager for their members to be tested. In this way, they know their members are receiving proper and adequate training.

Conclusion

The use of training consultants is really no different in concept than the use of any other consultant or contractor. You should first determine what you need, monitor what you are getting, and, finally, evaluate what you got.

TRAINING PROGRAM SETUP—A STEP-BY-STEP METHOD

Introduction

Among the many difficulties faced by employers wishing to establish a viable electrical safety program is the need to validate that workers have the proper background, skills, and training as required by OSHA and NFPA 70E. Unfortunately, the existing standards do not address detailed methods that allow employers to validate that an employee has truly "received training in and has demonstrated skills and knowledge in the construction and operation of electric equipment and installations and the hazards involved." The program described in this chapter uses long-recognized industrial

TABLE 14.1 Explanatory Notes for the Definition of a Qualified Person (From 29CFR 1910.331, .332, and .339)

Location in 29CFR	Text
1910.331	The provisions of 1910.331 through 1910.335 cover electrical safety work practices for both qualified persons (those who have training in avoiding the electrical hazards of working on or near exposed energized parts) . . .
1910.332(b)(3)	Additional requirements for qualified persons. Qualified persons (i.e., those permitted to work on or near exposed energized parts) shall, at a minimum, be trained in and familiar with the following: • The skills and techniques necessary to distinguish exposed live parts from other parts of electric equipment. • The skills and techniques necessary to determine the nominal voltage of exposed live parts, and • The clearance distances specified in 1910.333(c) and the corresponding voltages to which the qualified person will be exposed. Note 1: For the purposes of 1910.331 through 1910.335, a person must have the training required by paragraph (b)(3) of this section in order to be considered a qualified person. Note 2: Qualified persons whose work on energized equipment involves either direct contact or contact by means of tools or materials must also have the training needed to meet 1910.333(C)(2).
1910.399	Note 1: Whether an employee is considered to be a "qualified person" will depend upon various circumstances in the workplace. It is possible and, in fact, likely for an individual to be considered qualified" with regard to certain equipment in the workplace, but "unqualified" as to other equipment. (See 1910.332(b)(3) for training requirements that specifically apply to qualified persons.)

education procedures such as job and task analysis, skills assessment, and training needs analysis to develop and validate a program that will provide the highest level of compliance with existing regulatory and consensus standards.

Background

All of the modern regulatory and consensus standards allow only "qualified persons" to perform work that requires approaching energized conductors and circuit parts. OSHA defines a qualified person as *one who has received training in and has demonstrated skills and knowledge in the construction and operation of electric equipment and installations and the hazards involved.* (29CFR1910.399)

OSHA also offers some additional clarification in several locations as shown in Table 14.1.

Unfortunately, none of the regulations give clear methods for certifying that personnel are truly qualified in the OSHA sense. This can lead to situations in which the employer and employee are caught in a "no-win" scenario.

For example, several years ago a worker was injured in an arc-flash incident. This individual was a journeyman electrician with almost 25 years of direct, hands-on experience and training. He was working in an electrical switch that was identical to hundreds of others he had serviced over the years. The arc occurred when the worker allowed a grounded lead to contact exposed, energized conductors. These conductors were exposed via mounting hardware that was in a very difficult-to-reach location in the equipment.

Through subsequent investigation, it was determined that the worker did not know that the mounting hardware was directly connected to the energized bus through a plastic mounting piece. The design of the equipment predated the dead-front requirement of modern equipment. The injured worker had never realized that the exposed hardware was energized even when the switch was turned off.

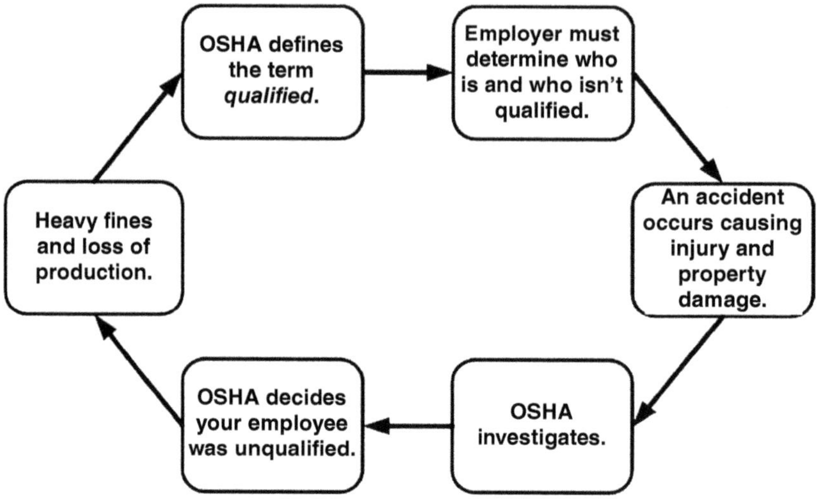

FIGURE 14.2 The regulatory training problem.

The experience and the training history of the worker did not sway the OSHA investigator, who correctly determined that the worker was, in fact, unqualified. He had not recognized and had not avoided the hazard.

This puts employers in the difficult position of declaring that their employees are qualified for certain work, only to find out later that, in fact, OSHA disagrees. Short of OSHA certifying employees (a highly unlikely scenario), there seems to be no way to validate that employees are qualified.

Figure 14.2 illustrates the problem.

The regulatory standards define the term *qualified* and give definitions and explanations as to what the term means. The standards also task the employer with the responsibility of determining who is and who is not qualified.

An accident involving a "qualified" employee occurs, which OSHA investigates. As a part of the evaluation, the OSHA investigator determines that one or more of the employees involved in the accident were not qualified. OSHA then issues fines and requires that the employer modify procedures and training.

At this point, the cycle starts again. The employer must again use the OSHA definitions to determine who is and who is not qualified, and so on.

In addition to the OSHA definition of a *qualified person* noted previously, OSHA provides some clarifying notes to this definition as follows:

> **Note 1 to the definition of "*qualified person*:"** Whether an employee is considered to be a "qualified person" will depend on various circumstances in the workplace. For example, it is possible and, in fact, likely for an individual to be considered "qualified" with regard to certain equipment in the workplace, but "unqualified" as to other equipment. (See 1910.332(b)(3) for training requirements that specifically apply to qualified persons.)
>
> **Note 2 to the definition of "*qualified person*:"** An employee who is undergoing on-the-job training and who, in the course of such training, has demonstrated an ability to perform duties safely at his or her level of training and who is under the direct supervision of a qualified person is considered to be a qualified person for the performance of those duties.

A Plan

The following paragraphs present a program that can be used to validate the training of electrical workers as well as organize the overall electrical safety program. This program has been successfully

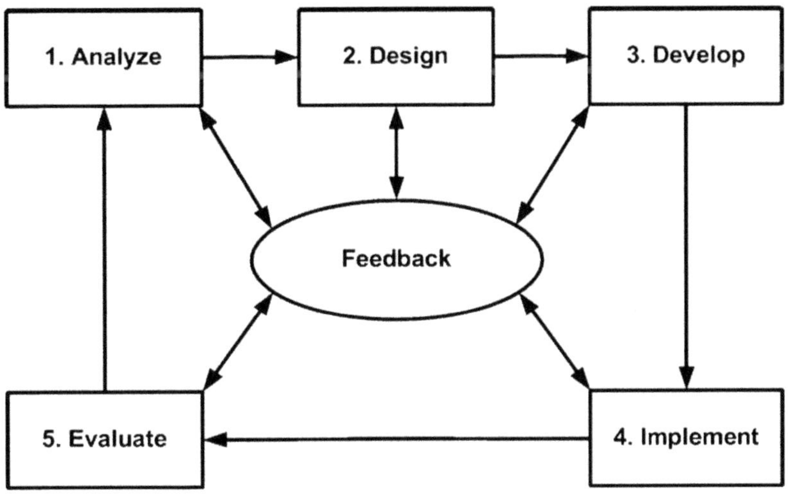

FIGURE 14.3 The instruction systems design (ISD) model.

deployed in many locations throughout the United States and continues to be introduced in a number of other plants.

This validation program uses a holistic approach by integrating well-established engineering, technical, and educational methods. At its core, the overall strategy is based on the Instructional Systems Design (ISD) model, which is shown in Fig. 14.3.

Note that this model uses a "scientific" format that should be familiar to those in the engineering and technical disciplines. Note also that sufficient time and effort must be made in each stage to ensure that subsequent steps will be completed effectively. For example, shortening the analysis stage will create a disproportional increase in the time required for the design and development stages.

While the overall approach is designed to be a comprehensive methodology, it is flexible enough to be adjusted to fit the specific needs and situations found in a variety of different industrial, commercial, and utility cultures. The six major sections of the program are described in the following paragraphs.

Analyze

Needs Analysis. Not all organizations are in need of a comprehensive training effort. Some, for example, may need only special skills training in a few isolated areas. A needs analysis is performed to determine what, if any, training is needed in the organization.

You should consult three sources to assess your needs—employees, supervisors and managers, and company performance.

1. *Employees.* Formally poll your employees for training requirements. Usually they know their shortcomings better than anyone. The poll should be concise and sincere. Be prepared for a revelation because experience has shown that, more often than not, employers and employees have very different views about the state of their training.
2. *Supervisors and managers.* Extend the training poll to your management and supervision. Give them a chance to tell you where training deficiencies are causing problems. Their input should be solicited for their employees' and their own training needs. Do not forget to include yourself in this analysis. Employees at all levels may need training.
3. *Company performance.* An honest evaluation of company performance can be one of the best sources of training needs assessment. The most obvious example of this element is in the area of safety.

Do some departments have a consistent record of safety problems? Such a department is an obvious target for safety training. Remember to involve your employees and supervisors in the review and analysis of the material. The people who are going to receive the training should be involved at all stages. An electrical safety audit, as described in Chap. 13, can be used to satisfy all the requirements of the needs analysis. This comprehensive audit requires a minimum of two days for a small facility and a maximum of several days for a large facility. A combination of interviews, questionnaires, and site inspections creates a thorough assessment of the existing conditions. An exit presentation is held with key plant personnel to review critical findings. In most cases, severe hazards are discovered during the audit that can be corrected on the spot. A comprehensive report follows the audit within two weeks. This report details the findings and recommendations of the audit and is the first concrete step in the process.

Job and Task Analysis. After the needs analysis is complete, you should have a well-defined set of training needs. The next step is the performance of a job and task analysis (JTA) for each of the areas that requires training. This critical step defines the actual composition of a given job. The JTA involves the following three steps:

1. *Employee interviews.* Each employee is interviewed, and a listing is made of the jobs and tasks that he or she performs. The listing should be very detailed to allow for the development of performance objectives.
2. *Supervisor and manager interviews.* The department supervisors are interviewed to obtain their input regarding the jobs and tasks performed by their employees. This listing is compared to the one developed by the employee interviews, and modifications are made as required. Many training problems disappear when this step is performed since supervisors and employees develop a common understanding of the task at hand.
3. *Job performance observation.* Employees are observed in the performance of their jobs. The job and task listings developed during the interviews are reviewed and refined to be consistent with actual performance. Remember that the actual performance of the given job is under question, so do not modify a JTA based on performance observation unless you are certain that the job is being performed correctly.

Arc-Energy Analysis. An arc-energy analysis that includes a short-circuit study, coordination study, and flash-hazard analysis is a critical part of the analysis stage and should be performed if it is not already available. Some companies may prefer to perform a "mini-study" before any of the other steps to determine the severity of the flash hazard at their location. The results of this study will determine the minimum PPE requirements for workers, provide the required flash-hazard warning labels for each location, and introduce overall electrical safety requirements to plant personnel.

Because of the perceived cost of such a study, many companies opt for the use of the so-called simplified method as described in NFPA 70E and discussed in Chap. 3. The authors of this book discourage this practice for several reasons, including the following:

1. The simplified method requires knowledge of much of the same baseline data including short-circuit duties and clearing times.
2. The simplified method frequently results in an overdressed workforce. In fact, if done conservatively, it should.
3. We have documented situations in which the simplified method actually results in underdressed personnel. That is, even when correctly applied, the simplified method may give results that will not provide adequate protection. While this occurrence is rare, even one time could result in severe injuries or death.
4. While most will agree that the current arc-energy calculation methods are a work in progress, they still provide the best available information of the degree of exposure. We owe our workers our best possible effort in determining the risk and attendant protection methods.

Design

Performance Objective Design. Based on the results of the JTA, a set of measurable performance objectives is developed. You may wish to develop training objectives in two tiers—terminal objectives and enabling objectives. Terminal objectives are broad-based objectives that represent the actual required behavior. Enabling objectives are "subobjectives" that define skills required to meet a given terminal objective. The following give examples.

- Terminal objectives. The employee will be able to test a Westinghouse induction disk overcurrent relay. The tests will be performed using the manufacturer's instruction leaflet, a transformer-loaded overcurrent relay test set and associated instructions, and the company-supplied test sheet. Test results shall be as defined in the relay instruction leaflet.
- Enabling objectives

1. The employee will be able to describe the purpose and use of each of the controls on the transformer-loaded test set.
2. The employee will be able to describe the pickup and timing adjustments on the Westinghouse induction disk overcurrent relay. (Add other objectives as defined from the JTA.)

Note the specificity of the objectives. The regulations and standards are very clear in the requirement that employees must be trained on the equipment with which they work.

Structure Design. The information generated during the needs analysis, JTA, and performance objectives development is reviewed, and an overall structure for the training effort is defined. Of course, not all companies are going to need a large, expensive training center with a staff of trainers.

- *In-house training.* If you define an extensive training need with many ongoing course requirements, you may need to consider the development and use of an in-house training system. The materials for such training can include both specifically developed and/or vendor-supplied materials. On-the-job training is usually a very important part of in-house training.
- *Vendor-supplied training.* Specific skills such as cable testing, relay testing, breaker testing, and instrument repair may best be supplied by training companies that specialize in these areas. In fact, in some cases an entire program may be implemented through commercially purchased training.
- *Self-study.* Many skills can be learned effectively through one of the many self-study programs that are commercially available. Even if a program of this type is not purchased, employees should be encouraged (monetarily if necessary) to put in some study time on their own.

Infrastructure Design

1. Training department organization
   a. The training department should report to the executive level of management. This is especially important for technical training departments.
   b. Training department supervisors should be technically qualified personnel. Furthermore, training supervisors should maintain a teaching schedule.
   c. Salaries for training department personnel should be equal to or exceed the salaries of equivalent level personnel in other departments.
2. Training facilities. The selection of a training facility will depend on the quantity and nature of the training that you wish to conduct. If you plan to use a lot of outside vendors, you may be able to use their facilities. Conference rooms, local meeting centers, and hotel facilities may also be viable alternatives. Specialized laboratory needs may require the construction of facilities. As an alternative to this, you may consider using your company work areas for laboratories; however, this is not recommended as a general practice.

Develop

Overall Training Structure. In some cases, the required training will be minimal and perhaps limited only to the specific qualification exercises necessary to ensure that all workers have the requisite technical and safety skills to perform their work safely. However, it should always be recalled that OSHA requires skilled personnel to be familiar with the hazards as well as the operation of the equipment, meaning that every worker must have a solid grounding in the hazards of electricity along with the various specific job requirements. The final training program will be a mix of several different training scenarios, including the following:

1. Prerequisite skills training from apprenticeship programs, public education, and/or validated on-the-job programs. The proof of such training may be a prerequisite for obtaining a job.
2. Formal classroom training to be attended by all personnel as they progress through their job structure. Such training may comprise one or a combination of the following:
   a. Internally developed and presented classes
   b. External training from vendors, vocational schools, or community colleges.
   c. Formal on-the-job training requiring demonstration of skills in the presence of a supervisor with an appropriate qualification form.

Whatever the mix of training, there is only one way to ensure that workers are truly qualified. They must demonstrate their competence in a controlled environment, under supervision. To be fully qualified, workers must be trained on a switch of the exact type that they will encounter in the field. Trainees must demonstrate to a supervisor where all of the hazardous locations are and what steps/precautions should be taken to prevent an accident. Figure 14.1 is a section of an employee on-the-job training qualification form for technical skills. Such forms should be used along with examinations and field demonstrations to ensure that workers are capable of performing all of the tasks to which they may be assigned.

Course Development. Courses must be developed or purchased to fulfill the objectives defined during the setup step.

1. *Purchased materials.* Both self-study and classroom materials are available for purchase from a variety of training vendors. Make certain that the purchased materials meet the objectives defined during the setup step.
2. *Developed materials.* Although somewhat more expensive, developing company-specific course materials has some distinct advantages. In the first place, your employees will tend to identify with materials that are obviously site-specific; moreover, presentation time is optimized since all material is company-specific. Note that these materials may be developed by either in-house personnel or consultants.

Implement

The format of training implementation depends to a large extent on the structure of the facility and the training needs of the workers as determined in previous sections. However, the correct implementation is absolutely essential in the validation process. At each step, the workers must demonstrate their knowledge of the operation and the hazards of the equipment that they work on; moreover, they must demonstrate their ability to plan ahead and avoid those hazards by using appropriate safe work practices, PPE, and other precautions as required.

Of course, examinations and hands-on exercises will be developed based on the job and task analysis and other assessments discussed previously. Any missed questions should be reviewed with students until they understand why they missed them and have demonstrated their ability to answer similar questions correctly. In other words, where safety is at risk, 70 percent is not a passing grade. In fact, anything less than 100 percent success is insufficient when safety is involved.

The actual presentation of the training program will depend upon the extent and type of training being implemented. Some courses may be presented on an ongoing basis. For example, skills updates such as cable testing, motor and generator maintenance, and the like may be repeated immediately before maintenance intervals that will require those skills. Safety training should be repeated frequently, especially in areas such as lifesaving and CPR.

Training in areas such as basic electricity, mathematics, and other such topics is usually performed only once per employee. Large companies may need to schedule such training on a regular basis to allow for turnover. Smaller companies, however, may opt to use self-training or correspondence courses for such topics.

Evaluate

Although evaluation is identified as a separate step, it should be an ongoing effort employed at every level of the training program. Five basic elements of the training program may be used to analyze its effectiveness—*test review*, *course critiques*, *employee interviews*, *course content*, and *needs analysis*.

Test Review. Examinations evaluate the student, the instructor, and the course. They should be administered for every training course. Tests should be reviewed for a variety of problems such as the following:

- *Commonly missed questions.* If one question or group of questions is consistently missed by all students, you have a problem with the instructor and/or the course.
- *Consistently high or low grades.* If the average grades on a test are consistently high or low, the course is probably too easy or difficult, respectively. Another possible cause of such a problem is that the course material does not fulfill the course objectives.

Course Critiques. Carefully review the course critiques for trouble spots. Consistently high or low evaluations may be an indication of a problem. Take the occasional "sour grapes" critique with a grain of salt.

Employee Interviews. Periodically interview employees to keep your finger on their pulse. Make certain that the employees are learning and that the training program is accomplishing what you intended.

Course Content. Electrical technology is changing rapidly. Although basic electrical principles remain the same, testing technology changes rapidly. Periodically you must have your course materials reviewed for technical accuracy and relevance. Nothing frustrates students more than learning yesterday's technology.

Needs Analysis. Periodically update the performance of the needs analysis. Make certain that the desired behavioral objectives are being met. Ask the following questions:

Has company performance improved as desired?

Is employee and supervisor morale at the levels you wished to achieve?

Is your company safety record acceptable?

Tracking the Training Program. Keeping track of the feedback received during the various portions of the program is a monumental task, especially for very large companies. A system for tracking the training program should be implemented to document and verify all of the various training steps.

For example, during training implementation, records are stored in an electronic database. Forms, queries, and reports are structured to allow the user to search for information relating to employees, hazards, or departments. Individual status reports are quickly and easily accessible and show each hazard for which the individual has completed training. Such record keeping is critical to optimize safety and to serve as proof of training if an OSHA investigation occurs.

Whatever criteria were used to establish the original need for the training program must be evaluated. If the desired changes have occurred, then the program is meeting its goals.

Modify

Update and modify your training program based on your evaluation. Although this may seem obvious, a surprising number of companies evaluate their programs and then make no changes. An evaluation is useless if no changes are implemented.

GLOSSARY

A

AFCI See *arc-fault circuit interrupter.*

AHJ See *authority having jurisdiction.*

arc (electric) The electrical breakdown of and subsequent electrical discharge through an electrical insulator, such as air.

arc energy input The total amount of energy delivered by the power system to the arc. This energy will be manifested in many forms, including light, heat, and mechanical (pressure) energy.

arc eye A medical condition of the eye caused when ultraviolet light kills cells in the cornea. After they are destroyed, the cells will fall off the surface of the cornea and expose sensitive tissue. When the eye blinks, it feels as though there is sand in the eye. Very painful. Resting in a dark room and/or applying certain types of anesthetic drops will relieve the pain.

arc-fault circuit interrupter (AFCI) A circuit interrupter found in the form of a circuit breaker that provides both overcurrent protection and arc-fault protection. It has internal electronic circuitry that analyzes the waveform of the current for the presence of an arc-fault. It is intended primarily as a fire prevention device. It offers only minimal personnel shock protection unless it is accompanied by a GFCI.

arc-flash The visible light released in association with an electric arc, often accompanied by radiated heat including invisible ultraviolet and infrared radiation.

arc-flash boundary See *flash boundary.*

arc incident energy The amount of radiated thermal energy delivered by an electric arc to the surface clothing or body of a worker. This amount of energy will be less than the total arc energy. The amount of arc incident energy will depend on factors in the arc event and the workplace's electrical system.

arcing fault A short-circuit in an electrical component accompanied by an electrical arc.

arc rating See *Arc Thermal Performance Value.*

arc-resistant material A material that, when exposed to an electrical arc, attenuates the energy passed through to the other side of the material or garment.

arc-resistant switchgear Metal-clad switchgear that features strengthened mechanical construction as well as pressure relief systems. Arc-resistant switchgear is designed and tested to minimize the probability that the energy of an electric arc in the switchgear will be released in a destructive manner to nearby people and facilities.

Arc Thermal Performance Value (ATPV) The incident thermal energy on a fabric or material that results in sufficient heat transfer through the fabric or material to cause the onset of a second-degree burn based on the Stoll curve.

authority having jurisdiction (AHJ) An organization, office, or individual responsible for enforcing the requirements of a code or standard, or for approving equipment, materials, an installation, or a procedure.

B

blast (electric) The explosive effect caused by the rapid expansion of air and other vaporized materials that are superheated by the sudden presence of an electric arc.

bonding/bonded The permanent joining of metallic parts to form an electrically conductive path that will ensure electrical continuity and the capacity to conduct safely any current likely to be imposed.

bonding jumper A reliable conductor used to ensure the required electrical conductivity between metal parts that must be electrically connected.

bonding jumper, equipment Used to connect (bond) two or more equipment grounding conductors together.

bonding jumper, main Used to connect (bond) the equipment grounding conductor and the grounded circuit conductor together at the service.

bypassing control See *control bypass.*

C

capacitance The ability of a body to hold an electric charge. The capacitance is calculated by $C = \frac{q}{V}$ where C = capacitance in farads, V = voltage drop across the charge, and q = net charge stored on each conductor across which the voltage is applied.

capacitive reactance The opposition to current flow by a capacitor when the capacitor voltage changes. The capacitive reactance ($\bar{X}_C$) is measured in ohms and is calculated by $\bar{X}_C = \frac{-j}{2\pi fC}$ where $j = \sqrt{-1}$, f = frequency in hertz, and C = capacitance in farads. The magnitude of the capacitive reactance is given by $\left|\bar{X}_C\right| = \frac{1}{2\pi fC}$.

capacitor An electronic device that is used to add capacitance to an electrical circuit.

CBM A commonly used acronym for condition-based maintenance.

clearing See *de-energize.*

cold-flow The tendency of materials to change shape as they expand and contract due to temperature changes. This is a significant problem when aluminum is improperly installed. The shape change creates air gaps that will oxidize. (Aluminum oxide is a very poor conductor.)

complex number A number of the form $a + jb$ where $j = \sqrt{-1}$. Complex numbers are sometimes informally referred to as *rotating vectors.*

condition-based maintenance A maintenance and testing program in which efforts and expenditures are based on the actual condition of the maintained equipment. CBM is a subset of reliability-centered maintenance (RCM).

conductivity The reciprocal of resistivity. $\sigma = \frac{1}{\rho}$ where ρ = resistivity in ohm-meters and σ = conductivity in siemens per meter.

conductor (electrical) A material that presents very little resistance to current flow. Conductors have resistivities within the range of 10^{-4} to 10^{-2} ohm-meters. Conductors may be made of silver, copper, aluminum, gold, or many other metals. Silver is the best naturally occurring conductor.

contact tester A voltage measuring device that requires physical, metal-to-metal contact with the electrical conductors. A contact tester has two probes. The tester measures the voltage between the two conductors.

continuous current rating The maximum current that a fuse or circuit breaker is designed to carry continuously without overheating (circuit breaker) or opening (fuse).

control bypass An often-used term to describe a situation in which someone is required to remove the lock and/or tag of the person who placed it.

CT See *current transformer.*

current The rate of charge flow past a given point. Current flow is given the symbol I and is measured in coulombs per second (C/s). C/s = amperes.

current transformer (CT) A device that transforms the current in one of its windings (the primary) to a different current in its other winding (the secondary). These devices are applied to reduce high primary currents into lower secondary currents suitable for instrumentation.

D

de-energize 1. In operations: Removing the voltage from a circuit.

2. Safety-related: Removing power, locking, tagging, trying, and testing. Grounding as required.

dermis The inner layer of the skin.

ductility The ability of a material to be reshaped when in its solid form. Thinning a material by hammering or extruding a metal into wire are both examples of processes that require metal to be ductile.

E

earthing See *ground.*

electrically safe See *electrically safe work condition.*

electrically safe work condition A state in which an electrical conductor or circuit part has been disconnected from energized parts, locked/tagged in accordance with established standards, tested to verify the absence of voltage, and, if necessary, temporarily grounded for personnel protection.*

electrocution Death caused by the passage of electricity through the body. Death caused by electric shock.

electromagnetic spectrum The range of all possible frequencies of electromagnetic radiation.

*NFPA 70E, *Standard for Electrical Safety in the Workplace,* National Fire Protection Association (NFPA), Quincy, MA, p. 70E-10, 2018.

electromotive force (EMF) The electrostatic force that causes electrons to flow in a closed circuit. EMF is also referred to as *voltage*. It is measured in volts (V).

electroporation Cell death resulting from the enlargement of cell membrane pores due to high-intensity electric fields.

energized Electrically connected to a source of potential difference.

energy break-through (E_{BT}) The average of the five highest incident energy exposure values below the Stoll curve where the specimens do not exhibit breakopen.†

epidermis The outer layer of the skin.

equipment A general term including material, fittings, devices, appliances, fixtures, apparatus, and the like used as a part of, or in connection with, an electrical installation.

equipotential zone A work area in which a worker cannot be subjected to dangerous differences in electrical potential. Equipotential zones are created by the proper installation of electrical safety grounds.

extensor muscle A muscle whose contraction increases the angle of a joint. The triceps is an extensor muscle.

F

fault Although frequently used interchangeably with *short circuit*, the broader definition of *fault* is *any abnormality in an electrical power system*. Examples include:

- A loss of a single phase in a three-phase power system
- An unintentional connection between two current-carrying wires of different voltage and/or phase angle (see *short circuit*)
- Failure of a ground (earth) connection causing abnormal overvoltages

ferroresonance An unstable high voltage that can occur on three-phase electrical systems under specific conditions, potentially causing the failure of equipment.

fibrillation Rapid and inefficient contraction of muscle fibers of the heart caused by disruption of nerve impulses.

flame-resistant material A fabric or material that burns very slowly and self-extinguishes upon the removal of an external flame.

flame retardant A chemical applied to or incorporated into a fabric at the time of production that significantly reduces a fabric's flammability.

flammable Easily ignited and capable of burning rapidly.

flash See *arc-flash.*

flash boundary The radial distance from an energized conductor at which the heat energy will equal 1.2 cal/cm^2 in the event that the worst-case arc-flash event occurs.

flash protection boundary See *flash boundary.*

---

†ASTM Std F 1959/F 1959M B 99, "Standard Test Method for Determining the Arc Thermal Performance Value of Materials for Clothing," American Society for Testing and Materials (ASTM), Conshohocken, PA.

flexor muscle A muscle whose contraction reduces the angle of a joint. The biceps is a flexor muscle.

frame size (circuit breakers) See *continuous current rating.*

fuse rating See *continuous current rating.*

fuzzy logic A mathematical analysis method that is nonbinary. Rather than classic 0 or 1 truth statements, fuzzy logic will allow multiple truth data between 0 and 1.

G

ganging A slang term sometimes used to mean the placing of multiple locks and tags on electrical equipment.

GFCI See *ground-fault circuit interrupter.*

ground A conducting connection, whether intentional or accidental, between an electrical circuit or equipment and the earth, or to some conducting body that serves in place of the earth. (The NEC now simply states that "ground" is "the earth.")

ground fault An unintentional electrical connection between an energized ungrounded conductor and a grounded or non-current-carrying conductor, enclosure, or surface, or the earth.

ground-fault circuit interrupter (GFCI) A device intended for the protection of personnel that functions to de-energize a circuit or portion thereof within an established period of time when a current to ground exceeds some predetermined value that is less than that required to operate the overcurrent protective device of the supply circuit.

grounded Connected to ground (earth) or to some conducting body that serves in place of the earth or that extends the ground conductor or connection.

grounded conductor A system or circuit conductor that is intentionally grounded.

grounded, effectively Intentionally connected to earth through a ground connection or connections of sufficiently low impedance and having sufficient current-carrying capacity to prevent the buildup of voltages that may result in undue hazards to connected equipment or to persons.

ground-fault protection of equipment A system or method that provides line-to-ground fault current protection for electrical equipment by sensing the fault and causing the disconnecting means to open all ungrounded conductors in the faulted circuit. The ground-fault protection sensing device is set at lower current levels than the circuit overcurrent protective device, which is set to protect the conductors from damaging overcurrent.

grounding conductor The conductor used to connect the non-current-carrying metal parts of equipment, raceways, and other enclosures to the system grounded conductor, the grounding electrode conductor, or both, at the service equipment or at the source of a separately derived system.

grounding electrode A conducting object through which a direct connection to earth is established.

grounding electrode conductor The conductor used to connect the grounding electrode to the equipment grounding conductor, to the grounded conductor, or to both, of the circuit at the service equipment or at the source of a separately derived system.

grounding electrode system Two or more grounding electrodes that are effectively bonded together shall be considered a single grounding electrode system.

H

horny layer The commonly applied name for the stratum corneum layer of the epidermis. The stratum corneum is called the horny layer because its cells are toughened like an animal's horn.

I

i Symbol used in mathematics for $\sqrt{-1}$. The symbol *j* is used in electrical circuits. (See *j*.)

impedance The total opposition to current flow in an alternating circuit. It is measured in ohms and comprises the sum of the resistance and the reactance of the circuit. It is calculated as $Z = R + jX$ where R is the circuit resistance and X is the circuit reactance.

incident energy The calculated heat energy to which workers are exposed when they are located at the *working distance* away from the arc source. Incident energy is measured in cal/cm^2 or joules/cm^2.

incipient failure A power system fault that is imminent. An incipient fault may not be detectable other than through sophisticated testing procedures.

inductance The ability of a conductor to store energy in a magnetic field. The inductance of a coil is given by $L = \frac{N\phi}{i}$ where L = the inductance in henrys, ϕ = the flux that passes through each turn of the coil in webers, and N = the number of turns in the coil.

inductive reactance The opposition to current flow by an inductor when the current changes through the inductor. The inductive reactance is measured in ohms and is calculated by the formula $X_L = j2\pi fL$ where $j = \sqrt{-1}$, π = 3.14159, f = frequency in hertz, and L = the inductance in henrys.

inductor An electronic device that is used to add inductance to an electrical circuit. Inductors are usually in the form of a coil of wire such as a motor or transformer winding.

instantaneous tripping An engineering concept in which a protective device trips as quickly as possible with no intentional time delay.

insulator (electrical) A material that presents a very high resistance to current flow. Insulators have resistivities in the range of 10^{10} to 10^{20} ohm-meters. Good insulators include glass, plastic, and rubber.

interrupting capacity The maximum current that a circuit breaker can interrupt at the specified voltage level and system *X/R* ratio. Note that for any given installation, this may be a different value than the *interrupting rating* depending on the type of testing that is performed. This should be taken into account, especially in circuit breaker installations.

interrupting rating The maximum short-circuit current that an overcurrent protective device can safely interrupt under standard test conditions.

intrinsically safe Inherently safe by design. A battery-powered hand tool is considered to be electrically, intrinsically safe because it is supplied by a low-voltage battery instead of the power system. This eliminates any arc-flash and drastically reduces the possibility of electric shock.

ionizing radiation Radiation that is capable of ionizing substances through which it passes. X-rays, alpha rays, beta rays, and neutron bombardment are all examples of ionizing radiation.

J

j $j = \sqrt{-1} = 1\angle 90°$. In electrical circuits the letter *i* or *I* is used as the symbol for current flow. Thus, using *j* prevents any confusion.

job briefing A safety meeting held to inform workers of the job requirements and safety hazards associated with the job. Job briefings should be held before the beginning of each job, at the beginning of each new work shift, and again any time conditions on the job site change.

just curable burn A term used by engineers to refer to the medical circumstances of a second-degree burn; it does not have a specific treatment meaning.

L

limited approach boundary The closest that an unqualified person is allowed to approach an exposed energized conductor.

loading a circuit A condition that exists when a voltmeter's internal impedance is very low compared to the system's impedance. This will cause the meter to read a lower voltage than is actually present.

M

make electrically safe See *de-energize* definition No. 2.

maximum voltage rating The maximum allowable voltage of the circuit in which a piece of equipment may be applied.

minimum trip rating The maximum current that a circuit breaker or fuse can carry without *tripping*. Note that currents that are only slightly higher than the maximum trip rating usually require great amounts of time before the device opens.

O

oil-impregnated paper An excellent insulator used for many years for electrical cable insulation. This type of cable was often sheathed in lead to form a cable that was relatively easy to install and remained impervious to chemical attacks.

P

PdM A commonly used acronym for *predictive maintenance*.

permeability The measure of the ability of a material to form or support a magnetic field within itself.

phase vector See *phasor*

phasor A complex number that is used to represent the magnitude and phase angle of a time varying waveform such as ac voltage or current. Phasors are used in polyphase power systems to show the relationships among ac waveforms.

plasma A high-temperature, electrically ionized gas. Because of the high temperatures and electrical characteristics of plasma, it is often identified as a fourth state of matter. The others include solid, liquid, and gas. Some plasmas are referred to as *cold* plasma. The plasma above a candle flame is a cold plasma. Clearly, all plasmas are very hot compared to standard ambient conditions.

PM A commonly used acronym for *preventive maintenance.*

predictive maintenance (PdM) A maintenance program that uses off-line test data to analyze and predict the future condition of the equipment. PdM may also employ test data gathered while equipment is operating. Infrared scan is an example of off-line PdM. PdM is a subset of CBM.

pre-job briefing See *job briefing*

preventive maintenance (PM) A program that schedules and performs periodic maintenance based on the calendar. PM is a subset of RCM.

prohibited approach boundary A distance from an energized conductor that is considered to be the same as physically contacting the conductor.

proximity tester A type of voltage indicator that does not require electrical contact with a conductor to indicate the presence of voltage. The tips of proximity testers are insulated. The device works by sensing the electric field created by the voltage.

Q

qualified person One who has received training in and has demonstrated skills and knowledge in the construction and operation of electric equipment and installations and the hazards involved (from 29 CFR 1910.399).

R

racking The physical act of removing (*racking out*) or installing (*racking in*) a circuit breaker into its operating position and connecting or disconnecting it from the electrical bus. The process may involve an electrical mechanism that uses an electric motor to move the breaker, or it may be a manual system in which the worker moves the breaker by hand.

RCM A commonly used acronym for reliability-centered maintenance.

reactance The opposition to current flow of a circuit element caused by a change in the current or the voltage in the circuit. The symbol used for reactance is *X*. See *capacitive reactance* and *inductive reactance.*

real-time monitoring A system that collects operating data from electrical equipment on a real-time basis. The rich dataset that is collected allows for very accurate predictions of incipient failures. RTM is a subset of CBM.

reliability-centered maintenance A holistic program that schedules and performs maintenance to optimize system reliability.

resistance One of the values that quantifies the amount of opposition to electric current. Its magnitude is given in ohms (Ω). The current flow through a pure (noncomplex) resistance causes the dissipation of real power in watts (W). An object's resistance can be calculated using $R = \frac{\rho L}{A}$ where ρ = resistivity in ohm-meters, L = the length of the object through which current is passing in meters, and A = the cross-sectional area through which the current is passing in square meters.

resistivity A measure of how strongly a material opposes the flow of electrical current. It is defined as $\rho = \frac{E}{I/m^2}$ where E = magnitude of the electric field in volts/meter (V/m) and $\frac{I}{m^2}$ = the current density in amperes per square meter. Conductors have very low values of ρ.

restricted approach boundary The approach distance at which qualified persons are required to wear insulating protective equipment suitable for the voltage.

RTF A commonly used acronym for *run-to-failure.*

RTM A commonly used acronym for *real-time monitoring.*

run-to-failure Inexpensive and/or noncritical equipment is allowed to operate until it fails. This is a legitimate maintenance philosophy that must be used with caution. RTF is a subset of RCM.

S

safety ground The intentional connection of one or more de-energized (but normally energized) circuit conductors to the earth or some conductive plane in place of the earth. The purpose of a safety ground is to create a zone of equalized potential for the safety of workers. See Chaps. 4 and 5 for specific details.

selective tripping An engineering criterion used in the calculation of settings for electrical protective devices. Only the closest upstream protective device to the fault must trip.

shock (electric) The physical stimulation or trauma that occurs as a result of an applied electric current from an external source passing through the body.

shock circuit The path or paths that electric current takes through the body. If the shock circuit includes critical organs, severe trauma is more likely than if it does not.

short circuit The unintentional connection of two electrical conductors that normally have a difference in voltage between them. If the short circuit includes an electrical arc, it is called an *arcing fault.*

Stoll curve A tabular or graphical relationship established in the 20th century that quantitatively describes the relationship between heat energy and human skin.

subcutaneous layer Body tissue below the skin layers.

T

three-step process A procedure that calls for testing a voltage measurement instrument on a known live circuit both before and after the actual circuit is checked. The purpose of this procedure is to make sure that the instrument is working and that the zero voltage measurement is accurate.

tripping Opening a circuit breaker. Most often applied to opening by automatic protective devices.

V

varnished cambric An insulating material made by impregnating cotton (or linen) with varnish or insulating oil. After impregnation, it is baked.

vector Any quantity that is defined by both its magnitude and its direction. Special types of vectors called phasors are used extensively in electrical power engineering.

ventricular fibrillation *Fibrillation* of the ventricles.

voltage The electric potential energy per unit charge, measured in joules per coulomb (J/C). Voltage is given the symbol V, E, or U. J/C = volts. Voltage is measured as the potential difference between two points in a circuit.

W

working distance The distance at which the incident energy is calculated in determining the required arc rating of arc protective clothing. Default working distances vary from 0.46 m (18 in) at lower voltages to 1.1 m (3 ft 6 in) at higher voltages. Other values can be used when conditions warrant.

X

***X/R* ratio** The ratio of the reactance of an electrical circuit (X) in ohms to the resistance (R) of the same circuit in ohms. The magnitude of the X/R ratio has a significant effect on the way the system behaves. For example, power systems with very high X/R ratios tend to have much higher fault currents for the first few cycles of a short circuit; therefore, the ratio must be taken into account when applying circuit breakers that may have to interrupt that current.

Z

zone of equalized potential See *equipotential zone.*

INDEX

www.ingramcontent.com/pod-product-compliance
Ingram Content Group UK Ltd.
Pitfield, Milton Keynes, MK11 3LW, UK
UKHW050816280526
471566UK00003B/18